Lecture Notes in Comput

Edited by G. Goos, J. Hartmanis an

Advisory Board: W. Brauer   D. G

C000021855

Springer

Berlin
Heidelberg
New York
Barcelona
Budapest
Hong Kong
London
Milan
Paris
Tokyo

Carlo Braccini   Leila DeFloriani
Gianni Vernazza (Eds.)

Image Analysis
and Processing

8th International Conference, ICIAP '95
San Remo, Italy, September 13-15, 1995
Proceedings

 Springer

Series Editors

Gerhard Goos, Karlsruhe University, Germany

Juris Hartmanis, Cornell University, NY, USA

Jan van Leeuwen, Utrecht University, The Netherlands

Volume Editors

Carlo Braccini
DIST, University of Genova
Via Opera Pia,13, I-16145 Genova, Italy

Leila DeFloriani
DISI, University of Genova
Via Dodecaneso, 35, I-16146 Genova, Italy

Gianni Vernazza
DIEE, University of Cagliari
Piazza d'Armi, I-09123 Cagliari, Italy

Cataloging-in-Publication data applied for

Die Deutsche Bibliothek - CIP-Einheitsaufnahme

Image analysis and processing : 8th international conference ;
proceedings / ICIAP '95, Sanremo, Italy, September 1995. Carlo
Braccini ... (ed.). [IAPR]. - Berlin ; Heidelberg ; New York ;
Barcelona ; Budapest ; Hong Kong ; London ; Milan ; Paris ;
Tokyo : Springer, 1995
  (Lecture notes in computer science ; Vol. 974)
  ISBN 3-540-60298-4
NE: Braccini, Carlo [Hrsg.]; ICIAP <8, 1995, San Remo>; International
  Association for Pattern Recognition; GT

CR Subject Classification (1991): I.4, I.5, I.3.3, I.3.5, I.3.7, I.2.10

ISBN 3-540-60298-4 Springer-Verlag Berlin Heidelberg New York

© Springer-Verlag Berlin Heidelberg 1995
Printed in Germany

Typesetting: Camera-ready by author
SPIN 10485511   06/3142 – 5 4 3 2 1 0   Printed on acid-free paper

Foreword

This volume collects the papers accepted for presentation at the eighth International Conference on Image Analysis and Processing (ICIAP), to be held in Sanremo, Italy, September 13-15, 1995. ICIAP'95 is the eighth of a series of international conferences promoted since 1981 by the Italian Chapter of the International Association for Pattern Recognition (IAPR) as an international forum for presentation and discussion of the most recent advances in the fields of image analysis and processing, and of pattern recognition.

The 8th ICIAP is organized jointly by the Department of Biophysical and Electronic Engineering (DIBE), by the Department of Computer and Information Sciences (DISI), by the Department of Communication, Computer and System Science (DIST) of the University of Genoa, and by Elsag Bailey.

An international call for papers, mainly sent by electronic mail, resulted in over 180 papers submitted. The submissions spanned over both theoretical and applicative aspects in low-level image processing, image coding, range and acoustic images, statistical and structural pattern recognition, shape recognition, texture analysis, motion analysis, scene understanding, neural networks, and architectures. The application domains covered include document analysis, biomedicine, remote sensing, robotics and industrial automation.

Each paper underwent a careful refereeing process, at the end of which 108 contributions were selected for inclusion in this volume. The selection of the papers has been extremely difficult because of the many good-quality submissions. We wish to thank all the members of the Program Committee and the other referees for their invaluable cooperation in reviewing the papers in a very restricted amount of time. We would like to thank all the authors for the quality of their work and for their collaboration in keeping the length of their papers within the page limit.

The technical program consists of 34 oral presentations and of 74 posters, and is enhanced by six invited talks given by the A. Jain (University of Michigan), H. Bunke (University of Bern), L. Shapiro (University of Washington), M.D. Levine (McGill University), G. Nagy (Rensselaer Polytechnic Institute) and A.N. Netravali (AT&T Bell Labs.). The lectures survey established approaches, recent results and directions of future work on different topics of primary importance within the field, namely texture analysis, graph-based shape recognition, CAD-based vision, three-dimensional object description, document analysis and image communication.

We wish to express our appreciation to all those who have cooperated for the success of the 8th ICIAP. First of all, we want to thank Giovanni Garibotto for his invaluable contribution to continuously monitoring the scientific level of the conference. Special thanks are due to Enrico Puppo for his invaluable cooperation in handling the refereeing process and putting together the technical program. We are also grateful to Fabio Cocurullo for his fundamental assistance in preparing these proceedings and to Sebastiano Serpico for coordinating the local organization. It is a pleasure to thank Ombretta Arvigo, Michela Bertolotto, and Paola Magillo of DISI, and Franco Fontana, Alessandra Tesei, Paolo Pellegretti, and Cinthya Ottonello of DIBE for their friendly help with the conference organization.

Finally, we would like to stress that the conference could not take place without the support of all the different organizations, companies and local institutions listed in this volume.

Genova, June 1995

Carlo Braccini   Leila De Floriani   Gianni Vernazza

General Chairs

C. Braccini	DIST, University of Genova
L. De Floriani	DISI, University of Genova
G. Vernazza	DIBE, University of Genova
	DIEE, University of Cagliari

Scientific Chair

G. Garibotto	Elsag Bailey - A Finmeccanica Company

Scientific Coordinators

F. Cocurullo	DIST, University of Genova
E. Puppo	IMA - CNR, Genova
S.B. Serpico	DIBE, University of Genova

Program Committee

J.K. Aggarwal, U.S.A.	C. Arcelli, Italy
V. Cantoni, Italy	V. Cappellini, Italy
L.P. Cordella, Italy	J. Desachy, France
V. Di Gesù, Italy	O. Faugeras, France
M. Ferretti, Italy	H. Freeman, U.S.A.
C. Guerra, Italy	T. Huang, U.S.A.
S. Impedovo, Italy	J. Kittler, U.K.
W. Kropatsch, Austria	O. Kübler, Switzerland
M. Kunt, Switzerland	S. Levialdi, Italy
P. Mussio, Italy	T. Pavlidis, U.S.A.
S. Peleg, Israel	V. Roberto, Italy
H. Samet, U.S.A.	A. Sanfeliu, Spain
R. Stefanelli, Italy	S. Tanimoto, U.S.A.
A. Venetsanopoulos, Canada	S. Vitulano, Italy
P. Zamperoni, Germany	B. Zavidovique, France

Scientific Secretariat

O. Arvigo	DISI, University of Genova

Local Arrangements

S. Vaccaro	DIBE, University of Genova

Financial

F. Bagnus	AEI - Ligurian Chapter

Publications

F. Cocurullo	DIST, University of Genova

Registration

F. Ravaschio	Elsag Bailey - A Finmeccanica Company

Additional Referees

Albanesi, M.G.
Ayer, S.
Bigun, J.
Borgefors, G.
Bruzzone, E.
Cocurullo, F.
Curinga, S.
De Natale, F.
Dellepiane, S.
Dickinson, S.
Duc, B.
Euriemdy, G.
Fischer, S.
Gong, X.
Lancini, R.
Magillo, P.
Marcelli, A.
Menard, C.
Murino, V.
Nicchiotti, G.
Ottonello, C.
Ramella, G.
Reusens, E.
Rosenfeld, A.
Sansone, C.
Scagliola, C.
Tesei, A.
Tubaro, S.
Werman, M.
Zappatore, S.

Alessandri, A.
Bertolotto, M.
Bischof, H.
Brelstaff, G.
Castello, P.
Colla, A.
De Micheli, E.
De Stefano, C.
Di Zenzo, S.
Doermann, D.
Egger, O.
Ferenberg, G.
Foresti, G.L.
Grattarola, A.
Lavagetto, F.
Makarov, A.
Masciangelo, S.
Moccagatta, I.
Nagy, G.
Ottaviani, E.
Puppo, E
Regazzoni, C.
Roli, F.
Sanniti di Baja, G.
Sazaklis, G.
Soffer, A.
Trucco, A.
Vento, M.
Yacoob, Y.

Organization and Support

Organized by:

- University of Genova
  DIBE - Dept. of Biophysical and Electronic Engineering
  DISI - Dept. of Computer and Information Sciences
  DIST - Dept. of Communication, Computer and System Science
- Elsag Bailey - A Finmeccanica Company

Sponsored by:

- IAPR - International Association for Pattern Recognition
- AEI - Electrical and Electronics Engineers Association (Ligurian Chapter)
- EURASIP - European Association for Signal Processing
- IEEE - Institute of Electrical and Electronics Engineers (North Italy Section)

Supported by:

- APT Riviera dei Fiori
- Banca CARIGE - Cassa di Risparmio di Genova ed Imperia.
- Casinò Municipale di Sanremo
- Città di Sanremo
- CNR - Consiglio Nazionale delle Ricerche
- Consorzio Sanremo Congressi Turismo
- Elsag Bailey - A Finmeccanica Company
- Gruppo Marazzi
- IRST - Istituto per la Ricerca e lo Sviluppo del Trentino
- Orsi Impianti
- Provincia di Imperia
- Telecom Italia - Direzione Regionale Liguria
- University of Cagliari
- 3M Italia

Contents

Software and Hardware Architectures for Image Processing

Neural Networks

Image Coding I

Biomedical Applications I

Scene Understanding

Robot Vision

Digital Topology and Morphology

Range and Acoustic Images

Low-Level Image Processing

Pattern Recognition and Document Processing

Optical Character Recognition

Biomedical Applications II

Image Coding II

Motion

Miscellaneous Applications

Textures

Texture Analysis: Representation and Matching

Anil K. Jain and Kalle Karu

Department of Computer Science
Michigan State University
East Lansing, MI 48824, USA
jain@cps.msu.edu, karukall@cps.msu.edu

Abstract. Texture has found many applications in computer vision. Examples where texture analysis methods are being used include: (i) classifying images and browsing images based on their texture; (ii) segmenting an input image into regions of homogeneous texture; (iii) extracting surface shape information from 'texture gradient'; and (iv) synthesizing textures that resemble natural images for various computer graphics applications. Image texture is characterized by the gray value or color 'pattern' in a neighborhood surrounding the pixel. Different methods of texture analysis capture this gray-level pattern by extracting textural features in a localized input region. Practical texture-based image processing methods define texture in a manner that is most appropriate for achieving a given goal and ignore the issue whether the input image really contains any texture. This paper describes attempts to learn 'optimal' texture discrimination masks using neural networks.

1 Introduction

Texture characterizes local variations of image color or intensity. Although texture-based methods have been widely used in computer vision and graphics, there is no single commonly accepted definition of texture. Each texture analysis method defines texture according to its own model. Texture is often described to consist of primitives that are arranged according to some placement rule. This definition gives rise to structural methods of texture analysis which explicitly attempt to recover the primitives and the replacement rules. Statistical methods, on the other hand, consider texture as a random phenomenon with identifiable local statistics. What is common in all descriptions of texture is that texture is a neighborhood property. Image regions that are perceived to have a similar texture have a similar pattern of color or intensity variation, which may be independent of scale, rotation, or illumination.

The following four areas of texture analysis have been studied most extensively: texture segmentation, classification, synthesis, and 'shape from texture' [23]. The purpose of texture segmentation is to divide the input image into homogeneously textured regions, without knowing *a priori* what the textures are. Texture classification methods, on the other hand, attempt to assign a known texture class to each image region. Figure 1(a) shows an image which can be segmented into five homogeneous textured regions. Closely related to texture

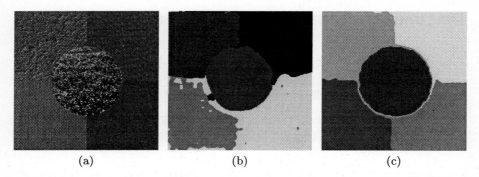

(a)   (b)   (c)

Fig. 1. Example of textured image: (a) a composition of natural images; (b) segmentation using Gabor filtering; (c) segmentation using masks from a neural network.

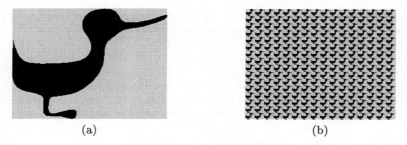

(a)   (b)

Fig. 2. Does the image contain texture? (a) an image without texture; (b) a textured image with the same texture primitive as in (a).

classification is the newly emerging area of image database retrieval based on texture [5, 19]. Elaborate classification methods are often simplified in image database applications to achieve higher processing speed on a large number of images. Texture synthesis methods are used in computer graphics to generate naturally looking surfaces from a few parameters. Changes in the granularity of texture, or in the size of its primitives, give cues about the 3D shape of a textured surface. Shape from texture algorithms use these cues to reconstruct the 3D shape, often with the help of other information, such as stereopsis or shading [17].

Texture analysis methods usually assume that the input images really have texture. In a practical application, it may not be known if the image has any texture and if texture-based methods are appropriate for processing it. Figure 2 shows two images composed of the same primitive. Texture analysis algorithms should not be applied to the first image which contains no texture [14].

In this paper we consider the texture segmentation and classification problems. Section 2 gives background information on different texture analysis methods. Section 3 describes a neural network model to optimize the feature extraction and classification tasks. Section 4 presents experimental results, and Section 5 concludes the paper.

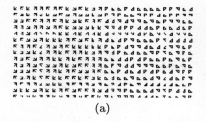
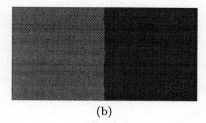

(a)   (b)

Fig. 3. A texture pair with identical second-order statistics: (a) input texture pair; (b) its segmentation using masks learned by a neural network.

2 Texture Analysis Methods

Texture is a popular image attribute which has been successfully used in a variety of applications (remote sensing, medical imaging, industrial inspection). The main goal of texture analysis is to replicate human ability to segment textures. Psychological and psychophysical studies have been conducted to understand human perception of texture and its limits [1]. Tamura *et al.* [22] selected six perceptual characteristics of texture – coarseness, contrast, directionality, linelikeness, regularity, roughness, and proposed algorithms for automatically determining numerical values for these features from given input images. The same texture features have recently been used in image database retrieval [5].

Figure 3 (a) contains a part of synthetic texture images used by Julesz and other researchers to establish the limits of our ability to segment textures preattentively. Julesz [13] conjectured that people cannot segment textures that agree in second-order statistics, that is, probabilities of observing gray values at points that are a certain displacement from each other. The image in Figure 3 (a) is a counterexample to this conjecture. Co-occurrence matrices estimate the second-order statistics by counting the frequencies for all pairs of gray values and all displacements in the input image. Since the number of co-occurrence matrices is very large, they are usually combined in a more compact representation. Haralick has proposed several textural features that can be extracted from co-occurrence matrices – uniformity of energy, entropy, maximum probability, contrast, inverse difference moments, correlation, and probability of run lengths [7]. When a large number of textural features are available, feature selection becomes an important issue. Schistad Solberg and Jain [21] have studied feature selection and combination for SAR image segmentation.

Structural texture analysis methods consider texture as a composition of primitive elements arranged according to a placement rule. Recovering these primitives from natural images is often very difficult, and this severely limits the applicability of structural methods. Model-based texture analysis methods are used both in computer vision and computer graphics. In texture recognition, a model (Markov random field [2], autoregressive [15], fractal [18], or other) is fitted to the image data. The estimated model parameters can then be used to segment or classify the image, or to synthesize new images.

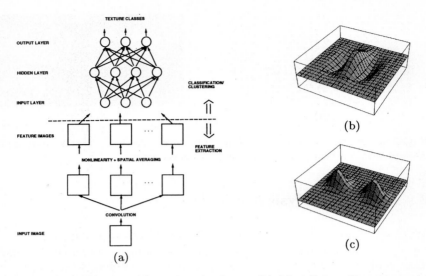

Fig. 4. Multichannel filtering: (a) a general scheem; (b),(c) a Gabor filter in spatial and spatial frequency domains, respectively

The statistical methods described above have the common property that processing is divided into two parts: (i) feature extraction and (ii) classification/clustering. Similar two-phase computation can be found in multichannel filtering methods (Figure 4 (a)) where features are extracted by convolving the input image with a bank of filters. Filters with different band-pass characteristics extract different properties of input texture. The resulting feature values are often called textural energies. The second stage of the multichannel filtering method classifies or clusters pixels based on the extracted features. Statistical classification and clustering methods (nearest neighbor classifier or k-means clustering, for instance) can be used to segment the feature vectors. In Figure 4 (a), we have drawn a neural network as a universal classification or clustering device.

Gabor features have been used to extract textural features corresponding to different orientations and frequencies [3, 11]. In the experiments by Jain and Farrokhnia [11], 20 even-symmetric Gabor filters were used to form the filter bank, $tanh(\beta x^2)$ nonlinearity was applied to each output pixel, and the results were averaged with a Gaussian filter. Figures 4 (b) and (c) show an even symmetric Gabor filter in spatial and spatial frequency domains. The other 19 filters can be obtained by rotating and stretching this kernel. The advantages of Gabor filtering include its multi-scale representation from which invariant features could be computed. Complex Gabor filters have been shown to minimize the joint localization in spatial and spatial frequency domains [6, 3]. Figure 1 (b) shows segmentation of the textured image in Figure 1 (a) using Gabor filters. Sixteen Gabor filters (with four orientations and four radial frequencies) were used for filtering. The clustering program CLUSTER [10] was used to segment the image. The number of clusters was set to five. The classification is quite

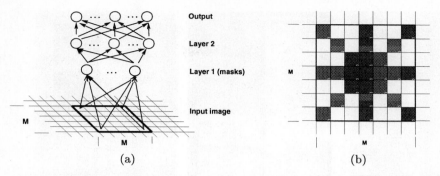

Fig. 5. Neural network for texture classification: (a) neural network architecture; (b) input configuration; shaded pixels are used as inputs to the network for the classification of the center pixel.

accurate except near border areas where the averaged features from different textures cause confusion.

3 Neural Network Classifier

A texture classifier can be obtained by specifying a set of filters and the non-linearity in Figure 4 (a). A multi-layer perceptron can then be trained in a supervised mode to classify the extracted feature vectors. For an optimal classifier, not only the best set of filters has to be found, but the network has to be trained to classify with minimal error. When considering the convolution and nonlinearity operations as part of the neural network (Figure 5), the two tasks – filter specification and classifier training – can be performed at the same time. The input to the multi-layer perceptron in Figure 5 (a) comes directly from a $M \times M$ window in the image. In this architecture, the weights in the first layer correspond to the mask coefficients of a filter bank in the general multichannel filtering scheme. Rather than use the entire $M \times M$ window for input values, we have constrained the network's input to the shaded pixels in Figure 5(b). This configuration helps to reduce the number of network parameters when the mask size M grows larger.

The neural network is trained with the back propagation method [20]. We have used a constant learning rate and momentum term. About one million steps were sufficient for convergence. During training the network weights are updated in all the layers, including the first one, whereas in Figure 4 (a) only the network classifier is trained with fixed filter coefficients. Therefore, training the unified neural network involves performing two optimization tasks at the same time – finding the best mask coefficients and the best classifier, so that the classification error is minimized. The selection of a suitable number of filters can also be included in the network training by growing or pruning the neural network. We have used the node pruning method by Mao *et al.* [16]. Since we are interested in the selection of masks, we only prune nodes in the first layer.

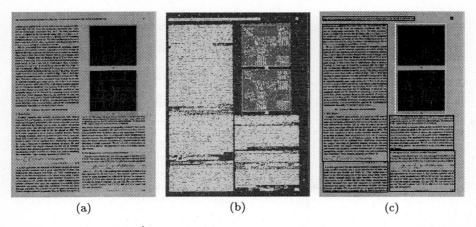

(a)   (b)   (c)

Fig. 6. Page layout segmentation: (a) a 1000×800 document image scanned at 100 dpi; (b) segmentation into text, graphics, and background; (c) the post-processed image with bounding boxes surrounding different regions.

4 Experimental Results

In this section, we show classification results from several applications. Figure 3 (b) shows a classification result of segmenting the synthetic image in Figure 3 (a) by a two-layer network with only three masks of size 5×5. The masks learned in a supervised neural network can also be used in unsupervised texture segmentation. A two-layer neural network with sixteen 11×11 masks was trained on nine different textures, including four textures from the image in Figure 1 (a). The learned sixteen masks were then separated from the network and used to perform unsupervised segmentation of the five-textured image in Figure 1 (a). The input image was convolved with the masks, output pixel values were squared and smoothed by a Gaussian function, and the resulting feature vectors were clustered using the CLUSTER algorithm [10]. Figure 1(c) shows the unsupervised segmentation result.

In a document processing system, a scanned page needs to be segmented into regions of text, graphics and background, after which character recognition or other image processing algorithms are used with the segmented parts. This segmentation can be done based on the texture of the different regions [8]. Figure 6(a) shows a digitized image of a page containing both text and graphics. A three-layer neural network with 20 nodes in each hidden layer was trained to classify the input pixels into three classes – background, text, and graphics. The result of applying the network with ten masks (pruning ten nodes in the first layer did not decrease the performance) to the page is shown in Figure 6(b). Due to the small mask sizes (11×11 in this experiment), the network is accurate in locating the texture boundaries, and finding even such small regions as the page numbers. Figure 6(c) shows the image after applying the post-processing steps described in [12].

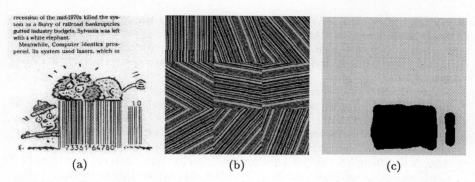

(a)   (b)   (c)

Fig. 7. Locating barcode: (a) a barcode image; (b) training patterns; (c) segmentation.

The problem of barcode localization is to find the barcode in an input image, which may be present in any orientation, position, and scale [9]. In the experiments shown in Figure 7, a three-layer network with ten hidden nodes was trained to classify input image regions into three categories: (i) barcode, (ii) text, and (iii) graphics and uniform gray-value combined. The training patterns for barcode were taken from a different image, and they were rotated to ensure rotational invariance of the classifier. The result of applying the trained network to the input image and median filtering the output class labels is shown in Figure 7 (c).

5 Conclusions

We have shown how a neural network can be trained for supervised as well as unsupervised texture segmentation. Practical problems, such as barcode localization and page layout segmentation, can be solved more efficiently with specifically designed masks than with a general filter set. The use of neural network makes the mask selection automatic. For a new problem, instead of manually choosing a system and testing its performance, one can simply train a neural network (or several of them) on known images. The neural network approach is particularly suitable in texture classification applications where the input textures are known. Compared with general filtering methods, a neural network that is optimized for the specific input textures is able to achieve higher classification accuracy and processing speed.

References

1. Caelli, T.: Visual Perception: Theory and Practice. Pergamon Press. (1981)
2. Cross, G.C, Jain, A.K.: Markov Random Field Texture Models. IEEE Trans. on PAMI. **PAMI-5** (1983) 25-39
3. Daugman, J.G.: Uncertainty relation for resolution in space, spatial-frequency, and orientation optimized by two-dimensional visual cortical filters. J. Opt. Soc. Amer. **2(7)** (1985) 1160-1169

4. Elfadel, I.M., Picard, R.W.: Gibbs random fields, cooccurrences, and texture modeling. IEEE Trans. on PAMI. **PAMI-16** (1994) 24-37

5. Faloutsos, C., Barber, R., Flickner, M., Hafner, J., Niblack, W., Petkovic, D., Equitz, W.: Efficient and Effective Querying by Image Content. JIIS. **3** (1994) 231-262

6. Gabor, D.: Theory of communication. J. IEE (London). **93** (1946) 429-457

7. Haralick, R.M.: Statistical and Structural Approaches to Texture. Proc. of the IEEE **5** (1979) 786-804

8. Jain, A.K., Bhattacharjee, S.: Text Segmentation Using Gabor Filters for Automatic Document Processing. Machine Vision and Appl. **5(3)** (1992) 169-184

9. Jain, A.K., Chen, Y.: Barcode Localization Using Texture Analysis. Proc. Sec. Int'l Conf. on Document Anal. and Rec., Tsukuba city, Japan. (1993) 41-44

10. Jain, A.K., Dubes, R.C.: Algorithms for Clustering Data. Prentice Hall, New Jersey. (1988)

11. Jain, A.K., Farrokhnia, F.: Unsupervised Texture Segmentation Using Gabor Filters. Pat. Rec. **12** (1991) 1167-1186

12. Jain, A.K., Zhong, Y.: Page Layout Segmentation based on Texture Analysis. (under review)

13. Julesz, B., Gilbert, E.N., Shepp, L.A., Frisch, H.L.: Inability of humans to discriminate between textures that agree in second-order statistics – revisited. Perception. **2** (1973) 391-405

14. Karu, K., Jain, A.K., Bolle, R.M.: Is there any texture on the image? Tech. report, Michigan State University. (1995)

15. Mao, J., Jain, A.K.: Texture Classification and Segmentation Using Multiresolution Simultaneous Autoregressive Models. Pat. Rec. **2** (1992) 173-188

16. Mao, J., Mohiuddin, K., Jain, A.K.: Minimal Network Design and Feature Selection Through Node Pruning. Proc. 12th ICPR, Jerusalem. **2** (1994) 622-624

17. Pankanti, S., Jain, A.K.: Integrating Vision Modules: Stereo, Shading, Grouping, and Line Labeling. IEEE Trans. on PAMI. (to appear)

18. Pentland, A.: Fractal-based description of natural scenes. IEEE Trans. on PAMI. **PAMI-9** (1984) 661-674

19. Picard, R.W., Minka, T.P.: Vision texture for annotation. Multimedia Systems. **3** (1995) 3-14

20. Rumelhart, D.E., Hinton, G.E., Williams, R.J.: Learning Internal Representations by Error Propagation. In D.E. Rumelhart and J.L. McClelland, editors, Parallel Distributed Processing: Exploration in the Microstructure of Cognition, Cambridge, MA: MIT Press. **1** (1986) 318-362

21. Schistad Solberg, A.H., Jain, A.K.: Texture Analysis of SAR Images: A comparative Study. IEEE Trans. on Geoscience and Remote Sensing. (under review)

22. Tamura, H., Mori, S., Yamawaki, Y.: Textural Features Corresponding to Visual Perception. IEEE Trans. on SMC. **SMC-8** (1978) 460-473

23. Tuceryan, M., Jain, A.K.: Texture Analysis. Chapter 2.1 in Handbook of Pattern Recognition and Computer Vision, C.H. Chen, L.F. Pau, P.S.P. Wang (eds.) World Scientific Publishing Co. (1993) 235-276

Region-Based Segmentation of Textured Images

Catherine Rouquet and Pierre Bonton

LAboratoire des Sciences et Matériaux pour l'Electronique, et d'Automatique
-URA 1793 CNRS- Université B. Pascal F-63177 Aubière Cedex

Abstract. This paper presents a region-based segmentation algorithm which can be applied to various problems since it does not require a priori knowledge concerning the kind of processed images. This algorithm, based on a split and merge method, gives results both on homogeneous grey level images and on textured images. We modeled exploited fields by Markov Random Fields (MRF), the segmentation is then optimally determined using the Iterated Conditional Modes (ICM). Results from road scenes without white lines are presented.

1 Introduction

The creation of an autonomous vehicle has been the subject of various research programs. To answer this question a vision system can be used from road tracking, obstacle detection, or domain recognition. All these questions require the same initial step which consists in reducing the data contained in images. This step is called segmentation.

When a road does not include white lines, detection is a particularly difficult problem because contrasts are generally not strong enough to use an edge detection algorithm. In such case, images require the use of a region-based segmentation, where we are not looking for discontinuity, but, on the contrary, for homogeneous zones in the sense of one or several given criteria. Segmentation is achieved using a split and merge technique, and regions are represented by grey level and texture features. The originality of this algorithm is that texture is used only to improve on the results of the first detection using grey levels.

In the first part of this paper, the splitting step is presented which is achieved by studying a co-occurrence matrix shape. Merging will be explained in the second part. Markov Random Fields are used to model exploited fields, and the segmentation is optimally determined using the Iterated Conditional Modes (ICM). Experimental results will be presented on road images without white lines.

2 Our Approach

2.1 Splitting

The splitting algorithm described here is based on the use of co-occurrence matrix [5] defined by R. Haralick [4]. We consider a window of interest (or sector) containing three regions with uniform grey level a, b, c. The co-occurrence matrix, and its diagonal can be represented by Fig. 1, where point size is proportional to the number of transitions (grey level - grey level) present in the window of interest. We notice that each region is characterized by a peak of the co-occurrence matrix diagonal.

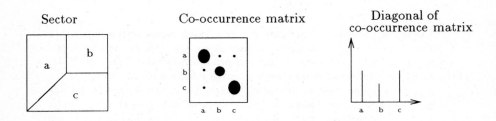

Fig. 1. Co-occurrence matrix

The splitting algorithm based on this particularity, can be broken down into five parts :

- Dividing the image into sectors : this is done independently of data, and consists in putting a grid on the image. We will discuss later the size of the grid elements.
- Computation of the co-occurrence matrix : for each sector, we compute the co-occurrence matrix diagonal. The matrix will be computed from a distance of one pixel, a neighborhood of eight pixels, and orientations of 0, and 90 degrees.
- Peak extraction of the co-occurrence matrix diagonal : different classes of the diagonal are extracted, a class being defined as a peak and its two adjacent valleys.
- Image labeling : we create an image of labels in which different labels are given to each set of connected pixels, belonging to the same class. Such pixel sets are called regions.
- The removal of small regions : small, non-significant regions (whose number of pixels is lower than a threshold which we discuss later) are removed by merging them with the neighboring region which has the closest grey levels average.

The results of the splitting step on road images without white lines are shown in Fig. 2. The initial size of the sectors is 32x32 pixels, and we simply work with a frame. Thus, we find sectors of 32x64 pixels on the result images.

Figure 2 presents the splitting results for two images, each region being represented by its outlines. When two neighboring regions have very different grey levels, their common outline is very well detected, and the splitting results are

 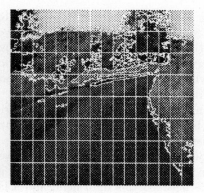

a - Splitting result on image 1   b - Splitting result on image 2

Fig. 2. Results.

sufficient to initiate the merging process. On the other hand, when some textured regions are presented, with a similar grey levels average, the splitting algorithm, which use only the diagonal of the co-occurrence matrix does not separate the two regions. So, we must use texture features in the segmentation algorithm. These features are only introduced during the merging step. The reasons for this choice are explained in the following section.

2.2 Merging

As seen above (see Fig. 2-b), it is not sufficient to use grey level data to obtain robust image segmentation algorithm (in order to find road edges). In this case, we will include texture parameters. Among all extraction methods of texture parameters, we have chosen to use texture features which are computed from the co-occurrence matrix [4], i.e. homogeneity and energy. Several authors such as M. Derras [2] have split original images using a grid, and computed texture parameters on sectors obtained in this way. The greatest disadvantage of this method is that it leads to a strong resolution.

For these reasons we compute texture coefficients from regions produced during the splitting step. Segmentation will then provide one pixel resolution edges, when the computed image presents homogeneous grey level. For textured images, resolution will be identical to the initial grid. Some initial sectors of 32x32 pixels are chosen. The strongest resolution will also be 32 pixels.

The final step of the splitting process is the removal of small regions. A small region is defined by a number of pixels lower than a threshold. This threshold is fixed at the minimal number of pixels to be considered for texture computation. Here, it is fixed at fifty pixels.

In order to obtain an autonomous merging algorithm, Markov Random Fields are used. They have been widely quoted in literature. D.W. Murray et B.F. Buxton [7] have obtained a spatio-temporal segmentation algorithm by modeling apparent motion with Markov Random Fields. H. Derin, et H. Elliot [1] and S. Geman et C. Graffigne [3] have obtained textured image segmentation algorithms. M. Derras [2] used Markov Random Fields to achieve natural UN-supervised image segmentation. We should also mention C. Kervrann et F. Heiz [6] who propose a segmentation method of textured images with a Markov re-laxation.

Input data in our possession are regions produced by splitting. They are mod-eled by a features vector with four components : the grey levels average, the peak of the co-occurrence matrix diagonal, the homogeneity, and the energy. These regions cannot be split further and we call them elementary sites, considering them as data fields (denoted Y). The problem is then to establish a correspon-dence between this data field, and a label field X, for which a label represents a set of homogeneous regions. We model exploited fields using Markov Fields which will be integrated into a Bayesian formalism in order to use constraints derived from a priori knowledge.

The algorithm we present was inspired by the ICM (Iterated Conditional Mode) algorithm : at each iteration, all sites are observed, and, for each of them, we compute the different possible values of an a posteriori energy function with respect to different labels. Only the label with minimal energy function is re-tained for this site. An a posteriori energy function is modeled by :

$$U(X \backslash Y) = U(Y \backslash X) + U(X) \ . \tag{1}$$

Or, for a given site,

$$U(X_s \backslash Y_s) = U(Y_s \backslash X_s) + U(X_s) \ . \tag{2}$$

where :

– $U(X_s)$ is the a priori energy function related to the label field. We also denote it $U(e_s)$
– $U(Y_s \backslash X_s)$ is the energy function representing the probability of observation, given the pixel label X_s. We denote it $U(O_s \backslash e_s)$.

With these new notations, we obtained :

$$U(e_s \backslash O_s) = U(O_s \backslash e_s) + U(e_s) \ . \tag{3}$$

Then, we defined a neighborhood system $V = \{V_s, s \in S\}$ on S such that $s_1 \in V_{s2}$ if, and only if, the sites s_1 and s_2 are adjacent in the sense of 8-connexity.

We model an a priori energy function $U(e_s)$ so that two neighboring sites will probably belong to the same label. The energy function can be noted :

$$U(e_s) = \sum_{r \in V_s} V_c(e_r) \ . \tag{4}$$

where V_c is the energy potential so that :

$$V_c(e_s) = 0 \ \text{if} \ e_s = e_t \ . \tag{5}$$

$$= 1 \ \text{else} \ . \tag{6}$$

where e_t is the second site of the clique c. Here, a clique is a set of two elements : the considered site, and one of the neighbor sites.

The transposition energy will be modeled by the Euclidean distance between feature vectors of the site s, and those of the label e :

$$U(O_s \backslash e_s) = (moy_s - moy_e)^2 + (pic_s - pic_e)^2 + (hom_s - hom_e)^2 + (ene_s - ene_e)^2 . \tag{7}$$

3 Experimental Results

We present here, segmentation results obtained from the previously used images (Fig. 3-a and 3-b).

a- Segmentation result on image 1   b- Segmentation result on image 2

Fig. 3. Results.

Generally, the *road* area is well detected. In the worst case, the resolution is the same as in the initial grid. So the segmentation algorithm provides elements which are necessary to a possible interpretation process. For some 512x512 images, this segmentation algorithm takes approximately 2.5 seconds on a HP9000/735.

4 Conclusion and Perspectives

The aim of this work is to segment road images without white lines. We were confronted with a difficulty which is that grey levels were not sufficient in these images, so we decided to use texture. We also abandoned all methods based on edge detection and instead we used a region-based segmentation algorithm. Our first approach was to divide images into sectors, and to compute texture on each sector. A disadvantage of this method is that the resolution of the segmented image is that of the initial grid. On the one hand we have segmentations using grey level whose resolution is only a one pixel resolution, but the *road* area cannot be found in some images. On the other hand we have segmentations using texture, which finds the *road* area, but has a high resolution. So, we decided to use both of this two major class algorithm. First, we split the images using a homogeneity criterion based on grey levels. Then, we compute texture features on the resulting regions. Both parameters (texture and grey level) are used to achieve merging. This segmentation algorithm provides results which have a one pixel resolution when grey levels are sufficient to extract the *road* area. When the *road* area and the background have the same grey levels average, and different textures, resolution is the same than in initial grid. This algorithm, which is entirely UN-supervised, gives very satisfying results for images which do not include objects which are too small relative to the initial grid size. It can be applied both on textured images and on UN-textured images. Furthermore, the segmentation algorithm provides elements which are necessary to a possible interpretation process.

References

1. Derin, H.,Elliot, H.: Modeling and Segmenting of Noisy and Textured Images using Gibbs Random Fields. IEEE trans. on pattern analysis and machine intelligence PAMI 9 n 1 (1987)
2. DERRAS, M.: Segmentation non Supervisée d'Images Texturées par Champs de Markov : Application à l'Automatisation de l'Entretien des Espaces Naturels. Thèse de docteur es Sciences LASMEA Clermont-Ferrand (dec. 1993) 174 p
3. GEMAN, S.,GRAFFIGNE, C.: Markov random fields image and their applications to computer vision. Proceeding of the International Congress of Mathematicians Ed. A.M. Gleason American Mathematical Society Providence (1987)
4. HARALICK, R., SHANMUGAN, K.,DINSTEIN, H.: Textural Features for Image Classification. IEEE Transactions on Systems and cybernetics vol. SMC-3 (1973) 610–621
5. HOUZELLE, S.,GIRAUDON, G.: Segmentation Région par Modélisation de Matrices de Cooccurrence. AFCET 8eme congrès Reconnaissance des formes et intelligence artificielle Lyon Villeurbanne **vol 3** (Nov 1991)
6. KERVRANN, C.,HEITZ, F.: A Markov Random Field Model-Based Approach to Unsupervised Texture Segmentation Using Local and Global Spatial Statistics. INRIA n **2062** (1993)
7. MURRAY, D.W.,DUXTON, B.F.: Scene Segmentation from Visual Motion using Global Optimization. IEEE transactions on pattern analysis and machine intelligence vol PAMI 9 n 2 (1987) 220–228

A New Approach to Image Segmentation

Antonio Succi and Vincent Torre
Dipartimento di Fisica dell'Università di Genova
Via Dodecaneso 33, 16146 Genova, Italy
e-mail: succi@genova.infn.it

Introduction

The segmentation of an image ([1], [3], [2]) is a major problem in Machine Vision, usually representing the initial step for high-level processing stages such as recognition. Its purpose is to partition the image into distinct regions, each corresponding to an area which is homogeneous in some sense; these regions are usually assumed to represent distinct relevant "objects" in the scene. This problem has been addressed in several ways, including very specialized and *ad hoc* procedures.

In this paper a general scheme based on a "mechanical" (deterministic) approach is outlined; segmentation is defined as a solution to the problem of finding a configuration of static equilibrium for an ensemble of "bubbles" interacting with external fields in various ways. In this aspect, the method has points in common with a version of the Munford & Shah approach ([4]) and with *snake* models ([7], [6]).

Minimization of the energy functional for the bubble ensemble is achieved by means of a deterministic Iterated Conditional Mode (ICM) algorithm in a multi-scale (pyramidal) scheme ([5]); this choice was motivated by its good properties in terms of convergence speed and relative insensitivity to the presence of local minima of the functional.

The segmentation is achieved in two steps. In the first one a coarse segmentation is produced and the maximum number of different bubbles is determined; in the second stage the segmentation is refined and a precise localization of the bubble boundaries is obtained.

The proposed method has been shown to be able to segment a large variety of images, both static and dynamic. Experiments have been carried out with good results on synthetic and natural textures, aerial images, biomedical images, road images and road sequences.

1 The Segmentation Model

1.1 Energy Functional

Let us consider an elastic 2D bubble whose generic boundary element is assumed to experience three kinds of forces: a hydrostatic force directed along the normal, an external force with an arbitrary direction (referred to as an "electrostatic" force) and finally a pair of tangential forces caused by surface tension and acting on the element endpoints.

The newtonian equation for the equilibrium of the bubble boundary can be obtained as the Euler-Lagrange equation associated to the following energy functional:

$$U[\phi] = -p \int_\Omega d^2\mathbf{x} + \int_\mathcal{P} \mathcal{F}(\|\phi'(\xi)\|)\, d\xi + \int_\mathcal{P} V(\phi(\xi))\, d\xi \tag{1}$$

where $\mathcal{P}$ is the parameter domain, Ω is the region whose boundary is represented by the curve $\phi(\xi)$, V is a potential for the external force field $\mathbf{f}$, p is the hydrostatic pressure (positive if inflating), and $\mathcal{F}$ is the potential for the tension forces.

The second term (surface tension) is a regularizing term, which promotes smooth boundaries. The third (electrostatic) term will be used to take account of intensity edges, by making V an attractive potential generated by intensity edges. The first (hydrostatic) term represents the internal energy of the bubble: its effect (with $p > 0$) is to promote configurations such that the bubble covers the largest surface available. In the absence of other terms the bubble inflates indefinitely. In order to obtain a meaningful image segmentation, the internal energy term must be appropriately modified so as to promote configurations in which the bubble doesn't cover just the largest area, but rather the largest *homogeneous* area. The internal energy takes the following form:

$$U_{int} = \Phi_\Omega[\psi] \tag{2}$$

where Φ_Ω is a functional with a support in the region inside the closed curve $\partial\Omega$, and ψ is the image intensity field. The choice of this functional defines the criteria of homogeneity of an area of the image.

The energy functional for a many–bubble system becomes:

$$U = \sum_k \Phi_{\Omega_k}[\psi] + \frac{1}{2}\int_{\mathcal{P}_k} \mathcal{F}(\|\phi'_k(\xi)\|)\, d\xi + \frac{1}{2}\sum_k \int_{\mathcal{P}_k} V(\phi_k(\xi))\, d\xi \tag{3}$$

where k index runs over the set of bubbles.

1.2   External Forces And Surface Tension

The external force term can be defined by assuming the following form for the external potential:

$$V(\mathbf{x}) = -\epsilon(G_\sigma * E)(\mathbf{x}) \tag{4}$$

where E is the edge map extracted from the original image, ϵ is a weight coefficient, and G_σ is a bidimensional gaussian function with width σ and unitary amplitude:

$$G_\sigma(\mathbf{x}) = \exp\left(-\frac{1}{2\sigma^2}\|\mathbf{x}\|^2\right) \tag{5}$$

Expression (4) represents a short-range attractive potential field generated by the intensity edges; it causes the bubble boundaries to be attracted by intensity edges which are sufficiently close.

A suitable choice for the regularizing term (surface tension) is obtained by taking the function $\mathcal{F}$, describing the elastic properties of the boundary, to be linear:

$$U_{s.t.} = \lambda \int_{\mathcal{P}} \|\phi'(\xi)\| d\xi \qquad (6)$$

This is the form corresponding to the surface tension energy of a physical bubble of fluid.

Other useful choices for the function $\mathcal{F}$ are quadratic or higher-degree polynomials; the corresponding energy term is the continuum limit of a chain of linear (Hooke) or higher order springs. As with snakes, a curvature-dependent term could be added, thus obtaining a higher degree of regularity for the boundaries.

2   Implementation Details

Segmentation of an image is achieved by minimizing the energy functional for the bubble ensemble with respect to bubble creation/destruction, merging of adjacent bubbles, deformation of bubble boundaries, and variation of the internal parameters (if *a priori* models are employed). Minimization is performed by means of a pyramidal deterministic scheme as in [5]; at a generic resolution, the bubble ensemble domain is partitioned in a set of pixel blocks of side n, each bubble being represented by a subset of these blocks.

In the case in which no *a priori* model is employed, the computation of internal energy requires the subdivision of each bubble into windows where image attributes are evaluated. The window size is held fixed at a value that allows a meaningful computation of attributes.

The multiscale optimization algorithm is organized as two distinct modules operating at different resolutions. The first one terminates when the block size is the same as the window size, thus generating a coarse approximation of the segmented image. The second one takes this coarse segmentation as its input state and carries on the minimization process up to a resolution in which the block size is 2×2 pixels. The coarse segmentation defines an upper limit for the number of regions appearing in the final result, since no new regions are generated by the high–resolution stage. The two stages differ in some points, as explained in the following.

In the coarse segmentation stage, energy minimization is performed with respect to the complete set of possible variations (deformation, creation, merging). The minimization procedure is structured as follows.

First, a raster scan is performed over the set of blocks in which the image is partitioned at a given resolution. For each block, "virtual" updates are performed on its label and the consequent energy variations are computed; the following updates are considered:

- the new label is one of the labels of the adjacent blocks (local deformation of the boundary);
- the new label is one not appearing in the current configuration (creation of a new bubble);
- the new label is the same as the old one (no operation).

The update corresponding to the lower (most negative) energy variation is actually performed. The scan is iterated until the number of blocks whose label is actually updated becomes lower than a given percentage of the total block number; being deterministic, the process is convergent.

After the block–wise minimization, a region merging is performed: first, the energy variations induced by the merging of adjacent pairs of bubbles are computed. Bubble pairs are sorted in order of the associated energy variations, and then merged in this order; at each merge the energy variations are recomputed and the list is updated. The process terminates when no bubble pair gives a negative energy variation when virtually merged.

The energy functional is represented in this stage by the internal energy term alone, the other terms being considered as accessory at low resolution.

In the high–resolution stage the energy functional is written in its complete form. Region creation is not allowed: energy minimization is performed only with respect to local updates of the bubble boundaries and possibly region merging, that can be optionally inhibited. The minimization procedure operates along the same lines as before, the main differences being that the block scan in the block–wise minimization is performed only on the bubble boundaries, and no new labels are considered in the virtual updates.

When making use of *a priori* models, each bubble is characterized not only by its shape but also by the value of its model–related parameters, which become involved in the minimization problem together with the bubble boundaries. This is true in particular for segmentation in the sense of motion.

It is difficult to perform the minimization with respect to the complete set of variables of the problem, starting from an arbitrary state. The task is easier if an approximate state is already available, under the form of a low–resolution segmentation. In this case an approximate value θ^* for the bubble parameters is first computed by minimizing only the internal energy term with respect to these variables, having fixed the bubble configuration to the one identified by the approximate segmentation:

$$\theta_k^* = \arg \min_{\theta_k} \Phi_{\Omega_k}[\psi, \theta_k] \tag{7}$$

After completing the initial parameter computation, the minimization with respect to bubble deformations is performed with the parameters being fixed to the computed approximate value. When the convergence is reached, the parameters are recomputed and the next step in the multiresolution process is started; this defines a sub–optimal algorithm, whose convergence properties are strongly conditioned by the choice of the starting state (presegmentation). The algorithm terminates when convergence is reached at the resolution of 2×2 pixels.

The approximate segmentation representing the starting state can be obtained in several ways; a possibility is to use the low–resolution stage with no *a priori* models, but other alternatives are possible. In the experiments, simple thresholding methods have often been employed.

3 Experimental Results

Only a little number of the results of the experiments performed with our segmentation method are here reported, due to lack of space.

Figure 1-A shows an echocardiographic image; for this image it was useful to have a binary segmentation into cavities and tissues. This was achieved by employing a binary presegmentation as input of the high–resolution stage. The presegmentation was obtained by a thresholding method at low resolution; the threshold was determined as the main valley of the bimodal histogram of the whole image.

Figure 1-B shows the final segmentation at 2×2 resolution. The boundaries of the detected regions are shown superimposed to the original image. The image attributes used were the first three moments of the histogram computed over 8×8 windows. Parameters were set to $\lambda = 2$, $\epsilon = 1$. Figure 2-A shows a detail

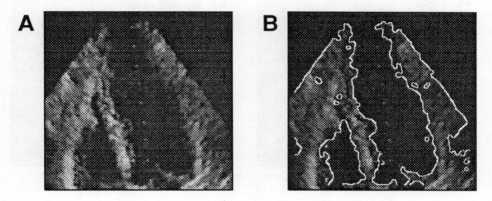

Fig. 1. Echocardiographic image and its segmentation.

of an aerial image obtained by a SAR image after a suitable preprocessing to reduce the amount of speckle noise due to coherent illumination.

The simple two–parameters model proved to be useful to discriminate the different regions present in this image. Figures 2-B shows the final segmentation at 2×2 resolution. Parameters were set to $\lambda = 3$, $\epsilon = 1$.

Figure 3 shows an example of motion segmentation. Figure 3-A shows a frame of a road sequence taken by a camera mounted on a vehicle moving toward the horizon; the car on the left is approaching. Segmentation was performed on a limited square zone of the image surrounding the approaching car. The presegmented image (not shown) was obtained by a thresholding method applied on a sparse optical flow field: a threshold was put on the modulus of the difference between the computed optical flow and the quadratic theoretical flow corresponding to the planar motion on the tarred road.

Figure 3-B shows the result of the final segmentation; parameters were set to $\lambda = 3$, $\epsilon = 2$.

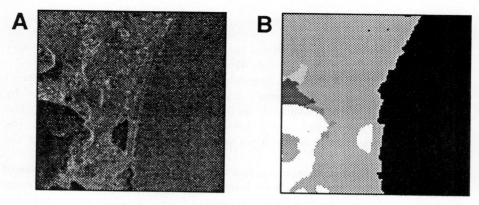

Fig. 2. SAR aerial image and its segmentation.

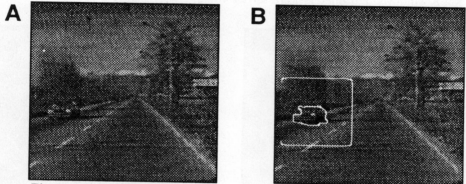

Fig. 3. A frame of a road sequence and the result of motion segmentation.

References

1. R. M. Haralick and L. G. Shapiro, "Image segmentation techniques", *Comput. Vis. Graph. Image Process.*, vol. 29, 1985
2. E. Riseman and M. Arbib, "Segmentation of static scenes", *Comput. Graphics Image Process.*, vol. 6, 1977
3. S. Zucker, "Region growing: Childhood and adolescence", *Comput. Graphics Image Process.*, vol. 5, 1976
4. D. Munford and J. Shah, "Optimal approximation by piecewise smooth functions and variational problems", *Communication on Pure and Applied Mathemathics*, vol. 42, No 5, 1988
5. A. Ackah-Miezan and A. Gagalowicz, "Discrete models for energy-minimizing segmentation", *Proceedings of the 4th International Conference on Computer Vision*, 1993
6. L. D. Cohen and I. Cohen, "A finite element method applied to new active contour models and 3D reconstruction from cross sections", *Technical report 1245*, INRIA, June 1990
7. A. Kass, A. Witkin and D. Terzopoulos, "Snakes: active contour models", *Int. J. Computer Vision,*, vol. 1, No. 3, 1988

Segmentation

Using Hopfield Networks to Segment Color Images

P. Campadelli*, D. Medici*, R. Schettini°

*Dip. di Scienze dell'Informazione, Universita' degli Studi di Milano,
Via Comelico 39, Milano, Italy

°Istituto Tecnologie Informatiche Multimediali, Consiglio Nazionale delle Ricerche
Via Ampere 56, 20131 Milano, Italy

Abstract. In this paper we present two algorithms for color image segmentation based on Huang's idea of describing the segmentation problem as the one of minimizing a suitable energy function for a Hopfield network. The first algorithm builds three different networks (one for each color feature) and then combine the results. The second builds a unique network according to the number of clusters obtained by histogram analysis. Experimental results, heuristically and quantitatively evaluated, are encouraging.

1. Introduction

The segmentation of color images is frequently based on supervised pixel classification when prior knowledge about the object colors is available, while clustering and histogram thresholding are the most widely used unsupervised methods. The algorithms employed assume that homogeneous regions correspond to distinct clusters in the feature space without taking into account spatial information, and may, therefore, generate noisy results. Relaxation algorithms have been proposed to modify the initial cluster labels of the segmented image on the basis of their spatial context [4]. In 1992 Huang [3] described the segmentation problem for grey-level images as the problem of minimizing a suitable energy function for Hopfield networks. We present here two algorithms for color image segmentation based on Huang's idea and evaluate their performance quantitatively.

2. Histogram analysis

We can, by applying one-dimensional histogram analysis to each color component, partition the color space into several hexahedra corresponding to prominent clusters in the color space. A segmented image can be obtained by assigning the mean color of each cluster to the corresponding pixels of the image. The method assumes that homogenous regions in the image correspond to color clusters which when projected on the feature axis generate at least a visible histogram peak, and that these clusters lie apart from each other in the feature space (i.e. the number of clusters can be reliably found). Segmentation results depend largely on the methods adopted to detect "significant" peaks and their respective cut-off values in the histograms. Since in image segmentation the number of clusters and their statistical descriptions are usually not known a priori, we can not use statistical methods to obtain optimal decision boundaries for the peaks in the histograms. We adopt, instead, a technique based on scale-space histogram filtering [7].

3. The Hopfield model

Symmetric networks of non linear graded-response neurons [2] are composed of units whose input, output and state assume values within the interval $[0,1]$ (or $[-1,1]$). The input u_i to neuron i $(1 \leq i \leq n$, where n is the number of network units) is the weighted sum of the outputs of the units connected with neuron i: $u_i = \sum_j w_{ij} V_j$, where w_{ij} is the weight of the connection between neuron j and neuron i, V_j is the output of neuron j $(1 \leq j \leq n)$. The input u_i is converted into the output value V_i by a continuous monotone-increasing function g_i: $V_i = g_i(u_i) = g_i(\sum_j w_{ij} V_j)$. When u_i and V_i vary within the range $[0,1]$, a commonly used function is $1/2[1 + tanh(\beta\ u_i)]$.

The following set of coupled differential equations describes how the state variable u_i $(1 \leq i \leq n)$ of the neurons changes with time:

$$\frac{du_i}{dt} = -\frac{u_i}{\tau_i} + \sum_j w_{ij} V_j \tag{1}$$

where τ_i are suitable time constants.

As Hopfield [2] has shown, the evolution in time of a symmetric network of analog neurons is a motion in the state space that seeks out minima in the function:

$$E_a = -\frac{1}{2}\sum_i \sum_j w_{ij} V_i V_j + \int_0^{V_i} g_i^{-1}(V) dV \tag{2}$$

When β (the parameter responsible for the steepness of the sigmoid activation function g_i) is large, the second term in expression (2) is negligible, and the network minimizes locally the function :

$$E = -\frac{1}{2}\sum_{i \neq j} w_{ij} V_i V_j \tag{3}$$

This is the Lyapunov function for networks of binary threshold neurons with a symmetric connection matrix (of non-negative diagonal elements) and random sequential updating [1]. Takefuij [6] has proved that a network of analog neurons can always be made to converge in the local minima of expression (3), when its dynamical behavior is described by the following set of differential equations:

$$\frac{du_i}{dt} = -\frac{\partial E}{\partial V_i} \tag{4}$$

that is, the term u_i/τ_i in equation (1) has been eliminated.

4. Segmentation algorithms

In 1992 Huang [3] described the segmentation problem for grey-level images as one of minimizing a suitable energy function for Hopfield networks: Let $f(x,y)$ be an

NxM grey-level image to be segmented and S (estimated by histogram analysis) the maximum number of classes to be obtained by the segmentation process. Consider a Hopfield network made of $NxMxS$ analog neurons organized in S layers of NxM neurons each. Each neuron in a layer represents an image pixel and is denoted by the triple (x,y,i), where (x,y) $(1 \le x \le N, 1 \le y \le M)$ are the pixel coordinates and i $(1 \le i \le S)$ denotes one of the S classes. The output $V_i(x,y)$ $(V_i(x,y) \in [0,1])$ of the neuron (x,y,i) is interpreted as the probability that pixel (x,y) will be assigned to class i. The function E, from which the network architecture is derived, expresses the constraints of the segmentation problem and is the sum of three terms: $E = AE_1 + BE_2 + CE_3$, where A, B, and C are suitable, positive constants and E_1, E_2, E_3 have the following expressions:

$$E_1 = \sum_{x=1}^{N} \sum_{y=1}^{M} \sum_{i=1}^{S} \sum_{l=-1}^{1} \sum_{k=-1}^{1} (V_i(x,y) - V_i(x+l, y+k))^2 \qquad (5)$$

$$E_2 = -\sum_{x=1}^{N} \sum_{y=1}^{M} \sum_{i=1}^{S} \sum_{J=1, J \ne i}^{S} \sum_{l=-1}^{1} \sum_{k=-1}^{1} (V_i(x,y) - V_j(x+l, y+k))^2 \qquad (6)$$

$$E_3 = \sum_{x=1}^{N} \sum_{y=1}^{M} (\sum_{i=1}^{S} V_i(x,y) - 1)^2 \qquad (7)$$

Observe that the term E_1 favors the assignment of adjacent pixels to the same class; the term E_2 is minimum when a pixel and its neighbors are assigned to only one class, and, finally, the term E_3 is zero if, and only if, the sum of the outputs of the neurons corresponding to the same pixel is equal to one. In his algorithm Huang initialized all the neurons, corresponding to the pixels belonging to a given peak, at the same values and let the network evolve according to a simulated annealing algorithm.

We have devised two different algorithms, the first one, which resembles Huang's algorithm for grey-level images, builds three different networks (one for each color feature considered) and then combine the results. The second builds a unique network according to the number of clusters obtained by histogram analysis. In both of these algorithms we have changed both the network initialization and its dynamic evolution with respect to the original proposition.

ALGORITHM 1

Input: Color Image
FOR each color component DO:
- Apply histogram analysis (Section 2) to locate significant peaks;
- Build a network with a number of layers equal to the number of peaks;
- Initialize each network neuron (x,y,i) according to the following rule:

$$V_i(x, y) = \frac{1/d_i(x, y)}{\sum_j 1/d_j(x, y)} \tag{8}$$

where $d_i(x, y) = |f(x, y) - p_i|$ and p_i is the location of the i-th peak;

- Let the network evolve according to equation (4) until each neuron output changes less than a predefined threshold T.

END

Combine the segmentation results (union of edges).

Output: Segmented image

ALGORITHM 2

Input: Color Image

- Partition the color space into several hexahedra corresponding to prominent clusters in the color space;
- Calculate the centroids of the selected clusters;
- Build a network with a number of layers equal to the number of selected clusters;
- Calculate for each pixel *(x,y)* the Euclidean distance d_{c_i} between the pixel color and each cluster centroid C_i;
- Initialize each network neuron *(x,y,i)* according to the following rule:

$$V_i(x, y) = \frac{1/d_{c_i}(x, y)}{\sum_j 1/d_{c_j}(x, y)} \tag{9}$$

- Let the network evolve according to equation (4) until each neuron output changes less than a predefined threshold T.

Output: Segmented image

5. Evaluation of the algorithms

We have tested Algorithm 1 and Algorithm 2 on many images in several different domains. In both algorithms, parameters A, B, C, and T must be set. We have found experimentally that the choice of the parameters A and B is not critical (and have set both of them at 0.04). The parameter C has been increased by steps of 0.1 during network evolution from an initial value of 1 to a value equal to the number of layers. We have observed that varying C according to this rule makes the network converge more quickly to solutions satisfying equation (7). The threshold T for the termination condition has been set at 0.0000001. In order to evaluate the performance of the algorithms more objectively we have adopted the evaluation function recently proposed by Yang and Liu [5]. This function is defined as

$$F(I) = \frac{1}{1000(N \times M)} \sqrt{R} \sum_{i=1}^{R} \frac{e_i^2}{\sqrt{A_i}} \tag{10}$$

where I is the image to be segmented, NxM, the image size, and R, the number of regions of the segmented image, while A_i and e_i are the area and the average color error of the i-th region respectively.

As an example, the segmentation of a 256x256 24-bit pixels test image is reported here. The image in figure 1a represents a aircraft photograph digitalized by a flat bed scanner. Figure 1b shows the segmentation obtained by the histogram thresholding of each RGB color feature independently, without taking into account spatial correlations among pixels. Figures 1c shows the segmentation results obtained using the same color features and smoothing parameters as above, and applying to the network the Huang's initialization and the deterministic evolution described in section 3 (instead of simulated annealing). The results obtained using Algoritm 1 and Algorithm 2 are shown in figures 1d and 1e. The experimental results are summarized in Table I for a quantitative comparison of the performance of the different segmentation algorithms. The performance of the devised algorithms are comparable for all the images analyzed, however the computational cost of Algorithm 2 is lower.

Segmented image	number of regions	average area	Average color error	Evaluation measure (f)
1b	30711	17.65	6.14	181.70
1c	269	243.62	10.38	225.94
1d	414	158.29	8.74	121.75
1e	246	266.40	8.83	109.47

Table I: Summary of experimental results.

6. References

1. J.J. Hopfield "Neural networks and physical systems with emergent collective computational abilities. Proc. Natl. Acad. Sci. USA 79: 2554-2558 (1982).

2. J.J. Hopfield "Neurons with graded response have collective computational properties like those of two-states neurons". Proc. Natl. Acad. Sci. USA 81: 3088-3092 (1984).

3. C-L. Huang "Parallel image segmentation using modified Hopfield model" *Pattern Recognition Letters* 13: 345-353 (1992).

4. A. Rosenfeld, A.C. Kak "Digital picture processing" Academic Press Inc, Orlando, Florida, 1982.

5. Y-H Yang, J. Liu "Multiresolution image segmentation" *IEEE Trans. on Pattern Analysis and Machine Intelligence* 16: 689-700 (1994).

6. Y. Takefuji "Neural network parallel computing", Kluwer Academic Publishers, 1992.

7. A.P. Witkin "Scale-Space Filtering: A New Approach to Multi-scale Description" *Proc. IEEE Int. Conf. on Acoustics, Speech, and Signal Processing,* 3: 39A.1.1-39A.1.4 (1984).

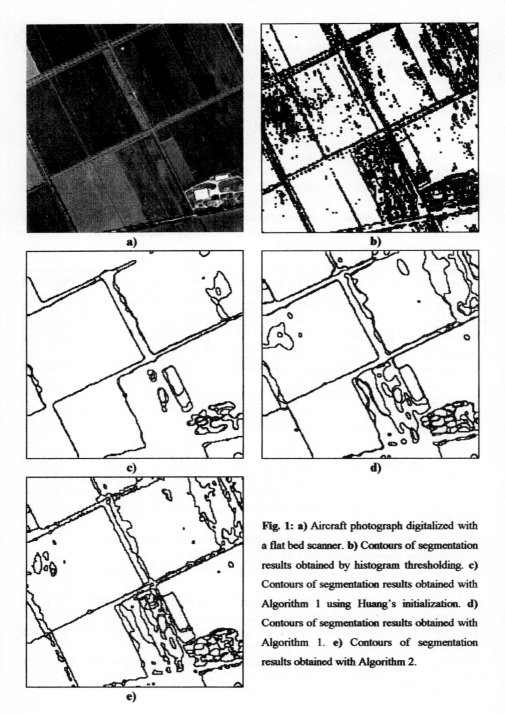

Fig. 1: a) Aircraft photograph digitalized with a flat bed scanner. b) Contours of segmentation results obtained by histogram thresholding. c) Contours of segmentation results obtained with Algorithm 1 using Huang's initialization. d) Contours of segmentation results obtained with Algorithm 1. e) Contours of segmentation results obtained with Algorithm 2.

The Watershed Transformation for Multiresolution Image Segmentation

S. Wegner , T. Harms, J. H. Builtjes, H. Oswald, E. Fleck

Deutsches Herzzentrum Berlin, Augustenburger Platz 1,
D-13353 Berlin, Germany

Abstract. For a hyperthermia planning environment the main requirements for the following numerical computation and visualization of the temperature field are the separation of several anatomical objects. Since different contrast objects cannot be seen at the same resolution, a segmentation technique using multiresolution segmentation planes is described here. The computation of these planes is based on the technique of region growing using watersheds in digital images. First, the watershed transformation is applied on the gradient image. The resulting image is then transformed into a graph on which a region growing process is performed. The iteration of these two actions results in hierarchical segmentation planes which differ in region amount and region size. Consequently low contrast objects can be separated in the lower planes whereas high contrast objects can be extracted in the higher planes.

1 Introduction

The main aspect of hyperthermia planning in cancer therapy is the patient-specific computation of the physical relevant temperature field parameters. Moreover, these computations are the basis for the optimization of the temperature treatment determining the optimal values for the variable technical parameters. This planning requires the following steps: acquisition of the Computer Tomography images, segmentation of these images, computation and visualization of the temperature field and finally the optimization of this field. Obviously, one of the main requirements for the succeeding steps is the separation of several anatomical objects, for instance tumor, skin, bones, different organs, fat, muscle. In particular the extraction of low contrast or even diffuse objects is required. Since these objects are best seen at different resolutions, multiresolution segmentation planes are computed. In the lower planes low contrast objects can be separated, whereas in the higher planes higher contrast objects can be extracted.

The construction of the multiresolution segmentation planes is based on the concept of region growing using watersheds in digital images. A preprocessing segmentation is obtained using the watershed transformation on the gradient image. The resulting image is then transformed into a graph on which the region growing is performed. This can be seen as the watershed transformation on graphs. The gradient of the graph is computed and regions with similar graylevels are merged to one region. The iteration of this procedure leads to the multiresolution segmentation planes.

These planes form the basis of an interactive segmentation tool for the hyperthermia planning environment. The user can select the high and low contrast objects in different planes for a final image segmentation. If an object is separated into several regions the user has the possibility of merging these regions. For most images a sufficient segmentation result is achieved when applying these two actions. However, for some diffuse objects interactively obtained contours are necessary.

2 Watershed Transformation

Watersheds were introduced in the field of mathematical morphology where grayscale pictures are often considered as topographic reliefs. In the topographic representation of an image, the numerical value of each pixel represents the elevation at this point. A drop of water falling on the topographic relief flows down until it reaches a regional minimum. A region or catchment basin associated with a minimum is given by the amount of points from which a drop of water flows down until this minimum. A watershed is defined as a side from where a drop of water can flow down to two minimas. The optimal contours of the image are these watersheds.

Employing watersheds to image segmentation the gradient of the image has to be interpreted as a relief. The original graylevel image is inappropriate because the amount of the graylevel elevation does not allow the unmistakable classification of valleys to regions and elevations to edges. The gradient of an image provides the absolute amount of inhomogeneities in an image. Homogeneous regions of the image are characterized by low contrast and hence a low gradient, whereas the high values of the gradient image are created by the high contrast contours of the original image.

The algorithmic computation of the watershed approach can be interpreted as an immersion process [Beu 90] [Vin 91]. Drilling holes in each regional minimum of the relief, the surface is immersed into a lake with uniform vertical speed. It is supposed that the immersion speed is slow enough to ensure a constant water level in all the basins. Starting from the minima of the lowest altitude, the water will fill up the different catchment basins. To avoid the confluence of the flood, dams are built along the lines where the water of two different catchment basins would merge. At the end of this immersion procedure each minimum is completely surrounded by dams, which delimit its associated catchment basins. These dams correspond to the watersheds of the image and hence the contours of the image.

3 Multiresolution Image Planes

The described technique results in regions with good located closed contours, but it leads to a strong oversegmentation, i.e., the interesting contours can not be seen clearly because of many irrelevant contours. A solution is the application of the watershed transformation on graphs. Instead of using the numerical value of each pixel for the topographic representation here presegmented regions are considered. This region merging can be interpreted as a region growing process on graphs where the regions are represented as nodes and the relations between them as arcs:

Let G = (V, E) be a decimal non-oriented, non-reflexive graph, where V is the set of nodes and E is the set of arcs.

For a region graph the nodes describe regions and the arcs represent the neighborhood relationship between these regions (Figure 1).

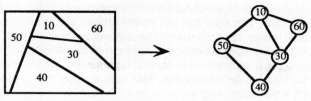

Fig. 1. Mosaic image and region graph

The regions are obtained using the watershed transformation on the gradient image. Each region is then characterized by only one graylevel achieving a so called mosaic image. The selected graylevels should be similar to the graylevels of the original image. Meyer and Beucher [Beu 90] chose the graylevel of the accompanying gradient minimum since this should represent the homogenest part of the region. We recognized that this is not always the best characterization of a region according to the segmentation result. Hence, in our environment the user has the possibility to select an appropriate criterion: homogenest part, minimum, maximum, average or median of the region.

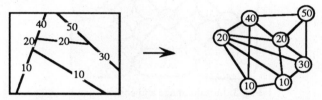

Fig. 2. Mosaic image - Figure 1 - with edge values and contour graph

Representing the regions as nodes in a graph leads to the problem that the watershed transformation cannot be performed directly on the graph. The reason is that this graph includes no information about the graylevel differences of neighbor regions. A solution could be an approximation, for instance the average, of all graylevel neighbor differences [Zeh 93]. However, since this is only an approximation of the graylevel differences, the finally obtained contours are often inexact. Zehetbauer [Zeh 93] suggested a so called contour graph as a model for the gradient. Every arc in the region graph - edge in the mosaic image - is represented as a node in the contour graph. The value of the node is the absolute difference of the two neighbor regions. Then all nodes belonging to one region are linked by arcs (Figure 2).

Considering this model some additional difficulties appear because in this representation a node is constructed of two regions. For instance, if the upper and lower region contours have been labeled as watersheds and the left and right as x or y, then the region could be labeled x or y (Figure 3). A solution is obtained by a combination of the region graph and the contour graph. On the basis of the labeled contour graph the nodes of the region graph are labeled with the most often contour label. If there are equally occurring labels the lowest label is selected.

The multiresolution segmentation planes are obtained by computing iteratively the mosaic image and applying the watershed transformation on the contour graph of this mosaic image. This process is continued until only one region remains. This action results in mosaic images of order n representing the n-th segmentation plane.

These planes are the basis of an interactive segmentation tool for the hyperthermia planning environment. The user has the possibility to select objects in different planes for a final image segmentation. The advantage here is that the different contrast objects that have been separated in different planes are unified in one image. Moreover, if an object is separated into several regions the user has the possibility of merging these regions for the final image. This is again possible in different planes. For most images a sufficient segmentation result is achieved when applying these two actions. However, diffuse objects are nearly impossible to separate even at different resolution planes. Hence, the user has additionally the possibility to obtain contours interactively.

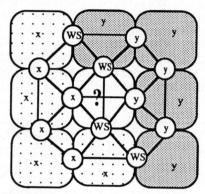

Fig. 3. Mosaic image and contour graph

4 Results and Conclusion

The upper image of Figure 4 shows a typical CT image slice acquired for a patient-specific hyperthermia planning, whereas the lower image shows the segmentation result of this slice. The different anatomical object are merged together from different resolution planes without interactively obtained contours. Only those objects are selected that are relevant for the following numerical computation of the temperature field parameters. The segmentation result is sufficient for most anatomical objects of the image but some contour corrections have to be made for the very low contrast objects. Despite these small corrections a user of this segmentation environment has the advantage to obtain most of the relevant contours automatically.

In order to prove the generality of this segmentation approach Figure 5 shows an example from a different application domain. The upper image presents the original X-ray ventriculogram whereas the lower image shows the obtained segmentation result. Current efforts focus on the model based selection of the high and low contrast objects in different planes. Furthermore, the approach should be extended to a 3D segmentation tool.

References

[Beu 90] S. Beucher, F. Meyer: Morphological Segmentation, Journal of Visual Communication and Image Representation, Vol. 1, No. 1, pp 21-46, Sept. 1990

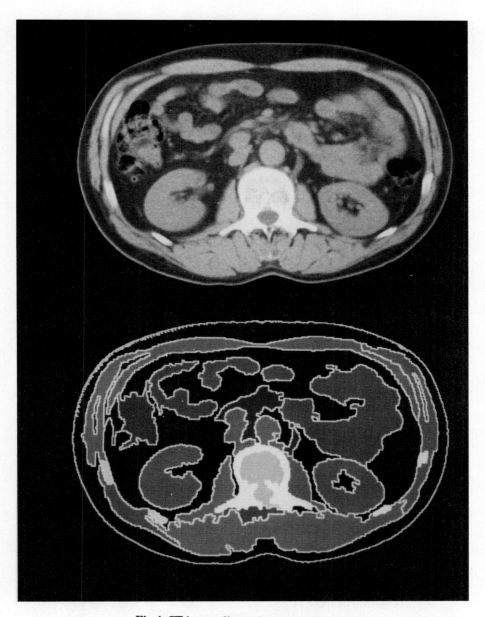

Fig 4. CT image slice and segmentation result

[Vin 89]   L. Vincent: Graphs and Mathematical Morphology, Signal Processing 16, pp 365-388, 1989

[Vin 91]   L. Vincent, P. Soille: Watersheds in Digital Spaces: An Effcent Algorithm Based on Immersion Simulation, IEEE Transactions on Pattern Analysis and Machine Intelligence, Vol.13, No.6, pp 583-598, Jun. 1991

[Zeh 93]   S. Zehetbauer, U. Meyer-Gruhl: Segmentierung und Analyse drei- und vierdimensionaler Ultraschalldatensätze, S. J. Pöppl, H. Handels, Mustererkennung 1993, pp 118-125

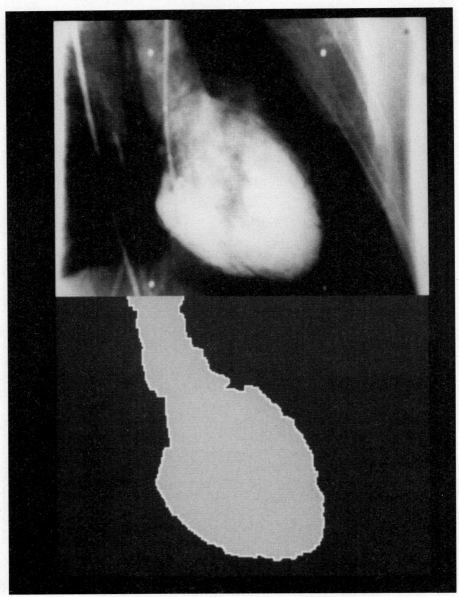

Fig. 5. X-ray ventriculogram and segmentation result

Bounds on the Optimal Elasticity Parameters for a Snake

Ole V. Larsen[1], Petia Radeva[2] and Enric Martí[2]

[1] Laboratory of Image Analysis, Aalborg University, Denmark
[2] Computer Vision Center, Autonomous University of Barcelona, Spain

Abstract. This paper develops a formalism by which an estimate for the upper and lower bounds for the elasticity parameters for a snake can be obtained. Objects different in size and shape give rise to different bounds. The bounds can be obtained based on an analysis of the shape of the object of interest. Experiments on synthetic images show a good correlation between the estimated behaviour of the snake and the one actually observed. Experiments on real X-ray images show that the parameters for optimal segmentation lie within the estimated bounds.

Keywords: snakes, elasticity parameters, segmentation.

1 Introduction

Snakes have several advantages compared to traditional segmentation techniques especially in case of noisy images and partly occluded objects [1, 2]. Snakes provide a model-based technique since ways exist to incorporate information about the object of interest into the snake formalism. An initial snake can be defined specifying the snake in accordance with the object of interest. Besides, it is possible by controlling the internal energy to constrain the smoothness of the snake. The smoothness required for an accurate segmentation is controlled by the elasticity parameters and is closely related to the shape of the object.

Despite the great interest in snakes, estimation of the elasticity parameters has not been given much attention previously. Samadani [5] dynamically estimates and adjusts the parameters to avoid instability in the deformation process. In contrast with his work our purpose is, based on knowledge about the size and shape of the object, to establish bounds on the parameters of elasticity that provide optimal smoothness of the snake.

Here we present a new formalism by which the upper and lower bounds on the elasticity parameters can be obtained through an analysis of a given sample of the object of interest, hereafter the object model. First, the theoretical foundation for the existence of bounds on the parameters is developed independently of the chosen snake implementation. Based on an implementation using the Finite Difference Method (FDM) formulas are then derived for calculating the bounds based on the shape of the object. At last, experimental results are presented to determine the correlation between the estimated behaviour of the snake and the one actually observed.

2 Parameters of elasticity

A snake is an elastic curve $u(s) = (x(s), y(s))$ for which an energy function E_{snake} is defined based on an internal energy E_{int} and an external energy E_{ext},

$$E_{snake} = \int_0^1 E_{int}(u(s)) + E_{ext}(u(s)) \; ds.$$

The internal energy is given by the sum of the membrane energy and the thin-plate energy: $E_{int}(u) = \alpha E_{membrane} + \beta E_{thin-plate} = \alpha u'(s) + \beta u''(s)$. The parameters α and β are the parameters of elasticity. α controls the stretching and β the bending of the snake curve.

The external energy is obtained from a potential field derived as the image gradient [2] or as a distance map of the edge points [1]. Without loss of generality we use the latter definition leading to the following expression, $E_{ext}(u(s)) = P(x, y) = d(x, y)$, where $d(x, y)$ is the distance between pixel (x, y) and the closest edge point. Minimizing the external energy the snake is attracted towards the edge points of the image. Minimizing the internal energy the shape of the snake is smoothed. The result of the segmentation is obtained when the snake detects a minimum of the total energy.

High values of the parameters of elasticity put great weight on the internal energies and consequently give a smooth curve. In general, we want the parameters to be as high as possible. It will decrease the risk of the snake being attracted to spurious noisy edges during its iterations [1]. However, if the parameters become too high, problems in terms of contour surpassing occur.

Let us consider a snake, u_c, of length n placed in the potential valley corresponding to the contour of the object we are looking for. And let ϵ be some predetermined constant. The following definitions can be stated,

Definition 1 *The minimum area around the contour valley containing all possible deformed versions, u_t, of snake u_c in a distance less than ϵ to u_c ($\|u_t - u_c\| < \epsilon n$) is called the ϵ-neighbourhood of the contour valley.*

Definition 2 **Local surpassing** *is when the snake u_c in the process of energy minimization moves locally out of the contour valley.* **Global surpassing** *is when the snake u_c deforms to a snake outside an ϵ-neighbourhood of the contour valley given a constant ϵ.*

An illustration of local and global surpassing is given in figure 1.

Definition 3 *The set of maximum values of the elasticity parameters for a snake that does not give rise to any local surpassing is called* **local (surpassing) parameters.** *The set of maximum values that, given some ϵ, does not give rise to any global surpassing is called* **global (surpassing) parameters.**

The fact that the local parameters are the maximum parameters that do not give rise to any local surpassing from the contour means that the deforming

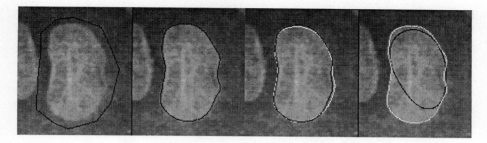

Fig. 1. Different situations for segmentation of a bone structure in a X-ray image; from left to the right: an initial snake, the snake shrunk to the contour, a local surpassing of the contour shown in white by the snake shown in black and a global surpassing by the snake.

snake by these parameters of elasticity will not get shapes more stretched neither bended than the represented in the object model. Due to the monotony property of the parameters of elasticity, stated in Lemma 1 and 2, each pair of parameters greater than the local parameters will not allow discontinuities in the snake not proper to the object model. Hence, the local parameters can be thought of as lower bounds on the parameters giving an optimal segmentation. The global parameters are the maximum values for which the displacement of the snake does not exceed some ϵ-neighbourhood of the contour. The ϵ-neighbourhood is to be determined by the user as the area around the contour where the snake does not lose its influence. The global parameters therefore determine the upper bounds on the parameters of elasticity.

Smoothing its shape the snake has a tendency to shrink itself. Based on the properties of the global parameters of elasticity it can be shown that the initial snake placed around the object is able to fall into an ϵ-neighbourhood of the object contour. This paper therefore proposes an approach to determine the bounds by an analysis of the internal and external energies when the snake, from a position on the contour, deforms to surpass the contour.

2.1 Fundamentals

Before being able to determine formulas for calculating the bounds it is necessary to see some fundamental relations concerning certain properties of the parameters of elasticity. The fundamentals are stated without proofs, for more details see [3]. It should be emphasized that relations described in this section do not depend on any specific snake implementation, neither on any specific calculation of the potential field.

Let us consider the initial snake u_0, the snake u_c shrunk to the contour valley and the snake u_t converged from the snake u_c. Let $E^c_{membrane}$, $E^c_{thin-plate}$, P^c, $E^t_{membrane}$, $E^t_{thin-plate}$ and P^t be the membrane energy, the thin-plate energy and the potential energy of the snakes u_c and u_t, respectively.

Lemma 1. *Let the parameters of elasticity (α', β') be such that the snake leaves the ϵ-neighbourhood of the contour valley given some ϵ. Then for any greater parameters of elasticity $\alpha \geq \alpha'$ and $\beta \geq \beta'$ the snake leaves the ϵ-neighbourhood.*

Lemma 2. *Let the parameters of elasticity (α', β') be such that the snake remains in the ϵ-neighbourhood of the contour valley given some ϵ. Then for any smaller parameters of elasticity $\alpha \leq \alpha'$ and $\beta \leq \beta'$ the snake remains in the ϵ-neighbourhood.*

These lemmas define a kind of monotony of the parameters. They show that for a given parameter setting which is found to give local or global surpassing, any parameter value greater than the ones found will also cause surpassing.

Looking at differences in E_{int} and E_{ext} between u_c on the contour and u_t outside an ϵ-neighbourhood we get the following theorem,

Theorem 3. *The maximum pair of parameters of elasticity that retains the snake u_c in an ϵ-neighbourhood of the contour is given by the formulas,*

$$\alpha = \frac{P^t - P^c}{2(E^c_{membrane} - E^t_{membrane})}, \quad \beta = \frac{P^t - P^c}{2(E^c_{thin-plate} - E^t_{thin-plate})} \tag{1}$$

where the snake u_t is obtained from the snake u_c and $\|u_t - u_c\| = \epsilon n$.

2.2 Calculating the bounds of the optimal parameters of elasticity

Given a specific shape model, the task now becomes one of determining the parameters not giving rise to local and global surpassing with respect to some constant ϵ. They can be obtained from formulas (1) substituting the estimations of the energies of the snake u_c placed in the contour valley, and its deformation u_t in a distance ϵ from the contour. Without loss of generality we compute the parameters for the snake implementation by FDM and use the fact that the potential is generated as a distance map of the edge points of the original image. In order to simplify our calculus we ensure that the snake pixels are evenly distributed on the curve in each iteration of the snake movement.

The energies of u_c can be estimated by the initial snake u_0 and the object model [3]. In order to calculate the global parameters we have to estimate the deformation of the snake u_c in a distance ϵ. In [3] we have shown that the deformation of a snake in a given distance can be approximated by the deformation of a circle with the same length. Substituting in (1) the energies of the circle and its deformation after global surpassing in distance ϵ, we get,

$$\alpha_{gl} = \frac{kn^2}{8\pi(n - \epsilon k\pi)}, \quad \beta_{gl} = \frac{kn^4}{32\pi^3(n - \epsilon k\pi)} \tag{2}$$

where $k = \frac{u_0}{u_c}$ is the accumulation rate of the initial snake in the contour valley.

For the case of the local parameters we use the fact that local surpassing firstly occurs in the most curved segment of the snake. From formulas (1) we get,

$$\alpha_{loc} = \frac{\theta\lceil\frac{m}{2}\rceil k^2}{2m(1 - 2k^2(r+\epsilon)^2(1 - \cos\frac{\mu}{m}))}, \quad \beta_{loc} = \frac{\theta\lceil\frac{m}{2}\rceil r^2}{2m(1 - 4r^2(r+\epsilon)^2(1 - \cos\frac{\mu}{m})^2)}$$

$$\theta = \sqrt{r^2 + (r+\epsilon)^2 - 2r(r+\epsilon)\cos\frac{\nu-\mu}{2}} - \epsilon, \quad \mu = \arccos\left(\frac{\epsilon(2r+\epsilon) + r^2\cos(m\arccos(1-\frac{1}{2r^2}))}{(r+\epsilon)^2}\right)$$

where m, ν and r are the length, the angle, and the radius of the most curved segment of the snake u_c.

3   Results and discussion

Two lines of experiments have been conducted. One based on synthetic images and one based on real X-ray images of bone structures.

The test on synthetic images investigated to what degree the calculated local and global surpassing parameters gave the expected surpassings if a snake was initialised with the values. Six different types of objects have been used: circle, octagon, hexagon, square, rectangle and triangle, each represented with scaled versions having length $100, 200, ..., 500$ pixels. For each object the values of α and β from formulas (1) were computed given $\epsilon = 1.5$. The experiments showed that the precision by which the estimated behaviour fitted the observed behaviour did not depend on the length of the snake. Furthermore, it showed that for 5 out of 6 objects the converged snake had a distance from the contour in the interval $[1.5; 2.0]$, with ϵ being 1.5. Only for the triangle the distance was as high as 3.5. From the experiments it can be concluded that the surpassings estimated based on parameters calculated from the object model are similar to the surpassings observed when a snake is used to converge around the object, with a slight tendency to underestimate the actual surpassings. Objects with sharp corners tend to have less similar surpassings. By letting snakes converge with parameters from the interval $[0.5; 2.0]$ times the estimated (α, β), it was observed that the distance between the contour and the converged snake increased as the values increased. This supports the monotony estimated in Lemma 1 and 2.

The test on real images comprised 15 images of hand radiographs. For each bone we used as an initial snake a model with similar structure but larger in size. Using local surpassing parameters it was observed that depending on the difference between the initial snake and the object contour in some cases the snake moved correctly to the contour (Fig. 3a (left)). In other cases the snake remained outside the contour due to attraction by near edge points (Fig. 3b(left)).

In accordance to our observations the global parameters assure us that the snake will surpass the contour going out in some distance with predetermined magnitude (Fig. 3a (right) and Fig. 3b (right)). In this way any snake detention in a local energy minimum before arriving to the contour will be avoided. Due to the monotony property of the elasticity parameters we can state that the parameters of the optimal segmentation belong to the intervals determined by the local and global parameters.

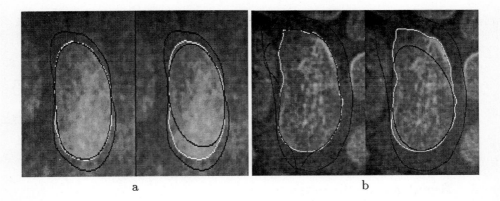

a   b

Fig. 2. Two different situations of using local (left) and global (right) parameters. The white curve is the contour of the objects. The outer curve represents the initial snake and the inner curve the resulting converged snake.

4   Conclusion

Based on the concept of local and global surpassing we have defined some local and global surpassing parameters which serve as lower and upper bounds on the elasticity parameters of a snake in terms of segmentation of a given object. Formulas are presented by which an estimate of the parameters bounds can be calculated for a given object model. Experiments show a good correlation between the estimate and results obtained on synthetic images. Experiments on real X-ray images show that optimal segmentation is only possible by parameters within the bounds defined in this paper. By establishing bounds on the parameters the search space of the values giving the best segmentation is limited. In [4] we propose guidelines for non-snake experts to determine the optimal parameters of elasticity based on an automatic inspection of the object models.

References

1. I. Cohen, L. D. Cohen, and N. Ayache. Using deformable surfaces to segment 3-d images and infer differential structures. *CVGIP: Image Understanding*, 56(2):242–263, 1992.
2. M. Kass, A. Witkin, and D. Terzopolous. Snakes: Active contour models. In *International Conference on Computer Vision, London*, pages 259–268, 1987.
3. O.V. Larsen and P. Radeva. Calculating the bounds on the optimal parameters of elasticity for a snake. Technical report, Laboratory of Image Analysis, Aalborg University, Denmark, December 1994.
4. O.V. Larsen, P. Radeva, and E. Martí. Guidelines for choosing optimal parameters of elasticity for snakes. In *Proceedings from CAIP'95 - International Conference on Computer Analysis and Image Processing (Accepted)*, 1995.
5. R. Samadani. Adaptive snakes: Control of damping and material parameters. In *SPIE Geometric Methods in Computer Vision*, volume 1570, 1991.

Matching

Efficient Attributed Graph Matching and its Application to Image Analysis

H. Bunke and B.T. Messmer

Institut für Informatik und angewandte Mathematik, University of Bern,
Neubrückstr. 10, CH-3012 Bern, Switzerland, bunke@iam.unibe.ch

Abstract. Graphs are a very powerful data structure for many tasks in
image analysis. If both known models and unknown objects are represented by graphs, the detection or recognition problem becomes a problem of graph matching. In this paper, we first review different methods
for graph matching. Then we introduce a new family of exact and error-
tolerant graph matching algorithms that have a number of interesting
properties. The algorithms are particularly efficient if there is a large
number of model graphs to be matched with an unknown input graph.
Moreover, they allow the incremental updating of a database of known
models. This property supports the application of graph matching in a
machine learning context. As an example, we show a 2-D shape recognition system based on graph matching that is able to learn new shapes.

1 Introduction

Graph structures are a powerful and universal tool with applications in various
subfields of science and engineering. In pattern recognition and image analysis,
graphs are often used for the representation of structured objects. For example,
if the problem is to recognize instances of known objects in an image, then often
models, or prototypes, of the known objects are represented by means of graphs
and stored in a database. The unkown objects in the input image are extracted by
means of suitable preprocessing and segmentation algorithms, and represented
by graphs that are analogous to the model graphs. Thus the problem of object
recognition is transformed into a graph matching problem.

Generally, the term *graph matching* refers to the process of comparing two
(or more) graphs with each other. There are several classes of graph matching
problems. In the *graph isomorphism* problem, we want to determine if two given
graphs are isomorphic to each other. An isomorphism is a bijective mapping
between the nodes of the two graphs such that the structure of the edges is
preserved. Informally speaking, two graphs are isomorphic to each other if they
are structurally identical. In the *subgraph isomorphism* problem, we are given
two graphs g_1 and g_2, and want to find out if g_2 contains a subgraph that is
isomorphic to g_1. More generally, a *bidirectional subgraph isomorphism* between
g_1 and g_2 means the existence of subgraphs g_1' and g_2' of g_1 and g_2, respectively,
such that g_1' and g_2' are isomorphic to each other. Finally, in *error-tolerant graph
matching*, we want to establish a graph, subgraph, or bidirectional subgraph

isomorphism that may include some distortions. The admissable distortions are often problem dependent. A general distortion model may include, for example, the deletion, insertion, and substitution of both nodes and edges. These distortions are also called *edit operations*. In order to model the fact that certain distortions are more frequent than others, one can assign a cost to each individual edit operation. Error-tolerant graph matching can be used to calculate a measure of similarity, or dissimilarity, for a given pair of graphs. This measure of similarity is based on the sequence of edit operations that has the minimum cost among all possible sequences that transform one of the given graphs into the other. Thus, approximate graph matching is a generalization of string edit distance computation [25]. For a more comprehensive introduction to graph matching see [1, 21].

It is still an open question whether the graph isomorphism problem is in the complexity class P or NP [7]. In this paper, we will consider only subgraph, bidirectional subgraph and error-tolerant graph matching as these problems are more important with respect to applications in image analysis. All these problems are known to be in NP. This means that all available methods have an exponential time complexity. Consequently, graph matching algorithms that are guaranteed to yield the correct solution are applicable only if the underlying graphs are relatively small. Graph matching for large graphs becomes computationally intractable. The only choice to deal with large graphs are approximate algorithms. These have usually a computational complexity that is lower than exponential, but they are no longer guaranteed to find the correct solution for a given problem.

The best known algorithm for subgraph isomorphism detection is that of Ullman [24]. Is is based on tree search with backtracking. In order to speed up the search, a particular lookahead technique is used, which allows to detect and prune dead ends in the search tree early. Another well-known method for subgraph and bidirectional subgraph isomorphism detection is based on maximal clique detection in a compatibility graph [10]. The algorithms that have been proposed for error-tolerant matching are based on tree search, similar to Ullman's algorithm [2, 5, 19, 23]. As the search space in error-tolerant matching is even larger than in regular subgraph or bidirectional subgraph isomorphism detection, the use of good heuristics together with A^*-like search techniqus [18] becomes indispensable.

Ullman's method, the technique based on maximal clique detection, and tree search based error-tolerant graph matching are optimal algorithms in the sense that they are guaranteed to yield the correct solution to a given problem. In the area of approximate algorithms, on the other hand, methods like simulated annealing [9], neural networks [6], genetic algorithms [11], continous optimization[13] and probabilistic relaxation [12] have been proposed. The idea common to all these methods is to iteratively minimize an objective function that represents the distance of the current solution to the correct solution. The most serious problem of these approaches is that the minimization procedure may either not converge or get trapped in a local minimum.

Numerous image analysis applications of graph matching have been described in the literature. These include character classification [15], schematic diagram and 2-D shape analysis [14], 3-D object recognition [8, 26], stereo vision [10], dynamic scene analysis [3] and muscle tissue classification [20].

2 Efficient subgraph and error-tolerant subgraph matching

In this section, we introduce a new family of optimal algorithms for subgraph and error-tolerant subgraph isomorphism detection. These algorithms have been developed particularly for the case where an input graph g representing some unknown object in an image is to be matched against a database of prototype graphs $p_1, \ldots, p_M$ in order to find each p_i that is a subgraph of g, or – in the case of error-tolerant matching – to find the p_i that is most similar to g. All graphs under consideration may have directed edges and any number of symbolic labels or numeric attributes attached to their nodes and edges. Given g and $p_1, \ldots, p_M$, any of the known algorithms, for example that by Ullman or the method based on maximal clique detection, would sequentially match each p_i against g. In many applications, however, the prototypes p_i will not be completely dissimilar. Instead, there will be graphs s_j that occur as subgraphs simultaneously in several of the p_i's. These s_j's will be matched multiple times against g by any of the known algorithms. This clearly leads to some redundancy.

In the new approach to graph matching described in this paper, the prototype graphs $p_1, \ldots, p_M$ are preprocessed, generating a symbolic data structure, the so-called *network* of prototypes. This network is a compact representation of the prototypes $p_1, \ldots, p_M$ in the sense that a graph s that occurs as a subgraph multiple times within the same or different prototypes is represented only once in the network. Consequently, such a graph will be matched once and only once with the input graph g. Thus the computational effort will be reduced. For the case of error-tolerant subgraph isomorphism detection, the new algorithm can be combined with a very efficient lookahead procedure. For a detailed description of the new method, see [4, 16, 17]. In this paper, we will only briefly sketch the main ideas of the matching algorithm, give an example, and show some results.

The new algorithm follows the divide-and-conquer paradigm. That is, if we want to check if there exists a subgraph isomorphism from one graph G to another graph g, we divide G into two disjoint parts, G_1 and G_2, and check if there exist two subgraph isomorphisms, one from G_1 to g, and another from G_2 to g. If there are two such subgraph isomorphisms and, additionally, the structure of the edges between G_1 and G_2 is preserved in g, we can conclude that there is a subgraph isomorphism from G to g. This observation can be utilized by successively dividing all prototypes p_i - they correspond to G in the description above - into smaller subgraphs until we eventually reach the level of single nodes. The subgraphs resulting from such a recursive division of the prototypes can be arranged in a network, where identical subgraphs resulting from different prototypes are represented only once.

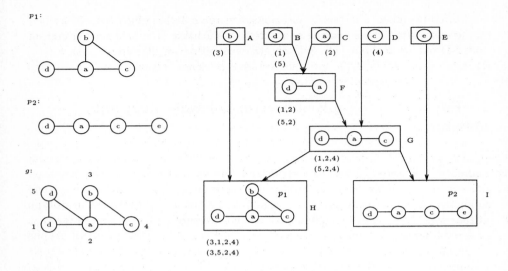

Fig. 1. Network for the graph models p_1 and p_2 and the instances found by the network units after the graph g was processed.

Two prototype graphs, p_1 and p_2, and the network that represents the recursive subdivision of p_1 and p_2 into smaller graphs are shown in Fig. 1. (Note that the network is shown "upside down" with the smallest units, i.e. network nodes representing single nodes of the p_i's, on top and full prototypes at the bottom.) A network like the one shown in Fig. 1 is not uniquely defined, in general. However, this property doesn't influence the matching performance. The generation of a network like the one given in Fig. 1 from a set of prototypes is described in more detail in [16].

A network like the one shown in Fig. 1 can be used for subgraph isomorphism detection in a straightforward manner. Each unit in the network representing some subgraph s that occurs once or multiple times in one or several prototypes has a procedure attached to it that finds all instances of s in the input graph g. Suppose that s has been split into two disjoint parts s_1 and s_2 in the recursive prototype subdivision procedure. Then in order to find all instances of s, we consider all pairs of instances of s_1 and s_2. Every such disjoint pair for which the structure of the edges between s_1 and s_2 is preserved in g is an instance of s. Therefore, in order to find all instances of the prototype graphs in the input, we start with the individual nodes at the top of the network and determine all their instances in g. Then we descend into the network and successively determine all instances of the units at the lower network levels. The procedure terminates once we have reached units at the bottom of the network, which represent the prototypes.

To illustrate the new graph matching procedure, let's consider an example.

Assume we want to determine all occurrences of p_1 and p_2 in the input graph g shown in Fig. 1. Note that the small letters denote node labels, while the number $1, 2, \ldots, 5$ in g are used to uniquely identify the nodes in g. Furthermore, capital letters $A, B, \ldots, I$ are used as identifiers for the network units. Units A through E check for instances of single nodes in the input graph, H and I check for the prototypes p_1 and p_2, respectively, and units F and G check for proper subgraphs of p_1 and p_2 consisting of more than one node. In Fig. 1 the instances that were found by the network units in the input graph g are printed in round brackets below each unit. For example, the nodes 1 and 5 are instances of network unit B in g, the pairs $(1, 2)$ and $(5, 2)$ are instances of the subgraph represented by network unit F, a.s.o. Finally, two instances of p_1 and no instance of p_2 are found. It can be easily verified that this is the correct result.

It is possible to generalize the algorithm to error-tolerant subgraph isomorphism detection. The static network, an example of which is shown in Fig. 1, as well as the algorithm that compiles a network from a set of prototype graphs remain the same. But the dynamic procedure that finds instances of network units in the input graph has to be extended. In the error-tolerant version, each network unit also accepts distorted subgraphs in the input graph g. Together with each distorted instance of a network unit, the corresponding edit costs are stored. The network nodes are no longer activated strictly from top to bottom, i.e. from smaller to larger units, but in a best-first manner where the subgraph with the smallest edit cost is considered first. It is possible to integrate a lookahead procedure that takes into regard a lower bound estimate of the future cost when selecting the unit with the smallest edit cost. This lookahead procedure can be implemented such that almost no additional computational overhead arises. Thus, we get a very efficient error-tolerant graph matching procedure. For more details of this method see [16, 17].

In a theoretical computational complexity analysis, the best and the worst case behavior of the new family of algorithms were studied and compared to Ullman's method for the case of subgraph isomorphism detection, and A^*-like search techniques for the error-tolerant case [18]. The main result is that for one prototype or completely disjoint prototypes – these are the most unfavorable scenarios for the network based algorithms – the worst case time complexity of both approaches is the same, while in the best case the network-based method is better by a factor of P, where P is the number of nodes in one prototype. Moreover, for the case of $M > 1$ prototypes $p_1, \ldots, p_M$ it can be shown that the complexity of the new method is only sublinearly dependent on M if there are common parts that are shared by different p_i's. In the limit with all p_i's being identical, the complexity is no longer dependent on M. By contrast, the complexity of any traditional, non-network-based matching procedure is always linear in M, no matter how similar or dissimilar the individual p_i's are.

The results of the theoretical complexity analysis could be confirmed in a series of practical experiments where network-based subgraph isomorphism detection was compared to Ullman's algorithm. All graphs in these experiments were randomly generated.

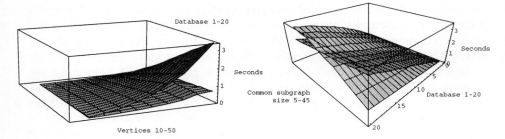

Fig. 2. Computation time for increasing the number of vertices and the number of prototypes in the database (the lower plane denotes the network algorithm, while the upper plane corresponds to the traditional algorithm).

Fig. 3. Computation time for increasing the number of vertices in the common subgraph and the size of the database (the lower plane at the front corner denotes the network algorithm, while the upper plane corresponds to the traditional algorithm).

In the first experiment shown in Fig. 2 we increased the number of vertices in the prototype graphs from 10 to 50 and the number of prototypes in the database from 1 to 20. The labels of the vertices were randomly chosen from a set of 10 different labels. For each prototype graph a corresponding, isomorphic input graph was generated by reordering the adjacency matrix. The lower plane in Fig. 2 denotes the time of the new algorithm while the upper plane denotes the time of the Ullman's algorithm. Fig. 2 confirms that for a growing number of prototypes the time for the new algorithm grows only sublinearly while the traditional algorithm is linearly dependent on the size of the database. Additionally, we observe that for larger prototype graphs the new algorithm performs remarkably better than the traditional algorithm. This can be explained by the fact that for larger graphs, the number of identical subgraphs within a single prototype and among different prototypes also grows. In this first experiment, no common subgraphs of the different prototypes was explicitly defined. However, due to the limited number of labels, common subgraphs evolved naturally.

In the second experiment, the effect of sharing common substructures among prototypes was examined more closely. For this purpose, we generated prototype graphs consisting of 50 vertices and approximately 60 edges and increased the number of prototypes from 1 to 20 and the size of the common subgraph from 5 to 45 vertices. Except for the common subgraph, all prototypes were disjoint. The results of the second experiment are given in Fig 3 where the lower plane at the front corner denotes the times of the new algorithm while the upper plane represents the traditional algorithm. Clearly, the intersection of the two planes indicates that for a small or no common subgraph in the different prototypes the performace of the new algorithm is slightly worse than that of the traditional algorithm. But for an increasing size of the common subgraph, the time needed by the new algorithm decreases and becomes much less than that required by **the traditional algorithm.**

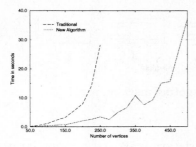 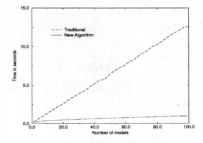

Fig. 4. Computation time for an increasing number of vertices in the prototype graphs.

Fig. 5. Computation time for an increasing number of prototypes in the database.

Finally, in the last two experiments we repeated the first experiment for large graphs and a large number of prototypes, respectively. In Fig. 4 we kept the number of prototypes at one and steadily increased the size of the prototype from 50 to 500 vertices. In Fig. 5 the size of the prototypes was fixed at 50 vertices and the number of prototypes was increased from 1 to 100. It can be conluded that under the selected scenario network-based graph matching clearly outperforms the traditional algorithm. Furtermore, even relatively large graphs and databases are computationally tractable using the proposed approach.

3 Recognition and learning of 2-D shapes using network-based graph matching

The graph matching methods introduced in the last section have the property that any network of prototypes can be incrementally built and updated. Given a network that represents prototypes $p_1, \ldots, p_M$, $M \geq 0$, we can successively add new prototypes $p_{M+1}, \ldots, p_{M+L}$ to the network without the need to recompile the whole network from scratch. This property makes network-based graph matching very interesting in the context of machine learning. While research on machine learning in the general artificial intelligence field has produced interesting results [22], machine learning in the domain of computer vision is still in its infancy. In this section, we consider the problem of symbol recognition in engineering diagrams as an example of an application of network-based error-tolerant graph matching. Input data to our system are line drawings that may contain both known and unknown symbols. All known symbols are stored in a database. The output produced by the system is a list of instances of the known symbols found in the diagrams. Furthermore, unknown symbols are identified and added to the database. Thus the system is able to learn new symbols and continuously update its database.

Symbols and drawings are represented by attributed relational graphs. The line segments of a drawing and their spatial relationships are encoded in the

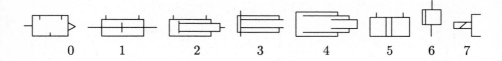

Fig. 6. DIN ISO symbols denoting machine parts.

graph such that the representation is translation, scale and rotation invariant. The process of recognizing a particular symbol in a drawing can then be formulated as the search for an error tolerant subgraph isomorphism between the symbol graph and the drawing graph. As the database of the known symbols usually contains more than one element, the recognition task for a given drawing consists of finding all symbol graphs for which a subgraph isomorphism to the drawing graph with an error less than a certain threshold exists.

The system starts with a number of a priori known symbols in the database. (This number may be zero.) Our proposed learning procedure has to satisfy two objectives. First, all symbols that are unknown, i.e. not contained in the database, should be added to the database in order to guarantee the complete interpretation of the input diagram. Secondly, if the same symbol occurs multiple times in the same input diagram (perhaps with some distortions), only one representative instance of it should be added to the database. This prevents the database from growing unnecessarily large.

In order to satisfy these two objectives, our learning scheme works in two steps. First, after all known symbols have been recognized and removed from the input diagram, all line segments that do not belong to a known symbol are collected and grouped into possible new symbols. A potential new symbol must satisfy certain constraints, such as the presence of a minimal number of line segments or at least one closed loop. These constraints are user-defined and vary from one application to the other. In the second step, a hierarchical clustering procedure is applied to the set of potential new symbols. Initially, the graph distance between each pair of potential symbols is calculated by the error tolerant graph matching method. Then clusters of symbols are formed depending on these distances. Finally, a single symbol is taken as the representative for each cluster and incrementally added to the database of known symbols.

We illustrate the process of recognizing and learning symbols with an example dealing with machine layouts. Notice that our proposed procedure can be very easily adapted to other types of engineering drawings. In Fig. 6 the set of symbols a priori known to the system is given. Based on these symbols, a first interpretation of the drawing in Fig. 7 is attempted. Instances of the symbols 0, 1 and 3 are detected and removed from the drawing; see Fig. 8. The learning scheme is then applied to the remaining symbols in Fig. 8 and the symbols given in Fig. 9 are learned and added to the database of symbols. With these new symbols it is now possible to completely interpret the drawing in Fig. 7.

It is obvious from the symbols displayed in Fig. 6 that this application is well

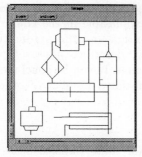 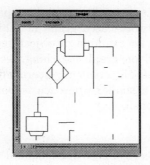 

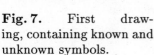

Fig. 7. First drawing, containing known and unknown symbols.

Fig. 8. First drawing after all known symbols have been removed.

Fig. 9. Two symbols learned from the drawing in Fig. 8.

suited for the new algorithm. There is a large number of common substructures in the machine part symbols such that the detection process can be efficiently done with the network based algorithm. Furthermore, the new algorithm supports the learning scheme by allowing the network structure to be incrementally updated with new symbols.

4 Conclusions

In this paper we described a new method for exact and error-tolerant subgraph isomorphism detection, which is based on a compact network representation of model graphs. With several models in the database, the network represents identical subgraphs of the different model graphs only once, thus reducing the number of computational steps necessary in order to detect exact and error-tolerant subgraph isomorphisms from a model graph to an input graph. In addition to sharing identical graph structures, the network can be combined with a fast lookahead procedure. The efficiency of the new algorithm was analytically shown and practically demonstrated in experiments.

The graph matching algorithm described in this paper is very general and powerful. As a matter of fact, there are no problem dependent assumptions included in the algorithm. Our distortion model consists of the deletion, insertion and substitution of both nodes and edges, which is powerful enough to model any type of error that may be introduced into an input graph. Adapting the matching algorithm to a particular application requires the solution of two concrete tasks. First, a suitable graph representation of the objects in the problem domain has to be developed. Secondly, appropriate costs of the graph edit operations have to be found. It is anticipated that there are many applications where both tasks are not really difficult if a set of sample pattern is given.

One of the most challenging problems in computer vision and image analysis is the automatic learning of object models. The network of prototypes that was proposed in this paper can be incrementally expanded by new model graphs. This property makes network-based graph matching applicable in the context of machine learning. As an example of object recognition and the automatic learning of models based on graph matching, we have described a prototype system that deals with 2-D objects in engineering drawings. In can be concluded from this example that a similar approach is possible for a number of other machine vision applications.

Acknowledgement

This work is part of project No 5003-34285 funded by the Swiss National Science Foundation under the "Schwerpunktprogramm Informatikforschung".

References

1. H. Bunke. Structural and syntactic pattern recognition. In C.H. Chen, L.F. Pau, and P. Wang, editors, *Handbook of Pattern Recognition and Computer Vision*, pages 163–209. World Scientific Publ. Co. Singapore, 1993.
2. H. Bunke and G. Allerman. Inexact graph matching for structural pattern recognition. *Pattern Recognition Letters 1*, 4:245–253, 1983.
3. H. Bunke, T. Glauser, and T.-H. Tran. Efficient matching of dynamically changing graphs. In P. Johansen and O.Olsen, editors, *Theory and Applications of Image Analysis*, pages 110–124. World Scientific Publ., 1992.
4. H. Bunke and B.T. Messmer. Similarity measures for structured representations. In M. M. Richter, S. Wess, K.-D. Althoff, and F. Maurer, editors, *Proceedings EWCBR-93 Lecture Notes on Artifical Intelligence*. Springer Verlag, 1994.
5. M.A. Eshera and K.S. Fu. A graph distance measure for image analysis. *IEEE Transactions on Systems, Man, and Cybernetics*, 14(3):398–408, May 1984.
6. J. Feng, M. Laumy, and M. Dhome. Inexact matching using neural networks. In E.S. Gelsema and L.N. Kanal, editors, *Pattern Recognition in Practice IV: Multiple Paradigms, Comparative Studies and Hybrid Systems*, pages 177–184. North-Holland, 1994.
7. M.R. Garey and D.S. Johnson. *Computers and Intractability: A Guide to the Theory of NP-Completeness.* Freeman and Company, 1979.
8. E. Gmuer and H. Bunke. 3-D object recognition based on subgraph matching in polynomial time. In R. Mohr, T. Pavlidis, and A. Sanfeliu, editors, *Structural Pattern Analysis*, pages 131–147. World Scientific, 1990.
9. L. Herault, R. Horaud, F. Veillon, and J.J. Niez. Symbolic image matching by simulated annealing. In *Proc. British Machine Vision Conference*, pages 319–324. Oxford, 1990.
10. R. Horaud and T. Skordas. Stereo correspondence through feature grouping and maximal cliques. *IEEE Transactions on Pattern Analysis and Machine Intelligence PAMI*, 11(11):1168–1180, 1989.
11. K.A. De Jong and W. M. Spears. Using genetic algorithms to solve NP-complete problems. In J.D. Schaffer, editor, *Genetic Algorithms*, pages 124–132. Morgan Kaufmann, 1989.
12. J. Kittler, W. J. Christmas, and M. Petrou. Probabilistic relaxation for matching of symbolic structures. In H. Bunke, editor, *Advances in Structural and Syntactic Pattern Recognition*, pages 471–480. World Scientific, 1992.

55

13. P. Kuner and B. Ueberreiter. Pattern recognition by graph matching - combinatorial versus continous optimization. *International Journal of Pattern Recognition and Artificial Intelligence*, 2(3):527–542, 1988.
14. S.W. Lee, J.H. Kim, and F.C.A. Groen. Translation- rotation- and scale invariant recognition of hand-drawn symbols in schematic diagrams. *International Journal of Pattern Recognition and Artificial Intelligence*, 4(1):1–15, 1990.
15. S.W. Lu, Y. Ren, and C.Y. Suen. Hierarchical attributed graph representation and recognition of handwritten Chinese characters. *Pattern Recognition*, 24:617–632, 1991.
16. B. T. Messmer and H. Bunke. A network based approach to exact and inexact graph matching. Technical Report IAM-93-021, Universität Bern, September 1993.
17. B.T. Messmer and H. Bunke. A new method for efficient error-correcting subgraph isomorphism. In D. Dori and A. Bruckstein, editors, *Syntactic and Structural Pattern Recognition*. World Scientific Publishers, Singapore, to appear in 1995.
18. N.J. Nilsson. *Principles of Artificial Intelligence*. Tioga, Palo Alto, 1980.
19. A. Sanfeliu and K.S. Fu. A distance measure between attributed relational graphs for pattern recognition. *IEEE Transactions on Systems, Man, and Cybernetics*, 13:353–363, 1983.
20. A. Sanfeliu, K.S. Fu, and J.M.S. Prewitt. An application of a graph distance measure to the classification of muscle tissue patterns. *Int. Journal of Pattern Recognition and Artificial Intelligence*, 1(1):17–42, 1987.
21. L. Shapiro. Relational matching. In T.Y. Young, editor, *Handbook of Pattern Recognition and Image Processing: Computer Vision*, pages 475–496. Academic Press, 1993.
22. J.W. Shavlick and T.G. Dietterich, editors. *Readings in Machine Learning*. Morgan Kaufman, San Mateo, 1990.
23. W.H. Tsai and K.S Fu. Error-correcting isomorphisms of attributed relational graphs for pattern recognition. *IEEE Transactions on Systems, Man, and Cybernetics*, 9:757–768, 1979.
24. J.R. Ullman. An algorithm for subgraph isomorphism. *Journal of the Association for Computing Machinery*, 23(1):31–42, 1976.
25. R.A. Wagner and M.J. Fischer. The string-to-string correction problem. *Journal of the Association for Computing Machinery*, 21(1):168–173, 1974.
26. E. K. Wong. Three-dimensional object recognition by attributed graphs. In H. Bunke and A. Sanfeliu, editors, *Syntactic and Structural Pattern Recognition- Theory and Applications*, pages 381–414. World Scientific, 1990.

Matching Delaunay Graphs

Andrew M. Finch and Edwin R. Hancock

Department of Computer Science,
University of York, U.K.

Abstract. This paper describes a Bayesian framework for matching Delaunay graphs. Our matching process is realised in terms of probabilistic relaxation. The novelty of our method stems from its use of a support function specified in terms of triangular face-units of the graphs under match. In this way we draw on more expressive constraints than is possible at the level of edge-units alone. We present a particularly simple face compatibility model that is entirely devoid of free parameters. It requires only knowledge of the numbers of nodes, edges and faces in the model graph. The resulting matching scheme is evaluated on radar images and compared with its edge-based counterpart.

1   Introduction

Inexact graph matching is a critical process for effective intermediate level scene interpretation [2, 4, 5, 6]. The technique allows a corrupted relational description of noisy image entities to be matched against an idealised model graph. As a consequence of its pivotal role in practical vision applications, the graph matching problem has been the focus of sustained activity in a number of diverse methodological areas. Early efforts focussed on the search for subgraph isomorphisms [5] and the definition of a meaningful relational distance metric [5]. More recently, optimisation ideas have provided a powerful alternative framework for relational matching [2, 4, 6].

Basic to these inexact methods is the aim of optimising a measure of matching consistency [4]. Ideally, this measure should be capable of simultaneously capturing both the constraints conveyed by the model graph and the corruption processes at play in the data graph. The majority of matching algorithms are satisfied with exploiting only the pairwise constraints provided by graph edges [4]. This restricted use of available relational constraints not only limits the achievable performance, it may also actually lead to internal inconsistencies in the specification of the matching process. In effect it corresponds to discarding a considerable body of detailed information concerning the local topology of the relational graphs under match.

Delaunay graphs provide an important example where the relational structure is not adequately represented at the edge-level [1]. These graphs are composed of triangular faces and arise in the representation of Voronoi tesselations of planes or surfaces. The intrinsic relational structure of Delaunay graphs is invariably overlooked when it comes to matching. Our aim in this paper is to develop an internally consistent matching process for Delaunay graphs by drawing directly upon their local triangulated topology. In so doing we aim to exploit more of the available relational structure and hence improve the fidelity of match beyond that achievable with edge-based constraints alone [4]. Our framework for this study is provided by probabilistic relaxation. Application of this optimisation approach to the matching problem, requires that we develop not only a support-function that draws on triangulated graph units, but also an associated model of compatibility expressed at the level of consistent graph faces. Both of these ingredients are prescribed in terms of a Bayesian framework.

2 Probabilistic Relaxation

We denote Delaunay graphs by the triple $\mathcal{G} = (\mathcal{V}, \mathcal{E}, \mathcal{F})$. Here $\mathcal{V}$ is the set of nodes and $\mathcal{E} \subset \mathcal{V} \times \mathcal{V}$ is the set of edges. The novel ingredient of the work reported here is to exploit constraints provided by the set of triangular faces $\mathcal{F}$ in the matching process. We use the term *face* to mean a Cartesian triple of of node labels, i.e. $\mathcal{F} \subset \mathcal{V} \times \mathcal{V} \times \mathcal{V}$, such that $(i, j, k) \in \mathcal{F} \Leftrightarrow (i, j) \in \mathcal{E} \wedge (j, k) \in \mathcal{E} \wedge (k, i) \in \mathcal{E}$.

Previous work on relational matching has concentrated on exploiting constraints provided by the edge-set $\mathcal{E}_m$ and is not well suited to the matching of Delaunay graphs. It has been shown that edge-based strategies are best suited to matching graphs which are tree-like in form (i.e. graphs of low edge density in which $|\mathcal{V}| \approx |\mathcal{E}|$) containing few first order cycles (triangular faces). In this paper we aim to overcome this shortcoming by developing a matching scheme which draws on constraints represented by the face-sets of the data and model graphs. In so doing, we aim take advantage of the powerful constraints provided by the cyclic ordering of the nodes in the graphs under match. This ordering concept is central to the ideas presented here and is illustrated in Fig 1. Such an ordering exists if the graph is planar.

We define a graph $\mathcal{G}_d$ on the labels J_α of the objects u_α, $\alpha = 1 \ldots N$ to be: $\mathcal{G}_d = (\mathcal{V}_d, \mathcal{E}_d, \mathcal{F}_d)$, where $\mathcal{V}_d = \{J_1, \ldots, J_N\}$. The graph $G_m = (\mathcal{V}_m, \mathcal{E}_m, \mathcal{F}_m)$, where $\mathcal{V}_m = \{j_1, \ldots, j_m\}$ is defined on the values j_α attributable to these labels. In order to accommodate entries in the data graph that have no physical match in the model graph, we augment the indexed set of data nodes $\mathcal{V}_m$ by a null category ϕ. This symbol us used to label unmatchable data entities which represent noise or extraneous clutter. We denote the index set of the contextual neighbours of

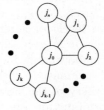

Fig. 1. Label value j_0 together with its contextual neighbours $j_1 \cdots j_n$

object u_α as I_α. This set includes the index α. For the purpose of notational simplicity and without loss of generality, we re-index the objects in the network. This ensures that the object under consideration has index 0 and its neighbours are indexed from $1 \ldots n$ (as in Fig. 1).

The starting point for developing the matching strategy is the general purpose Bayesian evidence combining formula of Kittler and Hancock [3]. According to this framework, the probability at iteration i for the match $J_0 \mapsto j_0$ on the centre-node of the contextual neighbourhood is updated as follows:-

$$P^{(i+1)}(J_0 \mapsto j_0) \leftarrow \frac{P^{(i)}(J_0 \mapsto j_0) Q^{(i)}(J_0 \mapsto j_0)}{\sum_{j_\alpha \in \mathcal{V}_m \cup \phi} P^{(i)}(J_0 \mapsto j_\alpha) Q^{(i)}(J_0 \mapsto j_\alpha)} \tag{1}$$

The critical ingredient in this formula is the support function $Q^{(i)}(J_0 \mapsto j_0)$ which combines evidence for the match $J_0 \mapsto j_0$ over the reindexed contextual neighbourhood of object u_0. For the triangulated neighbourhood shown in Fig 1 the support function may be evaluated recursively over the constituent faces faces commencing at the node indexed 1:-

$$Q^{(i)}(J_0 \mapsto j_\gamma) = \frac{1}{P(J_0 \mapsto j_\gamma)} \sum_{j_1 \in \mathcal{V}_m \cup \phi} \frac{P^{(i)}(J_1 \mapsto j_1)}{P(J_1 \mapsto j_1)}$$

$$\times \sum_{j_n \in \mathcal{V}_m \cup \phi} P(J_1 \mapsto j_1 | J_n \mapsto j_n, J_0 \mapsto j_0) \epsilon(J_1 \mapsto j_1, J_n \mapsto j_n) \tag{2}$$

We initialise this recursion by setting:-

$$\epsilon(J_{n-1} \mapsto j_{n-1}, J_n \mapsto j_n) = \frac{P^{(i)}(J_n \mapsto j_n)}{P(J_n \mapsto j_n)} P(J_n \mapsto j_n | J_{n-1} \mapsto j_{n-1}, J_0 \mapsto j_0) \tag{3}$$

The basic recursion-kernel employed in evaluating support is:-

$$\epsilon(J_{k-1} \mapsto j_{k-1}, J_n \mapsto j_n) = \sum_{j_k \in \mathcal{V}_m \cup \phi} \frac{P^{(i)}(J_k \mapsto j_k)}{P(J_k \mapsto j_k)}$$

$$\times P(J_k \mapsto j_k | J_{k-1} \mapsto j_{k-1}, J_0 \mapsto j_0) \epsilon(J_k \mapsto j_k, J_n \mapsto j_n) \tag{4}$$

In the recursion formula (4), the index $J_k \mapsto j_k$ of the quantity $\epsilon(J_k \mapsto j_k, J_n \mapsto j_n)$ is needed because the label assigned to each object enters two levels of summation in a cyclic manner; the index $J_n \mapsto j_n$ is required since it is necessary to break the cyclicity of indices at some object in the neighbourhood, in order to perform a recursive evaluation. This twofold level of summation effectively propagates the face-based constraints around the perifery of the Delaunay neighbourhood. The Bayesian ingredients of the support function are the single node prior $P(J_k \mapsto j_k)$ and the conditional prior $P(J_k \mapsto j_k | J_{k-1} \mapsto j_{k-1}, J_0 \mapsto j_0)$. It is the conditional prior that measures the consistency between the label match $J_k \mapsto j_k$ and the matches $J_{k-1} \mapsto j_{k-1}$ and $J_0 \mapsto j_0$ on the remaining nodes of the face $\{J_k, J_{k-1}, J_0\}$ belonging to the data graph. According to our Bayesian framework, the mutual information measure $R(j_k, j_{k-1}, j_0) = \frac{P(J_k \mapsto j_k | J_{k-1} \mapsto j_{k-1}, J_0 \mapsto j_0)}{P(J_k \mapsto j_k)}$ may be viewed a plying the role of a compatibility coefficient. In constructing a model of the compatibility coefficients $R(j_k, j_{k-1}, j_0)$ we would like to capture some of the constraint violations introduced by different classes of segmentation error. These errors include noise contamination, fragmentation due to over segmentation and accidental merging due to under segmentation.

3 Compatibility Model

Our adopted modelling philosophy is that the nodes in graph $\mathcal{G}_d$ represent data that must be matched in the light of constraints provided by graph $\mathcal{G}_m$. In order to cope with inconsistencies between the model and data graphs, we employ a system of constraint softening in which it is assumed that segmentation errors occur with uniform probability p. There are four classes of constraint corruption that can occur at the face level. Firstly, faces of the data graph with all nodes matched to a face in the model graph are uncorrupted and occur with total probability mass $(1 - p)^3$. Secondly, there are faces with two nodes matched to an an edge in the model graph and one-node matched to the null label; these have total probability mass $p(1 - p)^2$. The third class consists of faces in which two nodes are null-matched have a probability of $p^2(1 - p)$. Finally, there are configurations in which all nodes in the face are null matched take the remaining probability mass, i.e. p^3. Matches involving non-null label triplets which are inconsistent with the above configurations are completely forbidden and therefore account for zero total probability mass. In each of the four cases listed above, the available mass of probability is distributed uniformly among the label configurations falling into the relevant constraint class. Using this we arrive at a rule which yields the joint probabilities for face configurations:-

$$P(J_k \mapsto j_k, J_{k-1} \mapsto j_{k-1}, J_0 \mapsto j_0) = \begin{cases} \frac{(1-p)^3}{|\mathcal{F}_m|} & \text{if } \{j_k, j_{k-1}, j_0\} \in \mathcal{F}_m \\ \frac{p(1-p)^2}{|\mathcal{E}_m|} & \text{if } \{j_k, j_{k-1}\} \in \mathcal{E}_m \wedge j_0 = \phi \\ \frac{p^2(1-p)}{|\mathcal{V}_m|} & \text{if } j_k \in \mathcal{V}_m \wedge j_{k-1} = \phi \wedge j_0 = \phi \\ p^3 & \text{if } j_k = \phi \wedge j_{k-1} = \phi \wedge j_0 = \phi \\ 0 & \text{otherwise} \end{cases} \tag{5}$$

The edge-priors and single-node priors required in the computation of compatibility coefficients are obtained by summing the joint probabilities in the axiomatic way. The resulting compatibility coefficients are given below:-

$$R(j_k, j_{k-1}, j_0) = \begin{cases} \frac{|\mathcal{E}_m||\mathcal{V}_m|}{|\mathcal{F}_m|} & \text{if } \{j_k, j_{k-1}, j_0\} \in \mathcal{F}_m \\ \frac{|\mathcal{V}_m|^2}{|\mathcal{E}_m|} & \text{if } \{j_{k-1}, j_0\} \in \mathcal{E}_m \wedge j_k = \phi \\ & \text{if } \{j_{k-1}, j_k\} \in \mathcal{E}_m \wedge j_0 = \phi \\ 1 & \text{if at least 2 of } \{j_k, j_{k-1}, j_0\} \text{ are null} \\ & \text{or } \{j_k, j_0\} \in \mathcal{E}_m \wedge j_{k-1} = \phi \\ 0 & \text{otherwise} \end{cases} \tag{6}$$

The pattern of compatibilities grades the different face-constraints according to their overall consistency. Fully consistent faces have higher compatibility values than partially matched faces which feature isolated edges or nodes. These compatibilities also discourage the violation of the neighbourhood ordering relation. Partially matched faces containing an edge which could could potentially disrupt the ordering of the contextual neighbourhood (i.e. $\{j_k, j_0\} \in \mathcal{E}_m$) receive a lower compatibility than the others which could not. In other words, the compatibility pattern favours null matched nodes that are surrounded by a consistently ordered neighbourhood over matches that are connected to a plethora of incorrectly ordered yet consistent edges. This ability to impose ordering relations enhances the internal consistency of our face-based relaxation scheme and represents the main advantage over edge-based compatibility models [4].

4   Experiments

In order to demonstrate some of the performance advantages of our face-based matching scheme, we have taken an application involving synthetic aperture radar data. We are interested in matching linear hedge structures from radar images against their cartographic representation in a digital map. We establish Delaunay graph representations of data and model, by seeding Voronoi tessellations from the midpoints of the linear segments. To illustrate the results of applying our face-based matching process to the Delaunay graphs, Fig 2 shows the correctly matched segments overlaid on the original radar image.

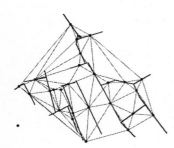 

Fig. 2. The model graph (left) and correctly matched SAR line segments (right)

To give some idea of the effectiveness of our compatibility model, there 28 nodes, 46 faces and 142 edges in the model graph. In consequence, the face compatibility $\frac{|\mathcal{E}_m||\mathcal{V}_m|}{|\mathcal{F}_m|} = 86.43$, while the edge compatibility $\frac{|\mathcal{V}_m|^2}{\mathcal{E}_m} = 5.52$. The graph in Fig 3 shows the accuracy of match (defined here to be the number of correctly matched line segments, minus the number of incorrectly matched line segments) plotted against the edge and face compatibility values. The graph shows a broad plateau around the theoretically optimal parameter values. In this region (3-8 edge compatibility and 30-100 face-compatibility) the correct/incorrect surplus is consistently high (i.e. 7-9). This is an encouraging observation, it means that our very simple compatibility model is adequate in describing the matching errors in a complex experimental example without excessive parameter tuning. Moreover, the results obtained with the face-based technique offer a number of tangible advantages over their edge-based counterpart. In the case of the face-based method 18 nodes are correct, 64 are null matched and only 5 nodes match in error. For the edge-based method, on the other hand, although 19 nodes are correctly matched, only 22 are null matched with 46 residual nodes matched in error. In other words, the face-based method is more effective in identifying unmatchable elements and assigning them to the null category.

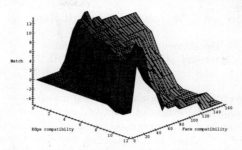

Fig. 3. Matching accuracy vs. face and edge compatibility

5 Conclusions

We have described a novel Bayesian framework for matching Delaunay triangulations. Our matching process is realised in terms of probabilistic relaxation. The novelty of our method stems from its use of a support function specified in terms of face-units of the graphs under match. In this way we draw on more expressive constraints than is possible at the level of edge-units alone. Viewed from the perspective of information processing utility, our method represents two basic advances over edge-based matching schemes. Firstly, we draw on the use of more powerful constraints since we are able to take advantage of neighbourhood ordering relations. Secondly, we are capable of matching graphs of high edge density containing many first-order cycles.

References

1. N. Ahuja, B. An and B. Schachter, "Image Representation using Voronoi Tessellation", *CVGIP*, **29**, pp 286–295, 1985.
2. K. Boyer and A. Kak, "Structural Stereopsis for 3D Vision", *IEEE PAMI*, **10**, pp 144-166, 1988.
3. J. Kittler and E.R. Hancock, "Contextual Decision Rule for Image Analysis", *Image Vision Computing*, **5**, pp, 145–154, 1987.
4. J. Kittler, W.J. Christmas and M. Petrou, "Probabilistic Relaxation for Matching Problems in Machine Vision", *Proceedings of the Fourth International Conference on Computer Vision*, pp. 666-674, 1993.
5. L. Shapiro and R.M. Haralick, "Structural Description and Inexact Matching", *IEEE PAMI*, **3**, pp 504–519, 1981.
6. R.C. Wilson and E.R Hancock, "Graph Matching by Discrete Relaxation", *Pattern Recognition in Practice IV: Multiple Paradigms, Comparative Studies and Hybrid Systems*, North Holland, pp. 165–177, 1994.

An Integrated Approach to Grouping and Matching

Richard C Wilson and Edwin R Hancock

Department of Computer Science,
University of York, York Y01 5DD,UK.

Abstract. Perceptual grouping and relational matching are conventionally viewed as sequential stages of effective intermediate level scene interpretation. Relational structures established on the basis of perceptual grouping criteria are utilised in a bottom-up control strategy and hence form the input data representation for subsequent matching. Our standpoint in this paper is that the two processes should be tightly coupled to one-another so that the relational model can prevail upon the extraction of perceptual groupings, providing additional constraints on an otherwise potentially fragile processing operation. We realise this objective by casting the grouping and matching operations as an iterative discrete relaxation process. The dual operations of re-organising the perceptual relation graph and subsequent matching against a model, both optimise a single objective function in the *maximum a posteriori probability* sense. Grouping and matching are therefore cast into an integrated optimisation framework that is realised by tightly coupled update processes.

1   Introduction

Perceptual grouping [4] and relational matching [1, 3, 5] are processes of central importance in the established hierarchy of computer vision operations. Conventionally perceptual grouping is used to establish relational data structures for later matching. Viewed from this perspective grouping is a process of structural filtering aimed at removing relational clutter. When applied effectively, the grouping process may significantly enhance the fidelity of subsequent matching operations [4].

Although component parts of a clearly established hierarchy, grouping and matching are invariably viewed in a bottom-up perspective [4]. Little interaction or feedback exists other than by virtue of the coupling induced by the vertical propagation of a common data abstraction. Moreover, this abstraction is usually devoid of any evidential index associated with the significance of perceptual tokens. Our viewpoint in this paper is that there are important symbiotic advantages to be gained from the tight integration and feedback between the two processes. We aim to demonstrate that relational constraints provided by a model can prevail upon the effective extraction of meaningful perceptual groupings from otherwise heavily corrupted and distorted data through the propagation of an index of evidential significance. By deleting redundant perceptual tokens we aim to reconstruct corrupted relations and hence improve the quality of final match.

Our framework for realising this objective of an integrated approach to grouping and matching is provided by *maximum a posteriori probability* (MAP) estimation [2]. We assess the effects of both modifying the set of perceptual tokens and updating their matched realisation using a Bayesian measure of evidential significance [5]. Simultaneous optimisation of relational structure and consistency of match is realised by two distinct classes of update operations. The first of these involves deleting and reinstating nodes from the relational graphs representing the salient perceptual structures [4]. This structural grouping process is one of dynamic graph reconfiguration which takes place by direct reference to the raw image data. Once a graph node has been identified as being relationally redundant, the corresponding perceptual token is deleted from the data and the grouping is recomputed. The second class of updating operations maintain a consistent match between the prevailing grouping of perceptual entities in the data and the available relational model. In this way the processes of perceptual grouping and matching interact through a common objective function. Effectively, the dual update processes are integrated together in a unified optimisation framework.

The outline of this paper is follows. Section 2 introduces the formal ingredients and computational framework underlying our method. We present an experimental evaluation of our method in Section 3. Finally, Section 4 offers some conclusions.

2   Computational Framework

We abstract the matching of perceptual structures in terms of attributed relational graphs [1]. According to this representation the nodes are entities to be matched. The arcs represent perceptual relations operating between the nodes. We use the triple $G = (V, E, \mathcal{A})$ to denote the graphs under match, where V is the set of nodes, $E \subset V \times V$ is the set of arcs and $\mathcal{A} = \{\mathbf{x}_i, \forall i \in V\}$ is a set of unary measurements associated with the nodes. Our aim in matching is locate correspondences between the nodes in a graph $G_1 = (V_1, E_1, \mathcal{A}_1)$ representing the perceptual organisation of the data and those in a graph $G_2 = (V_2, E_2, \mathcal{A}_2)$ representing the available relational model. Formally, the matching is represented by a function $f : V_1 \rightarrow V_2$ from the nodes in the data graph G_1 to those in the model graph G_2.

In performing the matches of the nodes in the data graph G_1 we will be interested in exploiting structural constraints provided by the model graph G_2. In order to strike a compromise between exploiting constraints of maximum power, on the one hand, and initiating an uncontrolled explosion of computational complexity, on the other, we will use subgraphs that consist of neighbourhoods of nodes interconnected by arcs; for convenience we refer to these structural subunits or N-ary relations as supercliques. The superclique of the node indexed j in the graph G_1 with arc-set E_1 is denoted by the set of nodes $C_j = j \cup \{i | (i, j) \in E_1\}$. We use the notation $R_j = (u_1, u_2, ..., u_{|C_j|})$ to denote the N-ary symbolic relation represented by the nodes of the superclique $C_j \subset V_1$

in the data graph G_1. The matched realisation of this superclique is denoted by the relation $\Gamma_j = (f(u_1), f(u_2), ..., f(u_{|C_j|}))$. The critical ingredient in developing our matching scheme is the set of structure preserving mappings between each superclique of graph G_1 and those of graph G_2. The set of mappings, or dictionary, for the superclique C_j is denoted by $\Theta(C_j) = \{S_i\}$ where each element S_i of $\Theta(C_j)$, is a relation formed on the nodes of the model graph. We denote such consistent symbolic relations by $S_i = (v_1, v_2, ...)$.

Our aim is to cast the processes of perceptual organisation and relational matching into a unified optimisation framework. This combined optimisation process involves two classes of update operation. The first of these is the iterative reconstruction of the relational graph G_1 by node deletions and reinsertions to improve the overall grouping of perceptual entities. The second class of update operations are concerned with modifying the match so as to maintain consistency with the relational model G_2.

At the computational level, these dual update operations involve maintaining two separate representations of the perceptual relation graph. Suppose that $G_1 = (E_1, V_1, \mathcal{A}_1)$ is the current representation of the perceptual organisation of the data. Graph $G_1' = (E_1', V_1', \mathcal{A}_1')$, on the other hand, is the reconfigured relational structure formed by excluding the single perceptual entity represented by node u in G_1 and reassigning it to the null or outlier set Φ. The node-set $V_1' = V_1 - u$ and the attribute-set $\mathcal{A}_1' = \mathcal{A}_1 - \mathbf{x}_u^1$ are trivially recomputed by deleting the entry associated with node u. Determination of the edge-set E_1' is more complex and must be undertaken by recomputing the relational affinity of the nodes when the perceptual entity associated with node u is excluded from the raw image data. The match between the reconfigured graph G_1' and the model graph G_2 is represented by the function $f' : V_1' \to V_2$.

2.1 MAP Criterion

From the standpoint of information theory, we therefore seek the grouped and matched configuration of nodes that has *maximum a posteriori probability* with respect to the available unary measurement information. Computationally, the objective is to formulate iterative local decision schemes that are capable of both grouping and matching nodes. More formally, these iterative reassignment processes are aimed at partitioning the set of entities into those that are grouped to form a consistently matched graph f and and those that are assigned to a set of relational outliers Φ. In other words, we seek the grouped and matched configuration of nodes that optimise the quantity $P(f, \Phi | \mathcal{A}_1, \mathcal{A}_2)$. Our development of a detailed MAP update process proceeds from the starting point described in [2] and centres around dual assumptions concerning the matching process. Firstly, we assume conditional independence over the unary measurements. Of greater significance to our unified grouping and matching process is the assumption that we can factorise the joint prior $P(f, \Phi)$ over the set of matched nodes and the set of outliers, i.e. $P(f, \Phi) = P(f).P(\Phi)$.

In order to realise the dual processes of grouping and matching, we require a model of the joint prior $P(f)$. In the context of relational matching, we have recently described a Bayesian model of the joint prior which draws on the structure preserving relational mappings contained within the dictionary [5]. Central to this process is the idea of gauging relational consistency using the Hamming distance between an assigned match Γ_j and a structure preserving mapping S_i by the Hamming distance $H(\Gamma_j, S_i) = \sum_{i \in C_j}(1 - \delta_{f(u_i), v_i})$. With this ingredient, the joint prior for the matching configuration Γ_j is equal to

$$P(\Gamma_j) = \frac{K_{C_j}}{|\Theta(C_j)|} \sum_{S_i \in \Theta(C_j)} \exp[-k_e H(\Gamma_j, S_i)] \tag{1}$$

where $K_{C_j} = (1 - P_e)^{|C_j|}$ and the exponential constant is related to the probability of memoryless matching errors P_e by $k_e = \ln \frac{1 - P_e}{P_e}$. The joint prior $P(f)$ is computed by averaging the clique priors $P(\Gamma_j)$ over the nodes of the data graph. Our model of the joint prior for the outlier set is based on the assumption that is contains no meaningful relational structure and that $P(\Phi) = P_\phi^{|\Phi|}$, where P_ϕ is the uniform probability of relational outliers.

2.2 Coupled Updates

In elucidating our grouping process, we make the observation that in order to gauge the net effect of deleting a node using the consistency measure given by equation (12) we need only examine those contributions that arise from the modified superclique set. This set is constructed by identifying those nodes that form a superclique with node u in graph G_1, i.e. $C_u - \{u\}$, and determining the new superclique set for these nodes in the reconfigured graph G_1'. We let χ_u^+ denote the superclique set of object u in graph G_1 and χ_u^- denote the corresponding superclique set in the reconfigured graph G_1'. With this notation the change in the MAP criterion caused by the deletion of the node u is proportional to

$$\Delta_u^- = P_\phi \sum_{j \in \chi_u^-} P(\Gamma_j) \tag{2}$$

By contrast, when considering the change in the MAP criterion due to the re-insertion of node u it is the superclique set χ_u^+ to which we turn our attention. The quantity of interest depends not only upon the joint matching priors $P(\Gamma_j)$ but also upon the a posteriori matching probabilities $P(u, v|\mathbf{x}_u^1, \mathbf{x}_v^2)$ and single node priors $P(u, v)$

$$\Delta_u^+ = \frac{P(u, v|\mathbf{x}_u^1, \mathbf{x}_v^2)}{P(u, v)} \sum_{j \in \chi_u^+} P(\Gamma_j) \tag{3}$$

With these two measures to hand, we can both delete and reinstate nodes in such a way as to monotonically increase the MAP criterion. We therefore delete node u provided $\Delta_u^+ < \Delta_u^-$ and reinstate the node if $\Delta_u^+ > \Delta_u^-$.

The graph reconfiguration process described above is aimed at restoring meaningful relational groupings using constraints provided by a model-graph.

With this enhanced relational structure to hand, a more consistent match may be recovered. In commencing the grouping process, we have assumed that a match of maximum achievable consistency had already been established for the mapping f. A consistent match may be maintained in the MAP sense provided that the mapping f is continually updated as follows

$$f(u) = \arg \max_{v \in V_2} \frac{P(u, v | \mathbf{x}_u^1, \mathbf{x}_v^2)}{P(u, v)} \sum_{i \in \chi_u^+} P(\Gamma_i) \tag{4}$$

Relational consistency is therefore incrementally maintained provided that we update those matches for the nodes in supercliques modified by the deletion or insertion processes.

3 Experiments

Evaluation of our unified matching process is concerned with matching hedge structures segmented from synthetic aperture radar (SAR) images against their cartographic representation in a digital Ordnance Survey map. The initial matches between the linear segments extracted from the SAR data and their map representation are established on the basis of the affinity between the vectors of unary node attributes in the two graphs. The matching probabilities are computed from exponential distributions of the Mahalanobis distance between attribute-vector pairs computed using an estimate of the variance-covariance matrix Σ, i.e.

$$P(u, v | \mathbf{x}_u^1, \mathbf{x}_v^2) = (1 - P_\phi) \frac{\exp[-\frac{1}{2}(\mathbf{x}_u^1 - \mathbf{x}_v^2)^T \Sigma^{-1}(\mathbf{x}_u^1 - \mathbf{x}_v^2)]}{\sum_{w \in V_2} \exp[-\frac{1}{2}(\mathbf{x}_u^1 - \mathbf{x}_w^2)^T \Sigma^{-1}(\mathbf{x}_u^1 - \mathbf{x}_w^2)]} \tag{5}$$

Our assumed model for the matching priors is one of uniformity over the set of matches i.e. $P(u, v) = (|V_1|.|V_2|)^{-1}$.

The experimental matching study is based on 95 linear segments in the SAR data and 30 segments contained in the map. However only 23 of the SAR segments have feasible matches within the map representation. Figure 1 illustrates the results of applying the unified grouping and matching technique. Figure 1a. shows the original SAR image. Figure 1b. shows the line segments extracted from the digital map. Figure 1c shows the initial matches for the corresponding line segments in the SAR data; the black lines are correct matches while the grey lines are matching errors. With the same coding scheme Figure 1d. shows the final result once the iterative grouping and matching process has converged. Here those lines assigned to the set Φ have been deleted from the figure. Comparing Figures 1c. and 1d. it is clear that the main effect of simultaneous grouping and matching operations, has been to delete the majority of the clutter segments from the SAR data graph. To give some idea of relative performance merit, in the case of the initial matching configuration 20 of the 23 matchable segments are correctly identified with 75 incorrect matches, while after application of the unified grouping method the final graph contains 19 correct matches, only 17 residual clutter nodes and 59 deleted nodes.

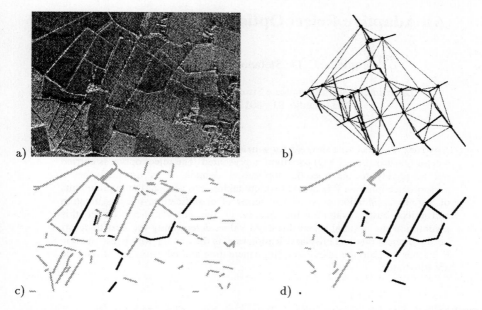

Fig. 1. Integrated grouping-matching process: a) Original image, b) Extracted line segments, c) Initial match, d) Final grouped match.

4 Conclusions

We have described a novel approach to relational matching. Rather than performing sequential grouping and matching operations, we accomplish the two processes simultaneously in unified optimisation framework. The dual operations of re-organising the perceptual relation graph and matching it against a model, both optimise a single objective function in the *maximum a posteriori probability* sense. We have demonstrated the utility of the resulting technique both on the matching of synthetic aperture radar images and on simulated data.

References

1. K Boyer and A Kak, "Structural Stereopsis for 3D Vision", *IEEE PAMI*, 10, pp 144-166, 1988.
2. E.R. Hancock and J. Kittler, "Discrete Relaxation," *Pattern Recognition*, **23**, pp.711–733, 1990.
3. R. Horaud and T. Skordas, "Stereo Correspondence through Feature Grouping and Maximal Cliques", *IEEE PAMI*, **11**, pp. 1168–1180, 1989.
4. S. Sarkar and K.L. Boyer, "Perceptual Organisation in Computer Vision: A Review and Proposal for a Classificatory Structure", *IEEE SMC*, **23**, pp 382–399, 1993.
5. R.C. Wilson and E.R Hancock, "Graph Matching by Discrete Relaxation", *Pattern Recognition in Practice IV*, North Holland pp. 165–177, 1994.

An Adaptive Reject Option for LVQ Classifiers

L.P. Cordella, C. De Stefano, C. Sansone, M. Vento

Dipartimento di Informatica e Sistemistica, Università di Napoli
Via Claudio, 21   80125 Napoli (Italy)

Abstract. A reject rule devised for a neural classifier based on the Learning Vector Quantization (LVQ) paradigm is presented. The reject option is carried out adaptively to the specific application domain. It is assumed that a performance function P is defined which, taking into account the requirements of a given application expressed in terms of classification, misclassification and reject costs, evaluates the quality of the classification. Under this assumption the optimal reject threshold value, determining the best trade-off between reject rate and misclassification rate, is the one for which the function P reaches its absolute maximum. Implementation and performance of the rule are illustrated.

1   Introduction

Neural networks have been widely used in the recent past in many application areas. In Pattern Recognition, neural networks revealed to be very interesting for building classifiers with good performance and trainable in a flexible way. To this concern, various learning algorithms have been widely studied and criteria for selecting and sorting the training set have been defined [1-4]. Mostly investigated topics concern techniques for obtaining better convergence rates and criteria for stopping the learning phase when an acceptable trade-off between generalization power and specialization degree of the net has been achieved.

Anyway, in real recognition problems, the samples belonging to the real world (data set) can be affected by distortions that make them quite different from the ones belonging to the training set; therefore, even a well trained network, when attempting the classification of a distorted sample, risks to misclassify it. This problem has been addressed for other kinds of classifiers by introducing a reject option essentially based on the identification of not reliable classifications [5]; the reject decision is made by evaluating the advantage of rejecting a sample instead of running the risk to misclassify it. It appears desirable that this advantage is measured taking into account the requirements of the specific application domain. In fact, there are applications for which the cost of a misclassification is very high, so that a high reject rate is acceptable just to keep misclassification rate as low as possible; a typical example could be the classification of medical images in the framework of a prescreening for early cancer detection. In other applications it may be desirable to assign every sample to a class even at the risk of a high misclassification rate; let us consider for instance the case of a character classifier used in applications in which a text has to be successively widely edited by man.

Between these extremes, a number of applications can be characterized by intermediate requirements.

In [6] we illustrate the rationale of a method for determining the optimal reject threshold value to be used with a neural classifier in a given application domain, in order to get the best trade-off between reject and misclassification rates. In this paper, the specialization of the method in case of an LVQ classifier is discussed. We suggest to measure the performance of a classifier by means of a performance function $P=P(R_c,R_r,R_m)$ defined in terms of recognition rate (R_c), reject rate (R_r) and misclassification rate (R_m). No hypotheses are in principle necessary on the form of P, but the one that it effectively represents the quality of the classification. For several applications it can be assumed that the cost of a correct classification, of a misclassification and of a reject doesn't vary with R_c, R_r and R_m. Therefore $\partial P/\partial R_c$, $\partial P/\partial R_r$ and $\partial P/\partial R_m$ can be considered constant, implying that P is a linear function that can be written in the form:

$$P = R_c - C_r R_r - C_m R_m$$

where C_r is the cost of a reject and C_m is the cost of a misclassification, both normalized, for the sake of simplicity, with respect to the cost (actually a gain!) of a correct classification. The values of the cost coefficients can be assigned by taking into account the specific application domain in which the classifier is used.

The proposed approach implies the introduction of a rule according to which an input sample is rejected if the value of a suitable parameter, effectively representing the classification reliability (*reliability parameter*), is lower than a given threshold. The selection of the threshold is made in such a way to maximize the performance function P in the considered context. This last aspect is especially important since, for different applications, both the form of P and the values to be assigned to the cost coefficients may be different. Moreover, the implementation of the method depends on the classifier architecture.

To evaluate the improvement of the performance function P obtained by applying the reject rule, we consider the occurrence densities of both correctly classified and misclassified patterns (in the following quoted as D_c and D_m respectively), obtained without using any reject option, as a function of the chosen reliability parameter.

It is worth noting that the distributions are computed after the training of the classifier, on a set S of labeled samples.

2   Reject Rules for LVQ Classifiers

In case of LVQ neural net based classifiers [7], the output vector is made of the distance values between the input sample and the prototypes of every class (corresponding to neurons belonging to the Kohonen layer): classification could thus be performed by selecting the class whose representative prototype has the smallest distance from the input sample (Winner-Takes-All rule). However, in this way, no samples would be rejected and consequently it would be impossible to affect the reliability of the classification.

In order to introduce the reject option, the reliability parameter can be determined by taking into account the characteristics of the output vector provided by the network. We have experimented with three different parameters which revealed to be particularly appropriate for different situations in the feature space.

The selection of the best parameter for a given situation is carried out a-posteriori: first the threshold value maximizing P is determined for each parameter and then the parameter corresponding to the highest value of P is selected.

Let us now consider an LVQ classifier, and let C the number of classes to be recognized, N the number of neurons per class and w_{ij} the j-th neuron standing for the j-th prototype belonging to the i-th class. The distance between the input sample x and the class i is given by:

$$\delta_i = \min_j (d(w_{ij}, x)) \qquad j = 1..N$$

where $d(w_{ij}, x)$ is the Euclidean distance between w_{ij} and x.

According to the Winner-Takes-All rule (W-rule) applied to the net output vector, the input sample x would be assigned to the class k if:

$$\delta_k = \min_i (\delta_i) = \delta_{WIN} \qquad i = 1..C.$$

Relatively small values of δ_{WIN} mean that the input sample is very close to the prototype, and consequently a high classification reliability has been achieved. On the basis of this simple consideration the first reliability parameter has been defined as:

$$\rho_1 = 1 - \delta_{WIN}/\delta_{MAX}$$

where δ_{MAX} represents the maximum value assumed by δ_{WIN} on the set S. Even if ρ_1 showed to work well in a set of cases, it fails when the input sample is close to the winner prototype, but placed in an overlapping region between two classes. In this case, comparing δ_{WIN} with the distance δ'_{WIN} of the input sample from the second winner class, can give more useful information. On this basis, we have introduced two more parameters:

$$\rho_2 = (\delta'_{WIN} - \delta_{WIN})/(\max (\delta'_{WIN} - \delta_{WIN})) \quad \text{and} \qquad \rho_3 = 1 - \delta_{WIN}/\delta'_{WIN}$$

where the maximum has been evaluated on the set S and

$$\delta'_{WIN} = \min_{i, i \neq k} (\delta_i) \qquad i = 1..C.$$

Using each of the above parameters, we can introduce a reject option by means of a classification rule (WR-rule), which rejects a sample if the corresponding value of the considered reliability parameter is lower than a given threshold s. In the following discussion, we will indicate the generic reliability parameter with the symbol ρ, since all the general considerations are independent on the specific parameter used. Note that, by definition, $0 \leq \rho \leq 1$.

Let us define the occurrence density curves $D_c(\rho)$ and $D_m(\rho)$ so that

$$R_c = \int_0^1 D_c(\rho)d\rho \qquad \text{and} \qquad R_m = \int_0^1 D_m(\rho)d\rho$$

provide recognition rate and misclassification rate respectively, according to the W-rule. The introduction of the threshold s has two opposite effects on the performance function P. On the one hand, the WR-rule classifies only that percentage R'_c of the samples correctly classified with the W-rule for which the value of P is greater than s and rejects the remaining R'_{rc} determining a decrease of P whose amount is $[R'_{rc}(1+C_r)]$. On the other hand, the WR-rule misclassifies only that subset of the samples misclassified with the W-rule for which the value of P is greater than s (their percentage be R'_m), while the remaining R'_{rm} are rejected; this implies an increase of P whose amount is $[R'_{rm} (C_m-C_r)]$.

Thus, the use of the WR-rule gives rise to a positive variation of P if:

$$R'_{rc}/R'_{rm} < (C_m-C_r)/(1+C_r)$$

It is clear that the greater is C_m with respect to C_r, the more convenient is the WR-rule with respect to the W-rule.

The performance function P, when using the WR-rule, can be written in the form:

$$P_{WR}(s) = R'_c - C_r R'_r - C_m R'_m \qquad (1) \qquad \text{with:} \quad R'_r = R'_{rm} + R'_{rc}$$

Note that P_{WR} depends on s through R'_{rc} and R'_{rm} which are integral functions of s. The method for determining the optimal value of s by maximizing relation (1), is not discussed here. On the analogy of (1), in the following P_W will denote the performance function when using the W-rule.

3 Experimental Results and Conclusions

The classification rule and the reliability parameters discussed above have been experimented in the framework of two tests. In Test 1 we used the same type of data chosen by Xu [8], grouped in four clusters in a bidimensional feature space, each representing a different class. The clusters, all having the same a-priori probability, were generated by two independent Gaussian distributions with variance 0.1, and are respectively centered in $(0,1),(0,-1),(1,0),(-1,0)$. In Test 2, in order to simulate a real situation, a noisy component made of samples of the four classes uniformly distributed within the range $[-2.5,2.5][-2.5,2.5]$ was added to the clusters. In Test 1, the training set was made of 2000 samples (500 for each cluster) while, in Test 2, 2223 samples (including about 10% of noise) were used. Two test sets, with the same number of samples, were generated according to the same criteria.

As the number of clusters per class is 1, we chose N = 1, but also carried out a test in which N was overdimensioned (N = 5). The networks were trained with the basic LVQ1 algorithm [7]. The algorithm basic version was used, instead of its improvements, e.g. [8], because our main goal was just to demonstrate that P can be improved by introducing the reject option.

In Table 1, the recognition rates achieved on the considered training and test sets by a Bayesian classifier (computed from the known probability density functions) and by the LVQ net without reject option are compared. In Table 2, the percentage increment I of P, achieved after introducing the reject option, is shown for Test 1 and Test 2. The value of $I = ((P_{WR} - P_W) / (\max(P_{WR}) - P_W)) \cdot 100$ is reported for each of the three considered reliability parameters and for different values of C_r and C_m. The quantity $(\max(P_{WR})-P_W)$ is the maximum increment of P, corresponding to the optimal case in which all the misclassified samples are rejected while the recognition rate remains unchanged. The values considered for C_r and C_m were chosen within the sets $\{3,4,5\}$ and $\{7,9,11,13,15\}$ respectively. This choice seemed adequate to include a bunch of possible real situations and correspond to the hypothesis that, in practical cases, C_r is at least three times greater than the cost of a correct classification, while C_m is at least twice C_r. Referring to Table 2, note that the use of

	TEST 1		TEST 2		
	Bayes	**LVQ (N=1)**	**Bayes**	**LVQ (N=1)**	**LVQ (N=5)**
Training set	97.5	97.5	89.7	89.6	88.6
Test set	97.8	97.7	90.1	90.1	89.1

Table 1. The recognition rates achieved by the Bayesian and LVQ classifiers on training set and test set, for the two considered experiments.

TEST 1		Reject on ρ_1				Reject on ρ_2				Reject on ρ_3			
C_m	C_r	Clas.	Mis.	Rej.	I	Clas.	Mis.	Rej.	I	Clas.	Mis.	Rej.	I
7	3	97.7	2.3	0	*0*	97.2	1.7	1.1	*2.2*	97.1	1.5	1.4	*4.3*
9	3	97.7	2.3	0	*0*	96.6	1.2	2.2	*13.8*	97.1	1.5	1.4	*13.8*
11	3	97.7	2.3	0	*0*	96.6	1.2	2.2	*21.7*	96.8	1.3	1.9	*20.6*
13	3	97.7	2.3	0	*0*	96.6	1.2	2.2	*26.5*	96.8	1.3	1.9	*24.8*
15	3	97.7	2.3	0	*0*	96.6	1.2	2.2	*29.7*	96.8	1.3	1.9	*27.5*
9	4	97.7	2.3	0	*0*	97.2	1.7	1.1	*1.7*	97.1	1.5	1.4	*4.3*
11	4	97.7	2.3	0	*0*	97.2	1.7	1.1	*8.7*	97.1	1.5	1.4	*12.4*
13	4	97.7	2.3	0	*0*	96.6	1.2	2.2	*18.8*	96.8	1.3	1.9	*18.4*
15	4	97.7	2.3	0	*0*	96.6	1.2	2.2	*23.7*	96.8	1.3	1.9	*22.5*
11	5	97.7	2.3	0	*0*	97.2	1.7	1.1	*2.2*	97.1	1.5	1.4	*4.3*
13	5	97.7	2.3	0	*0*	97.2	1.7	1.1	*8.1*	97.1	1.5	1.4	*11.4*
15	5	97.7	2.3	0	*0*	96.6	1.2	2.2	*17.0*	97.1	1.5	1.4	*15.6*

TEST 2		Reject on ρ_1				Reject on ρ_2				Reject on ρ_3			
C_m	C_r	Clas.	Mis.	Rej.	I	Clas.	Mis.	Rej.	I	Clas.	Mis.	Rej.	I
7	3	88.7	5.5	5.8	*30.2*	88.7	8.0	3.3	*6.8*	88.5	7.2	4.3	*12.6*
9	3	88.3	5.1	6.6	*36.7*	88.2	7.7	4.1	*10.7*	86.0	5.4	8.6	*18.3*
11	3	88.3	5.1	6.6	*39.9*	84.7	6.2	9.1	*11.3*	84.1	4.7	11.2	*22.6*
13	3	87.8	4.7	7.5	*42.9*	84.7	6.2	9.1	*16.7*	84.1	4.7	11.2	*28.6*
15	3	83.9	3.6	12.5	*43.2*	84.7	6.2	9.1	*20.2*	79.1	3.2	17.7	*31.1*
9	4	88.7	5.5	5.8	*30.4*	88.7	8.0	3.3	*6.8*	88.5	7.2	4.3	*12.7*
11	4	88.3	5.1	6.6	*35.9*	88.2	7.7	4.1	*9.8*	86.0	5.4	8.6	*16.2*
13	4	88.3	5.1	6.6	*38.8*	84.7	6.2	9.1	*8.4*	84.8	4.9	10.3	*22.0*
15	4	87.9	4.8	7.3	*42.4*	84.7	6.2	9.1	*13.8*	84.1	4.7	11.2	*25.3*
11	5	88.7	5.5	5.8	*30.3*	88.7	8.0	3.3	*6.9*	88.5	7.2	4.3	*12.7*
13	5	88.3	5.1	6.6	*35.3*	88.2	7.7	4.1	*9.2*	86.0	5.4	8.6	*14.8*
15	5	88.3	5.1	6.6	*38.0*	84.7	6.2	9.1	*6.0*	86.0	5.4	8.6	*21.0*

Table 2. Values of I for the three considered parameters in case of Test 1 and Test 2, for an LVQ net with N = 1.

ρ_1 with the data set of Test 1, doesn't allow to locate any threshold determining an improvement of P. This can be attributed to the symmetry of the chosen data, for which the misclassified samples have almost always a distance from the winner prototype comparable with the distance of the correctly classified samples. Conversely ρ_2 and ρ_3 perform well and, for $C_m/C_r > 3$, ρ_2 makes the increment of P greater than that generated by ρ_3. On the contrary, on Test 2, ρ_1 gives the greatest increment of P: in this case the distance of a sample from the prototype is a good indicator of the classification reliability. The above results show that, as expected, we cannot select a parameter performing better than the others independently of the distribution of the samples and/or of the cost values in the specific domain. As for $N = 5$, in Test 2 the recognition rate obtained with the W-rule decreases, as expected and shown in Table 1, but P still improves: the parameter ρ_1 performs better than the other parameters, with I ranging from 25.6% to 38.3%.

In conclusion, the method confirmed to be especially useful in recognition problems characterized by high variability among the samples belonging to a same class and by partial overlaps between the regions pertaining to different classes.

References

1. S. Becker, Y. Le Cun: Improving The Convergence of Back-Propagation Learning With Second Order Methods, in Proc. of the 1988 Connectionist Models Summer School, D. Touretzky, G. Hinton, and T. Sejnowsky Eds., San Mateo, CA: Morgan Kauffman, pp. 29-37, 1989.

2. J. Y. Han, M. R. Sayeh, J.Zhang: Convergence and Limit Points of Neural Network and its Application to Pattern Recognition, IEEE Transactions on Systems, Man, and Cybernetics, Vol. 19, No. 5, pp. 1217-1222, 1989.

3. R. P. Brent: Fast Training Algorithms for MultiLayer Neural Nets, IEEE Transactions on Neural Networks, Vol. 2, No. 3, pp. 346-354, 1991.

4. S. E. Fahlman: Faster-Learning Variations on Back-Propagation: An Empirical Study, in Proc. of the 1988 Connectionist Models Summer School, D. Touretzky, G. Hinton, and T. Sejnowsky Eds., San Mateo, CA: Morgan Kauffman, pp. 38-51, 1989.

5. M. E. Hellman: The Nearest Neighbor Classification Rule with a Reject Option, IEEE Transactions on Systems, Science and Cybernetics, Vol. 6, No. 3, pp. 179-185, 1970.

6. L.P. Cordella, C. De Stefano, F. Tortorella, M. Vento: A Method for Improving Classification Reliability of Multi-layer Perceptrons, to appear in IEEE Transactions on Neural Networks, Vol. 6, No. 5, 1995.

7. T. Kohonen: The Self-Organizing Map, Proceedings of the IEEE, Vol. 78, No. 9, pp. 1464-1480, 1990.

8. L. Xu, A. Krzyzak, E. Oja: Rival Penalized Competitive Learning for Clustering Analysis, RBF net and Curve Detection, IEEE Transactions on Neural Networks, Vol. 4, No. 4, pp. 636-649, 1991.

Shape Features

A Model Based Method for Characterization and Location of Curved Image Features

Thierry Blaszka and Rachid Deriche

INRIA Sophia-Antipolis, BP 93, 06902 Sophia-Antipolis Cedex, France

Abstract. This paper deals with the development of a parametric model based method to locate and characterize accurately important curved features such as ellipses and B-splines based curves. The method uses all the grey level information of the pixels contained within a window around the feature of interest and produces a complete parametric model that best approximates in a mean-square sense the observed grey level image intensities within the working area. Promising experimental results have been obtained on real data.

1 Introduction

This paper presents an approach which is a natural extension and generalization of the work presented in [3]. It deals with the localization and characterization of curved image features.

After this introduction, a first section is devoted to the modelization of the image features, then the next one will present the evaluation of the parameters of our models, the third section will be devoted to the experimental results and the perspectives and the applications of this work will conclude.

2 Characterization of Image Features

The linear models, defined in [3], are very useful for indoor scenes because of their polygonal environment. But for more general processes primitives not limited to lines are required, and to this end features delimited by curves will be considered.

The motivation is to have a complete characterization of curved features and to propose an approach that allows us to detect them with a sub-pixellic accuracy. Following the ideas used in the case of the linear features, the considered curve features are of the same global type: n regions with constant intensity defined by lines or by curve boundary in the working area. This type of attributes can be defined by the use of the function of Heaviside U. This function allows us to define features with sharp edges, but in the real images such attributes don't appear because of the blur introduced by the acquisition system. Then, a convolution operation with some smoothing kernel S is used to characterize this blur; these functions are defined as:

$$U(x) = \begin{cases} 1 \text{ if } x \geq 0 \\ 0 \text{ if else} \end{cases} \quad I \otimes S(x,y) = \int_{\mathbb{R}} \int_{\mathbb{R}} S(\alpha,\beta) I(x-\alpha, y-\beta) d\beta d\alpha \ . \quad (1)$$

The considered smoothing kernels are the Gaussian and the exponential filters introduced in [1]:

$$g(x) = \frac{1}{\sqrt{2\pi}\sigma} e^{-\frac{x^2}{2\sigma^2}} \quad G(x,y) = g(x)g(y) \quad e(x) = \frac{\alpha}{4}\left(\alpha|x| + 1\right)e^{-\alpha|x|} \quad E(x,y) = e(x)e(y) \ .$$

These filters lead us to define the models which will be denoted, in the rest of the paper, as Gaussian model and exponential model depending of the smoothing kernel used.

2.1 Ellipse Models

In our context, the simplest way to define ellipses, is to consider their analytical formulation: $N_l(x,y) = \frac{x^2}{a^2} + \frac{y^2}{b^2} - 1 = 0$, where a and b denotes the lengths of the ellipse axis. The combined use of this equation and the Heaviside function yields to the model of a sharp ellipse: $U(N_l(x,y))$. In order to consider more general ellipses, a frame change is done to take into account the orientation θ of its axis, and the position (x_0, y_0) of its center. Considering the new coordinates (x', y') and adding the grey-level intensities inside (A) and outside (B) the ellipse, the expression of the model becomes: $N_l'(x, y, x_0, y_0, \theta, a, b, A, B) = (A - B)U(N_l(x', y')) + B$. Convolving this model with one of the smoothing kernels (Gaussian or exponential), leads to the general model of ellipse, defined by eight parameters, $M_l(x, y, x_0, y_0, \theta, a, b, \sigma, A, B)$.

At this stage, one can note the difference between the approach of Lipson et al in [4] which first computes the ellipse parameters, and then evaluates the mean grey-level inside it, while our model intrinsically includes the radiometric (grey-level and blur) and geometric informations.

2.2 Closed B-Spline Models

In order to deal with a larger class of curved features, another type of curve model based on B-Spline closed curves is defined.

Within the large set of possible B-spline curves, only the subset of the smooth closed curves defined by their degree d and their control vertices of multiplicity one are used. The points $\mathbf{M}$ of such a B-Spline curve are defined as:

$$\mathbf{M}(t) = \sum_{i=1}^{n} \mathbf{V}_i \mathbf{B}_i^d(t) + \sum_{i=1}^{d} \mathbf{V}_i \mathbf{B}_{n+i}^d(t) \tag{2}$$

where t is the parameter varying along the curve, n the number of control vertices $\mathbf{V}_i$, and $\mathbf{B}_k^d$ the basis functions of degree d.

For our application, the selected curve has a fixed degree and a fixed number of control vertices, then its parameters are only the position of the control vertices. This leads to a model with $2n + 3$ parameters: the coordinates of the control vertices, the grey-level intensities inside and outside the curve (A and B), and the blur coefficient (σ).

This representation of curve prohibits the use of the Heaviside function in its analytical form, and consequently a close form for the smoothed model can't be derived. To solve this problem, an algorithmical solution is used. First, a synthetic image of the curve is created and it is filled by the use of one classic algorithm. At this stage, the model of a B-Spline curve $N_b(V_1, \ldots, V_n, A, B)$ including the grey-level intensities, but without smoothing is defined. The next step is to smooth this image, using a discrete convolution. As for the previous features, two models of B-Spline curve are considered depending on the smoothing filter (f) used, Gaussian or exponential:

$$M_b^f(V_1, \ldots, V_n, \sigma, A, B) = N_b(V_1, \ldots, V_n, A, B) \otimes F(\sigma) \tag{3}$$

Due to the CPU time needed by a direct convolution operation, the recursive implementation of the Gaussian and the exponential smoothing described in [2] and [1] are used. These approaches lead us to reduce the computation time twice, at least, without any lost of precision.

3   Approximation of the Data

To characterize the features from the images, using the previous models, an iterative method called the model based approach is used. This method supposes to have a region of interest around the feature to characterize and a feature type selected. But an iterative method needs a first vector to initiate the process, and even if the method has been proven to be robust (see the experimental part), starting with a parameter vector far from the solution leads to a great CPU time. To tackle this problem, a fast method called *variance descent approach* has been developed. This method is designed to fastly produce a close initial parameter vector which is a rough solution to the minimization process of the model based approach.

3.1   Variance Descent Approach

This method is based on the remark that the considered curves define two iso-intensity regions in the working area. If the parameters of this feature are known and if there is no blur then the sum of the standard deviations within each region will be null. Therefore this method consist to define an energy criterion Σ which is the sum of the standard deviations of each region defined by the curve in the working area.

Case of Ellipses. Considering a first ellipse given inter-actively by the user or by a previous process, the energy Σ corresponding to this initial ellipse is evaluated. Then, each geometric parameter of the ellipse is moved from its initial position, keeping the others unchanged, and the energy Σ is computed. The parameter set corresponding to the minimal energy term computed is retained and the process iterates until the energy stops to decrease. The way the parameters are moved depends on the considered parameter: the center of the ellipse is moved in the eight directions corresponding to its eight neighbors, the axis lengths and orientation are increased and decreased separately. The initialization vector is composed of the founded geometric parameters, the grey-level intensities are set to the means of each region and the blur coefficient is set to 1.

Case of B-Spline Curves. the method used in this case is derived directly from the ellipse algorithm: the initial energy term is calculated; each control point is moved in the eight directions of its eight neighbors, while keeping the other control points invariants; the energy term corresponding to this new set of control points is calculated; the set of control points corresponding to the lower energy term is retained and the process iterates until the energy term stop to decrease.

As expected due to the number of control points of the curve, this direct method, denoted direct *vda* approach in the rest of the paper, is computationally very expensive. Then, a more efficient method in term of CPU have been developed: the following steepest gradient method, denoted gradient *vda* method. This approach corresponds to evaluate the initial energy, to compute the gradient of the energy function using finite difference, to find the best step in order to minimize the energy function in the gradient direction and to iterate until the process stop to decrease.

The initialization parameter vector is composed of the set of control points founded with one of the previous *vda* approaches and the grey-level intensities are set to the mean grey-levels of each region and the blur coefficient is set to 1.

3.2 Model Based Approach

The final step is now to start from the close initial conditions provided by the previous approaches and use the following method to evaluate the solution parameters with sub-pixellic precision while taking into account the blur introduced by the acquisition system. The refinement of the parameters is done by a numerical method which is intended to minimize the error function:

$$F(\mathbf{P}) = \frac{1}{m} \sum_{i=1}^{m} \left(M_a\left(i, \mathbf{P}\right) - I(i) \right)^2 \tag{4}$$

where M_a denotes the model of the considered feature, $I(i)$ the intensity of the pixel i, m the number of pixels of the working area, and $\mathbf{P}$ the vector of parameters of the considered feature model. The minimization of this function which is a sum of squares of non-linear functions is done by the routine *lmdif* of the *Minpack* library which implements the Levenberg-Marquardt algorithm.

4   Experimental Results

To test our models and the robustness of our method a lot of experimentations on noisy synthetic data and on real images have been done. But only results on real images are presented here.

Fig. 1. Application of ellipse models.   **Fig. 2.** Application of B-Spline models.

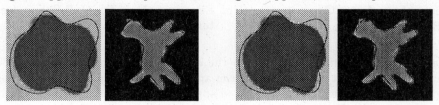

Fig. 3. The direct *vda* method.   **Fig. 4.** The gradient *vda* method.

Figure 1 presents the application of the ellipse model. On the left image the manual initialization is drawn in black, and the convergence of the *vda* approach in white. In the right image lies the result of the model based approach in white initialized by the *vda* in black. Following the same scheme, Fig. 2 presents the application of the B-Spline curve model on the same image. One can note that the final results are the same in the two types of model, but the B-spline based approach is two-times faster.

In the case of B-Spline curves Fig. 3 and 4 show the results of the application of the direct *vda* method and of the gradient *vda* method respectively. The CPU time required for the first approach is about 200 seconds for the cloud image (left) and about 650 seconds for the dog image (right). But the gradient *vda* method takes roughly just 30 seconds. However, it is worthwhile to note that the initial conditions provided by these two approaches both leads to the same result when applying the model based approach.

The robustness of the method is illustrated by Fig. 5 where the initialization (in black in the left image) was given far from the solution. The result of the gradient *vda* method is shown in white of the left image and the result of the Gaussian model of B-spline initialized by the previous process corresponds in a satisfactory way to the solution (in white in the right image). This illustrates the fact that the model of B-Spline has a good convergence on images of smooth curves and the use of the *vda* allows to reduce drastically the CPU time of the convergence.

The application of our B-Spline models on real images is presented in Fig. 6. The black curves show the results of the *vda* approaches (Left direct *vda* and right gradient *vda*) and the white curves show the results of the application of the exponential model initialized by the black curves. In term of CPU time the direct *vda* approach is very long, up to ten times longer than the gradient *vda* approach. The CPU time required by the model based approach initialized by

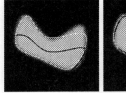

Fig. 5. Combination of the gradient *vda* and the Gaussian model of B-Spline.

Fig. 6. The *vda* approaches and the Exponential model of B-Spline (see text).

the direct *vda* output was 250 sec on the image representing the dog, and 300 sec using the output of the gradient *vda* approach (right Fig. 6). On the image representing the cloud, the model based approach has converged in 140 sec with the output of the direct *vda* approach as initial condition (left Fig. 6) and in 150 sec using the output of the gradient *vda* approach.

The direct *vda* method allows to be slightly faster but globally the sum of the CPU times of the two steps is smaller in the case of the use of the gradient *vda* method. Due to the fact that the accuracy of the model based approach is the same in the two cases of initialization, one can consider the *vda* gradient approach as the good way to produce a close initialization.

5 Conclusion

An efficient model based method has been developed to locate and characterize precisely curved image features. Two different models have been developed to describe efficiently these features and a minimization process has been proposed to find the parameters that best approximate locally the observed grey level image intensities. Among the directions in which the approach presented in this paper can be extended, one can consider the generalization of the models to take into account non planar intensity regions, and the application of the estimation of the blurring parameter to the problem of recovering depth from focus.

References

1. R. Deriche. Fast Algorithms For Low-Level Vision. *IEEE Transactions on Pattern Analysis and Machine Intelligence*, 12(1):78–88, January 1990.
2. R. Deriche. Recursively Implementing the Gaussian and Its Derivatives. In *Proc. Second International Conference On Image Processing*, pages 263–267, Singapore, September 7-11 1992.
3. R. Deriche and T. Blaszka. Recovering and Characterizing Image Features Using An Efficient Model Based Approach. In *Computer Vision And Pattern Recognition*, pages 530–535, New-York, June 14-17 1993.
4. P. Lipson, A.L. Yuille, D. O'Keeffe, J. Cavanaugh, J. Taaffe, and D. Rosenthal. Deformable Templates for Feature Extraction from Medical Images. In O.D. Faugeras, editor, *First European Conference on Computer Vision*, pages 413–417, Antibes France, April 1990.

Estimation of Curvature and Tangent Direction by Median Filtered Differencing

J. Matas, Z. Shao, and J. Kittler

University of Surrey, Guildford, Surrey GU2 5XH, United Kingdom

Abstract. We present a new method, *median filtered differencing*, for estimation of tangent direction and curvature of digitised curves. On three synthetic examples and two images we show the algorithm performs successfully on both straight and curved segments even in the neighbourhood of discontinuities.

1 Introduction

Curvature estimation is closely linked to a number of problems studied in connection with object recognition, eg. curve partitioning [3], corner detection [10] or extraction of salient points [15]. It is therefore not surprising that a number of curvature estimation methods have been proposed in literature; see [18] [4] [8] for recent surveys.

Worring [17] recognises three classes of approaches to curvature estimation: orientation based [13] [1] [2], path based [9] [11] [10] [14] and osculating circle based [16]. The classification is based on disparate definitions of discrete curvature. The formulations widely differ, but the central underlying assumption remains similar: the discrete curve, typically defined as a sequence of 8-connected pixels, represents a set of samples of the original continuous curve corrupted by *uncorrelated gaussian noise*. In a subsequent step the least squares machinery is invoked to fit a spline [11], low order polynomial [7] or circle[16]. Kalman filtering [14] requires a similar assumption.

In our opinion sampling of a continuous curve followed by corruption by additive independent gaussian noise is a very poor model of digital curve formation. First, the very fact that edge detection, boundary tracking etc. produces a connected sequence shows the dependence of noise at neighbouring pixels. Rasterisation noise, which can be defined as the distance from a smooth continuous curve to the the nearest point on a discrete grid, is strongly spatially correlated (see fig.4b) and not at all gaussian. The poor match between standard assumption and the actual process of formation and detection of discrete curves might be one of the reasons why Flynn [4] had to conclude that the curvature estimation methods he studied give reasonable results for images with noiseless, real-valued measurements, but none of the five methods gave good estimates after truncation even in the absence of additive noise. Another significant problem related to least square approaches stems from the lack of robustness. Usually, as in images depicted in figures 4a and 7a, the analysed curve has only *piecewise* smooth derivatives. Fits computed in the neighbourhood of a discontinuity are distorted as they are based on data belonging to two different smooth segments.

In the paper we propose a new method for estimating tangent direction and curvature based on two assumptions: 1. errors due to noise and rasterisation are symmetrically distributed and 2. at any location along a curve, at least 50% of neighbouring points belong to the same smooth segment. These assumptions lead naturally to the *median filtered differencing* method described in section 2. Experiments on curvature and tangent direction estimation on curves linear and curved segments separated by discontinuities in orientation and curvature are reported in section 3. Results are summarised in section 4.

2 Median filtered differencing

In the proposed method curvature estimates are obtained by two differencing steps. First, to estimate the tangent direction, median filtered differencing is applied to a sequence of points representing the digital curve. Next, the same computation is performed on the sequence of tangent direction estimates. Will present the algorithm in the form used for tangent direction estimation because of its intuitive geometric interpretation.

Algorithm 1: **Median filtered differencing**

1. Let $p_i, i = 1, \ldots, N$ be a set of 8-connected pixel locations. At each p_i define a set of $2M$ difference vectors $d_{i,i+j}, j = -M, \ldots, -1, 1, \ldots, M$, such that

$$d_{i,i+j} = \begin{cases} p_{i+j} - p_i & j = 1, \ldots, M \\ p_i - p_{i+j} & j = -1, \ldots, -M \end{cases}$$

2. Represent $d_{i,.}$ in polar coordinates. Sort $d_{i,.}$ according to the polar angle θ.
3. Let θ_i be the i-th angle in the sorted sequence. The direction of tangent dir_t is estimated as a median of the $2M$ angles:

$$dir_t = (\theta_M + \theta_{M+1})/2$$

Examples of the application of algorithm 1 shown in figures 1-3 represent prototypical situations. Figure 1 depicts a noisy rasterised straight segment. The difference vectors $d_{i,.}$ are shown in fig. 1a, sorted and translated in fig.1b. Vectors marked '2x' appear twice. In this example, M equals four. The median angle, ie. the estimated tangent direction, of the eight difference vectors is denoted dir_t. Note that for a straight line, the differences from points farthest from p_i are in the *center* of the distribution because they are least influenced by rasterisation noise.

Figure 2 demonstrates behaviour of alg. 1 near a discontinuity in orientation. The distribution of $d_{i,.}$ consists of two components: the directions belonging to the segment with p_i and outlier directions from the other side of the discontinuity. As p_i approaches the discontinuity, the proportion of outliers will grow, but will remain just under 50%. The median is therefore not grossly influenced [5]. Unlike methods based on smoothing the median differencing performs well near corners, *regardless* of the size M of the neighbourhood used to obtain the estimate.

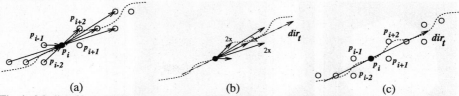

Fig. 1.: Median filtered differencing on noisy rasterised straight line. (a) Sequence of points around p_i and the $2M$ difference vectors, $M = 4$ (b) Difference vectors translated to p_i. (c) The estimated tangent direction dir_t superimposed on the original curve.

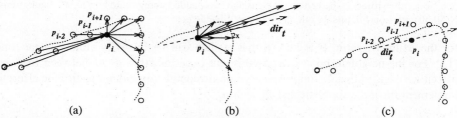

Fig. 2.: Median filtered differencing. (a) Sequence of points around p_i and the $2M$ difference vectors, $M = 6$ (b) (c) see fig. 1

Figure 3 show performance of alg. 1 on a general smooth curve. In this case difference vectors far from p_i are strongly biased, vectors from immediate neighbourhood of p_i are significantly influenced by discretisation. The median is most likely drawn from difference vectors with average distance from p_i (see figs. 3a-c). To summarise, we can conclude that the algorithm behaved intuitively correctly in all three cases. On a straight segment it estimated the tangent direction from points as far apart as possible. Near a breakpoint, outliers were rejected. On a smooth curve with non-zero curvature the algorithm finds a compromise between systematic bias and errors due to noise.

The same algorithm was used for curvature estimation with $d_{i,i+j}$ replaced with $(\delta\theta, j)$. The algorithm effectively selects a median slope of the tangent direction as function of arc length.

As implemented the efficiency of the method is $O(NM \log M)$, ie. it is linear in the number of points at which estimates are sought. The sorting of directions is responsible

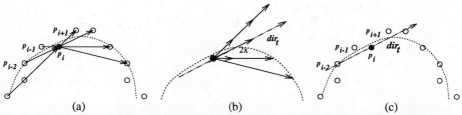

Fig. 3.: Median filtered differencing on a general smooth curve. (a) Sequence of points around p_i and the $2M$ difference vectors, $M = 4$. (b)(c) see fig. 1

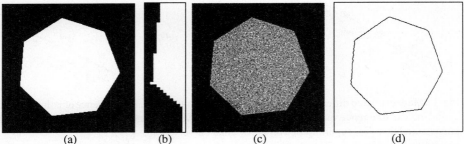

Fig. 4.: Heptagon. (a) Rasterised, noiseless. (b) Magnification of the area around the leftmost corner. Notice the artifact at the vertex. (c) Gaussian noise added; foreground N(200,30), background N(50,30). (d) noisy boundary output by Canny edge detector.

for the $M \log M$ term. It is known that median can be found in almost linear time [12]. For the neighbourhood sizes used in our experiments with M between five and fifteen we decided that the improved asymptotic complexity did not justify the effort to implement the more sophisticated algorithm.

3 Experiments

Due to lack of space we will describe only experiments on two 2D shapes. The heptagon depicted in fig. 4a, was chosen because it demonstrates performance of our algorithm on linear segments with different directions with respect to the discrete grid. The silhouette of scissors shown in fig.7 comprises a number of smooth segments as well as a number of discontinuities.

Figures 5 and 6 summarise performance of median filtered differencing on the heptagonal shape. Fig. 5a shows the angle estimate computed on the boundary of the noiseless rasterised image 4a. The results is close to the 'ideal' step function showing that the algorithm is a capable of filtering the highly correlate rasterization noise. In a

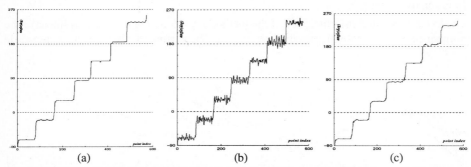

Fig. 5.: Heptagon. Tangent angle estimation. (a) Median differencing of noiseless rasterised image 4(a), $M = 13$. (b) Canny edge detection, gradient direction (c) median differencing on edge string 4d, $M = 13$

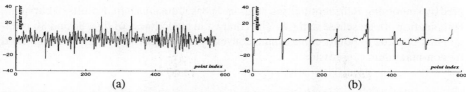

(a)   (b)

Fig. 6.: Heptagon. Error of tangent angle estimation for the Canny edge detector(a)(see fig. 5b) and the median filtered difference on the edge string (see fig. 5c).

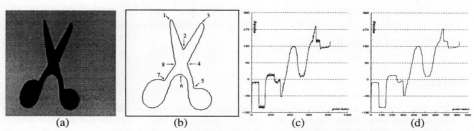

(a)   (b)   (c)   (d)

Fig. 7.: Scissors. (a) Original image. (b) Boundary output by Canny edge detector. (c) Canny edge detector, gradient direction. (d) Tangent angle direction estimated by median differencing, $M = 15$

following experiment, gaussian noise was added to the heptagon image (fig. 4c). Next, we run Canny edge detector to obtain boundary of the heptagon (fig. 4d). Edge direction is a part of the edge detector output (fig. 4b). We applied algorithm 1 to the edge string. Figures 5b-c and 6a-b compare the results. On the straight section, the standard deviation of the error in angle estimate (from the known ground truth) is reduced by a factor of 10. To achieve similar results by linear filtering, significant smoothing would take place at corners. When interpreting the comparatively large errors of median differencing near the corners we have to bear in mind the corners were rounded off by edge detection, not by smoothing of the estimation process.

The first part of the 'scissors' experiment, summarised in figure 7, shows that the median differencing produces reliable estimates of the tangent direction on a complex contour. Comparing figs. 7c and 7d it is clear that the noise reduction is *not* accompa-

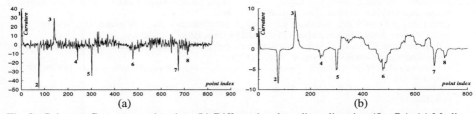

(a)   (b)

Fig. 8.: Scissors. Curvature estimation. (b) Differentiated gradient direction (fig. 7c). (c) Median filtered differencing of angle estimate (fig. 7d). Numbers mark points of maximum convexity and concavity (above a threshold). median

nied by smoothing at discontinuities; steps and peaks remain sharp. Results of curvature computation are presented in fig. 8. The contrast in smoothness of figs. 8a and 8b is apparent. Selection of points of maximum convexity and concavity is easily accomplished by non-maximum suppression.

4   Conclusions

We have presented a new method, *median filtered differencing*, for estimation of tangent direction and curvature. On three synthetic examples and two images we show the algorithm performs successfully on both straight and curved segments even in the neighbourhood of discontinuities.

References

1. I.M. Anderson and J.C. Bezdek. Curvature and tangential deflection of discrete arcs: a theory based on the commutator of scatter matrix pairs and its application to vertex detection in planar shape data,. *IEEE Transactions on Pattern Analysis and Machine Intelligence*, 6:27–40, 1984.
2. H. Asada and M. Brady. The curvature primal sketch. *IEEE Transactions on Pattern Analysis and Machine Intelligence*, 8:2–14, 1986.
3. Martin A. Fishler and Helen C. Wolf. Locating perceptually salient points on planar curves. *IEEE Transactions on Pattern Analysis and Machine Intelligence*, 16:113–128, February 1994.
4. Patrick J. Flynn and Anil K. Jain. On reliable curvature estimation. In *CVPR'88 (IEEE Computer Society Conference on Computer Vision and Pattern Recognition, Ann Arbor, MI, June 5–9, 1988)*, pages 110–117, Washington, DC., June 1988. Computer Society Press.
5. F. R. Hampel, E. M. Ronchetti, P.J. Rouseseeuw, and W.A. Stahel. *Robust Statistics*. John Wiley, 1986.
6. *11th IAPR International Conference on Pattern Recognition (The Hague, The Netherlands, August 30–September 3, 1992)*, Washington, DC, 1992. IEEE Computer Society Press.
7. C. Lee, R. Haralick, and K. Deguchi. Estimation of curvature from sampled noisy data. In *CVPR'93 (IEEE Computer Society Conference on Computer Vision and Pattern Recognition, New York City, NY, June 15–17, 1993)*, pages 536–541, Washington, DC., Jun 1993. Computer Society Press.
8. R. Legault and C.Y. Suen. A comparison of methods of extracting curvature features. In ICPR1992 [6], pages 134–138.
9. D.G. Lowe. Organization of smooth image curves at multiple scales. In *Second International Conference on Computer Vision (Tampa, FL, December 5–8, 1988)*, pages 558–567, Washington, DC., 1988. Computer Society Press.
10. G. Medioni and Y. Yasumoto. Corner detection and curve representation using B-splines. *Computer Vision, Graphics and Image Processing*, 39:267–278, 1987.
11. T. Pavlidis. Application of splines to shape description. In *Visual Forms – Analysis and Recognition, Proceedings of International Workshop on Visual Form*, pages 431–441, 1991.
12. W.H. Press, B.P. Flannery, S.A. Teukolsky, and W.T. Vetterling. *Numerical Recipes in C*. Cambridge University Press, 1988.
13. A. Rosenfeld and E. Johnston. Angle detection in digital curves. *IEEE Trans. Computers*, C-24:875–878, September 1973.
14. P.T. Sander. Estimating curvature by kalman filters. In *Visual Forms – Analysis and Recognition, Proceedings of International Workshop on Visual Form*, pages 469–477, 1991.
15. Zhimin Shao and Josef Kittler. Estimating angles and curvature features in grey scale images. In E. Hancock, editor, *British Machine Vision Conference*, pages 115–124. BMVA Press, 1994.
16. S.M. Thomas and Y.T. Chan. A simple approach for the estimation of circular arc center and its radius. *Computer Vision, Graphics and Image Processing*, 45:362–370, 1989.
17. M. Worring and A. W. M. Smeulders. The accuracy and precision of curvature estimation methods. In ICPR1992 [6], pages 139–142.
18. M. Worring and A. W. M. Smeulders. Digital curvature estimation. *CVGIP: Image Understanding*, 58:366–382, November 1993.

Adaptative Elimination of False Edges for First Order Detectors

Djemel ZIOU and Salvatore TABBONE

Département de mathématiques et d'informatique, université de Sherbrooke, Qc, Canada, J1K 2R1
Crin/Cnrs & Inria-Lorraine, BP 239, 54506 Vandoeuvre-les-Nancy, France

Abstract. In this paper, we propose a new rule for the elimination of false edges produced by gradient detectors. This rule is based on the characteristics of the image and the properties of the detector used. The rule has been tested on realistic images using multi-scale edge detectors.

1   Introduction

In edge detection, the most commonly proposed schemes include three operations: differentiation, smoothing and labeling. Differentiation evaluates the derivatives of the intensity image. Smoothing consists in the reduction of noise included in the image. Labeling localizes edges and increases the signal-to-noise ratio by suppressing false edges. We are interested in this paper in the elimination of these false edges. We limit ourselves to gradient detectors. In this case false edges originate from noise. The commonly used classification rule of edges as being true or false is that the plausibility measure (i.e., the gradient modulus) of a true (false) edge is above (below) a given threshold. The threshold is the minimum acceptable gradient modulus. Due to the fluctuation of the gradient modulus, edges resulting from this binary decision rule are broken. So, this rule has been improved to take the continuity of edges into account. Indeed, the hysteresis algorithm uses two thresholds and a given edge is true if the gradient modulus of any edge point is above a low threshold and at least one is above a high threshold. Elsewhere, this edge is false.

Another aspect of the elimination of false edges concerns the threshold computation. Usually, a threshold is found using a trial-and-error process and the same threshold is used for all edges of an image. As we will show, the threshold is a function of edge characteristics and of the properties of the edge detector. Consequently, it is not easy to find a single value of threshold for a given image. We have proposed in [4, 7] a cleaning rule for multi-scale edge detection based on the behavior of the ideal step edge in scale space. Improvements for this algorithm are proposed in this paper to be used with any edge model and with a gradient operator combined with any smoothing filter. We have tested our algorithm with gradient and multi-scale detectors using various realistic images. In the next section, we show the influence of scale and edge orientation on the gradient modulus and on the gradient direction. By using the behavior of the gradient we propose two algorithms for the elimination of false edges in section 3. Finally, we discuss the results obtained.

2 Behavior of the Gradient

In this section, we consider the influences of both the scale of the smoothing filter and the edge orientation on the gradient modulus and direction.

Let us consider the influence of the edge orientation on the gradient modulus and direction. For the sake of simplicity, we consider here only the case of an ideal step edge. This edge model is sufficient to show that it is suitable to take the edge orientation into account in the false edge elimination process. Consequently, the consideration of blurred edges and double edges does not affect our study. We have carried out this study in the continuous domain and shown that [6]: 1) The gradient magnitude of rotationally symmetric detectors is unaffected by edge orientation. For rotationally dependent detectors the gradient magnitude is often affected by edge orientation. However, there are rotationally dependent detectors whose gradient magnitude is unaffected by edge orientation as it is in the case of rotationally symmetric detectors. This influence is symmetric at edge orientation equal to $\pi/4$ with an extremum at this orientation. 2) For rotationally symmetric detectors the estimated edge orientation is accurate ; that is, the edge orientation θ is equal to the gradient direction ψ. For rotationally dependent detectors the estimation of edge orientation is often biased (i.e., $\theta \neq \psi$), even if the signal is noise free. However, there are rotationally dependent detectors such that $\theta = \psi$, as is the case for rotationally symmetric detectors. Furthermore, we have shown that in practice the property of invariance to rotation is not preserved, due to the tessellation of the image plane and the numerical error approximation. Rotationally dependent detectors are also influenced by these discretization problems and therefore remain the most sensitive to edge orientation.

Since the gradient modulus may be affected by edge orientation, the suppression of false edges may be sensitive to the orientation of the edge. In fact, the best threshold for one edge may be bad for another edge. We propose to compute an appropriate threshold for each edge pixel using the influence of the edge orientation (see next section).

Now, we will consider the influence of the scale on the edge plausibility (i.e., the gradient modulus). For the sake of simplicity, we examine the convolution of an one-dimensional smooth filter (normalized first order derivative) with one dimensional edges to infer the behavior of edges in two dimensions. The filter considered here is a tempered function fulfilling the regularization requirements of Tikhonov (i.e., exponential filter, Gaussian filter). Let us consider four different noise-free models of step edges which are frequently encountered in indoor images and which can be easily modeled: the ideal, the blurred, the pulse and the staircase model. The ideal edge usually corresponds to the clean border of the objects in the scene. The blurred edge comes from changes in illumination of the scene. The pulse and staircase edges result from mutual illumination between objects that are adjacent or from thin objects over a background. Although the problem of noise is not considered in our study, experimentation shows that the basic theory presented here is not affected by the presence of noise. The influence of the scale on the estimated contrast has been studied in [4, 5, 7] and is summarized by: 1) ideal edge: the plausibility is independent of the scale; 2) blurred edge: the plausibility is an increasing function of the scale. When the scale is large it is equal to the edge contrast; 3) pulse edge: the plausibility decreases when the scale increases. For a large scale, the weaker edge disappears while the stronger one remains, its plausibility is equal to the difference of the two steps; 4) staircase edge: the plausibility of the two maxima increases with the scale. At large scale, the plausibility is equal to the sum of the two steps.

Except the ideal edge, the plausibility of edges increases or decreases with the scale. For a blurred edge, the plausibility is very low at fine scale and it is difficult to distinguish it from noisy edges using only the contrast as criterion. This is also true for a pulse edge at high scale. One practical solution for this problem is to associate to each edge point its own threshold and to bind the scale used. Figure 1.a shows a grey level image containing a pulse edge and a blurred edge. To this image, we have added a strong Gaussian white noise with variance equal to 225. The contrast of the pulse edge is $(50, 20)$ and the contrast of the blurred edge is 25 and its width is 6. Figure 1.b presents edges obtained using Canny's detector [1] at scale $\sigma = 1.5$. The gradient modulus of the pulse is $(41, 8.25)$ and the gradient modulus of the blurred is 17.5. Although the image has been smoothed, the amplitude of noise is greater than both the gradient modulus of the weak edge of the pulse and the gradient modulus of the blurred. Therefore, by using a single threshold it is difficult to suppress false edges while preserving true edges (see figure 1.c). It is necessary to use a specific threshold for each of those edges.

3 Elimination of False Edges

We have shown in the previous section that if a cleaning algorithm is used, the choice of the threshold must take into account the behavior of edges in scale space and the sensitivity of the plausibility measure to the edge orientation. This means that it is suitable to compute one threshold per pixel and therefore to associate for each edge model a specific computation rule. Given an edge model $m(x, y)$, thresholds are computed according to this rule:

$$t(x, y) = t_0 c_{m,\theta,s}(x, y) \tag{1}$$

t_0 is a given threshold which can be considered as the smallest acceptable contrast at a given scale s. The same value of t_0 is used for all edge models. There is no known rule which can be successfully used for choosing this threshold. The trial-and-error process remains a reliable means for choosing the appropriate t_0. The quantity $c_{m,\theta,s}$ (see eq. 2) represents the sensitivity of the detector to the edge orientation (θ) and the influence of the scale (s). The gradient modulus of the response of the filter $f(x, y, s)$ describes the influence of the scale and of the edge orientation. Since $c_{m,\theta,s}$ must be independent of the contrast, we define it as the ratio of the gradient modulus of the response of the filter $f(x, y, s)$ to the gradient modulus of the response of the filter $f(X, Y, s)$:

$$c_{m,\theta,s} = \sqrt{\frac{(I(x,y) * f'_x(x,y,s))^2 + (I(x,y) * f'_y(x,y,s))^2}{(I(x,y) * f'_x(X,Y,s))^2 + (I(x,y) * f'_y(X,Y,s))^2}} \tag{2}$$

where s is the scale used, $X = x\cos(\theta) + y\sin(\theta)$, $Y = -x\sin(\theta) + y\cos(\theta)$, $f(X, Y, s)$ is the rotation an angle θ about the origin of the filter $f(x, y, s)$. The quantity θ is the estimated edge orientation which approximates the gradient direction ψ. The gradient modulus of the response of $f(X, Y, s)$ (i.e., the denominator in eq. 2) is orientation independent. In fact, let $x = \rho\cos(\phi)$ and $y = \rho\sin(\phi)$ be the polar coordinate system, $m(X)$ a linear edge of any type with an orientation $\theta \in [0, \pi/2]$ and $f(x, y, s)$ a smoothing filter. It is easy to show that:

$$\int_0^{+\infty} \int_0^{2\pi} f(\rho\cos(\phi - \theta), \rho\sin(\phi - \theta))m(\rho\cos(\phi - \theta)) - f(\rho, \phi)m(\rho\cos(\phi))\rho d\rho d\phi = 0 \tag{3}$$

This result means that the convolution of an edge with orientation θ with a filter $f(x, y, s)$ taken in the direction θ is equivalent to the convolution of the filter taken in the horizontal direction with a vertical edge. As we have shown earlier, in this case, the value of the denominator in (eq. 2) is the largest or the smallest depending on the detector used. Therefore, the value of $c_{m,\theta,s}(x, y)$ is one if the gradient modulus is orientation independent as is the case with Canny's detector. Elsewhere, it can be smaller (greater) than one when the gradient modulus decreases (increases) depending on the edge orientation and the detector used. For instance, for Deriche's detector[3] $c_{m,\theta,s}(x, y)$ is greater than one. The convolution of the image and the filter $f(X, Y, s)$ can be computed immediately using convolution masks. The cleaning algorithm can be summarized as follow: given the first order derivatives of the smoothed image $I(x, y) * f_x'(x, y, s)$, $I(x, y) * f_y'(x, y, s)$, for each edge pixel compute $I(x, y) * f_x'(X, Y, s)$, $I(x, y) * f_y'(X, Y, s)$ and $c_{m,\theta,s}$ (eq. 2) and then compute t (eq. 1).

In a multi-scale edge detection scheme, one needs to suppress false edges in all recovered images by cleaning each of them before the edge combination process. In general, this method is tedious since the trial-and-error process is required at each scale. To automate this task, we have proposed in [4, 7] an algorithm based on threshold propagation. More precisely, it is possible to compute a threshold for given edge pixels obtained at a high scale and to propagate it automatically at lower scales. The choice of the high scale as a starting point is based on the fact that, at this scale, there are few false edges and they are usually easy to suppress. We give here some improvements of this algorithm taking into account the effect of edge orientation and extending it to any edge model and to any edge detector. The multi-thresholding algorithm is given as follows:

1. Threshold edge pixels obtained at a high scale using the algorithm mentioned above; that is $t_h(x, y) = t_0 c_{m,\theta,s_h}(x, y)$. t_0 represents the smallest acceptable contrast of horizontal and vertical edges at high scale. c_{m,θ,s_h} used in this case represents the sensitivity of the detector to the edge orientation at the same scale.

$$c_{m,\theta,s_h}(x, y) = \sqrt{\frac{(I(x, y) * f_x'(x, y, s_h))^2 + (I(x, y) * f_y'(x, y, s_h))^2}{(I(x, y) * f_x'(X, Y, s_h))^2 + (I(x, y) * f_y'(X, Y, s_h))^2}} \tag{4}$$

2. Threshold edges obtained at smaller scales. The threshold is computed automatically from the one obtained at a high scale; that is, $t(x, y) =$

$$t_h(x,y)c_{m,s_h,s}(x,y),$$

$$c_{m,s_h,s}(x,y) = \sqrt{\frac{(I(x,y) * f_x'(X,Y,s))^2 + (I(x,y) * f_y'(X,Y,s))^2}{(I(x,y) * f_x'(X,Y,s_h))^2 + (I(x,y) * f_y'(X,Y,s_h))^2}} \qquad (5)$$

The influence of the orientation is considered only at the high scale. This influence is transcribed in t_h and therefore it is propagated to the low scales.

4 Experimental Results and Conclusion

We have experimented our algorithm using various real images and various edge detectors. We present here results of Deriche's detector. It should be noted that Deriche's detector is not rotationally symmetric. Consequently, edges produced by this detector are not easy to classify. To eliminate false edges we have used only one threshold combined to the binary decision rule. The use of the hysteresis algorithm leads to better results. Figure 2.a presents an indoor image which contains various edge models. Figures 2.b, 2.c and 2.d are respectively edges obtained at scale $\alpha = 0.9$, true edges and false edges ($t_0 = 9$). Figure 2.e shows edges obtained at $\alpha = 1.2$. At this scale, false edges are suppressed by the propagation of the threshold used at $\alpha = 0.9$. Figures 2.f and 2.g display respectively true edges and false edges.

We have shown that the use of the same threshold for all edges in the image does not lead to sufficient results. Our method takes into account the following parameters: edge orientation, signal-to-noise ratio, edge model, behavior of edges in scale space, multi-scale processing, properties of both the smoothing filter and the differentiation operator. Experiments show the efficiency of the classification scheme to distinguish false edges from true ones and for the propagation of thresholds between scales.

References

1. J.F. Canny. *A Computational Approach to Edge Detection.* IEEE Trans. on (PAMI), 8(6):679–698, 1986.
2. J.J. Clark. *Authenticating Edges Produced by Zero-Crossing Algorithms.* IEEE Trans. on (PAMI), 11(1):43–57, 1989.
3. R. Deriche. *Using Canny's Criteria to Derive a Recursive Implemented Optimal Edge Detector.* Int. Jour. Comp. Vision, 1(2):167–187, 1987.
4. S. Tabbone. *Edge Detection, Subpixel and Junctions Using Multiple Scales.* PhD thesis, Institut National Polytechnique de Lorraine, France, (In french), 1994.
5. D.J. Williams and M. Shah. *Edge Characterization Using Normalized Edge Detector.* CVGIP, 55:311–318, 1993.
6. D. Ziou. *Edge Contrast and Orientation Estimation: What About Rotation Invariance?* IEEE Trans. on (PAMI), *Submitted*, 1994.
7. D. Ziou and S. Tabbone. *A Multi-Scale Edge Detector.* Pattern Recognition, 26(9):1305–1314, 1993.

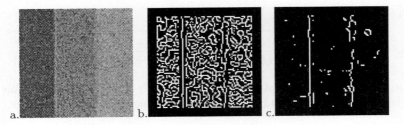

Fig. 1. *a) Original image:* 99×99 *of 170 grey levels, left edge is a pulse of a contrast (50, 20), right edge is a blurred of contrast 25 and width 6 pixels, the variance of noise is 255. b) Edges obtained using Canny's detector with* $\sigma = 1.5$. *c) Cleaned edges, the threshold used is 15.*

Fig. 2. *(a) Indoor image (256 $\times$ 256 pixels and 256 grey levels). (b) Edges resulting from Deriche's detector at the scale* $\alpha = 0.9$. *(c) and (d) True edges and noisy edges* $(t_0 = 9)$. *(e) Edges obtained at* $\alpha = 1.2$. *(f) and (g) True edges and noisy edges resulting from the procedure of threshold propagation.*

Software and Hardware Architectures for Image Processing

An Image Processing Library based on Abstract Image Data-types in C++*

Dennis Koelma and Arnold Smeulders

Faculty of Mathematics and Computer Science, University of Amsterdam
Kruislaan 403, 1098 SJ Amsterdam, The Netherlands
tel.: +31 20 525 7516, e-mail: koelma@fwi.uva.nl

Abstract. The paper presents a library for image processing based on abstract image data-types. The library finds a balance between programmability and efficiency of implementation of image processing operations. Operations are designed using abstract concepts for images and pixel manipulation to be independent of the image data-type. Operations may be optimized in an iterative way by replacing abstract concepts with data-type specific operations, reaching the level of efficiency of operations designed from scratch for a specific image data-type.

1 Introduction

The emerging of abstract data types and the object-oriented (OO) paradigm in good programmer's practice makes it worth while to reconsider the design of image processing libraries. In this paper we will review the design considerations of an image processing library and discuss an implementation of such a library in C++.

Over the years it has proven difficult to reach consensus on the functionality of an image processing library [1]. In fact, not more than 25 % of the 22 libraries surveyed in the framework of the ANSI proposed imaging standard coincide on the bulk of supported operations. The driving force behind the diversity are the conflicting design criteria to which every specific library is a compromise. The conflicting criteria include efficiency, applicability, extensibility, programmability, maintainability, and portability. As they are important in the design of any library we review them briefly.

Efficiency in image processing operations on standard hardware is achieved through decent programming and the use of specialized image data structures [2] or assembler and microcode [3]. In a library optimized for efficiency, images are typically represented in specific form, e.g. two-dimensional grey-value images in [2]. The focus on efficiency in the design reduces the possibility to extend the library with other image data-types. In general, designing a software library with efficiency as the primary target reduces its applicability, programmability, extensibility, and maintainability.

---

* This research was supported by STW project AWI92-1691

The lack of consensus in software libraries for image processing implicates the need for extensibility. Extension of a library with new operations may be facilitated through a high level of programmability. Math-like languages [4] and image algebras [5] use abstract specifications to facilitate implementation of operations. We feel that the high level of programmability in these approaches enhances extensibility but reduces efficiency. The loss of efficiency is due to neighbourhood addressing by indexing, requiring an additional function to translate a logical address to a physical address. Producing efficient code is even more difficult in case specialized data structures are used to store images, e.g. bitmaps for binary valued images [6].

Our aim in designing a software library is to find a balance between programmability and efficiency of image processing operations. The library is based on concepts designed to provide a high level of programmability and a simple optimization process. Operations implemented using the abstract concepts are valid for all existing and future image data-types. Once implemented, operations can be made more efficient in an iterative way by replacing the abstract concepts with data-type specific methods.

A remaining requirement to influence the design of our library is the need for portability. To assure portability, the library is implemented in a standard programming language.

In section 2 of the paper the design of the abstract concepts in the library is described. The implementation of some commonly used image processing operations is demonstrated in section 3. Section 4 describes the optimization of operations.

2 Concepts

Following the OO methodology we distinguish the following important concepts in the library: images, operations on images, and pixel manipulation.

2.1 Images

Image types are usually ordered in a class hierarchy to obtain a clear correspondence between operations and image types. A reasonable class hierarchy for image types is based on the following defining characteristics of images:

- number of dimensions of an image
- form of the pixel representation
- semantics of the contents of an image

The number of dimensions has a large influence on addressing of (neighbouring) pixels in an image. We consider images to be two- or three-dimensional observations of the real world. The number of dimensions of an image is therefore two or three. We do not represent scale spaces or time series in an equal dimension as they are essentially different from images. They are considered to be time or

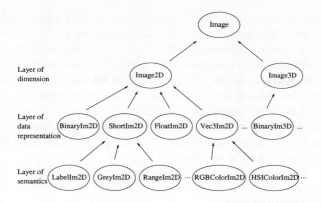

Fig. 1. Image class hierarchy

scale sets of two- or three-dimensional images with their own set of admissible operations in that dimension.

The representation of the pixel data is an important aspect in the definition of the class hierarchy as it determines how operations have to manipulate the actual pixel values. The pixel value representation is covered by introduction of a concept to facilitate future extensions with new pixel forms.

The semantics of the contents of an image gives an interpretation to the pixel values. In most operations, semantics will not influence manipulation of pixel values but in some operations it makes a difference.

We have adopted a hierarchical image class structure (see Fig. 1) avoiding multiple inheritance. Operations that are easily specified for all image types are implemented at the `Image` class level and inherited by all other classes. Operations can be optimized for a specific image type by re-implementing them at a lower level of abstraction in the hierarchy.

2.2 Operations on Images

Operations on images can be expressed in a number of ways. Consider the following list:

(i) `c=a+b`, (ii) `add(a, b, c)`, (iii) `a.add(b, c)`, and (iv) `c.assign(a).add(b)`.

The first way is a very elegant solution. However, the mathematical syntax is difficult to incorporate in a general purpose programming language. It also has an undesired side effect on memory allocation in that temporary objects are created in longer expressions: `c=a+b-d+7` has to be interpreted as `c=((a+b)-d)+7`.

The second approach is the classical way of expressing an operation and its arguments in a function-call. In this case, extension to larger expressions is not elegant. It can only be done through creation of temporary variables or introduction a new function for every new number of arguments.

Translation of the function-call approach to an OO approach results in the third expression. This approach has the same problem as the previous one when it is extended to larger expressions. Furthermore, it implies that objects create new objects which is not consistent with good OO practice [2].

The last approach suits the OO paradigm in the sense that each object hides its data from the outside world and functions are expressed as methods of a class. Also, creation of temporary objects or unnecessary copying is avoided. The approach yields an orthogonal function interface that is easily extensible for the specification of larger expressions: `c.assign(a).add(b).sub(d).add(7)`. The approach is adopted in our library.

2.3   Pixel Manipulation

In a general form, a pixel holds an N-dimensional data vector. Using a class to capture the concept of a pixel introduces a layer of abstraction that leads to a considerable loss of performance. The use of a separate pixel class also implies that storage of pixels cannot be optimized by the image class. For efficiency reasons we do not use a separate pixel class but defer storage of pixels to the image class.

There are two ways of addressing pixels in an image: by indices or by pointers. Indices imply a higher level of programmability than pointers because they offer more degrees of freedom in addressing pixels. The additional degrees of freedom also make operations using indices more difficult to optimize. We have chosen to use pointers for efficiency.

Pixel pointers are based on a common pixel data type. Pixel values expressed in the common pixel data type can be read-from or written-to the location indicated by a pixel pointer.

3   Image Processing Operations

To give an impression of the functionality of the concepts in the library we demonstrate the implementation of some typical image processing operations like point and neighbourhood operations. The example operations are given here without sanity checks to save text.

The first example is pixel by pixel addition of two images, called by `a.add(b)`:

```
Image &Image::add(const Image &op) {
  int npix = size();
  PixelPtr &p1 = pixel_ptr_clone(); PixelPtr &p2 = op.pixel_ptr_clone();
  while (--npix ≥ 0) { p1.write(p1.read() + p2.read()); p1++; p2++; }
  delete &p1; delete &p2; return *this;    //this image contains the result
}
```

Neighbourhood operations can be expressed on an abstract level through the generalized-template operation [5]:

```
Image &Image::g_template(const Image &templ, PixelIter &titer,
                                          PCDF g_mul, PCDF g_add) {
  Image &res = clone(0); PixelPtr &rptr = res.pixel_ptr_clone();
  PixelPtr &sptr= pixel_ptr_clone(); PixelPtr &tptr= templ.pixel_ptr_clone();
  PixelIter &siter = titer.clone(); siter.image(this); siter.check(1);
  for (k=0 ; k<depth() ; k++) {
    for (j=0 ; j<height() ; j++) {
      for (i=0 ; i<width() ; i++) {
        Common val; siter.center(Point(i, j, k)); titer.reset();
        titer.next(tptr); siter.next(sptr); val = g_mul(tptr.read(), sptr.read());
        while (titer.next(tptr)) {
            siter.next(sptr); val = g_add(val, g_mul(tptr.read(), sptr.read()));}
        rptr.write(val); rptr++; } } }
  delete &rptr; delete &tptr; delete &sptr; delete &siter;
  eat(res); return *this;          //cannibalism: steal data and destroy res
}
```

The operation combines weights of a template (templ) with pixel values in the neighbourhood (defined through an auxiliary class PixelIter) of a pixel by means of the g_mul operation. The resulting set of values is reduced to a single value by applying the g_add operation to the elements of the set in an iterative way. The operation will, for example, perform a convolution in case titer is set equal to IterArea2D (iterate over pixels in a rectangular, two-dimensional area in a row-wise order), g_mul is multiplication, and g_add is addition.

4 Optimization

The function add is optimized for the short-valued two-dimensional images by adding a function add to the interface of class ShortIm2D. The virtual function mechanism of C++ results in a call to ShortIm2D::add instead of Image::add whenever the object is an instance of the ShortIm2D class.

Optimization is straightforward in case pixels are actually stored in an array as assumed implicitly by the image concept. For example, optimizing the function add for the image class ShortIm2D consists of replacing the abstract PixelPtr's with pointers to short's and replacing the read and write functions with the pointer dereferencing operation.

Optimization is more difficult in case pixels are not stored in an array-like data structure. The efficiency of binary mathematical morphological operations on binary images can be improved significantly by representing images as bitmaps [6]. Direct translation of an operation defined for the Image class to an operation for the BinaryIm2D class will not produce the desired efficiency. In this case, operations must be redesigned on an algorithmic level to take optimal advantage of the bitmap structure.

To give an impression of the loss of efficiency caused by introducing abstract concepts we compare the execution speed of abstract operations with their optimized counterparts. In the abstract operations two different common pixel representations are used: a vector of three doubles (`Common`) and a single double (`Single`).

As expected, optimized point operations execute at the same speed as C-functions performing the same operations on continuous memory blocks. Abstract point operations using `Common` are typically 4-8 times slower. Using `Single` reduces the factor to 3-6.

Abstract neighbourhood operations have a larger overhead factor because they require more pixel operations. A Sobel operation using `Common` is typically 15-20 times slower, using `Single` yields a factor of 5-10.

5 Discussion

We have presented an image processing library based on abstract data-types that finds a new balance between programmability and efficiency. The library is based on abstract concepts for images, operations on images, and pixel manipulation. Operations based on the abstract concepts are valid for all existing and future image data-types, reducing the amount of software to be maintained. For a library with more than 10 different types of images, as exemplified in Fig. 1, the source code will collapse by a factor over 10. Even greater factors can be reached by using multi-functional operations like the generalized template operation but such only at the expense of a severe loss of performance. The high level of programmability provided by the concepts allows for fast prototyping of new image processing operations, making the library easily extensible.

The improved programmability, maintainability, and extensibility of the library is payed for by a significant loss of efficiency. However, with current workstations a typical operation will still provide interactive response and thus not hamper prototyping the operation. When needed, an operation may be optimized for efficiency upon completion of its development. Our approach of optimizing individual operations allows us to reach the same level of efficiency as operations designed from scratch for a specific image data-type.

References

1. T. Butler and P. Krolak. An overview of the programmer's imaging kernel (PIK) proposed standard. *Computers and Graphics*, 15(4):465–472, 1991.
2. J. Piper and D. Rutovitz. Data structures for image processing in a C language and Unix environment. *Pattern Recognition Letters*, 3:119–129, March 1985.
3. A. C. Sleigh and P. K. Bailey. DIPOD: an image understanding development and implementation system. *Pattern Recognition Letters*, 6:101–106, July 1987.
4. D. Hatfield, R. A. Miner, and C. T. Wilkes. A mathematical expression language for imaging applications. *Computers and Graphics*, 15(4):495–506, 1991.
5. G. X. Ritter, J. N. Wilson, and J. L. Davidson. Image algebra: an overview. *Computer Vision, Graphics, and Image Processing*, 49:297–331, 1990.
6. R. van den Boomgaard and R. van Balen. Methods for fast morphological image transforms using bitmapped binary images. *CVGIP: Graphical Models and Image Processing*, 54(3):252–258, May 1992.

Spatial-Temporal Data Management for a 3D Graphical Information System

Minoru Maruyama[1], Teruhiko Teraoka[1], Takashi Tamada[1], and Shogo Nishida[2]

[1] Central Research Laboratory, Mitsubishi Electric Corporation
Amagasaki, Hyogo 661, Japan
[2] Faculty of Engineering Science, Osaka University
Toyonaka, Osaka 560, Japan

Abstract. In this paper, we propose a spatio-temporal data management method for a 3D graphical facility management system. To provide highly interactive environment, we propose a 3D spatial data management method based on multiple 2D-trees. With the method, interactive operations such as smooth walkthrough and interactive information retrieval in 3D space are made possible. Then, we augment the data structure to deal with the temporal information as well as the spatial information. MP-tree is proposed for spatio-temporal data management. With the use of the data structure, efficient temporal search is made possible keeping the performance of spatial search. We also present a 3D electric facility management system using our data management method.

1 Introduction

In most of the geographical information systems (GIS), 3D objects are represented by 2D shapes. However, due to the growth of the computational power of workstations, the system that can directly handle 3D shapes themselves is coming to within the scope. The 3D graphical information system that enables users to interactively explore the virtual 3D environment is very useful for many tasks such as management, analysis and planning etc. In 3D systems, as conventional 2D systems, various interactive operations including interactive information retrieval, smooth walkthrough in 3D space are required.

In this paper, we describe the data management method for a 3D graphical facility management system. To realize interactive operations, rapid spatial search based on the hierarchical data structure is very helpful. Based on the hierarchical data structure, we propose a data management method for spatial information of city area to realize various kinds of interactive operations. Then, we augment the data structure to deal with the temporal information (e.g. the installation/update history of facilities) as well as the spatial information. MP-tree (Multidimensional Persistent tree) is proposed for spatio-temporal data management. With the use of the data structure, efficient temporal search is made possible keeping the performance of spatial search. We also present a 3D electric facility management system using our data management method.

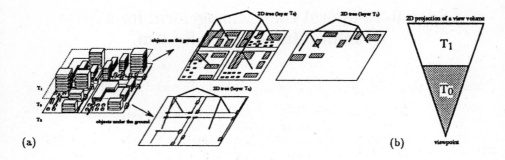

Fig. 1. (a)Management of 3D objects by multiple 2D trees (b)Division of the view volume and the corresponding layers

2 Spatial data management for interactive operations

We propose a spatial data management method for interactive operations in 3D space such as smooth walkthrough. To realize smooth walkthrough, the frame update rate around 10 frames/sec is required. During the walkthrough, usually the number of visible objects is very small compared to the total number of objects. Therefore, if the potentially visible objects are found very rapidly, it prunes unnecessary objects to render and can accelerate the rendering process. To find the potentially visible objects rapidly, efficient range search based on the hierarchical data structure is quite helpful[4]. In our method, each 3D object is managed by using its 2D projection onto the ground. The projected 2D shape is managed by the 2D kd-tree[1]. Potentially visible objects whose 2D projections intersect with the 2D projection of the view volume can be retrieved rapidly by the range search.

For the efficient display control, the objects on the ground and the objects under the ground are managed separately by distinct 2D trees (layers) as shown in Fig.1. Moreover, the space on the ground is further sliced into multiple layers depending on the height. As Fig.1 shows, all the objects on the ground are managed by the tree (layer) T_0. If the object is higher than the given threshold, it is also managed by another tree T_1. In the city area, if an object is far from the viewpoint, it is likely that the object is occluded by the other objects which are closer to the viewer. This implies that an object far from the viewer is visible only if it is tall enough. Based on the heuristics, the search area is divided into two regions (Fig.1(b)). During the walkthrough operation on the ground, in the region which is close to the viewpoint, the potentially visible objects are searched using the layer T_0. In the other region, which is far from the viewpoint, visible objects are searched using the *higher* layer T_1. With this technique, we can expect to reduce the number of objects to render without degrading the image quality.

3   Augmenting the data structure for spatio-temporal data management

3.1   Possible methods to manage spatio-temporal information

Time is an important aspect of real-world phenomena. In this section, the multidimensional data structure is augmented to handle the geometric objects with temporal information. In this paper, we consider spatio-temporal data which consists of location and its insertion/deletion version. It is required to design a data structure that manages temporal information efficiently keeping the performance of the spatial search. We consider several possible methods and discuss about their advantages and disadvantages. The methods considered are :
- **Single tree structure (ST)**
  Geometric objects are managed by a single tree. Version (temporal) information is dealt with as one of the attributes of each object.
- **Multiple trees (MT)**
  For each version, a distinct tree is constructed. Thus, to handle all the data, multiple trees are employed.
- **Multidimensional tree (3DT)**
  N-dimensional data with version information can be regarded as (N+1)-dimensional data. Multidimensional data structures (e.g. kd-tree) can be directly applied to the (N+1)-dimensional data.

Each of the above three methods has its own advantages and disadvantages with respect to the storage cost and the time (search) cost. For example, if a version number is specified, the search performance is very efficient with MT. However, it is not so efficient for spatio-temporal range search, where multiple versions are involved. Moreover, since long-lived data gives rise to many copies, MT requires huge storage cost. On the other hand, ST does not require any *additional* storage cost. However, since the structure of ST does not reflect any time information, ST can be inefficient for time involving search. A 2D point with its insertion/deletion version information can be regarded as a line segment in 3D space. If the data set is static, where time domain is bounded, then 3DT may

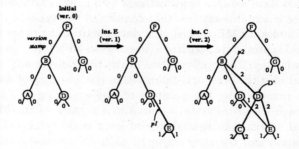

Fig. 2. Persistent binary search tree (node copying method)

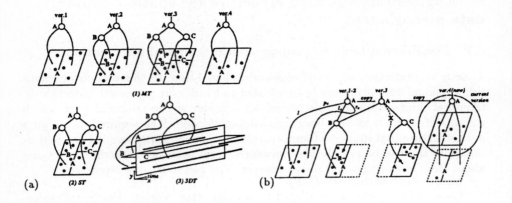

Fig. 3. (a)Data management by ST, MT and 3DT (b)Data management by MP-tree (In the figures, A,B,C show the discriminator values, that specify the partitions)

be a very good data management method. However, if the data set is dynamic, where time domain is unbounded, then 3DT cannot be directly applied.

Based on the above analysis, we try to design a new data structure that satisfies the following requirements: 1. Storage cost is not so expensive as MT, 2. Rapid spatio-temporal search can be performed with the data structure, 3. A dynamic data set can be handled without difficulty, 4. Spatial search for the current version can be performed as fast as in the case of MT.

3.2   The multidimensional persistent tree

To satisfy the above requirements, we use MP-tree (Multi-dimensional persistent tree)[5]. MP-tree is the multidimensional extension of the persistent search tree [3]. MP-tree is constructed by extending the conventional spatial data structures (e.g. kd-tree) using the node copying method[3] (see Fig.2) to save the changes. In Fig. 3 we show data management methods by ST, MT, 3DT and the MP-tree.

The MP-tree is a multi-way tree that consists of two types of nodes : internal nodes and leaf nodes. In MP-tree, spatial data partitioning procedure is same as the conventional spatial data structure (in our case kd-tree). Thus, each internal node has a spatial key. The configuration of the tree reflects not only the spatial information but also the temporal information (i.e. changes of data). Temporal information is managed by the pointers of nodes storing *the time stamp*. In the MP-tree, each historical state is stored in a subtree of the MP-tree. In the spatio-temporal search, the spatial key of each node is examined to get the spatial information and the time stamp of each pointer is examined to get the temporal information while traversing the tree. The features of MP-tree include : 1. A bucket method [2] is applicable, 2. At each version update, the number of

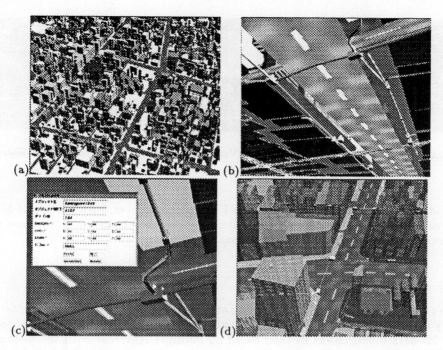

Fig. 4. (a)Entire region to be managed (b) A scene from the viewpoint under the ground (c)Interactive information retrieval by picking the 3D object (d)A scene using transparency control

data to be inserted or deleted is not limited to one, 3. A spatial search for the current version can be performed as fast as in the case of MT, 4. A multiversion range search for the short time interval can be performed faster than ST, MT, 5. The storage cost is much less expensive than MT.

4 A 3D facility management system

We have implemented our spatio-temporal data management method to a 3D facility management system. The system manages 3D objects on the ground such as buildings, houses, roads, and the electric facilities installed under the ground such as pipes, cables, man-holes. In Fig4 (a),(b), we show a scene of the buildings of the area, and a scene of the electric facilities under the ground. By our data management method, users can explore the 3D space freely and interactively retrieve the information of facilities (see Fig.4(c)). Due to the use of 3D models users can intuitively understand the spatial relations, such as proximity. In addition, exploiting some visualization techniques, such as transparency control, understanding of spatial relations between the ground facilities and the

Fig. 5. An example of temporal search. Visualization of the facility installation history

underground facilities gets easier (see Fig.4(d)). Moreover, with the use of MP-tree, our system provides efficient spatio-temporal search. In Fig. 5, we show an example of visualization of facility installation history.

5 Conclusion

We have presented an efficient data management method for a 3D graphical facility management system. To provide highly interactive environment, we propose a 3D data management method based on multiple 2D-trees. With the method, interactive operations such as smooth walkthrough and interactive information retrieval are made possible. For facility management, not only the spatial information but also the temporal information such as facility installation history is very important. We have proposed the MP-tree for efficient spatio-temporal data management. Using the MP-tree, efficient temporal search is carried out keeping the performance of spatial search. We have implemented the data management method for electric facility management system. The 3D graphical facility management system that enables users to explore the virtual 3D environment will become a strong tool for not only data management but analysis and planning.

References

1. J.L.Bentley, "Multidimensional binary search trees used for associative searching" , Comm. ACM, Vol.18, pp. 509 - 517 (1975).
2. H.Samet, "The design and analysis of spatial data structures" , Addison-Wesley (1989).
3. N.Sarnak and R.E.Tarjan, "Planar point location using persistent search trees" , Comm. ACM, Vol.29, pp. 669 - 679 (1986).
4. T.Tamada, Y.Nakamura and S.Takeda, "An efficient 3D object management and interactive walkthrough for the 3D facility management system", Proc. IECON '94, pp. 1937 - 1941 (1994).
5. T.Teraoka, M.Maruyama, and S.Nishida, "The MP-tree : A data structure for spatio-temporal data", Proc. International Phoenix Conference on Computers and Communications '95, pp. 326 - 333 (1995).

PhotoPix: an Object-Oriented Framework for Digital Image Processing Systems

Alisson Augusto Souza Sol
Arnaldo de Albuquerque Araújo

Departamento de Ciência da Computação
Universidade Federal de Minas Gerais
Caixa Postal 702
30.161-970 Belo Horizonte, MG, Brazil
{alisson | arnaldo}@dcc.ufmg.br

Abstract. This work describes how the use of object-oriented technology can help to separate the implementation of algorithms in digital image processing systems from the coding of "non-essential" functionality. The goal was the design and implementation of a system to allow researchers of digital image processing to keep their attention in the issues related to the specific algorithm they are implementing, without worrying about image reading/writing, image displaying/printing, user interface, etc. It describes a framework for digital image processing systems, named PhotoPix, which was designed using object-oriented methodology and coded in the C++ language, using the application programming interface of the Microsoft® Windows™ environment.

1 Introduction

The continuous evolution in microelectronics technology has made possible the production of microcomputers with greater computational power than most past decade mainframes. With the available personal computers, users expect to produce multimedia presentations instead of just plain text documents. So, there is an increasing interest in digital image processing (DIP) systems.

The purpose of DIP systems is the enhancement of desired attributes of digital images and the automation of procedures for information extraction, such as segmentation and pattern recognition [1,2].

Currently, two big problems pervade most implementations of DIP algorithms:
- the numerous formats for files storing digital images [3,4];
- the incompatible hardware platforms and operating environments available nowadays, making difficult the portability of user interface code.

Being almost impossible to code a system with support for every known file format, hardware platform or operating environment, developers must concentrate efforts on a reduced number of choices, based on available technology and resources. However, those choices can make the system useless after a few months, as technical and commercial issues have rendered unused several once popular hardware platforms, operating environments and file formats.

2 System Overview

The goal of the PhotoPix system is to provide a theoretical and practical basis for object-oriented DIP systems. The system should be portable not only to the several hardware platforms and operating environments of present time, but also to the future ones, making real the concept of reusable software [5,6].

Many efforts are usually duplicated when programmers implement DIP algorithms, because most available source code can not be reused. One of the main reasons for that is the difficulty to separate the algorithms code from user-interface code, file manipulation code or other code only needed due to restrictions of hardware technology. In the PhotoPix system, the code implementing DIP algorithms is separated from the code that deals with "non-essential" functionality.

Nowadays, the available hardware is closely approaching the perfect technology concept of Essential Analysis [7]. Therefore, for a system to be portable over time, it must not depend on any particular hardware characteristic -- or it must have minimum dependencies and they must be clearly isolated. Several available APIs (application programming interfaces) already make possible the coding of DIP systems without any references to the hardware. Among them, it was chosen the Microsoft Windows API [8,9] for the first implementation of the PhotoPix system.

Object-oriented programming is currently the best available option to code systems where one needs to encapsulate details and separate functionality in several modules. To code the PhotoPix system, it was chosen the most popular object-oriented programming language: the C++ language [10]. As there is not that same consensus about methodologies for object-oriented analysis and design, several techniques were used to specify and design the PhotoPix system, the main ones being Essential Analysis [5] and the Booch methodology [11].

A widespread solution for the problem of dealing with many different image file formats is the adoption of a unique internal image data representation -- encapsulated from algorithms by access functions of the image abstract data type. Converters between the internal format and some of the several popular image file formats allow the system to exchange data with other systems.

A major problem faced when choosing a data structure for image representation is the number of bits used to represent the quantized spectral information about each pixel. Most systems just adopt a standard spectral resolution, usually 8 or 24 bits/pixel. Those systems will not be able to read/write images from/to several popular file formats. Another common solution is to allow any -- or, at least, many -- spectral resolutions and color systems. However, the usual penalty in this approach is the obligation for the programmer to code many versions of the same DIP algorithm.

To avoid programmers having to code several versions of the same DIP algorithm or to use a generic and inefficient one, the PhotoPix system allows implementation of algorithms that might be executed only on specific spectral resolutions. The algorithm is queried about its ability to process the input images and, if it can not generate a reasonable result, it is not allowed to execute. A possible penalty in this approach is the need for the end-user to convert an image to another spectral resolution before being able to use a particular algorithm.

Several algorithms to convert the spectral information between images with different resolutions were studied. The most difficult issue is the reduction of the number of bits used to store spectral information for each pixel. To solve that problem, it is usual to map the original color space to a reduced one created by uniform quantization of the full spectrum, the popularity algorithm or the median cut algorithm [12]. Other considered issues were the rendering of images using a gray scale palette and local error diffusion [13,14].

3 System Specification

Essential systems analysis [7] was the main methodology used to specify the PhotoPix system. In that technique, the analyst must suppose that the system under specification will be executed by a machine built using "perfect technology" (perfect and infinite memory and an ideal processor with infinite execution speed). So, it is not necessary to list usual requirements that only deal with technological restrictions, as "read file into memory" (something not needed if there were "perfect memory") or "schedule process execution" (something that is usually needed just because a process is computationally so expensive that it can not be executed on-line with currently available hardware).

For a specification done with the assumption of execution under perfect technology, the inevitable requirements for a system lead to what is known as "system essence", formed by the union of the essential activities that the system must execute and the data that must be available for the system to execute them. However, it is important to recognize system borders, so that the analyst does not ignore usual limitations of humans and other systems with which the new one must communicate.

3.1 PhotoPix Essence

The essential memory of the PhotoPix system is a repository of images. Its fundamental essential activity is the execution of algorithms, transforming the images in the repository or generating new ones.

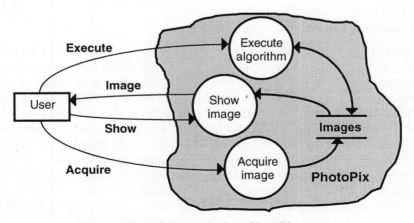

Figure 1 Essential model of the PhotoPix system

The "high-level" essential model of the PhotoPix system, shown in Figure 1, also presents the secondary essential activities of acquiring images from the external world to feed the image repository, and displaying those images in some output device. All those activities will be executed at user's discretion.

4 Design and Codification

The essential model of a software system can be implemented in several different ways. However, most procedural implementations lead to a result that barely resembles the specification. Additional potential problems for digital image processing systems are their high requirements for the hardware platform. Most programmers do not resist the temptation of getting a better performance by using low-level calls to functions available in just a particular video board or graphics accelerator. That code is rarely portable.

A very useful abstraction available today to help in the development of portable code is that of an API (application programming interface). An API can be seen as a documented set of function calls, their implementation for some hardware systems and the specification of a mechanism to use that functionality from code written in some computer language. Through the API, a "virtual machine" is made available to the programmer, helping in the preservation of the essential model in the final code.

After some research and evaluation of several alternatives, the Microsoft Windows API [8,9] was chosen to code the PhotoPix system. That graphical API is currently available to several different hardware platforms, being directly implemented or through emulation.

The use of an API can be a burden to a software system, specially in hardware systems where the API is emulated. Even in that case, the programmer can benefit from the similarity among the several available graphical APIs and make use of the most suitable to a particular platform. However, another solution is the use of class libraries, which are even higher level abstractions available in some object-oriented languages. The PhotoPix system makes use of the MFC (Microsoft Foundation Classes) library [7].

4.1 Identifying Classes

The design of an object-oriented system starts with the identification of abstract data types and their functionality. That can be done by using scenarios, which are supposed snapshots of the final software system when in use. In those snapshots, several objects (class instances) appear collaborating to achieve the system functionality, making possible to identify the semantics of classes and objects.

From the essential model presented in Figure 1, it is easily seen that there might be an image class in the PhotoPix system. That class would offer basic functionality for the execution of the digital image processing algorithms. An abstraction that is harder to be seen by people used to the development of procedural systems is that of an algorithm class. Instances of that class would be called to execute their functionality in some input images, modifying them or generating new ones. The basic functionality would be implemented in an abstract base class. By using inheritance, a new class would be defined for each particular DIP algorithm.

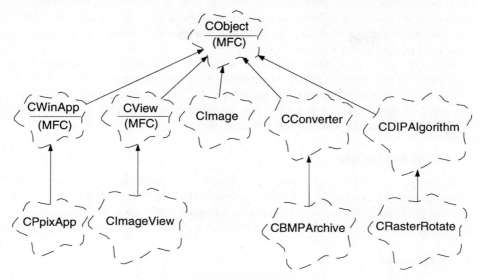

Figure 2 PhotoPix class hierarchy

The class hierarchy used for the PhotoPix system is shown in Figure 2, using the notation of [11] (only inheritance relationship is shown and some levels have been omitted to simplify the picture). The image and algorithm classes are the fundamental ones for the PhotoPix system. However, there is the need for classes that define how messages would flow between the user and the internal objects. That functionality is provided by the document-view architecture of the MFC library [9].

One object of the *CPpixApp* class is used to control the execution of the PhotoPix system. Objects of the *CImageView* class act like lens through which one can view the objects of class *CImage* in the system repository. The *CDIPAlgorithm* class is the basis for derivation of classes that would implement DIP algorithms.

The image repository would ideally be maintained in memory. However, that is not possible with current technology. That leads to the need for the *CConverter* class, which is the basis for classes that make the conversion between the internal representation of image data and image file formats, providing persistence for the image objects.

5 Conclusions

It was presented a simple structure for digital image processing systems, which helps to achieve reusability and efficient use of resources when doing research about DIP algorithms. Using object-oriented technology, it was designed a framework for DIP systems, which had a successful implementation for the Microsoft Windows API, using the C++ language and the MFC library.

Preliminary versions of the PhotoPix system were used in courses on digital image processing offered at the Computer Science Department of the UFMG. After a brief presentation of the interface and mechanisms of the image and algorithm classes, most students with basic understanding of the C language could add at least one DIP algorithm to the system in less than a week. This result indicates that a great level of abstraction was achieved.

The system is now being extended to enable the end-user to create new algorithms by composition of the basic ones. Also, the kernel subsystem is being extended to distribute processing among instances of the system running on different machines. In another future extension, the system will be made able to execute DIP algorithms over images in multimedia sequences.

6 Acknowledgments

The authors would like to thank the Conselho Nacional de Desenvolvimento Científico e Tecnológico - CNPq (grants 400.190/90-7, 500.908/91-5 and 130.990/90-6) and the Fundação de Amparo à Pesquisa do Estado de Minas Gerais - FAPEMIG (grant TEC 1113-90) for the financial support of this work.

7 References

1. Jain, A.K., *Fundamentals of Digital Image Processing*, Prentice-Hall, Inc., 1989
2. Pratt, W., *Digital Image Processing*, John Wiley & Sons, 1978
3. Kay, D.C. & Levine, J.R., *Graphics File Formats*, Windcrest Books, 1992
4. Murray, J. & van Ryper, W., *Encyclopedia of Graphics File Formats*, O'Reilly & Associates, 1994
5. Cox, B.J., *Object-Oriented Programming: An Evolutionary Approach*, Addison-Wesley, Inc., 1986
6. Martin, J., *Principles of Object-Oriented Analysis and Design*, Prentice-Hall, Inc., 1993
7. McMenamin, S.M. & Palmer, J.F., *Essential Systems Analysis*, Prentice-Hall, Inc., 1984
8. Petzold, C., *Programming Windows 3.1*, Microsoft Press, 1992
9. Microsoft Co., *Microsoft Visual C++ compiler manuals*, Microsoft Co., 1995
10. Ellis, M.A. & Stroustrup, B., *The annotated C++ reference manual*, Addison-Wesley, 1990
11. Booch, G., *Object-Oriented Analysis and Design - with Applications, 2nd. Ed.*, The Benjamin/Cummings Publishing Company, Inc., 1994
12. Heckbert, P., Color image quantization for frame buffer display, *Computer Graphics*, Vol. 16, No. 3, Jul. 1982
13. Ulichney, R., *Digital Halftoning*, MIT Press, 1988
14. Sol, A.A.S., *PhotoPix: an object-oriented framework for digital image processing systems*, Master thesis, DCC/UFMG, Belo Horizonte, MG, Brazil, 1993 (In portuguese)

Data Packing vs Processing Speed on Low-Cost Massively Parallel Systems: a Case Study *

Alberto Broggi

Dipartimento di Ingegneria dell'Informazione
Università di Parma
Parma, ITALY, I-43100
broggi@CE.UniPR.IT

Abstract. This work presents some considerations on the hardware organization of the external memory in low-cost massively parallel systems and discusses a few basic criteria for the development of efficient algorithms. These criteria are illustrated with the help of a case study.

1   Introduction

This work discusses the problem of data coding on *low-cost* SIMD massively parallel systems, namely where the Processor Array (PA) is composed of a set of extremely simple Processing Elements (PE) disposed on the nodes of a 2-dimensional mesh.

Since a 1:1 mapping between the data set and the PA is not feasible for generic Image Processing tasks, a processor virtualization mechanism is needed. When each PE can only handle a small amount of memory, the computation is serialized in windows: the PA is loaded with a sub-window of the data set, then the computation is performed until a special instruction is reached, and finally the results are stored back into the memory. These steps are iterated until all the sub-windows have been processed. Then the first sub-window is reloaded again into the PA and the computation is resumed until the next special instruction is found. Generally, due to the low amount of memory associated to a single PE, low-cost systems [1, 4, 5, 6] utilize this second solution (the so-called *external virtualization* [1]), which requires the use of an external memory for data storage.

Moreover, the specific choice of the hardware organization of the external memory is another crucial architectural key point. Two possible solutions can be devised: the first one corresponds to the one considered in this work, where each memory word contains the values of different binary layers belonging to the same image pixel. In the second one, each memory word contains the values of adjacent image pixels belonging to a single binary layer. The former has the disadvantage that for each data transfer a fixed number of binary layers (equal to the bus parallelism) are moved from the image memory to the PA. As a consequence, the data bus efficiency (η_{BUS}) seldom reaches high values,

* This work was partially supported by CNR Progetto Finalizzato Trasporti under contracts 93.01813.PF74, 93.04759.ST74 and 94.01371.PF74.

because the parallelism in the data transfer seldom matches the bus parallelism. Conversely, the latter has the advantage of moving through the data bus only the required amount of information, but it requires some additional hardware. In fact, a single data coming in parallel from a sensor (camera, VCR,...) carries the information of a single pixel, and this second solution would need a hardware *data shuffler* for the transposition of the data. The Connection Machine CM-2 [3] has a hardware extension (*Sprint-chip*) explicitly devoted to this purpose.

The architecture considered in this work, PAPRICA [1], is composed of a 16×16 PA, linked to the image memory through a 16-bit data bus, which transfers 16 bits of a single image pixel in parallel, and utilizes the *external* virtualization mechanism. The aim of this paper is the study of the classical trade-off between data packing and processing speed in the conditions described above.

The following Section presents an Image Processing case study which shows that, thanks to a plain data coding, it is possible to replace the evaluation of complex boolean functions with simple morphological operations. Section 3 ends the paper with some concluding remarks.

2 A morphological algorithm for slope detection

The aim of the algorithm is to label each foreground pixel of a binary image with a value representing the local slope s of the line the pixel belongs to. In the continuous case it is:

$$s(P) \triangleq \lim_{P' \to P} \frac{y' - y}{x' - x} , \tag{1}$$

where P and P' represent two foreground pixels with coordinates (x, y) and (x', y') respectively, belonging to the same line as shown in fig. 1.a.

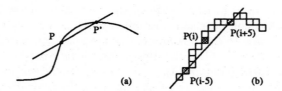

Fig. 1. Slope: continuous (a) and discrete (b) cases

Due to the specific architecture, the problem must be reduced to the evaluation of the slope value through the analysis of the line morphology contained into a finite-dimension neighborhood of pixel P. A 3×3 neighborhood is too small to characterize the line slope, and thus it is necessary to increment the neighborhood dimension.

As an example, fig. 1.b shows a discrete case corresponding to the analysis of the line morphology along an 11 steps long monodimensional neighborhood.

Called $P^{(1)}(x^{(1)}, y^{(1)})$, $P^{(2)}(x^{(2)}, y^{(2)})$, ..., $P^{(n)}(x^{(n)}, y^{(n)})$ the sequence of n foreground pixels, $s_{11}(P^{(i)})$ can be obtained as follows:

$$s_{11}(P^{(i)}) = \frac{y^{(i-5)} - y^{(i+5)}}{x^{(i-5)} - x^{(i+5)}} , \quad \text{with } 5 < i \leq n - 5 , \tag{2}$$

where the subscript "11" indicates the length of the monodimensional neighborhood analyzed. As shown in Eq. (2), the most intuitive solution, directly derived from the continuous case (Eq. (1)), is based on the iterative propagation of the pixels' x and y coordinates. Unfortunately its implementation on low-cost SIMD architectures is quite complex, due to the extremely simple PEs internal structure and interconnection network.

On the other hand, the algorithm presented in this work is based on an *iterative* analysis of a 3×3 neighborhood. The increment in the number of iterations corresponds to the widening of the dimension of the analyzed neighborhood.

The output quantization, namely the number of different slope values is a basic parameter of the algorithm. In the following case study, the final quantization is limited to 16 different slope values.

2.1 Slope coding in a 3 × 3 neighborhood

Assuming that the incoming binary image comes from a thinning filter, no more than two pixels can be set into the 3×3 neighborhood of each foreground pixel. Thus, excluding the case of line-ending pixels, there are only $\binom{8}{2} = 28$ possible neighborhood configurations (3×3 patterns), which are then reduced to 16: in fact, the 3 patterns shown in fig. 2.a and their 9 $\frac{\pi}{2}$-rotations are not possible, since the thinning filter would have deleted the redundant pixels, reducing them to the ones of fig. 2.b. A successive filtering can be added to transform the 4 configurations shown in fig. 2.c, in the ones shown in fig. 2.d. The 12 configurations left are shown in fig. 3. Let us now focus on the line configuration depicted

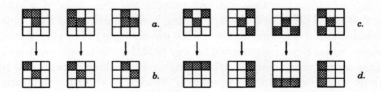

Fig. 2. (a) 3×3 patterns which do not come from a thinning filter; and (b) how they would be transformed by a thinning iteration. A prefiltering step transforms patterns (c) into patterns (d).

in fig. 4.a. For coding purposes, each pixel accumulates the whole neighborhood configuration in its own memory, following the scheme of fig. 4.b, where A,B,...,H are different binary locations in the PEs memory and $a, b, ...g$ represent different

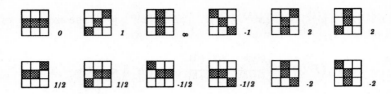

Fig. 3. The 12 valid configurations and their associated slope value

PEs associated to the corresponding foreground pixels of fig. 4.a. This operation requires 8 binary shifts, generally implemented through binary dilations [2]. In this way a 8-bit pattern, shown in fig. 4.d, is associated to each foreground pixel, using a 2-out-of-8 (2/8) coding technique[2].

Only 8 different slope values can be associated to the 12 patterns (as shown in fig. 3). The 8 values, whose coding is shown in fig. 4.c, are computed as the slope of the line connecting the two neighboring pixels in a 3×3 neighborhood. Thus, the number of bits needed to code the 8 different slopes is 3, but as explained in the next subsection, even if a 1/8 slope coding is more redundant, it reduces the computational complexity of the next processing step, thanks to its uncompressed format.

Figures 4.e shows the 1/8 coding, obtained by the application of 8 boolean functions (with an 8-bit input vector each) to the patterns of fig. 4.d and fig. 4.f shows the corresponding slope value.

2.2 Coding a 5×5 neighborhood

Due to the previous uncompressed coding, the logical union of the 1/8 codings of the two neighboring pixels provides a new 2/8 coding for each pixel, as shown in fig. 4.g. Figure 4.h shows the slope value computed on a 5 steps long monodimensional neighborhood: in this case the 2/8 coding corresponds to a 16-level quantization.

The logical union mentioned above reflects the definition of a binary morphological dilation [2][3]; thus, the 2/8 coding can be performed by 8 binary 3×3 neighborhood-based dilations (one for each binary layer). The use of morphological dilations, naturally implemented on any SIMD cellular system, together with a 1/8 plain coding simplifies the 2/8 coding process, since background pixels are transparent with respect to this operation. Moreover, if both neighboring pixels encode the same value, the resulting coding is 1/8, which simplifies furthermore the final step.

The advantage of the use of the 1/8 coding instead of a more compact one (for example using 3 bits only) is mainly due to the possibility to replace the synthesis of complex boolean functions with simple morphological operations.

---

[2] This means that only 2 bits out of 8 can be set at the same time.

[3] In fact a morphological dilation can be defined as the result of the logical union among the pixels belonging to a given neighborhood.

2.3 The iterative process

As shown in fig. 4, the slope values stored with a 2/8 coding is furthermore requantized to 8 levels (the quantization can be measured by comparing the values presented in fig. 4.h and fig. 4.j) by a set of boolean functions in order to obtain a new 1/8 coding (fig. 4.i). The process is iterated until the required neighborhood dimension has been reached.

The rationale underlying the whole process is the following:

- The first 2/8 coding (fig. 4.d) is used to get the neighborhood configuration.
- The 1/8 slope coding (fig. 4.e) encodes the following value:

$$s_3(p^{(i)}) = \frac{y^{(i-1)} - y^{(i+1)}}{x^{(i-1)} - x^{(i+1)}} \ . \tag{3}$$

Moreover, according to the previous assumptions, it follows that:

$$\max \left[\left| y^{(i-1)} - y^{(i+1)} \right|, \left| x^{(i-1)} - x^{(i+1)} \right| \right] = 2 \ . \tag{4}$$

From Eq. (3) and (4) it is now possible to derive: $\begin{cases} \Delta x = x^{(i-1)} - x^{(i+1)} \\ \Delta y = y^{(i-1)} - y^{(i+1)} \end{cases}$.

- The 2/8 coding (fig. 4.g) is obtained through a union of the 1/8 values encoded into each pixel's neighbors. From the 2/8 coding it is possible to extract $s_3(P^{(i-1)})$ and $s_3(P^{(i+1)})$, which encode the four following quantities:

$$\begin{cases} x^{(i-2)} - x^{(i)} \\ y^{(i-2)} - y^{(i)} \end{cases} \text{ and } \begin{cases} x^{(i)} - x^{(i+2)} \\ y^{(i)} - y^{(i+2)} \end{cases} \ . \tag{5}$$

After few algebraic manipulations, it follows that $\frac{x^{(i-2)} - x^{(i+2)}}{y^{(i-2)} - y^{(i+2)}} \triangleq s_5(P^{(i)})$, which represents the line slope along a 5 steps long 1D neighborhood.

It is thus proven that the 2/8 coding (fig. 4.g) encodes the value of $s_5(P^{(i)})$.

- The 8-levels quantization (fig. 4.i) is used to convert the 2/8 coding to a 1/8 coding to be used in the iterative process. It is important to note that in this operation there is a loss of information due to the quantization process.

3 Conclusions

The case study discussed in this paper has shown that the discrete version of the solution in the continuous case is quite complex to be implemented on low-cost SIMD systems. Conversely, due to the specific features of the hardware architecture, a different algorithm has been proposed in this work, based on local computations only and on a simple morphological processing. In fact with a plain data coding the synthesis of complex boolean functions has been reduced to the trivial application of morphological dilations.

The algorithm has been implemented directly in Assembly language on PA-PRICA architecture. Two iteration of this filter on a 256×256 image takes about 50 ms.

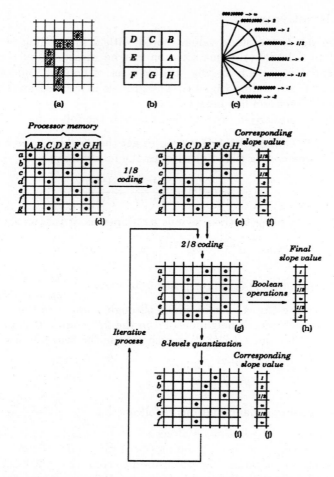

Fig. 4. Example of a slope coding process; the output values in (h) refer to the analysis of a 5 × 5 neighborhood.

References

1. A. Broggi, G. Conte, F. Gregoretti, C. Sansoè, and L. M. Reyneri. The PAPRICA Massively Parallel Processor. In *Proceedings MPCS - IEEE Intl. Conf. on Massively Parallel Computing Systems*, pages 16–30, 1994. IEEE CS - EuroMicro.
2. R. M. Haralick, S. R. Sternberg, and X. Zhuang. Image Analysis Using Mathematical Morphology. *IEEE Trans. on PAMI*, 9(4):532–550, 1987.
3. W. D. Hillis. *The Connection Machine*. MIT Press, Cambridge, Ma., 1985.
4. NCR Corporation, Dayton, Ohio. *Geometric Arithmetic Parallel Processor*, 1984.
5. S. Reddaway. DAP - A Distributed Array Processor. In *1st Annual Symposium on Computer Architectures*, pages 61–65, Florida, 1973.
6. L. A. Schmitt and S. S. Wilson. The AIS-5000 Parallel Processor. *IEEE Trans. on PAMI*, 10(3):320–330, May 1988.

A New Parallel Method Based on a Genetic Approach for Determination and Classification of Skin Spots

Andrea KUTICS   Munehiro DATE

The Institute of Physical and Chemical Research (RIKEN),
2-1, Hirosawa, Wako-shi, Saitama 351-01, JAPAN

Abstract: A low cost CCD camera system connected to a transputer network as a parallel processing device has been developed for the determination of human skin objects. Ultraviolet, visible and penetrative infrared images recorded by CCD cameras are used as the input data. A new genetic method based on mathematical morphology has been developed to detect objects and estimate shape properties of skin objects such as speckles and blood vessels on an arbitrarily chosen area. A very fast system could be achieved, and as no presumptions are used on the shape of the objects in question, the developed method is widely applicable to various kinds of images. The described transputer network was found to be very suitable for object analysis tasks due to its high performance and flexibility.

1 Introduction

Detection and investigation of external and internal skin objects originating from benign or malign skin deformities have a great significance not only in the field of dermatology but also for the cosmetic industry. However it is difficult the obtain appropriate data even by high cost systems like Computer Tomography and Magnetic Resonance Imaging on arbitrary skin areas in few mm depth. The reason for this is not only the low resolution but the difficulty and time requirement of obtaining cross sections in few mm wide intervals by these equipment. There are also instruments to provide pictures on a given spot or on a few mm wide skin area, like ultrasonic dermatology scanners etc. However, there is no system developed to provide automatic shape characteristics of skin objects of a large skin surface. It is not a simple task to determine objects and shapes within images, especially when the time requirements for this operation are severe. The developed techniques are rather specific for a given problem, and use computationally heavy methods for segmentation and pattern matching [1].

This paper presents a new method based on a genetic approach to detect objects and estimate shape properties (perimeter, area, mean diameter, shape factor) and distribution of skin objects like speckles and blood vessels on an arbitrarily chosen area. Our objective was to develop a fast system based on a parallel architecture that can provide global internal information about the human skin of the whole body or given parts of the body in 10 to 15 mm depth and to provide a fast method for the detection of the skin objects.

2 The optical and image acquisition system

For the above purpose infrared, visible and ultraviolet images have been used. The infrared light, even in the near infrared band, where the wavelength is between 700 and 1000 nm, can penetrate to 10 to 15 mm depth under the surface of the skin and the reflected part of this penetrating light can be recorded by CCD cameras. On the other hand, the ultraviolet light beam is totally reflected from the surface without any penetration into the subsurface layers of the skin. A schematic description of the light skin surface interaction is shown in Fig. 1.

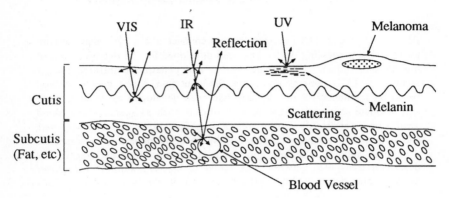

Fig. 1. Lightbeam and skin surface interaction

A CCD camera device, equipped with a mirror system has been developed. Our equipment consists of five CCD cameras, two for the ultraviolet band, filtered by 200-300 nm and 300-400 nm band filters, a visible camera for the 400-700 nm band and two CCD cameras for the infrared band extracted by 700-800 nm and 800-900 nm band filters. The light beam entering the camera system is first split into ultraviolet, visible and infrared bands on a mirror and filter system that is located in front of the cameras. The video signals produced by the camera system are digitized on an 8bit, 10 Msampling/s analog/digital converter and stored in the memory of the main transputer. The total hardware layout is shown in Fig. 2.

3 The transputer network

The transputer network consists of two pipeline connected lattice networks, each of them containing sixteen T800 transputers. This architecture supports both algorithmic and data parallelization at the same time. The main transputer is also a T800 transputer with 4 Mbyte extended memory, while each subnode in the network has 1 Mbyte local memory. The main transputer is connected to the root transputer board located in the host Sun workstation through an RS422 interface. This host machine provides the user interface comprising the user control of the focus and iris, the selection of the Look Up Tables, booting the application into the network and displaying and post processing of the calculated images.

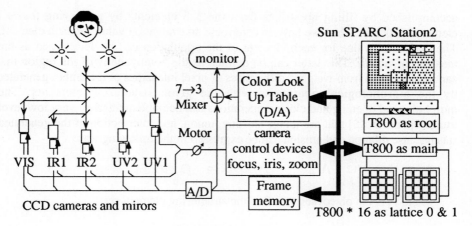

Fig. 2. Hardware layout

4 The imaging method

4.1 Initialization

For the detection of the skin objects, combinations of ultraviolet, visible and infrared band images are composed by using the camera device. First an arbitrary segmentation process is carried out by accomplishing separated two-dimensional Laplacian of Gaussian operation [2]. Object areas surrounded by edges are then decomposed for the determination of the morphological structuring element series representing the starting solution set for the genetic method. In this processing step, the separated consequent one dimensional calculation tasks are carried out parallel on the 32 transputer nodes by applying a careful data division, keeping overlapping data on the neighborhood nodes. Two horizontally adjacent nodes accomplish the separated LoG operation steps and then the resulting data are summed and saved on the appropriate transputer nodes.

4.2 Genetic method

Genetic algorithms were first introduced by Holland [3, 4] and implement evolutionary theories for solving nontrivial, mainly optimization problems. Here we use this approach to provide a new method for object detection. In our case a series of small size (3x3) structuring elements as shape composition blocks of the skin objects to be detected is acting as the element (chromosomes) of the genetic method. The starting structuring element series are determined on the basis of the arbitrarily segmented objects and used to create initial population. A series of morphological openings and closings is carried out by the elements of the operator sets. Then a binary encoding is accomplished for the generation of the initial population on the basis of the results obtained from the morphological operations. This can be accomplished by filling up strings from the 3x3 elements by preserving the 3x3 elements structure. Then we have to determine an evaluation value for each element. The evaluation value for each element of the population can be determined as the measure of fitness. This latter can be defined as the homogeneity of the region that can be calculated from weighed differences of pixel intensities of the umbra generated by the morphological operations with the given structuring elements. The morphological opening and closing operations can be implemented by very low level image operations[5] that are image shifting, union, and intersection of the generated sets because dilation and erosion can be expressed in the following way:

$$A \oplus B = \bigcup_{b \in B} A_b \qquad A \ominus B = \bigcap_{b \in B} A_{-b} \qquad (1)$$

where A is the object set, and B is the structuring element set.

The above algorithm is accomplished on the transputer network by dividing the image into segments, such as each segment includes one or a few skin spots. We use an image scanning and boundary tracing task running on the main transputer to determine the data segments, and another one also running on the main transputer to provide these data for the transputer nodes by multidirectional data loading. The dilation and erosion steps are independent tasks on the nodes. In order to avoid the unbalanced processing on the network, data can be provided from a computationally heavy node to an idling one. It can be accomplished by registering independent tasks for local communication on the four serial links. These communication tasks can also handle the problem of providing external data for the shift operations on a given segment. After calculating the fitness values of the determined objects, finding regions fitting better to the image data is started by carrying out crossovers and some mutations on selected elements of the population that have relatively higher fitness values. The crossovers and mutations are accomplished on randomly paired elements by a given probability that is relatively low especially for mutation (0.01). In our case, the crossovers are two dimensional operations. However, as they are restricted to horizontal and vertical splits, they can be implemented as crossovers of bitsequences. Tasks for carrying out the selection, crossover, and mutation can be accomplished parallel by alternating the processing and data getting and passing tasks by applying a proper timeout. A new offspring of the population can be obtained by the crossovers and mutations and the fitness values of the new elements have to be regenerated. Then a new processing loop can be started until either the required fitness is reached or no more progress can be measured. The average time required for the calculation on 512x512 size images is less than 350 ms applying 32 nodes for images containing 10 to 100 objects and using a population of 50 elements. However, this time is depending very much on the size and number of objects. If the number of objects is more than 100 the efficiency is significantly reduced due to the communication time required for the data reloading. The best fit shapes are then classified by determining the shape factors as the relation between the perimeter, the area, and the mean diameter of the regions. An actual implementations for a female face in Fig. 3.1. The pictures in the third line of this figure present the size and perimeter distributions and

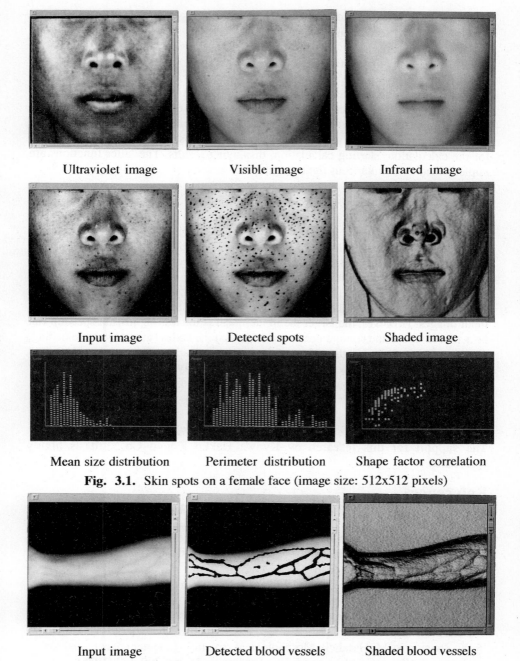

Ultraviolet image   Visible image   Infrared image

Input image   Detected spots   Shaded image

Mean size distribution   Perimeter distribution   Shape factor correlation

Fig. 3.1. Skin spots on a female face (image size: 512x512 pixels)

Input image   Detected blood vessels   Shaded blood vessels

Fig. 3.2. Blood vessels on a female arm (image size: 512x512 pixels)

the shape factor versus size correlation of the spots. Fig. 3.2. shows an application for blood vessel detection on a female arm.

4.3 Network programming and user interface

As a programming language, we have used OCCAM 2 for carrying out the algorithms on the transputer network as it generates an effective code [6]. The user interface is written in C on the host machine providing a convenient environment to the user in controlling and setting the cameras and LUTs, selecting the input picture for the calculation, starting calculation, displaying results. The latter functions are implemented as simultaneous light weight processes.

5 Conclusions

A new parallel system for low cost analysis of skin objects such as skin spots and blood vessels is discussed in this paper. A genetic algorithm has been developed on the basis of morphological structuring elements. A very fast system with about 15 times speedup can be achieved by the 32 transputer nodes as compared to the performance of the 28MIPS Sun workstation 2. Due to the flexibility and high performance of the transputers there are no constraints regarding on the size of the input images. However, the larger the input image is the less significant speedup can be achieved due to the communication overhead on the nodes caused by the reloading. Avoiding this problem is the object of further investigation. The developed method can be easily applied to wide areas of the image analysis field as no preliminary information is required about the objects to be determined.

Acknowledgement

The support for this research by Kyowa Hakko Kogyo Co., Ltd. is highly appreciated.

References

1. R. M. Haralick, L.G. Shapiro: Computer and robot vision. New York: Addison Wesley 1992
2. A. Huertas, G. Medioni: Detection of intensity changes with subpixel accuracy using Laplacian of Gaussian masks. IEEE Transactions on Pattern Analysis and Machine Intelligence 7, pp. 651-664 (1986)
3. J. Holland: Adaptation in natural and artificial systems. Ann Arbor: University of Michigan Press 1975
4. Z. Michailewicz: Genetic algorithms + data structures = evolution programs. Berlin: Springer 1992
5. X. Zhuang, R. M. Haralick: Morphological structuring element decomposition. Computer Vision Graphics and Image Processing 35, pp. 370-382 (1986)
6. INMOS Limited: The transputer applications notebook, architecture and software, London: Prentice Hall 1989.

A VLSI Scalable Processor Array for Motion Estimation

P. Baglietto, M. Maresca, A. Migliaro and M. Migliardi

DIST - University of Genoa
via Opera Pia 13 - 16145 Genova
tel +39-10-353.2983 fax +39-10-353.2948
e-mail prp@dist.unige.it

Abstract - In this paper we describe a parallel architecture for motion estimation based on the Full Search Block Matching Algorithm. The distinctive characteristic of the proposed architecture is its suitability to be implemented both in a high performance dedicated device for embedded systems (e.g. an ASIC) and on mesh connected SIMD massively parallel computers. The paper describes the first of these options in detail.

Keywords: MPEG, Motion Estimation, Processor Array, VLSI.

1 Introduction

Motion estimation is one of the issues in full motion video compression [1][2]. Motion estimation allows to reduce the amount of information needed to represent a frame in a video sequence by eliminating the temporal redundancy due to the fact that a frame is usually very similar to the previous ones.

A video coder implements motion estimation by encoding the current frame with respect to one of the preceding frames in the video sequence taken as a reference frame. The current frame is represented as a combination of the estimation function to be applied to the reference frame and of the error resulting from the application of such a function to the reference frame. The frame resulting from the application of the estimation function to the reference frame, called estimated frame, should be as similar as possible to the current frame and the error is the difference between the estimated frame and the current frame.

The computation of the estimation function is called motion estimation. The motion estimation in the MPEG and H.261 standards is computed following the Block Matching Algorithm (BMA) which consists of dividing the current frame into fixed size blocks of pixels and for each block finding the most similar block of pixels in the reference frame. As a consequence, each block of the current frame is estimated by means of a motion vector representing the difference between the position of the block in the current frame and the position of the corresponding block in the reference frame. The set of motion vectors associated to the blocks of the current frame are the estimation function.

As the amount of computation requested to identify the motion vectors is large and the degree of parallelism that can be exploited appears to be high, special purpose architectures for motion estimation have been proposed and are currently available [3][4][5].

In this paper we describe a metodology that allows to efficiently implement the Full Search BMA (FSBMA) in parallel both in a high performance dedicated device for embedded systems (tipically ASIC) and on mesh connected SIMD massively parallel computers (e.g. the MasPar MP-1).

The FSBMA for motion estimation presents three levels of potential parallelism:

1. at the frame level the computation of the motion vector of each block is totally independent from the computation of the motion vectors of the other blocks;
2. at the block level the computation of the error associated to each possible motion vector is totally independent from the computation of the errors associated to other possible motion vectors;
3. at the error level all the differences to be computed (see next section for details) can be carried out in parallel;

The implementation described in this paper, at the ASIC level, exploits the parallelism at the block level while an implementation on SIMD massively parallel computers may in addition exploit the parallelism at the frame level to reach higher performance. Parallelism at the error level cannot be easily exploited, especially in ASICs, as it requires the use of parallel adders, and gives less performance improvements compared to the other two types of parallelism. The available ICs for motion estimation prevalently exploit parallelism at the block level [3][4]. The VLSI implementation of our architecture follows the same technique and improves the performance of these ICs.

The paper is organized as follows. In the next section we present the BMA algorithm and a parallelization technique suitable for implementation both in a SIMD massively paralllle computer and in VLSI. In section 3 we present an abstract parallel architecture supporting such a parallelisation technique and in section 4 we describe a VLSI ASIC implementing such an abstract architecture and discuss its performance. Finally we give some concluding remarks.

2 The parallel algorithm for full block matching

The BMA estimates the amount of motion on a block by block basis. In a tipical BMA, a frame on $N\times M$ pixels is divided into blocks of $n\times n$ pixels (in MPEG and H.261 standard $n=16$). Each block of pixels of the frame, called reference block, is compared with a set of blocks of a previous frame, called candidate blocks, and the candidate block most similar to the reference block is identified. The blocks in the set are not taken from the whole frame, but, on the contrary, they are taken from a search area of size $(n+2p)\times(n+2p)$ around the position of the reference block, where p is the maximum displacement (see fig. 1). The motion vector of a reference block corresponds to the displacement of the best matching candidate block.

Among the possible search methods, the Full Search BMA, which searches all possible candidate blocks in a search area, is the optimal solution and has the lowest control overhead. This last feature makes the FSMBA easy to be implemented in a SIMD processor array and highly suitable for VLSI implementation.

Among the possible matching criteria the Mean Absolute Difference (MAD) is usually preferred for VLSI implemetation because it requires the repetition of simple operations. The MAD is defined as follows:

$$MAD(u, v) = \sum_{i=1}^{n} \sum_{j=1}^{n} \left| S(i+u, j+v) - R(i, j) \right|$$

where $R(i, j)$ is a pixel of the reference block and $S(i+u, j+v)$ is the corresponding pixel of the candidate block displaced of (u, v), which is also the corresponding motion vector. The motion vector associated with the reference block is the motion vector corresponding to the least MAD. According to this definition of the FBMA, in each search area there are $(2p+1)^2$ candidate blocks and the computation of the MAD of a candidate block requires n^2 subtractions, n^2 absolute value operations and n^2 accumulations. A total of $3n^2 \times (2p+1)^2$ operations for each reference block.

In order to be able to perform the motion estimation in real-time, a parallel architecture is required. The architecture we propose in the paper is based on the parallelization of the MAD calculation for a single reference block over the $(2p+1)^2$ possible displacements, taking advantage of the fact that the computation of the MADs corresponding to the $(2p+1)^2$ displacements are independent from each other. The same technique can be implemented on a massively parallel computer taking advantage also of the parallelization over the blocks, as the MADs of all the blocks can be calculated in parallel.

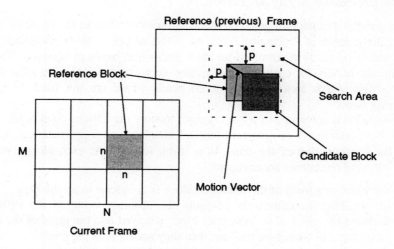

Figure 1 - Block matching for motion estimation.

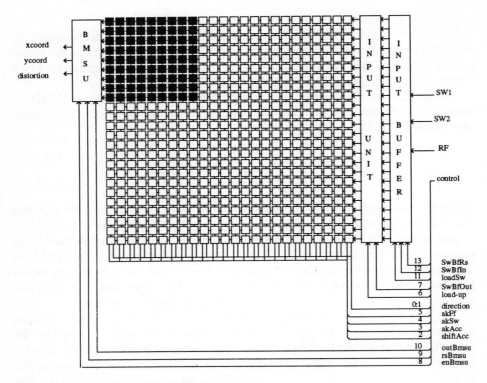

Figure 2 - Processor array architecture.

3 The processor array architecture

The parallel architecture based on the technique described in the previous section is basically a mesh of Processing Elements (PEs) and of Memory Elements (MEs) which are respectively represented as black and white boxes in figure 2. The PEs compute the MADs for each possible motion vector and the MEs to store all the pixels of the search area including those pixels which are not used at a given processing step.

A single PE is composed of a 8-bit register to store the current pixel of the search area, a 8-bit adder, a 1-complement unit for absolute value computation, a 16-bit adder for accumulation of the result in a 16-bit register. At each clock cycle the three following operations are executed:

- the value of a pixel of the reference block is broadcast to all the PEs;
- all the PEs simultaneously compute the absolute value of the difference between the pixel of the reference block received and the pixel of the search area stored in their 8-bit register; then they accumulate the result;
- the mesh of registers (which include the 8-bit registers inside each PE) shift its content along one of the four directions (NEWS).

The pixels of the search area (contained in the 8-bit registers, i.e. the ME and the 8-bits registers of the PEs) shift over the mesh of PEs according a snake shaped path, while the values of the pixels of the reference block are sequentially broadcast to all the PEs. At each clock cycle the $(2p+1)^2$ PEs compute the difference between a pixel of the reference block and all the pixels of the search area onto which the reference block pixel can be mappedi n parallel. Then they compute the absolute values of the results and accumulate them in the 16-bit registers.

A total of $n^2+4p(n-1)-1$ registers are needed to store the pixels of the search area in addition to the $(2p+1)^2$ PEs. A total of n^2 clock cycles are required to compute the MADs while a few more clock cycles are needed to extract the MADs from the PEs, to compare them and to initialize the mesh with a new search area.

In addition to the processor and register array, the architecture is composed of:

- a Best Match Selection Unit compares all the MADs computed by the PEs and extracts the minimum among them along with the associated motion vector;
- an Input Buffer Unit which allows to overlap the input of the next search area with the computation of the MADs for the current search area;
- an Input Unit which distributes the pixels stored in the Input Buffer according to the pattern required by the current step of computation.

4 The VLSI implementation

The VLSI implementation of the architecture defined in the previous section has been realized with the Epoch Silicon Compiler using a standard cell technology. The characteristics of the chip are:

Technology:	0.5 μ, 3 metals, 3V	clock speed:	50 MHz
Package:	PLCC 84 pins	block rate	183000 blocks/sec
Core area:	8.8×7.3 mm.	on chip memory	1062 Bytes
# of PEs:	81	# of MOPS:	12150

The chip was designed for the special case of blocks of size $n=16$ (as reccomended by the MPEG and H.261 standards) and a search area of size $p=4$. As a consequence the ASIC hosts a mesh of 576 elements (81 PEs and 495 MEs) and is able to deliver the motion vector associated to a reference block every 273 clock cycles (256 cycles for MADs computation, 16 cycles for best match selection and initialization with the next search area pixels, 1 cycle for input buffer reset). However the chip can be used with blocks and search areas of different size.

The chip can operate at a maximum clock frequency of 50 MHz which corresponds to a maximum block rate of 183 Kblock/sec and to a maximum pixel rate of 46,9 Mpixels/sec.

The use of this chip as a part of a board, or more in general of an embedded system, is very flexible as can be seen from the folowing table which lists the number of chips and the clock speed required to process various video streams at two level of accuracy (video streams at a rate of 30 frame/sec., blocks of size $n=16$, search area of size $p=4$ and $p=8$):

Frame size	# of blocks	Search Area 24×24 pixels	Search Area 32×32 pixels
640×480	36.000	1 ASIC at 10 MHz	1 ASIC at 30 Mhz with time sharing
800×600	57.000	1 ASIC at 16 MHz	1 ASIC at 48 Mhz with time sharing
1280×1024	153.600	1 ASIC at 42 MHz	4 ASIC at 42 Mhz in parallel
1500×1024	180.480	1 ASIC at 50 MHz	4 ASIC at 50 Mhz in parallel

5 Concluding Remarks

We have presented a general approach to the problem of the parallelization of the Full Search Block Matching Algorithm for motion estimation suitable to be adopted both in SIMD massively parallel computers and in ASIC. In particular we have discussed an ASIC which can be used in embedded systems to compress video sequences according the MPEG and H.261 standards since the degree of parallelism exploited allows to obtain real time performance even at a low clock rate. The paper demonstrates that the VLSI implementation of the proposed approach is feasible and convenient using available technology. Finally the proposed approach can be used both in emebdded systems and in general purpose computers since its performance is fully scalable and proportional to the size of the parallel system used.

References

[1] *Video codec for audio visual services at p× 64 kb/s*, CCITT Reccomendation H.261, 1990.

[2] *Coding of moving pictures and associated audio*, Committee Draft of standard ISO11172: ISO/MPEG/90/176, Dec. 1991.

[3] P. Ruetz, P. Tong, D. Bailey, D. Luthi and P. Ang, *A high performance full-motion video compression chip set*, IEEE Trans. on Circuits and systems for video technology, Vol. 2, N. 2, June 1992, pp. 111-122.

[4] H. Fujiwara, M. Liou, M. Sun, K. Yang, M. Maruyama, K. Shomura and K. Ohyama, *An all-ASIC implementation of a low bit-rate video codec*, IEEE Trans. on Circuits and systems for video technology, Vol. 2, N. 2, June 1992, pp. 123-134.

[5] C. Hsieh and T. Lin, *VLSI architecture for block-matching motion estimation algorithm*, IEEE Trans. on Circuits and systems for video technology, Vol. 2, N. 2, June 1992, pp. 169-175.

Evaluating Digital Angles by a Parallel Diffusion Process

V. Cantoni[+], L. Cinque[°], S. Levialdi[°], L. Lombardi[+]

[°]Dipartimento di Scienze dell'Informazione, Universita' "La Sapienza" di Roma
Via Salaria 113, 00198 Roma, Italy
[+]Dipartimento di Informatica e Sistemistica, Universita' di Pavia
Via Abbiategrasso 209, 27100 Pavia, Italy

Abstract. This paper describes a new method that uses the diffusion-like process to classify angles in a image. The proposed method exploits the curvature value of a contour line to characterize the correspondent angle. The effectiveness of the new method is illustrated with the results of experiments for a number different angle types.

1   Introduction

The purpose of low-level vision is to extract useful information from the images. To date, differential edge detectors have proven the best tools for this purpose. Although primitive features such as edge and lines are vital visual clues, intersection of edges such as corners (angles) and junctions, which are commonly refereed to as 2D features, provide richer informations for examining frame-to-frame displacement characteristics of images. In general, edge detectors do not make good corner detectors because they give reduced output at corners, and because of this, considerable research has been specifically directed towards isolating corners points.

Our interest in the problem of corner detection arises from the desire to use the curvature value of a contour line to classify the correspondent angle. Many methods have been proposed in the literature for corner detection. Some of these are applied on digital curves or region boundaries after a prior segmentation of the image [1, ?]. More recent approaches are focused on developing methods which could detect corners by operating directly on the gray level images [3, 4]. A detailed review and a comparative study of gray level corner detectors can be found in [5, 6] .

In this paper we describe a new method the uses the diffusion-like process to classify angles. The proposed method exploits the curvature value of a contour line to characterize the correspondent angle. The diffusion approach, our implementation and how we use this method to classify angles is described in the next sextion. Results of experiments for selecting a safe set of thresholds for angles classification are carried out in section 3, and we draw conclusions in section 4.

2 Description of Basic Method

Our method for the characterization of the shape [7], is based on the analogy with a diffusion process computing numerical values which reflect the object contour curvature.

In our approach the diffusion process acts on a physical object having the same shape as the digital object on which the contour must be analyzed. The contour of the object is heated, instantaneously at time $t = 0$, at a given energy value. After a time $t = t_f$ the temperature of the contour is measured for each pixel; if we assume the object is thermally insulated from the background, the value of contour pixels can be used to generate a shape-related code.

In our implementation the procedure starts by assigning to all the contour elements of the digital object to be labeled, an arbitrary (non-zero) value which initializes the diffusion process, while all the other elements are set to zero.

At the end of the process, the contour elements that preserve high values correspond to local convexities and those in which a sharp decrement is produced correspond to local concavities.

To detect these features it is necessary to stop the diffusion process at a certain stage. When the simulated diffusion process is stopped, the values are distributed along the contour depending on the local shape configuration; these values will be used for contour labeling.

2.1 Angle characterization

We have applied the proposed method to classify angles. For a given number of iterations a set of thresholds may be chosen so as to associate values exceeding such thresholds with angle labels. As an example the method has been applied on a square with a hole (figure 1), the result is presented in figure 2. A uniform gray scale was chosen to show the different values obtained on the border: white represents a high value and black a low one.

3 Experimental results

The purpose of this section is to show that we can use the proposed method for classifying angles. We have performed experiments on images with known characteristics.

The experimental protocol involved the application of the procedure to each image with different values, for diffusion coefficient D and the number of steps N, as shown in Table 1. Test images are circular sectors with different angles θ ($\theta = 30, 60, 90, \ldots, 330$) at different spatial orientations.

The results of our experiments are shown in a set of tables (one for each value of D and N) in which for each angle the minimum, the medium and the maximum values are given (see Table 2,3). These values describe the range associate to the angle.

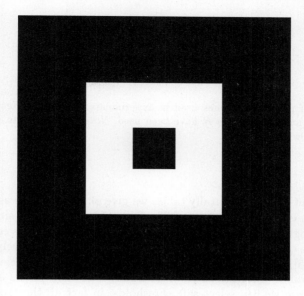

Fig. 1. Test sample object.

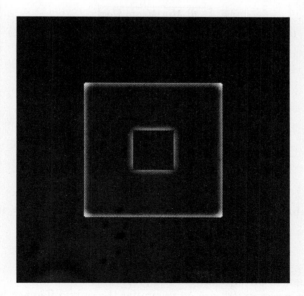

Fig. 2. Result of the diffusion procedure on the object of figure 1. White represents a high value and black a low one.

D	N				
0.05	20	30	40	50	60
0.025	40	60	80	100	120
0.0125	80	120	160	200	240

Table 1. Set of parameter values used in experiments. For each diffusion coefficient D five different number of steps N have been used. Note that prodocts $N \times D$ are the same on each row.

From the set of experimentally tables we give only two tables as an example, these represent two meaningful cases. The first table represents a situation where we cannot recognize safely classes of angles because some ranges of values overlap. This is the case of the ranges for angles 60° and 90° respectively [1927, 2201] and [1980, 2285]; these values are obtained with a diffusion coefficient $D = 0.025$ and with a number of iterations $N = 40$. The second table shows instead a positive result where we can select safe thresholds; in fact all the range values are disjoint. These values have been computed with $D = 0.025$ and $N = 120$.

θ	Min	Mean	Max
30	1730	1764	1815
60	1927	2013	2201
90	1980	2166	2285
120	2305	2411	2548
150	2641	2743	2833
180	2936	3301	3596
210	3766	3966	4362
240	4082	4369	4379
270	4969	5083	5246
300	5607	5931	6363
330	6524	6992	7950

Table 2. Experimental results with $D = 0.025$ and $N = 40$.

A global view of all experiments are plotted in figures 3,4.

The graphs, as expected, are quite similar the only difference is in the width of errorbars; infact growing the number of steps the range of values associate with each angle becomes smaller.

Our final scope is to determine a set of thresholds that allow us to safely recognize eleven different angles (30, 60, ..., 330), having disjoint ranges. As

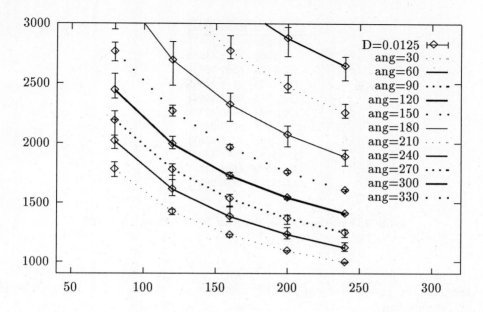

Fig. 3. Results of our procedure for different numbers of steps and D=0.0125.

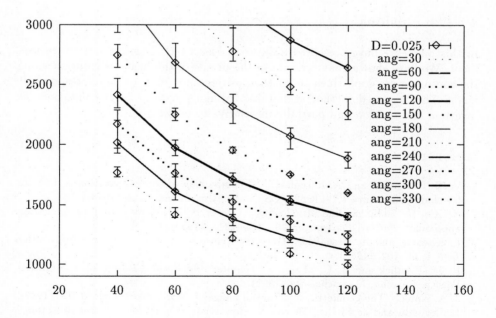

Fig. 4. Results of our procedure for different numbers of steps and D=0.025.

θ	Min	Mean	Max
30	978	992	1037
60	1083	1118	1168
90	1173	1240	1281
120	1375	1398	1436
150	1593	1597	1607
180	1805	1885	1939
210	2208	2259	2380
240	2506	2639	2767
270	3188	3241	3301
300	4074	4180	4315
330	5203	5638	6655

Table 3. Experimental results with $D = 0.025$ and $N = 120$.

shown in the graphs a set of possible solutions are $D = 0.025$ and $N \geq 120$. Safe results can be obtained using a high number of steps, the drawback is a higher computational cost. A good compromise is represented by the couple $D = 0.0125, N = 240$.

4 Conclusions

In this paper we have presented a diffusion based technique for angle measurement. We have shown that extrema of the profile of the heat distribution along the contour gives good clues about the position and width of the correspondent angle. The proposed method is suitable to highly parallel implementations, is invariant for rotation and partially insensitive to noise.

References

1. W. S. Rutkowski and A. Rosenfeld , "A comparison of corner detection techniques for chain-coded curves," TR-623, CS Center, Uni. of Maryland (1978).
2. X. Xie, R. Sudhakar and H. Zhuang, "Corner detection by a cost minimization approach," *Pattern Recognition* **26**, 1235–1243, (1993).
3. L. Kitchen, and A. Rosenfeld, "Gray level corner detection," *Pattern Recognition Lett.* **1**, 95–102, (1982).
4. R. M. Haralick and L. G. Shapiro, *Computer and Robot Vision*, Vol. I, Addison-Wesley. Reading, Massachusetts (1992).
5. J. A. Noble, "The geometric structure of images," *MSc Report.* Oxford UK (1987).
6. F. Eryurtlu and J. Kittler, "A comparative study of gray level corner detectors," *Signal processing VI:Theories and Appications.* Elsevier Sci. Publ., (1992).
7. L. Cinque and L. Lombardi, "Shape description and recognition by a multiresolution approach," *Image and Vision Computing.* (to appear).

A Pyramidal Approach to Convex Hull and Filling Algorithms

M.G. Albanesi, M. Ferretti, L. Zangrandi
Dipartimento di Informatica e Sistemistica, University of Pavia,
Via Abbiategrasso 209, I-27100 Italy

Abstract   In the paper, a class of algorithms for filling concavities in binary images is presented. The paradigm of computation is a serial, multi-resolution approach. Among the algorithms which have been implemented, the most significant one is fully described, while for the others, some hints are given on the computational complexity. The algorithm for the approximation of the convex hull has a complexity which is linear in the image dimensions, provided that the multi-resolution is already available. The performance of the algorithms has been measured on a broad set of images and experimental results are reported. Final considerations about parallelization are also given.

1 Preliminary considerations

In binary images, the problem of describing objects often is formulated in terms of an efficient characterization of silhouettes, which are identified by their convex contour. Therefore, in the analysis of the image, the determination of the *convex hull* of the shape or the filling of concavities are used in many applications. It is useful, for the following discussion, to distinguish between two classes of approach; *planar* and *hierarchical*. In the first class, the data structure involved in the computation is a bi-dimensional array, in which operations may occur *locally* or *globally*. On the contrary, in the hierarchical approach, a multi-resolution version of the input image is built, and the computation occurring on a generic level of resolution usually exploits information provided by other levels.

Planar versions of algorithms for the computation of filling have been proposed [2] [3], in which computation occurs locally and iteratively in a 3 x 3 neighbourhood. A hierarchical approach [4] applies iteratively the planar algorithm [3] in each level, followed by a projection operation to pass to the higher resolution level.

The new methods for convex hull and filling here proposed belong to the serial paradigm of computation, performed in multi-resolution. In particular, three algorithms have been implemented for the filling task (named *Fill1*, *Fill2* and *Fill3*, respectively), and in the following section the one with best performance (*Fill3*) is described carefully. For the other two (*Fill1* and *Fill2*), which have been preliminary in our study of hierarchical approaches, we will give some hints on their computation complexity.

2 Hierarchical approaches to convex hull and filling computation

The class of the three algorithms is characterized by the computation of the OR-pyramid on the input binary image; the first algorithm (*Fill1*) applies the classical planar method [2] on a generic level k (where k is an input parameter of the procedure). The computation is followed by a set of projections on the lower levels, until the base is reached. Projection is associated to a refinement operation, which is necessary to maintain the convexity of the shapes.

We have computed [5] the computational complexity of the algorithm, which is of the order $O(N^2)$, for an image of NxN pixels.

This algorithm gives an approximated version of the filling; the error, defined as the number of pixels beyond the boundary of the filled concavity due to the project operation, can be evaluated [5] as $2^{k+1} / \sqrt{2}$. In order to remove this error, a second algorithm can be proposed (*Fill2*), in which the refinement is performed directly on the basis of the pyramid at the end of the projection phase.

In order to understand the strategy of the third algorithm for filling (*Fill3*), it is necessary to introduce a new preliminary algorithm; it aims at finding a convex hull approximation of an object [6] to be used as a starting point in the strategy for filling.

Let us consider a digital image $\Im$ defined on N x N pixels. The approximation of the convex hull is done by considering n straight lines which are tangent to the edges of the object and which envelope the internal concavity of the object. The approximation is done under the following assumptions:

1)   The image is binary and an object is defined as the set of black pixels which are contiguous, according to a specified topology (see point two).

2)   The algorithm uses the 8-connected topology.

3)   The algorithm produces one convex hull approximation for the whole image (see figure 1); therefore, if more than one object are present, the convex hull is the one which includes all of them.

4)   The algorithm can approximate the convex hull in n directions ($n \geq 3$); in the following, only the 8 directions are considered ($0°$, $\pm 45°$, $\pm 90°$, $\pm 135°$, $180°$). This is not a strict limitation, because this choice is meaningful to show the effectiveness of the algorithm in most of the considered cases (see section 3).

The algorithm for the convex hull (*CH1*) is formulated in a multi-resolution frame, and it can be described by the following procedure:

Algorithm CH1:

Step 1:   On the given image $\Im$, an OR-Pyramid is built up to the 2x2 level. It consists of $\log_2 N$ levels ($l = 0, 1, \ldots \log_2 N - 1$). Level $l = 0$ is the original image $\Im$ level $l = \log_2 N - 1$ is a 2 x 2 pixel image.

Step 2:   Starting from level $l = \log_2 N - 1$, n searches are performed, one for each direction D_j ($j = 0, \pm 45, \pm 90, \pm 135, 180$) of the straight lines which approximate the convex hull. In our case, 8 searches are performed in a serial way. The search consists of finding the first straight line at a given direction D_j which crosses the object, namely its tangent at that direction D_j. This assumes that the image is scanned along a direction which is normal to D_j. When the line is found, the tangency point is marked as P_j and its coordinate (x_j, y_j) are stored. The step ends when all the 8 lines and the corresponding tangency points are determined.

 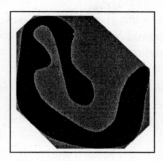

Figure 1. An example of the convex hull approximation: original image (left) and result (right). In the result, the concavity has also been filled.

Step 3: Step 2 is repeated at level l-1, but the search of the 8 lines is spatially restricted by the results of the search at the previous level. In fact, for a given direction D_j the scan of the image at level l-1 starts from the line which passes through the coordinates $(2x_j, 2y_j)$

Step 4: Step 3 is iterated until level $l = 0$ is reached. The result in level $l = 0$ is the output of the algorithm, namely the coordinates of eight points in the image: each point is associated to one direction D_j, and the edges of the convex hull are approximated by the eight straight lines which pass through the resulting points.

The advantage of using a multi-resolution approach is the possibility of restricting the search at a generic level l by using information (i.e. the coordinates of the tangency points) provided by the algorithm at level l-1. This gain can be quantitatively evaluated in the following way: a digital straight line L_j represented by a couple (D_j, P_j), namely passing through point P_j with direction D_j, computed at level l has obviously a thickness of one pixel. The search at level l-1 is restricted to a set of lines, which results from the project of line L_j. The number of lines of this set is 2, because of the topology of an OR-pyramid. This means that the search for each level and each direction requires at most 2 trials. Each trial requires at most $\sqrt{2}N / 2^l$ computations, because there is one possible computation of a new coordinate pair for each pixel of the corresponding straight line of the image at resolution level l. For this reason, the whole complexity of the algorithm can be evaluated [5] as of $O(n\ N)$; the cost of convex hull approximation is linear with respect to the image dimension N. This overcomes the square complexity $O(N^2)$ of a planar approach. Obviously, in a serial computation one should consider also the cost of the construction of the OR-pyramid (Step 1). For this reason, the whole complexity of *CHI* is asymptotically $O(N^2)$. However, software implementation of the algorithm shows that the execution times are preferable in the hierarchical approach, if compared to the planar one, due to smaller constants (see section 3).

The algorithm *CH1* can be modified into algorithm *Fill3* in order to produce the filling of the concavities of the object (see figure 1):

Algorithm Fill3:

Step 1: On the given image $\mathfrak{I}$, an OR-Pyramid is built as in Step 1 of *CH1*. A second pyramid Ψ of the same dimensions is created and it is filled with ones. This second pyramid will contain the result of the filling in its level $l = 0$.

Step 2: On a generic level l, starting form $l = \log_2 N - 1$ Step 2 of *CH1* is performed, with a little overhead: it consists of filling with zeros the pyramid Ψ in every pixel belonging to each straight line which does not cross the object. Before executing step 3 of *CH1*, a projection of zeros is performed on level $l - 1$ in pyramid Ψ .

Step 3: Step 2 is iterated until $l = 0$. Pyramid Ψ contains the result of filling.

The overhead due to the building of pyramid Ψ, essentially in the projection phase, raises the algorithm asymptotic complexity to $O(N^2)$, comparable to the planar approach. Experiments show that the execution times are preferable in the hierarchical approach, even if the gain is not high (see next section for details).

3 Experimental Results

This section analyzes the results of software implementation of the hierarchical approaches *Fill1*, *Fill2* and *Fill3*. The test has been done on a series of binary images, realized by thresholding naturalistic gray level images (see figures 2a-2b).

| (a) | (b) | (c) |

Figure 2. An instance of one of the test:: (a) gray level image "blackbird", (b) binarized image, (c) the result of filling algorithm *Fill3* on image "blackbird".

In the test, CPU times have been measured; the algorithm have been implemented in C++ on a HP 9000/735 workstation. In figure 3 a plot of the CPU time, expressed in seconds as a function of the image linear dimension N, is shown for a comparison between algorithms *Fill1* and *Fill2*, namely the first two hierarchical algorithms. CPU time has been computed running the algorithms on synthetic images of different size. Experiments confirm the theoretic evaluation of time complexities: time growth is quadratic, but the constants for algorithm *Fill1* are smaller.

Figure 4 shows a plot of the CPU time, expressed in seconds as a function of the image linear dimension N, for algorithm *Fill3*, compared with a corresponding planar serial version (*PL*). In this version, the simple search for the eight straight lines which envelope the concavity is performed at the maximum resolution (N). Experiments confirm the validity of algorithm *Fill3*, expecially for higher image dimensions (N > 800) with a maximum gain of about 30% (N = 1600). Moreover, algorithm *Fill3* is preferable if compared with *Fill1* and *Fill2*: this can be easily verified by comparing plots of figure 3 and 4. For example, for an image dimension of 1024 x 1024, *Fill1* and *Fill2* take over 18 and 20 secs, respectively, while *Fill3* takes 3.4 secs, with a maximum gain of 83%.

4 Conclusions: towards parallelization

In the paper, a hierarchical approach to filling concavities in binary images is introduced. The proposed algorithms are serial, but few consideration can be done for what concerns parallelization. In a parallel pyramidal machine, the problem of filling objects can be solved in constant time if an operation of propagation is available in hardware ([1],[7]).

In a standard SIMD pyramid of $\log_2 N$ +1 levels, the n equations defining the approximation of the convex hull can be obtained in $n\log_2 N$ steps. In fact, the algorithm considers in turn each direction D_j ($1 \leq j \leq n$): each processor computes its distance from the line oriented along D_j and passing through one of the four corners of the basis. The minimum among such distances is associated to the point of tangency or direction D_j; such minimum is extracted using the pyramid by well known algorithm [6] in $\log_2 N$ steps. At this point, the n lines of the approximation of the convex hull are detected, so the filling can be built in a constant time.

References

1. V. Cantoni, V. Di Gesù, M. Ferretti, S. Levialdi, R. Negrini, R: Stefanelli: The Papia System, Journal of VLSI Signal Processing, 2, 1991, pp. 195-217.
2. J. Sklansky, L. Cordella, S. Levialdi: Parallel detection of concavities in cellular blobs, IEEE Trans. Computers, Vol. C-25, Feb. 1976, pp. 187-196.
3. G. Borgefors, G. S. di Baja: Filling and analysing concavities of digital patterns parallelwise, Visual Form - Analysis and Recognition, Plenum ed., 1994, pp. 57-66.
4. G. Borgefors, G. S. di Baja: Methods for hierarchical analysis of concavities, Proc. Int. Conf. 11th IAPR, Vol C, 1992, pp. 171-175.
5. L. Zangrandi: Strategie gerarchiche di raffinamento in algoritmi per l'elaborazione di immagini, Thesis, Dipartimento di Informatica e Sistemistica, Università di Pavia, 1994.

6. R. Klette: On the Approximation of Convex Hulls of finite Grid Point Set, PRL; vol. 2, 1983, pp. 19-22.

7. M. J. Duff: Propagation in cellular logic arrays, Proc. Workshop on Picture Data Description and Management, 1980, pp. 259-262

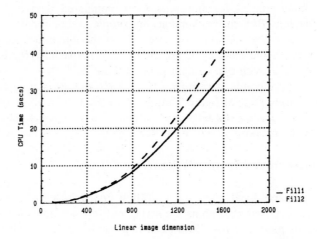

Figure 3. CPU time, expressed in seconds, as a function of the image linear dimension N, for a comparison between algorithms *Fill1* and *Fill2*.

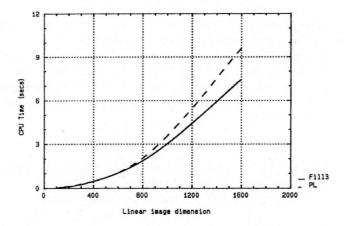

Figure 4. CPU time, expressed in seconds as a function of the image linear dimension N, for algorithm *Fill3* and its corresponding planar version (*PL*).

A Code Based Approach to Image-Processing Teaching

A. Biancardi and M. Pini

Università di Pavia
Via Abbiategrasso 209 - 27100 Pavia - Italy
e-mail: alberto@ipvvis.unipv.it
maxi@ipvvis.unipv.it

Abstract. The Pacco Teaching Tools library extends Pacco, an object-oriented data-processing environment, and offers a new approach to i.p. teaching where the source code is placed at the centre of the educational experience, providing a common framework between theory and practice and accomplishing students' learning of the development environment while improving their knowledge of i.p. techniques. Thanks to the automatic monitoring of processed objects and to the interactivity and robustness of the tools, the environment seems very suited to the targeted task.

Introduction

The increasing diffusion of image processing (i.p.) courses at the (uder)graduate level makes even more glaring the almost total absence of tools to support and complement i.p. teaching.

The accepted evolution of a course can be roughly divided into four main stages, going in students' increasing involvement:

1 exposition of ip techniques and strategies
2 learning of the development system
3 experimental verification of theoretical results
4 project implementation

The big problem with i.p. teaching is not really the fact that examples and results require images and graphical items and that they cannot be easily described by means of words; the difficulty lies in the unavoidable dynamicity of those images that complement a lecture. Moreover it is easier to keep user's attention focused on the topic if a set of evolving pictures is presented (e.g. choosing a data file, performing an image transform or a special task, showing effects and results, choosing another file, and so on). This way theory becomes more interesting and not boring.

The almost interactive nature of this process means that some kind of active-textbook should be used. If this guideline is coupled with the importance of teaching not only the theoretical techniques but also the way to perform them, it follows that the programming development environment is the best candidate to act both as an active-textbook and as a workbench.

Anyway the limited time students have to get aquainted to the development environment plays a major role in the academic context. The ideal environment should:

- offer a smooth learning curve
- get students productive as soon as possible
- do not encumber them with eccessive details

Finally, another issue worth considering is the ability to provide autonomous or self–paced learning, which is strictly connected to the management of on–line (and off–line) documentation together with some kind of *help* facility.

To cope with these requirements, the Pacco Teaching Tools (PTT) library was developed to let Pacco, an object-oriented data-processing environment[1], become even more effective as an authoring system and a development environment for non–experienced programmers.

Before peeking into the details of the PTT library, a brief overview of Pacco will be given. Features, use and implementation of PTT will follow, while the discussion of related work will conclude the paper.

Compound Objects and Active Components

Pacco stands for "Pavia Active-Component Compound Objects" and is an extension of a general purpose command language (Tcl) and its graphical toolkit (Tk)[2]. Pacco design is characterized by a new approach to object-oriented programming where run–time extensions, coded in C language or Tcl, are used to enhance the behaviour of objects and to supply unique services, such as transparent feedback to data changes.

In order to allow the programmer capture data dependent structures and relationships, Pacco defines a new object model by introducing the concept of *compound object*, which is called box, of *container class*, named Cbox as child-box, and of *unknown class*, while reserving the usual concept of *object* [3] to standard classes.

Pacco objects are tree structures made up of *components*; each component belongs to a *class* and Cbox components are both embedded (being components) and embedding (like boxes), thus allowing the creation of the tree hierarchy.

Boxes are the atomic items as far as object creation and destruction is concerned, while components are the atomic information holders as far as the tcl API is concerned; finer grained data inside a component can only be accessed by means of class functions, named *actions* and written in C–language.

The clear-cut division between structuring and information processing, which is enforced by the box/component model, allows Pacco to handle most of the object–management chores disregarding completely the data-types of the components being managed. Therefore whichever task can be performed in a type-independent way is assigned to the kernel. It deals with computation dispatching and result gathering. It is in charge of data input/output handling at the box or Cbox level. It performs loading of unknown classes and actions on demand. It centrally manages the retrieval of defaults for any Pacco feature that can be given a default value.

The most peculiar kernel service is the bind action. It allows the Tcl programmer to add a list of commands to be executed whenever the data of a component change (or the box is destroyed when the command is bound to the whole box). This capability allows the easy implementation of data-driven applications. Moreover, the ability to freely intermix data–generated and user–generated events opens the way to an effective realization of *reactive* scripts.

Pacco defines a new tcl command named after the box anytime a new box is created. Each component in the box gets a tcl command, too. These commands are used to invoke actions to be performed over the component or box. This is whence *active component* comes from. Component command act as component identifiers as well and may be used to uniquely reference components.

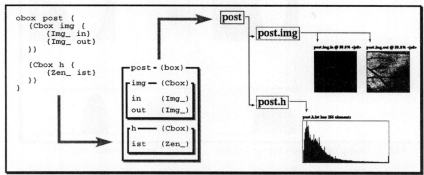

Fig. 1. Structure and dump of the box associated to the command example.

A graphic library extends Tk, the tcl-based X11[4] toolkit, with a set of new widgets and canvas items that can be linked directly to Pacco components. A number of utility procedures further enhance the environment with a polymorphic display command, region of interest handling, colormap animation support and so on. Bindings show their usefulness by allowing the automatic update of component visualization.

Pacco is available by anonymous ftp at `iride.unipv.it` in the directory `/pub/Pacco`.

Pacco Teaching Tools

Being a command language, tcl is simple, has a minimal syntax and is similar to commonly used unix-shells; therefore learning tcl is almost immediate (at least as far as basic concepts and command set are concerned).

Moreover the powerful combination of tcl, tk and Pacco creates a functionally rich environment making straightforward the implementation of interactive tools that keep track of data changes.

Thus, box visualization and monitoring can be supplied transparently through the redefinition of the box–creating command. (Fig. 1 and 2 show two dumps of such windows into directly created PostScript[5] files)

The coesistence of this two factors opens the way to a new approach to i.p. teaching which is embodied by the PTT library. The idea is that the code, implementing an i.p. task, can act both as a reference of the task and as the way to perform the task. In short, it makes possible to provide a common framework between theory and practice and to accomplish students' learning of the development environment while improving their knowledge of i.p. techniques.

In addition to the monitoring support, the library deals with the authoring system and a special command tool to ease code creation and development.

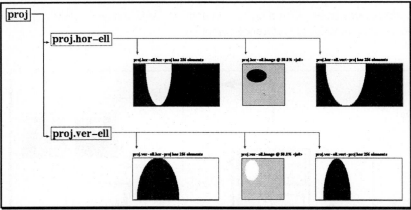

Fig. 2. Dump of the monitoring window associated to the tutorial example.

"Hands On" Tutorials

Each PTT tutorials is structured into *sections*, one for each specific sub-topic. Every section has the following common structure: a global description of its goals and algorithm, the section source code, subdivided into groups of instructions called *steps*, and a set of *captions*, which are synchronized to and explain the code steps.

Thanks to the code-based approach, the authoring tool (Fig. 3) is a specially tailored tcl–source debugger. Navigation controls are the natural translation of the debugger execution commands into the new context, extended by a kind of *undo* feature where the user can restart the current step. Synchronization is achieved by tagging the code according to the step it is related to, thus allowing the user to edit the code without disrupting existing links.

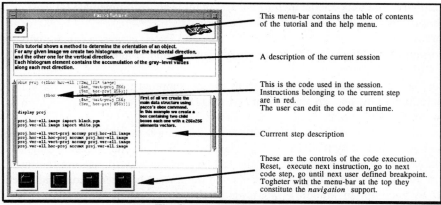

Fig. 3. The tutorial window.

The resulting environment is highly interactive and robust; it should attract users' exploration and therefore be very suited to novices, as experience proved [6].

One problem, currently being addressed, is the implementation of a tool to speed up tutorial creation and caption translation into another language (Pacco structured objects already offer an ideal way to store tutorials).

Project Implementation Support

The command tool (Fig. 4) is intended as a first attempt at giving a better visibility and history handling to pacco programmers. Just like visual programming systems, the command tool tries to expose all the available commands, to make it easier to track a working–session evolution and to retrieve past commands or command results. For example if the first item of the command line is a pacco's component, pressing a mouse button in the command line builds up a pop-up menu showing all the actions available for that component; selecting an action pastes its name into the command line.

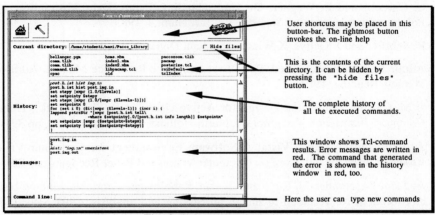

Fig. 4. The command window.

Related Work

Many packages try to be simple and accessible (like Vista[7] or LaboImage[8]) so that they can be used in research and didactic environments. Anyway, up to now, only Khoros[9] explicitly addressed the problem.

The DIP Toolbox[10] of Khoros 2.0 relies on a www-client and cantata, Khoros visual programming (VP) tool. The author's main claim states that visual programming is the only means to effectively hide the huge complexity of the system.

While a deeper discussion on merits and pitfalls of VP is beyond the scope of this paper, a few comments may be sketched out:
- if students already know how to program and are used to a unix shell, extending their programming skills with a very-high-level command language is not an issue, at all;
- flow-control is sometimes difficult to express in a VP tool [11];

- the density of information in a VP layout cannot be higher than in a textual program;
- it is possible to supply a VP layer to tcl-based environments [12].

Conclusions

The PTT library offers a new point of view to i.p. teaching by placing the source code at the centre of the educational experience. Thanks to the automatic monitoring of processed objects and to the interactivity and robustness of the tools, the environment seems very suited to the targeted task.

References

1. A. Biancardi and A. Rubini, Pacco – An Effective Approach To I.P. Programming, Proc. 12th ICPR, IEEE Press, Los Alamos, 1994

2. J.K. Ousterhout,Tcl and the Tk Toolkit, Addison-Wesley, Reading, 1994.

3. L. Cardelli and P. Wegner, On Understanding Types, Data Abstraction, and Polymorphism, *Computer Surveys*, Vol. 17, Num. 4, 1985.

4. R. Schifler and J. Gettys, X Window System, Digital Press, 1990.

5. Adobe System Incorporated, PostScript Language Reference Manual, Addison-Wesley, Reading, 1985.

6. J. M. Carroll, Minimalist Design for Active Users, in B. Shackel(ed.), *Human Computer Interaction – Interact'84*, North-Holland, Amsterdam, 1985, pp 39-44.

7. A. Pope and D. Lowe, Vista: A Software Environment for Computer Vision Research, Proc. of Intl. Conf on Computer Vision and Patter Recognition, 1994.

8. A. Jacot-Descombes M. Rupp and T. Pun, LaboImage: a portable window-based environment for research in image processing and analysis, 1992 SPIE/IS&T Symposium on Electronic Imaging, SPIE Proceedings Vol. 1659, 1992.

9. J. Rasure and S. Kubica, The Khoros Application Development Environment, in H.I Christensen and J.L Crowley (ed.), *Experimental Environments for Computer Vision and Image Processing,* World Scientific 1994.

10. R. Jordán and R. Lotufo, Digital Image Processing with Khoros 2.0, available at the URL http://www.khoros.unm.edu/dip/html/dip.html (and following pages).

11. C. Balasubramanian, Dataflow image processing, Computer, November 1994, pp 82-84.

12. M. Bernini and M. Mosconi, VIPERS: a Dataflow Visual Programming Environment Based on the Tcl Language, Proc. of the Workshop on Advanced Visual Interfaces AVI'94, ACM Press, 1994.

Neural Networks

A Visual Speech Model based on Fuzzy-neuro Methods

Hans H. Bothe

Department of Electronics, Technical University of Berlin
Einsteinufer 17, D-10587 Berlin

Abstract. This paper describes a new approach of modeling visual speech movements, based on a codebook of characteristic *key-pictures* and a complex fuzzy neural network (FNN). Goal is the development of a computer animation program as a training aid for learning lip-reading. The network architecture makes possible a fusion of linguistic expert knowledge into the FNN. The current PC version allows a synchronization of the animation program with a special stand-alone speech synthesis computer via a Centronics parallel interface.

1 Introduction

From the experimental work of Menzerath, together with de Lacerda [1] it is known that the movements of the speech organs are structurally interrelated within the spoken context. The speech organs needed for the formation of upcoming phones, even though currently not engaged, take up position relatively early to their actual use. They produce sound in the course of a fully overlapping phonal coarticulation. The visual reflections of these movements on the speakers face may be seen as *visual speech.*

Whereas the smallest speaker-independent units derived from the acoustic signal being semantically distinguishable are the *phonemes*, there also exist smallest perceptible visual speech units , the *visemes*. The articulatory positions and transitions needed for the production of a phoneme find their visual expression in the related viseme. Knowledge about relationship and structure has been investigated in extensive experiments, especially with hearing-impaired persons. The key-work for German language is the dissertation of Alich [2]. According to these investigations, the 40 phonemes of German can be related to 12 visemes, whereas the phonemes /t, d, n/ and / , / have to be related to two different visemes, depending on the spoken context. Thus, for use of this knowledge in a computer-based facial animation system, the viseme scheme should be modified by isolation of those phonemes and defining two extra visemes. The following table 1 shows this modified scheme with eight consonant and six vowel visemes. The extra visemes V_{C5} and V_{V2} could be perceived as V_{C3}-V_{C5} or V_{V1}-V_{V3}, respectively.

Consonant Visemes				Vowel Visemes			
V_{C1}	/p, b, m/	V_{C5}	/l, r/	V_{V1}	/a, a: ë,/	V_{V5}	/o, œ/
V_{C2}	/f, v/	V_{C6}	/k, g, x, N, /	V_{V2}	/ä, Ä/	V_{V6}	/R/
V_{C3}	/s, z/	V_{C7}	/ç, j/	V_{V3}	/ i, i:, e:/		
V_{C4}	/t, d, n/	V_{C8}	/S , Z	V_{V4}	/o:, ø:, u, u:, y, y:/		

Table 1. Scheme of modified consonant and vowel visemes V_{C1}-V_{C8} and V_{V1}-V_{V6} of German.

Visual speech movements mostly contain sufficient information to enable hearing-impaired persons to lip-read a spoken text. Since the visual recognition is largely focussed on the speakers mouth region, especially on the lips, and the lip movements contain most of the visually perceptible information, this paper proposes the modeling of face movements based on the corresponding lip shapes.

2 Data Acquisition

In order to model visual speech movements, the acoustic speech signal and movement data of prototype speakers were recorded on videotape and analyzed with a multi-media workstation. For automatic visual feature extraction in the speaker's face several points on nose and forehead, as well as the lip contours, were marked with a contrasting fluorescent color. To increase the contrast even more and to smooth the contours, the face was lightly UV-radiated so that shady parts of the lips as wrinkles or grooves were self-radiating instead of only reflecting light. The set points and contours were then localized with the help of an automatic contrast search program.

The two marked reference points and the set point on the nose refer to the head coordinate system and are used for correction of global head movements during the recording. This was necessary since an artificial fixing of the speakers head was felt to have influence on the naturalness of the articulatory movements. Three example frames of the video film, together with the feature extraction scheme, are shown in figure 1. Applying this scheme, each frame is represented by a five dimensional visual feature vector.

   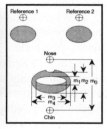

Fig.1. Characteristic video frames and extraction of primary visual features $<m_i>$.

Those frames fitting best with the subjective impressions for a well pronounced sound were interactively indicated with the help of both the acoustic and visual material by different experts in lipreading; the acoustic phone boundaries - determined with the help of oscillogram, sonagram and playback - limit the scanning range of each wanted frame [3]. If in certain cases as, for instance, for the phonemes /h,g,k/, no characteristic frame could be determined, this information was used for the later facial animation.

The determined phone characteristic frames of the video film were classified in this feature space with respect to lip shape and position. The cluster centers compose a set of representative visual feature vectors and define a codebook of key-frames.

3   Generation of the Codebook of Key-frames

The feature vectors were classified with the help of of the fuzzy c-means algorithm as described in [3]. The iteration algorithm generates optimum location of the clusters automatically with respect to a given number of clusters. Thus, a desired number of key-frames can be defined by proposing the number of clusters before starting the classification process.

The fuzzy c-means clustering algorithm does iteratively calculate the cluster centers and with them the new membership grades of the objects. It may be processed mainly in the following four steps, with N being the number of sample vectors $\underline{m}_i$:

Step 1:   Choose
- the desired number of clusters n with $2 \leq n \leq N$,
- the initial condition of the fuzzy membership matrix $\underline{U}^{(0)} = (\mu_{ij})^{(0)}$ of the sample vectors $\underline{m}_i$ to the clusters Q_j.

Step 2:   Adjust the n cluster centers $\underline{v}_j^{(s)}$ of iteration step (s).

Step 3:   Calculate the membership values $(\mu_{ij})^{(s)}$ of the sample vectors for the new cluster centers.

Step 4:   Stop iteration, if $\| \underline{U}^{(s+1)} - \underline{U}^{(s)} \| \leq \varepsilon$ with $\|...\|$ as a suitable matrix norm.

After the classification process, any straightforward phoneme-to-key-frame mapping is lost. Exemplary results with $n_1 = 15$ and $n_2 = 25$ clusters is shown in figure 2. The diagram draws the features width $<m_3>$ over height $<m_1>$ of the outer lip contour in arbitrary units, i.e. screen pixels. The centers of the circuits determine the cluster centers, and the radii the amount of sample vectors in that cluster.

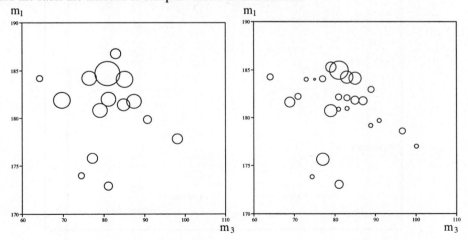

Fig.2. Exemplary classification results projected on a 2D space of the visual features $<m_1>$ over $<m_3>$ (units in screen pixels) with 15 or 25 clusters.

4   Key-frame Selection by a Fuzzy Neural Network

The subject of modeling visual speech and coarticulation effects has been addressed by several authors for different languages (e.g., [4-8]), whereas the movements are either directly controlled by certain visual features or calculated by a target and interpolation algorithm using a codebook of key-frames. The latter has the advantage of an easy implementation on computers with a relative small calculation power as, for instance, PCs; but a larger memory is needed for the storage of the codebook.

A first order approximation for modeling backward and forward coarticulatory effects takes into account the immediate next neighboring phonemes. For this purpose, the phonetic text can be split into overlapping diphones, diphthongs be represented by two closely connected single phones. The frame of the second is classified with respect to the first one [4]. This process leads to a deterministic diphone related phoneme-to-key-frame mapping. The frames may simply be depicted from a look-up table.

In reality, coarticulation effects extend often far beyond the immediate next neighboring phonemes. A proposed area of influence has strong limits by the need of a finite text corpus. Thus, in this work a complex fuzzy neural network (FNN) was trained to map the given phoneme sequence onto a corresponding sequence of key-frames. It relates the single phonemes together with the surrounding next 3+3 neighbor phonemes to the frames of the codebook. For the training of the FNN, half of the text corpus was used, the other half being reserved for evaluation of the FNN generalization quality.

In the later speech synthesis, the same FNN is used for key-frame depiction, and the film is generated by calculating interim frames. The general design of the FNN is shown in figure 3.

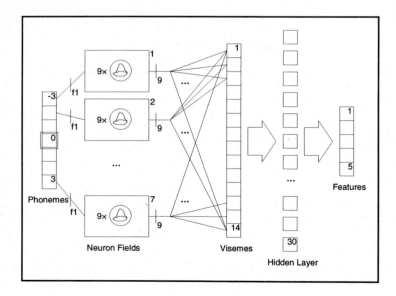

Fig. 3. Design of the ANN for key-frame selection.

The input coding is carried out by a self-organizing Kohonen map that represents the visual feature vectors correlated with the phonemes [9]. After the training process of the Kohonen map, each phoneme corresponds with a specific weight distribution over the map. The localization of the centers of gravity of these distributions, as shown in figure 4, may be taken as a complex similarity measure of the visually represented phonemes. The two dimensions x and y have no physical meaning, but rather serve for ordering the phonemes.

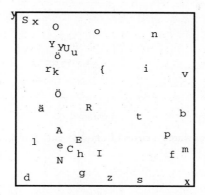

Fig. 4. Localization of the phonemes on a trained Kohonen map.

The FNN maps the input sequence of 7 phonemes - each out of the set of 41 phonemes - on a set of 14 output neurons by a radial basis function network (RBFN) with fixed Gaussian distribution functions. These overlapping functions cover the universe of the map. They are interpreted as fuzzy IF...THEN...-rules by detecting similarity of the phoneme locations. For instance, the phonemes /y, y:, u, u:, œ/ have about the same coarticulatory influence on the key-frame selection, but most probably a different one than /p, b, m, f/.

Each output neuron of the RBFN represents one viseme. This means that a certain phoneme in its context of 3+3 neighboring phonemes is represented by 14 membership values of the set of visemes. Thus, the existing linguistic knowledge on the perception of articulatory movements can be used in before adjusting the RBFN weights. The single neuron fields of the RBFN are interpreted as fuzzy rules, i.e. .

The viseme neurons are taken as input neurons for a subsequent multi-layer perceptron (MLP) with 30 neurons in one hidden layer. The 5D output vector is pointing to the proposed corresponding feature values. The actual key-frame is selected by using the nearest neighbor method and the Euclidian distance measure.

The network is trained in three steps: i) the phoneme-to-viseme mapping with respect to the visematic system (e.g., since /p,b,m/ belong to the same viseme, a crisp mapping on the /p,b,m/-viseme neuron is proposed when /p,b,m/ are in the center position of the input sequence), ii) the viseme-to-feature-vector mapping with respect to the corresponding training sets, iii) the connected FNN with respect to the given phoneme-to-feature-vector mapping of the training sentences.

The FNN approach allows to i) forecast the course of features for any given input text and ii) refine the so far in the literature crisp phoneme-to-viseme mapping by taking contextual influences into account. The forecast of the short phoneme sequence /a:b/, together with the measured course of the features $<m_1>$-$<m_4>$, are shown in figure 5. The timing of the courses of features is calculated with the help of averaged key-frame distances.

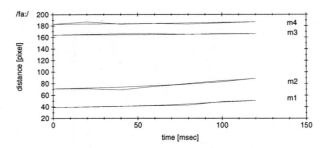

Fig. 5. Forecasted and measured courses of $<m_1>$ to $<m_4>$ for the phoneme sequence /a:b/.

References

1.   P. Menzerath and A. de Lacerda: Koartikulation, Steuerung und Lautabgrenzung, Berlin, (1933).
2.   G. Alich: Zur Erkennbarkeit von Sprachgestalten beim Ablesen vom Munde (Dissertation), Bonn, (1961).
3.   H.H. Bothe and N. v. Bötticher: Key-frame selection for the analysis of visual speech with fuzzy-c-means algorithm. In: B. Bouchon-Meunier & R. Yager & L.A. Zadeh (Eds.), Advances in Intelligent Computing, Springer-Verlag, Berlin-Heidelberg (to appear, 1995).
4.   H.H. Bothe, G. Lindner and F. Rieger: The Development of a Computer Animation Program for the Teaching of Lipreading, In: E. Ballabio, I. Placencia-Porrero and R. Puig de la Bellacasa (Eds.), Technology and Informatics 9, Rehabilitation Technology: Strategies for the European Union, Amsterdam, (1993), 45-49.
5.   D. Storey and M. Roberts: Reading the Speech of Digital Lips: Motives an Methods for Audio-visual Speech Synthesis, Visible Language 22 (1989), 112-127.
6.   M.M. Cohen and D.W. Massaro: Synthesis of Visible Speech, Behaviour Research Methods, Instruments & Computers, (1990), 260-263.
7.   M. Saintourens, M.H. Tramus, H. Huitric, and M. Nahas: Creation of a Synthetic Face Speaking in Real Time with a Synthetic Voice, Proceedings of the ESCA Workshop on Speech Synthesis, Autrance, (1990), 381-393.
8.   F. Lavagetto: Converting Speech into Lip Movements: A Multimedia Telephone for Hard of Hearing People. Trans. Rehabilitation Engineering, (to appear; 1995).
9.   Bothe, H.H.: Fuzzy input coding for an artificial neural network. (ACM / SAC'95), Nashville, (1995).

Edge Detection Filters Based on Artificial Neural Networks

Armando J. Pinho[1] and Luís B. Almeida[2]

[1] Dep. Electrónica e Telecomunicações / INESC
Universidade de Aveiro, 3800 Aveiro, Portugal
(Fax +351–34–370545, Email ap@inesca.pt)
[2] INESC / Inst. Superior Técnico
R. Alves Redol, 9, 1000 Lisboa, Portugal
(Fax +351–1–525843, Email lba@inesc.pt)

Abstract. This paper presents quantitative results on the problem of edge detection using neural network filters. These results are compared with the results provided by the derivative of the Gaussian edge detection filter. A new figure of merit for edge quality, based on Pratt's figure of merit, is introduced. The results displayed in this paper give evidence that neural network edge detection filters can perform better than the linear "optimal" filters.

1 Introduction

In [1] some qualitative results were presented on the subject of edge enhancement using artificial neural networks. It was shown that this type of non-linear filters can provide good localization and low distortion of edges and also a good response to corners. In this paper we further extend that work and we present a more systematic and quantitative evaluation of the properties of these filters.

As in [1] the representation adopted for the edge elements is based on the inter-pixel *cracks* [2]. This type of edge representation offers an unambiguous placement of edge elements, an effective region separation and also a potentially greater capacity to retain information.

The evaluation of the quality of edges is often a subjective operation due to the lack of knowledge of the correct answer. The work reported here is based on synthetic images which allow us to access the respective edge maps. Therefore, in this case we can use an objective measure to evaluate the quality of the edges generated by the detectors. Pratt proposed a figure of merit [3] defined as

$$F_{\text{Pratt}} = \frac{1}{\max(E_{\text{ID}}, E_{\text{AC}})} \sum_{i=1}^{E_{\text{AC}}} \frac{1}{1 + \alpha d^2} \tag{1}$$

where E_{ID} and E_{AC} represent the number of ideal and actual edge points, α is a scaling parameter which penalizes offset edges, and d is the distance from the actual edge to its correct location. This figure of merit although useful does not correctly penalize spurious responses, due for example to noise or blur. As stated in (1) every actual edge is matched with one ideal edge. This means that

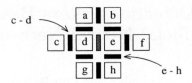

Fig. 1. The neural network input values are the first differences of the context pixels. As can be seen, the eight context pixels (represented as squares) generate nine differences (the rectangles), which form the input of the neural network used to process the central crack (the gray rectangle). The horizontal cracks are processed in a similar way, due to symmetry properties.

if E_{AC} is greater than E_{ID} there are some ideal edge elements that have to match more than one actual edge element. This results in a lack of explicit and effective accounting of false edges.

To improve Pratt's figure of merit we propose the following formulation:

$$ F = \left(\frac{1}{E_{ID}} \sum_{i=1}^{E_{ID}} \frac{1}{1 + \alpha d^2} \right) \times \left(\frac{1}{1 + \beta \frac{E_{FA}}{E_{ID}}} \right) \tag{2} $$

The first term in parentheses is a modified version of Pratt's figure of merit, differing only in the direction of matching. While in (1) each actual edge element is matched with one ideal edge element, in the modified version each ideal edge looks for one, and only one, actual edge element. Also, an actual edge element can, at most, be allocated to an ideal edge element. All un-allocated actual edges are considered false edge elements and are handled by the second term in parentheses of (2). E_{FA} denotes the number of false (unmatched) edge elements.

The two parameters included in (2) (α and β) are responsible for a good balance between the missing and misplaced edge error types (α) and false edge error type (β). For the first one (α) we adopted the value $1/9$, suggested by Pratt. For the second parameter (β) we used the value 1, which seems to be a reasonable choice. This means that if the number of false edge elements is equal to the number of ideal edges the second term in parentheses will be 0.5.

2   Neural Network Filters for Edge Detection

In this work we used the same type of neural network topology as described in [1], i.e. feed-forward multi-layer perceptrons trained with back-propagation [4], improved with the acceleration technique proposed by Silva and Almeida [5].

The input of the neural networks is formed by the nine first differences calculated using adjacent pixels, as shown in Fig. 1. This figure also displays the topology of the eight pixels used as the input context (i.e. the support of the filter), which was inspired on early work of Hanson and Riseman on image segmentation based on relaxation labeling [6].

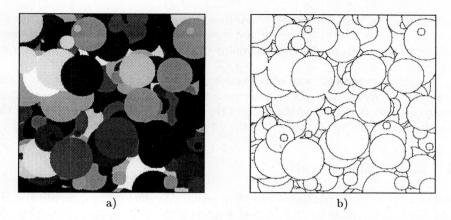

Fig. 2. a) Synthetic image used to train the neural networks and b) its desired contours.

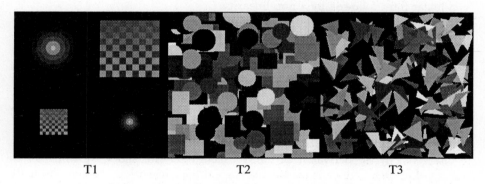

Fig. 3. Synthetic test images (T1, T2, T3).

The neural networks were trained and tested using synthetic images. On one hand this provides a better control of the training and test data and also an easy access to the desired edge maps. On the other hand the variability of the synthetic data can be considered poor when compared to data obtained from real images.

Figure 2a shows an image of 256×256 pixels, with 8 bits per pixel, composed of disks of diameters randomly drawn from the set $\{10, 20, \cdots, 60\}$. The gray level inside each disk is constant and was obtained randomly from the set $\{0, 10, 20, \cdots, 250\}$. Figure 2b depicts the desired contours, obtained by construction. To test the behavior of the neural network filters we used three synthetic images which are displayed in Fig. 3.

All images (training and test) were blurred using the 3×3 low pass mask

$$\begin{bmatrix} a & a & a \\ a & b & a \\ a & a & a \end{bmatrix}$$

with $a = 0.075$ and $b = 0.4$. This provides more realistic data since edges are rarely step like. Also, four levels of additive Gaussian noise were used to degrade the images: $\sigma = 3$, $\sigma = 5$, $\sigma = 10$ and $\sigma = 20$. The objective is to obtain filters robust to noise (from the training point of view) and also to test their noise rejection capabilities.

To provide a simple way of referencing the images (training and test) subjected to the different levels of noise we adopt the following notation: nameSnn, where name is the name of the image (Tr, T1, T2 and T3) and nn refers to the standard deviation of the noise (03, 05, 10 and 20).

3   Simulation Results

Four neural network configurations were tested. We denote them as H0, H2, H4 and H8, which stands for 0, 2, 4, and 8 units in the hidden layer. Note that the H0 network degenerates into a single layer neural network, which can be viewed as a linear filter followed by a non-linearity.

All networks were trained using 10000 examples, randomly extracted from the training image. Each training was performed during 2000 epochs[3], and was repeated with three different random initializations. The training process was controlled periodically (every 100 epochs) through the processing and evaluation of the training image. This procedure can be considered a kind of cross-validation[4], which is used to ensure that the optimal training point is not missed. The best of the three neural networks (which resulted from the three random initializations) was then retrained until the optimal point of the cross-validation curve.

We compared the results obtained with the neural network filters with the results given by the derivative of the Gaussian filter (for short, we will refer to it as the Ln filter), which is considered a good approximation of some optimally derived linear filters for step edges [7] [8]. The Ln filter is defined as

$$
\mathrm{Ln}(x) = -\frac{x}{\sigma_f^2} \exp\left(-\frac{x^2}{2\sigma_f^2}\right),
$$

where σ_f is the scale of the filter. For each image we determined the σ_f that provided the largest figure of merit. Also, since we need binary edge maps to compute the figure of merit, we always used the best possible threshold, both for the Ln and neural network filters.

Table 1 shows the figure of merit of the test images, after processing by the neural networks obtained from the various network topology and training data combinations. In addition, it also shows the results obtained with the Ln filter, using the best σ_f for each image. Note that the test images were never used during the process of neural network training, including the cross-validation operation. Only the training image was used for that purpose.

Figure 4 shows image T3S05 processed by the Ln filter ($F = 0.53$) and also processed by the H4 neural network trained with image TrS05 ($F = 0.67$).

---

[3] One epoch is a complete pass through the training set.

[4] Note that only a small part of the training image is actually used for training.

Table 1. Comparison of the figures of merit calculated for T1, T2 and T3 after processing by the Ln filter and the filters resulting from the several combinations of neural network topology and training sets. The σ_f line displays the scale of the Ln filters that offered the best result (maximum F) for each image.

	T1S03	T1S05	T1S10	T1S20	T2S03	T2S05	T2S10	T2S20	T3S03	T3S05	T3S10	T3S20
Ln	0.43	0.39	0.33	0.29	0.47	0.47	0.44	0.38	0.54	0.53	0.50	0.42
σ_f	0.35	0.55	0.65	0.70	0.10	0.20	0.30	0.60	0.05	0.05	0.30	0.55
Neural networks trained with image TrS03												
H0	0.44	0.36	0.27	0.19	0.55	0.53	0.41	0.25	0.64	0.61	0.49	0.33
H2	0.49	0.41	0.26	0.17	0.66	0.59	0.38	0.20	0.71	0.65	0.47	0.29
H4	0.63	0.46	0.28	0.19	0.70	0.60	0.40	0.22	0.74	0.68	0.49	0.31
H8	0.63	0.45	0.27	0.19	0.69	0.59	0.38	0.21	0.74	0.67	0.48	0.31
Neural networks trained with image TrS05												
H0	0.43	0.38	0.30	0.23	0.47	0.47	0.44	0.31	0.54	0.53	0.50	0.39
H2	0.41	0.40	0.29	0.18	0.64	0.59	0.45	0.22	0.67	0.63	0.51	0.31
H4	0.57	0.49	0.34	0.24	0.64	0.62	0.48	0.29	0.70	0.67	0.56	0.38
H8	0.57	0.50	0.35	0.24	0.66	0.63	0.51	0.35	0.71	0.67	0.57	0.42
Neural networks trained with image TrS10												
H0	0.37	0.37	0.32	0.27	0.39	0.39	0.39	0.37	0.42	0.41	0.42	0.41
H2	0.36	0.36	0.33	0.25	0.49	0.48	0.45	0.34	0.51	0.50	0.47	0.40
H4	0.52	0.46	0.37	0.25	0.59	0.58	0.50	0.33	0.63	0.61	0.56	0.42
H8	0.46	0.42	0.36	0.26	0.61	0.59	0.52	0.37	0.65	0.63	0.57	0.44

4   Discussion and Conclusions

Analyzing globally the results of Table 1, it is evident that neural network edge detection filters can outperform linear filters. Some other important observations can also be drawn from that table.

As a first observation we note the degradation of performance of the neural networks for images of high level of noise. This should be expected since the input support of these filters is small (only 4 pixels in the direction orthogonal to the edge) when compared to the support of at least 7 pixels used for scales $\sigma_f \geq 0.5$. Image T3, which needs smaller scales, is the least affected by this behavior (compare the T1S20, T2S20 and T3S20 columns of Table 1).

Another important observation is the variation of performance with the choice of the noise level in the training image. Low noise during training offers a good performance on low noise images, as would be expected, but at the expense of poor performance on medium and high noise images. If the noise level of the training image is increased then we obtain a somewhat lower performance on low noise images but the medium and high noise images are improved.

The variation of performance due to the complexity of the neural networks (i.e., with the number of hidden units) is more or less as expected. It increases, in general, with the increase in network complexity. However, we could not explain why it decreases when we pass from a H0 to a Hn, $n \neq 0$ topology, for high noise images (see the T1S20, T2S20 and T3S20 columns of Table 1).

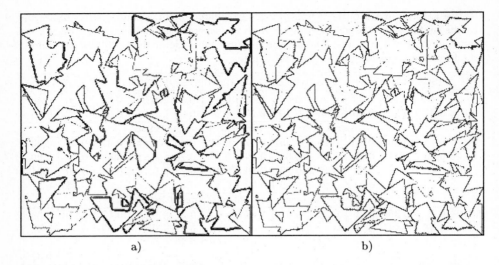

a)   b)

Fig. 4. a) Image T3S05 processed by the Ln filter ($F = 0.53$); b) Image T3S05 processed by the H4 neural network trained with image TrS05 ($F = 0.67$).

The study that we present in this paper is focused on small input support neural networks. Using only these results we are not able to extrapolate about the behavior of this kind of filters for large input supports. However, we believe that the improvement in performance that we observed here will also show for large input supports.

References

1. A. J. Pinho and L. B. Almeida. Some results on edge enhancement with neural networks. In *Proc. of the 1st IEEE Int. Conf. on Image Processing (ICIP'94), Austin, Texas, U.S.A.*, 1994.
2. D. H. Ballard and C. M. Brown. *Computer vision.* Prentice-Hall, Inc., 1982.
3. W. K. Pratt. *Digital image processing.* Wiley-Interscience, 1978.
4. D. E. Rumelhart, J. L. McClelland, and PDP Research Group. *Parallel Distributed Processing - Explorations in the Microstructure of Cognition: Foundations.* Volume 1, MIT Press / Bradford Books, 1986.
5. F. M. Silva and L. B. Almeida. Acceleration techniques for the backpropagation algorithm. In L. B. Almeida and C. J. Wellekens, eds., *Neural Networks, Proc. EURASIP Workshop, Sesimbra, Portugal,* Springer-Verlag, 1990.
6. A. R. Hanson and E. M. Riseman. Segmentation of natural scenes. In A. R. Hanson and E. M. Riseman, eds., *Computer Vision Systems,* pp. 129–163, Academic Press, 1978.
7. J. Canny. A computational approach to edge detection. *IEEE Trans. on Pattern Analysis and Machine Intelligence,* 1986, **8**, pp. 679–698.
8. S. Sarkar and K. L. Boyer. On optimal infinite impulse response edge detection filters. *IEEE Trans. on Pattern Analysis an Machine Intelligence,* 1991, **13**, pp. 1154–1171.

Use of Cortical Filters and Neural Networks in a Self-Organising Image Classification System

Nikolay Petkov

Centre for High Performance Computing and
Institute of Mathematics and Computing Science
University of Groningen
P.O.Box 800, 9700 AV Groningen, The Netherlands
petkov@cs.rug.nl

Abstract. A preprocessor based on a computational model of simple cells in the mammalian primary visual cortex is combined with a self-organising artificial neural network in order to to set up an image classification system. After learning with a sequence of input images, the output units of the system turn out to correspond to classes of input images. Notably, a group of output units which are selective for images of human faces emerges. These units mimic the behaviour of face selective cells that have been found in the inferior temporal cortex of primates.

1 Introduction

The system described in this paper was inspired by two facts which are known from neurophysiological research on the visual cortex of primates. The first fact is that the majority of neurons in the primary visual cortex react strongly, in one way or another, to oriented lines, line and curve segments, bars and edges [2, 3]. The second fact is that in certain hierarchically high areas of the visual cortex of monkeys, more precisely in the inferior temporal cortex, cells have been found which react selectively to very complex visual patterns, such as faces [5].

As to the first above mentioned fact, the extensive neurophysiological studies carried out in the past three and a half decades have provided models of primary cortical cells based on so-called receptive field functions. On the basis of such models, one can compute quantities which correspond to the activities of primary visual cortex cells when an arbitrary input image is projected on the retina, i.e. one can compute (an approximation to) the primary cortical representation of that image.

As to the second above mentioned fact, at present there is no computational model of face selective and, more generally, class selective cells. The facts known about the cortical areas between the primary cortex and the inferior temporal cortex are not sufficient to complete the picture and become able to conceive a full computational model which would make possible automatic classification according to and with the efficacy of the mechanisms employed by natural visual systems. In this study, a self-organising artificial neural network is used to bridge this gap in our knowledge of the visual system in order to set up an artificial image classification system.

2 Preprocessor based on cortical filters

The following model is used to compute the response r of a visual neuron characterised by a receptive field function (impulse response) $g(x, y)$ to a composite input visual signal $s(x, y)$, $(x, y) \in \Omega$ (Ω denotes visual field):

An integral

$$\tilde{s} = \iint_{\Omega} s(x, y) g(x, y) \, dx dy \tag{1}$$

is evaluated and the result $\tilde{s}$ is submitted to thresholding

$$r = 0 \quad \text{if} \quad \tilde{s} \leq T, \tag{2}$$

where T is a threshold value, and non-linear local contrast normalisation:

$$r = \frac{r_\infty \tilde{s}}{c_{\frac{1}{2}} L + \tilde{s}} \quad \text{if} \quad \tilde{s} > T, \tag{3}$$

where L is the mean illuminance within the receptive field of the concerned neuron and r_∞ and $c_{\frac{1}{2}}$ are constants which specify the saturation response and the value $(\tilde{s}_{\frac{1}{2}}/L)$ of the contrast for which half-saturation is reached, respectively.

In the following, a family of Gabor functions proposed by Daugman [1] to model the responses of so-called simple cells in the mammalian primary visual cortex is used in a slightly modified parametrisation form:

$$g_{\xi, \eta, \sigma, \gamma, \Theta, \lambda, \varphi}(x, y) = e^{-\frac{(x'^2 + \gamma^2 y'^2)}{2\sigma^2}} \cos\left(2\pi \frac{x'}{\lambda} + \varphi\right) \tag{4}$$

$$x' = (x - \xi)\cos\Theta - (y - \eta)\sin\Theta$$

$$y' = (x - \xi)\sin\Theta + (y - \eta)\cos\Theta$$

The arguments x and y specify the position of a light spot in the visual field and ξ, η, σ, γ, Θ, λ and φ are parameters as follows:

The pair (ξ, η), which has the same domain Ω as the pair (x, y), specifies the *center of a receptive field* within the visual field. The standard deviation σ the Gaussian factor determines the (linear) *size of the receptive field*. The eccentricity of the Gaussian factor and herewith the eccentricity of the receptive field ellipse is determined by the parameter γ which is referred to as the *spatial aspect ratio*. Its value has been found to vary in a very limited range of $0.23 < \gamma < 0.92$. One constant value $\gamma = 0.5$ is used in this study.

The angle parameter Θ ($\Theta \in [0, \pi)$) specifies the *orientation* of the normal to the parallel excitatory and inhibitory stripe zones (this normal is the axis x' in eq.4). The parameter λ is the *wavelength* of the harmonic factor $\cos(2\pi \frac{x'}{\lambda} + \varphi)$. The ratio σ/λ determines the bandwidth of a cell in the spatial frequency domain and the number of parallel excitatory and inhibitory zones which can be observed in its receptive field. The value $\sigma/\lambda = 0.5$ which has biological relevance [1] is used in this study.

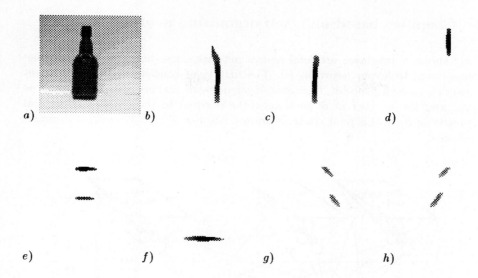

Fig. 1. An input image (a) and cortical images $(b-h)$ computed from it using receptive field functions of different orientations and symmetries: $b)$ $\Theta = 0$, $\varphi = -\frac{1}{2}\pi$; $c)$ $\Theta = 0$, $\varphi = \frac{1}{2}\pi$; $d)$ $\Theta = 0$, $\varphi = \pi$; $e)$ $\Theta = \frac{1}{2}\pi$, $\varphi = -\frac{1}{2}\pi$; $f)$ $\Theta = \frac{1}{2}\pi$, $\varphi = \frac{1}{2}\pi$; $g)$ $\Theta = \frac{1}{4}\pi$, $\varphi = -\frac{1}{2}\pi$; $h)$ $\Theta = \frac{3}{4}\pi$, $\varphi = -\frac{1}{2}\pi$.

Finally, for the phase parameter φ we use the following values: 0 for symmetric receptive fields with an excitatory central lobe, π for symmetric receptive fields with an inhibitory central lobe and $-\frac{1}{2}\pi$ and $\frac{1}{2}\pi$ for antisymmetric receptive fields with opposite polarity.

Substituting a receptive field function $g_{\xi,\eta,\sigma,\gamma,\Theta,\lambda,\varphi}(x,y)$ in eqs.1-3, one can compute the response $r_{\xi,\eta,\sigma,\gamma,\Theta,\lambda,\varphi}$ of a simple visual cortical cell modelled by this function to an input image $s(x,y)$. Fig.1a shows an input image for which a number of such quantities are computed and shown in Fig.1b-h, grouped together in so-called *cortical images*. The quantities grouped in one such image are computed with receptive field functions of the same values for all parameters but (ξ, η); the later specify the coordinates of a pixel to which the value of such a quantity is assigned. The filters which generate such images are referred to as *cortical filters* or *channels*.

Roughly speaking, the effect of such a filter is to enhance luminance transitions of a given orientation and at a give scale. The filters with antisymmetric receptive field functions are more selective for edges and those with symmetric receptive field functions are more selective for bars of a given width. These selectivity properties of the cortical filters can be further enhanced by additional non-linear mechanisms in which the four filters responsible for the same orientation but having different phase (symmetry) interact with each other to produce cortical images in which either edges or bars (but not both at the same time) are enhanced in a cortical image [6].

3   Classifier based on a self-organising neural network

Fig.2 shows a two-layer artificial neural network constructed of a set of one-dimensional Kohonen networks [4]. The first layer consists of m independent networks, each of n nodes. Each first-layer network accepts a cortical image as input and the number m of such networks is equal to the number of cortical channels used. In this pilot study, a limited number of $m = 32$ cortical channels were used.

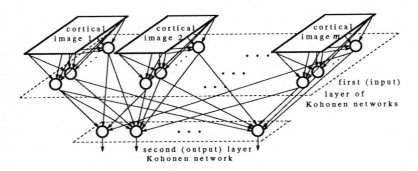

Fig. 2. A separate one-dimensional Kohonen network is associated with each cortical filter channel, accepting the corresponding cortical image as an input. The two-dimensional activity pattern produced by the set of networks in the first layer is used as an input pattern to a second-layer (output) Kohonen network.

As to the number of nodes n, it is chosen depending on the target application taking into account the following considerations: (i) n has to be greater than the number of classes which have to be discriminated and (ii) n should not be very large to prevent a very sparse assignment of units to classes and a large ensuing number of learning cycles needed to achieve convergence. In this study, the number of units used for each first-layer network was taken to be $n = 20$. This number was chosen according to the above mentioned considerations and a particular test application in which five classes of objects (images of faces, bottles, armchairs, tea cups and triangles) had to be discriminated.

For each first-layer network, the learning sequence consisted of 30 cortical images computed from 30 corresponding input images (six images per class for each of the above mentioned classes). Convergence was achieved amazingly very quickly — typically in less than ten epochs (presentations of the learning sequence). The fast convergence can be explained by the fact that single cortical images are simple patterns containing a small number of features (oriented elongated connected activity regions). In most of the cases these patterns are disjoint from class to class so that, after such a pattern is assigned to a unit, it cannot be modified by patterns which arise from images of a different class. As to self-organisation, as expected the units which are activated by images of the same class tend to build clusters, but interleaved units of different classes or units which are never activated as well as mixed-class units, i.e. units which react to image patterns that belong to different classes, can also be observed.

The existence of mixed-class units means that, if one would use just one single-orientation cortical channel with one associated Kohonen network, one can get misclassifications. This is not amazing, since for complex visual input patterns, individual cortical images are not necessarily characteristic of the classes of the corresponding input images. However, *combinations of such images* are characteristic of the class to which an input image belongs.

The proposed two-layer network structure is based on this assumption and its function can intuitively be explained as follows: Each of the first-layer networks associated with the corresponding cortical channels makes its own classification of the cortical image it receives as input. One can think of the first-layer networks associated with different cortical channels as voting for different classes. These votes are counted by another network arranged in the second layer and the class which collects the largest number of votes wins.

The second-layer network has a structure which is similar to the structure of the individual networks in the first layer. It is also one-dimensional and has the same number of units n and the same neighbourhood relations. The only difference is that, while the inputs to the first-layer networks are cortical images of size $k \times k$ ($k = 32$ in this particular case), the inputs to the second-layer network are $m \times n$ binary activity patterns produced by the m first-layer networks, each of n units ($n = 20$, $m = 32$ in this case).

The learning process for the second layer is started after the learning in the first layer is completed. The learning sequence for the second-layer network consisted of 30 activity patterns induced in the first-layer networks for the corresponding 30 input images. Similar to the convergence behaviour of the first-layer networks, the learning process for the second layer took not more than ten epochs. The classification of the learned input patterns was always correct, i.e., images from different classes always activated different units whereby the units corresponding to a class tend to form a cluster. The units of one of these clusters are activated only by input images of faces and in that they mimic the function of face selective cells in the inferior temporal cortex. As to test patterns, the system was tested only for the classification of images taken under similar conditions as the learning images. The system succeeded to classify all test images correctly.

4 Implementation and conclusions

Computing one cortical image on a powerful contemporary workstation takes approximately 5 seconds. In this study only 32 cortical images were computed for each input image. In an ongoing extension, a set of 320 cortical images is computed for each input image. This computation takes more than half an hour on a powerful workstation. In order to accelerate computations, the set of cortical filters has been implemented on a Connection Machine CM-5 scale 3 parallel supercomputer (16 nodes, 64 vector units, 2 Gflop/s, 512 Mbyte) and effective acceleration by a factor of nearly forty was achieved.

The computational effort connected with the neural network part of the system depends on the number of units used. In our first experiments, this number was relatively small, so that computing time did not present a major problem both for learning and classification. The real time learning and classification for a very large number of cortical channels and thousands of different image classes exceed the power of the currently available parallel supercomputers.

The presented system does not incorporate any a priory model of the visual patterns which have to be learned and classified. No knowledge of the visual world, such as the fact that a face has two eyes, a nose, a mouth, etc., or that a bottle has a body and a neck, is used. The system computes its own internal representations of visual patterns, memorises them and uses them to classify new patterns. The output units of the classifier subsystem correspond to classes of visual patterns and not just to individual patterns (generalisation property). This in a way corresponds to the neurophysiological observation that face selective cells are broadly tuned, in that such a cell would react to different faces instead of just one individual face.

As demonstrated by the above example, studying and simulating the principles and mechanisms employed by natural vision systems can lead to new image analysis and object recognition techniques which may have the potential to outperform traditional machine vision approaches. For further details and discussion the reader is referred to [6, 7]. The current paper is a short version of [7].

References

1. Daugman J.G.: "Uncertainty relations for resolution in space, satial frequency, and orientation optimized by two-dimensional visual cortical filters", *Journal of the Optical Society of America A*, Vol. 2 (1985) No. 7, pp.1160-1169.
2. Hubel D.H. and Wiesel T.: "Receptive fields, binocular interaction, and functional architecture in the cat's visual cortex", *J. Physiol. (London)*, Vol. 160 (1962), pp.106-154.
3. Hubel D.H.: "Explorations of the primary visual cortex, 1955- 1978" (1981 Nobel Prise lecture), *Nature*, Vol. 299 (1982) pp.515-524.
4. Kohonen T.: *Self Organization and Associative Memory* (Berlin: Springer-Verlag, 1990, third edition)
5. Perrett D.I., Hietanen J.K., Oram M.W., and Benson P.J.: "Organization and functions of cells responsive to faces in the temporal cortex", *Phil. Trans. Roy. Soc. Lond. B: Biolog. Sci.*, Vol. 335 (1992) pp.23-30.
6. Petkov N.: "Biologically motivated computationally intensive approaches to image pattern recognition", *Future Generation Computer Systems*, special issue *HPCN'94*, ed. W. Gentzsch, 1995, in print.
7. Petkov N.: "Image classification system based on cortical representations and unsupervised neural network learning", Proc. Int. Conf. on *Computer Architectures for Machine Perception, CAMP'95*, Villa Olmo, Como, Italy, September 18-20, 1995 (IEEE Press, 1995, in print).

Occluded Objects Recognition Using Multiscale Features and Hopfield Neural Networks

Jiann-Shu Lee,* Chin-Hsing Chen,* Yung-Nien Sun,**
and Guan-Shu Tseng*

*Department of Electrical Engineering, **Institute of
Information Engineering,
National Cheng Kung University, Tainan, Taiwan, R.O.C.

Abstract. A new method to recognize partially visible two-dimensional objects by means of multiscale features and Hopfield neural network is proposed. The Hopfield network is employed to perform global feature matching. Since the network only guarantee to converge to a local optimal state, the matching results heavily depend on the initial network state determined by the extracted features. To acquire more satisfactory initial matching results, a new feature vector, consisting of the multiscale evolution of the extremal position and magnitude of the wavelet transformed contour orientation, is developed. These features contain the contour primitives information in a multiscale manner, hence good initial states can be obtained. The good initiation enables the method to recognize objects of even heavily occluded, that can not be achieved by using the Nasrabadi-Li's method. In addition, to make the matching results more insensitive to the threshold value selection of the network, we replace the step-like thresholding function by a ramp-like one. Experimental results have shown that our method is effective even for noisy occluded objects.

1 Introduction

The problem of recognizing partially visible objects is of considerable interest in the field of industrial automation. This problem has attracted many researchers [1-3] in the field of computer vision. In this paper, we propose a new approach to deal with this problem. To match features in a global manner, we employ the Hopfield neural network. The Hopfield neural network only guarantees to converge to a local optimal state. If the number of false matches is much larger than the number of correct matches in the initial matching, that is the initial state is away from the optimal state, the network possibly converges to an undesirable state. Therefore, an algorithm with bad initial states, like Nasrabadi-Li [4], tends to fail in the case of heavily occluded objects because the visible feature points are few. To cope with this drawback, features of more effectiveness should be employed. In this paper, we develop a new feature set consisting of the multiscale evolution of the wavelet transformed extremal position and magnitude of the contour orientation. In our previous paper [5], we have demonstrated that the evolution at larger scales contains the information about the neighboring primitives. In this paper, we will show that the evolution at smaller scales can even be used to recognize the single primitives including corners and arcs. This means that the evolution conveys the information of the contour primitives in a multiscale manner. In addition, to make the matching results by the Hopfield network less sensitive to the threshold value selection, a ramp-like thresholding function was used to replace the conventional thresholding function which is step-like. Experimental results have shown that our method is effective even for heavily occluded noisy objects.

2 The Adopted Mother Wavelet

In this paper, we choose the first derivative of the normalized Gaussian function as the mother wavelet $\psi(x)$,

$$\psi(x) = -x\exp(-x^2/2).\tag{1}$$

The orthogonal wavelets are not used because they usually do not possess mathematical close forms such that analytically mathematical derivation is difficult. In addition, the adopted mother wavelet is the first derivative of a smoothing function, hence its local extrema, namely, the curvature extrema, correspond to the sharper changes of a contour, e.g. corners and arcs.

3 Single Primitives Recognition Using Wavelet Transformation

3.1 Smoothed Orientation Function

Let $C(t) = (X(t), Y(t))$ represents a regular planar curve where t is the arc length. To improve the orientation resolution, the orientation at point p_i is defined as

$$\phi(i) = \tan^{-1}((Y_{i+q} - Y_{i-q})/(X_{i+q} - X_{i-q})),\tag{2}$$

for some $q > 1$ to obtain a smoothed version of orientation. In this paper, q is equal to three. This choice causes the orientation profile of a corner to be a ramp-like with ramp-width equal to the smoothing length $SL = 2q + 1 = 7$ instead of a step. As to an arc, the smoothed orientation at p_i in Eq. (2) is equivalent to calculating the tangential angle at p_i that varies linearly along the arc. Therefore, the smoothed arc orientation profile is still a ramp with the ramp-width equal to $l + SL - 1$, where l is the length of the arc. Assuming that w_c and w_a denote the ramp-width of a corner and an arc with $w_c = SL = 7$ and $w_a = l + SL - 1$, respectively. From above, we know that the smoothed orientation profiles of corners and arcs are ramps, and the only difference between them is the ramp-width w.

3.2 Single Primitives Recognition

In this section, we will show that the wavelet transform ratio can be used to determine the ramp-width and even angle changes. We define a ramp function $R(x, w)$ as

$$R(x,w) = \begin{cases} c & if \ x < -\frac{w}{2} \\ c+m(x+\frac{w}{2}) & if \ -\frac{w}{2} \leq x \leq \frac{w}{2} \\ c+mw & if \ x > \frac{w}{2} \end{cases}. \tag{3}$$

Performing wavelet transform on $R(x, w)$, we obtain

$$WR(s,x,w) = m \int_{x-w/2}^{x+w/2} \exp(-\frac{u^2}{2s^2}) du . \tag{4}$$

The modulus maximum of $WR(s,x,w)$ is at $x = 0$. The extrema ratio of two scales s_a, s_b, namely, $R(s_a, s_b, w)$ is given by

$$R(s_a, s_b, w) = WR(s_a, 0, w) / WR(s_b, 0, w) . \tag{5}$$

Differentiating $R(s_a, s_b, w)$ with respective to w, we obtain

$$R'(s_a, s_b, w) = (\exp(-\frac{y^2}{2})\int_0^{ky} \exp(-\frac{u^2}{2}) du - k \exp(-\frac{k^2 y^2}{2}) \int_0^y \exp(-\frac{u^2}{2}) du) / (2(\int_0^{ky} \exp(-\frac{u^2}{2}) du)^2) \tag{6}$$

where $y = w/(2s_a) > 0$ and $0 < k = s_a / s_b < 1$ (i.e. the scale ratio). Let $N(k, y)$ denote the numerator. By simple computation, we know that $N(k, y)$ is negative, that is, $R'(s_a, s_b, w)$ is negative, if $w \geq 2s_a \sqrt{(2\ln(s_b / s_a)) / (1 - s_a^2 / s_b^2)}$. This result implies that $R(s_a, s_b, w)$ is strictly monotonic decreasing for $w \geq 2s_a \sqrt{(2\ln(s_b / s_a)) / (1 - s_a^2 / s_b^2)}$. In other words, if a w satisfies this condition a certain R corresponding to a certain w can be used to determine the ramp-width. Because the only difference between a corner and an arc is the value of w, this means that the extrema ratio R can not only discriminate a corner from an arc but also can determine the arc length. Since the w is acquired, the other parameter--angle change mw also can be determined easily by applying Eq. (5). This shows that a single primitive can be recognized by using two scales wavelet transform extrema.

4 Feature Locating and Feature Selection

Observing the wavelet transform orientation, we find that most of the local extrema corresponding to the significant structures can appear at least three consecutive scales s_1, s_2 and s_3 in the test images. Thus, an extremum at s_1 is regarded as a feature point if it can also appear at s_2 and s_3. The feature position is decided at the first scale because the locating capability is better at a smaller scale. We select $s_1 = 2$ for the occluded objects recognition. This selection is not sensitive to noise. In addition, this selection makes the wavelet transform extrema at s_1 and s_2 can analyze the single primitive with orientation ramp-width $w \geq 6$. In other words, we can analyze single primitives by using the wavelet transform extrema at s_1 and s_2 because the orientation ramp-width of every single primitive is longer than six for a smoothed contour. Hence, we employ the evolution, $\left[W\phi(s_1,p_1), W\phi(s_2,p_2), W\phi(s_3,p_3), p_2 - p_1, p_3 - p_2\right]$ as the contour features, where $W\phi$ is the wavelet transform of the smoothed orientation function $\phi(\cdot)$, and p_i denotes the local extremum position at the scale $s_i = is_1$. We call the evolution the Integrated Multiscale Feature (IMF).

5 Feature Matching

The feature matching task is performed by the Hopfield neural network. The location of the object is then estimated by a least squares fit among the matched feature points. We represent each object by a set of features consisting of IMFs. The energy function used in our method is the same as the Nasrabadi-Lis' [4]

$$E = -\sum_{i=1}^{M}\sum_{k=1}^{N}\sum_{j=1}^{M}\sum_{l=1}^{N} C_{ikjl} V_{ik} V_{jl} + \sum_{i=1}^{M}(1-\sum_{k=1}^{N}V_{ik})^2 + \sum_{k=1}^{N}(1-\sum_{i=1}^{M}V_{ik})^2. \tag{7}$$

The major component of the compatibility measure C_{ikjl} (the synaptic strength) between a neuron in row i column k and in row j column l is defined by

$$C_{ikjl} = W_1 F(\sum_{y=1}^{5}|f_{i,y} - f_{k,y}|) + W_2 F(\sum_{y=1}^{5}|f_{j,y} - f_{l,y}|) + W_3 F(|d_{i,j} - d_{k,l}|) \tag{8}$$

where W_i for i = 1, 2, 3 are the weighting factors, $f_{i,y}$ and $f_{j,y}$ are the yth elements of the IMFs of the ith and the jth feature points of the model object, $f_{k,l}$ and $f_{k,l}$ are the yth elements of the IMFs of the kth and the lth feature points of the scene, $F(x)$ is the thresholding function, and d_{xy} denotes the distance between the xth and the yth features. The network is updated in the same way as the Nasrabadi-Lis' [4]. The adopted thresholding function $F(x,y)$ in Nasrabadi-Li's algorithm is a step function. There exists an abrupt change at the threshold value in the function. Such a discontinuity causes the feature matching sensitive to the threshold value. To avoid such a drawback, we select a ramp-like function $F(x)$ as the thresholding function. Since there is no abrupt change in $F(x)$, the feature matching results of our algorithm are more stable.

6 Experimental Results

To test the effectiveness of our method, a noisy image shown in Fig. 1(a) is adopted. The superimposed result of all the recognized objects is shown in Fig. 1(b). The result shows that our method is effective even for heavily occluded objects.

7 Conclusions

In this paper, we have proposed a new occluded object recognition method using the multiscale feature vector and the Hopfield neural network. The Hopfield network is used to perform global feature matching. The feature vector consists of the multiscale extrema evolution of position and magnitude of the wavelet transformed contour orientation. The vector contains the information of the contour primitives in a multiscale manner, hence a good initial state of the Hopfield network can be obtained. Such good initiation enables the network to converge to a state which is closer to the global optimum. To make the converging results more insensitive to the threshold value selection of the network, the step-like thresholding function is replaced by a ramp-like one. Experimental results have shown that our method is effective even for noisy objects with heavy occlusion.

References

1. J. L. Turney, T. N. Mudge and R. A. Voltz, "Recognizing partially occluded parts," *IEEE Trans. Pattern Analysis Mach. Intell.* **PAMI-7**, pp. 410-421, 1985.
2. H. C. Liu and M. D. Srinath, "Partial shape classification using contour matching in distance transformation," *IEEE Trans. Pattern Analysis Mach. Intell.* **PAMI-12**, pp. 1072-1079, 1990.
3. N. Ansari and E. J. Delp, "Partial shape recognition: a landmark-based approach," *IEEE Trans. Pattern Anal. Mach. Intell.* **PAMI-12**, pp. 470-483, 1990.
4. Nasser M. Nasrabadi, and Wei Li, "Object recognition by a Hopfield neural network," *IEEE Trans. on Syst. Man Cybern.* **SMC-21**, pp. 1523-1535, 1991.
5. J. S. Lee, Y. N. Sun, C. H. Chen, and C. T. Tsai, "Wavelet based corner detection," Pattern Recognition, Vol. 26, No. 6, pp. 853-865, 1993.

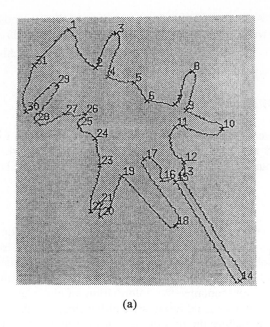

(a)

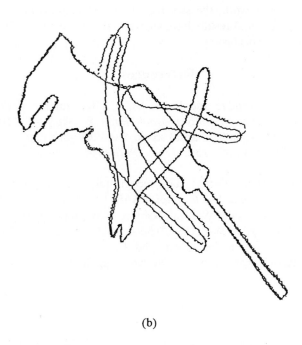

(b)

Fig.1 (a) The noisy objects and (b) the superimposed result of all the recognized objects.

Image Coding I

Automatic Video Segmentation through Editing Analysis

J. M. Corridoni[1] and A. Del Bimbo[12]

[1] Dipartimento di Sistemi e Informatica, Università degli Studi di Firenze,
via S. Marta 3, 50139 Firenze, Italy.
[2] Dipartimento di Elettronica per l'Automazione, Università degli Studi di Brescia,
via Branze, 38, 25133 Brescia, Italy

Abstract. In the present paper the problem of video segmentation for indexing and retrieval in multimedia databases is addressed. This work proposes a statistically based new algorithm to detect cuts between two shots in a video sequence. Such algorithm is data driven, since no a priori knowledge about the shots is available. A mathematical model is formulated for detecting fades, dissolves and mattes. All the algorithms are inserted into a global framework, in order to generate a low level parsing and segmentation of video streams.

1 Introduction

With the emergence of multimedia applications, a growing interest is being focused on databases of digital images, image sequences and video, concerning the way in which contents of these pictorial data are described, indexed and retrieved. The central problem for the effective access and usage of a video database lies in representing video content meaning, which is determined not purely by the visual content of the video stream, but also by the way in which frame sequences are ordered and organized into the entire stream. Visual content of frame sequences and video may be expressed either by detecting the spatial occurrences of objects in the individual frames,[3], and encoding through appropriate representation languages the temporal change of the spatial relationships between imaged objects, [1],[4], or at a higher level, by describing spatio-temporal interactions among imaged objects, through statements taken from descriptive languages [2]. Detection of cuts between clips, based on data analysis, is developed in [3]. There, the *Template Matching* technique and the χ^2 *Test* are applied to the color histograms of two subsequent frames. Segmentation obtained not purely considering data analysis, is presented in [5], where a different cut technique is presented (*Intensity Average Difference*) and Different editing procedures like page translate, fades and dissolves are modeled through mathematical functions. The film editing is to be considered not only the "glue" between two shots, but also an essential contribute to the meaning conveyed by the video. Where film becomes art, editing techniques create a *meta–language*, through which a part of the visual message is communicated to the viewer. The detection of the editing features in a film can be therefore a useful help for technical analysis on films as forms of art. In this paper we present a new prototypal system for automatic segmentation of video into its syntactic units through the detection of the editing operations that are present in the film. Experimental evidence of the system operation is provided by presenting results selected from the set of experiments carried out over critical video sequences. A critical comparison with the performance of systems discussed in the literature is also presented.

2 Editing Analysis

The parsing procedure herein introduced is aimed to detect all the editing operations that are present in a film. The stream of frames recorded between the time in which a video camera is turned on and turned off is called *shot*. A *scene* is a sequence of shots related by semantic features. The content of a single scene must have the four Aristotelian properties of unity of *space, place, time* and *action*. All the shots sharing these properties are part of the same scene. The shot is the syntactic unit of a video, while the scene constitutes the semantic atom upon which the visual speech is based. In the following we will address both the detection of cuts, and of the other editing techniques.

Cut Detection: A cut between two shots hasn't undergone any editing process and has therefore no mathematical model, since no information about the shots connected is available. The basic idea that lies under the design of a *cut detector* is that consecutive frames belonging to the same shot are in some way *more similar* than frames belonging to different shots. Measures for the cut detection in a gray-level or a RGB video proposed in [5], [3] show poor performance whenever there is large motion in the scene, for example in presence of rapid moving objects or camera zoom. Even in such cases the global image background does not change significantly, while, on the contrary, a cut implies a global change in the scene. As a result, we expect a cut to be present when the difference between two frames is much larger than the *standard* difference between frames belonging to the same shot. Therefore a relative difference between frames should be introduced, which is expressed as an incremental ratio: $\mathbf{CD} = \mathbf{D}(f_{t+1}, f_t)/\mathbf{D}(f_t, f_{t-1})$, where $\mathbf{D}(f_{t+1}, f_t)$ is a suitable difference measure of two frames. Previous equation states that a cut is detected when the difference between two frames is larger with respect to the former. This statement eliminates the difference as an absolute value, but states its relativeness. In order to increase robustness to local changes in the image, such as the appearance of an object in the visual field, each frame has been divided into $n \times n$ rectangular regions. Instead of comparing two whole frames, as in [5], we compare every pair of subframes between two frames, to obtain n^2 difference values. Discarding the largest values, we make $\mathbf{D}$ more robust against local changes between consecutive frames of the same shot. The problem of defining a meaningful measure of difference has been addressed, using some statistical parameters of the color histograms. Instead of encoding the color levels, as in [3] and [5], the three color histograms H_r, H_g, H_b, are collected for each subframe $r_t(i)$ of frame f_t. Such histograms can be parameterized well enough by the three statistic moments of first (m_1), second (m_2) and third order (m_3). Denoting with $\mathbf{D}_i^{red}(f_t, f_{t+1})$ the difference between the couple of subframes i, with reference to the red color, we define:

$$\mathbf{D}_i^{red}(f_t, f_{t+1}) = \sum_{i=1}^{3} a_i |m_i^{red}(t+1) - m_i^{red}(t)| \tag{1}$$

being $\mathbf{a} = [a_1, a_2, a_3]^T$ a set of parameters experimentally tuned. These parameters can be modulated, whenever more information is available about the shots. For the generic subframe i we define the difference measure in the following way.

$$\mathbf{D}_i(f_t, f_{t+1}) = \mathbf{D}_i^{red}(f_t, f_{t+1}) + \mathbf{D}_i^{green}(f_t, f_{t+1}) + \mathbf{D}_i^{blue}(f_t, f_{t+1}). \tag{2}$$

Finally, the global measure $\mathbf{D}(f_t, f_{t+1})$ of the difference between frames f_t and f_{t+1} is obtained, discarding the k largest values among the n^2 that have been computed and averaging the remaining values. A shot-cut in a video is detected by thresholding: whenever $\mathbf{CD}$ is over a predefined threshold, then a cut is inferred. Since there is no mathematical model underlying the production of a shot, nor an *a priori* knowledge about its characteristics, then no conceptual reasoning is applicable to determine the threshold value, which has to be stated experimentally.

Chromatic Editing Detection: While the *cut* has no model, a mathematical approximation can be proposed for the fades, dissolves and mattes, since these effects are obtained in laboratory, according to a specific technique. Let $\mathbf{c} = (r, g, b)$ represent the intensity level of an arbitrary pixel in a video frame, we can identify the frames belonging to a shot $\mathbf{S}$ as $\mathbf{S} = \{(x, y, t) | \mathbf{c} = \mathbf{S}(x, y, t)\}$. Let $\mathbf{S}_1(x, y, t)$ and $\mathbf{S}_2(x, y, t)$ be two shots that are being edited, and $\mathbf{S}(x, y, t)$ the edited shot. All the chromatic processes can be described as a linear pixel intensity manipulation, since the machines used (*truka*) follow a linear law. Therefore, the equation that expresses $\mathbf{S}$ has the following form [5]:

$$\mathbf{S}(x, y, t) = \mathbf{S}_1(x, y, t)\left(1 - \frac{t}{l_{out}}\right)\Big|_{(t_1, t_1 + l_{out})} + \mathbf{S}_2(x, y, t)\left(\frac{t}{l_{in}}\right)\Big|_{(t_2, t_2 + l_{in})} \tag{3}$$

The fade–in and fade–out are peculiar cases, in which $\mathbf{S}_1 = 0$ or $\mathbf{S}_2 = 0$, respectively. In the matte case, $\mathbf{S}_2$ is zero outside a specified pixel area expressed by bidimensional function of shape $f(x, y)$. To design a chromatic feature detector, let us consider the case of a fade–in. Let $\mathbf{S}_{fi}$ be the fade–in sequence; deriving it with respect to time we obtain:

$$\frac{\partial \mathbf{S}_{fi}(x, y, t)}{\partial t} = \frac{\partial \mathbf{S}_2(x, y, t)}{\partial t}\left(\frac{t}{l_{in}}\right) + \frac{\mathbf{S}_2(x, y, t)}{l_{in}} \tag{4}$$

If it is supposed that the scene does not change much during the fade–in, which is often verified in films, the first addend can be skipped. Dividing (3) by $\mathbf{S}_2$, we obtain a constant value. This means that in a fade–in (out) sequence the frame ratio $\frac{f(t+1) - f(t)}{f(t+1)}$ is approximately constant. The verification of the constancy of this ratio can be assumed as an efficient measure for detecting fades. In a typical dissolve sequence there is always a phase, where the chromatic dynamics of one of the two shots being edited prevails over the other, in such a way that at least two phases of "dominant fade–in" and "dominant fade–out" can be separated. During each of these two phases the model in equation 3 applies and the value of the chromatic feature is meaningful. In [5], this measure is obtained only based on local analysis. This approach contrasts with the way in which these artifices are actually devised in the editing of a video, since dissolves, fades and mattes have no local meaning, being obtained as an optical process, which is performed on a sequence of frames, not on a single frame like the cut. Moreover, such editing processes are performed using the "truka" machine, which can operate only on sequences of standard duration (16, 24, 34, 48 or 96 frames). Therefore, we analyze a sequence over a wide temporal window which is approximately of the size of the minimal chromatic sequence: 16 frames. Whenever the temporal average over such a window overcomes a threshold, then a chromatic editing is detected. Differently from [5], the algorithm allows for a discrimination among fade–in, fade–out, dissolve and matte. By extracting the color histograms of the first and last frames in the edit sequence detected, a distinction between fade-in and fade-out can be easily made, since the average luminance of the first or last frame, respectively, is approximately zero. The dissolve, has first and last frame with non zero luminance and generates typically a local minimum in the feature histogram, as it will be clear in section 3. As to the mattes, they look like fades, apart from the fact that luminance varies over the frames following a geometrical law. Once a fade has been detected, a statistical analysis over one or two frames, typically the central ones, has to be made. If the luminance has a big statistical discontinuity, due to the presence of a black mask the partially covers the frame, then a matte is detected.

3   Experimental Results

The techniques described in previous sections have been applied to movies and to television commercials. The cut and chromatic features detection has been tested on about four hours of video sequences, and a comparison with editing other detection techniques has been made. Results obtained with two sample scenes are reported below. The scenes presented show specific criticality for cut and dissolve detection respectively. **Sequence 1:** Figure 1 shows a sequence taken from a TV commercial, where 6 cuts are present. The longest shot in this sequence is characterized by a large amount of motion, due to the rapid panning of the camera which follows the arrow during its run to the target. This large motion condition could deeply impair the cut feature. The histogram in figure 2a, shows the feature response detected. Its performance is compared with the *Average Intensity* measure [5] (fig. 2b), the *Template Matching* (fig. 2c) and the χ^2 *Test* techniques (fig. 2d), [3]. The comparison among the histograms shows that, while the first measure detects only the cuts, the others give some peak response during the camera panning. **Sequence 2:** The histogram in fig. 3a shows the chromatic feature response computed on a typical dissolve sequence [3], according to the algorithm described in Sec. 2. In the figure the three phases of *dominant fade-in* (**A**), equal fade–in and out (**B**) and *dominant fade-out* (**C**), detected by the measure are put in evidence. The diagram in fig.3b shows the output response for the technique proposed in [5]. The cut feature response shows no peaks and its diagram is therefore here omitted. While histogram in fig. 3b has peaks outside the dissolve zone and low peaks even in the dissolve region, diagram in 3a has a large region with high values only in correspondence with the dissolved frames. In that region two relative maxima and a local minimum can be detected. They describe the three phases of dominating fade–in, equal fade–in and fade–out and dominating fade–out, that have been highlighted previously. The local minimum in phase (**B**) is to to the fact that the error made approximating a dissolve with a simple fade as in equation 4 becomes larger during that phase. Finally, table 1 summarizes the global results obtained on four hours video. The result summary indicates a 97% correct segmentation.

Edit Type	Correct	False	Total	Error Rate
Cut	798	15	813	2%
Fade–in	23	1	24	4%
Fade–out	34	2	36	5%
Dissolve	16	2	18	11%
Matte	7	0	7	0%
TOTAL	878	20	898	3%

Table 1. Error rates for the video segmentation algorithm, tested on 4 hour video.

References

1. T.Arndt, S.K.Chang, "Image Sequence Compression by Iconic Indexing", *IEEE VL '89 Workshop on Visual Languages*, Roma, Italy, Sept.1989.
2. M.Davis, "Media Streams, an Iconic Visual language for Video Annotation", *Telektronik*, No.4, 1993, (also appeared in reduced version in *Proc.IEEE VL'93 Workshop on Visual Languages*, Bergen, Norway, Aug.1993).
3. A.Nagasaka, Y.Tanaka, "Automatic Video Indexing and Full Video Search for Object Appearances," in *IFIP Transactions, Visual Database Systems II*, Knuth, Wegner (Eds.), Elsevier Pub. 1992.
4. A.Del Bimbo, E. Vicario, D. Zingoni, "Symbolic Description of Image Sequences with Spatio Temporal Logic, *IEEE Transactions on Knowledge and Data Engineering*, to appear.
5. A. Hampapur, R. Jain, T. Weymouth "Digital Video Indexing in Multimedia Systems" in *Proc. of AAAI-94 Workshop on Indexing and Reuse in Multimedia Systems* Seattle, Wa, Aug. 1994.

---

[3] Beginning sequence from the episode by Martin Scorsese in *"New York Stories"*.

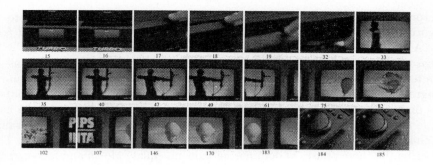

Fig. 1. Frames sampled from a TV spot, characterized by rapid camera panning

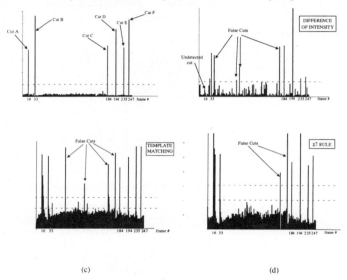

(c)   (d)

Fig. 2. Comparison among cut detectors. (a) Algorithm described in sect. 2. (b) *Average Intensity Difference.* (c) *Template Matching.* (d) χ^2 *Test.*

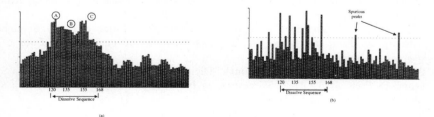

Fig. 3. (a) The response diagram for the dissolve detector presented in section 2. (b) The response diagram for the dissolve detector introduced in [5].

Image Retrieval by Elastic Matching of User Sketches

A. Del Bimbo[1] and P. Pala[2]

[1] Dipartimento di Elettronica per l'Automazione,
Università degli Studi di Brescia, via Branze, 38, 25133 Brescia, Italy
[2] Dipartimento di Sistemi e Informatica,
Università degli Studi di Firenze, via S. Marta 3, 50139 Firenze, Italy

Abstract. Image retrieval by contents from database is a major research subject in advanced multimedia systems. The intrinsic visuality associated with pictorial data suggests the use of iconic indexes and visual techniques to perform retrieval effectively. In this paper we present a method for image retrieval based on sketches of object shapes. In our method, the shape drawn by the user is deformed to match as well as possible the objects in the images. The degree of match achieved, and the elastic deformation energy spent to achieve such a match are used as a measure of the similarity between the template and the image object. The elastic matching is integrated with arrangements to provide for scale invariance, to take into account rotations and spatial relationships between objects for multiple-object queries.

1 Introduction

The intrinsic visuality of the information contents associated with pictorial data advises against the use of indexing and retrieval based on textual keywords as traditionally used in text documents. Iconic indexes have been proposed in [8] to effectively support image retrieval by contents. Iconic indexes are in the form of symbolic descriptions of pictorial data or pictorial data relationships but may also include the actual values of object features, or be in the form of abstract images taking the salient features of the original image. The use of iconic indexes naturally fits with the accomplishment of image retrieval according to visual querying by-example. Once iconic indexes have been built, the user reproduces on the screen an approximated visual representation of pictorial contents of images to be retrieved. Retrieval is reduced to the matching of the user visual representation against iconic indexes in the database. Visual queries by-example for pictorial data exploit human natural capabilities in picture analysis and interpretation and largely reduce the cognitive effort of the user in the access to the database. A number of techniques have appeared in the literature which deal with iconic indexing and visual querying by example of single images; different approaches in performing queries are related to the type of facets of pictorial data that are taken into account. Indexing and querying based on spatial relationships have been proposed in [2], [1], [4], [3]. Spatial relationships are represented symbolically through *2D strings* according to the *symbolic projection* approach.

2D-strings encode the positional relationships between the projections of the objects on two reference coordinate axes. Indexing and querying based on picture color distribution or object texture organization has been proposed in [7] and [6], respectively: images are requested that contain object colors and textures similar to those selected from a menu; matching is performed by comparing color histograms or the Euclidean distance in the texture space. Retrieval by-contents based on similarity between imaged object shapes and user-drawn sketches has been proposed by a few authors [6], [5]. Unlike indexing and retrieval by colors or textures, or spatial relationships, here the problem is complicated by the fact that shape does not have a mathematical definition that exactly matches what the user feels as a shape. In this paper, we present a system for image retrieval-by-contents based on shape matching with elastic deformations. In our system the user sketches a drawing on the computer screen and this is deformed to adapt to shapes of objects in the images. For queries including multiple objects, spatial relationships between the different objects are taken into account.

2 The elastic approach to shape matching

Suppose we have a one-dimensional template, modeled by a first order spline $\tau : \mathbf{R} \mapsto \mathbf{R}^2$. We will always assume that the template is parametrized with respect to arclength, and normalized so as to result of length 1. We have an image $I : \mathbf{R}^2 \mapsto [0,1]$ – we suppose the luminance at every point normalized in $[0,1]$ – that we search for a contour with a shape *similar* to that of τ. To make a robust match even in the presence of deformations, we must allow the template to warp. If $\theta : \mathbf{R} \mapsto \mathbf{R}^2$ is the deformation, then the two components of the deformed template (also parametrized with respect to arclength) are given by: $\phi_j(s) = \tau_j(s) + \theta_j(s)$. The template must warp taking into account two opposite requirements. First, it must follow as closely as possible the *edges* of the image. The second requirement to take into account is the deformation of the template. Allowing arbitrary deformations every template matches every image, and results in a mathematically ill-posed problem. In order to discover similarity between the original shape of the template and the shape of the edge areas on the image, we must set some constraint on deformation. In this way our goal is to minimize the compound functional:

$$\int_0^1 \alpha \left[\left(\frac{d\theta_x}{ds} \right)^2 + \left(\frac{d\theta_y}{ds} \right)^2 \right] + \beta \left[\left(\frac{d^2\theta_x}{ds^2} \right)^2 + \left(\frac{d^2\theta_y}{ds^2} \right)^2 \right] - |\nabla I\left(\phi(s)\right)|^2 \, ds. \quad (1)$$

The quantity depending on the first derivative is a measure of how the template τ has been *locally stretched* by the deformation θ, while the quantity depending on the second derivative is an approximate measure of the energy spent to *locally bend* the template. The match between the deformed template and the edges in the image can be measured through the third integral of the compound functional (1). Using the gradient descending technique, the solution θ can be approximated as a third order spline. To support template deformation the edge image is blurred with a gaussian filter. The elastic energy depends only on the first and second derivatives of the deformation θ. This allows to do not penalize discontinuities and sharp angles that are already present in the template, but to penalize only the degree by which we depart from those discontinuities or angles. Also, since the energy depends only on the derivatives of θ, pure translation of the template, for which θ is constant, does not result in additive cost. This makes our scheme inherently translation invariant.

2.1 Template Matching

After a template reached convergence over an image shape, we need to measure how much the two are *similar*. Again, the similarity is a fuzzy concept, and to measure it we need to take into account a number of things. A first thing to take into account is, of course, the degree of overlapping between the deformed template and the gradient of the image. This can easily be measured as:

$$\mathcal{M} = \int_0^1 \left[\nabla I(\tau(s) + \theta(s)) \right]^2 ds \tag{2}$$

Another factor to consider is how much the template had to warp to achieve that match. We use two different measures of deformation, corresponding to the two energy terms used in the compound functional (1): *Strain energy* ($\mathcal{S}$) and *Bend energy* ($\mathcal{B}$) defined as:

$$\mathcal{S} = \int_0^1 \left(\frac{d\theta_j}{ds} \right)^2 ds \qquad \mathcal{B} = \int_0^1 \left(\frac{d^2\theta_j}{ds^2} \right)^2 ds \tag{3}$$

Coefficients $\mathcal{M}$, $\mathcal{S}$, $\mathcal{B}$ alone are not enough to operate a good discrimination between different shapes. In our approach constraints are imposed by considering the changes of the number $\mathcal{C}$ of zeroes of the curvature function. So another factor to take into account is the variation of the number of zeroes of the curvature function during the deformation process; that is $\mathcal{C} - \mathcal{NC}$. All these 5 parameters $(\mathcal{M}, \mathcal{S}, \mathcal{B}, \mathcal{C}, \mathcal{C} - \mathcal{NC})$ are classified by a back-propagation neural network suitable trained. The neural classifier gives one output value ranging from 0 to 1, which represents the similarity between the shape in the image and the template.

3 Spatial Relationships

Another important source of information about the scene represented in an image is the spatial disposition of objects. In our approach each object is represented by its minimum enclosing rectangle (MER); we use spatial relationships between rectangles both as a mean to filter the database and as a mean to make a more precise multi-objects query. We do this by a slight modification of a method developed in [1]. Projection of such rectangle on the two coordinate axes determine begin and end boundaries of the object. In this way each object is represented by four boundaries: begin (bb) and end (eb) boundaries in the x-axis direction and begin (bb) and end (eb) boundaries in the y-axis direction.

Boundaries can then be sorted introducing two precedence operators: "$<$" (left-right, below-above) and "$=$" (same location). All possible relations between two objects may be ranked in five categories based on relations among object's boundaries:

$$A \; disjoint \; B = \{[bb_x(A) < eb_x(B)] \vee [bb_x(B) < eb_x(A)] \vee$$
$$[bb_y(A) < eb_y(B)] \vee [bb_y(B) < eb_y(A)]\}$$
$$A \; meet \; B = \{[eb_x(A) = bb_x(B)] \vee [bb_x(A) = eb_x(B)] \vee$$
$$[eb_y(B) = bb_y(A)] \vee [bb_y(A) = eb_y(B)]\}$$
$$\wedge \tilde{\;}(A \; disjoint \; B)$$
$$A \; contain \; B = \{[eb_x(A) \leq eb_x(B)] \wedge [bb_x(B) \leq bb_x(A)] \wedge$$
$$[eb_y(A) \leq eb_y(B)] \wedge [bb_y(B) \leq bb_y(A)]\}$$
$$A \; inside \; B = \{[eb_x(B) \leq eb_x(A)] \wedge [bb_x(A) \leq bb_x(B)] \wedge$$
$$[eb_y(B) \leq eb_y(A)] \wedge [bb_y(A) \leq bb_y(B)]\}$$
$$A \; partly \; overlap \; B = \tilde{\;}(A \; disjoint \; B) \wedge \tilde{\;}(A \; meet \; B)$$
$$\wedge \tilde{\;}(A \; contain \; B) \wedge \tilde{\;}(A \; inside \; B)$$

To make the system's description coherent with that operated by our visual perception, four orientation parameters O_1, O_2, O_3, O_4 are introduced that represent relations as NORTH-SOUTH EAST-WEST. While in [1] these four parameters are defined based on precedence relationships among object's boundaries, we found that a more appropriate definition of these parameters for the relation of object A with respect to object B considers the position of centroid of A with respect to boundaries of object B. The spatial relation of object A with respect to object B is represented by a symbolic 5-tuple: $R(A, B) = [C, O_1, O_2, O_3, O_4]$, where C is the category (*disjoint, meet, contain, inside, partly overlap*) which the spatial relation belongs to. In order to speed up the spatial matching process a binary codeword *signature file* is associated with each image encoding mutual spatial relationships, and matching is performed through a hash function.

4 Image Retrieval System

Based on shapes and spatial relations representations previously discussed, a system has been developed for image retrieval by contents. In the databases raw images are passed through a Canny edge detector and blurred through a gaussian filter. To reduce the computational effort in the retrieval phase, objects which are considered interesting for retrieval purposes are bounded with their minimum enclosing rectangle. Templates are deformed over the shapes included in these rectangles. Based upon the technique proposed in Sect. 3, spatial relationships among objects are analyzed and recorded in a symbolic *description file* associated to the image. Then, a *signature file* is built which is used as an index for fast access to spatial relationships informations. In this way to each raw image of the database the following structures are associated: *edge image, image description file, signature file*. The system can be requested to retrieve images representing one or more shapes. A query is composed by drawing a sketch of one or more shapes on a graphic screen. If the sketch is composed of N templates, the system first searches an image where N objects are represented in the same spatial relation of drawn templates. Once it has been found, the system warps each template over the shape located in the same relative position in the image, computing a coefficient $S_i \in [0,1]$, (as the output of the neural classifier) to measure similarity between the shape and the template. A similarity coefficient for the whole image is derived as $S = \sum_{i=1}^{N} S_i$. Once all the images in the database have been processed, they are sorted depending on the value of S and presented to the user. The technique explained in the previous sections has been applied for retrieval of images by sketch from a database of 20th century pictures from the Morandi catalogue, in the context of a joint project with industry for artwork musem database management systems. Fig. 1 shows the sketch of a Morandi's bottle, with a roughly rounded body sketched on the blackboard. Retrieval results are shown in Fig. 4 where the six more similar bottles are presented. The deformed template is shown superimposed over the original image. Pictures retrieved are ordered by decreasing similarity rank; rounded body bottles are ranked in the first positions, followed by those with more strained shape. A query for a different bottle shape over the same database and results obtained ordered according to their similarity rankings are shown in Figs. 2, 5. Examples of retrieval based on shapes and relative spatial relationships are shown in Figs. 3, 6. The user is allowed to draw object sketches and arrange their positions in the blackboard. Images are analyzed for matching of spatial relations between object including rectangles; only those rectangles which have passed the sieve are hence subjected to the template elastic deformation process.

References

1. S.Y. Lee and F.J. Hsu. "Spatial Reasoning and Similarity Retrieval of Images using 2D-C String Knowledge Representation". *Pattern Recognition*, 25(3), 1992.
2. S.K.Chang, Q.Y.Shi, C.W.Yan, "Iconic Indexing by 2-D Strings". *IEEE Transactions on Pattern Analysis and Machine* Vol.9, No.3, July 1987.
3. S.K.Chang, C.W.Yan, D.C.Dimitroff, T.Arndt, "An Intelligent Image Database System". *IEEE Transactions on Software Engineering*, Vol.14, No.5, May 1988.
4. A.Del Bimbo, E.Vicario, D.Zingoni, "A Spatial Logic for Symbolic Description of Image Contents". to appear on *Journal on Visual Languages and Computing*.
5. K.Hirata, T.Kato, "Query by Visual Example: Content-Based Image Retrieval". In *Advances in Database Technology - EDBT'92*, A.Pirotte, C.Delobel, G.Gottlob (Eds.), Lecture Notes on Computer Science, Vol.580,
6. W.Niblack et alii, "The QBIC Project: Querying Images by Content Using Color, Texture and Shape". Res.Report 9203, IBM Res.Div. Almaden Res.Center, Feb.1993.
7. M.J.Swain, D.H.Gallard, "Color Indexing". *Int.Journal of Computer Vision*, Vol.7, No.1, 1991.
8. S.L.Tanimoto, "An Iconic/Symbolic Data Structuring Scheme". in *Pattern Recognition and Artificial Intelligence*, C.H.Chen (Ed.), New York Academic, 1976.

Fig. 1. A rounded body Morandi's bottle.

Fig. 2. Sketch made to retrieve a thin bottle.

Fig. 3. Sketch made to retrieve a bottle with a fruit dish at its right hand side.

Fig. 4. Retrieval results for the sketch of Fig. 1.

Fig. 5. Retrieval results for the sketch of Fig. 2.

Fig. 6. Only one image in the DB matches both in shape and in spatial relation the sketch representing the bottle with the fruit dish (Fig. 3).

Finding Facial Features Using an HLS Colour Space

Petter Ranefall, Bo Nordin, Ewert Bengtsson

Centre for Image Analysis,
Uppsala University, Sweden

Abstract. A method for finding facial features using information from an HLS colour space is presented. This method attempts to find the face with simple single band operations and in case of failure a more advanced method is used. Using this approach the nice photos which are the majority will be handled quickly, more difficult photos will be handled with more advanced methods and only a few images will be reported as failures.

1 Introduction

The general image segmentation and interpretation problem is still largely unsolved. But for constrained images, where substantial a priori information about the allowable image contents is available, rather complex segmentation and interpretation tasks can readily be solved. The recognition of facial features in images of human faces is an example where a priori information can be used to advantage. One of the first attempts to extract facial features was made by Kanade [1]. There are many possible applications for finding facial features e.g. face recognition [2,3,4], tracking the motions in the face for high compression image coding [5,6] or controlling a computer cursor simply by moving the head [7], to mention a few examples. Virtually all recent international conferences in image analysis or pattern recognition includes a few papers about face image analysis.

Our application is a rather simple one. We need to find the outline of the head as well as the position of the eyes and mouth in order to position and scale the image properly when producing driver's licenses or ID cards automatically from colour portrait images glued onto standardized application forms. The method has to be simple and robust as well as having the potential of being implemented at high speed in standard image array processing hardware.

2 Method

As a preprocessing step we convert the RGB image to HLS. Then the algorithm works in three major steps. If any of the three steps fails completely the processing reports a failure for the current image.

In the first step we first attempt to find the head contour as easily as possible by thresholding the Lightness image. If that does not work we use a Maximum Likelihood (ML) classifier on the RGB image. Secondly we search for the eyes in the Saturation image since the eyes will have very low saturation. The search area is restricted to regions surrounded by skin. The skin is found by an ML classifier on the RGB image. Thirdly, when the eyes are found we use the natural geometry of faces and search in an area under the eyes for the lips which differs from skin in the Hue image. A problem with the Hue image is that highlighted parts of the face (e.g. the nose) may also differ from the normal skin. The mouth is found at the broadest part of this region.

In the following description of the different steps in our method the *italicised* terms will be described in the Tools section (3).

2.1 Finding the Head Contour

1. Transform the image to HLS (see figure 1 and figures 2a, b, c). There are many ways of converting RGB images to HLS proposed in the literature. In our case Hue is defined such that 0° is between blue and green and Saturation is defined as the difference between the highest and the lowest RGB-value. Lightness is defined as the average of the highest and lowest RGB-value. This definition is useful for our purposes since the white and black in the eyes gets low saturation values and the skin colour is in the middle part of the hue scale.

2. *Automatic threshold* of the Lightness image.

3. *If the head contour is not OK*:

   ML train & classify the original colour image. This is done by using a pattern image (see figure 3) which marks some areas where background, clothes and head are expected.

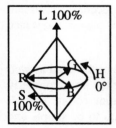   

Fig. 1. HLS colour space.   Fig. 2a. Light-ness.   Fig. 2b. Hue.   Fig. 2c. Satura-tion.

2.2 Finding the Eyes

The saturation image is used since the eyes differ from the rest of the face in this image. In some cases (e.g figure 2b) the hue image could serve equally well.

Narrow the search area by creating an *ML-classifier for skin and hair* (figure 4) and extract the skin within the head contour.

Line-by-line search within this region in the Saturation image.

193

2.3 Finding the Mouth

1. Extract a sub image under the eyes in the Hue image (see figure 5a). The sub image is a square with the side equal to the distance between the eyes and it is placed one half of the side length straight below the eyes.

2. *Automatic threshold* of this mouth region (see the result in figure 5b).

3. Fill holes in the marked area.

4. Find the maximum horizontal width within this marked area.

Fig. 3. Pattern.

Skin   Hair

Fig. 4. Training regions.

Fig. 5a. Contrast enhanced mouth region in the Hue image of figure 4.

Fig. 5b. Thresh-olded mouth region.

3 Tools

This is a more detailed description of the tools used in the previous section.

3.1 Automatic Threshold

Because of variations in skin and hair colour, lighting, background etc. we use a dynamic thresholding method which works as follows:

Split the histogram at the point S where the maximum split effect is obtained. The split effect for $S \in [x_L, x_U]$ is defined as follows [8]:

$$\sum_{k=x_L}^{x_U} (k-\bar{k})^2 - \left(\sum_{m=x_L}^{S} (m-\bar{m})^2 + \sum_{n=S+1}^{x_U} (n-\bar{n})^2 \right) \quad [eq.1]$$

where x_L and x_U are the lower and upper limits of the histogram, and $\bar{k}$, $\bar{m}$ and $\bar{n}$ are the mean values for the histogram and the sub histograms, respectively. If this does not separate the background from the object then the sub histogram which contains both the object and the background is split. This procedure continues until the object is separated from the background. The upper corners of the image are assumed to contain background and the centre of the image is assumed to contain object and the average in these areas defines background and object, respectively.

3.2 Test of Head Contour

Broken head contours can be detected using the classical shape measure:

$$P^2/A = \frac{\text{Lenght of contour}^2}{4\pi \cdot \text{Area}} \qquad [\text{eq.2}]$$

where the contour in our case is measured on the upper half of the portrait and the area is the area within the same contour. We have empirically found that heads with normal contours have P^2/A values in a range from approximately 1.2 to 1.8.

3.3 ML Classifier for Skin and Hair

For each column a number (we have used 1/6 of the Y-dimension) of pixels closest to the upper contour of the head are selected as training region for hair. Only pixels in the upper half of the image are used. Around the average of the x-values within the contour a rectangle of size X-dim/3, X-dim/6 is placed standing on the horizontal centre line (figure 4). This rectangle defines the training region for skin.

3.4 Line-by-line Search for Eyes

Search for the lowest threshold value for which darker points are found on both the left and the right side of the middle within the skin area on the same horizontal line. This search also tests if conditions for the distance between the eyes and the distance to the edge of the face are fulfilled.

4 Application

ID KORT AB which produces all driver's licences and most of the ID cards in Sweden uses digital image processing based on a previous method designed by two of the authors. This has been used to produce several million licences and cards so far. The old method does however rely on rather rigid assumptions about the size and position of the photos under the camera. Since these assumptions frequently are violated many manual interventions are necessary. The company now wants to be able to process their cards with a higher degree of automation. That includes moving and resizing the photos based on the positions of the eyes and mouth.

4.1 Finding the Photo

As a preprocessing step the photo has to be found on the form. This is because many people have problems in placing the photo at the correct place in spite of rather detailed instructions. An automatic threshold applied to the colour band with the highest split effect (eq. 1) with the value for background selected as 255 and the value for object selected as 0 does the work. This often finds the contour of the head, which could make that special first operation superfluous.

The photo is then cut out from the larger digital image which has been scanned and the rest of the processing is done on this sub image. The images are subsampled to approximately 200x250 pixels to speed up the calculations.

4.2 Finding the Facial Feature Positions

Using the algorithm described above (section 2-3).

4.3 Smoothing the Background

The colour and intensity of the background is smoothed linearly from the contour of the head to the edge of the photo in order to blend the image into the ID card without disturbing effects from the background or edges of the photo.

5 Results

Figure 6 shows the result of facial feature detection of the face in figure 2. For this image the more advanced method was used to find the contour because the face was illuminated mainly from the right. Features are found in faces with darker hair and complexion or glasses (e.g figure 7) or facial hair (e.g figure 8), but glasses, facial hair as well as no hair at all could cause some problems. We will continue to study these kinds of problems.

The average runtime were about 15 seconds on a DECstation 5000/120. On a more advanced workstation this would decrease to a few seconds which would be acceptable for production.

Fig. 6. Face with difficult illumination.   Fig. 7. Face with glasses.   Fig. 8. Face with facial hair.

6 Discussion

Computer interpretation of facial images has recently been discussed in many papers as pointed out in the introduction. Depending on the application the images can be assumed to be more or less constrained. In our case we have a substantial amount of a priori information about the images. If they do not fulfil these conditions they are not accepted for making ID-cards. Still there is significant variation in the way people may look and a few images are relatively hard to process. This has been utilized in our algorithms which attempts to use simple and quick methods when possible but reverts to more demanding processing to deal with the more difficult cases. This increases the robustness of the method without a corresponding penalty in processing time.

The algorithm has so far been tested on a limited number of cases. It will be installed in the production line of ID KORT AB and used to process about a million images per year. This will provide reliable statistics about its performance in the near future.

Another aspect of colour photo processing is to normalize the colour, so that very dark/ light images will be improved. This must be done without removing the variance due to skin and hair colour. The lightness compensation must also be done without changing the perceived hue. Perhaps this should be done in an (approximately) perceptually uniform colour space (e.g CIELAB or CIELUV). We will investigate these problems in the near future.

Acknowledgement

We are very grateful to ID KORT AB for their financial support and cooperation.

References

1.   T. Kanade: Picture processing system by computer complex recognition. PhD thesis, Department of Information Science, Kyoto University, Japan, 1973.

2.   Chung-Lin Huang, Ching-Wen Chen: Human facial feature extraction for face interpretation and recognition. Proc. of the 11th IAPR Int. Conf. on Pattern Recognition, The Hague, The Netherlands, August 1992, vol. 2, pp. 204-207.

3.   Alex Pentland, Baback Moghaddam, Thad Starner: View-based and modular eigenspaces for face recognition. Proc. of the 1994 IEEE Computer Society Conf. on Computer Vision and Pattern Recognition, Seattle, Washington, USA, June 1994, pp. 84-91.

4.   Guangzheng Yang, Thomas S. Huang: Human face detection in a complex background. Pattern Recognition, Vol. 27, No 1, pp.53-63, 1994

5.   K. Aizawa, H. Harashima, T. Saito: Model-based analysis synthesis image coding (MBASIC) system for a person's face. Proc. of GLOOBECOM-87, November 1987, pp. 45-49.

6.   Thomas S. Huang, Subhash C. Reddy: Human face motion analysis. Proc. of the Int. Workshop on Visual Form, Capri, Italy, May 1991, pp 287-292.

7.   Thomas C. Chang, Thomas S. Huang, Carol Novak: Facial feature extraction from color images. Proc. of the 12th IAPR Int. Conf. on Pattern Recognition, Jerusalem, Israel, October 1994, vol. 2, pp 39-43.

8.   Otto Milvang: An adaptive algorithm for color image quantization. Proc. of the 5th Scandinavian Conf. on Image Analysis, Stockholm, Sweden, June 1987, vol. 1, pp. 43-47.

A New and Fast Real-time Implementation of 2-D DCT

Jianmin Jiang

School of Engineering, Bolton Institute, United Kingdom

ABSTRACT: A new and fast 4×4 DCT algorithm is proposed in this paper. This algorithm classifies the input 2-D pixel data into four groups, each of which is then further rearranged into 1-D DCT transform. Therefore, the computation of 2-D DCT can be implemented by four 1-D DCTs. As a result, the efficiency of 2-D DCT algorithm is dependent on the 1-D DCT algorithm adopted, and all the existing fast algorithms for 1-D DCT can be directly applied to further optimise the algorithm design. The proposed algorithm can also be extended to compute general 2-D DCT by a recursive procedure. Specific algorithm flowchart and design is also included and described.

1 Introduction

Ever since the discrete cosine transform (DCT) was discovered in 1970's, it has been widely accepted as one of the major means for image signal processing and filtering. Its application is particularly widespread on image coding in communication technology. In the image compression standards developed by JPEG and MPEG, DCT is used as one of the most important stages to process the input pixel data.

Research on DCT of 4×4, 8×8, 16×16 blocks has been very fruitful [1-3]. All the developments can be summarized into two categories. One is to address the two dimensional DCT directly by investigating the algebraic structure of DCT's and reducing the arithmetic complexity of its computations[4]. The second category features the development of various fast algorithms for one dimensional DCT. Thus, the two dimensional DCT can be addressed by considering its rows and columns individually. The work covered in this paper, however, put forward a novel idea in designing the fast 2-D DCT algorithm by one dimensional approach.

2 1-D Transformation of 2-D DCT

In this section, a one dimensional approach is developed to tackle the computation of 4×4 DCT's. This development can also be used to compute $N \times N$ DCT recursively. The $2^m \times 2^m$ DCT, for example, can be computed by four $2^{m-1} \times 2^{m-1}$ DCT's. This procedure goes on until we get to the basic 4×4 DCT.

Let [g] be an 4×4 matrix representing the two-dimensional input pixel data and [G] be the 2-D DCT of [g]. Then the uv-element of [G] is given by[1]:

$$G(u,v) = \frac{2c(u)c(v)}{\sqrt{(4\times4)}} \sum_{m=0}^{3} \sum_{n=0}^{3} g_{mn} \cos\left\{\frac{(2m+1)u\pi}{8}\right\} \times \cos\left\{\frac{(2n+1)v\pi}{8}\right\} \qquad (1)$$

where:

$$c(k) = \begin{cases} \dfrac{1}{\sqrt{2}} & k=0 \\ \\ 1 & otherwise \end{cases}$$

The above equation can be further rearranged into:

$$G(u,v) = \frac{c(u)c(v)}{4} \sum_{0}^{3} \sum_{0}^{3} g_{mn} \left\{ \cos\frac{(2m+1)u\pi+(2n+1)v\pi}{8} + \cos\frac{(2m+1)u\pi-(2n+1)v\pi}{8} \right\} \quad (2)$$

If one variable in the two dimensional transform over m and n can be represented by another variable, the two dimensional DCT can be changed into 1-D DCT. From equation (2), it can be seen that if $(2n+1)$ can be replaced by $p(2m+1)$, p is any integer, the transform becomes two 1-D DCT over m for (u+pv) and (u-pv). To obtain a suitable relationship between m and n, let n=m+x, hence:

$$2n+1 = 2m+2x+1 = 2m+1+2x$$

As $(2n+1)$ should be replaced by $p(2m+1)$, x can be determined by:

$$x = \frac{p(2m+1)-(2m+1)}{2} = \frac{(p-1)(2m+1)}{2}$$

If p is further replaced by 2k+1, x is of the form:

$$x = k(2m+1)$$

where k = 0, ±1, ±2, ...

Thus, n is represented by:

$$n = m+x = m+k(2m+1) = (2k+1)m+k = pm+k$$

As m varies from 0 to 3 in 4×4 DCT, the accurate expression of n should be:

$$n = pm + k \quad (modulo \; 4)$$

The relationship between n and m can be summarized in Table I. From Table I, it can be observed that among all the groups, only four groups are independent. In other words, no matter how the values of k and p varies, all the groups obtained can be described by only four groups: x_0, x_1, x_2 and x_3.

Table I 1-D Transform of Input Pixel

k	p	n	$n \; (mod \; 4)$	x_k
0	1	m	m	$x_0 = \{ \; g_{00}, \; g_{11}, \; g_{22}, \; g_{33} \; \}$
1	3	3m+1	3m+1(mod 4)	$x_1 = \{ \; g_{01}, \; g_{10}, \; g_{23}, \; g_{32} \; \}$
2	5	5m+2	m+2 (mod 4)	$x_2 = \{ \; g_{02}, \; g_{13}, \; g_{20}, \; g_{31} \; \}$
3	7	7m+3	3m+3(mod 4)	$x_3 = \{ \; g_{03}, \; g_{12}, \; g_{21}, \; g_{30} \; \}$
4	9	9m+4	m	$x_0 = \{ \; g_{00}, \; g_{11}, \; g_{22}, \; g_{33} \; \}$
5	11	11m+5	3m+1(mod 4)	$x_1 = \{ \; g_{01}, \; g_{10}, \; g_{23}, \; g_{32} \; \}$
...	...	...	...	...
-1	-1	-m-1	-m-1	$x_3 = \{ \; g_{03}, \; g_{12}, \; g_{21}, \; g_{30} \; \}$
-2	-3	-3m-2	-3m-2(mod 4)	$x_2 = \{ \; g_{02}, \; g_{13}, \; g_{20}, \; g_{31} \; \}$
...	...	...	...	...

Therefore, in terms of the four independent groups, the 4×4 2-D DCT can be further rearranged as:

$$G(m, n) = \frac{c(u)c(v)}{4} \sum_{0}^{3} \sum_{0}^{3} g_{mn} \left\{ \cos \frac{(2m+1)u\pi + (2n+1)v\pi}{8} + \cos \frac{(2m+1)u\pi - (2n+1)v\pi}{8} \right\}$$

$$= \frac{c(u)c(v)}{4} \sum_{k=0}^{3} (X_k + Y_k)$$
(3)

where:

$$X_k = \sum_{g_{mn} \in x_k} g_{mn} \cos \frac{(2m+1)u\pi + (2n+1)v\pi}{8}$$
(4)

$$Y_k = \sum_{g_{mn} \in x_k} g_{mn} \cos \frac{(2m+1)u\pi - (2n+1)v\pi}{8}$$
(5)

and x_k is the k'th group shown in the fifth column of Table I.

To find the best way of transforming the 2-D DCT into 1-D DCT, we represent the value of n by m, $(3m+1)(\text{mod } 4)$, $(-3m-2)(\text{mod } 4)$ and $(3m+3)(\text{mod } 4)$ respectively. Corresponding to the four relationships, all the 4×4 input pixels are arranged into the following:

$$x_0 = \{ g_{00}\ g_{11}\ g_{22}\ g_{33} \} \qquad\qquad x_1 = \{ g_{01}\ g_{10}\ g_{23}\ g_{32} \}$$
$$x_2 = \{ g_{02}\ g_{13}\ g_{20}\ g_{31} \} \qquad\qquad x_3 = \{ g_{03}\ g_{12}\ g_{21}\ g_{30} \}$$

By examining each group individually, the equations (4) and (5) can be rearranged into one dimensional DCTs for all the four groups as follows:

$$
\begin{aligned}
X_0 &= D_{(x_0)}(u+v) & Y_0 &= D_{(x_0)}(u-v); \\
X_1 &= D_{((-1)^{a_m}x_1)}(u+3v) & Y_1 &= D_{((-1)^{a_m}x_2)}(u+3v); \\
X_2 &= D_{(-1)^{a_m}x_2)}(u-3v) & Y_2 &= D_{((-1)^{a_m}x_2)}(u+3v); \\
X_3 &= D_{((-1)^{y}x_3)}(u-v) & Y_3 &= D_{((-1)^{y}x_3)}(u+v);
\end{aligned}
$$

$$(6)$$

where:

$$
a_m = \begin{cases} 0 & m=0 \\ 1 & m=1 \\ 1 & m=2 \\ 2 & m=3 \end{cases}
\qquad\qquad
b_m = \begin{cases} 1 & m=0 \\ 2 & m=1 \\ 2 & m=2 \\ 3 & m=3 \end{cases}
$$

and $D_{(\ )}(L)$ stands for the L'th 1-D DCT of the input elements $\{\ \}$.

As an example, since n= 3m+3 (modulo 4) is selected for group x_3, the modulo operation of n can be eliminated by specifying:

$$n = 3m + 3 - 4b_m$$

The value of b_m can be determined the same as m, in accordance with the modulo operation and all the pixels in group x_3. Hence:

$$2n + 1 = 6m + 6 - 8m + 1 = -(2m+1) + 8$$

Therefore X_3 and Y_3 can be obtained from equations (4) and (5) as follows:

$$X_3 = \sum_{m=0}^{3} (-1)^y g_{mn} \cos\left\{ \frac{(2m+1)(u-v)\pi}{8} \right\} = D_{((-1)^y x_3)}(u-v)$$

$$(7)$$

$$Y_3 = \sum_{m=0}^{3} (-1)^v g_{mn} \cos\left\{\frac{(2m+1)(u+v)\pi}{8}\right\} = D_{((-1)^v x_3)}(u+v) \qquad (8)$$

By putting the equations (6) into (3), the 2-D DCT can finally be arranged into the following form:

$$G(m, n) = \frac{c(u)c(v)}{4} \sum_{k=0}^{3} (X_k + Y_k) = \frac{c(u)c(v)}{4} \{D_{(x_0 + (-1)^v x_2)}(u+v) + D_{(x_0 + (-1)^v x_2)}(u-v)$$
$$+ D_{((-1)^{c-v} x_1 + (-1)^{c-v} c_3)}(u-3v) + D_{((-1)^{c-v} x_1 + (-1)^{c-v} x_3)}(u+3v)\} \qquad (9)$$

Table II Computation of G(u, v)

(u, v)	G(u, v) (I)	G(u, v) (II)
0	$\{D_1(0)+D_1(0)+D_3(0)+D_3(0)\}$	$\{2D_1(0)+2D_3(0)\}$
1	$\{D_2(1)+D_2(1)+D_4(3)+D_4(3)\}$	$\{2D_2(1)-2D_4(3)\}$
2	$\{D_1(2)+D_1(2)+D_3(6)+D_3(6)\}$	$\{2D_1(2)+2D_3(6)\}$
3	$\{D_2(3)+D_2(3)+D_4(9)+D_4(9)\}$	$\{2D_2(3)-2D_4(1)\}$
10	$\{D_1(1)+D_1(1)+D_3(1)+D_3(1)\}$	$\{2D_1(1)+2D_3(1)\}$
11	$\{D_2(2)+D_2(0)+D_4(1)+D_4(1)\}$	$\{D_2(2)+D_2(0)+2D_4(1)\}$
12	$\{D_1(3)+D_1(1)+D_3(7)+D_3(5)\}$	$\{D_1(3)+D_1(1)-D_3(1)-D_3(3)\}$
13	$\{D_2(4)+D_2(2)+D_4(10)+D_4(8)\}$	$\{D_2(2)-D_4(2)-D_4(0)\}$
20	$\{D_1(2)+D_1(2)+D_3(2)+D_3(2)\}$	$\{2D_1(2)+2D_3(2)\}$
21	$\{D_2(3)+D_2(1)+D_4(5)+D_4(1)\}$	$\{D_2(3)+D_2(1)-D_4(3)+D_4(1)\}$
22	$\{D_1(4)+D_1(0)+D_3(8)+D_3(4)\}$	$\{D_1(0)-D_3(0)\}$
23	$\{D_2(5)+D_2(1)+D_4(11)+D_4(7)\}$	$\{-D_2(3)+D_2(1)-D_4(3)-D_4(1)\}$
30	$\{D_1(3)+D_1(3)+D_3(3)+D_3(3)\}$	$\{2D_1(3)+2D_3(3)\}$
31	$\{D_2(4)+D_2(2)+D_4(6)+D_4(0)\}$	$\{D_2(2)-D_4(2)+D_4(0)\}$
32	$\{D_1(5)+D_1(1)+D_3(9)+D_3(3)\}$	$\{-D_1(3)+D_1(1)-D_3(1)+D_3(3)\}$
33	$\{D_2(6)+D_2(0)+D_4(12)+D_4(6)\}$	$\{-D_2(2)+D_2(0)-D_4(2)\}$

3. A Complete Fast 4 × 4 DCT Algorithm

When the value of v is considered as odd and even respectively, the two combinations in equation (9), $\{x_0+(-1)^v x_2\}$ and $\{(-1)^{a \cdot v} x_1+(-1)^{c \cdot v} x_3\}$, can be further simplified into four combinations: $\{x_0+x_2\}$, $\{x_0-x_2\}$, $\{x_1+x_3\}$ and $\{(-1)^{a \cdot v} x_1+(-1)^{c \cdot v} x_3\}$.

Therefore, as the values of u and v vary from 0 to 3, the complete 4 × 4 DCT can be computed, according to equation (9), as shown in Table II where $D_1(L)$, $D_2(L)$, $D_3(L)$ and $D_4(L)$ represent the 1-D DCT of the above four combinations respectively, such as $D_1(L)=D_{\{x_0+x_2\}}(L)$ etc. From the final simplified results given in the second column in Table II, the complete flow chart for the algorithm to compute the 2-D DCT can be worked out as shown in Figure 1.

4. Conclusions

A fast and efficient 2-D 4 × 4 DCT algorithm has been developed and presented in this paper. The algorithm features one dimensional transformation and simple additions for the computation of 2-D DCT. This algorithm can also be applied to compute general 2-D $2^m \times 2^m$ DCT by recursive procedures. As the one dimensional DCT is the basic core for this algorithm, its efficiency in terms of computing complexity, such as the number of multiplications, additions, is dependent on the 1-D DCT algorithm adopted. Therefore, all existing fast algorithm development for one dimensional DCT can be directly utilized to implement the 2-D DCT algorithm.

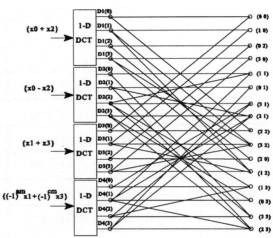

Figure 1 Flowchart of 4 × 4 DCT

References

1.   Rao, K.R. and Yip, P. 'Discrete cosine transform algorithms, advantages, and applications', Academic Press, INC. 1990, ISBN: 0-12-580203-x.

2.   Hou, H.S. 'A fast recursive algorithm for computing the discrete cosine transform', IEEE Trans. Acoust., Speech, Signal Processing, Vol. ASSP-35, No. 10, 1455-1461, 1987.

3.   Lee, B.G. 'A new algorithm to compute the discrete cosine transform' IEEE Trans. Acoust., Speech, Signal Processing, Vol. ASSP-32, No.6, 1243-1245, Dec. 1984.

4.   Feig, Ephraim, 'Fast algorithms for the discrete cosine transform'. IEEE Trans. on Signal Processing, Vol. 40, No 9, 2174-2193, Sept. 1992.

A Hybrid Lossless Compression of Still Images Using Markov Model and Linear Prediction

Seishi TAKAMURA and Mikio TAKAGI

Institute of Industrial Science, The University of Tokyo

Abstract. Markov model approximation of digital images gives very low entropy, but requires a huge amount of storage to maintain the frequency table for encoding and decoding. Therefore, it has only been used for binary picture compression so far.

We have developed a high performance DPCM-based predictive lossless image coder, and have combined this technique with a Markov model, which provides a better compression ratio for gray scale images.

The basic idea is, encode the first part (this size is variable) of an image using the normal DPCM-based encoder and concurrently use these data as a Markov model training set. Then the latter half will be encoded using the Markov model with the assistance of the DPCM coder.

1 Introduction

Lossless still image compression techniques have recently been highlighted. The most popular method for lossless compression of still images with more than 8 bits/pixel is DPCM-based predictive coding like JPEG lossless mode, but performance limitations are beginning to appear.

On the other hand, the Markov model entropies of the images are attractively lower than the bit rate of predictive coders (e.g. table 1 in section 4), but a frequency table with a huge amount of data is necessary for encoding and decoding, therefore it has been unrealistic to apply the Markov model for image compression up to the present.

In this paper, we propose using a Markov model for image compression and show that the performance of current lossless image compression techniques can be improved.

Our method consists of Markov model encoding and classical DPCM-based encoding. It does not require the transmission of a huge frequency table. In addition, our DPCM-based coder itself outperforms the JPEG lossless coder.

2 Markov Model Image Representation for Image Coding

If the symbol occurrence probability depends only on the m preceding symbols, the source is called an *"m-th order Markov source"*. The series of m preceding symbols is called the *"state"*. If the source has N different kinds of symbols, the m-th order Markov model can be divided into N^m memoryless sources, the same as the number of possible states.

To encode such a Markov source, N^m different occurrence frequency tables are necessary. As each table has N words, total table size amounts to N^{m+1} words. At the moment, current techniques using the Markov model only treat binary ($N = 2$) images for $m \simeq 10$. Needless to say, it is impractical to transmit the whole table for gray images ($N \simeq 2^8 = 256$) because of its astronomical size.

We use a 2-dimensional second order Markov model whose state S is represented as $S(X_1, X_2)$, where X_1 and X_2 are as shown in figure 1. Although this model is very simple, the entropy of this model is much lower than the bit rate of former lossless compression scheme (see table 1), which implies further compression possibilities.

Fig. 1. Support of a second order Markov model

2.1 Learning Markov model

First of all, we encode the first image part (it is variable in size) of an image with a certain lossless coder, and use these image data for the training of the Markov model. After the training we try to encode the latter part with the Markov encoder (see the following section for details). The coder is also trained while it encodes. Therefore there is no need to transmit the frequency table of each Markov state.

Appearance of an unknown symbol In compensation of no table transmission, a key question is how to deal with in the coding process of the latter image part. It is "the appearance of an unknown symbol s", because the decoder has no means of detecting it. Therefore some means of informing the decoder must be provided.

To solve this problem, at the moment we prepare the special signal *"Unknown"*. Each frequency table prepares room for this signal. In encoding, when the coder detect the new symbol, "Unknown" is transmitted. Suppose A is the fixed frequency for this symbol, "theoretical code size for the model" H_A is

$$H_A = \sum_S \sum_{s \in S} -\#(X) \log_2 \frac{\#(s)}{\#(S) + A} \quad \text{(bit)}, \tag{1}$$

where S is a state, s is a symbol and $\#(\cdot)$ means the number of the set. Note that $H_A|_{A=0}$ is equal to the Markov model entropy. If the state S is unknown or the symbol is unknown, we use the normal prediction coding methods using the neighbor pixels (see section 3) to recover the value of current pixel.

We use arithmetic coder which is suited to encode dynamically-varying probability fractions with high coding efficiency. The evaluation of the other trial of letting the decoder know when a new symbol appears and the comparison with JBIG will also be shown in subsection 4.1.

3 DPCM-based Encoding Algorithm

In this section we explain our DPCM-based predictive coder which is used together with the Markov model image coder. The scheme consists of two main parts: pixel classification and source distribution fitting.

3.1 Pixel classification

We use two methods to classify each pixel. The first method is used to decide the *group* of each pixel and calculate the prediction coefficients respectively, and the second method calculates the *class* for each pixel, which is used to split the data stream of prediction error and to reduce entropy.

An important property of these methods is they can be done using only past data. Therefore the encoder have to transmit no information about *group* and *class* since the decoder can generate the same information independently.

Grouping according to local edge Figure 2 shows X's neighbor pixels $a \sim f$ used to determine the group it belongs to.

The group number of X is decided according to which number among ($|a - f|$, $|b - f|$, $|c - f|$, $|e-f|$, $|f+c-2b|$, $|b-c|$, $|d-c|$) is the biggest. This method enables X to be distinguished according to its local edge direction. We calculate the prediction coefficients using the least square error method for each group respectively.

a	b	c	d
e	f	X	

Fig. 2. Neighbor pixels $a \sim f$ to decide the group of pixel X

Classification according to the statistical property[4] It is also possible to classify a pixel according to the value $Q = |e-f|+|b-f|+|c-f|+|d-f|$, which physically means the *"edge intensity around X"*. After obtaining this Q value, it is quantized properly and used as *class* number. Since this procedure is independent from the former classification method, they can be used together.

Effect of pixel grouping For example, Gaussian sources with different variances should not be entropy coded together, but separately, because the latter always gives a shorter code. Therefore the source splitting improves the performance of compression if it is successful.

Our classification method is not directly related to the source property itself. But from the experimental results, the source is split into different statistical properties, which means an improvement in compression.

3.2 Improvement of coding efficiency

To encode signals effectively, a knowledge about its distribution is necessary. However sometimes the size of the table becomes significant compared to the code size. In our method, we find the best fit function of the error distribution and then transmit the parameters, which has negligible overhead so that the decoder can re-generate an approximation of the frequency table. Since the fitting curve is not perfect, the coder updates the internal table after encoding/decoding each symbol, which is described next section.

Error conversion Normally the prediction error is rounded off to an integer and encoded. In addition to this simple rounding, we also try to convert the prediction error in the following way and examine its entropy to obtain a smaller code.

First we obtain the upper and lower bound of a group (max, min). Then convert the actual pixel value into an non negative integer (figure 3). (In this figure, if the actual pixel is equal to 'max', $E = 9$.) This conversion is obviously reversible.

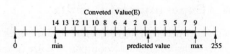

Fig. 3. Concept of error conversion

Criterion of fitting Giving consideration of subsequent entropy coding, the criterion of "better fitting" should be considered. We use "the theoretical information amount I" according to the probability function, i.e.:

$$I = \sum_s -\#(s)\log_2 f(s) \quad \text{(bits)},\tag{2}$$

where $\#(s)$ means the number of occurrences of symbol s and $f(s)$ is the fitting function which satisfies $\sum_s f(s) = 1$.

Error distribution fitting To find well approximated fitting, some different fitting curves, such as *(Generalized) Gaussian function* and *Spline function* (cubic natural spline) are attempted. Generalized Gaussian function with zero mean $f_{p,\sigma}(s)$ is represented as $p\exp\left(-|s/(\gamma\sigma)|^p\right)/\left(2\Gamma(1/p)\gamma\sigma\right)$, where $\gamma = \sqrt{\Gamma(1/p)/\Gamma(3/p)}$ and σ^2 is the variance of s. Sometimes this distribution do not fit the actual distribution well because of its asymmetricity. This is the reason why spline function is also tried.

Both simple prediction error distribution and E (converted value) distribution are tried to fit by these functions and the best one according to the criterion equation (2), which also considers the *overhead*, is chosen. This *overhead* is the number of the bits for each fitting function to represent its parameters such as node coordinates of spline, σ and p for $f_{p,\sigma}$. An example of fitting of a certain distribution is shown in figure 4. In this case, the generalized Gaussian distribution for E is chosen.

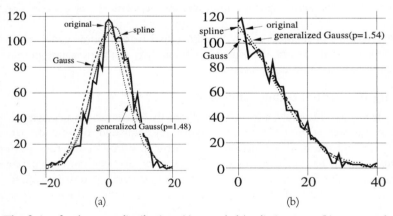

Fig. 4. The fitting for the error distribution. (a): rounded prediction error (b): converted error E.

4 Experimental Results

The compression performance of our coder with varying training sizes is tested on several gray-scale images and compared with other lossless compression schemes. We used 24 neighbor pixels for DPCM-based linear prediction, the results are shown in table 1. The percentage "0%" means trying to compress all the image with Markov encoder except the points where the new states or new symbols appear. "100%" means whole-DPCM encoding.

The smallest bit-rate for each image is underlined. The first 5 images are 720×576 pixels, "Washdc1..7" are multi-band LANDSAT images with 512×512 pixels.

Table 1. Entropy, bit rate of JPEG lossless coder and our method for each image(unit = bit). **0th E**: 0th order entropy, **M.M.E**: Second order Markov model entropy, **JPEG**: Bit rate of JPEG Lossless coder (by *'crush'* [6]). The numbers inside the parentheses is the predictor number, **percentage**: Ratio of the first DPCM-encoded trainig part. **overhd**: Overhead size (bytes) for noticing the occurrence of new symbols, **JBIG**: Code size (bytes) for informing the same information by the JBIG coder.

Image	0th E.	M.M.E.	JPEG	0% (overhd:JBIG)	50% (overhd:JBIG)	100%
girl	6.608	3.079	4.981(8)	3.653 (23396:25571)	4.136 (10589:11771)	4.861
boats	7.088	3.300	4.185(8)	4.465 (45978:35521)	4.371 (34060:22798)	4.107
goldhill	7.530	3.675	4.653(8)	4.943 (62949:46104)	4.720 (27073:20430)	4.450
zelda	6.741	2.990	4.768(8)	3.501 (20151:22735)	4.091 (7518:8551)	4.770
hotel	7.546	3.320	4.667(8)	4.719 (29953:41969)	4.582 (14960:18471)	4.352
washdc1	4.685	3.442	3.849(7)	3.748 (9074:10309)	3.652 (2989:3588)	3.543
washdc2	3.846	2.726	2.981(7)	2.881 (4107:5402)	2.817 (1249:1802)	2.738
washdc3	4.307	3.153	3.484(7)	3.390 (7105:8357)	3.286 (2330:2842)	3.183
washdc4	5.432	3.909	4.324(7)	4.434 (20084:18670)	4.206 (7315:7052)	4.007
washdc5	5.681	4.056	4.649(7)	4.816 (27274:22968)	4.567 (10226:9097)	4.351
washdc6	3.441	0.903	0.920(4)	0.948 (413:712)	0.894 (171:334)	0.938
washdc7	4.757	3.454	3.792(7)	3.747 (9652:10825)	3.617 (3141:3721)	3.511

4.1 Representation of new symbol appearance

To substantiate our method to deal with the occurrence of new symbols, the theoretical overhead of the extra information necessary for representing the occurrence of new symbol according to equation (1) is calculated. It is also possible to inform the decoder about the new symbol by encoding the binary figure like figure 5(b), since with such binary figures the decoder can know where the new symbol occurs. We adopt the JBIG coder to compress such data.

Figure 5(a) shows the original image "zelda" and the black pixels in figure 5(b) shows the occurrence points of the unknown symbols. The first half (50%) of the image is used for the initial training. There are 12551 occurrences within the latter half (207360 pixels). This image is compressed to 8551 bytes by JBIG coder while the calculated overhead of our method is 7518 bytes. Other results are shown within the parentheses in table 1. Sometimes the JBIG coder outperforms our method in regard to new symbol treatment.

(a)

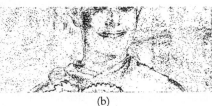
(b)

Fig. 5. (a) Sample image "zelda". (b): The positions of the new occurrence of the symbols in image (black dots).

5   Conclusion and Future Work

In this paper, the algorithm for continuous tone still image compression using a Markov model is proposed. The coder always outperforms JPEG lossless coding schemes by 0.2~1.3 bits/ pixel.

In our method, the predictive coding part and the Markov model coding part are independent, therefore it is possible to combine more powerful predictive coders.

Although it is clear for some images (girl, zelda) that the Markov model coding inproves the performance, there are still some images for which it does not work well. We will further investigate the following points:

 – efficient treatment of the appearance of the unknown symbol
 – investigate the Markov state reduction
 – application for the color image coding and near-lossless coding

References

1. S. Takamura et al.: "Lossless Image Compression with Lossy Image using Adaptive Prediction and Arithmetic Coding", Proc. Data Compression Conference '94, pp. 166–174, 1994
2. D. M. Abrahamson: "An Adaptive Dependency Source Model For Data Compression", Comm. ACM, Vol. 32, No. 1, pp. 77–83, Jan. 1989
3. P. G. Howard, J. Vitter, "New Methods for Lossless Image Compression Using Arithmetic Coding", Proc. Data Compression Conference, Apr. 1991, pp.257–266
4. T. Taniguchi et al.: "Variable-Length-Code-Selective Reversible Predictive Coding for Multi-Level Images", Trans. Institute of Electronics, Information and Communication Engineers (in Japanese), Vol.J70-B, pp.654–663, Jun. 1987
5. S. Utsui et al.: "Study on a Lossless Coding of Natural Images", Proc. 9th Picture Coding Symposium of Japan (in Japanese), pp.103–104, Oct. 1994
6. 'crush' package is available via anonymous ftp at `dftnic.gsfc.nasa.gov`:
   `disk$moe:[anonymous.files.software.unix.crushv3]:crush_v3_tar.Z`

The Impact of Aliasing and Quantization on Motion Compensation

Christoph Stiller

INRS-Telecommunications**, 16 Place du Commerce, Verdun, Québec
H3E 1H6 Canada, *Tel.:+1 (514) 761-8653, stiller@inrs-telecom.uquebec.ca*

Abstract. Motion estimation is among the key techniques for a variety of problems in image sequence processing. Even for ideal conditions, when all temporal changes in the image sequence are solely due to motion, the performance of motion compensated prediction is limited by aliasing and quantization noise. This paper provides a theoretical framework for these impact allowing quantification of the power spectral density of the prediction error signal. Both aliasing and quantization are shown to have important impact for typical image sequences. As an illustrative example for an application of the findings, an adaptive in-loop filter for motion compensated prediction is designed. Improvements of up to 1 dB over a standard Wiener filter are demonstrated when the filter adapts on the local motion as suggested by the model calculation.

1 Introduction

Motion information is an important element in analysis, processing, and compression of image sequences. Its basic underlying idea is that some local property of the image is 'nearly' preserved along motion trajectories. Most commonly, the intensity itself is assumed 'almost' constant in this direction. For this reason, most motion estimators include minimization of a norm of the error signal of motion compensated prediction [5].

The knowledge of properties of prediction error images is important for a variety of applications. For motion estimation, statistical models of the prediction error and the motion field allow a Bayesian solution of the estimation problem (e.g., [4, 7]). Motion compensated prediction can be improved by Wiener filtering when the power spectral density of the prediction error image and the original image are given [1]. Hybrid video coding as is employed in current standards (H.261,MPEG-1,MPEG-2) includes transmission of the prediction error image itself. Hence, near rate-distortion optimum encoding implies a good knowledge of the prediction error statistics.

This contribution extends fundamental work of [1, 2] on the analysis of the prediction error by not only accounting for the impact of quantization of motion information but also for the impact of aliasing noise, which is present in any digital imaging system. Experimental studies indicate that the latter impact is far from negligible. On one hand, the analysis gives insight in performance bounds of motion compensated prediction. On the other hand, it provides information

---

** Major parts of the underlying research were performed at: Institute for Communication Engineering, Aachen University of Technology, Germany
The author recently joined Bosch Research Communications, Hildesheim, Germany

for the design of some elements in video processing, such as motion estimators, motion vector quantizers or in-loop filters and prediction error coders in hybrid video coding.

The rest of this paper is organized as follows. The main part given in the next section is concerned with a theoretical analysis of the impact of aliasing and quantization on motion compensation. Model assumptions for the image sequence and the motion fields allow quantitative evaluation of these impact and give insight into some basic properties. Section 3 provides an illustrative example of an application of these findings for motion compensated prediction as employed in standardized video coding. Conclusions are presented in Section 4.

2 Theoretical Analysis of Motion Compensation

For the analysis, areas of the image sequence are considered, where all intensity changes are solely due to motion, i.e., it is assumed that the image acquisition system is ideal, all objects are diffuse reflecting, illumination is constant and no occlusions are present. Clearly, the performance of motion compensation in such regions will serve as an upper bound for real imagery. A system model for the prediction error is depicted in Figure 1. Input to the system is the intensity

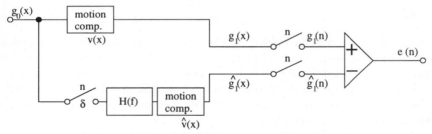

Fig. 1. System model for the prediction error

pattern $g_0(\mathbf{x})$ at time $t = 0$ with continuous spatial coordinates $\mathbf{x} = (x, y)^T \in \mathbb{R}^2$. In contrast, $\mathbf{n} = (n, m)^T \in \mathbf{Z}^2$ will denote discrete coordinates of sampled fields. The upper branch of the system models the formation process of the following image $g_1(\mathbf{n})$. Firstly, the pattern $g_0(\mathbf{x})$ is displaced according to some true motion field $\mathbf{v}(\mathbf{x}) = (v(\mathbf{x}), w(\mathbf{x}))^T$ to form the continuous pattern

$$g_1(\mathbf{x}) = g_0(\mathbf{x} - \mathbf{v}(\mathbf{x})) \tag{1}$$

at time $t = 1$. Secondly, it is subsampled to $g_1(\mathbf{n})$ which is the output of the image acquisition system at time $t = 1$. Note that the model assumes existence of a true motion field $\mathbf{v}(\mathbf{x})$.

The motion compensated prediction $\hat{g}_1(\mathbf{n})$ of $g_1(\mathbf{n})$ is composed in the lower branch. A subsampled version of $g_0(\mathbf{x})$ is interpolated by the filter with transfer function $H(\mathbf{f})$, where $\mathbf{f} = (f_x, f_y)^T \in \mathbb{R}^2$ denotes 2D frequency. Subsequently motion compensation employing some estimated motion field $\hat{\mathbf{v}}(\mathbf{x}) = (\hat{v}(\mathbf{x}), \hat{w}(\mathbf{x}))^T$ is performed. Finally, the motion compensated prediction $\hat{g}_1$ is subsampled onto the image lattice. The prediction error is defined as $e(\mathbf{n}) = g_1(\mathbf{n}) - \hat{g}_1(\mathbf{n})$.

Even for these ideal conditions, motion compensated prediction cannot be expected to perform perfectly, i.e., $e(\mathbf{n}) \neq 0$ for the following two reasons:

- The image sequence is subsampled in both temporal and spatial direction causing motion-dependent aliasing.
- Motion estimation is subject to estimation errors and transmission of motion estimation is subject to quantization noise, i.e., $\hat{\mathbf{v}} \neq \mathbf{v}$.

For a sufficiently small neighborhood, the motion field can be approximated by a first order expansion as

$$\mathbf{v}(\mathbf{x}) \approx (b_x x + c_x, b_y y + c_y)^T.$$

This comprises translation by $\mathbf{c} = (c_x, c_y)^T$ as well as linear dilation by $\mathbf{b} = (b_x, b_y)^T$. If the intensity image and the prediction error are at least weakly stationary, the power spectral density of the latter can be decomposed as

$$\phi_{ee}(\mathbf{f}) = \phi_{g_1 g_1}(\mathbf{f}) + \phi_{\hat{g}_1 \hat{g}_1}(\mathbf{f}) - 2\mathrm{Re}\{\phi_{\hat{g}_1 g_1}(\mathbf{f})\}. \tag{2}$$

For exact estimation of integer $\frac{1}{1+b_x}, \frac{1}{1+b_y}$ the prediction error $e(\mathbf{n})$ is derived to be at least weakly stationary in the appendix. Moreover the power spectral density of the prediction error (2) is shown to become

$$\phi_{ee}(\mathbf{f}) = a^2 \text{Ш}(\mathbf{f}) * \left\{ \frac{1}{a} \phi_{g_0 g_0}(\bar{\mathbf{f}}) + [\phi_{g_0 g_0}(\bar{\mathbf{f}}) * \text{Ш}(\bar{\mathbf{f}})] \cdot |H(\bar{\mathbf{f}})|^2 \right. \tag{3}$$

$$\left. - 2\mathrm{Re}\left\{ \phi_{g_0 g_0}(\bar{\mathbf{f}}) e^{-j 2\pi \mathbf{cf}} \left[\left(H^*(\bar{\mathbf{f}}) e^{j 2\pi \hat{\mathbf{c}} \mathbf{f}} \right) * \text{Ш}(\bar{\mathbf{f}}) \right] \right\} \right\}$$

where $a = (1+b_x)(1+b_y)$, $\bar{\mathbf{f}} = ((1+b_x)f_x, (1+b_y)f_y)^T$ and $\text{Ш}(\mathbf{f}) = \sum_{\mathbf{n} \in \mathbb{Z}^2} \delta(\mathbf{f} - \mathbf{n})$ denotes the two-dimensional Dirac-pulse sequence (*Shah function*) [6]. This finding gives insight into important properties of prediction error signals as will be discussed in the sequel.

In order to isolate the aliasing impact, a perfect motion estimator is assumed for the moment, i.e., $\hat{\mathbf{b}} = \mathbf{b}$, $\hat{\mathbf{c}} = \mathbf{c}$.

The autocorrelation function of the intensity pattern $g_0(\mathbf{x})$ is modeled isotropic with power 1 [3]

$$\varphi_{g_0 g_0}(\mathbf{x}) = \rho^{\|\mathbf{x}\|},$$

where a typical correlation coefficient is $\rho = 0.9$.

Figure 2 depicts examples of the prediction error energy due to aliasing for an ideal lowpass interpolation filter

$$H(\mathbf{f}) = \begin{cases} 1 \text{ if } |f_x|, |f_y| < \frac{1}{2} \\ 0 \text{ otherwise.} \end{cases}$$

The aliasing vanishes when all motion vectors connect pixel sites between the two images. It is worth noting that the distance between the original image and the aliasing can be less than 12 dB and hence aliasing is far from negligible in typical coding applications.

When the prediction error due to inaccurate motion vectors is assumed independent from aliasing, it can be derived from eq. (3) as

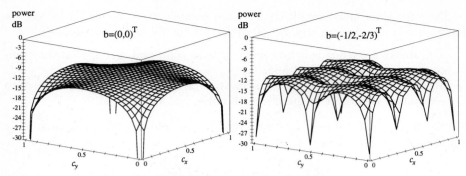

Fig. 2. Prediction error due to aliasing

$$E\{\phi_{ee}(\mathbf{f})\} = a \cdot \text{Ш}(\mathbf{f}) * \{\phi_{g_0 g_0}(\bar{\mathbf{f}}) \cdot [1 + |H(\bar{\mathbf{f}})|^2 - 2\text{Re}\{H(\bar{\mathbf{f}}) \cdot F_{\Delta c}(\mathbf{f})\}]\}, \quad (4)$$

where $F_{\Delta c}(\mathbf{f}) = E\{e^{j 2\pi \Delta c \mathbf{f}}\}$ is the characteristic function of the random variable $\Delta c = c - \hat{c}$ denoting the error of translatoric motion. For the special case of $\mathbf{b} = \mathbf{0}$, (4) becomes a result of [1] which was derived for rate-distortion optimum coding of gaussian distributed images at high data rate.

When motion inaccuracy is solely caused by linear scalar quantization with small step size Δ, it can be modeled as uniformly distributed with characteristic function $F_{\Delta c}(\mathbf{f}) = sinc(\pi \Delta f_x) sinc(\pi \Delta f_y)$. Figure 3 depicts the expected power of the prediction error according to eq. (4) for this situation.

A straightforward choice for the accuracy Δ in a high quality video coding scheme is to postulate that the error due to motion inaccuracies should at most be of the same magnitude as the error due to aliasing determined by the image acquisition system. As can be seen from Figure 3, this is already satisfied in average for $\Delta = 1$, i.e., pixel accurate motion compensation. A theoretical justification for the common choice of sub-pixel accurate motion vectors can be provided by a closer look at the power spectral density (3) of the prediction error depicted in Figure 4. Visual distortion in the motion compensated prediction occurs mainly in regions with a true motion vector pointing between pixel sites, i.e., $\mathbf{v}(\mathbf{x}) = (n + \frac{1}{2}, m + \frac{1}{2})^T$, where $n, m \in \mathbf{Z}$. In those regions the prediction

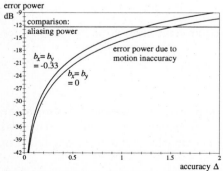

Fig. 3. Prediction error due to inaccurate motion

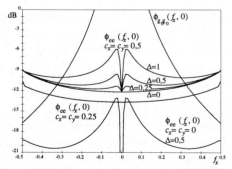

Fig. 4. Power spectral density of prediction error

error is dominated by aliasing at high frequencies. At lower frequencies which are significant to a human observer, an accuracy of $\Delta = \frac{1}{2}$ or even $\Delta = \frac{1}{4}$ is required if no significant distortion is to be added by motion inaccuracy. Hence sub-pixel accuracy is required to reduce the error due to motion inaccuracies to the magnitude of the aliasing error in all regions of the image.

3   An Application – Adaptive Wiener In-Loop Filter for Motion Compensated Prediction

This section sketches an example for an application of the above findings. In video codecs employing motion compensated prediction, a filter with transfer function $W(\mathbf{f})$ is placed behind the motion compensation in the lower branch in Figure 1, known as in-loop filter. Qualitatively it attenuates the prediction at high frequencies. A Wiener filter with transfer function $W_W(\mathbf{f}) = \frac{\phi_{g_1 \hat{g}_{1s}}}{\phi_{\hat{g}_{1s} \hat{g}_{1s}}}$ is the fixed filter that achieves optimum SNR, where $\hat{g}_{1s}$ denotes the motion compensated prediction before the filter. A fundamental fact that can be seen from Figure 4 is that the necessary accuracy of motion vectors depends on the true motion. Therefore, the in-loop filter ideally adapts to the local motion. For integer local motion, it should be close to an ideal allpass whereas for non-integer motion it should attenuate the prediction stronger. This suggestion has been verified for a simple block-based motion compensating predictor, as is employed by H.261 and MPEG1/2. Motion estimation was performed by full search block-matching maximizing PSNR of motion compensated prediction.

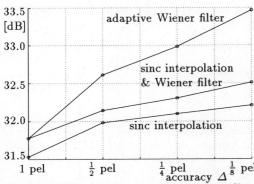

Fig. 5. PSNR of motion compensated prediction for the sequence 'salesman'

Figure 5 depicts the PSNR for the CCITT sequence 'salesman' for adaptive and non-adaptive Wiener in-loop filters, both with 4 tabs. The adaptive Wiener filter was optimized for each individual sub-pixel amount of motion. The improvement of adaptive Wiener filtering over standard Wiener filtering of up to 1 dB confirms the model calculations of the previous section. Since in hybrid video coding, motion information is transmitted to the receiver anyway, these improvements can be achieved without transmission of additional data.

4   Summary and Conclusions

A theoretical framework to investigate the influence of motion vector quantization and aliasing on motion compensated prediction has been presented. Based on a simple model, fundamental properties of the power spectral density of the prediction error signal were examined. It was shown that the impact of aliasing and quantization delimit the performance of motion compensation for typical situations. Interestingly, these distortions strongly depend on local motion. As an illustrative example for an application of these findings, an adaptive in-loop filter for motion compensated prediction has been designed. Significant improvements over a standard Wiener filter confirm the model calculation. Since in video coding, motion information is transmitted to the receiver anyway, these improvements can be achieved without transmission of additional data.

Appendix

Equation (3) is reached by evaluation of the three terms at the right side of equation (2). From (1) the autocorrelation of g_1 becomes

$$\varphi_{g_1 g_1}(\mathbf{n}) = \varphi_{g_0 g_0}\left(\frac{n}{1+b_x}, \frac{m}{1+b_y}\right)$$

with Fourier transform

$$\phi_{g_1 g_1}(\mathbf{f}) = a\phi_{g_0 g_0}(\bar{\mathbf{f}}) * \text{Ш}(\mathbf{f}). \tag{5}$$

Without loss of generality the filter H can be assumed a non-ideal lowpass $H(\mathbf{f}) = 0 \ \forall |\mathbf{f}| > 0.5$. Denoting its autocorrelation function by $\varphi_{hh}^E(\mathbf{x})$ one can write

$$\varphi_{\hat{g}_1 \hat{g}_1}(\mathbf{n}) = \sum_{i,j=-\infty,-\infty}^{\infty,\infty} \varphi_{g_0 g_0}(i,j)\varphi_{hh}^E\left(\frac{n}{1+b_x}-i, \frac{m}{1+b_y}-j\right)$$

with Fourier transform

$$\phi_{\hat{g}_1 \hat{g}_1}(\mathbf{f}) = a^2\left[\left(\phi_{g_0 g_0}(\bar{\mathbf{f}}) * \text{Ш}(\bar{\mathbf{f}})\right) \cdot \left|H(\bar{\mathbf{f}})\right|^2\right] * \text{Ш}(\mathbf{f}). \tag{6}$$

Finally, for integer $\frac{1}{1+b_x}, \frac{1}{1+b_y}$ the cross power spectral density $\phi_{\hat{g}_1 g_1}$ exists and after some calculations can be written as

$$\phi_{\hat{g}_1 g_1}(\mathbf{f}) = a^2\left[\phi_{g_0 g_0}(\bar{\mathbf{f}})e^{-j2\pi\mathbf{c}\mathbf{f}}\left[\left(H^*(\bar{\mathbf{f}})e^{j2\pi\hat{\mathbf{c}}\mathbf{f}}\right) * \text{Ш}(\bar{\mathbf{f}})\right]\right] * \text{Ш}(\mathbf{f}). \tag{7}$$

Insertion of (5), (6) and (7) into (2) yields (3).

References

1. B. Girod, "The efficiency of motion-compensating prediction for hybrid coding of video sequences," *IEEE J. Sel. Areas Commun.*, vol. 5, pp. 1140–1154, Aug. 1987.
2. B. Girod, "Motion-compensating prediction with fractional-pel accuracy," *IEEE Trans. Commun.*, vol. 41, pp. 604–612, Apr. 1993.
3. J. Jain and A. Jain, "Displacement measurement and its application in interframe image coding," *IEEE Trans. Commun.*, vol. COM-29, pp. 1799–1808, Dec. 1981.
4. J. Konrad and E. Dubois, "Bayesian estimation of motion vector fields," *IEEE Trans. Pattern Anal. Machine Intell.*, vol. PAMI-14, pp. 910–927, Sept. 1992.
5. H. Musmann, P. Pirsch, and H.-J. Grallert, "Advances in picture coding," *Proc. IEEE*, vol. 73, pp. 523–548, Apr. 1985.
6. C. Stiller, *Modellbasierte Bewegungsschätzung in Bildfolgen.* PhD thesis, Aachen University of Techn., Fortschr.-Ber., Ser. 10, No. 320, Düsseldorf: VDI-Verlag, 1994.
7. C. Stiller and B. Hürtgen, "Combined displacement estimation and segmentation in image sequences," in *Proc. SPIE/EUROPTO Video Commun. and PACS for Medical Appl.*, vol. 1977, pp. 276–287, Apr. 1993.

Biomedical Applications I

Rule–Based Method for Tumor Recognition in Liver Ultrasonic Images

Vassili A. Kovalev

Computer Center , Belarus Academy of Sciences, Kirova St., 32-A, 246652, Gomel, BELARUS   e-mail: goim@nauka.belpak.gomel.by

Abstract. Rule-based method is considered for recognition of arbitrary 64×64 pixel regions selected in liver ultrasound images. Recognition rules are based on parameters describing spatial distribution of different gradient levels and anisotropy of liver texture. High recognition accuracy has been obtained in case of the same image acquisition conditions.

1 Introduction

There are many approaches for texture characterization based on co-occurrence matrices [1,2], spatial and frequency domain texture features [3], statistical measures, which is not-based on a pre-defined formulation [4], etc. The main problem of the computer analysis and recognition of liver ultrasound texture concerned with great image inhomogeneity [5,6]. Inhomogeneity is due to both anatomical differences between regions of same image and "technical" differences between instruments used and conditions of image acquisition.

The elementary structure balance method [6] based on multidimensional co-occurrence matrices was used at preliminary stage of the work. However, good recognition has been gotten only for the case of one-type regions. It means that only region of interest (ROI) selected and appreciated by specialist in advance can be given in input of a recognition program. So, the following requirements were taken as the initial precondition.

- *A simple choice of the ROI.* The ROI must be as rectangle or square.
- *Arbitrary structure of the ROI.* Any section including liver texture, blood vessel, shadow region, and image edge can be selected in the image analyzed.
- *High automation.* User must not select any regimes and set control parameters.
- *Recognition stability.* It involves in low sensibility to intensity range, image contrast, spatial distortion, and other factors.

Obviously, the requirements listed can be satisfied only using rule-based approach with flexible multi-step scheme of image analysis and recognition.

2 Initial Image Data

Twenty-six liver images of 256×256 pixel size acquired by ultrasonic convex scanner (Fig. 1) were used for parameter selection, recognition rule construction, and testing. The ROI examples of 64×64 pixels are shown in Fig. 2. They present basic types of the liver regions. These examples will be used for illustration of different recognition aspects. Numbers pointed in Fig. 2 over the each ROI will be employed for simplifying references. Seven CT images of 512×512 pixels that present the normal liver were used additionally for estimation of " true " liver texture anisotropy.

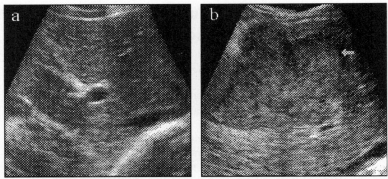

Fig. 1. Example of initial images of normal liver (a) and liver with tumor (b).

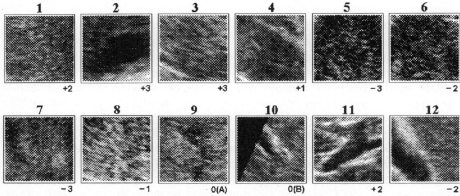

Fig. 2. Example of the liver ROIs: the norm(1-4), the tumor(5-8), and particular cases(9-12).

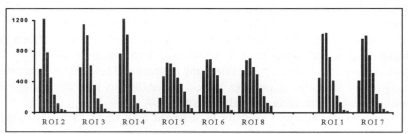

Fig. 3. Gradient histograms for ROIs of Fig. 2: (a) different shape for the norm (ROIs 2-4) and tumor (ROIs 5,6, and 8); (b) similar shape for the norm (ROI 1) and tumor (ROI 7).

3 Method

The parameter (feature) set selection is a key point of the work.

Gradient parameters. The Sobel differential operator with 3×3 mask and root-mean-square magnitude of gradient functions was used for gradient calculation [7]. Investigation of the gradient-based features showed that gradient histograms have as a rule various shapes for the ROI that present the norm and tumor and can be used for recognition (Fig. 3.a). However, in a few isolated cases for typical sections of the normal liver and tumor histogram shape is similar (Fig. 3.b). The cause is that the histogram shape is not sensitive to global scale-space structure of an image. So, the following procedure was used as a practical manner of the gradient utilizing on account of above mentioned problems.

(a) Dividing of gradient range into four unequal levels (the thresholds used are pointed in the brackets):

code L = 0: "nought" gradient (0-30);   code L = 1: small gradient (30-80);
code L = 2: middle gradient (80-300);   code L = 3: great gradient (300-800).

(b) Calculation of the quantized gradient image, whose the each pixel is the gradient level code for corresponding pixel of the analyzed ROI.

(c) Segmentation of the gradient image and calculation of every blob area.

(d) Calculation of the "solid" blob area, that is, area without external border pixels and border pixels of holes.

Thus, 8 parameters including area and its solid part as a percentage for each of 4 gradient levels as well as minimum, maximum, and average gradient were selected.

Orientational parameters. By the "orientational parameters" we mean here the quantitative parameters that describe direction and "stretching" degree (anisotropy) of the texture. High spatial image frequency of the liver texture permits to use of the gradient azimuth orientation into 3×3 window as the base of orientational analysis [8]. The orientational histogram was used for the quantitative description of the orientational image structure. It was calculated by the following steps.

(a) The number of intervals (histogram dimension) is chosen, i.e., the number of sectors on which all azimuth range of 0-180 degree are divided (N parameter).

(b) The azimuth orientation of the gradient is calculated for each image pixel with the gradient level code L=2.

(c) The orientational histogram is calculated: $W = (w(1), w(2), ... , w(N))$, where $w(i)$ is the number of the image pixels whose the gradient has the azimuth orientation within sector by "i" number. It is convenient to depict the orientational histogram as the circular diagram symmetric about the coordinate center.

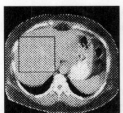 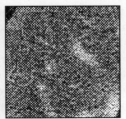 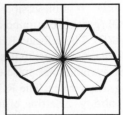

Fig. 4. "True" orientational histogram of liver texture (liver CT image, K_a =1.51).

The anisotropy coefficient $K_a \geq 1.0$ is used as an integral parameter which characterizes the ROI texture anisotropy. It is defined as maximum-to-minimum element ratio of the orientational histogram, i.e. $K_a = Max[w(i)] / Min[w(i)]$. In more details the texture orientational analysis technique discussed in [8]. Typical orientational histogram of normal liver texture are shown in Fig. 4. Investigation of the orientational properties of liver texture showed that the anisotropy decreasing is a symptom of the tumor (Fig. 5). So, the anisotropy coefficient K_a is selected as the important feature because it characterizes only plane (spatial) properties of the texture and is not sensitive to such transformations as multiplication of the pixel intensity by a constant. The "verticality" coefficient K_V is used as parameter which characterizes the orientation of the ROI structure relatively to image edges. It is determined at N=6 as $K_V = [w(3)+w(4)] / [w(1)+w(6)]$.

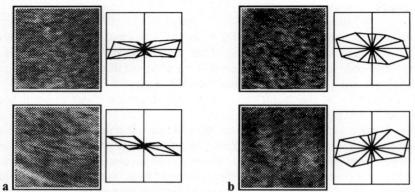

Fig. 5. Orientational histograms for normal and tumor liver regions: (a) normal liver (K_a =8.5 and K_a =15.3); (b) liver with tumor (K_a =2.3 and K_a =2.8).

Diagnostic scale. The ROI recognition result was presented as some diagnostic estimate DS. It indicates that the ROI is related to the norm (DS>0) or to the tumor (DS<0). The confidence of estimate DS is presented in the scale of seven possible levels (colors used for recognition result display are pointed in the brackets).

DS = +3, positive YES (bright green);   DS = +2, YES (green);

DS = +1, nonpositive YES (dark green);   DS = –3, positive NO (bright red);

DS = –2, NO (red);   DS = –1, nonpositive NO (dark red);

The ROI recognition results are ambiguous, i.e., DS = 0 (neutral color) in the cases:

- DS = 0(A). The ROI is some "boundary case" between the norm and tumor;
- DS = 0(B). It is impossible to solve the recognition task due to the poor select of the ROI (great background area, the ROI contains only blood vessel and others);
- DS = 0(C). The selected ROI is not image region.

The ROI diagnostic estimates DS are shown under the each ROI in Fig. 2. All diagnostic estimates are presented except DS=0(C).

Recognition rules. The ROI recognition consists of four stages. 1.Input check of the ROI validity. 2.Setting of initial magnitude for some internal diagnostic estimate IDS. 3.Application of the recognition rules to improving and refinement of internal estimate IDS. 4.Calculation of the final diagnostic estimate DS by the internal estimate IDS and result display.

The input check of the ROI is performed to test an opportunity of recognition task solution, that is, lack of causes for diagnostic estimates as DS=0(B) and DS=0(C). The following recognition rules can be presented as an example.

if (area at L=2 < 15%) *or* (average gradient of ROI < 90) *then* DS=0 (C);

if (area at L=0>10%) *and* (solid at L=0 > 6%) *then* set Edge-Vessel flag;

if (Edge-Vessel flag) *and* ((K_a < 7.0) *or* (K_V > 0.5)) *then* DS=0(B);

Internal variable of recognition algorithm is diagnostic estimate IDS. It shows closeness of the recognition ROI to the norm or tumor in some conditional scale (1.2, 120.0). The anisotropy coefficient K_a is the initial value for IDS. The following rules can be presented as an example of rules used in the third stage.

if (Edge-Vessel flag) *and* ($K_a > 7.0$) *then* increase IDS by 3.0;

if (area at L=3 > 2.5%) *and* (solid at L=3 > 50%) *and* (Edge-Vessel flag) *then* set Vein flag; *if* (Vein flag) *and* ($K_a < 6.8$) *then* increase IDS by 2.5;

if (area at L=1 > 20%) *and* (area at L=3 > 4%) *and* (solid at L=3<10%) *then* increase IDS by 1.5;

The diagnostic estimate DS is determined by the internal estimate IDS at the final fourth stage. The different DS corresponds to the different subintervals of IDS:

IDS:	<3.0	3.0–4.5	4.5–5.4	5.4–6.3	6.3–7.3	7.3–10.0	>10.0
DS:	−3	−2	−1	0(A)	+1	+2	+3

4 Results and conclusions

Sixteen test images of convex ultrasonic scanner acquired by same conditions were used for software checking and recognition quality testing. Eight of them presented the normal liver and eight other -- the liver with tumor. The recognition results were agreed with expert's opinion as a rule (see, for instance, Fig. 2).

The initial images were scanned in raster scan order by 64×64 pixels window with 20 pixel displacement for quantitative estimate of recognition accuracy. So, 100 ROIs were automatically selected in each image. They were differed by 31% pixels as a minimum. General recognition results are presented in Table 1.

Table 1. General recognition results

	ROIs distribution by diagnostic estimate DS								
	−3	−2	−1	0(A)	0(B)	0(C)	+1	+2	+3
Normal liver	0	2	0	4	134	18	26	130	486
Liver with tumor	51	164	66	45	147	21	52	103	151

Normal liver. Nrally, the results with the diagnostic estimate DS<0 (tumor) are qualified as errors in case of normal liver testing (see first row of Table 1). The estimate DS = −2 is taken in two cases (one of them is ROI 12 in Fig. 2). The results DS = −3 and DS = −1 are not taken. So, the recognition errors are 0.25%. The estimate DS=0(A) was taken for four ROIs, that is the ambiguous result are 0.5%. The ROI with number 9 shown in Fig. 2 is a case in point. The main part of rejected ROIs with estimate DS=0(B) and DS=0(C) were obtained from left and right top parts of initial images (see, for example, ROI 10 in Fig. 2).

Liver with tumor. Eight images of liver with tumor are tested by the similar way. Any DS values are possible in the given case because the ROI may include the tumor section, or the normal liver section or some combination of them. Recognition accuracy $R_a = 0,1,...6$ for every ROI was calculated as the difference between expert's estimate DSe and program estimate DS, i.e., $R_a = |DSe - DS|$. Disagreement magnitude in limits of 1-2 units of confidence scale was qualified as "satisfactory recognition" and more than 2 units -- as a recognition error. The results obtained are shown in Table 2.

Table 2. Recognition accuracy for liver with tumor (632 ROIs).

ROIs	Precise recognition	Satisfactory recognition	Errors
Number of ROIs	517	76	39
Percent	81.8	12.0	6.2

It has been established as a result of special test, the method and software can be applied for tumor recognition in images acquired only by the same scanners and same conditions (ultrasound frequency, spatial resolution, etc.)

Finally, the following recognition results were obtained for all images. 1280 ROIs (without 0(B) and 0(C) cases) were tested. 1163 ROIs (90.9 %) were recognized precisely and 41 ROIs (3.2 %) were recognized with error. Recognition results of 76 other cases (5.9 %) were satisfactory. Prototype software was developed in Pascal for IBM AT compatible computers. The recognition time for 64×64 ROI is about 0.3 second on PC AT 486/66. The following conclusions can be made.

1. The parameters which characterized spatial distribution of different gradient levels and texture anisotropy may be used as features for tumor recognition in liver ultrasound images.

2. Arbitrary liver sections were recognized with high accuracy by the suggested method in case of the same image acquisition conditions.

Acknowledgements - The author wishes to thank MEDISON Co. for providing the initial image data. Special thanks belong to MD S.I.Pimanov for medical verification of the result.

References

1. R.M.Haralick, K.Shanmugam, and I.Dinstein. Textural features for image classification. IEEE Transactions on Systems Man and Cybernetics, SMC-3, No. 6, 1973, pp. 610-621.

2. J.F.Haddon and J.F.Boyce. Co-occurrence matrices for image analysis. Electronics & Communication Engineering Journal, 4, 1993, pp. 71-83.

3. R.Muzzolini, Y.H.Yang, and R.Pierson. Texture characterization using robust statistics. Pattern Recognition, Vol. 27, No. 1, 1994, pp. 119-134.

4. D.Patel and T.J.Stonham. Texture image classification and segmentation using RANK-order clustering. In: Proceedings, 11th Int. Conf. on Pattern Recognition, Vol. 3, The Hague, The Netherlands, 1992, pp. 92-95.

5. J.S. DaPonte and P. Sherman. Classification of ultrasonic image texture by statistical discriminant analysis and neural networks. Computerized Medical Imaging and Graphics, 15(1), 1991, pp.3-9.

6. V.A.Kovalev. Feature extraction and visualization methods based on image class comparison. In: Proceedings, Medical Imaging '94 Int. Symp., Image Processing, M.Loew (ed.), Newport Beach, CA, 1994, pp. 691-701.

7. I.E.Abdou and W.K.Pratt. Quantitative design and evaluation of enhancement/ thresholding edge detectors. IEEE Proc. 67(5), 1979, pp. 753-763.

8. V.A.Kovalev and S.A.Chizhik. On the orientational structure of solid surfaces. Journal of Friction and Wear, Vol. 14, No. 2, Alerton Press, 1993, pp. 45-54.

An understanding of muscle fibre images

C.C. Taylor, M.R. Faghihi and I.L. Dryden

Department of Statistics, University of Leeds, Leeds LS2 9JT, U.K.

Abstract. Images of muscle biopsies reveal a mosaic pattern of two (slow-twitch and fast-twitch) fibre-types. An analysis of such images can indicate some neuromuscular disorder. We briefly review some methods which analyse the arrangement of the fibres (*e.g.* clustering of fibre type) and the fibre sizes. The proposed methodology uses the cell centres as a set of landmarks from which a Delaunay triangulation is created. The shapes of these (correlated) triangles are then used in a test statistic, to ascertain normality of a muscle. Our "normal muscle" model supposes that the fibres are hexagonal (so that the triangulation is made up of equilateral triangles) with a perturbation of specified isotropic variance of the fibre centres. We obtain the distribution of the test statistic as an approximate function of a χ^2 random variable, so that a formal test can be carried out.

1 Introduction

An important component in the early diagnosis of muscle diseases is the ability to (automatically) analyse muscle biopsies. Using either the needle biopsy technique (Edwards, 1971), the open surgical technique (Dubowitz, 1985) or the semi-open technique (Henriksson, 1979), part of several hundred fibres can be extracted for examination. These cross-sections are usually histochemically stained to reveal a light microscopic examination of two type of fibres: type 1 (lightly stained) and type 2 (heavily stained). Figure 1 shows two examples (at different magnifications) of a small part of images obtained from *whole* muscle cross-sections (obtained post-mortem), but similar images would also result from biopsies. There are important changes that occur with age, for example a decrease in muscle area and a decrease in fibre size; see Lexell *et al.* (1988) for details. Deviations from normality, and hence indicating some neuromuscular disorder, can occur in a variety of ways. For example, the formation of large groups of fibres of the same histochemical type is considered to be evidence of a denervation and reinervation process (Walton, 1981; Dubowitz, 1985), and a variety of objective methods have been suggested to detect such grouping: see Lexell, Downham & Sjöstrom (1987) for a review of statistical approaches.

An analysis of fibre size has been investigated from fibre outlines digitised by hand on a digitising tablet. This enables an investigation of the difference in the distribution between type I and type II fibres within a given region of the muscle, and differences in distributions between different parts of the muscle, for example superficial fibres tend to have less variability. This difference may be expressed in terms of the mean and/or variance, but more often there are important features in the distribution, such as evidence of "subgroups", which are most easily seen in a nonparametric density estimate. These can be readily superimposed, as an example shows in Figure 2.

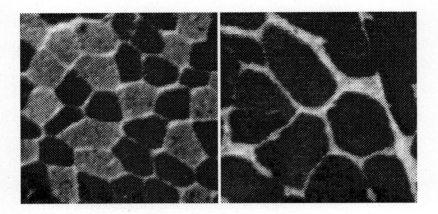

Fig. 1. Examples from a normal muscle. Left: the fibres are fairly hexagonal in shape and apparently quite random in type, Right: close-up shows some angular shaped fibres

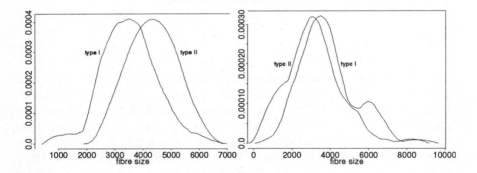

Fig. 2. Normal muscle examples. Note the sub-populations indicated by bimodality.

Hand digitisation is a laborious process, and a system which can segment each fibre and label it as type I or type II would capture all the essential information for an objective classification of each muscle biopsy. One difficulty in segmentation arises from grey-level trends in which the type of fibre can only be determined from the *relative* brightness. This could be dealt with by subtraction of a smoothed background. More importantly, detecting an edge between two abutting fibres of the same type can be very difficult. In principle, it should be possible to use prior information that fibres are usually convex in shape. A Bayesian approach, which attaches a probability distribution to each fibre shape (and size) could then give a more robust segmentation. A less time consuming digitisation would be to mark the "centres" (for example, the approximate centroid) of each fibre, together with the fibre type. This takes only a few minutes for each region studied (about 150 fibres).

2 Shape

Healthy muscles generally contain fibres which are roughly hexagonal in shape, so an analysis of fibre shape may detect an abnormality. However, in view of the above remarks on ease of digitisation we instead study the "shape" of the marked centres. The methodology that has been developed to date does not use the information in the fibre type, so we initially throw away the labelling.

2.1 Landmarks and Triangulation

Shape analysis is often based on a set of "landmarks" *i.e.* points of an object (or image) which are easily identified. This set of co-ordinates $\{(x_i, y_i), i = 1, 2, \ldots, n\}$ is then transformed to take out the effects of scale, rotation and translation. The information which remains is thus referred to as shape. In the current context we use the centres of the fibres as our set of landmarks, and we begin by forming a *Delaunay triangulation* (see Figure 3). We first associate, for each (x_i, y_i), the polygon $T(x_i, y_i)$ which is the set of points for which (x_i, y_i) is the nearest neighbour amongst all landmarks. These polygons, known as *Voronoi polygons* form a *Dirichlet tessellation* of the data, and they have the property that almost surely, except in the case of certain regular structures, every node is touched by exactly three cells. The Delaunay triangulation is constructed by connecting those landmarks whose cells share the same node. Note that an edge correction procedure is required in which cells which lie on the boundary of the image are removed in order to avoid flat "splinter like" triangles.

2.2 Summary statistic

The proposed model is that a normal muscle has fibres which are close to hexagonal in shape, in which case the Voronoi polygons associated with the fibre centres are just the fibres themselves, and the associated Delaunay triangles are equilateral. So we suppose that normal fibres could be simulated with centres (X_i, Y_i) with distribution

$$X_i \sim N(\alpha_i, \sigma^2), \qquad Y_i \sim N(\beta_i, \sigma^2), \qquad i = 1, 2, \ldots, n$$

all mutually independent, where the $(\alpha_i, \beta_i), i = 1, 2, \ldots, n$ are chosen such that their associated Delaunay triangulation contains equal sized equilateral triangles; see Mardia *at al.* (1977). The value of σ then gives a deviation from the regular (hexagonal) case; a large σ emulating a diseased muscle, and a small σ indicating a healthy muscle. Figure 3 has two realisations for different values of σ.

Let δ be the length of the edge of each equilateral triangle. As shape is independent of the scale of the image we examine whether or not $\tau = \sigma/\delta$ is large. Estimation of τ for some real data then gives a plausible test statistic which can be used to test $H_0 : \tau = \tau_0$ vs $H_1 : \tau > \tau_0$ if we have a given τ_0 which has been previously estimated from known healthy muscles. However, in practice, it is easier to work with the shape of the triangles, rather than estimate τ directly. Define the *shape distance* between two triangles $((x_i, y_i)$ and $(\alpha_i, \beta_i), i = 1, 2, 3)$ as ρ where

$$(\cos \rho)^2 = \frac{(S_{x\alpha} + S_{y\beta})^2 + (S_{x\beta} - S_{y\alpha})^2}{(S_{xx} + S_{yy})(S_{\alpha\alpha} + S_{\beta\beta})},$$

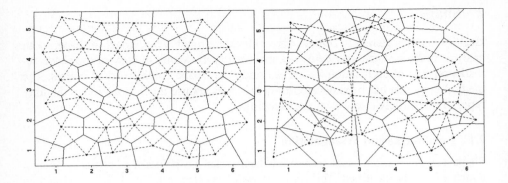

Fig. 3. Examples of Dirichlet tessellation (solid) and Delaunay triangulation (dashed). Left: $\sigma = 0.1$; Right: $\sigma = 0.3$

with

$$S_{xy} = \sum_{i=1}^{3}(x_i - \overline{x})(y_i - \overline{y}), \qquad \overline{x} = \frac{1}{3}\sum_{i=1}^{3}x_i, \qquad \overline{y} = \frac{1}{3}\sum_{i=1}^{3}y_i.$$

The distance is known as Riemannian distance between two triangle shapes and was introduced by Kendall (1984). We can regard ρ as a measure of similarity between two triangles, after removing location, rotation and scale effects.

2.3 Distribution

It can be shown (see Dryden *et al.*, 1995 for details) that, for small $\tau = \sigma/\delta$

$$\cos\rho \sim 1 - \frac{\tau^2}{2}\chi_2^2.$$

where χ_2^2 represents a Chi-squared random variable with 2 degrees of freedom. This work is related to previous work on perturbations in Procrustes methods; see Sibson (1978, 1979), Langron & Collins (1985), Goodall (1991) and Mardia (1989). Suppose now that we have n landmarks which will form m triangles. Note that the shape of neighbouring triangles are correlated, so the distribution of

$$\overline{\cos\rho} = \frac{1}{m}\sum_{l=1}^{m}\cos\rho_l$$

must take account of the neighbourhood structure. In principle it would be possible to use simulation to obtain results, for example, to give $p-$values for a certain null hypothesis. However, it is straightforward to obtain the approximate distribution

$$\overline{\cos\rho} \sim 1 - \frac{\tau^2}{4m}\sum_{i=1}^{n}\lambda_i\chi_2^2$$

where the $\lambda_i \geq 0$ are eigenvalues of a matrix associated with the contiguity structure of the (α_i, β_i). By matching the mean and variance we can further approximate this linear sum of χ^2 random variables to get

$$\overline{\cos \rho} \sim 1 - a\chi_\nu^2$$

with

$$\nu = \frac{8m^2}{\sum_{i=1}^{n} \lambda_i^2}, \qquad a = \frac{\tau^2}{\nu}$$

Exact values of $\sum \lambda_i^2$ can be determined for a given sample, but we can usually use the approximation that, for a rectangular array of $m_1 \times m_2$ triangles, we have

$$\sum_{i=1}^{n} \lambda_i^2 = \frac{4}{9}(21m_1m_2 - 8m_2 - 14m_1 + c)$$

where c can be 6, 8 or 10 according to the different arrangements of the triangles. It is of interest to examine how the variance of the statistic, and hence $\sum \lambda_i^2$ varies for different choices of (α_i, β_i), $i = 1, \ldots, n$; see Dryden et al. (1995) for details.

2.4 Hypothesis Testing

If we have only one image, then it is straightforward to compute $\overline{\cos \rho}$ which then gives an estimate of τ as

$$\hat{\tau} = \sqrt{1 - \overline{\cos \rho}}$$

We can then test $H_0 : \tau = \tau_0$ vs $H_1 : \tau > \tau_0$ by using the test statistic

$$T = \frac{\nu}{\tau_0^2}(1 - \overline{\cos \rho}) \sim \chi_\nu^2$$

if the null hypothesis is true.

If we have multiple images (from the same individual, or from a group of individuals from the same class) then we can estimate τ by maximum likelihood and proceed in the usual way. In principle, τ_0 can be estimated from *known* healthy muscles and used as a subsequent benchmark.

Example Suppose that the null hypothesis of regularity is $H_0 : \tau \leq 0.15$, and that the alternative hypothesis of irregularity is $H_1 : \tau > 0.15$. Suppose that the image of muscle fibres has $m = 176$ in a roughly rectangular region (with $m_1 = 11, m_2 = 16$), and that $\overline{\cos \rho} = 0.9711$. Then we have $\nu = 163$ and test statistic

$$T = \frac{163}{0.15^2}(0.0289) = 209.4 \sim \chi_{163}^2$$

if H_0 is true. For large ν we can use the Normal approximation $\sqrt{2T} - \sqrt{2 \times 163 - 1} \sim N(0, 1)$. This gives a p-value of about 0.0149, so we reject H_0.

3 Extensions

The above triangulation can also be used to describe the size of the fibres, s, given by

$$3s^2 = S_{xx} + S_{yy}$$

and in a similar manner we can obtain a test statistic for the variance of triangle sizes.

A further extension is to use the labels of the landmarks (type I or type II). We can then either consider the distribution of the 4 types of triangle, *i.e.* those which have 0, 1, 2 or 3 type I vertices, or we can obtain a separate triangulation for each label, and examine the size and shape of these triangles. Both of these approaches may usefully combine information in size, shape and arrangement of fibre type colours, and lead to powerful discrimination methods.

Acknowledgement: The authors would like to thank Jan Lexell, University of Lund for providing the muscle fibre images, and introducing us to this problem.

References

Goodall, C.: Procrustes methods in the statistical analysis of shape. Journal of Royal Statistical Society B, **53** 285–339 (1991)

Dubowitz, V.: Muscle biopsy. A practical. Baillière Tindal, London (1985)

Edwards, R.H.T.: Percunatenous needle biopsy of skeletal muscle in diagnosis and research. Lancet, **11** 593–596 (1971)

Dryden, I.L., Faghihi, M.R. and Taylor, C.C.: Shape analysis of spatial point patterns. In: K.V. Mardia, C.A. Gill (eds.): Proceedings in Current Issues in Statistical Shape Analysis. Leeds University Press 1995, pp. 40–48.

Henriksson, K.G.: "Semi-open" muscle biopsy technique. A simple outpatient procedure. Acta Neurol Scand **59** 317–323 (1979)

Kendall, D.G.: Shape - manifolds, procrustean metric and complex projective space. Bull. Lond. Math. Soc. **16** 81–121 (1984)

Langron, S.P. and Collins, A.J.: Perturbation theory for generalized Procrustes analysis Journal of the Royal Statistical Society B **47** 277–284 (1985)

Lexell, J., Downham, D. and Sjöstrom, M.: Morphological detection of neurogenic muscle disorders: how can statistical methods aid diagnosis? Acta Neuropathol **75** 109–115 (1987)

Lexell, J., Taylor, C.C. and Sjöstrom, M.: What is the cause of the ageing atrophy? Total number, size and proportion of different fiber types studied in whole vastus lateralis muscle from 15- to 83-year-old men. J. Neurol. Sci. **84** 275–294 (1988)

Mardia., K. V.: Shape analysis of triangles through directional techniques. Journal of Royal Statistical Society B **51** 449–458 (1989)

Mardia, K.V., Edwards, R. and Puri, M.L. Analysis of Central Place Theory. Bulletin of the International Statistical Institute **47** 93–110 (1977)

Sibson, R. (1978) Studies in the robustness of multidimensional scaling: Procrustes statistics.

Sibson, R.: Studies in the robustness of multidimensional scaling: Procrustes statistics. Journal of the Royal Statistical Society, B **40** 234–238 (1978)

Sibson, R.: Studies in the robustness of multidimensional scaling: perturbation analysis of classical scaling. Journal of the Royal Statistical Society, B **41** 217–229 (1979)

Walton, J.: Disorders of Voluntary Muscle. London: Churchill Livingston 1981

A New Plant Cell Image Segmentation Algorithm*

G. Fernàndez[1] and M. Kunt[1] and J-P. Zrÿd[2]

[1] Signal Processing Laboratory
Swiss Federal Institute of Technology
CH-1015 Lausanne, Switzerland
[2] Laboratory of Plant Cell Genetics
University of Lausanne
CH-1015 Lausanne, Switzerland

Abstract. A new plant cell image segmentation algorithm is presented in this paper. It is based on the morphological analysis of the cell shapes and on gradient information. The difficulty of the segmentation of plant cells lies in the complex shapes of the cells and their overlapping, often present due to recent cellular division. The algorithm presented tries to imitate the human procedure for segmenting overlapping and touching particles. It analyzes concavities in the shape of a group of cells as well as the existence of a coherent surface for the segmentation. The algorithm can be divided in two main parts, firstly a simple method for finding dominant concave points in shapes is introduced and secondly several parameters between concave points are calculated as a criterion for segmentation. It has been also shown that the algorithm produces good results when the output is applied as a marker for the morphological *watershed* algorithm. Results will be presented in real images of a cell suspension culture to show the validity of the chosen approach.

1 Introduction

In biology, a common problem is the segmentation of cells for counting and feature extraction purposes. Manual work is tedious and yields imprecise and subjective results. Computer assisted assessment has been shown to be a powerful tool due to its automatic, objective and fast measurement. A typical biological image analysis system starts with a digitizer (a video camera, scanner, etc..) followed by an enhancement of the image, segmentation of the cells and feature extraction for classification. The output provides statistical results concerning the state of the culture that can be interpreted by the biologists.

Image segmentation has been studied by many authors and it has been shown that it is often dependent on the type of elements to segment. A relatively new tool for segmentation is the *watershed* algorithm [1]. It has the ability to perform very accurate segmentations from image markers reducing the problem

---

* This work is supported by Swiss National Science Foundation (FNRS) grant N. 3136561.92

to a simple marker extraction. In this paper we present a new method to find markers of touching and overlapping cells that gives good results in combination with the *watershed* algorithm.

2 Project Background

Samples of *Beta vulgaris* cells are cultivated in vitro in liquid suspension. Previous work in this project yielded well enhanced images. Samples of the cells are conveniently diluted in order to have a reasonable spatial distribution. After digitation of the cells [2], the useful zone of the scene was obtained by an adaptive form of the *Hough transformation* [3] and the images were enhanced by the morphological *top hat* transformation [4]. The next step in the image analyzing process is the segmentation of the enhanced cells and will be discussed in the next sections.

3 Segmentation of Plant Cells

The enhanced cells present an even spatial distribution and some of them are touching or overlapping. They do not present a unique geometrical shape that could facilitate the marker extraction (i.e. *ultimate erosion* yields oversegmentation) the grey level is similar for all of them (the gradient information is often useless). Furthermore, the gradient image does not present a unique minimum per cell which could be used for the marker extraction. It is for all these reasons that classical segmentation algorithms either do not segment the touching cells or lead to oversegmentation. The new segmentation method presented here tries to solve this problem.

The main idea of the algorithm lies in the imitation of the human procedure in segmenting touching particles. A group of cells is segmented if it presents a shape with two concavities relatively near one to each other with respect to the perimeter of the group, and if a flat path exists between the concave points.

3.1 Contours and Their Concavities

After the enhancement by the *top hat* transformation, groups of cells are clearly differentiated from the background. The next step is to obtain the contours of the cells in order to study their concavities. This is done by the set difference between the binary image and the unitary erosion. The contours are extracted, labeled and analyzed to obtain the concave points.

3.2 Concave Points

Dominant points have been widely studied in Pattern Recognition literature, specially in polygonal approximation purposes, object and character recognition but not so much in segmentation [5, 6]. These algorithms are usually recursive and/or computationally demanding. Since the analysis of these points must be done for all the cells in the image and the number of cells can be well above several hundreds, the algorithm of concave points has to be simple and robust.

Each point of the contour will have a value of *concaveness* which is calculated in the following way:

$$concaveness(j) = \sum_{j-1}^{j+1} \sum_{x=1}^{5} \sum_{y=1}^{5} M_j \cap B \qquad (1)$$

where B is the binary image, j runs over the contour and M_j is a 5x5 mask. In other words, for the j_{th} point of the contour the value of *concaveness* is calculated as the number of points of a 5x5 mask centered on j that intersect the binary shape. Finally, in order to avoid uncertainties, the values of the two adjacent contour neighbors are added to the present contour value. Thresholding of these values yields the concave points of the shape. If two adjacent points are selected, the maximum of them is retained. Fig. 1 illustrates the method.

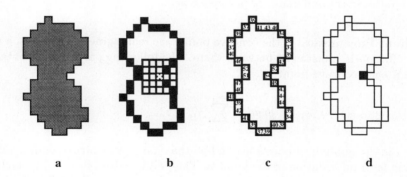

Fig. 1. a. Binary shape, b. Contour obtained by erosion and mask, c. Values of concaveness. d. Final concave points

These points can be due to the specific shape of the cells or the presence of touching or overlapping cells. In the next section the information of these concavities will be discussed to perform the segmentation.

3.3   Linking Concave Points for Segmentation

The main idea for segmenting is to cut the shapes by the places where the concave points are present. But the problem is that some concave points may be due to the particular features of the cells and, furthermore, the presence of more than two cells implies several pairs of concave points. The choice of the points for segmentation are based on two main features: The gray level information (flatness of the path between points) and spatial information (distance between points compared to the perimeter of the contour).

For each concave point the length and the gradient of the path to the other concave points is calculated. According to these parameters, each point selects its candidate as a partner. A concave point will choose the partner that minimizes the spatial and gradient parameters. Their description follows.

Spatial Parameter. The spatial parameter is the euclidean distance between two points. In order to prevent from segmenting cells that have concave points in their "real" shape, a condition has to be imposed: The euclidean distance between candidate segmenting points is compared to the perimeter of the shape (number of contour points) in the following way:

$$d < \frac{per}{\pi} \qquad (2)$$

where d is the distance between points and per the perimeter. The idea of the condition is to segment only when the distance d is smaller than the diameter of a hypothetic circumference of perimeter per.

Gradient Parameter. If the concave points are real segmenting points, a relatively "flat path" must exist between them. The following parameter is obtained between each concave point:

$$grad = \sum_{n=1}^{i} |Y_n - Y_{n-1}| \qquad (3)$$

where i is the number of points in the line that joins two concave points and Y_n the gray level information in the point n. The point that minimizes this factor is chosen as a candidate. The obtained value must not exceed $2 * N$ (N being the number of total grey levels) , which has been found to be a reasonable threshold of flatness.

The calculation of these parameters is done for all points and each point makes its choice. Finally, the selected points are compared, when the points select one to each other the cell is decided to be segmented by drawing a line between points. This line is dilated in order to let the *watershed* algorithm to find a more accurate segmentation line. In Fig. 2 where the algorithm is illustrated, each point selects its partner but only the mutual choice produces segmentation.

4   Results

For testing the performance of the algorithm, a collage image was built containing groups of cells (Fig. 3.a). Different stages of the algorithm are shown, contour image with concave points, segmenting lines and final segmented image after the *watershed* algorithm.

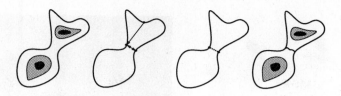

Fig. 2. Each concave point selects its candidate but only when a mutual choice occurs the segmentation is done.

The algorithm performs good segmentation in trivial cases where the shape has a pair of clear concavities close one to the other. For more complex combinations of cells, the results are also satisfactory, taking into account that sometimes even for the biologist it is difficult to say how many cells are in a group. It is important to remark the good selection of linking the points in the cases where even if they are fare one to each other, the algorithm recognizes them as the right path to chose.

After the segmentation of the cells, features such as length, width, and area are calculated for a final classification.

5 Conclusions

In this paper, a new plant cell image segmentation algorithm has been presented. It is based on the imitation of the human segmentation, that means segmentation of cells between the points that copes certain features of *shape and surface* within the cell. The algorithm has been shown to be robust in front of different cell shapes and groups of cells. Results in cultures in liquid suspension satisfy the segmentation needs for a complete image analysis system for plant cells.

Acknowledgments

The authors would like to thanks Mr P.Brigger for his useful suggestions in this work.

References

1. S. Beucher and F. Meyer. Morphological segmentation. *Journal of Visual Communication and Image Representation*, 1:21–46, 1990.
2. P. Brigger. Morphological plant cell analysis. In J. Serra and P. Salembier, editors, *Mathematical Morphology and its Applications to Signal Processing*, pages 101–106, Barcelona, May 1993.

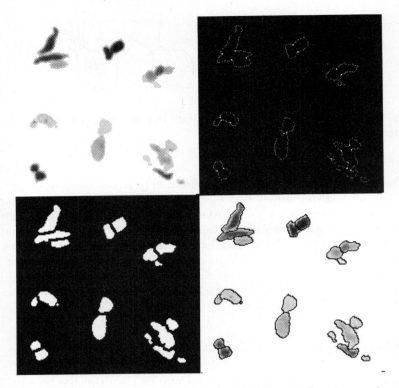

Fig. 3. From left to right and top to bottom: a.Original *collage* image, b. concave points of contours, c. segmenting lines, d. final segmented image after the *watershed* algorithm.

3. R. Duda and P. Hart. Use of the hough transformation to detect lines and curves in pictures. *Communications of the ACM*, 15:11–15, 1972.

4. F. Meyer. *Cytologie quantitative et morphologie mathématique.* PhD thesis, Ecole Nationale Supérieure des Mines de Paris, Fontainebleau, 1979.

5. A.K. Jain, S.P. Smith, and E. Backer. Segmentation of muscle cell pictures: a preliminary study. *IEEE Transactions on Pattern Analysis and Machine Intelligence*, 3:232–242, 1980.

6. A. Lerch and D. Chetverikov. Correction of line drawings for image segmentation in leather industry. In *Proc. 11th Int. Conf. on Pattern Recognition*, volume 1, pages 420–424, The Hague, 1992.

Extraction of Tumours from MR Images of the Brain by Texture and Clustering

João Batista* and Richard Kitney

Imperial College, Electrical Engineering Dept.
Exhibition Road, SW7 2BT, London - UK

Abstract. The characterisation of tumours from Magnetic Resonance (MR) images of the brain is still a challenging task. In this paper we present an approach based on a K–Means clustering algorithm combined with textural feature information as opposed to intrinsic MR paramenters T1,T2 and PD. This is due to the fact that MR parameters may exhibit significant alterations in the presence of pathological conditions and, therefore lead to incorrect classification. We also address two important aspects of clustering: the selection of the optimum number of classes (Cluster Validity) and the most effective features (reduction of the feature space).

Index Terms: Tissue Characterisation, Texture Analysis, Clustering, Feature Reduction, Cluster Validity.

1   Introduction

Considerable work has been done on the characterisation of normal and abnormal human brain tissue based on quantitative Magnetic Resonance Imaging (MRI), ie the analysis and interpretation of its intrinsic parameters Proton Density (PD) and the relaxation times T1 and T2. It has also been shown that these parameters vary from tissue to tissue [Fos 84, Tay 86] but may exhibit significant alterations in the presence of pathological conditions, not allowing classification to be carried out exclusively on PD, T1 and T2 values.

The approach presented in this paper regards the classification of MRI brain tissues as a pattern recognition task [Bez 93], which combines features extracted from the original image – textures – with a classification procedure – clustering – to devise the classes, or tissues.

Texture analysis plays an important role in the segmentation of images which do not exhibit a clear boundary between different objects within the image. This can be observed in MR images of the brain, especially if oedema or tumour is present.

The main reason for using clustering, as opposed to supervised classifiers, is that the former does not require a training data set. The imaged physical values

* Sponsored by CNPq - Conselho Nacional de Desenvolvimento Científico e Tecnológico, Brazil

in MRI vary from machine to machine due to different magnetic field values. Consequently, a training data set which is taken from a particular machine and then used to tune the classifier, may give misleading results if the classifier is applied to a different set of images – with the same physical characteristics as the training data set images – acquired from a different scanner. Moreover, an unsupervised classifier may reveal unnoticed structures within the image; also, no *a priori* number of classes needs to be defined.

We also discuss two important aspects of clustering: cluster validity and feature reduction. The former is concerned with the optimum number of clusters and the latter with redundant and/or undesirable features in the feature space.

2 Texture

When the discrimination of classes based on the grey level differences is ineffective, textural features can be considered. Texture is a neighbourhood property of an image and, therefore, it conveys properties such as fineness, coarseness, smoothness, granulation and randomness. Each of these nouns translates into some properties of the tonal primitives and the spatial interaction between pixels. As defined in [Goo 85], this interaction can be either deterministic or statistical.

For MR images, the nature of texture is typically statistical. The two methods described in this section aim at the characterisation of stochastic properties of the spatial distribution of the grey level in the image.

2.1 Spatial Grey Level Dependence Matrix (SGLDM)

The SGLDM describes the spatial distribution and spatial dependence among the grey tones in a local area based on the estimation of second order conditional probability density functions $f(i, j, d, \theta)$. Each of these functions are the probability of going from grey level i to grey level j separated by a distance d and aligned to the angle θ. The estimated values (which are grey–tone spatial–dependence frequencies) can be written in matrix form, the so–called co–occurrence matrices.

A number of texture features can be computed from the horizontal (0°), vertical (90°), left diagonal (45°) and right diagonal (135°) matrices. Haralick [Har 73] proposed 28 features extracted from 14 equations, but usually only 5 of them are used. They are angular second moment(ASM), contrast, correlation, inverse difference moment(IDM) and entropy.

2.2 Grey Level Run Length Method (GLRL)

The GLRL method is based on the calculation of the number of grey level runs with different lengths. A grey level run is a set of consecutive, co–linear pixels having the same grey level values. The length of the run is the number of pixels in the run.

Given an image and a direction (0°, 45°, 90°, 135°), a GLRL matrix can be computed. The GLRL matrix element (i, j) specifies the number of times the picture contains a run of length j, in the given direction, consisting of points having grey level values i, or within the grey level range i. Textural information can now be extracted from the GLRL matrices. A set of six functions used in this work can be seen in [Loh 88]. They are second moment with respect to run length, first and second moments with respect to grey level, grey level and run length non–uniformity and sum of variance.

3 Clustering

The objective of cluster analysis is to separate the features into groups so that the members of any one group differ from each other as little as possible, according to a chosen criterion. The clustering used in the work which underlies this paper is based on the K-Means clustering algorithm of Coleman and Andrews [Col 79] and comprises three major steps: (a) initialisation, (b) distribution of feature vectors among existing clusters, (c) cluster validity.

(a) **Initialisation**: consists of splitting the input data set into two initial clusters. The input is given by a set of m feature vectors $v = (f_1, f_2, \ldots, f_n)$ where m is the number of pixels in the image, n is the feature vector dimension and $f_i (i = 1, \ldots, n)$ is a texture feature. The two initial clusters are created by computing the mean and variance over the m feature vectors. Cluster centres are then calculated to be evenly spaced on the diagonal of positive correlation of $^{+}_{-}1$ standard deviation in the hyperspace of the feature set.

(b) **Distributing input among clusters**: Having established the new cluster centres, the M feature vectors must be assigned to the existing clusters until the algorithm converges. This is done by assigning each feature vector to the closest cluster centre through the Euclidean Distance [Fuk 72].

This step is repeated until all the cluster centres remain unchanged, ie until the algorithm has converged. At this point a new cluster can be devised by selecting amongst all clusters the feature vector which presents the largest distance from its cluster centre.

(c) **Cluster Validity**: One of the central points in clustering is to define the "correct" number of clusters for a given feature space. One possible approach is to obtain a measure of clustering quality represented by a parameter β derived from the within–cluster and the between–cluster scatter matrices [Fuk 72].

The within–cluster scatter matrix shows the scatter of samples about the cluster means and is expressed by:

$$S_w = \frac{1}{K} \sum_{k=1}^{K} \frac{1}{M_k} \sum_{x_i \in S_k} (x_i - \mu_k)(x_i - \mu_k)^T \tag{1}$$

where μ_k is the mean of the kth cluster; M_k is the number of elements in the kth cluster; x_i is an element in the kth cluster (the set of such elements given by S_k) and K is the total number of clusters.

The between–cluster scatter matrix shows the scatter of clusters in relation to the overall input set:

$$S_b = \frac{1}{K} \sum_{k=1}^{K} (\mu_k - \mu_o)(\mu_k - \mu_o)^T \tag{2}$$

where μ_o is the overall mean vector of the input set and μ_k is the mean vector of the kth cluster.

The β value which has produced the best results in this work is given by: $\beta_5 = \mathrm{tr}S_b \cdot \mathrm{tr}S_w$, where $\mathrm{tr}(\bullet)$ indicates trace (sum of the diagonal elements of a matrix). Note that when the number of clusters is equal to 1, $\mathrm{tr}S_w = \sigma^2$ (the variance of the input data set), $\mathrm{tr}S_b = 0$ and $\beta_5 = 0$. When the number of clusters equal M (M is the total number of feature vectors), $\mathrm{tr}S_w = 0$ and $\mathrm{tr}S_b = \sigma^2$, hence, $\beta_5 = 0$. This measure is 0 at the limiting points of the clustering and greater than 0 in the interval. Therefore, it must convey at least one maximum value somewhere in the interval.

4 Reduction of Feature Space

Another fundamental problem in clustering is the selection of features which are most effective in producing an optimum class or cluster separability. This also reduces the computation time while improving the quality of the classification by discarding noisy, redundant and less useful features. The Bhattacharyya Distance (B–distance) is a measure of how much each feature contributes to the cluster separability for a pair of clusters (S_1, S_2) and is given by

$$B_n(S_i, S_j) = \frac{1}{4} \ln \left\{ \frac{1}{4} \left(\frac{\sigma_i^2(n)}{\sigma_j^2(n)} + \frac{\sigma_j^2(n)}{\sigma_i^2(n)} + 2 \right) \right\} + \frac{1}{4} \left\{ \frac{(\mu_i(n) - \mu_j(n))^2}{\sigma_i^2(n) + \sigma_j^2(n)} \right\} \quad (3)$$

where n refers to the nth dimension of the feature space and σ_i and σ_j are variances of the ith and jth cluster data in dimension n, and μ_i and μ_j are their respective means.

The greater the differences in variance and mean for each pair, the larger the distances will be. The feature rejection criterion would be to retain those features which produce large B–distance measures. However, so that feature reduction can retain the best features, the input data set should be decorrelated (Hottelin Transform [Col 79]). That will rotate the features and orient them in the direction of those with higher eigenvector and, therefore, produce higher Bhattacharyya Distance. These features should then be retained and clustering performed again on the reduced feature set.

5 Application

We will now show the method applied to real MRI data. The data for this purpose is a pathological coronal T1-weighted image of the brain and is the 78th slice of a set of 124 images acquired from a 1.5 TESLA scanner, with TR=3500 and TE=5000. Seven textural measures have been selected using a 7×7 overlapping window: four from the GLRL method (first and second soment with respect to grey level, grey level non–uniformity and sum of variance) and three measures are functions from the SGLDM (contrast, correlation and inverse difference moment).

Figure 1 shows the normalised Bhattacharyya distance for each of the seven features. Both graphs show reasonably consistent behaviour of the distance measures as the number of clusters varies. Thus, selection based on this measure is a consistent procedure. The best of the rotated features (first moment with respect to grey level) was higher in Bhattacharyya distance than any other feature.

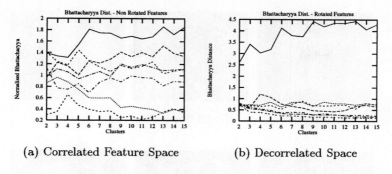

(a) Correlated Feature Space   (b) Decorrelated Space

Fig. 1. Bhattacharyya Distances - Feature Reduction

In order to identify the optimum number of clusters, the procedure is carried out with a fixed and excessive number of clusters: 15. Figure 2 plots the resulting β values against the number of clusters. Plot (a), shows 7 clusters as being the optimal value for all features. In plot (b), only the best feature has been selected and the optimum number of clusters was 5.

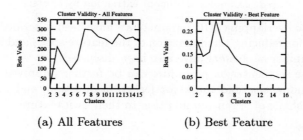

(a) All Features   (b) Best Feature

Fig. 2. Evaluation of the best number of clusters

At this point, the algorithm is again executed. It can be observed from figure 3 that the segmentation in (c) - although with misclassified regions - is more detailed than in (b). This shows that a reduction in features space has improved the classification. Also, it shows that a reduction in the number of clusters - from 7 in (b) to 5 in (c) - does not worsen the resulting classification. Note that in (d), the tumour has been partitioned in two cluster (a darker area, sorrounded by a lighter region). Finally, image (e) presents the classification for all seven features and 15 clusters. In (f), the isolated cluster corresponding to the extracted tumour is shown. Notice that the tumour is entirely extracted (with some misclassified regions though), despite the large pre–defined number of clusters.

6 Discussion and Conclusions

In this paper we have presented an approach to the extraction of tumours from MR images of the brain based on clustering and texture. The advantages of clustering and texture as features which convey important pathological information

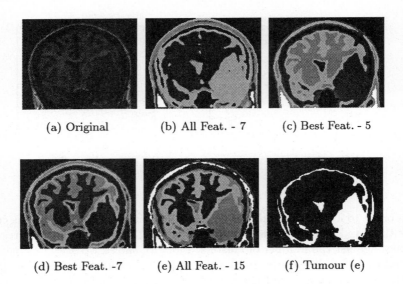

| (a) Original | (b) All Feat. - 7 | (c) Best Feat. - 5 |
| (d) Best Feat. -7 | (e) All Feat. - 15 | (f) Tumour (e) |

Fig. 3. Examples of tumour extraction by clustering

– as opposed to overlapping MRI parameters – have been described. The benefits of techniques for reducing the dimension of the feature space and estimating the appropiate number of clusters have also been discussed.

Finally, the segmentation procedure can be further improved by careful attention to window size and shape for feature acquisition, and the capacity to recognise the shape of regions in addition to their own texture.

References

[Bez 93] Bezdek, J.C.; Hall, L.O.; Clarke, L.P. - *Review of mr image segmentation tecniques using pattern recognition*. Medical Physics. 20(4):1033-1048, 1993.

[Col 79] Coleman,G.B.; Andrews,H.C. - *Image Segmentation by Clustering*. Proc. IEEE. 67(5):773-785, 1979.

[Fos 84] Foster,M.A. et al. - *Nuclear Magnetic Resonance Pulse Sequence and Discrimination of High- and Low-fat Tissues*. Magnetic Resonance Imaging. 2:187-192, 1984.

[Fuk 72] Fukunaga, K.. - *Introduction to Statistical Pattern Recognition*. Academic Press, London, 1972.

[Goo 85] Gool,L.V.; Dewaele,P; Oosterlink,A. - *Texture Analysis Anno 1983*. Computer Vision, Graphics and Image Processing. 29:336-357, 1985.

[Har 73] Haralick,R.M. at al. - *Textural Features for Image Classification*. IEEE Trans. on Systems, Man and Cybernetics. SMC-3(6):610-621, 1973.

[Loh 88] Loh,H.; Leu,J. - *The Analysis of Natural Textures Using Run Length Features*. IEEE Trans. on Industrial Electronics. 35(2):323-328, 1988.

[Tay 86] Taylor,D.G.; Bushell,M.C. - *NMR Characterisation of Healthy Rat Tissues in vivo*. Clin. Phys. Physiol. Meas. 7(1):57-62, 1986

[The 89] Therrien, C.W. - *Decision, Estimation and Classification: an introduction to pattern recognition and related topics*. John Wiley & Sons, New York, 1989.

Automatic Identification of Brain Contours in Magnetic Resonance Images of the Head

Fernando Bello[*] and Richard I Kitney

Biomedical Systems Group, Electrical Engineering Department,
Imperial College of Science, Technology and Medicine
London SW7 2BT
e-mail: f.bello@ic.ac.uk

Abstract— We present an automated procedure that can identify the contours of the brain from single-echo 3-D magnetic resonance (MR) images of the head. A first approximation to the desired contours is obtained by combining anatomical and MR imaging characteristics of the brain. A priori knowledge about the brain and its surrounding structures is then used to refine the original contours. This procedure has been successfully applied to various data sets from control patients, and a number of potential clinical applications are currently being considered.

1 Introduction

One of the present challenges in medical image processing is the segmentation of MR imaging data. Automated and semi-automated MR brain segmentation is relevant because of the large amount of data generated, and the need for reproducible brain analysis tools. A wide range of approaches has been proposed for the detection of various structures in the head [1-3].

In this paper we contribute a segmentation technique operating on single-echo 3-D MR images of the head. Being able to isolate the brain not only has the great advantage of reducing the amount of data to consider, but it is also a fundamental step to be carried out before any further classification or characterisation of brain tissues, in order to reduce computation time and complexity. Previous approaches include the use of different levels of histogram thresholding and morphological operations [5], a radial transform consisting of intensity profile analysis and the application of a series of heuristics [6], and a combination of 3D filters with watershed analysis [4]. The present approach is conceptually similar to that presented in [5], but it involves only one thresholding step and a smaller set of operations.

Our approach can be divided into three functional stages: Contrast Map generation, Contour Mask formation and Contour Mask refinement. These will be described in detail in the remainder of this paper.

---

[*] Sponsored by the Mexican National Research Council (CONACYT)

2 MR Data Sets

Our brain contour identification procedure was developed as a preliminary step for further identification of brain tissues and for clinical applications such as volumetric measurements of the brain. The data sets used for these test runs were acquired using a single echo pulse sequence on a 1.5T General Electric Signa scanner. They consist of 124 T_1-weighted coronal slices of dimension 256x256. Field of view was 240 mm minimised to encompass the head. Slice thickness was 1.5 mm and no inter-slice gap was used.

3 Contrast Map

In a MR image of the brain, the ring of high intensity seen around the boundary of the head corresponds mainly to adipose tissue (Fig. 1.(a)). By looking at horizontal profiles such as that of Fig. 1.(b), it can be seen that the adipose tissue is represented by the two largest peaks located on both sides of the profile. The width of such peaks is directly proportional to the amount of adipose tissue present. Due to the shape of the head, these peaks appear as high contrast pulse edges on a nearly semicircular fashion.

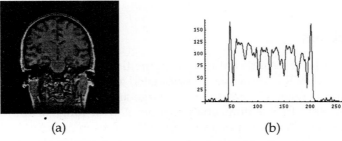

(a)   (b)

Fig. 1   (a) Brain MR mid-coronal slice and (b) horizontal profile at indicated location.

Based on the magnitude, width and location characteristics of these edges, it should be possible to apply a suitable edge detector to automatically map out these high contrast points of interest. Such an edge detector should be rotationally invariant to take into account the different orientations of the edges and have a response proportional to the edge contrast. It should also be possible to tune the detector to be sensitive to pulse edges of the desired width.

The Laplacian-of-Gaussian or LoG operator first proposed in [8] has these characteristics. It has been widely used in medical image analysis applications [3,4]. Although it tends to dislocate edges not complying with certain characteristics [9], we don't need to worry about this, as we are only interested in locating pulse-type edges of a certain width. Moreover, our concern is not the usual zero crossings detection, but analysing the actual values of the filtered image to extract the desired contrast information.

Consider the intensity profile of an ideal pulse edge of width $d = b - a$ such as that of Fig. 2.(a), and a LoG operator of size $M = 3w$, where w is the width of the excitatory region of the operator. The expected response obtained by performing the 1-D convolution of Eq. 1 and 2 is presented in Eq. 3 and graphically shown in Fig. 2.(b) for the case when $w \le d < M$. It has been shown [9] that for this case the resulting zero crossings will represent accurately the position of pulse edges.

$$\nabla^2 G(x) = \left(1 - \frac{x^2}{\sigma^2}\right) \exp\left(-\frac{x^2}{2\sigma^2}\right) \quad (1) \qquad I(x) = \begin{cases} 0 & \text{if } x > b, x < a \\ h & \text{if } a \le x \le b \end{cases} \quad (2)$$

$$C(x) = h \cdot \left. \frac{\alpha - x}{\exp\left(\dfrac{(x - \alpha)^2}{2\sigma^2}\right)} \right]_a^b = h \cdot \left(\frac{b - x}{\exp\left(\dfrac{(x - b)^2}{2\sigma^2}\right)} - \frac{a - x}{\exp\left(\dfrac{(x - a)^2}{2\sigma^2}\right)} \right) \quad (3)$$

As seen in Eq. 3, the magnitude of the peaks in Fig. 2.(b) is directly proportional to the edge contrast h. To detect pulse edges of the desired width d, we select a value for the scale constant σ large enough to guarantee only one positive peak at the mid-point of the pulse edge, but not so large as to dislocate the zero crossings from their expected positions. In other words, the magnitude of $C(x)$ at points a and b should be as close to zero as possible, and at point $d/2$ should have the largest possible magnitude. From our experiments, we know that d lies in the range of 10-15 pixels. Considering the minimum value of $d = 10$ and plotting the magnitude of $C(x)$ at the expected zero crossing locations of $x = a$ and $x = b$ (Fig. 2.(c)), and at the pulse mid-point of $x = d/2$ (Fig. 2.(d)), we see that a value of $\sigma = 3$ represents a good compromise. This value of σ is the one used in the present paper, where the separable property of the LoG operator has been exploited as in [10] to implement the LoG filtering operation with considerable computational savings.

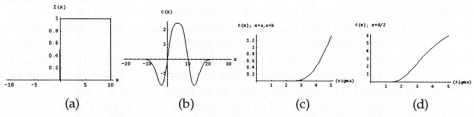

Fig. 2   (a) Edge profile with $a = 0$, $b = 10$, $h = 1$, (b) Response of a LoG operator with $w \le d$, $M > d$, and Magnitude of $C(x)$ at (c) expected zero crossing locations and (d) pulse edge mid-point.

By considering only the largest positive and negative peaks we concentrate on pulse edges having maximum contrast. Using the characteristic that pulse edges of interest will be located in a nearly semicircular fashion around the brain, and the fact that most of the brain is contained in the top half of a typical coronal slice, we can effectively generate a Contrast Map containing only those high contrast points of interest. An example of an absolute value LoG filtered image and its corresponding Contrast Map is illustrated in Fig. 3.(a) and (b).

4 Contour Mask

A Contrast Map shows locations in the original image where there are pulse edges of high contrast having a certain spatial configuration with respect to the brain. The expected positive peaks from the adipose tissue surrounding the brain almost form a closed path. It is not the same situation for the negative peaks that should indicate the skin/skull boundary (Fig. 3.(b)). Because these contours are not closed, they can not be directly used to identify the brain boundaries. Hence, two tracking/closing algorithms have been designed to obtain a binary Contour Mask that can be used to isolate the brain from the rest of the image.

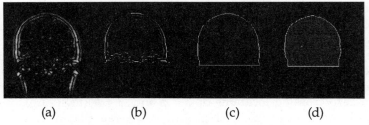

(a)	(b)	(c)	(d)

Fig. 3   (a) Absolute value LoG filtered image, (b) resulting Contrast Map, (c) Positive peaks Contour Mask and (d) negative peaks Contour Mask.

The first algorithm closes the positive and internal negative peaks paths for the top half. This procedure is repeated for the negative peaks, constraining its path to follow the same direction as the closed positive path, whenever a connection can not be made using a set of pre-defined templates.

The second algorithm approximates the brain contour for the bottom half. In this case there is no simple anatomical information to aid the tracking procedure. Therefore we make sure that no part of the brain is left out by finding its lowest point and using this as a landmark for the algorithm. An estimate for this lowest point is found by applying a peak differencing method to locate large bright to dark transitions in a number of equidistant vertical intensity profiles of the bottom half. An average location is obtained from transitions larger than a previously determined threshold, and it is used as the desired estimate. Fig. 3.(c) and (d) show the resulting Contour Masks for the positive and internal negative peaks of the Contrast Map in Fig. 3.(b).

5 Contour Mask Refinement

The brain is predominantly located in the superior portion of the head. By analysing the histogram of Fig. 4.(a) we can determine suitable thresholds for pixels located within the grey and white matter distributions. These thresholds are automatically obtained by using a parametric least squares (LS) fit of a sum of two Gaussians [11]. A bounded grey-value range limited on either side by the grey-value distance at one standard deviation from each peak is used to perform the fit. Upper and lower thresholds are calculated at two standard deviations away from each estimated mean. An example of the resulting LS curve fit is shown on Fig. 4.(b).

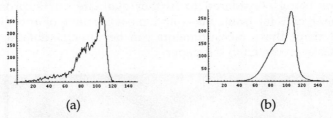

(a) (b)

Fig. 4   (a) Grey-value histogram of top half obtained using negative peaks Contour Mask and (b) LS fit of sum of two Gaussians.

This thresholding procedure results in binary threshold masks including mostly brain pixels, but also pixels from other structures. Moreover, low-intensity pixels in the brain appear as holes in the threshold masks (Fig. 5.(a)). Morphological operations that allow us to match information from the shape and location of the different structures, with knowledge about anatomy and the imaging process are used to solve these problems. Finally, the brain is isolated from the remaining structures by selecting the region with the largest area. The result of applying this procedure to the binary threshold mask of Fig. 5.(a) is shown in Fig. 5.(b). Fig. 5.(c) presents the extracted brain.

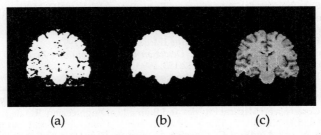

(a) (b) (c)

Fig. 5   (a) Binary threshold mask, (b) refined Contour Mask and (c) extracted brain.

6 Results and discussion

A total of 4 data sets were used for testing the identification procedure. Fig. 6 shows the final result for various slices of the different data sets.

Overall, the results of these test runs are encouraging and similar to those previously reported by other authors [4-6]. The procedure was successful in 80% of the slices. Most problems occurred in anterior and posterior slices, and in slices with large regions of visible bone marrow. Further improvements contemplated include using the third dimension by propagating the brain contours obtained from the middle slices, the use of a fixed distance to obtain the Contour Mask from the positive peaks contour in order to discard large bone marrow regions, and a region combination strategy that considers additional information to select brain regions from the labelled image

The procedure presented has been successfully used as a pre-segmentation step in a MR brain segmentation system [7]. Among the clinical applications being considered to further evaluate our approach are the measurement of total brain mass and measurement of total intracranial volume. Both of these measurements can be directly obtained from the Contour Masks presented in this paper.

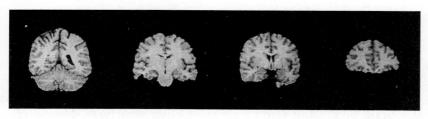

Fig. 6   Final contours of the brain obtained for various slices of the different data sets.

References

[1]   **Kapouleas I**, *Segmentation and Feature Extraction for Magnetic Resonance Brain Image Analysis*, Proc. of 10th Intl. Conf. on Pattern Recognition, Atlantic City, 1:583, 1990.

[2]   **Raya S**, *Low-level Segmentation of 3-D Magnetic Resonance Brain Images—A Rule–Based System*, IEEE Transactions on Medical Imaging 9:3, 1990.

[3]   **Ehricke H**, *Problems and Approaches for Tissue Segmentation in 3D–MR Imaging*, Proc. SPIE Med. Imaging IV, 1233:128, 1990.

[4]   **Katz W**, *et al*, *Segmentation of the Brain from 3–D Magnetic Resonance Images of the Head*, Proc. of 14th Annual Inter. Conf. of the IEEE EMBS, Paris, 1992.

[5]   **Brummer M E** *et al*, *Automatic Detection of Brain Contours in MRI Data Sets*, IEEE Transactions on Medical Imaging 12:153, 1993.

[6]   **Zijdenbos A P** *et al*, *Automatic Extraction of the Intracranial Cavity on Transvers MR Brain Images*, Proc. 11th Intl. Conf. on Pattern Recognition, The Hague, 3:430, 1992.

[7]   **Bello F** and **Kitney R I**, *The Use of Regional Features to Segment Brain MRI*, Proc. IFMBE World Congress on Medical Physics and Biomedical Engineering, Rio de Janeiro, 1994.

[8]   **Marr D** and **Hildreth E C**, *Theory of Edge Detection*, Proc. Royal Soc. London 207:187, 1980.

[9]   **Huertas A** and **Medioni G**, *Detection of Intensity Changes with Subpixel Accuracy Using Laplacian—Gaussian Masks*, IEEE Transactions on Pattern Analysis and Machine Intelligence 8:651, 1986.

[10]   **Sotak G E** and **Boyer K L**, *The Laplacian -of-Gaussian Kernel: A Formal Analysis and Design Procedure for Fast, Accurate Convolution and Full Frame Output*, Comput. Vision, Graphics and Image Processing 48:147, 1989.

[11]   **Press W H** *et al*, *Numerical Recipes in C*, 2nd Ed., Cambridge University Press, Cambridge UK, 1991.

Analysis of the Spatial Arrangement of Cells in the Proliferative Breast Lesions

Vincenzo Della Mea, Carlo Alberto Beltrami

Department of Pathology, University of Udine
33100 Udine - Italy

Abstract. A particularly difficult problem in the field of histopathology is the diagnosis among epitheliosis, atypical hyperplasia and carcinoma in situ.
This work is an attempt to reproduce the reasoning made by the pathologist when he/she analyzes the spatial arrangement of the cells in the duct, that is different depending on the grade of malignancy of the lesion. For this aim a method for the representation of the duct is needed, in order to identify the structures involved in the diagnosis.
A representation based on perceptual graphs is derived from an initial set of nodes that represents the cell nuclei of the lesion. On this representation the evaluation and quantification of structural features may be carried out.
A prototype was developed to evaluate the clinical significance of the approach. In a preliminary phase the statistical analysis of some features revealed significant differences among the three pathologies, encouraging the progression of the work.

1 Introduction

A difficult problem in the field of histopathology is the classification of proliferative breast lesions, i.e., the diagnosis among epitheliosis, atypical hyperplasia and carcinoma in situ. These three categories were first described by the College of American Pathologists [2] and by Page [9,11] and each of them was considered to have a different relative risk for the development of infiltrating breast cancer. In his work, Page presented a set of rules for the discrimination among the pathologies. These rules are divided in three different subsets, involving respectively cytologic features, architectural characteristics and anatomic extent of the lesion. The rules of the first subset are based on cytologic findings such as cytoplasmic aspect, nuclear shape, chromasia; these features have been investigated but the results are not satisfactory.

This paper focuses the attention on the rules belonging to the second subset, i.e., those capturing the way the cells occupy the space differently depending on the grade of malignancy of the lesion. The use of these rules implies a form of spatial reasoning on the lesion, with the aim of recognizing and quantifying spatial properties of the whole lesion or regions of it.
A modelization of the problem involving the use of perceptual graphs is proposed; more specifically, a hierarchy of graphs is used to represent the lesion.
The study of the structural characteristics of the lesions may offer some advantages: first of all, it guarantees the quantification of features hard to evaluate in an objective manner; then it offers the possibility to develop a system that avoids the classical problems related to morphometry and densitometry, and finally it gives the

opportunity to understand thoroughly the biological significance of proliferative lesions of the breast.

A system that implements some of the concepts presented here has been developed and tested for clinical significance, in order to build a diagnostic decision support system based on the structural features arising from an histologic image of a pathologic duct.

2 The Model of the Duct

Usually the pathologist comes to a diagnosis by viewing at microscope hematoxylin-eosin stained specimens of mammary tissue. The ducts referred by the diagnostic rules of Page appear as in figure 1a: the nuclei are blue, while cytoplasm is pink-red.

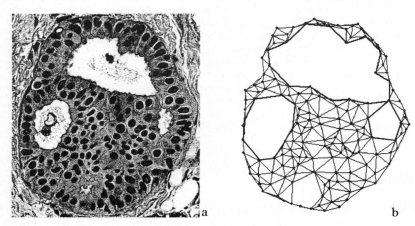

Fig. 1. a) mammary duct, 40x original magnification; b) graph representation

Our technique starts from an histologic image representing the transverse section of a single duct; in fact, the aforementioned rules are referred to a similar image.

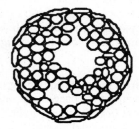

Fig. 2. Ducts

A normal duct is a "pipe" with the wall made by a double layer of cells; the internal layer (epithelial cells) is involved in the lesions (figure 2). The proliferative pathology causes a gradual filling of the internal cavity of the duct, and produces during the intermediate steps some structures, the so-called lumina and bridges; in addition, the filled ("solid") areas show a variable nuclear density. The completeness of the filling is not relevant for the diagnosis; the fundamental character is the way the proliferating cells occupy the internal space, generating some structures instead of others, or the same structures with different characteristics. In fact, benign cells behave similarly to the starting cells, while the malignant cells usually derive from clonal cells different in shape, dimension and behaviour from the epithelial cells.

The work presented in this paper is an attempt to reproduce the reasoning made by the pathologist when he/she analyzes the spatial arrangement of the nuclei in the duct, as stressed by the approach of Page. For this aim, it is necessary to represent the duct in a structured way so the architectural findings that characterize the different pathologies can be recognized, evaluated, measured and used for the decision support to the diagnosis.

3 The Structure Representation

The primary entities composing the duct are the cells, but the boundaries between them are very difficult to recognize. The cell nuclei are indeed clearly defined, although sometimes they are overlapping, so it is vantageous to use the nuclei as distinctive element for the cell identification. The nucleus has some properties traditionally analyzed with the tools of morphometry and densitometry, as cytoplasmic aspect, nuclear shape, chromasia, but these features - alone - have been demonstrated to be insufficient for a discrimination of the pathologies that have been taken into account.

A relatively new morphometric approach to the study of neoplastic epithelial lesions involves the use of graphs and was described in 1987-1992 by Kayser for lung carcinoma and soft tissue tumours [5-7], in 1992 by van Diest and Co. for breast cancer [12], in 1992 by Meijer and Co. for colorectal adenomatous polyps [8], in 1993 by Raymond and Co. for germinal center of lymph nodes [10], in 1993 by Darro and Co. for colorectal cancer cell lines [3].

These works use mainly the Voronoi tessellation approach, that seems to be suitable to describe space-filling mosaic-like structures resulting from growth processes [4]; also the minimum spanning tree is applied. Both the approaches seem not to be adequate to our need of structuration in the knowledge representation.

The basic idea of our approach is to represent the lesion by means of the neighbourhood relationship between nuclei as primary element that allows the recognition of the spatial properties at a more abstract level. A hierarchical representation based on three main graphs (neighbourhood, planar and dual graphs) is derived from an initial set of nodes that represents the cell nuclei of the lesion, as in the other works [3,5-8,10,12].

A graph G is a mathematical structure defined by a couple of sets N,A:
$$G = <N,A>, \; N \supset N, \; A = \{\{n_1,n_2\} \mid n_1 \in N, n_2 \in N\}$$
where N is the node set and A is the arc set; if there exists an arc $a=\{n_1,n_2\} \in A$, we say that n_1,n_2 are connected by a. By means of a semantic function v:
$$v: N \rightarrow D, \; \forall \, n \in N \; v(n) = d \;, \text{ where D is the information domain,}$$
some information can be associated to each node (in our case, the co-ordinates of the nucleus in a local system and eventually some cytological and morphological features of the cell). In the same way it is possible to associate information to the arcs.

We shall consider the neighbourhood graph $G=<N,A>$ ($N \supset N$) with threshold S defined in the following way:

i) N is the set of natural numbers;

ii) a semantic function v is associated with the node set N;

iii) for each $n \in N$, $v(n)$ represents the two coordinates of n, i.e. $<x,y>$. For convenience, we use two auxiliary functions $x(n)$ and $y(n)$ defined by: $x(n)= x$, $y(n) = y$ $\forall n \in N$, $v(n)=<x,y>$.

iv) $dist(n_1,n_2)$ $\forall <n_1,n_2> \in NxN$ is the usual Ecuclidean distance.

v) the set A consists of all arcs $a=\{n_1,n_2\}$ such that $dist(n_1,n_2) \leq S$.

vi) The length of an arc is defined on A by: $\forall a=\{n_1,n_2\} \in A$, $length(a)=dist(n_1,n_2)$.

The concept of neighbourhood, as expressed by this graph, is redundant because of the great number of intersecting arcs. Having a planar graph is fundamental for our approach because the dual graph, that represents the areas comprised within the arcs, can be generated from it.

The planar neighbourhood graph $P^*=<N^*,A^*>$ may be constructed from G as follows:

i) N^* coincides with N;

ii) A^* is recursively defined by:

$$A_0 = \{a\} \in M_0 = \{ m \in A \mid length(m) = \min_{x \in A} length(x)\}$$

$$A_k = A_{k-1} \cup \{a\},$$
$$a \in M_{k-1} = [\{ m \in A \backslash A_{k-1} \mid length(m) = \min_{x \in A} length(x)\} \cap$$
$$\{m \in A \backslash A_{k-1} \mid \forall m' \in A_{k-1}, m \text{ does not intersect}(m,m')\}]$$

$$A^* = A_k, k = \min_{k \in N} k : A_k = A_{k+1}$$

In this way, the complexity of the representation decreases, and from the planar graph another step can be made to obtain the dual graph. In the dual graph each node represents an area (called "face") comprised among arcs of the dual graph, and remains associated with the arcs of the planar graph that form the perimeter of the face. The last two graphs provide a core for the structural analysis of the lesion, adopting techniques as previously described elsewhere [13,14]. Each graph may be partitioned into subgraphs representing some structures of the lesion, or using criteria related to cytologic uniformity [7].

The main subgraphs represent respectively the lumina, the solid area and the external perimeter of the duct. The lumina are defined as faces of the dual graph with a relatively high number of sides and such that the corresponding region in the starting image contains a white area; the perimeter of the duct is the external face of the dual graph; the solid area comprehends the remaining faces.

These subgraphs may be analyzed in order to quantify the features described by Page or eventually discover other criteria; for example, the subgraph representing the solid area may give useful data related to the dimensions and regularity of the cells analyzing the distribution of the arc lengths or of the face areas, while the roundness and cribriformity of the lumina may be measured on the lumina subgraph by using some adequate shape factors.

This structural representation may be included in a decision support system able to generate a diagnostic suggestion starting directly from an histologic image. In such a system, it is necessary to study four modules: a perception module that realizes the segmentation tasks, a module for the generation and evaluation of the structural representation, a module for the management of the diagnostic knowledge (learning and use), and a man/machine interface.

4 Results

A prototype of the system for the structural analysis of lesions has been developed (SANE, Structural ANalysis Environment) [1]. At this stage of development, the prototype comprehends the perception module, the representation module capable to recognize only some structural features and a simple decision-support module based on decision trees. The algorithms for the derivation of the planar and dual graphs have been studied and tested.

We focused mainly on structure representation, so the perception module is actually implemented with a simple segmentation technique using filtering, thresholds and mathematical morphology procedures. For each region recognised as a nucleus, center of gravity, area, major axis and shape factor are computed and saved as attributes for the corresponding node.

For the preliminary analysis of the system, we studied images obtained from 70 cases. Each image represents a breast duct from hematoxylin-eosin stained specimens, acquired at 40x magnification by a videocamera connected to a motorized microscope, and processed by the system. Each image measured from 512x512 to 1536x1536 pixels, 24 bit depth. Only a limited subset of graph features was taken into account.

As an example, the planar/dual graph for the duct of figure 1a is shown in figure 1b.

To obtain a first evaluation of the clinical significance of this approach, the statistical analysis of the differences in some features of the ducts was carried out. Among those considered, we mention mainly the statistical characterization of solid area, obtained by using mean, standard deviation, minimum and maximum of node degree and arc lengths on neighbourhood and planar graphs, and of perimeters and areas for dual graph. The analysis of these features revealed some significant differences about cell dimensions and variability among the three diagnostic classes (table 1), validating our approach to the structure representation of the lesions as previously shown in [1]. In addition, the significant features were submitted to an automatic generator of decision trees; the results of this are yet to be evaluated.

feature	description	EPI	ATHYP	CIS	Tukey-Kramer HSD p value
GAMAX	Neighbourhood graph: max node degree	29.83±4.2	22.21±4.2	15.29±4.1	0.001
GAMED	Neighbourhood graph: mean node degree	16.48±2.8	12.56±1.8	7.90±2.4	0.001
GASD	Neighbourhood graph: SD node degree	5.252±1.0	3.874±0.8	2.851±0.7	0.01
PLMED	Planar graph: mean arcs length	40.00±3.2	44.43±2.9	51.79±4.7	0.05
PERSOL	Solid dual graph: mean face perimeter	119.3±9.4	132.7±9.0	154.5±14	0.05

Table 1. Significant features

5 Conclusions

The preliminary results obtained by the prototype system are encouraging. The possibilities of this technique are to be explored more deeply, with particular attention to the features not yet considered. Moreover, the knowledge based module that realizes the decision support to the diagnosis should be improved with the management of uncertainty.

This approach seems also to be applicable to other similar classification problems in the field of histopathology.

References

1. C.A. Beltrami, V. Della Mea, N. Finato: Structure Analysis of Breast Lesions using Neighbourhood Graphs. Anal Quant Cytol Histol, 17, 143-148 (1995)
2. College of American Pathologists, Consensus Meeting: Is "fibrocystic disease" of the breast precancerous? Arch Pathol Lab Med 110,171-173 (1986)
3. F. Darro, A. Krukzynski, C. Etievant, J. Martinez, J.L. Pasteels, R. Kiss: Characterization of the differentiation of human colorectal cancer cell lines by means of Voronoi diagrams. Cytometry 14, 783-792 (1993)
4. U. Hahn, U. Lorz: Stereological analysis of the spatial Poisson-Voronoi tesselation. J Microsc 175, 176-185 (1994)
5. K. Kayser, H. Stute: Minimum spanning tree, Voronoi's tesselation and Johnson-Mehl diagrams in human lung carcinoma. Path Res Pract 185, 729-734 (1989)
6. K. Kayser, K. Sandau, G. Bohm, K.D. Kunze, J. Paul: Analysis of soft tissue tumors by an attributed minimum spanning tree. Anal Quant Cytol Histol 13, 329-334 (1991)
7. K. Kayser, K. Sandau, J. Paul, G. Weisse: An approach based on two-dimensional graph theory for structural cluster detection and its histopathological application. J Microsc 165, 281-288 (1992)
8. G.A. Meijer, P.J. Van Diest, J.C. Fleege, J.A. Baak: Syntactic structure analysis of the arrangement of nuclei in dysplastic epithelium of colorectal adenomatous polyps. Anal Quant Cytol Histol 14, 491-498 (1992)
9. D.L. Page: Cancer risk assessment in benign breast biopsies. Hum Pathol 17, 871-874 (1986)
10. E. Raymond, M. Raphael, M. Grimaud, L. Vincent, J.L. Binet, F. Meyer: Germinal center analysis with the tools of mathematical morphology on graphs. Cytometry 14, 848-861 (1993)
11. S.J. Schnitt, J.L. Connolly, F.A. Tavassoli, R.E. Fechner, R.L. Kempson, R. Gelman, D.L. Page: Interobserver reproducibility in the diagnosis of ductal proliferative breast lesions using standardized criteria. Am J Surg Pathol 16, 1133-1143 (1992)
12. P.J. Van Diest, J.C. Fleege, J.A. Baak: Syntactic structure analysis in invasive breast cancer: analysis of reproducibility, biologic background, and prognostic value. Hum Pathol 23, 876-883 (1992)
13. L. Vincent: Graphs and Mathematical Morphology. Signal Processing 16, 365-388 (1989)
14. C.T. Zahn: Graph-Theoretical Methods for Detecting and Describing Gestalt Clusters. IEEE Trans. on Comp. c-20/1, 68-86 (1971)

Application of Image Processing in Neurobiology: Detection of Low Signals with High Spatial Resolution and a Non-Uniform Variance

Yasmina CHITTI

Unité de Neurocybernétique Cellulaire
280, Bd Ste Marguerite
13009 Marseille (FRANCE)
Tel: (33) 91 75 02 00 Fax: (33) 91 26 20 38
email: chitti@cptsu2.univ-mrs.fr

Abstract. Optical imaging of neuronal activity is a recent method for studying information processing by neuronal networks. It is known that neuronal activity was generated by the activity of membrane proteins, called "ion channels".

To see the spatial action of channels on a single neuron, we use an imaging system combining a CCD camera mounted on an inverted microscope and fluorescence voltage sensitive probe and taking images of fluorescence intensity with high spatial resolution and low noise.

To visualize the difference between an excited and a quiet state of a neuron we must compute the relative variation in fluorescence between both images representing both states. We obtained a very noisy resulting image of low signals. We have shown that its variance is non uniform and in inverse ratio to the square of raw data intensity and that the highest variances are located on the non-biological background.

We have developed a nonlinear filtering based on local image segmentation. The algorithm filters all the noise in the non-biological background and outlines the shape of the cell body. Tests of significance between groups of intensities have shown that the response of the neuron is patchy and biological experiments show that the patchyness is bound to electrical activities of the neuron.

1 Introduction

Several methods allows neurobiologists to study neurons. The most often used consists in introducing a microelectrode into a cell body of a single neuron. The electrode stimulates the neuron with an electrical impulse. The same microelectrode records the electrical response of the neuron, called action potential. It is known that the action potential was generated by the activity of membrane proteins, called "ion channels" but a single microelectrode can only measure the spatial sum of the potential fields created from each active channel of the whole membrane [5].

Optical recording with photodiode devices and voltage sensitive dyes have been previously used to detect changes in transmembrane voltage at high temporal resolution [4]. Bath-applied dye binds to the external membrane and acts as a molecular transducer that transforms changes in membrane potential into optical signals. It has been shown that the fluorescence of the stained neuron correlates linearly with its electrical activity. Taking advantage of both the linearity and the microsecond response time of the dye, the fluorescence changes in the neuronal membrane during recorded individual action potentials can be imaged. During the exposure time, the rapid transmembrane voltage changes produce local changes in the emitted fluorescence which are stored in the image. This image is the spatial counterpart of the time information conveyed by a microelectrode recording.

The mean diameter of a single cultured neuron is about $30\,\mu$m. The sources of action potential are about $1\,\mu$m large and the emitted signal is very low. Neurobiologists and physicists have cooperated to work out a high spatial resolution and low noise imaging system and an adequate image processing.

2 Optical Recording System

In order to check the response of the stained neuron, a microelectrode is introduced in its membrane. The electrode stimulates the neuron and records the action potential. The emission spectra of the probe linearly shifts with changes in membrane potential. A linear optical filter transforms the shift into changes in fluorescence intensity. Images of fluorescence are taken with a liquid nitrogen cooled charge-coupled device (CCD) camera mounted on an inverted microscope. The imaging system is a low noise (< 10 electrons r.m.s) system and it allows us to take 189×70 pixels images with a $1\,\mu$m^2 spatial resolution. Figure 1,A shows an image of the fluorescent neuron obtained with a such imaging system.

We want to compare two excitation states of a neuron: at rest -without stimulation- and during an action potential -after a stimulation. The changes in potential between this two excitation states is around $100\,$mV and correspond to a shift of the emission spectra of the probe around five percent of its initial value and a variation of fluorescence intensity around three percents: the variation does not allows us to see any difference between two images corresponding to the both states of excitation of the neuron.

3 Image Processing

In order to visualize the difference between the both studied states and to see the spatial distribution of active channels, it is necessary to compute the relative variation in fluorescence between the both images. Experimental noises like bleaching (the intensity of fluorescence decreases with the lighting of the field), cellular movements (images must be quickly taken)... have been estimated and removed. However, the result of the computation is very noisy as can be seen in Fig. 1,B. When the neuron is not stimulated the variation of fluorescence intensity of the cell is around 0% but the non-biological background is very noisy. When the neuron is stimulated we can see the edge of the cell and observe that the noise is higher on the background than on the cell.

As the noise provided by CCD is a Poisson-noise, the variance of every pixel intensity of the image of the fluorescent neuron is proportional to the pixel intensity and has a different value in each (x, y) pixel [1]. The study of the variance resulting of the relative variation computation shows that the variance is non uniform:

$$V\left(x, y\right) = V\left(\frac{f(x, y) - g(x, y)}{f(x, y)}\right) = \frac{g^2(x, y)}{f^4(x, y)} * \sigma_1^2(x, y) + \frac{1}{f^2(x, y)} * \sigma_2^2(x, y)$$

where

1. $f(x, y)$ is the spatial intensity of the (x, y) pixel of a control image of the neuron at rest.
2. $g(x, y)$ is the spatial intensity of the (x, y) pixel of a test image where the neuron was either stimulated or at rest.
3. $\sigma_1^2(x, y) = c * f(x, y)$ and $\sigma_2^2(x, y) = c * g(x, y)$ are their variance.

The computation give us a chart of the spatial variance (showed in Fig. 1,C):

$$V\left(x, y\right) = c * \frac{g(x, y)}{f^2(x, y)} * \left[1 + \frac{g(x, y)}{f(x, y)}\right]$$

The variance of the relative variation is in inverse ratio to the raw data intensity. In fact, as spots of high intensity of fluorescence in original images are located on biological membrane, the variance of the result image of relative variation is higher on the non-biological background than on biological membranes. The intensities of two pixels next to each other of the non-biological background of the raw image of fluorescence are similar but the intensities of these both pixels in the relative variation image really differ. Only a nonlinear processing based on local features will be able to filter this image.

The developed algorithm is a nonlinear filter based on local image segmentation[3]. We compute the variation between the intensity of each pixel and the mean intensity of the 4-neighbouring pixels. If the variation is greater than a local threshold which is proportionnal to the intensity of the centered pixel, we removed the centered pixel else the pixel is not removed. As Fig. 1,D shows, the algorithm filters all the noise in the non-biological background and outlines the shape of the cell body.

To localize activity areas, the result was filtered with a 3×3 median filter as Fig. 1,E presents. After the filtering by dynamic thresholding segmentation and by a median filter, active areas appear on the cell when the neuron was excited and do not appear when the neuron is at rest.

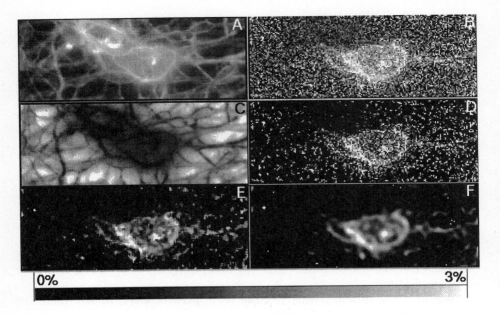

Fig. 1. Detection of active channels on a neuron membrane.
(A) Image in fluorescence after incorporation of the dye (RH237). (B) Relative variation of fluorescence (dF/F) between an image of the neuron at rest and an image of the excited neuron. (C) Map of local variances of the C image. (D) Filtering of image B by local segmentation removing the noise of the non-biological part of the image. (E) Filtering of the image D by a 3×3 median filter removing the spot of noise. (F) Detection of active channels by using the wavelet transform.(Scale used for D–F images.)

4   Results

The significance of the result was tested by several ways.

1. The first test consists to compare the mean intensity of two areas which have almost the same intensities. If two samples, A and B, are extracted, the result of the test give us the confidence of $p\,\%$ with which $mean(A)$ and $mean(B)$ differ.

   Generally, the intensity after filtering range from 0 % to 2 % of variation and four samples of 0.5 % large can be distinguished.

2. An experimental model which contains all experimental evaluated noises was evolved, filtered and tested with the previous test of significance. The same significance is obtained for the both experimental cases. These results show that the patchy response of the neuron is not due to the non uniform spatial distribution of fluorescence but due to differences in the emission wavelength of the probe and so to differences in local potential.

3. To increase the spatial resolution, images of relative variation will be filtered in Wavelets space with the *à trous* algorithm [6] by taking in account the spatial distribution of the variance. The evaluation of the intensity, the location and the size of each active areas on the membrane will be better evaluated.

Many biological experiments prove that the detected active channels are bound to the electrical activity of the membrane of the neuron. We have developped a new method to see the electrical excitation of a cultured neuron which add a new dimension to electrical reccordings of neurons [2].

References

1. F.J. Anscombe: The Transformation of Poisson, binomial and negative-binomial data. Biometrika 15,246–254 (1948)
2. P. Gogan, I. Schmiedel-Jakob, Y. Chitti and S. Tyč-Dumont: Fluorescence Imaging of Local Electric Fields During Excitation of Single Neurons in Culture. Biophys. J. (accepted)
3. R.C. Gonzalez and P. Wintz: Digital Image Processing. Addison Wesley 1987.
4. A Grinvald, R. Hildesheim, I.C. Farber, and L. Anglister: Improved Fluorescent Probes for the Measurement of Rapid Changes in Membrane Potential. Biophys. J. 39,301–308 (1982)
5. C. Hammond and D. Tritsch: Neurobiologie Cellulaire I. Doin 1990.
6. M. Holdschneider, R. Kroland-Martinet, J. Morlet, and P. Tchamitchian: A realtime Algorithm for Signal Analysis with the help of the Wavelet transform. In J.M. Combes et al.: Wavelets. Berlin: Springer 1989, pp 286–297

Computer assisted analysis of echocardiographic image sequences

Andrea Giachetti[1], Marco Cappello[2], Guido Gigli[3] and Vincent Torre[1]

[1] Dipartimento di Fisica Università di Genova
[2] Consorzio I.N.F.M. - Universita' di Genova
Via Dodecaneso 33, 16146 Genova
Tel: 39-10-3536311 Fax: 39-10-314218
[3] Servizio di Cardiologia, Ospedale di Rapallo, Piazza Molfino 10, 16035 Rapallo(GE)
Tel: 39-185-683231 Fax: 39-185-683277
E-mail: giachetti@genova.infn.it

Abstract. In this paper we present a semi-automatic system for the analysis of echocardiographic image sequences, able to provide useful information to cardiologists. The proposed approach combines well known techniques for the detection of left ventricular boundaries with the computation of optical flow. The initial detection of the cavity contour is based on an improved balloon model, the computation of optical flow is performed with a correlation technique and the contour tracking is obtained combining motion information provided by the optical flow with a snake-based regularization. The system is able to follow precisely the cavities motion, to provide several quantitative features of the heart beat and a dynamic representation of systolic and diastolic motion. Preliminary experimental results are presented and commented with particular attention to their clinical relevance.

Introduction

Echocardiography is the commonest technique for cardiac imaging and provides a large quantity of clinical information on cardiac structures and functions in an incruent manner. The evaluation of the left ventricle is one of the most valuable clinical application of echocardiography [1]; its volume and its ejection fraction have an extraordinary diagnostic and prognostic value [2]. In ischemic heart disease, echocardiography is useful in the diagnosis of myocardial infarction and transient ischemic episodes [3]. The analysis of echocardiographic image sequences is usually performed by visual inspection. However, some systems are able to produce image processing helpful to enhance the definition of the heart cavities and to provide some other useful parameters [4]. Computer Vision tools have been used in pilot studies to perform heart cavity detection: interpolations of calculated edges [5, 9], snakes [7, 9] and segmentation techniques based on Markov Random Fields [10] have already been exploited, while optical flow has been used to study heart motion [12, 13]. In this paper we describe some preliminary results of a system for the automatic analysis of echocardiographic images, based on the use of snakes [6, 7] and the computation of optical flow [11]. This system detects the left ventricular contour on final echocardiographic images (i.e. without any intervention in the processing cascade of ultrasonic signal), computes its area, tracks it and analyses its motion.

1   The detection of ventricular cavities

The analysis of the image sequence starts with the detection of the left ventricular cavity in the first image. This function is performed by an active contour model, inspired by the recent work on snakes [6] and balloons [7]. The contour is represented by a closed curve, parametrized by the arc length s: $\mathbf{p}(s) = (x(s), y(s))$. In our model, similar to the balloon model of Cohen & Cohen [7], the initial shape of the contour can be initialized by hand near the true boundaries or just giving the center of the cavity (see Section 4); in this case the initial contour is taken as a small ellipse centered in the given point. The curve undergoes an evolution driven by an internal force $\mathbf{F}_{int} = \alpha \mathbf{p}_{ss}$ tending to minimize the contour length, an inflating force $\mathbf{F}_g = f_g \gamma$ (where $\gamma(s)$ is the unit normal vector of the curve) tending to enlarge it, and a deflating force that stops the contour expansion near the cardiac boundaries. This force has been introduced because, due to noisiness and low contrast of echocardiographic images, the use of the usual edge force $\mathbf{F}_{edge}(\mathbf{p}) = -\nabla |\nabla E(\mathbf{p})|$. (where $E(\mathbf{p})$ is the grey level at the point $\mathbf{p}$) was not convenient. The new force is defined as follows:

$$\mathbf{F}_d = -f_d \left(1 - \left(exp\left(\frac{E(\mathbf{p}) - k}{T} \right) - 1 \right)^{-1} \right) \gamma \qquad (1)$$

This is a smooth step function where T controls the smoothing, k is the center of the small region where its value changes from zero to f_d. When the grey level is low (inside the cavity), the deflating force is approximately zero. When the grey level is above the threshold k, the force quickly reaches the maximum value f_d and if the bright region is sufficiently large, the contour is stopped at its border.

The implementation of the active contour is made substituting the curve $\mathbf{p}(s)$ with a closed chain $\mathbf{p}(i)$ of N points, replacing the derivatives of the internal force with finite differences and approximating the other forces with vectors applied to the points. Finally the points are simultaneously shifted of the quantity obtained giving unitary masses to the points and computing their approximated dynamics for a unitary time step. The procedure is then iterated until the growth of the contour is stopped.

2   Cardiac walls motion and contour tracking

The analysis of the motion of the ventricular contour can be useful to build a model of the complex motion of the heart and to detect the presence of ischemic regions from irregular motions during the contraction. Given the chain representing the initial ventricular boundaries, we simply compute the *optical flow* at each point of the chain. The procedure used for our system is the following: after a simple smoothing of the images with a gaussian filter $exp(-(x^2+y^2)/2\sigma^2)$ with a value of 1.5 pixels for σ, the point (x', y') at time $t+1$ corresponding to point (x, y) at time t is the one with the lowest value of the quantity:

$$\sum_{i=-n}^{n} \sum_{j=-n}^{n} (E(i + x', j + y', t + 1) - E(i + x, j + y, t))^2 \qquad (2)$$

where $E(x, y, t)$ is the grey level of pixel (x, y) of the frame labelled by the integer variable t. Usually the value of n was set equal to 15. The displacements computed at the chain points can be used to compute the new contour for the next image, by simply shifting the points according to their values. This procedure, however, is not accurate enough to produce a reliable detection of the new ventricular contour: the existence of erroneous displacements and the presence of the quantization noise can generate some problems. If the wrong matches are not eliminated, after a few frames the contour becomes irregular, with many points far from the cardiac walls. To overcome this problem, after shifting the contour points with the computed displacements, we applied a regularization procedure consisting in a few iterations of the forces described in Section 1. In this way, there is a fast convergence towards the true cardiac walls. To avoid that during the regularization even the points where the computed optical flow is correct undergo large displacements, the masses of the point are no longer unitary, but are taken inversely proportional to an appropriate confidence measure of the flow.

2.1 Area evolution and ejection fraction

A useful parameter for the analysis of heart functionality is the area S of the computed ventricular section. For each computed contour $p(i)$, an application of Gauss-Green's theorem gives the following formula for the area:

$$S = \frac{1}{2} \sum_{1}^{N} \mathbf{p}(i) \cdot \boldsymbol{\gamma}(i) |\mathbf{p}(i) - \mathbf{p}(i-1)| \tag{3}$$

The area of the cavity section can be used also to estimate ventricular volumes and important functional parameters such as the ejection fraction [2].

3 Experimental results

The proposed approach was tested on several image sequences of healthy and anomalously beating hearts acquired from a video used for cardiologists training. The aim of this experimentation was to establish whether the proposed approach was able to track the left ventricle reliably and to evaluate whether the description of cavity motion with the optical flow was useful. The performance of the system depended rather critically, as it might be expected, on the numerical values of the different parameters used. Contours chains were usually formed by 100 points. In order to have a strong but short range deflating force and an equilibrium position near the true boundaries of the cavity the values of the parameters T and f_r were both set equal to 1. The range of the inflating and internal forces was rather short, approximately of the order of 0.1 pixels and their ratio was tuned for each sequence to have the better results. The choice of the threshold k was very important and had to be differently made for each sequence; furthermore k should not be constant for all the points, because in the echo images the gray level corresponding to the cardiac membrane depends on the angle of incidence of the ultrasound and on the gain factors. This means that the gray value corresponding to the membrane is different for different contour points and can change with time. This do not affect too much the optical flow algorithm, but can create more problems for the image force driving the snake. To overcome this problem we introduced local values of the threshold k. If the curve was initialized by hand, the thresholds were automatically chosen with

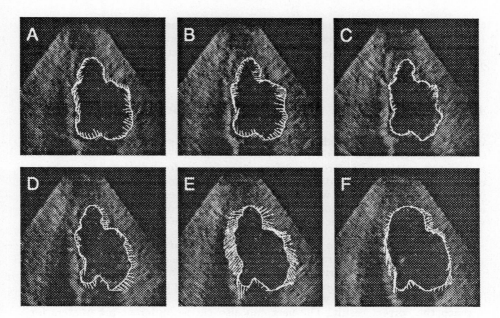

Fig. 1. A, B, C: Contour detection, tracking and optical flow computation for three consecutive frames of systolic motion of the left ventricle. D,E,F: The same for diastolic motion. Flow vectors are magnificated by a factor 2.

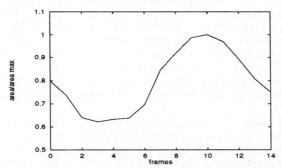

Fig. 2. A: The evolution of the area during the whole heart beat of Fig.1.

values close to the average gray level of the image outside the curve near each point. This adaptive choice was then repeated at each frame, so that the system was able to tolerate changes in the brightness of the cardiac boundaries. If the contour was initializated giving the center of the cavity, we left the possibility to choose four different thresholds for the upper, the left, the right and the lower part of the contour for the first contour detection.

The computation of optical flow of points lying on the cavity walls provided correct vector displacements for most of the points. Displacements were rather large because of the low acquisition rate (25 frame/s.). Fig. 1 shows the initial contour detection (A) and during a whole beat, where the computed boundaries of the ventricular cavity are superimposed to the corresponding images together with the optical flow vectors. The tracking of the contour was obtained shifting the chain points with the described technique. After an entire beat (in this case 14 frames), the contour was approximately in the original position.

The computed area of the cross section had a correct periodic behaviour (Fig. 2).

4   Interaction with medical doctors: the graphic tool

Once we had a way to track the contour and compute its motion we developed some elaborations and a user-friendly tool for an interactive data analysis. The aim of this work was to provide information for medical doctors, so we had to consider cardiologists' needs to build an interactive tool able to realize an easy initialization of the model with a correct tuning of the parameters, and to show as final results all the elaborations of the computed contours and of the motion vectors that can help medical diagnosis. These are the main features of the tool:

- **Contour initialization and tuning of the parameters** The user can either choose the initialization by hand, starting the contour by clicking with the mouse near the ventricle contour or the central initialization, just clicking with the mouse inside the ventricle and then tuning the parameters (number of contour points, force values ...) with the cursors on a dedicated panel to reach the correct result. The user can also choose different parameters for the contour following, turn on and off the optical flow computation, change the mask size and the search space for the correlation and so on. When the analysis is started, the graphic display shows the evolution of the contour and the flow vectors superimposed to the correspondent frames.
- **Contour envelope display** The envelope of the contour during a beat is traced during the evolution to show the entity of the motion of the ventricular boundaries with a colour code indicating the absolute value of the speed. When a region of the border is moving fast, the contour is traced with a bright colour (yellow, red), while when its motion is slow, the contour is traced with a dark colour. If a region of the contour where the quality of the image is sufficiently good is an ischemic one, this plot can evidentiate it (Fig. 3C).
- **Area/Volume display** The computed area of the ventricular cross-section is plotted in a dedicated window. The value of the area can be replaced by the volume estimated with an appropriate approximation. Values can also represent the true dimension of the cavity by setting the pixel/cm. ratio clicking on the reference points in the first image. The estimated ejection fraction is also given at the end of the analysis.
- **Anomalous motion** Regions where the motion of the contour is not regular are detected by studying the correlation between the computed evolution of the area and the motion of the contour points. Different values of this measure are represented with different colours on the representation of the contour envelope. If the computed flow is sufficiently reliable, this graphic can be used to detect diskinesia (Fig. 3D).

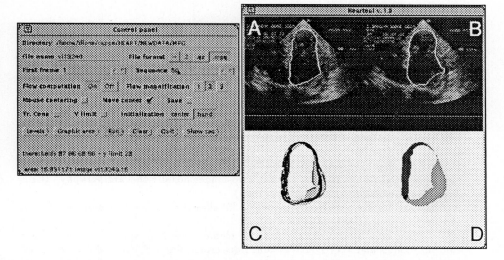

Fig. 3. The main panel and the graphic display of the tool used for the interactive analysis of ventricular boundaries motion. A shows the optical flows relative to the maxima of the expansion and the contraction; C represent the contour envelope with trace of speed (darker colour = slower motion), while in D the darker region represents the detection of anomalous motion. In this case, a slow and anomalous motion is detected on the upper left region, result comfirmed by the analysis of a cardiologist.

5 Discussion

The proposed approach for the analysis of echocardiographic image sequences is based on optical flow, that is used for three reasons: to improve the accuracy of the tracking of the cavity walls, to show the complex sequence of ventricular contractions, and as an independent test for contraction abnormalities, as shown in Fig. 3. Indeed, the computation of optical flow restricted on the points on the cavity walls, provides a good description of the dynamics of the heart and therefore of its functional state. In previous studies the computation of the optical flow on echocardiographic image sequences was impaired by several problems because of noise [12]. The computation of the flow over the entire image requires also a long computing time, which is not suitable for real-time clinical application. We solved these problems by computing motion only in the contour points and by using a robust correlation-based technique able to provide good results even in the presence of noise and even in the case of large shifts of the contour in successive frames. The preliminary results obtained with our system are promising: with good quality images the system identified wall motion abnormalities and the left ventricle area changes occurring through the cardiac cycle. The method is also sufficiently fast for clinical applications: the computation of flow vectors and the regularization of the contour are performed in about one second for each frame on a Sun 10 workstation.

References

1. S.J. Mason and N.J. Fortuin, "The use of echocardiography for quantitative evaluation of left ventricle function" Prog. Cardiovasc. Dis. **21**: 119 (1978)
2. M.A. Quinones et al. "A new simplified method for determining ejection fraction with two-dimensional echocardiography" Circulation **64**:744 (1981).
3. R. E. Kreber, M. L. Marcus, "Evaluation of regional miocardial function in ischemic heart disease by echocardiography". Prog. Cardiovasc. Dis. **20**:41 (1978).
4. V. F. Vanderberg et al. " Estimation of left ventricular cavity area with an on line, semiautomated echocardiographic edge detection system" Circulation **86**:159 (1992).
5. C. H. Chu, E.J. Delp and A.J. Buda, "Detecting Left Ventricular Endocardial and Epicardial Boundaries by Digital Two-Dimensional Echocardiography" IEEE Trans. on Medical Imaging **7**:2 266 (1988)
6. A. Kass, A. Witkin and D. Terzopoulos, "Snakes: Active contour models," Int. J. of Comp. Vision **1**, 321–331 (1988).
7. L.D. Cohen and Isaac Cohen, "A finite element method applied to new active contour models and 3D reconstructions from cross-sections" Proc. of 3rd Int. Conf. on Comp. Vision, pp. 587–591 (1990).
8. A. Blake et al. "Affine-invariant contour tracking with automatic control of spatiotemporal scale" Proc. 4th I.C.C.V. pp. 66–75 (1993).
9. I. L. Herlin and N. Ayache, "Features Extraction and Analysis Methods for sequences of Ultrasound Images" Proc. 2nd E.C.C.V., pp. 43–55 (1992).
10. I. L. Herlin, D. Breziat, G. Giraudin, C. Nguyen and C. Graffigne, "Segmentation of echocardiographic images with Markov random fields" Proc. 3rd E.C.C.V., **2** pp. 201-206 (1994).
11. B.K.P. Horn and B.G. Schunck, "Determining optical flow," Artificial Intelligence **17**, 185–203 (1981).
12. G. E. Mailloux, A. Bleau et al., "Computer Analysis of Heart Motion from Two-Dimensional Echocardiograms" IEEE Trans. Biom. Eng. **34**:5 356 (1987).
13. S. C. Armatur and H. J. Vesselle, "A New Approach to Study Cardiac Motion: The Optical Flow of Cine MR Images". Nuclear M. R. in Medicine **12**, 59–67.

Tissue Segmentation in MRI as an Informative Indicator of Disease Activity in the Brain

Simon Vinitski, Ph.D., Carlos Gonzalez, M.D., Claudio Burnett, M.D., Feroze Mohamed, M.S, Tad Iwanaga, M.S., Hector Ortega, M.D., Scott Faro, M.D.*

Department of Radiology, Thomas Jefferson University Hospital
132 S. 10th Street, Room 1098, Main Building, Philadelphia, PA   19107
215-955-7293   (FAX) 215-955-5329
* Department of Radiology, Medical College of Pennsylvania, Philadelphia, PA

Abstract. The presented tissue segmentation technique is based on a multispectral analysis approach. The input data were derived from high resolution MR images. Usually, only two inputs, proton density (PD) and T2-weighted images, are utilized to calculate the 2D feature map. In our method, we introduced a third input, T1-weighted MR image, for segmentation based on 3D feature map. k-Nearest Neighborhood segmentation algorithm was utilized. Tissue segmentation was performed in phantoms, normal humans and those with brain tumors and MS. Our technique utilizing all three inputs provided the best segmentation ($p<0.001$). The inclusion of T1 based images into segmentation produced dramatic improvement in tissue identification. Using our method, we identified the two distinctly different classes of tissue within the same MS plaque. We presume that these tissues represent the different stages involved in the evolution of the MS lesions. Further, our methodology for measuring MS lesion burden was also used to obtain its regional distribution as well as to follow its changes over time. The segmentation results were in full accord with neuropsychological findings.
  Key Words: MRI, Tissue segmentation, Multiple Sclerosis, brain tumor, 3D Feature Map

1   Introduction

The presented work deals with the application of tissue segmentation to characterize it both qualitatively and quantitatively using MR imaging. Previously, we utilized the method of Cline et al. who used probability and connectivity on two sets of images (proton density and T2-weighted) to segment brain tissues (1). We extended the segmentation algorithm into the third dimension. Many tissues (defined probabilistically) have overlapping MR signal intensities. We hypothesized that adding a third input image would make the separation of the clusters in a 3D feature space greater, with a resulting improvement in tissue segmentation. Initial results showed that segmentation image based on 3D feature map was better than or equal to the best segmented image based on 2D feature map (2). However, there was not enough data to perform quantitative analysis.

Since that time we have improved our methodology (e.g., move from probability model to k-Nearest Neighborhood). Initially, we tested this improved methodology in animals (experimental edema in cat brain). Results showed an inter-observer variability in the range of 7.6% - 9.3% and intra-observer variability range was 8.3% - 11.4%, indicating high stability of the technique (3).

To determine volumetric accuracy of our technique a phantom study was also performed (4). Five samples which represented CSF, white/gray tissues, blood and fat, were utilized as a phantom. To establish volumetric accuracy, the volume of each sample was accurately measured prior to MR imaging. Results of volumetric phantom measurements show accuracy between 3 - 5%.

In this study we had the following aims: a) to demonstrate the advantage of tissue segmentation based on 3D feature map over that based on 2D by quantitative means and in patients, particularly those with MS and, b) to investigate the effects of inclusion of T1-weighted MRI into the segmentation process of the brain tissues. Further, currently no reliable marker of the number of brain pathologies, such as MS activity, over time exists. Thus, the final aim of our study was to accurately measure regional lesion load as well as to follow its changes over time.

2 Methods and Materials

As we stated earlier, we have utilized a multispectral analysis approach (1). Three sets of MRI data were utilized in phantom and human experiments. Proton density and T2-weighted fast spin echo and, as a third input, T1-weighted spin echo were used. Image matrix was up to 256^2 and 3-5 mm slice thickness was used. A 1.5 T Signa GE scanner was utilized.

Humans were five normal, six with brain tumor, and thirteen patients with MS. From the brain MS patient population, three patients also undergo a second follow-up study one-half year later.

For segmentation, images were transferred to Sun Sparc Station 2. Usually, only two inputs, proton density (PD) and T2-weighted images, are utilized to calculate the 2D feature map. In our method, we introduced a third input, T1-weighted MR image, for segmentation based on 3D feature map. The qualified observer "seeded" tissue samples (40-50 samples/tissue) from three input images simultaneously. Later either three sets of image intensities (for 3D feature map) or any combination of two of them (for 2D feature map) were used for segmentation. Frequently, the most and least intense regions of the MS plaque were treated as two different tissues. Statistical filter narrowed sample distribution, thus, greatly reducing operator error in misclassified seeding. Anisotropic diffusion filter (5) was used for some correction of partial volume effects. Instead of probability model, as suggested by Clarke et al. (6), k-Nearest Neighborhood segmentation was utilized for both 2D and 3D feature map calculation. This data was used to create a stack of color-coded segmented images (one based on three inputs and three based on different combinations of two inputs) and up to eight tissues were classified. A connectivity algorithm (7) along with a dividing cube algorithm (8) constructs a surface of selected tissue(s). Finally, five board certified neuroradiologists ranked segmented images on a scale from 1 to 10. Statistical analysis was performed using the Wilcoxon Matched-Pairs Signed-Ranks Test (9).

3 Results

In MS, Table 1 demonstrates the results of ranking for one segmentation based on 3D feature map (PD, T1, and T2) and three based on 2D. Results show segmentation based on 3D feature map was found to be far superior than any combination of 2D feature map (average rank 7.3, p<0.001). We were surprised to find that the segmented image resulting from a combination of PD/T1 was superior to two other methods based on 2D feature map: combinations of PD/T2 and T1/T2 (average rank 5.1 vs 3.7 and 3.4 (p<0.01) respectively). Partial volume effects were visibly reduced. In addition,

Table 1: Evaluation of segmentation based on 3D
and three 2D feature maps in MS

Observers	3 D PD, T1,and T2	2 D PD and T2	T1 and T2	T1 and PD
#1	6.69	3.38	3.23	4.23
#2	7.38	2.38	3.08	5.23
#3	7.38	4.31	4.15	5.00
#4	7.62	3.23	3.69	5.77
#5	7.53	3.38	4.31	5.38
Average	7.32	3.34	3.69	5.12

PD, T1 and T2 = proton density, T1- and T2-weighted images, respectively.
Ranking ranges from 1 to 10, where 10 is the best in lesion identification.

accuracy of phantom volumetric measurements was improved by 54% as compared to the previous results (2).

Moreover, the inclusion of T1 based images into segmentation of MS patients produced dramatic improvement in tissue identification, particularly between different areas within the same plaque. There was no mix between color pixels belonging to both "tissues." The color (class) distributions closely resembled cluster pattern. While no statistical analysis of these two "tissues" was performed, this data, nonetheless, strongly indicates the possible existence of two different, at least from the MRI point of view, tissues within the plaque. We presume that these two different classes of tissue most likely represent the different stages involved in the evolution of the MS lesions, such as: edema, demyelination, as well as residual cystic formation.

Table 2. Region distribution of MS lesion (cc) in a patient

	Frontal	Parietal	Temporal	Occipital	Total
Skin and Fat	11.76	5.19	8.59	10.25	35.79
Gray Matter	196.27	159.36	127.47	88.70	571.80
White Matter	104.78	101.14	52.75	59.39	318.06
CSF	43.51	52.67	31.71	8.33	136.22
MS Lesion	27.78	30.16	16.61	9.67	84.22
Vascularity	31.72	15.61	25.00	12.34	84.67

Reproducible and accurate measurements of regional lesion load (temporal, parietal frontal, occipital) were obtained in each patient (Table 2). The majority of lesions were identified in the parietal occipital lobes. When the MS load increased it was associated with brain atrophy represented by an increase in CSF volume as well as a decrease in brain normal tissues volume. This is illustrated in Table 3. Importantly, these results of segmentation were consistent with the patients' clinical and neuropsychological evaluations of the disease over a six-month period.

Table 3. Changes in MS load (cc) with time

Tissue	Start	6 months	Change
Normal Brain	786.1	740.8	- 5.8%
CSF	145.0	171.8	+18.5%
MS	79.9	94.1	+17.7%
Total Volume	1011.0	1007.0	0.0%

4 Discussion

We do not have enough data to quantitatively evaluate the use of tissue segmentation in brain tumors. However, we gained some experience with investigation of brain MS activity using tissue segmentation based on 3D feature map.

Cerebral lesions in MS have been described as clinically relevant only in advanced stages of the disease. Unfortunately, their effects are not readily apparent on the standard neurologic examination, disability rating scales, or gross screening instrument. In contrast, MRI has shown lesions in the brain much earlier in the disease. Quantitative measurements were obtained originally by Rao et al. (10, 11) using a 2D "region of interest" method resulting in obtaining the area encompassed by each tracing. These measurements were then correlated with the patient's cognitive dysfunction. Their results, as well as ours (12, 13), demonstrated that there is a direct correlation between brain lesion load and neuropsychological deficit. In another study using specific rating of lesions involvement by frontal, temporal and parieto-occipital regions we quantified the topographic distributions of lesions and consequent effects upon cognitive function (14). In this context, quantitative MRI measurement of brain lesions provides an excellent opportunity to study altered brain behavior. All of these studies were performed using manual tracing of the lesions. Unfortunately, such studies are not very accurate. In a recent study of MS, a multi-institutional group (15) found that in manual measurements inter- and intra-observer variability can be as much as 100 - 200%; and effectively gave this task to one "selected" technologist.

Another important application of quantitative MRI measures is in the performance of therapeutic clinical trials. Due to the recurrent nature of MS lesions, and the fact that new lesions detected by MRI are neurologically silent, the results of therapeutical clinical trials are difficult to evaluate. Using MRI manual tracing of the lesions as an indicator of treatment effects, Paty et al. (15), studied successfully the effect of Interferon 1B in MS patients. Other parallel studies using MRI as an indicator for disease activity have also been performed. However, based on our results, we believe that tissue segmentation analysis and resulting volumetric measurements of the lesions provides a more sensitive, accurate and, most importantly, stable correlation with the results of the neuropsychological testing and can be relatively easily repeated and compared.

In conclusion, the presented work shows that segmentation based on 3D feature map is vastly superior to that based on 2D. Next, the results show that it is possible to apply a reasonably mature segmentation technique to the actual disease processes on the quantitative and useful level. The inclusion of T1-weighted information into our segmentation methodology allowed better characterization of MS plaque. The results also suggest that our method of tissue segmentation is a reliable technique to measure, at least, MS lesion burden. We have also shown that it can also be used as a surrogate marker for disease activity over time. Finally, this technique can also be a useful adjunct in the future performance of therapeutic clinical trials in MS and other diseases. Thus, further clinical evaluation of this technique in other brain abnormalities is warranted.

4 References

1. Cline HE, Lorensen WE, Kikinis R, Jolesz F. Three-dimensional segmentation of MR images of the head using probability and connectivity. J Comp Assist Tomogr 1990; 14:1037-1042.

2. Vinitski S, Seshagiri S, Mohamed FB, et al. Tissue characterization by MR: data segmentation using 3D feature map. In: Vernazza G, Venetsanopoulos AN, Braccini C (eds), Image processing theory and applications. Amsterdam:Elsevier Science Publishers B.V., 1993; 325-328.

3. Vinitski S, Gonzalez C, Burnett C, Seshagiri S, Mohamed FB, Lublin FD, Knobler RL, Frazer G. Tissue segmentation by high resolution MRI: improved accuracy and stability. Proc. IEEE Eng. Med. Biol. 1994; 16:577-578.

4. Vinitski S, Gonzalez C, et al. Improved Intracranial Lesion Characterization by Tissue Segmentation Based on 3D Feature Map. Radiology 1994; 193(P):253.

5. Perona P, Malik J. Scalespace and edge detection using anisotropic diffusion. Proc IEEE Workshop on Computer Vision, Miami, FL 1987; 6-22.

6. Clarke LP, Velthuizen RP, et al. MRI stability of three supervised segmentation techniques. Magn Reson Imaging 1993; 11:95-106.

7. Cline HE, Dumoulin Cl, Hart Jr HR, et al. 3D reconstruction of the brain from MRI using a connectivity algorithm. Magn Reson Imaging 1987; 345-349.

8. Cline HE, Lorensen WE, Ludke S, Crawford CR, Teeter BC. Two algorithms for the three-dimensional reconstruction of tomograms. Med Phys 1988; 15:320-327.

9. Dawson-Saunders B, Trapp RG. Basic and Clinical Biostatistics. Norwalk:Appleton & Langes, 1990; 79-99.

10. Rao SM. Neuropsychology of multiple sclerosis: a critical review. J Clin Exp Neuropsych 1986; 8:503-542.

11. Haughton VM, Yetkin FZ, Rao SM, et al. Quantitative MR in the diagnosis of multiple sclerosis. Magn Res Med 1992; 26:71-74.

12. Gonzalez CF, Mitchell DR, Sacchetti T, Seward JD, Knobler RL, Lublin FD. Correlation between structural brain lesions and emotional and cognitive function in patients with multiple sclerosis: an MRI study. Neuroradiology 1991 (suppl) 123-124.

13. Mitchell DR, Swirsky-Sacchetti T, Knobler RL, Gonzalez CF, Seward J, Field HL, Santiago RS, Lublin FD. Analysis and correlation of mood state with cerebral MRI and severity of illness in patients with multiple sclerosis. Neurology 1991; 41 (suppl 1):145.

14. Sacchetti T, Mitchell DR, Seward JD, Gonzalez CF, Lublin FD, Fnobler RL, Field H. Neuropsychological and structural brain lesions in multiple sclerosis: a regional analysis. Neurology 1992; 42:1291-1295.

15. Paty DW, Li DKB, The UBC MS/MRI Study Group, The IFNB MSII Study Group. MRI analysis results of a multicenter, randomized, double-blind, placebo-controlled trial. Neurology 1993; 43:655-661.

Scene Understanding

Analysis of Scenes Containing Multiple Non-polyhedral 3D Objects

Mauro S. Costa[1] and Linda G. Shapiro[1,2] *

[1] Department of Electrical Engineering, FT-10
[2] Department of Computer Science & Engineering, FR-35
University of Washington
Seattle WA 98195
U.S.A.

Abstract. Recognition of generic three-dimensional objects remains an unsolved problem. Scenes containing multiple nonpolyhedral 3D objects are particularly challenging. Conventional object models based on straight line segments and junctions are not suitable for this task. We have developed an appearanced-based 3D object model in which an object is represented by the features that can be most reliably detected in a training set of real images. For industrial objects with both flat and curved surfaces, holes, and threads, a set of useful features has been derived; and a recognition system utilizing these features and their interrelationships is being developed. The recognition system uses small relational subgraphs of features to index the database of models and to retrieve the appropriate 3D models in a hypothesize-and-test matching algorithm. This paper describes the new models, the matching algorithm, and our preliminary results.

1 Introduction

Recognition of general, non-polyhedral 3D objects remains an active area of research in computer vision. Many feature-based systems have been developed and proven useful in the recognition of polyhedral objects. However, due to the nature of the features they utilize, namely points and lines, these systems are not suitable for recognizing generic 3D objects. Our philosophy is that to accomplish this task, it is necessary: to divide the general-object case into classes of objects; to utilize the appropriate sensors for each object class; and to make use of the appropriate features that can be reliably extracted using those sensors. We are currently working towards accomplishing this kind of generic recognition in scenes containing multiple 3D objects. The following work is most related or important to our own.

* This research was supported by the National Science Foundation under grant number IRI-9023977, by the Boeing Commercial Airplane Group, and by the Washington Technology Center.

The work conceptually closest to ours is that of Gremban and Ikeuchi [6]. They introduce a paradigm called appearance-based vision, in which a new step in the recognition process is introduced. This step predicts and analyzes the appearances of the object models based on the CAD data and on the physical sensor models. The prediction can be either analytical or based on synthesized images of the objects in the model database. The predicted appearance is the set of features that are visible under a specific set of viewing conditions. The analysis of the predicted appearance allows for the generation of an object recognition program, to be used in the on-line phase of the recognition process. This method is also known as VAC (Vision Algorithm Compiler) because it takes a set of object and sensor models and outputs an executable object recognition program. The framework is general in the sense that it does not require any specific type of sensor. Their system has successfully recognized simple objects from range data in a bin-picking environment. The major drawbacks of this approach are: 1) analytical prediction is impractical in some domains; and 2) synthetic images are not yet realistic enough for general use.

The PREMIO system of Camps et. al. [3] utilizes artificially rendered images to predict object appearances under various environmental conditions (sensor, lighting and viewpoint location). The predictions generated by the system did not agree well enough with the real images acquired under the same set of conditions. In order to improve PREMIO's predictions, Pulli [10] developed the TRIBORS system. He initially attempted to improve the predictions by using a better ray tracer, but that was also insufficient. The solution he found was to bootstrap the prediction process with synthetic images and to train on real images. These new predictions led to better and faster object recognition.

Despite the fact that it only deals with two-dimensional objects, Bolles and Cain's Local-Feature-Focus Method [2] is very relevant to our work. This method automatically analyzes the object models and selects the best features for recognition. Typical features include holes and corners. The basic principle is to locate one relatively reliable feature and use it to partially define a coordinate system within which a group of other key features is located. If enough of these secondary features are located and if they can uniquely identify the focus feature, then the hypothesized position and orientation of the object (of which this feature is a part) is determined. A verification step that utilizes template matching is then performed to prove or disprove the hypothesis. The system has been proven to efficiently recognize and locate a large class of partially visible two-dimensional objects.

The work of Murase and Nayar [9] also involves appearance of objects. They argue that since the appearance of an object is dependent on its shape, its reflectance properties, its pose in the scene, and the illumination conditions, the problem of recognizing objects from brightness images is more a problem of appearance matching than of shape matching. They define a compact representation of object appearance that is parametrized by pose and illumination only, since shape and reflectance are intrinsic (constant) properties. This representation is obtained by acquiring a large set of real images of the objects under different lighting and pose configurations, and then compressing the set into an eigenspace. A hypersurface in this space represents a particular object. At recognition time, the image of an object is projected onto a point in the eigenspace and the object is recognized based on the hypersurface on which it lies. The exact location of the point determines the pose of the object. The major drawback of this method is that it cannot handle multiple-object scenes. Occlusion also adversely affects the performance of the system.

Though the work of Bergevin and Levine [1] on generic object recognition does not make use of the specific model-based paradigm, it is related to ours in philosophy. They utilize coarse, qualitative models that represent classes of objects. Their work is based on the recognition by component (RBC) theory of Biederman. The system is divided into three main subsystems: part segmentation, part labeling, and object model matching. The part segmentation algorithm is boundary-based and it is independent of the specific shape of the parts making up an object. The part (geon) labeling algorithm makes use of the concept of faces to further categorize the geons into generalized solids. At the matching stage, the labeled geons are used to index into the database of models. A measure of similarity is defined in order to discriminate between the models. An important observation made by the authors themselves is that it is not clear that suitable line drawings may eventually be obtained from real images. All their examples and tests have made use of ideal line drawings.

The evidence-based recognition technique proposed by Jain and Hoffman [8] defines an object representation and a recognition scheme based on salient features in range images. The objects are represented in terms of their surfaces, boundaries, and edges. The recognition scheme makes use of an evidence rulebase, which is a set of evidence conditions and their corresponding weights for various models in the database. The similarity between a set of observed image features and the set of evidence conditions for a given object determines whether there is enough evidence that the particular model is in the image. The model features must be carefully chosen in order to make possible the distinction between object classes.

2 Appearance-Based Models

The *appearance-based model* of an object is defined as the collection of the features that can be reliably detected from a training set of real images of the object. If a well-defined procedure exists through which a computer program can extract a given feature, this feature is said to be *detectable*. Even though appearance-based models can be *full-object models*, we choose to use *view-class models* in which an object is represented by a small set of characteristic views, each having its own distinct feature set [11]. Since we are currently only dealing with intensity images, all of our features are 2D features which may or may not directly correspond to a 3D feature, as in the case of limb edges.

Let $S_{V,M}$ be a set of training images for view class V of object model M. Each image $I \in S_{V,M}$ is processed to yield a set of features F_I. A feature f_n^I from image I is *equivalent* to another feature f_m^J from image J if they have the same type and are judged to have come from the same 3D source. The set of features that represent the view class is the set $F_{V,M}$ of equivalence classes of the union of the feature sets. The feature types we are investigating for use in our system are: coaxial circular arcs (two-cluster, three-cluster, and multi-cluster), ellipses, triples of line segments (U-shaped and Z-shaped), junctions (V-junction, T-junction, Y-junction, and Arrow), parallel line segments (close and far apart).

A natural extension to the use of features in a recognition task is the use of their properties and the relationships among them. In order to incorporate that, we define a view-class model by its *structural description* $D_{V,M} = (F_{V,M}, P_{V,M}, R_{V,M})$, where $P_{V,M}$ is a set of the properties of the features, and $R_{V,M}$ is a set of the relationships among the features.

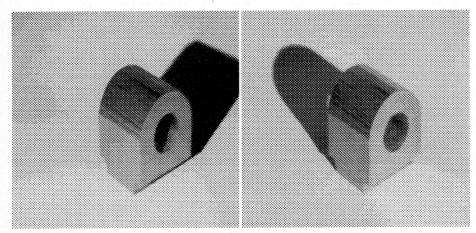

(a) Light source at the left of camera   (b) Light source at the right of camera

Fig. 1. Example of intensity image pair used by the system.

Our system works with pairs of intensity images. The two images are taken from the same viewpoint, but with two different lightings, one with the light source at the left and one with the light source at the right, as illustrated in Figure 1. By combining the two images, shadows can be eliminated and a more reliable edge image can be obtained. Figure 2 shows the features that were extracted by processing the sample image pair of Figure 1. Edges are detected using a Canny operator and the segmentation into lines and circular arcs is obtained using the Object Recognition Toolkit (ORT) package [5]. The line features (pair of parallel lines and the two V-junctions), the cluster of three circular arcs and the ellipse are detected by our system from the more primitive ORT features.

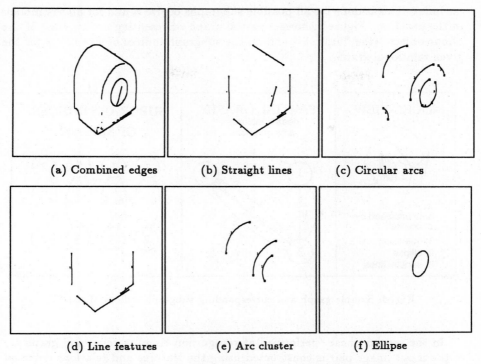

(a) Combined edges   (b) Straight lines   (c) Circular arcs

(d) Line features   (e) Arc cluster   (f) Ellipse

Fig. 2. A set of features extracted from the sample pair of Figure 1.

3 Scene Analysis Using Appearance-Based Models and Relational Indexing

We have created a database of appearance-based object models for a set of mechanical parts that have both flat and curved surfaces, holes, and threads. The structural descriptions $D_{V,M}$ of all the model-views were derived from a large set of training pairs of real images. We currently have 280 image pairs of 7 models. Our scene analysis paradigm makes use of the appearance-based models database and of a matching technique we call *relational indexing*.

The idea behind relational indexing is to utilize the structural description $D_{V,M}$ and represent each model-view as a relational graph $G_{V,M}$ of the features $F_{V,M}$ and relations $R_{V,M}$. The indexing principle is the same as in the original geometric hashing technique [7]. The method has two main phases: preprocessing and matching. The first one is an off-line phase in which the information contained in the entire database of models DB, is converted into a different representation that allows for a rapid retrieval. This is done in the following way: for each $G_{V,M}$ in DB, small relational subgraphs of size n are encoded and used as indices to access a hash table. The bin corresponding to a particular encoded subgraph stores information about which model-views gave rise to that particular index. This is done for all possible subgraphs of size n and for all the models in the database. Figure 3 shows a partial graph representing a view class of one of our objects (the "hexnut") and all the subgraph indices of size $n = 2$ for the given relational graph.

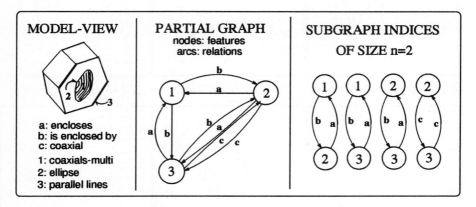

Fig. 3. Sample graph and corresponding subgraph indices of size 2.

In the second phase (performed at recognition time), a relational graph G_I of the input image pair is constructed using the features and relations detected in the scene. As in the off-line phase, all the subgraphs of size n are encoded and used to index into the hash table. Votes are cast for each model-view class stored in the bin indexed by each encoded subgraph. After all possible subgraphs have been used to index the table, the model-views with sufficiently high votes are taken as possible matching hypotheses. Details on the implementation of the hash table and the hashing scheme used can be found in [11].

Since some model-views share features and relations, it is expected that some of the hypotheses produced will be incorrect. This indicates that a subsequent verification phase is essential for the method to be successful. It is important to mention that the information stored in the hash table is actually more than just the identity of the model-view that gave rise to a particular subgraph index. It also contains information about which specific features (and their attributes) are part of the subgraph. This information is essential for hypothesis verification.

4   Multi-Level Indexing

In the case of single-object scenes, where there is no occlusion, one expects to extract most of the features and relations detected in the model-generation training phase. Therefore, the larger the subgraphs used, the more reliable and efficient the matching will be. However, in the case of multi-object scenes, only unoccluded objects will match to large subgraphs. Typically, in such scenes, features are missing or are only partially detected, and may even become different features due to occlusion. Consequently, their original relations are also greatly affected. Thus, it is more appropriate to use relational subgraphs of small size (a low level of indexing), which will include only a couple of features and relations, since these are more immune to the adverse effects occlusion has on both the features and the relations.

Taking the above into consideration, it seems natural to consider a multi-level indexing approach to matching. Without any knowledge of the degree of occlusion in the scene, the system starts at the largest subgraph level and goes down to lower levels as necessary to recognize all objects in the scene. Objects that are unoccluded are expected to be recognized at the higher levels of indexing while highly occluded objects may only be recognized at the lowest levels.

5 Results and Discussion

In order to illustrate our scene analysis methodology, we matched the image of a scene containing four objects to the database of model-views. The database of models was created by encoding all relational subgraphs of size $n = 2$ for each of the model-views. The test image pair was processed, features and relations were detected, the relational graph was built, and all subgraphs of size $n = 2$ were encoded. The relational indexing was then performed and the generated hypotheses were normalized and ranked in order of strength. Among the five significantly high-ranked hypotheses, four were correct and they are shown in Figure 4. These hypothesized models were taken through pose computation (affine correspondence of appearance-based model features and scene features) without verification. The fifth strong hypothesis (not shown) matched the object "hexnut" to an incorrect view of the corresponding object model. The subgraph indices shown in Figure 3 are among those that were used in the matching process.

As it can be seen, the method produces promising results. A verification procedure is being designed in order to effectively rule out incorrect hypotheses that may be generated. Future work includes the development of a Bayesian approach to the relational indexing paradigm, along the same lines as our previous work on indexing [4], and the exploration of the proposed multi-level indexing technique applied to the case of scenes with a large degree of occlusion.

References

1. R. Bergevin and M. D. Levine. Generic Object Recognition: Building and Matching Coarse Descriptions from Line Drawings. *IEEE Transactions on Pattern Analysis and Machine Intelligence*, 15(1):19–36, 1993.
2. R. C. Bolles and R. A. Cain. Recognizing and Locating Partially Visible Objects: The Local-Feature-Focus Method. *The International Journal of Robotics Research*, 1(3):57–82,1982.

Fig. 4. Right image of a test scene overlaid with the appearance-based features of the hypothesized model matches.

3. O. I. Camps, L. G. Shapiro, and R. M. Haralick. Image Prediction for Computer Vision. In *Three-dimensional Object Recognition Systems*, A. Jain and P. Flynn (eds). Elsevier Science Publishers BV, 1993.

4. M. S. Costa, R. M. Haralick and L. G. Shapiro. Optimal Affine Invariant Point Matching. In *Proceedings of 10th ICPR*, volume 1, pp. 233–236, 1990.

5. A. Etemadi. Robust segmentation of edge data. In *Proceedings of the IEE Image Processing Conference*, 1992.

6. K. D. Gremban and K. Ikeuchi. Appearance-Based Vision and the Automatic Generation of Object Recognition Programs. In *Three-dimensional Object Recognition Systems*, A. Jain and P. Flynn (eds). Elsevier Science Publishers BV, 1993.

7. R. Hummel and H. Wolfson. Affine Invariant Matching. *DARPA Image Understanding Workshop*, April, 1988.

8. A. K. Jain and R. Hoffman. Evidence-Based Recognition of 3-D Objects. *IEEE Transactions on Pattern Analysis and Machine Intelligence*, 10(6):783–802, 1988.

9. H. Murase and S. K. Nayar. Visual Learning of Object Models from Appearance. *International Journal of Computer Vision*, in press. Also Tech. Rep. CUCS-054-92.

10. K. Pulli. TRIBORS: A Triplet-Based Object Recognition System. *Technical Report 95-01-01*, Department of Computer Science and Engineering, University of Washington, January 1995.

11. L. G. Shapiro and M. S. Costa. Appearance-Based 3D Object Recognition. In *Proc. of the NSF/DARPA Workshop on 3D Object Representation for Computer Vision*, New York, NY, December 1994.

Fuzzy Segmentation and Astronomical Images Interpretation

Pierre Dhérété, Laurent Wendling, Jacky Desachy

Université Paul Sabatier - IRIT
118, route de Narbonne, 31062 Toulouse Cedex FRANCE

Abstract. This paper presents a fast segmentation method using fuzzy sets theory applied to astronomical images interpretation. It consists in separation in two classes (« stars » and « other objects ») with human-like decision. This separation constitutes the first step in astronomical images analysis, before spectrographic analysis and high level interpretation. A fuzzy segmentation isolates regions relative to objects and evaluates shape parameters for each region. Then, combined criteria are computed on the obtained fuzzy regions. The decision is founded on rules using these combined criteria. For each object we obtain the location, the orientation and the membership degrees to each class, then we build two images coding the membership degrees for each pixel to each class. With this method the use of fuzzy concepts in the first step allows to use fuzzy reasoning all along the classification process.

1. Introduction

The new large telescopes, with very sensitive and wide spectrum instruments, permit to explore regions of sky invisible until nowadays. They make possible deep space study and allow new horizons to astronomical research, mainly cosmology.

Remote sensing founded on deep sky images deals with different problems. The high number of objects excludes systematic study of each object. The bad signal/noise ratio makes the analysis difficult. So, it is very important to use fast classification as automatic as possible. The goal is to build catalogs, which regroup the main characteristics of objects, allowing to select them for a specific study.

A fast classification is only an approach, the result is tainted with errors. So, it is very important to take notice of these errors and to specify the reliance of the classification for each object. The notion of uncertainty needed by this type of classification may be expressed with fuzzy logic formalism. Thus, we can associate to each object the possibility to belong more or less to a specific class.

2. Astronomical images

An astronomical image, after an instrumental deconvolution, contains different object types, stars (of our galaxy), galaxies, gravitational bows, globular clusters, galaxy clusters, cosmic rays, etc.

A large field image contains a large amount of small objects. Generally the resolution does not allow to distinguish structure or details. Sometimes the sky background is not homogeneous or flat and may modify the object morphology.

Objects can be in interaction and belong to a group (e.g. clusters). In the neighborhood of clusters, a recognition process has to determine if objects belong to the cluster or not, and if they are before or behind it.

A visual analysis can take a long time and is only the first step in object analysis. Specific astronomical programs uses only a few object categories. So, before performing an analysis, we need to classify the objects. to select interesting objects.

3. Classification

After sky background correction and cosmic rays removal, classifications are generally performed step by step. The first step consists in separation of two classes, stars and other objets. Then the classification needs spectrographic criteria, redshift, red/blue ratio, etc., to separate kinds of stars and other objects. At last, we need more complex criteria (distribution, relationships, shapes) requiring high level concepts.

Current classifications are subjective. An astronomer can attribute different types to same object, so a category is always a fuzzy concept. To obtain fuzzy decisions it becomes interesting to use a fuzzy reasoning all along the classification process. Each step provides fuzzy data to the next step, allowing revision of former decisions and thus may provide a final result closer to reality.

Most of known automated astronomical images classifiers do not use fuzzy concepts, specifically in the first step. They are often founded on FFT and wavelets transforms. They use pattern matching with models or comparison with predetermined values. The parameters are conditioned by images and the adjustments are generally manual.

4. Selection of criteria

To separate stars and other objects (Galaxy) an astronomer uses rules and criteria to take a decision. The criteria are subjective and not precise. They are fuzzy concepts.

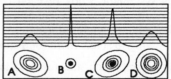

Shapes and levels of astronomical objects

The first criterion is the **ellipticity** : A star is circular (B) ; A galaxy is generally elliptic (A) (C) ; A galaxy may be circular (D) (like a star)
⇒ we need a second criterion,

The **sharpness** : A star is very sharp (B) ; A galaxy is generally stretched (A) (D) ; A galaxy with intense nucleus has a high sharpness (C).

5. Segmentation

After sky background rejection, segmentation is performed using a set of level-cuts. Because of high dynamic of images the successive values of α follow an exponential law by decade (1..9, 10..90, 100..900, etc.), to limit the number of level-cuts and to concentrate many cuts in image low levels.

For each level-cut we obtain a binary image, after applying a closing process to cancel the noise, we extract the contour for each region.

A single object profile is gaussian, see above figure, for a same object the successive level-cuts give contours included one into another. Several close objects may define only one region at low level α, and more with increasing α. The superimposed objects give a similar case.

So we create a tree describing inclusions of contours, then we regroup contours for each elementary object. The low level contour is the support for the object.

6. Fuzzy parameters

We need ellipticity and sharpness to classify objects. But we need to locate each object and its orientation too. We process these two parameters as fuzzy quantities.

6.1 Moments, barycenter, sharpness

Barycenter, ellipticity and orientation are deducted from moments of order 0 through order 2 [3]. For each object we have one support and higher level-cuts. For each level-cut A_α we compute the six moments, the barycenter and the sharpness.

Expressions of moments are :

$$M_{00_i} = \sum_x \sum_y I(x,y) \quad ; \quad M_{10_i} = \sum_x \sum_y x \cdot I(x,y) \quad \text{and} \quad M_{01_i} = \sum_x \sum_y y \cdot I(x,y)$$

$$\text{and} \quad M_{jk_i} = \sum_x \sum_y (x - B_x)^j \cdot (y - B_y)^k \cdot I(x,y) \quad \text{for } M_{02}, M_{11} \text{ and } M_{20}.$$

The barycenter coordinates are :   The sharpness is :

$$B_{x_i} = M_{10} / M_{00} \quad \text{and} \quad B_{y_i} = M_{01} / M_{00} \qquad S_i = -\ln\left(\alpha_i \middle/ I_{max} \right) \middle/ \text{Area}(A_{\alpha_i})$$

I_{max}: barycenter intensity ; α_i : level-cut level ; Area(A_{α_i}) : area of level-cut A_{α_i}.

6.2 Weight function

Images are not real membership functions, even if they approximate the belonging function to class « Object », thus we are not strictly in fuzzy logic formalism. It is important to normalize the function for each object.

The weight function privileges to much high level-cuts. In our images these regions might be very small (3 or 4 pixels), consequently computed parameters may be wrong. For instance, an elliptic object may have an high level-cut composed of 4 pixels (and then supposed to be circular). This will affects strongly the global ellipticity, and sharpness may not be defined. Moreover, we cannot use the same weight for all level-cuts of a same object. As low level-cuts (very noisy) are more numerous than high level-cuts the final result will be affected.

The chosen weight function minimizes low level-cuts and moderately privileges high level-cuts, and of course normalizes the function for each object.

Expression is : $m(A_{\alpha_i}) = \alpha_i \bigg/ \sum_{j=1}^{n} \alpha_j$ It respects the constraint : $\sum_{i=1}^{n} m(A_{\alpha_i}) = 1$

α_i is current level for the region, A_{α_i} are each level-cuts of the object.

6.3 Combined criteria

Then we define combined criteria for each object. The values of order 2 moments depend on level-cut area. Level-cuts with small area have low moments, and our high level level-cuts have low area. To minimize a bit more the influence of high level-cuts in combined ellipticity computation, the ellipticity is not computed for each level-cut and then combined. But we first combine order 2 moments, to get combined moments, and then we compute combined ellipticity using them. Thus medium regions with good resolution and good signal/noise ratio are privileged.

Combined moments are : $\quad CM_{jk} = \sum_{i=1}^{n} m(A_{\alpha_i}) . M_{jk_i}$

Combined barycenter is : $\quad CB_X = \sum_{i=1}^{n} m(A_{\alpha_i}) . B_{X_i} \quad ; \quad CB_Y = \sum_{i=1}^{n} m(A_{\alpha_i}) . B_{Y_i}$

Combined orientation is : $\quad CO = \dfrac{1}{2} \tan^{-1}\left(\dfrac{2 . FM_{11}}{FM_{20} - FM_{02}} \right)$ For circular objects CO is forced to 90° instead of 45°

Combined sharpness is : $\quad CS = \sum_{i=1}^{n} m(A_{\alpha_i}) . S_i$

Combined ellipticity is [3] : $\quad CE = \sqrt{\dfrac{(CM_{20} + CM_{02}) + \sqrt{(CM_{20} - CM_{02})^2 + 4 . CM_{11}^2}}{(CM_{20} + CM_{02}) - \sqrt{(CM_{20} - CM_{02})^2 + 4 . CM_{11}^2}}}$

7. Judgments associated to criteria

Ellipticity and sharpness have different ranges. A medium ellipticity has not the same value as medium sharpness. To judge this two values in a similar way we associate a fuzzy set to each criteria $\mu_E(CE)$ and $\mu_S(CS)$, coding them in [0, 1] domain. Thus the criteria are homogeneous.

7.1 Decision rules

The "Star" membership degree, SCF (Star Certainty Factor), depends on $\mu_E(CE)$ and $\mu_S(CS)$. The decision uses rules simulating human-like behavior. This rules are activated in the following order :

1. **If sharpness is very high (i.e. $\mu_S(CS) = 1$), then SCF is $\mu_E(CE)$.**
   If sharpness is high, then it is either a star or an object with sharp nucleus, we judge only with ellipticity.

2. **If object is very elliptic (i.e. $\mu_E(CE) = 0$), then it is not a star SCF = 0.**
   If ellipticity permits to assert it is not a star, then ignore sharpness. It squeezes doubt with sharp galaxies.

3. **If ellipticity is very low (i.e. μ_E(CE) = 1), then SCF is μ_S(CS).**
   If object is circular, then it is either a star or a stretched circular object, only sharpness is useful.

4. **If none of former rule is applied, then SCF is the medium value of μ_E(CE) and μ_S(CS).**
   And a contradiction between the two values gives 0.5, this is a fully doubt.

The certainty factor for class « Other object », OCF, is :   OCF = 1 - SCF.

We obtain SCF and OCF measures for each object. This is extended to each object pixel (x, y) to compute ISCF and IOCF images as follows :

$$ISCF(x,y) = \sum_{\alpha_i \,/\, (x,y)\, \in A_{\alpha_i}} m(A_{\alpha_i}) \cdot SCF \quad ; \quad ISCF(x,y) = \sum_{\alpha_i \,/\, (x,y)\, \in A_{\alpha_i}} m(A_{\alpha_i}) \cdot SCF$$

8. Results

We first test the process on synthesis images without noise with known catalog. Fully good answers represent 70% (CF=1 and is right). Fully bad answers represent 3% (CF=1 and is wrong). If till fully doubt answers (CF=0.5) are considered good we obtain 85% of success. Of course this result depends on the content of the image.

Then we test on noisy synthesis images. With a Poissonian additional noise (68±2). Fully good answers represent 60% . Fully bad answers 8%. Answers till fully doubt 85% too. Fully bad answers concern mainly very weak objects. The quality of classification depends only on dispersion of noise and not on mean value. Typically a good image has a dispersion between ±1 and ±3 with short exposure time, the noise increase with exposure time (i.e. sky background) following a Poisson's law.

The test with real noisy images gives a similar result, when sky background is uniform. One of the images, without catalog, has been processed without noise reduction, the noise dispersion is ±5. All very weak objects are classified in « Other » class. The problem is due to low quantization rate. A good classification should give 0.5 (fully doubt) on this objects.

Modifying ellipticity and sharpness functions (μ_E(CE) and μ_S(CS)), we record a problem. If classification of weak objects is good, classification of bright objects is less, and inversely. This is due to low quantization rate too. A weak star has a lower sharpness than a bright star. For some small objects, it is impossible to distinguish stars and other objects if they are circular or with a low ellipticity. We must force CF to 0.5 for this weak objects, only spectral analysis may improve the result.

Two other problems are the "seeing" value (blurring due to atmospheric perturbations during exposure), the star sharpness is different depending on images, and the telescope motion quality which modifies ellipticity of stars. When the seeing is bad it is more difficult to discriminate weak objects. When the motion of the telescope is bad stars are elliptic with the same value in the same direction.

The deep sky test image is provided with a catalog of another classic segmenter : the astronomical image processor IRAF. It does not specify classes but only shape parameters, area, compactness, order 2 moments, and location. The whole of these parameters is not absolutely right. The comparison between the catalogs give about 75% of success. This is a good result regarding errors of each segmenter and the disparity of criteria. Only the main bright objects are identified, and the classification is right for them. It is impossible to say if classification for weak objects is right or not. We do not have deep sky image with a known catalog (truth) to perform a good comparison.

8.1 Improvements

The quality of classification depends on noise, but modification of level-cuts (number and levels) may improve the results. Increase the doubt for very weak objects depending on noise dispersion.

When a galaxy with high sharpness nucleus is close to a star, it is recognized as a star. Image-adaptive level-cuts is possible to improve the classification.

It is important to take notice of seeing and star ellipticity values. Evaluation of medium sharpness and medium ellipticity of stars permits to adjust sharpness and ellipticity thresholds.

The sky background rejection used in the process is not sufficient, it is just a test operator. We need an operator able to remove non-uniform sky with interpolation method.

9. Conclusion

According to astronomers this is a good result regarding to classic processes and visual analysis. This process is only the first step in astronomical objects classification. The obtained membership degrees allow revision of decisions, and flexible human-like reasoning. Membership degree images may be used by neural networks in combination with spectrographic data.

This method is a new approach allowing the use of fuzzy concepts all along the classification process.

10. References

1. DUBOIS and H. PRADE : « Théorie des possibilités ». Ed. MASSON 1985.

2. DUBOIS and H. PRADE : « Possibility Theory, an approach to the computerized processing of uncertainty », Plenum Press, New-York, 1988.

3. REED TEAGUE : « Image analysis via the general theory of moments ». J. Optical Society of America, Vol.70, N°8, August 1980, p.920-921.

4. SERRA : « Image Analysis and Mathematical Morphology », Academic press, 1982.

5. SHAFER : « A Mathematical Theory of Evidence », Princetown University Press 1976.

6. THONNAT and A. BIJAOUI : « Knowledge Based Classification of Galaxies » from *Knowledge Based System in Astronomy*, A. Heck & F. Murtagh, Springer Verlag, p.121-159, Berlin, 1989.

7. A. ZADEH : *Fuzzy Sets. Information and Control* 8:338-353, 1965. Reprinted in « Fuzzy Models for Pattern Recognition », J.C. Bezdek and S.K. Pal eds., IEEE Press, 1992 p.35-45.

Learning How to Find Patterns or Objects in Complex Scenes

Walter F. Bischof[1] * and Terry Caelli[2]

[1] Department of Psychology, University of Alberta, Edmonton, Alberta T6G 2E9, Canada
[2] Department of Computer Science, Curtin University of Technology, Perth, WA 6001, Australia

Abstract. In this paper, we consider how machine learning can be used to help solve the problem of identifying objects or structures composed of parts as they occur in complex scenes. We first discuss an automatic conditional rule generation technique (CRG) that is designed to describe structures via part attributes and their relations. It does so by generating part-indexed decision trees where the branches define the types of pattern structures necessary to identify and to generalize from the different training examples. We then show how the resultant rules can be used for region labeling, and we examine grouping and constraint propagation techniques that are required for the identification of objects in complex scenes.

1 Introduction

Though the literature abounds with techniques for the recognition of *isolated* 2D patterns and 3D objects, the problem of efficiently detecting and recognizing such structures in complex scenes has not received as much attention. A number of authors have incorporated machine learning techniques to increase the robustness and efficiency to these methods, i.e. to improve their ability to generalize from training or known object data, and to improve their efficiency in searching scene data. For example, evidence-based methods have been used recently in 3D object recognition [4, 2] where object model views and parts are used to automatically generate rules (attribute bounds) which evidence different models. In such systems, generalizations to new views or to distortions are defined in terms of the rule attribute bounds and evidence weights. However, they typically have not encoded relational information and have not been adapted to recognition in complex scenes. This paper addresses both issues, the generation of rules encoding relational information and the application of rules to the interpretation of complex scenes.

---

* Supported by Grant OGP38521 from the Natural Sciences and Engineering Research Council of Canada.

2 Learning Structural Descriptions: CRG

Conditional Rule Generation (CRG) is a technique devised by the authors for learning structural descriptions of patterns as trees of hierarchically organized rules [1]. The rules are defined as clusters in conditional feature (attribute) spaces which correspond to either unary features of pattern parts or binary features of relation between parts. The clusters in a given attribute space are generated by splitting attributes (partitioning feature spaces) in such a way that each selected splitting operation creates new regions (defining rule bounds) which contribute better evidence for fewer classes. In our approach, such rules are generated conditionally through controlled decision tree expansion and a cluster refinement procedure.

More formally, each pattern sample (a 2D pattern or a view of a 3D object) is composed of a number of parts (pattern components). Each part $p_r, r = 1, \ldots, N$ is described by a set of unary features $\mathbf{u}(p_r)$, and pairs of parts (p_r, p_s) belonging to the same sample (but not necessarily all possible pairs) are described by a set of binary features $\mathbf{b}(p_r, p_s)$. Below, $S(p_r)$ denotes the sample to which a part p_r belongs to and H_i refers to the information, or cluster entropy statistic $H_i = -\sum_j q_{ij} \ln q_{ij}$ where q_{ij} defines the probability of elements of cluster i belonging to class j. We first construct the initial unary feature space for all parts over all samples and classes $U = \{\mathbf{u}(p_r), r = 1, .., N\}$ and partition this feature space into clusters U_i. Clusters that are unique with respect to class membership (with entropy $H_i = 0$) provide a simple classification rule for some patterns. Each non-unique (unresolved) cluster U_i is further analyzed with respect to binary features by constructing the (conditional) binary feature space $UB_i = \{\mathbf{b}(p_r, p_s) \mid \mathbf{u}(p_r) \in U_i$ and $S(p_r) = S(p_s)\}$. This feature space is partitioned with respect to binary features into clusters UB_{ij}. Again, clusters that are unique with respect to class membership provide classification rules for some objects. Each non-unique cluster UB_{ij} is then analyzed with respect to unary features of the second part and the resulting feature space $UBU_{ij} = \{\mathbf{u}(p_s) \mid \mathbf{b}(p_r, p_s) \in UB_{ij}\}$ is clustered into clusters UBU_{ijk}.

Uniqueness of pattern classification rules can be achieved either by repeated conditional clustering involving additional pattern parts or through cluster refinement. Refinement of a cluster C is achieved by finding the feature dimension F and the feature threshold T that minimizes the partition entropy $H_P(T)$:

$$H_\mathrm{P}(T) = n_1 H(P_1) + n_2 H(P_2) \ . \tag{1}$$

where P_1 and P_2 denote the two partitions obtained using the threshold T. In addition, rather than splitting only leaf clusters, one can split the cluster tree at any level, and the cluster minimizing (1) is considered optimal for refining the cluster tree.

It should be noted that each feature space in the cluster tree corresponds to a standard decision tree [5]. CRG thus produces a tree of decision trees that is indexed by sequences of pattern parts, i.e. it is "part-indexed", whereas decision trees are purely "attribute-indexed". The dynamic expansion of cluster trees constitutes a major advantage of CRG over decision trees: CRG can expand trees to the level optimized for a given data set whereas decision trees operate on fixed sets of features that have to be chosen a priori (see [1] for more details).

3 Scene Labeling and Recognition: SURE

Once CRG has generated rules from training samples, the problem of scene labeling reduces to that of instantiating rules in data, grouping labels and checking for their compatibilities. Indeed, the very purpose of the CRG method has been to "pre-compile" the types and number of parts, their attribute and relational attribute states that are necessary and sufficient for recognition. The problem remains, however, how to apply such rules to scenes composed of multiple objects. Of particular difficulty in such problems is the grouping of features or parts for rule evaluation. This problem has been studied by Grimson [3] and others in the context of model-based vision. Here, we discuss a solution in the context of a rule-based system that makes only weak and general assumptions about the structure of scene and objects. Our solution is based on the analysis of the relationships within (intra) and between (inter) instantiated rules. The solution method, termed SURE (Scene Understanding using Rule Evaluation), is based on the *sequential* evaluation of constraints described below.

Initial Rule Evaluation. The first stage in SURE involves direct activation of the CRG rules in a parallel, iterative deepening method. Starting from each scene part, all possible sequences of parts, termed *snakes*, are generated and classified using the CRG rules. Expansion of each snake $S = <s_1, s_2, \ldots, s_n>$ terminates if at least one of the following conditions occurs: 1) the part sequence $s_1, s_2, \ldots, s_n$ cannot be expanded without creating a cycle, 2) all CRG rules instantiated by S are completely resolved, or 3) the binary features $\mathbf{b}(s_n, s_{n+1})$ do not satisfy the features bounds of any CRG rule. If a snake S cannot be expanded, the evidence vectors of all rules instantiated by S are averaged to obtain the evidence vector $\mathbf{E}(S)$ of the snake S. Further, the set $\mathcal{S}_p$ of all snakes that start at p is used to obtain an initial evidence vector for part p:

$$\mathbf{E}(p) = \frac{1}{\#(\mathcal{S}_p)} \sum_{S \in \mathcal{S}_p} \mathbf{E}(S) \ . \tag{2}$$

where $\#(\mathcal{S})$ denotes the cardinality of the set $\mathcal{S}$. Classification of scene parts based on (2) has one major problem. Snakes that are contained completely within a single "object" are likely to be classified correctly, but snakes that "cross" two ore more objects are likely to be classified in an arbitrary way, and they therefore distort the classification in (2).

Snake Permutation Constraint. Every CRG rule encodes a set of model snakes $\{M_k = <m_{k1}, m_{k2}, \ldots, m_{kn}>, 1 \leq k \leq K\}$. When a snake $S = <s_1 s_2 \ldots s_n>$ instantiates such a rule each image part s_i indexes a set of model parts $\mathcal{M}(s_i) = \{m_{ki}, 1 \leq k \leq K\}$. The snake permutation constraint is based on the assumption that rule instantiations are invariant to permutations, i.e. if two snakes are permutations of each other, for example $S_1 = \, < A, B, C >$ and $S_2 = \, < B, A, C >$, their parts must index the same set of model parts, independent of snakes and independent of instantiated rules.

Single Classification Constraint. The single classification constraint is based on the assumption that *at least one* snake among all snakes starting at a scene part does not cross an object boundary and that at least one instantiated rule indexes the correct model parts. Given this, if there is any scene part that initiates a single snake S_i and this snake instantiates a single classification rule then the model parts indexed by S_i can be used to constrain all snakes that touch S_i.

These two deterministic constraints are very powerful in terms of eliminating inconsistent (crossing) snakes. Their usefulness breaks down, however, for cases where the assumptions formulated earlier are not met for a given training and test data set.

Inter-snake Compatibility Analysis. The idea of the inter-snake compatibility analysis is as follows. The less compatible the evidence vector of a snake S_i is with the evidence vectors of all snakes that S_i touches, the more likely it is that S_i crosses an object boundary. In this case S_i is given a low weight in the computation of (2). More formally, let $S_i = \, < s_{i1}, s_{i2}, \ldots, s_{in_i} >$ and $S_j = \, < s_{j1} s_{j2} \ldots s_{jn_j} >$ be touching snakes, and let T_{ij} be the set of common parts, i.e. $T_{ij} = \{p \mid \exists k \; p = s_{ik} \text{ and } \exists l \; p = s_{jl}\}$ with $\#(T_{ij}) > 0$. The compatibility of S_i and S_j, $C(S_i, S_j)$ is defined as

$$C(S_i, S_j) = \frac{1}{\#(T_{ij})} \sum_{p \in T_{ij}} \frac{\#(\mathcal{M}(p|S_i) \cap \mathcal{M}(p|S_j))}{\#(\mathcal{M}(p|S_i) \cup \mathcal{M}(p|S_j))} \; . \tag{3}$$

The overall compatibility of a snake S_i is then defined with respect to the set $\mathcal{S}_T$ of snakes that touch S_i, i.e. $\mathcal{S}_T = \{S_j \mid \#(T_{ij}) > 0\}$:

$$w_{\text{inter}}(S_i) = \frac{1}{\#(\mathcal{S}_T)} \sum_{S \in \mathcal{S}_T} C(S_i, S) \; . \tag{4}$$

Using the inter-snake compatibility, the averaging of the evidence vectors in (2) changes to

$$\mathbf{E}(p) = \frac{\sum_{S \in \mathcal{S}_p} w_{\text{inter}}(S) \mathbf{E}(S)}{\sum_{S \in \mathcal{S}_p} w_{\text{inter}}(S)} \; , \tag{5}$$

where $\mathcal{S}_p$ is defined as in (2).

Intra-snake Compatibility Analysis. The last rule for detecting boundary-crossing snakes is based on the following idea. If a snake $S_i = <s_{i1}, s_{i2}, \ldots, s_{in}>$ does not cross boundaries of objects then the evidence vectors $\mathbf{E}(s_{i1})$, $\mathbf{E}(s_{i2})$, $\ldots$, $\mathbf{E}(s_{in})$ computed by (5) are likely to be similar, and dissimilarity of the evidence vectors suggests that S_i may be a "crossing" snake. Similarity of any pair of evidence vectors can be measured by their dot product, and similarity of all intra-snake evidence vectors is captured by the following measure

$$w_{\text{intra}}(S) = \frac{1}{n(n-1)} \sum_{k=1}^{n} \sum_{\substack{l \neq k \\ l=1}}^{n} \mathbf{E}(s_{ik}) \cdot \mathbf{E}(s_{il}) \ . \tag{6}$$

With the incorporation of the intra-snake compatibility analysis, the part evidence vectors are computed using the following iterative (relaxation) scheme:

$$\mathbf{E}^{(t+1)}(p) = \Phi \left[\frac{1}{Z} \sum_{S \in \mathcal{S}_p} w_{\text{inter}}(S) w_{\text{intra}}^{(t)}(S) \mathbf{E}(S) \right] \ , \tag{7}$$

where Z is a normalizing factor and Φ a logistic function. Iterative computation of (7) is required since recomputation of $\mathbf{E}(p)$ affects the intra-snake compatibility (6). As indicated above, the four rules presented in this Section are evaluated sequentially, and the final part classification is given by the iterative scheme (7).

4 An Example

Due to space limitations, we present a single example involving the recognition of 2D patterns in complex scenes. The line configurations are simplified versions of patterns found in geomagnetic images that are used to infer the presence of different precious metals. The training set consisted of four classes, corresponding, for example, to the presence of different types of metals, with four training patterns each, and each pattern consisted of three lines (see Fig. 1a). Patterns were described by the unary features "length" and "orientation", and the binary features "distance of line centers" and "intersection angle". CRG was run with maximum rule length set to the *UBUBU*-form, and it produced 35 rules, 3 *U*-rules, 18 *UB*-rules, 2 *UBU*-rules, and 12 *UBUB*-rules.

The SURE procedure was then run on the montage of patterns shown in Fig. 1b. Unary features were extracted for all scene parts (lines) and binary features were extracted for all neighboring scene parts, i.e. for pairs of lines whose center distance did not exceed a given limit. Results of the classification procedure are shown in Fig. 1c where 35 out of 42 scene parts are classified correctly.

5 Discussion

In the present paper, we have addressed two major issues. First, given that CRG represents structural descriptions in terms of sets of independent pattern snakes,

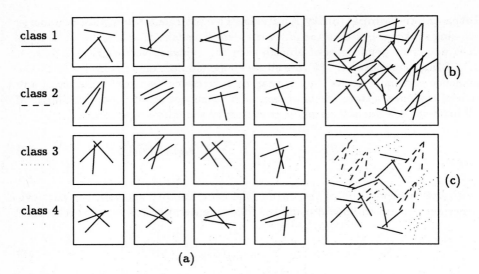

Fig. 1. (a) Four classes of patterns with four training patterns each. Each pattern is composed of three lines. (b) Montage of line triples. (c) Result of the pattern classification. Class labels for each line are shown on the left.

we have studied how interdependence of these snakes can be analyzed. Second, and more pertinent to this paper, we have studied how these interdependencies can be used to group pattern parts or image regions into groups that are likely to be associated with a single object.

References

1. Bischof, W. F., Caelli, T.: Learning structural descriptions of patterns: A new technique for conditional clustering and rule generation. Pattern Recognition **27** (1994) 689–698
2. Caelli, T., Dreier, A.: Variations on the evidenced-based object recognition theme. Pattern Recognition **27** (1994) 185–204
3. Grimson, W. E. L.: Object Recognition by Computer. Cambridge: MIT Press (1990)
4. Jain, A. K., Hoffman, D.: Evidence-based recognition of objects. IEEE Transactions on Pattern Analysis and Machine Intelligence **10** (1988) 783–802
5. Quinlan, J. R.: C4.5 Programs for Machine Learning. San Mateo: Morgan Kaufman (1993)

Adaptive Matching Using Object Modes Generated from Photometric Stereo Images

G. Bellaire, K. Schlüns, A. Mitritz, K. Gwinner

Institut für Tech. Informatik, TU Berlin, Sekr. FR 3-11
Franklinstr. 28-29, 10587 Berlin, GERMANY
{bellaire,karsten}@cs.tu-berlin.de

Abstract. We present a complete object recognition system for 3-D objects using a viewer-centered object description, so-called surface normal images (SNIs), recently introduced by Park et al. [3]. Based on this representation we utilize a weak active technique (the Photometric Stereo Method (PSM)) to extract 3-D features from the objects. We combine surface orientations with an approximated line drawing to build 2.5-D models.

Furthermore we develop an accumulator based matching method, which is adaptive and tolerant regarding the measurement errors. This includes a module to analyze the composition of the actual object library, that supports the construction of the index hierarchy. An effective technique is proposed that combines the results of the sequential feature matching of the rotated 2.5-D scene model set. Both the reconstruction level and the matching level of the object recognition system were tested successfully with synthetic and real object data bases.

1 Introduction

Viewer-centered approaches [2] develop models that represent the information visible from a certain view point. The projective relation between the 3-D object and the image is not considered during the matching process.

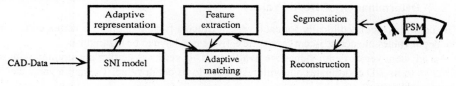

Fig. 1. Instances of our object recognition system.

The surface normal image (SNI) model [3] is a new and promising representative of this approach. It groups a set of views defined as follows: A view is represented as a 2-D line drawing. In the SNI set each view is a normal view. A normal view is defined by the alignment of the surface normal of a base face parallel to the line of sight. A SNI model of a 3-D object contains each object face as base face exactly once. In the scene domain a SNI is matched with so-called rotated input images (RII). The RIIs are generated from a 2.5-D scene model in the same manner as the SNIs. Following these descriptions the demand of an especially adapted matching strategy is obvious.

2 2.5-D model reconstruction

For generating the RIIs the first partial derivatives generated from PSM [5] have to be transformed to a height map or a geometrical 2.5-D model. This can be done for a large class of diffuse and hybrid reflecting surfaces [4]. We have developed a method, that combines dense gradient information and a line drawing of the object. The method consists of two steps. The projected object is taken apart in planar and curved patches. Then these parts are fitted together to a 2.5-D surface. First the gradient images are segmented with region growing. Using this kind of segmentation technique planar as well as curved patches can be extracted without additional effort. Curved patches are approximated by planar patches. These planar patches are attributed as belonging to a curved region. This is necessary for the treatment of occlusions and the elimination of approximation edges in the recognition part. Since region boundaries in the interior of curved patches are determined by the growing process, these patches are post-processed with a balancing algorithm. Pixels on region boundaries are reclassified, if the average orientation in an adjacent region has a smaller angular deviation than the original region. The reclassification process is done iteratively until an equilibrium is reached. Subsequently to this process the boundaries are approximated polygonaly. Consequently, for the next steps consistent region boundary information is available.

2.1   Part Assembler

Now, from the region and boundary data a winged edge model is generated. The model includes the vertices, edges, faces and the face orientations from the 2-D structure. This structure must be modified, if concave objects with partially occluded boundaries occur. In this case, we have to assign more than one depth value to some vertices. To prepare the depth calculation, such occluding edges have to be detected. Occluding boundaries are detected by using the face orientations. Since an orthogonal projection (denoted by op) is assumed, the expected edge orientation depends on the adjacent face orientations $\bar{n}_1$ and $\bar{n}_2$ as follows:

$$(x \; y)^T = (q_2 - q_1 \;\; p_1 - p_2)^T = op(\bar{n}_1 \times \bar{n}_2), \text{ with } \bar{n}_1 = (p_1 \;\; q_1 \;\; -1)^T, \; \bar{n}_2 = (p_2 \;\; q_2 \;\; -1)^T.$$

If this orientation is inconsistent with the line drawing, the edge becomes an occluding edge. Vertex splitting is done, if 1. both adjacent edges are occluding edges, or 2. one edge is occluding and the other is a 3-D boundary edge.

2.2   Determination of Depth Values

The depth is calculated locally for each vertex. Since a fixated depth of a vertex constrains the depth values of each adjacent face vertex, five cases have to be distinguished:

I. No adjacent vertex depth values are available: This vertex is neither constrained with respective to its 2-D coordinates nor with respective to its depth value. The depth value can be chosen arbitrarily.

II. For exactly one of the adjacent faces a fixation is made: The 2-D coordinates of the vertices can be substituted in the plane equation. This determines the depth values of these vertices.

III. For exactly two of the adjacent faces a fixation is made: 2-D inconsistencies of vertices can arise. Therefore an orthogonal projection onto the line of intersection of the two planes is determined. Thereafter the point of intersection is substituted into the plane equation.

IV. For exactly three of the adjacent faces a fixation is made: Consistency and the depth can be attained simultaneously by calculating the intersection of the planes.

V. More than three faces are fixed: If there is more than one point of intersection, than this inconsistency cannot be repaired. Therefore the order of depth calculation is made to be dependent on the number of adjacent faces. Such accidental events are less likely, especially for objects in general orientations and poses.

This procedure ensures that face orientations determined by the shape recovery method leave the assembling process unchanged. The edge structure in this process is variable. Surface slopes on and nearby edges are less reliable than in the interior of regions due to physical, non-ideal object edges. Accordingly it is appropriate to let the edge structure alterable. Surface orientations are measured over the whole region for each face, as a result the surface orientation associated with a face is sufficiently stable.

The above scheme is applied to each vertex in the described order. When the process is finished, the 2-D winged edge model derived from the line drawing is transformed to a 2.5-D model.

3   Accumulator based adaptive matching

Park et al compare sequentially all RII sets with the model data base. The sole recognizable optimization of this trivial matching strategy is the ordering of the RIIs' by base face attributes as face size and face type. They classify RII faces in not occluded and occluded to manage inexact data. This data is then especially treated during the matching procedure.

We introduce an adaptive inexact matching strategy. It gains an optimal matching scheme, that is adapted to the actual object data base, the a priori knowledge about the current sensor configuration, see [1], and the measured sensor errors. This article regards the processing of unavoidable sensor tolerances. The CAD-data processing modeling component computes a set of SNIs for each object. Weak perspective projection or scaled orthographic projection is assumed. The SNI sets of all model objects build the object data bases. Feature extracting functions generate feature vectors that describe the particular SNIs. For each SNI one feature vector exists. During the matching procedure the RII set is processed by comparing RIIs' feature vectors one by another with the complete SNI set feature vectors. The matching algorithm consists of an adaptive hierarchical indexing scheme, that contains an inexact matching module and that is enlarged by a learning module.

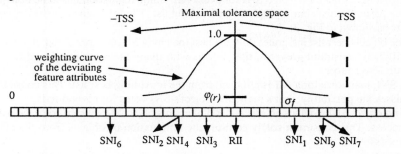

Fig. 2. Index layer showing the tolerance space dimesion and a Gaussian distributed voting function.

The indexing process consists of a relation from the area of the scene features into the area of the index space. If a set of n scene features exists, then a vector of g ($g \leq n$) feature attributes determine an index I_S. The values of the index depend on the content of the feature attributes. An index vector specifies exactly a point in the index space. The hierarchical system organizes g features of interest in g index layers. Each feature is represented by a certain layer in the layer model. The layer model is adapted for the data base and the used sensors. Layers delineate the index space of a feature. The layer indexes point to accumulators of those SNIs, that contain similar feature attributes.

In the case of inexact data a method has to be found, that relates the data to comparative models. The proposed inexact matching strategy increases the accumulators of several SNIs by the value $e^{-k^2/2}$, with $k = 0, ..., k_{max}$ denoting the deviation of a model feature to the measured data. That means the accumulator of an object with a feature value close to the measured data is increased more than an accumulator of an object view with a less similar feature value. This is called Gaussian distribution method. The exact method is denoted as spike distribution method. The accumulator values determine the ranking of the SNI best list.

The following conditions constrain the indexing procedure, see Fig. 2:

- The width of the Gaussian curve used to calculate the vote value is adapted for the standard deviation of the regarded feature σ_f multiplied by a scale factor s. For $s = 2.0$ it is well known, that ca. 95% of the deviations caused by measuring errors are covered by the curve. Pursuing this, the absolute amount of deviations remains relatively small.

- A tolerance space, that is larger than $s \cdot \sigma_f$, causes an assignment of the smallest possible vote $v(k) = \varphi(r)$ to the nodes outside the Gaussian curve. SNI candidates with larger errors in particular features are allowed to concur the matching race.

- A SNI is omitted from searching if a feature value drops out of the layer's tolerance space. Initially the tolerance space size (TSS) is 50%.

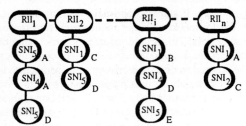

Fig. 3. The RII list matching algorithm concludes the resulting SNI lists: A SNI is marked by an object label (A, B, ...) and a number for the concerning SNI set. For each RII a list of SNI's is generated. This list is sorted with respect to the accumulated votes of the observed features.

In the proposed system a list of RII's is compared with the object data base. This process calculates a list of the most voted SNI for every RII. Some regulation has to be done to conclude the particular lists into a single object ranking list. The following constraints constrict SNI lists as shown in Fig. 3:

- The objects best list integrates only the best matching SNI of a particular object for each RII: This constraint prevents, that objects with many faces are privileged for objects with fewer faces.

- A SNI, that votes for a RII is removed from the remaining best list. This constraint dismisses the possibility, that a particular object's SNI votes for several RIIs.

- If no object SNI matches an especial RII, then this object is excluded from the matching process. This constraint solely controls the elimination of objects from the matching process.

3.1 Learning by integration of measured sensor accuracies

A technique to improve the performance of the matching is to adapt the tolerance space size (TSS) on the feature attribute distribution of the measured data. The system performs an Online adaptation by integrating the sensor accuracy: The matching strategy generates a list of the best matching SNI sets. A localization function verifies or discards the hypothesis of the matching procedure. Correctly selected view sets are compared with the RII sets.

The accuracy is measured by comparing of the correct result feature vector x_{SNI} and the vector x_{RII} of the corresponding RII. This process repeats for all feature vectors of the two view sets. For each feature attribute (element of the feature vectors) the maximal difference between the real and the measured feature value is stored in a learning map. This learning map supervises the mean standard deviation and maximal deviation value max_{dev} of the particular features. After t successfully processed recognition tasks the TSS and the σ_f of each index layers adapt. The learning map calculates the standard deviation for a feature g after the $n+1$

recognition task. The TSS is calculated by the max_{dev} of the features. Since the TSS excludes object views from the matching procedure, the learning algorithm allows the expansion of the TSS by incrementing max_{dev} by $max_{dev}./3$. The algorithm decreases the TSS by deleting the current max_{dev} after t recognition tasks and choosing the next smaller $max2_{dev}$ as new max_{dev}.

Fig. 4. Two data base objects and their reconstructed 2.5-D models.

3.2 The accuracy probability of results

In the following we estimate the accuracy of object propositions depending on the calculated votes of the SNI features. The sizes of the Gaussian curves of the several features rely on the standard deviations of the features' errors. Pursuing this, it is possible to regard votes as function values of the proportion z. This z represents the proportion between the actual feature difference of the RII and SNI and the standard deviation of the regarded feature. To get an estimation of the matching probability of a proposed model to an object, we calculate the inverse function $z = v^{-1}$ of the used Gaussian weight voting function: $z(vote) = \sqrt{-2\ln(vote)}$.
We use the results of z to calculate the probability of the accuracy of a model proposition, on the assumption of normal distributed feature errors. The probability, that a measured feature deviates at Δx from the correct feature value follows the Gaussian error integral:

$$p(z(vote)) = \sqrt{2/\pi} \bullet \int_0^z e^{-t^2/2} dt \, ; \quad t = \Delta x / \sigma_n \text{ and } z = a / \sigma_n$$

For a deviation Δx with $\Delta x / \sigma_n \geq z(vote)$ $(a \leq \Delta x)$ the value pi = 1 - $p(z(vote))$ delineates the probability for this Δx. For $z(vote) = 0$ it follows pi = 1.0.

Fig. 5. Shaded image of reconstructed real objects.

4 Results

We have examined the shape recovery scheme (PSM) and the proposed face assembling method with several synthetic and real objects. We have analyzed several PSM realizations, including analytic, LUT based and discrete solution finding approaches. Our 2-D-LUT technique has proved to give the best results with respective to calibration expenditure and accuracy. Fig. 4 shows single PSM input images of two objects in the data base. For each object two aspects are shown. Since only a 2.5-D model can be derived from the input images two faces are missing in the second aspect of the L-shaped object. Also, the F-shaped object has been reconstructed in a reliable manner, although there arise shadows and mutual illuminations. Fig. 5 shows the Lambertian shaded versions of the derived 2.5-D models of the real objects in the data base. It can be seen, that the 2.5-D models of some objects are not complete due to cast and self shadowing effects. To verify the recovered surface orientations the angle between right-angled faces has been measured. For most face pairs the angular difference was not higher than three degrees.

To test the recognition system we used a model data base with 14 objects. The objects are categorized in the classes convex, concave and smooth. We demonstrate six test series with varying parameters. The following features are calculated for the experiments:

- junction types, edge types, relations of the length of edge pairs, classes of edges with the same length, face number, face relations, angle types.

experiment #	adaptation steps	convex	curved	concave	sum
1	---	0 % ‖ ---	0 % ‖ ---	0 % ‖ ---	0 % ‖ ---
2	4	100 % ‖ 0.92	50 % ‖ 0.82	57 % ‖ 0.80	66 % ‖ 0.85
3	21	100 % ‖ 0.92	75 % ‖ 0.83	71 % ‖ 0.77	80 % ‖ 0.83
4	39	100 % ‖ 0.92	75 % ‖ 0.82	86 % ‖ 0.74	87 % ‖ 0.81
5	56	100 % ‖ 0.91	75 % ‖ 0.80	86 % ‖ 0.72	87 % ‖ 0.80
6	---	100 % ‖ 0.93	50 % ‖ 0.87	86 % ‖ 0.82	80 % ‖ 0.86

Tab. 1. Recognized real objects in percent and their recognition probability.

The experiment results shown in Tab. 1 depend on the variation of the tolerance parameters:

- Experiment 1: The TSS and the standard deviation for all features are zero. These settings correspond to the exact matching method.
- Experiment 2-5: The parameters are adapted stepwise. Distinct objects cause the generation of a distinct number of RIIs. This explains the non continuos growing of the adaptation steps. The recognition rate of the system grows continuously.
- Experiment 6: The system runs with extremely large TSS and standard deviation.

5   Conclusion

It has been shown, that a shading based shape recovery method (Photometric Stereo) leads to sufficient 2.5-D models for application in an elaborate object recognition system. Particularly this is appropriate, if the derived 3-D shape information is combined with an approximated line drawing. The line drawing is necessary, if occlusions have to take into consideration. We have shown, that good results are reached when combining PSM, the SNI model and the adaptive matching strategy. In the future, the shape recovery part of the project will be focused on making use of shadow information and mutual illuminations. The matching will be extended by an automatic verification module.

6   References

1.   G. Bellaire, "Feature-Based Computation of Hierarchical Aspect-Graphs", Machine GRAPHICS & VISION, Vol. 2, No. 2, pp. 105-122, 1993

2.   K. Ikeuchi, "Determining Linear Shape Change: Toward Automatic Generation of Object Recognition Programs", CVGIP, Vol. 53, pp. 154-170, 1991

3.   J. H. Park, T. G. Chang, J. S. Choi, "Three-Dimensional Object Representation and Recognition Based on Surface Normal Images", PR, 26, No. 6, pp. 913-921, 1993

4.   K. Schlüns: "Photometric Stereo for Non-Lambertian Surfaces Using Color Information", Proc. 5th Int. Conf. on Computer Analysis of Images and Patterns, Budapest, Hungary, Sept. 13-15, pp. 444-451, 1993

5.   R.J. Woodham R.J.: "Photometric Method for Determining Surface Orientations from Multiple Images", Optical Engineering, Vol. 19, No. 1, pp. 139-144, 1980

Robot Vision

Sensor Planning Techniques and Active Visual Inspection

Vito Roberto and Emanuele Trucco

Machine Vision Laboratory,
Department of Mathematics and Informatics
University of Udine, Italy

Department of Computing and Electrical Engineering
Heriot-Watt University, Edinburgh, Scotland

Abstract. This paper addresses the problem of computing the sequence of positions in space *(inspection script)* from which a robot-mounted sensor can perform a given inspection task optimally. We present the design and partial realisation of a sensor planning system, GASP, capable of generating inspection scripts for a variety of tasks, objects and sensors. The treatment of detection uncertainty is discussed with reference to the problem of detecting line features; the *generalised sensor* is introduced, and novel results in a simulated environment are reported.

1. Introduction. Sensor Planning and Inspection

A key feature of *active vision systems* is the ability to reconfigure themselves in order to improve their performance at given visual tasks. This ability can be exploited in advanced inspection, for instance by directing robot-mounted sensors to positions in space from which an inspection task can be performed optimally, e.g. the features to inspect are maximally visible, or most robustly detected by the vision system, or both. In general, several parts (features) of an object need inspecting, and each one is best observed from a different viewpoint. Therefore, a sequence of sensor positioning actions (an *inspection script*) must be planned to perform the task [5, 6, 7]. Visual inspection strategies and their automatic generation depend on the type and number of sensors and object features to be inspected: therefore, models of both sensors and objects must be available.

Typical inspection tasks that a planner can afford are the following. *Single-feature, single-sensor:* find the position from which a single imaging sensor (e.g., intensity or range) can inspect a single object features optimally. *Single-feature, multiple-sensor:* find the position from which a multiple sensor (e.g., a stereo head) can inspect a feature optimally. *Multiple-feature, single-sensor:* (a) find the position from which a sensor can simultaneously inspect a set of features optimally; (b) find the best path for a sensor to inspect a set of features from optimal positions. *Multiple-feature, multiple-sensor:* find the best path taking a stereo head to inspect a set of features from optimal positions.

This paper is a brief overview of GASP (General Automatic Sensor Planning), a system capable of generating inspection scripts for a variety of tasks, objects and sensors. The features currently examined by GASP are surface patches (planes, cylinders, cones, delimited by their possible intersections) and segments (straight lines and curves defined by the intersections of patches, as above). Using a feature-based CAD model of the object to be inspected, GASP identifies the viewpoints from which each feature is visible, how much, and how reliably it can be detected. We present the inspection tasks currently considered by GASP (section 2); section 3 concerns the treatment of uncertainty based on the concept of *generaised sensor* and its novel results.

2. Inspection Scripts

We have designed a representation, the FIR - *Feature Inspection Representation*, used by GASP to plan inspection scripts and carry them out in a simulated environment.

The (discrete) set of all the viewpoints accessible to the sensor is called *visibility space*. A FIR partitions the visibility space into *visibility regions*, each formed by all the viewpoints from which a given feature is visible. Viewpoints are weighted by two coefficients, *visibility* and *reliability*. The former indicates the size of the feature in the image (the larger the image, the larger the coefficient); the latter expresses the expected robustness with which the feature can be detected. The two coefficients are linearly combined into an *optimality* coefficient, in such a way that the relative importance of visibility and reliability can be adjusted.

In GASP, the visibility space is modelled by a *geodesic dome* [3] centered on the object; the viewpoints are the centers of the dome's facets. The algorithm to compute the FIR [7] generates a geodesic dome around a CAD model of the object to inspect, then raytraces the model from each viewpoint on the dome. A FIR for the widget shown in Figure 1 (left), observed from 320 viewpoints and with image resolution of 64x64 pixels, occupies about 80 Kbytes. The widget is about 250 mm in length.

2.1 Single-sensor Scripts

Single-feature Inspection. In a FIR, the viewpoints belonging to a visibility region are arranged in a list ordered by optimality. Figure 1 (right) shows the visibility region of the planar patch in Figure 1 (left) given a single intensity camera of focal length about 50mm; the widget has been magnified for clarity. The object-camera distance (dome radius) was determined by GASP as the minimum one such that the widget is visible from all viewpoints; other choices are possible. Viewpoints have been shaded according to their optimality (the darker the better), assuming equal weights for visibility and reliability.
Simultaneous Inspection of Several Features. In this case, a sensor must inspect several features simultaneously. The region of the visibility space from which a set of features is visible is called *covisibility region*, and is obtained as the intersection of the visibility regions of the individual features. This is done

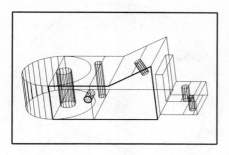
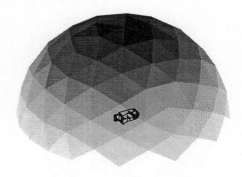

Figure 1: CAD model of a widget (left) and visibility region for highlighted planar patch (right).

simply and efficiently thanks to the FIR's structure. Notice that the merit (optimality) of a viewpoint for inspecting the set of features must be defined as a function of the optimality of the viewpoint for inspecting each individual feature.

Sequential Inspection of Several Features. Given a set of features, we now want to find a path in space which takes the sensor through the optimal inspection positions associated to all the features in the set; the path must also be as short as possible since we wish to minimise the number of sensor repositioning actions. Planning the shortest 3-D path through a given set of points is a NP-complete problem - the *Travelling Salesman Problem, TSP* - and only approximate solutions can be found. Three TSP algorithms have been designed and tested [8]: simulated annealing, CCAO [4], and the elastic net [2]. We found that, with up to 100 sites, CCAO outperformes the other two in terms of path length and distance from the overall optimal solution. An example of inspection path is given in Figure 2 (right) in the context of multiple-sensor, multiple-feature scripts.

2.2 Multiple-sensor Scripts

Single-feature Inspection. We want to determine the position from which a stereo head can observe at best a given feature. Using a FIR and information about the head geometry, GASP selects efficiently the most promising positions of the head inside the region, and picks the best one. Notice that the optimality must be defined as a combination of those of the (distinct!) viewpoints in which the two cameras are placed [8]. Figure 2 (left) shows the optimal placement for a stereo head found by GASP in the large (compared with the head's camera-to-camera distance) visibility region of one of the widget's back planar surfaces.

Multiple-feature Inspection. This class of scripts finds the shortest path for a

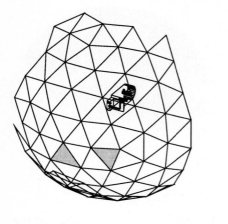 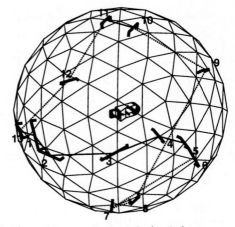

Figure 2: Stereo head inspection (left) and inspection path (right).

stereo head. Thanks to the uniform representation of viewpoint optimality in the FIR (i.e., referred to the viewpoint, not to the sensor), it is possible to apply the same path-finding module used for the single-sensor case. An example of optimal shortest 3-D inspection path is shown in Figure 2 (right). It has been computed assuming two intensity cameras of focal length about 50 mm and image resolution 64x64. The dome radius (object-camera distance) was fixed by GASP at 2400 mm. The camera-to-camera distance is 500 mm. 13 surfaces of the widget are visible to the cameras (others are too small or mostly occluded, e.g. the bottom surfaces of holes); the 13 viewpoints are the sites visited by the stereo head on the shortest path computed.

3. Modelling Sensor Reliability

A FIR records how much of a feature is visible from a viewpoint, and how reliably the feature can be detected by a given module. Reliability coefficients refer therefore to specific feature extractors - e.g., a line detector, a module computing the length of a linear feature, and so forth. Since the whole complexity of sensing and measuring is embedded in the two FIR coefficients - visibility and reliability - we call *generalised sensor* the compound formed by the physical sensor and the measuring or detection module.

The reliability has been estimated by simulation. To explain this we shall refer to features detected by an intensity camera and a line extractor. In this case our generalised sensor comprised a pinhole camera (perspective, no distortion), a Canny edge extractor, and a Hough line detector. We generated synthetic images of linear features across a wide range of the relevant parameters: line length and strength, noise variance. The images were corrupted with additive Gaussian noise, of variance in a range centeredon the values estimated from real images acquired by a real camera. For each image, we identified and located

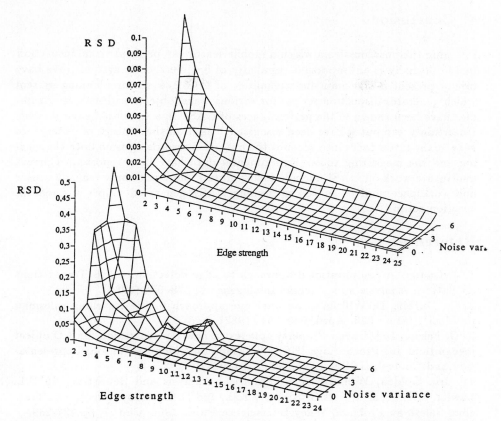

Figure 3: Uncertainty estimates via Root Square Distance (RSD). Experimental data (lower left) and interpolated values (upper right).

line using the target edge and line detectors. We then defined the uncertainty for the whole linear feature as a weighted sum of the observed variances of three measures: the *positional uncertainty* of each line point [1], the *root square distance* of each edge point from the estimated line (with subpixel accuracy) and *the strength* of each edge point. Second-order *uncertainty surfaces* interpolating the experimental values express the uncertainty of the detection for all possible occurrences of the line in an image. As an example, Figure 3 shows the measured uncertainties for the root square distance (left) and the corresponding interpolated surface (right), as functions of edge strength and noise.

The latter surfaces are used when building a FIR to enter appropriate reliability values for each feature and viewpoint. Since a FIR is generated by raytracing CAD models, the values of all parameters necessary to index the surface are always known, and a reliability coefficient is associated to each pair viewpoint-feature simply as the complement to one of uncertainty. At the moment, reliability data are available for linear features extracted from intensity images and surface patches extracted from range images.

4. Conclusions

Planning the positions from which a mobile sensor can perform visual inspection tasks optimally is an important capability of flexible robotic systems. We have briefly presented the main functionalities of GASP, a sensor planning system which generates inspection scripts for various tasks, objects and sensors. Examples have been shown of the performances of GASP in a simulated environment. Uncertainty estimates have been discussed, based on the concept of *generalised sensor*; the latter takes into account the uncertainties arising from both the sensing and the measuring tools. Experimental data have been reported. Current and future work on GASP include incorporating more and more realistic sensor and workspace models, as well as running GASP-generated scripts in a real inspection setup.

References

1. J. Canny: A computational approach to edge detection. IEEE Transactions on Pattern Analysis and Machine Intelligence 6, 679–698 (1986).
2. R. Durbin, D. Willshaw: an analogue approach to the travelling salesman problem. Nature 326, April, 689–691 (1987).
3. G. Fekete, L.S. Davis: Property spheres: a new representation for 3-D object recognition. In: Proceedings IEEE Workshop on Computer Vision, Representation and Control, (IEEE, Inc.), 192–201 (1984).
4. B.L. Golden, W.R. Stewart: Empirical Analysis and Heuristics. In E.L. Lawler, J.K. Lenstra, A.H.G. Rinnooy Kan, and D.B. Shmoys (eds.), The Traveling Salesman Problem, Wiley Interscience Pubs. 1985, Cap. 7, pp.227-239.
5. K.D. Gremban, K. Ikeuchi: Planning multiple observations for object recognition. International Journal of Computer Vision 12(2/3), 137–172 (1994).
6. K. Tarabanis, R.Y. Tsai, P.K. Allen: Analytical characterization of feature detectability constraints. Computer Vision, Graphics and Image Processing: Image Understanding 59(3), 340–358 (1994).
7. E. Trucco, E. Thirion, M. Umasuthan, A.M. Wallace: Visibility scripts for active feature-based inspection. In *Proceedings British Machine Vision Conference*, Springer-Verlag 1991, 538–547.
8. E. Trucco, M. Diprima, V. Roberto: Visibility scripts for active feature-based inspection. Pattern Recognition Letters 15, 1151–1164 (1994).

Experimental Self-calibration from Four Views

Reyes Enciso and Thierry Viéville

Projet Robotvis, INRIA Sophia-Antipolis, B.P. 93
F-06902 Sophia-Antipolis,France.

Abstract. The main goal of self-calibration [2, 3, 8] is to compute the intrinsic and extrinsic parameters of a camera without using a known pattern. In this paper we focus on the calibration of a binocular head-eye system from four views. The only information provided to the algorithm is the fundamental matrices [3] and the point correspondences between the 4 views. We exploit the information of the cross-correspondences to improve the Euclidean reconstruction.

1 Introduction

In this paper we address the problem of computing the intrinsic and extrinsic calibration parameters, in a binocular image sequence, given a set of point correspondences. Most authors [2, 5, 8] studied the case of point correspondences, but have restricted their approach to the case where the intrinsic parameters of the camera are constant, while only 2 or 3 views have been taken into account, or studied the monocular case for the long sequences as in [7].

The generalization to the case where intrinsic parameters are non-constant has already been addressed. But usually the analysis is restricted to the recover of affine or projective structure of the scene.

This paper extends these previous works to the case of non-constant intrinsic parameters and non-constant relative positioning of the cameras of the stereoscopic system. In particular, in the case of active vision, the extrinsic and intrinsic parameters of the visual sensor are modified dynamically. For instance, when tuning the zoom and focus of a lens, these parameters are modified and must be considered as dynamic parameters. It is thus necessary to attempt to determine dynamic calibration parameters by a simple observation of an unknown stationary scene, when performing a rigid motion.

2 The Camera Model

We use the well-known pinhole camera model (see Fig. 1) [3]. We assume that there will be a perfect perspective projection with center $\mathbf{C}$ (the *optical center*) at a distance $\mathbf{f}$ (*focal distance*) from the *retinal plane* $\mathcal{R}$. The plane containing the optical center is called the *focal plane*.

Three main coordinate frames are defined: the **world coordinate frame**, the **camera frame** defined by its origin $\mathbf{C}$ and the axis (X_c, Y_c, Z_c), and the

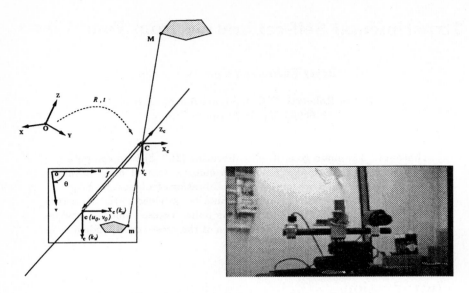

Fig. 1. The pinhole model and our binocular head.

normalized frame at focal distance equal to 1, which origin is called the *principal point* **c**, and axis (X_c, Y_c).

Using projective coordinates, the 3D world point $\mathbf{M} = [X, Y, Z]$ and its retinal normalized image $\mathbf{m} = [u, v]$, are related by

$$
\begin{pmatrix} su \\ sv \\ s \end{pmatrix} = \mathbf{A} \underbrace{\begin{pmatrix} 1\,0\,0\,0 \\ 0\,1\,0\,0 \\ 0\,0\,1\,0 \end{pmatrix} \mathbf{D}}_{\mathbf{P}} \begin{pmatrix} X \\ Y \\ Z \\ 1 \end{pmatrix} , \tag{1}
$$

where $\mathbf{A}$ is the matrix of the *intrinsic parameters* and $\mathbf{D}$ the matrix of the *extrinsic parameters*:

$$
\mathbf{A} = \begin{pmatrix} \alpha_u & 0 & u_0 \\ 0 & \alpha_v & v_0 \\ 0 & 0 & 1 \end{pmatrix} , \mathbf{D} = \begin{pmatrix} \mathbf{R}\ \mathbf{t} \\ \mathbf{0}\ 1 \end{pmatrix} .
$$

We have chosen to represent the 3x1 translation vector $\mathbf{t}$ in spherical coordinates ($\mathbf{t}$ is defined up to a scale factor and chosen to be unitary), the rotation matrix $\mathbf{R}$ with the Rodrigues formulation [8] and $\mathbf{A}$ depending only on four parameters α_u, α_v, u_0 and v_0. Other authors use a fifth parameter θ, which measures the non-orthogonality of the pixels, but its value is always less than noise [1].

3   The Fundamental Matrix.

As already demonstrated by Faugeras in [2] the equation of Longuet-Higgins relating the fundamental matrix $\mathbf{F}$ and the point correspondences $\mathbf{q}$ and $\mathbf{q}'$

between two views can be written:

$$q'^t F q = 0 , \qquad (2)$$

with:

$$\boxed{F = A'^{-1t} E A^{-1}} , \qquad (3)$$

where $E = TR$ is the **essential matrix**. T is a skew symmetrical matrix defined by the translation vector t such that $Tx = t \wedge x$ for all 3D vector x ($\wedge$ denotes the cross-product). Of course, $E = A^t F A$ when both cameras are identical.

For any pair of views we are trying to compute: 3 parameters for the rotation, 2 for the translation (defined up to a scale factor), and 4 intrinsic parameters. Since the essential 3x3 matrix $E = TR$ is of rank two, and because A and A' are invertible, F is of rank two. Then, we can compute 7 parameters from each fundamental matrix: the motion between the two views and 2 intrinsic parameters. The principal point (u_0, v_0) is fixed, and only the factors α_u and α_v need to be estimated.

4   Calibrating the Binocular Head-Eye System

We use a binocular head-eye system (Fig. 1). A description of this active visual system can be found in [6]. To calibrate the head-eye we need to know the intrinsic parameters of each camera at any position (in this study A_i with $i = 1..4$), the displacement of each camera (right D_r and left D_l), the displacement between both cameras (stereo displacements D and D'), and finally the cross-displacements (D^o, D^*). From Fig. 2 we can easily write: $D' D_l = D_r D$, $D^o = D_r D$ and $D_r = D' D^*$. This means that we can compute all rotation matrices and translation vectors from, for instance, R_r, t_r, R, t, R' and t'.

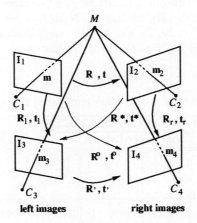

Fig. 2. Our stereoscopic system

4.1 The Non-Linear Minimization Algorithm

The **input** is the six fundamental matrices relating the four views (or at least the first four): $\mathbf{F}_{12}$, $\mathbf{F}_{13}$, $\mathbf{F}_{24}$, $\mathbf{F}_{34}$, $\mathbf{F}_{14}$, $\mathbf{F}_{23}$, and the correspondences. We denote $\mathbf{m}_{ij}^{ik}$ as the j-th correspondence in image i, and between the images i and k. As explained in Sect. 3, we obtain from each $\mathbf{F}$ an estimation of $\mathbf{R}$ and $\mathbf{t}$ once the intrinsic parameters are known. We compute the **output** intrinsic and extrinsic parameters replacing in the next **criterion** each $\mathbf{F}$ by the equation 3, and minimizing:

$$\min \left[\sum_{i=1}^{p} ((\mathbf{m}_{2i}^{21})^t \mathbf{F}_{12} \mathbf{m}_{1i}^{12})^2 + \sum_{j=1}^{q} ((\mathbf{m}_{3j}^{31})^t \mathbf{F}_{13} \mathbf{m}_{1j}^{13})^2 + \sum_{k=1}^{r} ((\mathbf{m}_{4k}^{42})^t \mathbf{F}_{24} \mathbf{m}_{2k}^{24})^2 + \right.$$

$$\left. \sum_{l=1}^{s} ((\mathbf{m}_{4l}^{43})^t \mathbf{F}_{34} \mathbf{m}_{3l}^{34})^2 + \sum_{m=1}^{t} ((\mathbf{m}_{4m}^{41})^t \mathbf{F}_{14} \mathbf{m}_{1m}^{14})^2 + \sum_{n=1}^{u} ((\mathbf{m}_{3n}^{32})^t \mathbf{F}_{23} \mathbf{m}_{2n}^{23})^2 \right]$$

4.2 The Parameters Computed depending on the Model Used.

The **Model 0**, or the simplest model, assumes that $\mathbf{D} = \mathbf{D}'$ and the intrinsic parameters are constant. This is the case when the vergence of the right camera, the zoom and the focus are not changed. The principal point is fixed to the center of the image: (255, 255) pixels. To reduce the number of intrinsic parameters (**Model 1**), we know [1] that the quotient $C_0 = \frac{\alpha_u}{\alpha_v} = 0.7$ is constant, so we can write the intrinsic parameter matrix for each camera position i:

$$\mathbf{A} = \begin{pmatrix} C_0.\alpha_v & 0 & u_0 \\ 0 & \alpha_v & v_0 \\ 0 & 0 & 1 \end{pmatrix} .$$

Using the **Model 2**, we do not a priori set any parameter as being constant. If $\mathbf{N}$ is the number of views [7], we have 11N-15 independent parameters, which is in our case 11*4-15 = 29. In the table 1 we have represented the different models. We denote i=1, ..., N as the index of the view.

Table 1. The Euclidean parameters computed depending on the model

EUCLIDEAN PARAMETERS					
Model	intrin	rotation	trans	TOTAL	N = 4
Model 0 u_0, v_0 fixed A_i fixed, $\mathbf{D} = \mathbf{D}'$	4	$\frac{3N}{2}$	$\frac{3N}{2}- 1$	3N+3	15
Model 1 $\frac{\alpha_u}{\alpha_v} = 0.7$	3N	3(N-1)	3(N-1)-1	9N-7	29
Model 2	4N	3(N-1)	3(N-1)-1	10N-7	33

4.3 Euclidean Reconstruction Results Using Real Images

Two **minimization routines** have been used: *e04fcf()* of NAG and *frprmn()* of Numerical Recipes (NR). The parameters are initialized to their values by default: (u_0, v_0) are initialized to (255, 255) and (α_u, α_v) are set to (800, 800). The extrinsic parameters are computed by developing (3), once the intrinsic parameters are set. In the Fig. 3, 4 we show different views of the **3D Euclidean reconstructed** segments of the scene.

SEQUENCE 1: Intrinsic and stereo parameters are constant. Between the two pairs only the elevation of the stereo frame changes. The correspondences are extracted automatically using the grid.

Fig. 3. From the left to the right: *Camera View* of the superimposed 3D Euclidean reconstruction using model 1 and the library NAG when the parameters are initialized to their default values, *Top View* of 3D Euclidean reconstruction result of our algorithm when the parameters are initialized to their default values, initialized by the result of "hard calibration" [4], and the result of the "hard calibration" method, respectively.

SEQUENCE 2: Intrinsic and stereo parameters change between the two views. We focused, zoomed and changed the vergence's value. The reconstruction using model 0 and 1 is insatisfactory. The parameters are initialized to their default values. The correspondences are extracted by hand using the sub-pixel accuracy.

Fig. 4. *Top, Camera and Front View* of the 3D Euclidean reconstruction using the model 2 with the library NR (the results with NAG are similar).

5 Conclusions

We want to point out that although we are using a calibration grid for a convenient feature detection, only sequence 1 uses the 3D model of the grid to increase the number of correspondences and to compare the results with a "hard calibration" method. The other sequence use 50 correspondences. In every case we are using a self-calibration method, that is to say, without using the 3D model of a known pattern in order to compute the perspective projection matrix. The segments are reconstructed automatically.

The most relevant conclusions are: a) The simplified model 0 is not suitable at all when extrinsic or intrinsic parameters change. To set the principal point to the center of the image when using a camera with a zoom is not realistic at all. b) The model 1 is convenient when we do not use the zoom, then the intrinsic parameters $\alpha_u = 0.7 * \alpha_v$. It works better than the model 2 because there are 29 unknowns (the same number of independent parameters).

The gradient conjugate method's results are worse than Newton's modified method results for real images.

In the case the **stereo displacement does not change** between the two frames, if we initialized the first frame's parameters with the calibration hard method, then the results are very good. In the case in which the **stereo parameters change**, the algorithm minimizes the epipolar distance but increases the reconstruction error.

References

1. R. Enciso, T. Viéville, and O. Faugeras. Approximation du changement de focale et de mise au point par une transformation affine à trois paramètres. *Traitement du Signal*, 11(5), 1994.
2. O. Faugeras, Q.-T. Luong, and S. Maybank. Camera self-calibration: theory and experiments. In *2nd ECCV*, pages 321–334, Santa-Margerita, Italy, 1992.
3. T. Luong. *Matrice Fondamentale et Calibration Visuelle sur l'Environnement*. PhD thesis, Université de Paris-Sud, Orsay, 1992. PhD thesis.
4. L. Robert. *Perception Stéréoscopique de Courbes et de Surfaces Tridimensionnelles, Application à la Robotique Mobile*. PhD thesis, Ecole Polytechnique, Palaiseau. France, 1992. PhD thesis.
5. T. Viéville. Autocalibration of visual sensor parameters on a robotic head. *Image and Vision Computing*, 12, 1994.
6. T. Viéville, E. Clergue, R. Enciso, and H. Mathieu. Experimenting 3d vision on a robotic head. In *The 12th Int. Conf. on Pattern Recognition*, pages 739–743, 1994.
7. T. Viéville, Q. Luong, and O. Faugeras. Motion of points and lines in the uncalibrated case. *International Journal of Computer Vision*, 1994. To appear.
8. Z. Zhang, Q. Luong, and O. Faugeras. Motion of an uncalibrated stereo rig: Self-calibration and metric reconstruction. Technical Report 2079, INRIA, 1993.

Vision-based Navigation in Service Robotics

G.Garibotto, M.Ilic, S.Masciangelo

TELEROBOT
via Hermada 6, Genova, Italy

Abstract. The paper describes an integrated mobile robot which has been designed to perform autonomous mission in service robotics. Computer Vision represents the main sensory system for autonomous navigation by artificial landmark recognition and 3D location. The orientation control through vanishing point analysis is performed by Kalman filtering. The system is proved to successfully work in a crowded scenario, it requires minimal impact on the environment, with a friendly user interface.

1. Introduction

Service Robotics is an emerging field of applications aimed to provide the necessary flexibility in performing repetitive tasks, as light material transportations (mail, documents, medicines, clinical data), and for heavy transportations (warehouse management and material handling). There are prospects [1] of wide expansion in various different fields, from hostile environments (nuclear, chemical) in surveillance and patrolling, entertainment, up to future domestic applications as cleaning and assistance.

A basic requirements in all these applications is autonomous navigation with capabilities to adapt to the different operating conditions. In order to demonstrate the maturity of vision-based navigation technology, also using low-cost PC-based processing platform, we have decided to realize a prototype of a mobile robot, named SAM (Autonomous Mobile System), suitable to address a sufficiently representative class of autonomous navigation and transport tasks, in an indoor environment, in presence of people.

Moreover, we have also experienced a novel application of service robotics, as a museum guide. The robot, equipped with a sound/voice generation system, was installed at Palazzo Ducale, Genova, an historical building where temporal expositions are periodically held. The robot was able to navigate between *points of interests* where it could stop and generate a vocal message or a sound. Moreover other messages were generated according to the particular navigation condition; for instance, the presence of people along the navigation path was managed by the by means of a proper message or sound. This paper does not include a description of the vocal interface, which is available in [6], but is mainly focused on the platform navigation capabilities. In the next section an overview of the prototype is provided, then, some technical aspects of the prototype are described in greater detail with particular reference to vision based global positioning subsystem. Finally some experimental results are referred in terms of reliability and easy of use of the system.

2.   Description of the mobile robot SAM

The logic architecture is decomposed in almost independent layers of competencies, each one in charge of a single, well defined task, such as obstacle avoidance or global position maintenance. The obstacle avoidance strategy is *reflexive*, that is the trajectory is heuristically determined on the basis of sensor readings rather than accurately planned starting from a reconstructed local environmental map. The suboptimality of the obtained trajectory is largely compensated by the fast response time which allows to navigate safely at an acceptable speed also in presence of cluttered environments. Moreover map making and planning capabilities are limited as far as possible in order to spare computational power and software development costs.

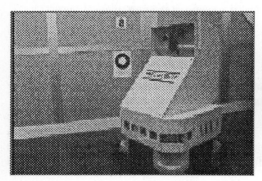

Figure 1: The mobile robot SAM   Figure 2: The artificial landmark

The hardware solution is based on a PC platform as the main computational infrastructure to reduce costs, minimise the development time and take advantage of the wide choice among a great variety of add-on boards which can be integrated to improve the system functionality.

3.   Computer Vision for Autonomous Navigation

3.1   High level planning and pose computation

The navigation system needs a periodic position and orientation estimation coming from an external sensor in order to reset drifts of odometry.
This is provided through a Vision system able to detect and recognise *navigation landmarks* placed in known position along the robot routes [4], and to recover the robot position and orientation with respect to them. A navigation landmark may an object already present in the environment *(natural landmarks)* or a feature placed on purpose along the robot routes *(artificial landmarks)*.
In our approach, each artificial landmark consists of a black annulus on a white background, visible in Fig.2. The technique that allows to recover the 3D position

and attitude of the camera from a single image is known as *model based perspective inversion* [2].

Although the use of artificial landmark is a reliable way to solve the problem of self-positioning, it requires an intervention on the environment, which should be minimized, as far as possible. Anyway, when moving along hallways, it is possible to recover the robot absolute orientation from the analysis of the vanishing points. They are natural landmarks, since they are already present in the world and may be extracted reliably in indoor environments, where the contours of visible surfaces are mainly formed by either parallel or mutually orthogonal straight lines. The vanishing point is invariant to camera translation movements, therefore its tracking along an image sequence can give an estimate of the camera orientation [3].

Which of the two Vision algorithms is invoked is specified in the global map and therefore depends on the robot current position. The global map is not a complete 3D map but rather an abstract description based on *points of interest* located in strategic positions for the navigation.

Neighbouring points of interest are connected by vectors, whose module represents the length of the path and the vector orientation is the robot's heading direction. A valid navigation plan is a sequence of points of interest, or subgoals, to be reached.

When the robot reckons to be on a point of interest, it can take either of the following actions:

- Stop and verify its position by observing the relative artificial landmark or vanishing point (this is specified in an attribute of the point of interest), reset odometric errors and turn towards the following subgoal;
- Execute a motion command such as an absolute turn.

Landmarks differ each other for the diameter of the inner circle. This parameter, that can be measured very accurately on the image plane, is a *perspective invariant,* i.e., does not change with respect to the viewpoint. That property, together with the perspective inversion, allows to determine the absolute location of the robot in the global map and to replan the mission, that is the next path segment towards the given final goal.

The same *search* strategy allows the robot to start the mission from any point in which at least one landmark is visible, without any a-priori information on the starting position.

3.2 Heading control from visually detected vanishing point direction

In order to improve the reliability of visual navigation a continuous visual control algorithm which employs together vision and odometry has been developed and integrated into the navigation system. The module controls the robot's heading angle by exploiting again the vanishing point generated by the 3D worls structure typical of indoor environment. In fact horizontal parallel lines generate two bundles, each one meeting the horizon line at infinity in a vanishing point, as shown in Fig. 3.b. By extension, the direction of the straight line that links the vanishing point to the camera pinhole is also called *vanishing point*. Thus the heading of the vehicle can be recovered from one of the two vanishing points, by adding the fixed angle between the forward vehicle direction, known with respect to the odometry coordinate system, and the projection of the camera optical axis onto the horizon, estimated by calibration.

In order to extract 3D geometric information, as the orientation of a vanishing point, from the 2D features of an image, it is necessary to compute, off-line, the camera intrinsic parameters as well as its orientation and position with respect both to the vehicle odometry coordinate system and the horizontal navigation plane. All these parameters are computed jointly by moving the vehicle in different positions and using the corresponding odometric information.

3.3   Sensor fusion through Kalman filtering

The integration with the odometry has been carried out through Kalman filtering, to deal with different confidence values (variance data) of the input measures, which are acquired at different time scales, according to their computational requirements. The process equation for the vehicle heading angle θ_i at time t_i for $i > 0$ may be written as

$$\theta_i = \theta_{i-1} + \Delta\theta_{\text{odo}_{i-1}} + w_{i-1} \quad \text{where} \quad \begin{cases} w_i = \Delta\theta_i - \Delta\theta_{\text{odo}_i} \\ \Delta\theta_i = \theta_{i+1} - \theta_i \\ \Delta\theta_{\text{odo}_i} = \theta_{\text{odo}_{i+1}} - \theta_{\text{odo}_i} \end{cases}$$

and θ_{odo_i} is the value of the odometry heading angle register on the vehicle as read at time t_i.

While performing explicit Kalman filter steps at vision measurement rate only, this approach exploits the odometry system embedded in the platform controller, according to the block diagram of Fig. 3.a, which polls encoders at higher rate. It would be interesting to work out a statistical model also for the odometry computation process in order to better integrate it in the data fusion schema provided by the Kalman filter. More details on the Kalman filter implementation can be found in [7]

Experiments on the performance and use of the fusion with the odometry of the heading angle measurements from vanishing point detection have been carried out on the SAM mobile system. Commands were issued to make the vehicle navigate straight ahead for 5000 mm at 250 mm/s constant speed.

In order to emphasise the effect of vision measurements updates with respect to a poor odometry performance, the path was suitably arranged to make the vehicle navigate with one wheel on the floor while the other was slipping on a thick carpet. Without vision this resulted in a great error of the final lateral displacement, about 0.5 m, and of heading, about 15 degrees, which was nonetheless recovered as soon as the vision-based heading measures have been introduced. In order to allow a motion feedback control loop based not only on wheel encoder readings, but also on the current odometry heading register, a high-level control loop was added to generate, while advancing, a corrective rotational velocity proportional to the heading angle error as reported by the odometry heading register.

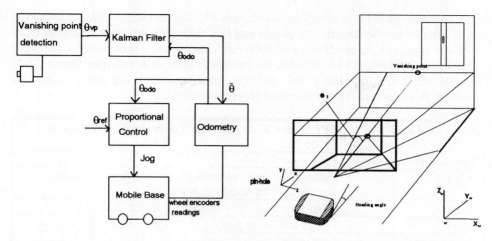

Figure 3: a) Schema of the vehicle direction feedback control system. b).The vanishing point of the scene and its mapping onto the image plane. The vehicle Heading angle is also shown.

The rate of this control was four times higher than the vision-based heading measurement procedure, whose average execution time was 0.7 s for a scene showing about 350 segments, on a PC486 33MHz. The final lateral displacement error was reduced to 0.25 m and the heading error to 1.5°. Even if the lateral displacement was not explicitly controlled, the vehicle heading feedback loop kept the vehicle from drifting aside too much, showing once again that the most important navigation control variable is heading.

4. Experimental results

It is important to evaluate the system as a whole more than the single modules. Experiments consisted in making the robot move on a fixed route first in our office environment, usually full of people wandering around. Later the robot has been equipped with a Soundblaster© board for sound generation and a radio modem providing a link to an host computer at a remote control station. The remote computer is able to select the most appropriate sound or voice file according to the robot position or the navigation status (presence of obstacles, landmark search, and so on) as communicated by the robot navigation system.

The robot in such a configuration was installed in an historical building during the Christmas '93 exhibitions, as reported in Table 1. Environments were not artificially structured, apart from the application of few circular landmarks (20 cm^2) on the walls. General public, mainly composed by children, was not prepared to any interaction with the robot, therefore it was not possible to foresee the reaction and the behaviour in presence of it.

The results were very encouraging and showed that the obstacle avoidance system is quite reliable for the detection of people and that the vision system was able to work robustly in every light condition. Problems came from the excessive number of people surrounding the robot, who occluded the camera field of view and from the very bad vehicle odometry, depending on the floor conditions, that caused occasionally mission aborts in between two points of self-positioning.

Navigation map	a corridor 30 m long and 7 m wide, plus a large area 20 X 15 m
Artificial landmarks	5
Function	attraction and entertainment, explanation about the exhibitions by vocal interface
Duration	four days including rehearsals, about 25 hours
Public	high density of people, great interest among children
Cruise speed	60 cm/s
Anomalies	a user intervention every 10 min in average due to: landmark occlusion for the presence of people (70 %), loss of orientation because of slippage on irregular floor patches (30\%).

Table 1: The main characteristics of the experiments carried out at Palazzo Ducale (Genova) during Christmas exhibitions 1993.

References

[1] J.Engelberger, "Robotics in Service", London, Kogan Page Ltd, 1989.

[2] M. Ferri, F. Mangili, G. Viano, "Projective Pose Estimation and Quadratic Primitives in Monocular Computer Vision", CVGIP: Image Understanding,vol. 58, N. 1, July 1993.

[3] Straforini, C. Coelho and M. Campani, "Extraction of vanishing points from images of indoor and outdoor scenes", Image and vision computing, vol. 11, no 2, March 1993.

[4] G.B. Garibotto and S. Masciangelo, "3D Computer Vision for Navigation/Control of Mobile Robots", in Machine Perception, AGARD Lecture Series 185, 1992.

[5] X. Lebegue and J.K.Aggarwal, "Extraction and Interpretation of Semantically Significant Line Segment for a Mobile Robot ", Proc. 1992 International Conference on Robotics and Automation, Nice, France, May 1992.

[6] G. B. Garibotto, S. Masciangelo, M Ilic, A Camurri, G.Vercellli, "A Mobile Robotic Museum Guide Demonstrating Intelligent Interaction with the Environment and the Public", ISATA'94, Aachen (Germany), October 1994.

[7] G.B. Garibotto, M. Ilic, S. Masciangelo, "Multisensor Integration for Autonomous Mobile Robotics", Proc. of the Workshop of the Italian Association of Artificial Intelligence, Parma, Sept. 1994

Digital Topology and Morphology

Segmenting 3D Objects into Geons

Kenong Wu and Martin D. Levine

Center for Intelligent Machines & Dept. of Electrical Engineering
McGill University, Montréal, Québec, Canada, H3A 2A7

Abstract. We propose a new approach to 3D object segmentation and description. Beginning with multiview range images of a 3D object, we segment the object into parts at deep surface concavities. Motivated by physics, we detect these concavities by locating surface points where the simulated electrical charge density achieves a local minimum. The individual parts are then described by parametric geons. The latter are defined as seven distinctive volumetric shapes characterized by constrained superellipsoids, with deformation parameters controlling the tapering and bending. We obtain a unique part model by fitting all parametric geons to each part and classifying the fitting residuals. The advantage of our classification approach is its ability to approximate the shape of an object, even one not composed of perfect geon-like parts. The resulting 3D shape description is a prerequisite for generic object recognition.

1 Introduction

Biederman has proposed a theory of Recognition-by-Components(RBC) [4] based on psychophysical experiments which involves a modest set of volumetric primitives, called *geons*, as qualitative descriptions of object parts. The theory requires that an arrangement of a few geons be recovered from an object by segmenting the object at deep surface concavities. Then, having the geon-based description of the object, it was postulated that it could be quickly recognized with great tolerance to variations in viewpoint and shape. Since 1985, significant computational efforts have been made to create geon-based descriptions of 3D objects [13, 11, 17, 3, 24, 29, 28, 20, 36, 23, 10].

The paradigm of geon-based representation consists of two interrelated stages, object decomposition and geon identification. Object decomposition is concerned with segmenting an object into parts, for each of which a geon model needs to be identified. Decomposition algorithms for geon-based representation can be categorized as being region(shape or primitive)-based or boundary-based approaches. Region-based approaches [11, 29, 12] first find image regions that correspond to object surface patches, and then group these patches into individual parts based on particular geon surface configurations. This scheme relies on the limited number of surface shape types of *perfect* geons. Therefore each surface corresponding to a geon can be inferred by (conceptually) examining all possible combinations of surface types. The disadvantage of this approach is that if an object is not composed of perfect geons or mistakes are made in surface segmentation, the surface type label might be in error. This would produce an incorrect object decomposition. Boundary-based approaches [2, 24] are directly inspired by the theory of Hoffman and Richards [16] and locate part boundaries at deep surface concavities. Since they concentrate on part boundaries rather than part shape, they can segment an object regardless of whether it is composed of perfect geons. However, there is a limitation since the part boundary points must be explicitly defined. This may not always be the case [16]. We also note that the literature mentions other part segmentation algorithms which were not specifically intended for geon-based representations but could well serve this purpose [26, 25, 8, 14, 15, 22, 30].

The early work on geon identification focused on the recovery of geon models from complete line drawings which depicted objects consisting of *perfect* geon-like parts [3, 11, 17]. Generally, however, such "clean" or complete line drawings of objects cannot be obtained easily due to the color and texture of object surfaces and complex illumination configurations. Because of this, and also for practical reasons, some research has focussed recently on data obtained from laser rangefinders [24, 29]. In both cases, part descriptions are determined in a bottom-up fashion, inferring global properties by first aggregating local features. This approach is not very effective when object features do not fully satisfy the exact definitions of the geons. Clearly, any computer vision system which successfully recovers qualitative descriptions needs to address the problem of *shape approximation*. Two approaches for geon model recovery have been proposed to achieve this. Both essentially amount to using *quantitative* methods. Raja and Jain [28] identified a shape as one of 12 geon types. They fitted range data to a deformed superellipsoid model [26, 32] and then classified its shape by comparing the object's superellipsoid parameters with a database of geon parameter models. However, these superellipsoid parameters are very sensitive to even minor shape variations. By comparison, we have defined only seven parametric geon models [35]. We characterize object shape by fitting all of the models to multi-view range data and classifying the fitting residuals [36, 37]. This method uses parametric geons as explicit shape constraints, thereby enabling shape verification by directly comparing the goodness of fit between the seven geons and the part shape. There are other reasons for preferring the quantification of parameters over purely qualitative models. For example, in the case where qualitative shape information is the same for two objects, discrimination must rely on quantitative information, such as relative size of object parts or the specific curvature of the part axis. Moreover, object pose is often required for certain tasks, such as manipulation, for example. Thus Dickinson et al. have used qualitative, followed by quantitative, methods to recover ten geon models from both intensity images [10] and range data [12].

The key issue for geon-based representation is how to reliably produce *shape approximations* as object descriptions. This implies two steps: object decomposition and part identification. In this paper, we propose a novel approach to object decomposition into parts which is motivated by physics [38]. Assuming that the object to be segmented is a charged conductor, we simulate electrical charge over the object surface. Then we detect part boundaries by locating deep surface concavities where local charge densities reach their minima. The object is then segmented into parts at these points. Our strategy is consistent with boundary-based approaches [16]. However, unlike previous work [14, 30], we do not compute surface curvature and also do not directly address the scale problem when computing surface properties. Other work has relied on surface curvature estimation. These approaches suffered because o f the need to select a proper scale for the window size to the compute surface curvature. Following the object decomposition stage we recover a parametric geon model for each part by simultaneously fitting all parametric geon models to a putative part and selecting the model with the minimum fitting residual. The advantage of this approach is that we can recover part models which *approximately* fit the data. Thus we can tolerate noisy data, object surfaces which are slightly rough, and parts whose shape deviates from perfect geons. By contrast, methods based on local feature aggregation would probably encounter difficulties using imperfect data.

2 Object Decomposition

2.1 Physics

When a charged conductor with an arbitrary shape is in electrostatic equilibrium, all charge resides unevenly on the outer surface of the conductor [7]. The charge density is very high at sharp convex edges and corners. Conversely, almost no charge accumulates at sharp concavities. Therefore, deep surface concavities which have been shown to delineate part boundaries [16] can be detected by significant charge density minima.

Electrical charge densities at sharp edges and corners have been carefully studied by Jackson [19]. An object edge or corner is defined as a C^1 discontinuity of an object surface. Fig. 1 depicts examples of these. By ignoring secondary global effects, Jackson has derived an approximate relationship governing the charge density ρ at an edge formed by two conducting planes, as shown in Fig. 2.

In Fig. 2 (a), β is the angle between two planes defining an edge and η is the distance from the edge to a point P, where the charge density is measured. Fig. 2 (b) shows ρ as a function of β and η. The larger β and the smaller η, the greater the charge density. We observe that all sections of the function are monotonic for constant η. This relationship(see [19]) also reveals a theoretical singular behavior of charge densities at edges (for $\eta = 0$) as follows:

$$\rho = \begin{cases} \infty & if \quad \beta > \pi \\ constant & if \quad \beta = \pi \\ 0 & if \quad \beta < \pi \end{cases} \tag{1}$$

This means that the charge density is infinite, constant and zero when the angle defined by the two planes is convex, flat and concave, respectively. The singular behavior of charge densities at corners, which is similar to that at edges, has also been investigated [19].

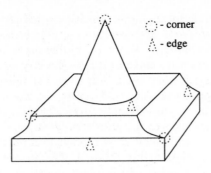

Fig. 1. Examples of edges and corners.

The physical model we have used is the charge density distribution on a perfect conductor in free space, where there is no other conductor or charge. Let $\mathbf{r}$ define an observation point and $\mathbf{r}'$ a source point charge q on the surface, as shown in Fig. 3. It is customary to select the reference point of the electric potential ϕ at infinity. Thus ϕ at $\mathbf{r}$ caused by q at $\mathbf{r}'$ is equal to [5]:

$$\phi(\mathbf{r}) = \frac{q}{4\pi\epsilon_0} \frac{1}{|\mathbf{r} - \mathbf{r}'|} \tag{2}$$

Here ϵ_0 is a constant, known as the permitivity of free space. Suppose the charge is continuously distributed over the object surface S (see Fig. 3). Then the electrical potential at $\mathbf{r}$ is contributed by *all of the charge on S*. This can be expressed as follows:

$$\phi(\mathbf{r}) = \frac{1}{4\pi\epsilon_0} \int_S \frac{\rho(\mathbf{r}')}{|\mathbf{r} - \mathbf{r}'|} dS' \tag{3}$$

Here $dq = \rho(\mathbf{r}')dS'$, $\rho(\mathbf{r}')$ is the charge density at $\mathbf{r}'$ and S' represents a unit area on S.

According to physics [7], all points on a charged conductor in electric equilibrium are at the same electric potential. If we restrict $\mathbf{r}$ in Equation (3) to the conductor surface, then $\phi(\mathbf{r})$ is constant. Accordingly, (3) may be rewritten as follows:

$$V = \int_S \frac{\rho(\mathbf{r}')}{|\mathbf{r} - \mathbf{r}'|} dS' \tag{4}$$

Here $V = 4\pi\epsilon_0\phi(\mathbf{r})$ is a constant.

We note that the object being considered here is located in a real 3D space. A diffusion-based method to compute surface properties of such objects has been reported in the literature [39]. Working on a voxel-based coordinate system, this approach simulates the propagation of a specific number of particles among the object voxels. At a certain stage of the diffusion process particle accumulations at sharp surface concavities and convexities become significant. Therefore sharp concave and convex surface points can be detected by diffusion. There are two

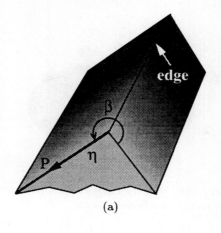

(a)

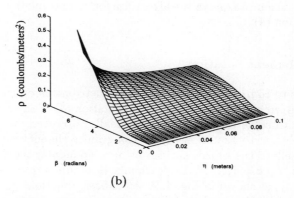

(b)

Fig. 2. Charge densities near edges. (a) An edge formed by two planes with an angle β. (b) The charge density at $P(\eta, \beta)$.

problems associated with this approach. First, it is difficult to decide when the diffusion process should be stopped. At diffusion equilibrium, the particle density is uniform everywhere within the object and therefore cannot indicate any specific geometrical property of the object surface. Thus the diffusion process must be stopped before equilibrium. However, if diffusion is stopped at an inappropriate time, the distinction between particle densities at concave and convex surface points will not be strong. In contrast, our approach deals with electrical equilibrium, where charge densities at concave and convex surface points are very distinct. Second, a diffusion-based approach must work in a voxel-based coordinate frame. This is very time-consuming if the object is large. However, our approach concentrates only on the surface of the object and does not need to perform computation within its interior. This reduction in dimensionality reduces the cost of both computational memory and time.

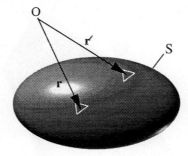

Fig. 3. The observation point **r** and the charge source point **r**′ on the surface S of an ellipsoid. O is the origin of the coordinate system.

In the next section, we derive the algorithm for computing the charge density based on Equation (4).

2.2 Finite Element Solution

Our objective is to compute the charge density distribution on the surface of an object having an irregular shape. Since S in Equation (4) is an arbitrary surface, it is impossible to solve the equation analytically. However, we can obtain an approximate solution to Equation (4) by using finite element methods [31]. To do this, we tessellate the object surface using a triangular mesh which has N planar triangles, $T_k, k = 1, ..., N$. Each triangle is assumed to possess a constant charge density, ρ_k, as shown in Fig. 4. A set of basis functions $f_k, k = 1, ..., N$ is defined on this triangular mesh as follows:

$$f_k(\mathbf{r}') = \begin{cases} 1 & if \quad \mathbf{r}' \in T_k \\ 0 & otherwise \end{cases} \tag{5}$$

The basis function, f_k, is nonzero only when $\mathbf{r}'$ is on the triangle T_k, as shown in Fig. 4. Therefore, the charge density $\rho(\mathbf{r}')$ can be approximated by a piecewise constant charge density function as follows:

$$\rho(\mathbf{r}') \approx \sum_{k=1}^{N} \rho_k f_k(\mathbf{r}') \tag{6}$$

Substituting (6) into Equation (4), we have

$$V = \sum_{k=1}^{N} \rho_k \int_{T_k} \frac{1}{|\mathbf{r} - \mathbf{r}'|} dS' \tag{7}$$

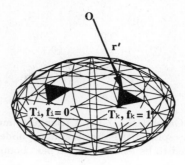

Fig. 4. Triangular mesh on the surface of an ellipsoid.

Since the charge density is assumed to be constant on each T_k, we may take $\mathbf{r}_i$ as the observation point on each T_i and rewrite Equation (7) as:

$$V = \sum_{k=1}^{N} \rho_k \int_{T_k} \frac{1}{|\mathbf{r}_i - \mathbf{r}'|} dS' \quad i = 1, \ldots, N. \tag{8}$$

Because of charge conservation, the sum of the charges on each triangle equals the total charge on the surface of the conductor. Let Q be the total charge on the conductor and S_k be the area of T_k. Then we have

$$Q = \sum_{k=1}^{N} \rho_k S_k \tag{9}$$

Assuming Q is known, and using (8) and (9), we obtain a set of linear equations with $N + 1$ unknowns, $\rho_1, \ldots, \rho_N$ and V. Since the integral in (8) can be evaluated analytically [34], the charge density distribution ρ_k and the constant V can be obtained by solving a set of linear equations.

2.3 Implementation

Our data are obtained using a laser rangefinder to scan objects from multiple views. Transformations between successive views are computed by a view registration algorithm [6]. Surface triangulation is performed by a method developed by DeCarlo and Metaxas [9][1]. The observation point $\mathbf{r}_i$ on each triangular patch is selected at the centroid of the triangle. The set of linear equations is solved by a standard LU decomposition method [27].

---

[1] Triangular mesh data used in this paper have been kindly provided by D. DeCarlo and D. Metaxas at the University of Pennsylvania.

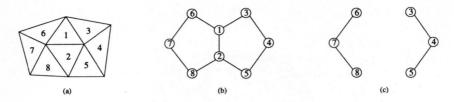

Fig. 5. Direct Connection Graph (DCG). (a) A triangular mesh. (b) DCG of the triangular mesh in (a). (c) Subgraphs of (b) after boundary node deletion. Here triangular patches 1 and 2 are assumed to be located on the part boundary.

2.4 Part Segmentation

After obtaining the charge densities on the object surface, we segment an object into parts by detecting and then deleting points on the part boundaries. The method is based on a *Direct Connection Graph*(DCG) defined on the triangular mesh, as shown in Fig. 5. Here the nodes represent the triangular patches in the mesh and the branches represent the connections between direct neighbors. By direct neighbors we imply that two triangles share two vertices. For example in Fig. 5 (a), triangles 1 and 2 are direct neighbors while 2 and 3 are not. Thus, the DCG provides a convenient coordinate system on the object surface.

We have assumed that a part boundary forms a *closed* 3D curve, explicitly defined by deep surface concavities. This ensures that the decomposition algorithm will be able to segment a part from the rest of the object. The assumption also provides a stopping criterion for the boundary tracing procedure.

The algorithm examines the charge density on all triangles to find a starting triangle for tracing each boundary. A starting triangle must satisfy the following conditions:

1. It must be a *concave* extremum; that is, its charge density must be a local minimum.
2. It must be located at a *deep* concavity. Thus the charge density on the triangle must be lower than a preselected threshold[2].
3. It and its neighbors must not have been visited before. This ensures that the same boundary will not be traced again.

Beginning at the starting triangle, the algorithm proceeds to the neighbor with the lowest charge density. During the tracing procedure, all triangles detected on the boundary are marked. The marked ones will not be checked again and

---

[2] This threshold determines when an object should not be decomposed further. If the charge density at a starting triangle is greater than this threshold, we assume that all boundary points have been found. The selection of the threshold depends on a priori knowledge of the surface concavity and there is no universal rule to determine it. Currently we choose 1.05 times the lowest charge density on the object surface as the threshold.

eventually will be deleted from the DCG. The process continues until it returns to the starting triangle. As a result of the above assumption, this means that all triangles on this part boundary have been visited. Next the algorithm finds a new starting triangle and traces another boundary. It recursively repeats the same tracing procedure, and then stops when the charge density at a starting triangle is higher than the preselected threshold. After all triangles on part boundaries have been found, the nodes of the DCG representing these triangles are deleted. Thus the original DCG is now divided into a set of disconnected subgraphs, as shown in Fig. 5. Physically the object has been broken into parts. Each subgraph represents one part of the object which is ready for geon identification.

3   Geon Identification

This section is an overview of our work on parametric geon recovery. Details can be found in [35, 36, 37].

Motivated by the art of sculpture, we have defined seven parametric geons (see Fig. 6). These shapes are commonly used by sculptors as basic volumetric shapes. They are regular, simple, symmetrical, distinctive and can be expressed by compact analytic functions. Defined by

$$f(\mathbf{x}, \mathbf{a}_i) = 0, \qquad i = 1, ..., 7 \tag{10}$$

parametric geons are derived from superellipsoids [1] by specifying appropriate shape parameters and applying tapering and bending deformations [32]. Here $\mathbf{x} \in R^3$, $\mathbf{a}_i$ is the nine- to eleven-dimensional vector of model parameters, and i is the model shape type. Parametric geons provide both qualitative shape and quantitative size and deformation information. Compared to the geon shapes proposed by Biederman [4], certain qualitative properties have been simplified for the sake of simplicity. The detailed derivation of the implicit and normal vector equations of parametric geons can be found in [35].

3.1   The Objective Function

The major step in parametric geon identification is to fit these parametric models to object data by searching a specific parameter set $\mathbf{a}$ such that a two-term objective function

$$\mathcal{E}(\mathbf{a}) = d_1(\mathbf{a}) + \lambda\gamma d_2(\mathbf{a}) \tag{11}$$

is minimized. This objective function reflects the similarity of size and shape between the object and fitted models. The first term of the objective function is given by

$$d_1 = \frac{1}{N} \sum_{i=1}^{N} |e(D_i, \mathbf{a})| \tag{12}$$

Here N is the number of data points denoted by $\{D_i \in R^3, i = 1, ..., N\}$.

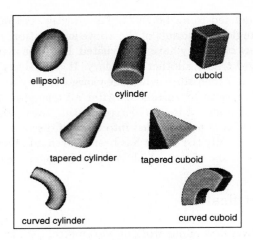

Fig. 6. The seven parametric geons.

For the three regular primitives (ellipsoid, cylinder and cuboid), $e(D_i, \mathbf{a})$ is defined as the Euclidean distance from a data point to the model surface along a line passing through the origin of the model and the data point [33]. Since tapering or bending significantly complicates the implicit primitive equations, we cannot obtain a closed-form solution for $e(D_i, \mathbf{a})$, as can be done for the regular primitives. Thus, in order to reduce the amount of computation, we have derived an approximate distance measure for the tapered and curved models [36].

We define the second term $(\lambda \gamma d_2)$ of the objective function by measuring a squared difference between the surface normal vectors $\mathbf{n}_d$ of objects and the surface normal vectors $\mathbf{n}_m$ of models at each corresponding position. Thus (d_2) is given by:

$$d_2 = \frac{1}{N} \sum_{i=1}^{N} e_n(i) \tag{13}$$

Here $e_n(i) = \|\mathbf{n}_d(i) - \mathbf{n}_m(i)\|^2$.

In (11) γ is taken as the average of the model size parameters. This tends to adapt the second term to the size of the object. λ is a weighting factor which controls the contribution of the second term to the objective function. In this paper, we choose $\lambda = 5$ according to a heuristic based on the shape difference between each pair of parametric geons [36]. A similar objective function [40] has been reported which used L_2 norms for both the first and second terms. In contrast, we use the L_1 norm for the first term in order to speed up the fitting process [36].

3.2 Very Fast Simulated Re-annealing

Fitting parametric geons to range data involves a search for a particular set of parametric geon parameters which minimize the objective function in (11).

Fig. 7. Recovery of parametric geons. (a) triangular mesh of the object; (b) Charge density over object surface. (c) Two separated parts; (d) object description in terms of parametric geons.

This function has a few deep local minima caused by an inappropriate orientation of the model, and many shallow local minima, caused by noise and minor changes in object shape. In order to find the model parameters corresponding to the global minimum of the objective function, we employ a stochastic optimization algorithm, Very Fast Simulated Re-annealing (VFSR) [18]. This algorithm permits an annealing schedule which decreases exponentially in annealing time and is dramatically faster than traditional simulated (Boltzmann) annealing [21] whose annealing schedule decreases logarithmically.

The search space can be limited somewhat by appropriately choosing constraints.seven A constraint on the model centroid and size can be estimated from the 3D range data. Constraints on the rotation, tapering and bending parameters can be specified by theoretical considerations [36].

After parametric geons are fitted to the 3D data, the best model for the object is selected according to the minimum fitting residual.

4 Experiments

4.1 Simulated Charge Distribution

We have applied our method to a simple vase in order to illustrate parametric geon recovery. Fig. 7 (a) shows the triangular mesh for this 3D object. Fig. 7 (b) shows the computed charge density distribution over its surface. The gray levels indicate charge densities, which are normalized in the range between 0 (darkest intensity) and 250 (white). It can be clearly seen that the lowest charge densities are located at surface concavities, which are at the intersection of the spherical and cylindrical portions of the vase. Conversely, since the edge on the top of the vase is sharply convex, the charge density at these points reaches a maximum.

4.2 Object Decomposition

Given that the charge density distribution over the object surface shown in Fig. 7 (b), our algorithm finds the part boundary shown in black. It then segments the object at the these boundary points, breaking the object into two parts as illustrated in Fig. 7 (c).

4.3 Parametric Geon Recovery

After the object data are segmented into parts, our algorithm fits all of the parametric geon models to the data of each part. The model for a particular part is selected based on the minimum fitting residual. An ellipsoid and a cylinder were found as models of the lower and upper parts of the vase, respectively. The parametric geon description of the vase is shown in Fig. 7 (d).

5 Conclusions

We have presented a new approach to 3D object segmentation and representation. The general strategy is to segment an object into parts at deep surface concavities and approximate the shape of each part in terms of finite volumetric primitives. We compute and analyse a simulated charge density distribution over the object surface to find the deep surface concavities. The object is broken into parts at these points. Unlike previous curvature-based approaches, this method computes local surface information without the frustration of having to choose a particular scale for analyzing local data. Although our algorithm works in a full 3D space, it does not deal with the interior of objects as do voxel-based approaches. This saves both memory and time by restricting the computations to the object surface. We recover a qualitative part description by comparing the data with all seven geon models. A particular one is selected based on a measured similarity function which accounts for both shape and size. Using an experimental example we have illustrated how our approach derives a parametric geon-based description from multiview range images. This kind of description is a prerequisite for generic object recognition.

Acknowledgements

We wish to thank Professors Lester Ingber, Demitri Metaxas, and Jonathan Webb, Gerard Blais, Douglas DeCarlo and Gilbert Soucy for their kind help. MDL would like to thank the Canadian Institute for Advanced Research and PRECARN for its support. This work was partially supported by a Natural Sciences and Engineering Research Council of Canada Strategic Grant and an FCAR Grant from the Province of Quebec.

References

1. A. H. Barr. Superquadrics and angle-preserving transformations. *IEEE Computer Graphics Applications*, 1:11–23, 1981.
2. R. Bergevin and M. D. Levine. Part decomposition of objects from single view line drawings. *CVGIP: IMAGE UNDERSTANDING*, 55(1):73–83, January 1992.
3. R. Bergevin and M. D. Levine. Generic object recognition: Building and matching coarse descriptions from line drawings. *IEEE Transactions on Pattern Analysis and Machine Intelligence*, 15(1):19–36, January 1993.

4. I. Biederman. Human image understanding: Recent research and a theory. *Computer Vision, Graphics, and Image Processing*, 32:29–73, 1985.

5. J. Van Bladel. *Electromagnetic Fields*. Hemisphere Publishing Corporation, revised printing edition, 1985.

6. G. Blais and M. D. Levine. Registering multiview range data to create 3D computer objects. *IEEE Transactions on Pattern Analysis and Machine Intelligence*, 1994, Accepted for publication.

7. F. J. Bueche. *Introduction to Physics for Scientists and Engineers*. McGraw-Hill Book Company, New York, 3rd edition, 1980.

8. T. Darrell, S. Sclaroff, and A. Pentland. Segmentation by minimal description. In *Proceedings of Third International Conference on Computer Vision*, pages 112–116, Osaka, Japan, December 1990. IEEE.

9. D. DeCarlo and D. Metaxas. Adaptive shape evolution using blending. In *IEEE Proceedings of International Conference on Computer Vision (to appear)*, 1995.

10. S. J. Dickinson and D. Metaxas. Integrating qualitative and quantitative shape recovery. *International Jounral of Computer Vision*, 13(3):311–330, 1994.

11. S. J. Dickinson, A. P. Pentland, and A. Rosenfeld. 3D shape recovery using distributed aspect matching. *IEEE Transactions on Pattern Analysis and Machine Intelligence*, 14(2):174–198, February 1992.

12. S. J. Dickson, D. Metaxas, and A. Pentland. Constrained recovery of deformable models from range data. In *Proceedings of Internaitional Workshop on Visual Form*, Capri, Italy, May 1994.

13. R. Fairwood. Recognition of generic components using logic-program relations of images contours. *Image and Vision Computing*, 9:113–122, 1991.

14. F. Ferrie and M. D. Levine. Deriving coarse 3D models of objects. In *IEEE Computer Society Conference on Computer Vision and Pattern Recognition*, pages 345–353, Ann Arbor, Michigan, June 1988.

15. A. Gupta and R. Bajcsy. Volumetric segmentation of range images of 3D objects using superquadric models. *CVGIP: Image Understanding*, 58(3):302–326, November 1993.

16. D. Hoffman and W. Richards. Parts of recognition. *Cognition*, 18:65–96, 1984.

17. J. Hummel and I. Biederman. Dynamic binding in a neural network for shape recognition. *Psychological Review*, 99(3):480–517, July 1992.

18. L. Ingber. Very fast simulated re-annealing. *Mathematical and Computer Modelling*, 12(8):967–973, 1989.

19. J. D. Jackson. *Classical electrodynamics*. Wiley, New York, 1975. Secs. 2.11 and 3.4.

20. A. Jacot-Descombes and T. Pun. A probabilistic approach to 3-D inference of geons from a 2-D view. In K. W. Bowyer, editor, *Applications of Artificial Intelligence X: Machine Vision and Robotics, Proceedings of SPIE*, volume 1708, pages 579–588, Orlando, Florida, April 1992.

21. S. Kirkpatrick, Jr. C. D. Gelatt, and M. P. Vecchi. Optimization by simulated annealing. *Science*, 220(4598):671–680, May 1983.

22. A. Lejeune and F. Ferrie. Partioning range images using curvature and scale. In *Proceedings of 1993 IEEE Computer Society Coonference on Computer Vision and Pattern Recognition*, pages 800–801, New York City, NY, June 1993. IEEE Computer Society Press.

23. R. C. Munck-Fairwood and L. Du. Shape using volumetric primitives. *Image & Vision Computing*, 11(6):364–371, July 1993.

24. Q. L. Nguyen and M. D. Levine. 3D object representation in range images using geons. In *11th International Conference on Pattern Recognition*, Hague, Netherlands, August 1992.

25. A. Pentland. Part segmentation for object recognition. *Neural Computation*, 1:82–91, 1989.

26. A. P. Pentland. Recognition by parts. In *The First International Conference on Computer Vision*, pages 8–11, London, June 1987.

27. W. H. Press, S. A. Teukolsky, W. T. Vetterling, et al. *Numerical Recipes in C*. Cambridge University Press, Cambridge, UK, 2nd edition, 1992.

28. N. S. Raja and A. K. Jain. Recognizing geons from superquadrics fitted to range data. *Image and Vision Computing*, 10(3):179–190, April 1992.

29. N. S. Raja and A. K. Jain. Obtaining generic parts from range data using a multi-view representation. *CVGIP: Image Understanding*, 60(1):44–64, July 1994.

30. H. Rom and G. Medioni. Part decomposition and description of 3D shapes. In *Proceedings of the 12th International Conference on Pattern Recognition*, volume I, pages 629–632, Jerusalem, Israel, October 1994. IEEE computer Society, IEEE Computer Society Press.

31. P. P. Silvester and R. L. Ferrari. *Finite Elements for Electrical Engineering*. Cambridge University Press, Cambridge, 2nd edition, 1990.

32. F. Solina and R. Bajcsy. Recovery of parametric models from range images: the case for superquadrics with global deformations. *IEEE Transactions on Pattern Analysis and Machine Intelligence*, 12(2):131–147, 1990.

33. P. Whaite and F. P. Ferrie. From uncertainty to visual exploration. *IEEE Transactions on Pattern Analysis and Machine Intelligence*, 13(10):1038–1049, October 1991.

34. D. Wilton, S. M. Rao, A. W. Glisson, et al. Potential integrals for uniform and linear source distributions on polygonal and polyhedral domains. *IEEE Transactions on Antennas and Propagation*, AP-32(3):276–281, 1984.

35. K. Wu and M. D. Levine. Parametric geons: A discrete set of shapes with parameterized attributes. In *SPIE International Symposium on Optical Engineering and Photonics in Aerospace Sensing: Visual Information Processing III*, volume 2239, pages 14–26, Orlando, FL, April 1994. The International Society for Optical Engineering.

36. K. Wu and M. D. Levine. Recovering parametric geons from multiview range data. In *IEEE Conference on Computer Vision & Pattern Recognition*, pages 159–166, Seattle, June 1994. IEEE Computer Society.

37. K. Wu and M. D. Levine. Shape approximation: From multiview range images to parametric geons. In *Proceedings of 12th International Conference on Pattern Recognition*, volume 1, pages 622–625, Jerusalem, Israel, October 1994. IEEE Computer Society Press.

38. K. Wu and M. D. Levine. 2D shape segmentation: A new approach. Technical Report TR-CIM-95-01, Centre for Intelligent Machines, McGill University, Montreal, Quebec, Canada, Feburary 1995.

39. Y. Yacoob and L. S. Davis. Labeling of human face components from range data. *CVGIP: Image Understanding*, 60(2):168–178, September 1994.

40. N. Yokoya, M. Kaneta, and K. Yamamoto. Recovery of superquadric primitives from a range image using simulated annealing. In *Proceedings of International joint Conference on Pattern Recognition*, volume 1, pages 168–172, 1992.

Delineation of Elongated Sub-Patterns
in a Piecewise Constant Foreground

Carlo Arcelli and Giuliana Ramella

Istituto di Cibernetica, CNR
80072 Arco Felice, Napoli, Italy

Abstract. A procedure useful to give evidence to the perceived linear structure of a gray-tone pattern is presented, which allows one to delineate its locally higher intensity regions with a connected set of simple digital lines, qualitatively analogous to the skeleton representation computed in the case of binary images. The pattern is regarded as constituted by a number of regions with constant gray-value, and the skeletonization is based on the detection of suitable pixels on its Distance Transform, computed according to the city-block distance. The set delineating the pattern is found by reducing the set of the skeletal pixels to a one-pixel-thick set, and by pruning part or all of its peripheral branches.

1 Introduction

A stylized version consisting of digital lines has often been used to represent a digital pattern in the case of binary images. However, since the binarization process may introduce shape distorsions that irremediably affect the presence of significant features, there is an increasing interest on the extraction of linear features directly from gray-tone images. We are interested in gray-tone images where the regions that mainly attract the attention of an observer are elongated and appear as having constant gray-value, greater than the gray-value of the neighboring regions. In the global context, the regions with locally higher gray-value could be perceived as lines whose spatial arrangement gives evidence to a still meaningful sketch of the image. By delineation of these elongated regions we mean the extraction of the set of such representative lines, that we call skeleton because of its resemblance with a similarly named representation often used in the case of binary images [1].

A gray-tone pattern has often been interpreted as a Digital Elevation Model (DEM), and the identification on it of the set of digital lines has been related to the detection of topographical features such as ridges, peaks, saddles and so on. For instance, the skeleton can be found by considering the image as a continuous surface, and by using the first and second partial derivatives of this surface to identify the skeletal pixels [2]. Alternatively, one may resort to the repeated application of local operations removing the pixels whose presence is not necessary either to ensure local pattern connectedness, or to mark a ridge configuration. As a result, the background propagates over the pattern and the erosion proceeds from lower gray levels towards higher gray levels until the skeleton is found as a one-pixel-thick subset, possibly in correspondence with locally higher intensity regions [3].

We find the skeleton by looking for specific configurations in the Distance Transform of the pattern. In the case of binary images, the structure of the Distance Transform shows the mode in which the background propagates over the pattern, and gives evidence to the regions where interaction occurs between wave fronts coming from different parts of the background. According to Blum [4], these regions are perceptually meaningful linear sets.

In our case, we regard the pattern as piecewise constant [5], and compute for each region with constant gray-value the Distance Transform, which is of the constrained type [6] whenever there exist adjacent regions with higher gray-value. Computation of the Distance Transform is accomplished according to the city-block distance, by ordered propagation over regions with increasing gray-value.

With respect to the traditional approach to skeletonization based on the Distance Transform, e.g., [7], additional conditions are devised to detect the pixels placed in the saddle configurations of the gray-tone pattern, and to ensure skeleton connectedness.

2 Preliminaries

Let us consider a gray-tone digital image on the square grid, where the pixels are assigned one out of a finite number of integer values, increasing from value 0. The set of pixels with value 0 constitutes the background B, while every connected set of pixels with positive value constitutes a component of the foreground F. Let g_k, k=1,N, denote the gray-value of the pixels of F, and g(x) denote the gray-value of the entity x, where x is either a pixel or a set of pixels. The 8-connectedness and the 4-connectedness are respectively chosen for F and for B. Pixels are denoted by letters, which may also indicate either the corresponding gray-value in the image or the corresponding distance label in the Distance Transform.

The neighbors of a pixel p in F are its 8-adjacent pixels. They constitute the neighborhood N(p) of p and are denoted by $n_1, n_2, ..., n_8$, where the subindexes increase clockwise from the pixel n_1 placed to the left of p.

A path between two pixels p and q in F is a sequence of pixels $p_0, p_1, ..., p_{s-1}, p_s$, such that $p_0 = p$, $p_s = q$, and p_i is a neighbor of p_{i-1}, i= 1,s. A 4-path is a path where p_i is a 4-neighbor of p_{i-1}, i= 1,s.

A connected set of pixels with gray-value k is said a region at elevation k.

A 4-connected region at elevation k from which any 4-path to pixels not in the region necessarily includes pixels with gray-value greater than k is termed a hollow at elevation k. It is denoted by H_k.

The depth of H_k is the difference in gray-value (h-k), where h is the minimal among the gray-values of a pixel p_r in any 4-path from H_k to pixels not in the region, such that $p_0 \leq p_1 \leq ... < p_s \leq ... \leq p_{r-1} \leq p_r > p_{r+1}$, for p_0 in H_k and p_s not in H_k.

An 8-connected region at elevation k from which any path to pixels not in the region necessarily includes pixels with gray-value smaller than k is termed a plateau at elevation k. It is denoted by PL_k.

The height of PL_k is the difference in gray-value (k-h), where h is the maximal among the gray-values of a pixel p_r in any i-th path from PL_k to pixels not in the region, such that $p_0 \geq p_1 \geq ... > p_s \geq .... \geq p_{r-1} \geq p_r < p_{r+1}$, for p_0 in PL_k and p_s not in PL_k.

Let X and X^c respectively denote a region of F and its complement, moreover let $X^c \supseteq R$, with R a set adjacent to the whole border of X. The Distance Transform of X with respect to the reference set R is the multi-valued set DT(X,R), which differs from X in having each pixel labeled with its distance from R, computed according a chosen distance function. If R is not adjacent to the whole border of X, the transform is called the constrained Distance Transform of X. Examples of Distance Transforms are shown in Figure 1.

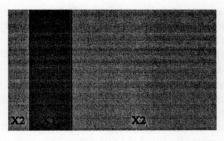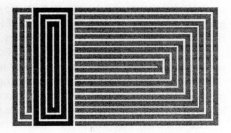

Fig. 1. Distance Transforms of the sets X1 and X2, where g(X1)>g(X2), computed with respect to the sets of pixels with gray-value less than g(X1) and less than g(X2), respectively. For a set for which the Distance Transform is constrained, as in the case of X2, the interaction between different wave fronts is as larger as more protruding is the set with respect to the adjacent region(s) with greater gray-value.

We regard the gray-tone pattern as the union of a number of connected single-valued regions, on each of which we compute the Distance Transform with respect to a reference set constituted by the pixels adjacent to the region and belonging to the regions with lower gray-value. We distinguish three types of regions, depending on the gray-value of the regions that are adjacent to the region taken into account:

type 1. all the adjacent regions have smaller gray-value.

type 2. only some of the adjacent regions have smaller gray-value, and with
    respect to the set of the adjacent regions with greater gray-value either

    2a. the region is a protruding extension of that set, or

    2b. the region is not a protruding extension of that set.

type 3. all the adjacent regions have greater gray-value.

It is straightforward to observe that for the regions of *type 1*, the Distance Transform is unconstrained since there are no adjacent regions with higher gray-value. These regions are characterized by locally higher intensity and will certainly include a skeleton branch.

For any region of *type 2*, the Distance Transform is constrained and its computation leads to a pattern of propagating wave fronts where the extent of interaction between the wave fronts is as greater as more protruding is the region itself. See Figure 1. Skeleton subsets will be found where such an interaction occurs.

Let X and Y denote two regions of F, with $g(X)<g(Y)$, and let L_X and L_{XY} be the lengths of the border of X and of the part of the border of X adjacent to Y, respectively. We say that X strongly protrudes from Y if $(L_X-L_{XY}) > 3L_{XY}$. Any region of *type 2* appears as perceptually dominated by the adjacent regions with higher gray-value, and the strength of this dominance may be defined as greater as less protruding is the region itself. We expect to find significant skeleton subsets only in correspondence with the regions which are not strongly dominated. In fact, a protruding region of *type 2a* is naturally perceived as having a locally higher gray-value since in turn, contrary to a region of *type 2b* , it dominates some large enough surroundings.

Finally, for the regions of *type 3* the Distance Transform cannot be computed since the reference set is empty.

Any region of *type 3* is a hollow in the DEM, and we require that its presence be reflected by the skeleton structure only if its depth is significant, i.e., is greater than a threshold value D* depending on the problem domain. When the depth is significant, we regard the hollow as a component of the background. When it is not significant, we change its gray-value to that of the adjacent region with lowest gray-value, and repeat this merging with the neighboring regions until the hollow is no longer detected. In the first case, we will have the hollow represented by a loop of the skeleton; in the second case, the hollow is filled in and does not influence the topology of the skeleton .

Hollow Detection. Any pixel p such that $p \leq n_i$, i odd, is first identified by a suitable marker. Then, the pixels not belonging to any hollow (i.e., greater than at least a 4-neighbor) are used as seeds from which propagation is accomplished through successive 4-neighbors so as to remove the marker from the pixels of the region with the same gray-level. Every connected set of marked pixels still remaining in F is a hollow.

The hollows with depth not smaller than D* are regarded as components of the background, while the remaining hollows are repeatedly filled in until their depth decreases to zero. To this purpose, the hollows' depth is decreased D*-1 times by increasing D*-1 times the gray-value of the pixels in each hollow. Hollow detection is repeated after each unit increase, to check whether not significant hollows still survive.

Plateau Detection. To ensure perceptual significance to the delineating digital lines, these should be detected in correspondence of sufficiently high plateaux (regions of *type 1*). To this purpose, using a technique similar to the filling in of not significant hollows, a preliminary step of the procedure is devoted to diminish, for an *a priori* given amount H*, the height of the plateaux so that the low ones disappear. The initial gray-value of the pixels of the survived plateaux is then restored by adding H*-1 units to their current gray-value.

3 Method

To avoid finding a skeleton structure too busy, it is convenient that the gray-tone pattern be cleaned of narrow regions of *type 1* and *type 3*. To this purpose a preprocessing, similar to "salt and pepper" noise removal performed when skeletonizing binary images, is accomplished by using shrink and expand operations [8]. Then, after the change of the status of the hollows and of the plateaux, the successive steps are concerned with the distance transformation and the skeletonization. They are briefly described in what follows.

Distance Transformation. The computation is accomplished on the array A, where the image is stored, and is performed according to the city-block distance, by ordered propagation over regions with ascending gray-value. Queues, i. e., first-in-first-out data structures, are employed to this purpose. As a result, the pixels in each region receive a distance label equal to their geodesic distance from a reference set constituted by the adjacent pixels belonging to the regions with lower gray-value. The obtained labeling of the gray-tone pattern originates its Distance Transform, which is the union of the (constrained and unconstrained) Distance Transforms of the regions constituting the pattern itself.

The gray-values g_k (k=1,...,N) of the pattern are detected and stored in increasing order in the 1-dimensional array G. Then, during an initialization pass the pixels interior in the pattern have their value increased of the large positive integer *untrans* ($untrans \gg g_N$). Moreover, the border points of the pattern are traced and stored in a queue Q constituted by a set of queues, a level of priority being attached to each of them (priority queue [9]).

The priority levels correspond to the increasing gray-values in the pattern, and the queue at priority level g_k contains only the pixels with gray value g_k currently under examination.

Gray-values are taken into account in increasing order. Let g_k denote the gray-value under consideration, and $dist_k$ the distance label to be updated of one unit at every propagation step. The initial value of $dist_k$ is $dist_k=0$ for k=1, and is $dist_k= dist_{k-1}$ for k>1. The w_k pixels stored in the queue at the priority level g_k are processed as follows.

For every pixel p, the corresponding value A(p) in array A is set to $dist_k+1$. The 4-neighbors of p are checked, and each neighbor p' such that: A(p')>*untrans* is inserted in the queue according to its gray-value A(p')-*untrans*. When all the initial w_k elements have been exhausted, $dist_k$ is increased by 1 and, if new pixels have been added to the queue at the current level, the propagation is applied again to these new elements, otherwise the process is applied to the elements of the queue at the next priority level g_{k+1}.

The presence of new elements (which may be added to the queue at the current level) requires a 1-dimensional array of counters. Each elements of this array contains the updated number of pixels for the corresponding gray-value.

Skeletonization. The skeletonization of a single-valued region based on the city-block Distance Transform consists in finding the intrinsic skeletal pixels, defined by certain neighborhood configurations [7], and then growing from some of them monotonically increasing paths terminating on pixels with locally higher distance labels. For the gray-tone pattern, we follow substantially the same procedure, by taking advantage of the monotonic evolution of the distance labels towards regions with higher gray-value.

Fig. 2. A gray-tone pattern (left) delineated by its skeleton (right).

However, it is necessary to introduce new conditions in $N(p)$ to ensure the connectedness of the set of the skeletal pixels. In fact, the morphology of the Distance Transform (regarded as a DEM) is no longer as simple and regular as that of the Distance Transform of a single-valued pattern, where the constraints on the distance label distribution in $N(p)$ are well defined. As in the binary case, the set of skeletal pixels is not generally one-pixel-thick, so that removal operations are applied to it in order to obtain a set consisting of simple digital lines.

The conditions we use to detect a skeletal pixel p on the Distance Transform are:

C1. Neither the horizontal nor the vertical pair of 4-neighbors of p are such that one 4-neighbor is greater than p and the other is smaller than p.

C2. There is not exactly one pair of successive 4-neighbors equal to p.

C3. There is more than one 4-component of neighbors with value less than p.

C4. There is at least a triple of successive neighbors, starting from a 4-neighbor, which are all equal to p.

C5. There is more than one 8-component of neighbors with value greater than p.

The procedure to find the set of the skeletal pixels consists of two phases, respectively devoted to the detection of the intrinsic skeletal pixels and of the induced skeletal pixels. For the latter type of pixels, the quality of being skeletal pixels is induced by the presence in their neighborhood of at least one pixel already classified as skeletal.

The intrinsic skeletal pixels are checked parallelwise on the Distance Transform, and a pixel p is detected if the Boolean expression "(*C1* AND *C2*) OR *C3* OR *C4* OR *C5*" is true.

Any induced skeletal pixels is found as the tip of a path originated from an intrinsic skeletal pixel. The paths are grown sequentially, one after the other, and stop as soon as their tip meet some other skeletal pixel, i.e., when in the neighborhood of the tip there is more than one 8-component of skeletal pixels. A pixel p is classified as induced skeletal pixel whenever the Boolean expression "$C1$ OR $C3$" is true. When computing this expression, note that the distance label assigned to the pixels already classified as skeletal in $N(p)$ should be understood as equal to the distance label of p.

Skeleton branches which are not detected in correspondence with elongated sub-patterns with locally higher intensity are clearly superfluous and should be removed. To this purpose, we obtained satisfactory results by employing a simple criterion based on the elongation property. This causes pruning of skeleton branches by removing successive pixels with not decreasing distance labels, until a given number of successive intrinsic skeletal pixel is found. The example in Figure 2 shows a pattern taken from a set of gray-tone images digitized at 16 levels, used to test the performance of the algorithm. Darker pixels are in correspondence with higher intensity regions.

References

1. M.D. Levine: Vision in Man and Machine. New York: McGraw-Hill 1985
2. L. Wang, T. Pavlidis: Detection of curved and straight segments from gray scale topography. CVGIP:Image Understanding 58, 352-365 (1993)
3. K. Abe, F. Mizutani, C. Wang: Thinning of gray-scale images with combined sequential and parallel conditions for pixel removal. In: C. Arcelli, L.P. Cordella, G. Sanniti di Baja (eds.):Visual Form Analysis and Recognition. New York: Plenum 1992, pp 1-10
4. H. Blum: Biological shape and visual science. J. Theor. Biol. 38, 205-287 (1973)
5. A. Rosenfeld: On connectivity properties of grayscale pictures. Pattern Recognition 16, 47-50 (1983)
6. J. Piper, E. Granum: Computing distance transformations in convex and non-convex domains. Pattern Recognition 20, 599-615 (1987)
7. C. Arcelli, G. Sanniti di Baja: A one-pass two-operations process to detect the skeletal pixels on the 4-distance transform. IEEE Trans. PAMI 11, 411-414 (1989)
8. Y. Nakagawa, A. Rosenfeld: A note on the use of local min and max operations in digital picture processing. IEEE Trans. Systems, Man, and Cybernetics 8, 632-635 (1978)
9. D. E. Knuth: The art of computer programming.Vol.3. London: Addison Wesley 1975

On a New Basic Concept and Topological Invariant

Franck XIA

Department of Information Systems
University of Macao, P.O. Box 3001 Macao
E-mail: fbafx@uealab.umac.mo

Abstract: Component is one of the most basic topological concepts for digital images. However, there is still few simple method, to date, permitting to distinguish holes from external borders. Such theoretic difficulty on components leads to quite complex and unnatural algorithms on component analysis. In this paper, the author proves that there exists, in 2D images, a basic topological invariant called total curvature. The total curvature on a border is the sum of local curvature representing the external angle of polygon vertices. We have discovered an important binary feature of this invariant: it equals to 2π for external contours and -2π for hole contours, independent of the orientation of contour tracing. The binary characteristics of total curvature reveals that external contours are global convex despite of local concavities and hole contours are global concave despite of local convexities. Therefore hole and external contours can be unified as total convex and total concave borders. Classifying all the borders of images by total curvature provides directly the number of both components and holes and it can be calculated easily in digital images.

I Introduction

Digital topology has been extensively explored in the past [1]. The effort was concentrated on a graph-theory approach which is widely accepted in computer science community. Digital topology provides a theoretic basis for various elementary operations in image processing such as thinning, border following, contour filling, object counting and extraction and the like.

Two concepts are fundamental in digital topology: connectedness and component. The problem of connectedness has been extensively studied in the past through skeletonization. Our interest in this paper is focused on the properties of two dimensional connected components, which is not well explored, according to the author, comparing to the abundant literature on connectedness. Undoubtedly, component analysis is an important issue in image processing, since components are basic elements of images and they represent generally objects or parts of object. The research on this issue, basically on parallel component labelling, is quite active [2].

Regions or connected components are entirely enclosed by their external contour in single connection case or with included holes in case of multiply connection. Intuitively, if we can detect all the borders and distinguish outer contours from hole contours in images, then the description of components is straightforward. In digital topology, people generally use Euler characteristics to describe component since Euler number equals to the number of components minus the number of holes.

The problem is that from computational viewpoint, Euler number is not really helpful for component analysis. This is because even to determine the number of components in images, we must determine first the number of holes and then deduce the number of components from Euler number. But detecting holes in an images is an unsolved problem, in practice, although the concept of hole is intuitive Z^2.

Until very recently, there was no satisfactory method presented in literature detecting directly inner contours. Suzuki and Abe proposed an algorithm being able to describe the nested structure of borders and hence differentiate outer contour from holes indirectly [3]. The differentiation of components from holes is not based on border's characteristics but on the nested structure of whole image: background surrounds components, components may include holes and holes may, in turn, surround objects, etc. [4]. Even to extract a small component, the whole image must be analysed first. Another tentative was based on mathematical morphology which has a serious drawback to change the topological features of images [5].

The only theoretical work on holes and their detection was carried out by Lee, Poston and Rosenfeld in 1991 [6]. The idea consists of counting the transition number of normal vector on contours in a given direction to distinguish holes from components. Their work provides the first theoretic method for component analysis. The problem revealed is that the normal vector of boundary is intrinsically discontinuous. Therefore they suggested to use a continuous smoothing function replacing the digital lattice and normal vectors are calculated with the smoothing function. Counting the winding number cannot be realized directly with digital images, unfortunately.

In general, due to the barrier to distinguish external borders from hole borders, component analysis algorithms remain generally unnatural and complex. A typical example is sequential component labelling through raster scanning, which requires either time consuming iterative label propagation or using complex and memory consuming equivalence relationship [7].

II Proposition

In this communication, we prove that total curvature - the sum of local curvature on borders representing the complement of internal angles at border vertices - is a very attractive and basic topological invariant. The real important discovery is that, in fact, the total curvature of an arbitrary borders has only two constant values 2π or -2π, relying on whether the border is external or internal.

The binary characteristics of total curvature reveals the intrinsic property of contours: external contours are global convex despite of local concavity and hole contours are global concave despite of local convexity. Therefore external contours and hole contours can be unified as total convex and total concave borders based on the concept of total curvature. Classifying all the borders of images by total curvature provides directly the number of both components and holes. The computation of total curvature allows a very simple separation of outer contours from hole contours. Applying this method to digital images is extremely simple.

We start to examine total curvature of polygons and then apply the invariant to digital images. Computation results are given to illustrate the simplicity and usefulness of the invariant.

III Topological Invariant of Polygon

We will describe in this section the invariant property of single closed polygons, since all digital images can be described by single closed polygon (we denote simply by polygon in this paper). Discussing the invariant of polygons provides an universal theory for 2D digital images which is completely independent of neighborhood used. This is a major advantage. The proof of the following results can be found in [8].

Definition 3.1: Local curvature $\Delta\alpha_i$ at vertex A_i of a polygon equals to $\pi - \beta_i$, with β_i the interior angle of the polygon at A_i (Fig. 1).

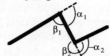

Figure 1: Local Curvature in R^2

Definition 3.2: For any polygon, the total curvature is the sum of local curvatures on all the vertices $\sum_i \Delta\alpha_i$.

Remark: the local curvature is independent of the direction in which all the vertices of the polygon are traversed. In fact the local curvature at A_i is positive if the internal angle A_i is convex and negative if concave.

Proposition 3.3: For any convex polygon of M edges, the sum of interior angles

$$\sum_i \beta_i(P) = (M\text{-}2)\pi.$$

Denote $\beta_i(P)$ the ith interior angle of P and $\beta_j^c(P)$ the jth concave interior angle of P, with $\beta_j^c(P) > \pi$.

Lemma 3.4: Any polygon P of N concave interior angle ($\sum_i \beta_i^c(P_1) = N \geq 1$) can be divided into two parts P_1 & P_2, by a segment starting from a vertex of P, such that
1) the number of concave interior angles in each polygon reduced at least 1:

$$\sum_i \beta_i^c(P_1) \leq N\text{-}1 \text{ and } \sum_i \beta_i^c(P_2) \leq N\text{-}1;$$

2) the sum of interior angle of P_1, P_2 and P have the following relation:

$$\sum_i \beta_i(P_1) + \sum_i \beta_i(P_2) - \gamma\pi = \sum_i \beta_i(P) \text{ where}$$

$\gamma = 0$ if the segment does not cut any edge of P and no new vertex is generated and $\gamma = 1$ if the segment cuts an edge of P and a new vertex is generated.

Lemma 3.4 is essential for the proof of theorem 3.5.

Theorem 3.5: For any polygon of M edges, the sum of its interior angles is $(M\text{-}2)\pi$.

Theorem 3.6: For any polygon P, its total curvature is 2π.

This theorem is fundamental for our method and evident by theorem 3.5.

Corollary 3.7: For a special sort of polygon in which two different polygon paths coincide, its total curvature is 2π.

This theorem is very useful for discussing holes. It allows us to cut a region with a open polygon inside the region and discuss the two sides of the same open polygon.

Proposition 3.8: For any open polygon $p_0p_1p_2\cdots p_{n-1}p_n$, the sum of vertex curvatures calculated from both sides of the open polygon excluding the two extremities is zero.

Corollary 3.7 and proposition 3.8 are useful for theorem 3.9.

Theorem 3.9: For any multiply connected region bounded by two polygons, the total curvature of its hole polygon equals to -2π.

In topology, vertices, sides and surfaces are major concerns as they have practical meaning in many application fields. In contrary, the property of boundary has not attracted the interest of topologists. Here we provide, by total curvature, a simple way to distinguish external borders from holes of polygon. Note that the distinction of hole borders from external borders is not at all related to the direction of border tracing. It is based on the intrinsic global convexity and global concavity of borders.

IV Contour Invariant in Digital Images

When applying the new property of polygon invariance to digital images, we follow basic definitions on discrete geometry described in [1].

Definition 4.1: Let P_{i-1}, P_i and P_{i+1} be three consecutive contour pixels and separate its j-neighbors into object side and background side. Let $\angle P_{i-1}P_iP_{i+1}$ be the interior angle (in object side) at P_i. *Local curvature* at P_i $LC(P_i)$ represents the complementary angle of $\angle P_{i-1}P_iP_{i+1}$: $\pi - \angle P_{i-1}P_iP_{i+1}$ (Fig. 2). Note shaded pixels are interior points.

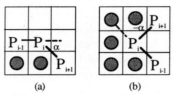

(a) (b)

Figure 2: Local curvature in 8-neighbors.

The value of local curvature $LC(P_i) \geq 0$ if $\angle P_{i-1}P_iP_{i+1}$ is convex (Fig.2.a) or $LC(P_i)$ < 0 when $\angle P_{i-1}P_iP_{i+1}$ is concave. (Fig. 2.b) As local curvature is a new concept in digital images, we illustrate in Figure 3 examples of local curvature in 8-neighbors. By the definition of contour, **LC** may have discrete values {-2, -1, 0, 1, 2, 3, 4}. In the figure, dark shaded circles designate object/contour pixels, clear shaded circles stand for the current edge points, and the numeric in center represents **LC** at these points. For simplicity, we omit those that can be obtained by rotation of multiples of 90°.

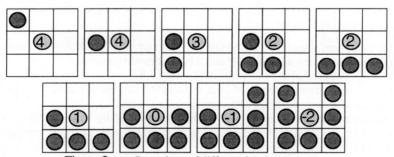

Figure 3: configurations of different local curvatures.

Definition 4.2: *Total curvature* **TC**(∂C) is the sum of local curvature **LC** on a contour ∂C and denoted by $\sum_i LC(P_i|P_i \in \partial C)$.

A special case is an isolated object point, since the definition of local curvature requires at least two connected points. We define the total curvature of an isolated point is 2π. The problem does not exist for hole contours since the minimum perimeter of hole contours is 4 corresponding to a hole of isolated pixel.

Corollary 4.3: In digital images, $TC(\partial C^+) = -TC(\partial C^-) = N$, where N equals 8 for rectangular tessellation and 6 for hexagonal tessellation .

V Result

Without loss of generality for sequential processing, we adopt the left-hand convention to trace all the borders of multiple connected regions. But in fact, the definitions of local curvature and total curvature are independent of contour tracing orientation.

As computational illustration, we present results of external and hole contour distinction through a very simple algorithm. The algorithm scans once the whole image to find out all the contours - both outer contours and hole contours, by a classical contour following procedure. Contours are encoded with Freeman direction codes and the difference of two consecutive codes is calculated during contour tracing which provides directly the local curvature of contour pixels. The total curvature is obtained simply by summing local curvatures. It is then used to classify outer contours and hole contours. The algorithm is extremely simple.

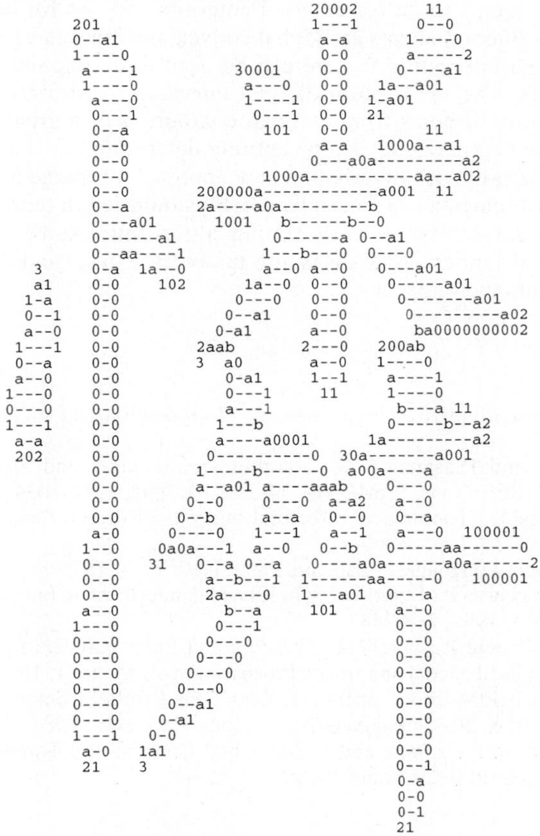

```
                                        20002       11
                      201               1---1       0--0
                      0--a1             a-a         1----0
                      1----0            0-0         a-----2
                       a----1    30001  0-0         0----a1
                      1----0     a---0  0-0         1a--a01
                       a---0     1----1 0-0         1-a01
                      0---1      0---1  0-0         21
                      0--a       101    0-0
                      0--0               a-a        1000a--a1
                      0--0               0---a0a---------a2
                      0--0          1000a-----------aa--a02
                      0--0    200000a-----------a001  11
                      0--a    2a----a0a---------b
                      0---a01 1001   b------b--0
                      0------a1       0------a 0--a1
                      0--aa----1      1---b---0  0---0
          3           0-a  1a--0      a--0 a--0   0---a01
          a1          0-0   102       1a--0 0--0   0-----a01
          1-a         0-0             0---0 0--0    0-------a01
          0--1        0-0             0--a1 0--0     0--------a02
          a--0        0-0             0-a1  a--0     ba0000000002
          1---1       0-0             2aab    2---0  200ab
          0--a        0-0          3  a0      a--0   1----0
          a--0        0-0             0-a1    1--1   a----1
          1---0       0-0             0---1   11     1----0
          0---0       0-0             a---1          b---a 11
          1---1       0-0             1---b          0-----b--a2
          a-a         0-0             a----a0001     1a---------a2
          202         0-0           0----------0  30a-------a001
                      0-0           1-----b------0  a00a----a
                      0-0           a--a01 a----aaab  0---0
                      0-0    0--0    0---0 a-a2   a--0
                      0-0    0--b    a--a  0--0   0--a
                      a-0    0----0  1---1  a--1   a---0 100001
                      1--0   0a0a---1 a--0  0--b  0-----aa------0
                      0--0   31  0--a 0--a  0----a0a------a0a------2
                      0--0       a--b---1 1------aa-----0  100001
                      0--0       2a----0  1---a01 0---a
                      a--0        b--a     101     a--0
                      1---0       0---1             0--0
                      0---0       0---0             0--0
                      0---0       0---0             0--0
                      0---0    0---0                0--0
                      0---0    0---0                0--0
                      0---0       0--a1             0--0
                      0---0    0-a1                 0--0
                      1---1    0-0                  0--0
                      a-0      1a1                  0--1
                      21       3                    0-a
                                                    0-0
                                                    0-1
                                                    21
```

Contour No.1: Perimeter 255, Total Curvature = 8 => outer contour
Contour No.2: Perimeter 22, Total Curvature = 8 => outer contour
Contour No.3: Perimeter 111, Total Curvature = 8 => outer contour
Contour No.4: Perimeter 14, Total Curvature = 8 => outer contour
Contour No.5: Perimeter 67, Total Curvature =-8 => hole contour
Contour No.6: Perimeter 26, Total Curvature = 8 => outer contour
Contour No.7: Perimeter 16, Total Curvature =-8 => hole contour
Contour No.8: Perimeter 18, Total Curvature =-8 => hole contour

Figure 4: Components with distinguished outer and hole contours

Figure 4 illustrates the resulting images in which a and b represent local curvature value -1 and -2 on contour pixels.

A practical advantage of the proposed method to evaluate topological property is that it may insensitive to noise. This is because the perimeter of contours provides the information on the size of components or holes. Depending on applications, small components and holes can be ignored during processing to provide robust topological property.

VI Conclusion

In digital space, there is still few simple method, to date, allowing to distinguish holes from other contours. Consequently, the difficulty leads to quite complex and unnatural algorithms in image analysis. In this paper, a novel invariant called total curvature has been presented. The total curvature of borders has an important binary characteristics : It equals to 2π for external contours and -2π for hole contours. This means that borders of components are global convex and borders surrounding holes are global concave. The interesting feature can be used to distinguish external contours from hole contours. The application of total curvature is straightforward in digital space. The distinction of holes from external contours is of a great importance, since regions or connected components can be entirely determined by external contours and holes. Thanks to the proposed invariant, a new approach of image analysis is available and methodology of image analysis can be greatly simplified in terms of computational and algorithmic complexity. A very simple algorithm can be derived using the invariant property of border - total curvature to detect, extract and describe connected components without labelling.

Reference

1. Kong T.Y. and Rosenfeld A., Digital topology: Introduction and survey, CVGIP, vol. 48, 1989, 357-393
2. Alnuweiri H.M. and Prasanna V.K., Parallel architectures and algorithms for image component labeling, IEEE Trans. PAMI, Vol. 14, No. 10, 1992, 1014-1034
3. Suzuki S. and Abe K., Topological structural analysis of digital binary images by border following, CVGIP, vol. 30, 1985, 32-46
4. Rosenfeld A., Picture Languages, Academic Press, New York, 1979
5. Qian K. and Bhattacharya P., Determining holes and connectivity in binay images, Comput. & Graphics, vol. 16, no. 3, 1992, 283-288
6. Lee C.N., Poston T. and Rosenfeld A., Winding and Euler numbers for 2D and 3D digital images, CVGIP: Graphical Model and Image Processing, vol. 55, no. 1, 1993, 20-47
7. Haralick R., Some neighborhood operators, Real-Time Parallel Computing Image Analysis, Onoe M., Preston K. Jr. & Rosenfeld A. (eds.), Plenum, New York, 1981
8. Xia F., Total Curvature - a New and Fundamental Concept and Topological Invariant for Digital Images, to appear in IEEE Trans. PAMI

Line moments and invariants for real time processing of vectorized contour data

Georg Lambert[1], Hua Gao[2]

[1] Technical University of Darmstadt, Control System Theory and Robotics Dept., Landgraf-Georg-Str.4, 64283 Darmstadt, Germany
[2] Technical University of Darmstadt, Institute of Production Engineering and Machine Tools, Petersenstr. 30, 64287 Darmstadt, Germany

Abstract. In this article a new concept for real time analysis of images based on the vectorized contours is presented. Given the polygone approximation of contours the geometric moments can be formulated for line contours. The resulting line moments (LM's) can be calculated for an arbitrary collection of contour fragments. Neither the existence of an area nor a closed contour is prerequisite to characterize an object by line moments. Due to these properties of the LM's time consuming preprocessing of noisy object contours is not required. Even objects with structured surfaces can be characterized by LM's. The formalism to apply the geometric moments to contour patterns is derived. A direct and a recursive algorithm to efficiently compute the LM's is given. As an application of the LM's an analysis system for structured textures is presented and results are discussed.

1 Indroduction

The application of image processing systems in industrial environments gains more and more in significance. One reason why image processing has been unattractive for many applications in the past was the lack of computation power to meet real time requirements.

In this paper we present an approach to real time analysis of images based on the contour data of the scene. In this context we consider the extraction of contours as a standard operation, which is universally applicable to many image processing and analysis tasks. Special hardware processors or implementations on signal–processor basis can realize contour extraction in video real time.

For this work we used a hardware contour vectorizer called *VectEx*[3] which delivers a polygone approximation of the gray level edges of images in video real time. (For a description of the processor see [HERR90].) As an example figure 1a shows the vectorized contours of a structured texture.

Based on the contour data we developed a fast recursive algorithm to compute the line moments (LM's) of patterns. Therefore the extension of the conventional moment definition (e.g. [HU62]) to LM's is one main aspect of this work. Due to this interpretation the computation of moments is no longer restricted

---

[3] Distributed by ELTEC Elektronik GmbH, D–55071 Mainz, Germany

to elements with a definite area and a closed contour. An element is considered to be composed of an arbitrary collection of line fragments.

Based on LM's the entire theory of moments and moment invariants can be applied for element recognition and analysis.

2 Line moments of contour patterns

Applying a contour vectorization on objects of a real scene yields in object contours broken down into lots of fragments. To extract objects from the contour data logically related fragments have to be recognized. This might be done by some rule based algorithms or by statistical analysis and segmentation. In [AMEL94] an algorithm is described, which assigns all contour fragments situated in local neighbourhoods of edge points recursively to the same object. Non–overlapping and not directly neighbouring objects like in figures 1a and 5a can be segmented efficiently through this algorithm.

Given the contour fragments C_j^* of an object B (figure 1b), the contour pattern C of B can be expressed as

$$C = \bigcup_j C_j^*. \tag{1}$$

If the contour fragments are approximated by polygones, the contour can be expressed in terms of line fragments c_{ji}

$$C_j^* = \bigcup_i c_{ji} \quad , \qquad C = \bigcup_j \bigcup_i c_{ji}. \tag{2}$$

A contour pattern C of an object B is consequently defined as an arbitrary collection of line fragments. There is no need to define either the area or the perimeter of an object.

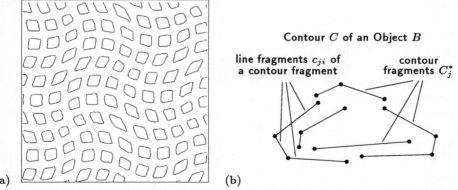

(a) (b)

Fig. 1. (a) Example of a vectorized texture scene. **(b)** Definition of a contour pattern

After the definition of contour patterns we will focus on the derivation of line moments. Let us first consider the well known geometric moments of the form $m_{pq}^{(2)} = \int_{-\infty}^{+\infty} \int_{-\infty}^{+\infty} x^p y^q f(x,y)\, dx\, dy$. Given a gray level image we can understand $f(x,y)$ as the gray level values. If the domain of definition $D(B)$ of an object

B is known, e.g. through segmentation, the object is uniquely determined by its moments

$$m_{pq}^{(2)} = \int \int_{(x,y) \in D(B)} x^p y^q f(x,y) \, dA \qquad p,q = [0,1,\dots,\infty] \qquad (3)$$

according to the Uniqueness Theorem ([HU62]). If we consider line contours, the line moments along a contour C are given by

$$m_{pq}^{(1)} = \int_C x(s)^p y(s)^q f(s) \, ds \qquad (4)$$

where $f(s)$ is called the linear density and s stands for the arc length. In the case of describing the contour shape, $f(s)$ can be set to $f(s) = 1$. Equation (4) can also be expressed in terms of the cartesian coordinates x or y

$$m_{pq}^{(1)} = \int_C x^p y(x)^q \sqrt{1 + y'(x)^2} \, dx = \int_C x(y)^p y^q \sqrt{1 + x'(y)^2} \, dy. \qquad (5)$$

If the contour pattern is approximated by a polygone with edge–points (x_i, y_i), $i = 1, 2, \dots, n+1$, each line segment can be parameterized to

$$c_i: \quad y = a_i x + y_i - a_i x_i, \quad x_i \le x \le x_{i+1}, \quad i = 1, 2, \dots, n \qquad (6)$$

whith $a_i = (y_{i+1} - y_i)/(x_{i+1} - x_i)$ as the slope of segment i. (The index j is omitted in the following.) Let D_i be the contribution of c_i to the line integral of eq.(5) the moments are given by

$$m_{pq}^{(1)} = \sum_{i=1}^n D_i, \qquad D_i = \int_{c_i} x^p y^q \sqrt{1 + a_i^2} \, dx. \qquad (7)$$

Inserting eq.(6) in (7) we get

$$D_i = \int_{x_i}^{x_{i+1}} x^p (a_i x + y_i - a_i x_i)^q \sqrt{1 + a_i^2} \, dx$$

$$= \sqrt{1 + a_i^2} \cdot \sum_{k=0}^q \left\{ \binom{q}{k} a_i^k (y_i - a_i x_i)^{q-k} \cdot \frac{x_{i+1}^{p+k+1} - x_i^{p+k+1}}{p+k+1} \right\}. \qquad (8)$$

If c_i is vertical, we take the alternative parameterization

$$c_i: \quad x = x_i, \quad y_i \le y \le y_{i+1}, \quad i = 1, 2, \dots, n \qquad (9)$$

which leads to

$$D_i = \int_{y_i}^{y_{i+1}} x^p y^q \sqrt{1 + x'(y)^2} \, dy = \int_{y_i}^{y_{i+1}} x_i^p y^q \, dy = x_i^p \cdot \frac{y_{i+1}^{q+1} - y_i^{q+1}}{q+1}. \qquad (10)$$

To obtain invariance under translation the central moments

$$\mu_{pq}^{(1)} = \int_C (x(s) - \overline{x})^p (y(s) - \overline{y})^q f(s) \, ds \qquad (11)$$

have to be calculated where $\overline{x} = m_{10}^{(1)}/m_{00}^{(1)}$, $\overline{y} = m_{01}^{(1)}/m_{00}^{(1)}$. To obtain scaling invariance the following normalisation by the contour length l_C leads to the result

$$\eta_{pq}^{(1)} = \int_C \left(\frac{x(s) - \overline{x}}{l_C} \right)^p \left(\frac{y(s) - \overline{y}}{l_C} \right)^q \frac{ds}{l_C} = \frac{\int_C (x(s) - \overline{x})^p (y(s) - \overline{y})^q \, ds}{l_C^{p+q+1}} = \frac{\mu_{pq}^{(1)}}{(\mu_{00}^{(1)})^\gamma} \qquad (12)$$

where $\gamma = p + q + 1$. It has to be stated, that in the case of LM's γ differs from the exponent, Hu derived for area based moments in [HU62]. The normalisation of eq.(12) is also given in [SARD94].

3 Algorithms

In this section we will give a direct and a fast, recursive algorithm to compute the LM's. The algorithm is similar to those given in [JIAN91]. Nevertheless it has to be emphasized that Jiang and Bunke computed area based moments out of the closed contours of objects through the Green's formula which is totally different from the approach to LM's presented here.

Figure 2 gives the direct algorithm to compute the LM's of polygonal contour patterns.

forall (pq) **do** $m_{pq}^{(1)} = 0$;
for $i = 1$ **to** n **do begin**
    if vertical(c_i) **then**
        forall (pq) **do** $m_{pq}^{(1)} = m_{pq}^{(1)} + D_i$;   (* D_i by eq.(10) *)
    else
        forall (pq) **do** $m_{pq}^{(1)} = m_{pq}^{(1)} + D_i$;   (* D_i by eq.(8) *)
end

Fig. 2. Direct algorithm to compute the line moments

The direct algorithm does not take into account relationships between the moments of higher order to those of lower order. To derive a recursive algorithm we define

$$A_i(p, q) = \int_{x_i}^{x_{i+1}} x^p \, (a_i x + y_i - a_i x_i)^q \, dx \tag{13}$$

(see eq.(8)). For $q \geq 1$ we get

$$A_i(p, q) = \int_{x_i}^{x_{i+1}} x^p \, (a_i x + y_i - a_i x_i)^{q-1}(a_i x + y_i - a_i x_i) \, dx$$

$$= a_i \int_{x_i}^{x_{i+1}} x^{p+1} \, (a_i x + y_i - a_i x_i)^{q-1} \, dx \;\; +$$

$$(y_i - a_i x_i) \int_{x_i}^{x_{i+1}} x^p \, (a_i x + y_i - a_i x_i)^{q-1} \, dx$$

$$= a_i A_i(p+1, q-1) + (y_i - a_i x_i)A_i(p, q-1) \; . \tag{14}$$

To start the recursion we have

$$A_i(p, 0) = \int_{x_i}^{x_{i+1}} x^p \, dx = \frac{x_{i+1}^{p+1} - x_i^{p+1}}{p+1} \; . \tag{15}$$

The recursive dependencies of the terms $A_i(p, q)$ are illustrated in figure 3a for an example with $(p, q) \leq 4$.

The notation $(p_1, q_1) \rightarrow (p_2, q_2)$ expresses that for the calculation of $A_i(p_2, q_2)$ the value of $A_i(p_1, q_1)$ has to be known. Based on these dependencies the order of calculation for all $A_i(p, q)$ with $(p, q) \leq 4$ can be derived as shown in figure 3b. The variables $B[0] \ldots B[4]$ (see also fig. 4) store temporary results during

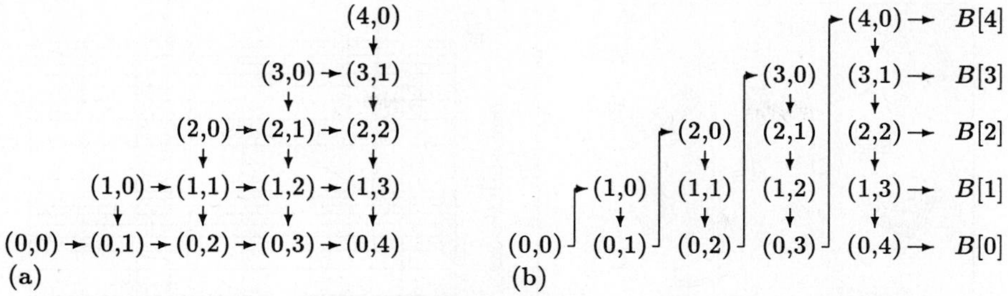

$$
\begin{array}{c}
(4,0) \\
\downarrow \\
(3,0) \rightarrow (3,1) \\
\downarrow \qquad \downarrow \\
(2,0) \rightarrow (2,1) \rightarrow (2,2) \\
\downarrow \qquad \downarrow \qquad \downarrow \\
(1,0) \rightarrow (1,1) \rightarrow (1,2) \rightarrow (1,3) \\
\downarrow \qquad \downarrow \qquad \downarrow \qquad \downarrow \\
(0,0) \rightarrow (0,1) \rightarrow (0,2) \rightarrow (0,3) \rightarrow (0,4)
\end{array}
$$

(a)

$$
\begin{array}{l}
\qquad\qquad\qquad\qquad\qquad\qquad (4,0) \rightarrow B[4] \\
\qquad\qquad\qquad\qquad\qquad\qquad\; \downarrow \\
\qquad\qquad\qquad\quad (3,0)\; (3,1) \rightarrow B[3] \\
\qquad\qquad\qquad\qquad\;\; \downarrow \qquad \downarrow \\
\qquad\quad (2,0)\; (2,1)\; (2,2) \rightarrow B[2] \\
\qquad\qquad\quad\;\; \downarrow \qquad \downarrow \qquad \downarrow \\
(1,0)\; (1,1)\; (1,2)\; (1,3) \rightarrow B[1] \\
\;\; \downarrow \qquad \downarrow \qquad \downarrow \qquad \downarrow \\
(0,0)\; (0,1)\; (0,2)\; (0,3)\; (0,4) \rightarrow B[0]
\end{array}
$$

(b)

Fig. 3. (a) recursive dependencies of $A_i(p,q)$ **(b)** order of computation

```
for order = 0 to max_order do begin
    B[order] = A_i(order, 0);                    (* by eq.(15) *)
    m_{order,0}^{(1)} = m_{order,0}^{(1)} + √(1 + a_i²) · B[order];
    for p = order − 1 downto 0 do begin
        q = order − p;
        B[p] = a_i B[p+1] + (y_i − a_i x_i) B[p];    (* eq.(14) *)
        m_{p,q}^{(1)} = m_{p,q}^{(1)} + √(1 + a_i²) · B[p];
    end;
end;
```

Fig. 4. Recursive algorithm to compute the line moments

computation. To evaluate the LM's according to eq.(7) we get (compare (8) and (13))

$$
D_i = \sqrt{1 + a_i^2} \cdot A_i(p,q) \ . \tag{16}
$$

Hence the recursive algorithm shown in figure 4 is appropriate to calculate the LM's in a very efficient way.

4 Experimental results

As an application of the LM's we present an analysis system for quality control of structured textures in this section.

It has to be emphasized, that the whole theory of moment invariants is applicable to LM's. Since the various types of moments, e.g. Legendre Moments, Zernike Moments, Complex Moments and others can be expressed through the geometric moments ([TEH88]) their formulation as line moments is possible.

In our experiments we calculated line moment invariants (LMI's) to characterize the extracted texture primitives. We implemented the 7 well known Hu moment invariants ([HU62]), as they are still a benchmark for other invariants. To extract texture primitives from the vectorized contour data a rule based algorithm described in [AMEL94] was applied. In a learning phase the LMI's of the different texels are calculated from texture samples and stored as feature vectors. During the defect detection phase the following tasks have to be applied to each texture scene: texel segmentation, calculation of LMI's and classification with a Mahalanobis distance classifier. The processing is based on the output of the hardware vectorizer *VectEx*.

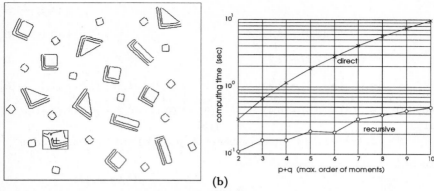

(a)   (b)

Fig. 5. (a) Detection of the corrupted texture element marked by a frame. **(b)** Computing time for the direct and the recursive algorithm (80486DX2-66MHz processor).

Figure 5a shows the result of a texture analysis with 4 different texel classes. The corrupted texture primitive is detected by the system. The computation has been done on a PC with a 80486 processor, 66 Mhz clock rate and VL bus. To calculate the HU invariants all moments up to the maximum order of $(p + q) \leq 3$ have to be determined. Using the iterative algorithm, the moment computation takes 0.16 sec for the above scene. Computation of all other tasks as texture primitive generation, invariants calculation and classification takes 0.15 sec. Hence the whole analysis requires 0.31 sec. on a scene with 1505 polygone points in 213 contours and 29 texture elements.

To point out the effectiveness of the iterative algorithm compared with the direct one, figure 5b depicts computation times for all the line moments up to an order $(p + q) \leq N$. For $N = 10$ the iterative algorithm is 20 times faster than the direct one.

References

[AMEL94] Jörg Amelung, Georg Lambert und Jörg Pfister: Ein vektorbasiertes Verfahren zur schnellen Fehlererkennung in strukturierten Texturen; Proc.16. DAGM Symp. Mustererkennung, Wien, 1994, pp. 666–675.

[HERR90] E.Herre und R.Massen: Symbolische konturorientierte Bildverarbeitung durch Echtzeit-Polygon-Approximation; Technisches Messen TM Vol.57 Nr.10, 1990, pp.384-388.

[HU62] Ming–Kuei Hu: Visual Pattern Recognition by Moment Invariants; IRE Trans. on Information Theorie, Vol. 8, 1962, pp.179-187.

[JIAN91] X. Y. Jiang, H. Bunke: Ein konturbasierter Ansatz zur Berechnung von Momenten; Proc.13. DAGM Symp. Mustererkennung, Munich, 1991, pp. 143-150.

[SARD94] H. K. Sardana, M. F. Daemi and M.K. Ibrahim: Global description of edge patterns using moments; Pattern Recognition, Vol. 27, No. 1, 1994, pp. 109-118.

[TEH88] Cho-Huak Teh, Roland T. Chin: On Image Analysis by the Methods of Moments; Trans. on Pattern Analysis and Machine Intelligence, Vol. 10, No. 4, 1988, pp.496-513.

Range and Acoustic Images

Restoration of Noisy Underwater Acoustic Images Using Markov Random Fields

Vittorio Murino, Enrico Frumento, Flavio Gabino

Department of Biophysical and Electronic Engineering (DIBE)
University of Genoa
Via all'Opera Pia 11A, 16145 Genova, Italy

Abstract. This paper describes a method for the restoration of noisy underwater 3D acoustic images. A beamforming technique is used to build a 3D map associated with corresponding map representing the confidence (i.e., reliability) of measures. Backscattered echoes coming from a scene previously insonified are received by a two-dimensional array antenna and an image is formed, where each pixel represents the distance from the sensor in a specified direction. In the proposed algorithm, depth and confidence images are modelled as Markov Random Fields whose associated probability distributions are specified by a single energy functional, that embeds a-priori knowledge on the images and on the noise affecting measurements. Confidence measures are also considered to improve the restoration. Simulated Annealing is used to find the optimal MAP estimate of both images. Experimental results show the improvements in the resulting images as compared with initial data.

1. Introduction

Depth-map extraction is one of the most important tasks accomplished by active sonar systems to determine the range of an observed scene. Imaging systems using a single non-selective insonification of a scene and the beamforming of echoes backscattered and received by an array antenna are to be preferred in order to obtain a high frame rate (see [1,2]). The beamforming process provides a time signal representing the backscattered echoes generated by the scene in a particular direction. From the detection of the temporal instant corresponding to the highest peak (and assuming known sound velocity in water), it is possible to compute the distance of the scene in that direction. Iterating this mechanism for a set of regular directions forms a 3D image. The amplitude of the envelope of the output signal coming from the beamforming process also provides a reliability measure (we call "confidence value") of the depth estimate. The simplest method for extracting confidence measures from an envelope is to detect the maximum peak; however other statistical analyses can be made to detect the right peak of a received signal [3]. Speckle noise and other types of effects (e.g., reverberation) may affect backscattered echoes, thus modifying and attenuating the signal received by the antenna. As a consequence, it is not an easy task to detect the right time instants of peaks, then depth estimation becomes critical. Noise is generated by constructive and destructive interferences of the echoes diffused by a scene and can be approximated by Additive White Gaussian Noise [4]. Moreover, halos may form around objects, thus producing high responses even in directions in which no object is present. This is due to the directivity characteristics of the antenna (i.e., side-lobe effects [5]). Eventually, these disturbing factors affect the estimation of the scene depth and give rise to images of the type shown in Fig. 2 (the ideal images are shown in Fig 1), where one can observe a non-null response where no object is present and null responses inside objects.

The goal of the proposed algorithm is to restore a range image modelled as a Markov Random Field (MRF) [6] using information available from a confidence map. The paper is organized as follows: in Section 2, the significance of the confidence map is pointed out, and, in Section 3, the MRF approach is described, with special emphasis on the energy functional. In Section 4, results are presented and discussed. Finally in section 5 conclusions are drawn.

2. The Confidence Map

To perform a correct estimation of a depth image, the envelope should have only one very high peak, representing the signal backscattered by a point of an object in a certain direction and received by an array antenna. The amplitude of this peak also provides a confidence value indicating the reliability of each depth estimate. The higher the peak, the higher the confidence. Denoting the envelope signal by V_0, the depth image by Z, and the confidence image by S, different situations may occur when noise affects V_0 more or less strongly:

For points inside objects:
if V_0 presents a single high peak then the confidence S is high and the depth estimate Z is always correct. If V_0 presents a single weak peak and some fluctuations near zero the confidence is low but the depth estimate is equally correct because low noise make possible to detect the single peak of the envelope. If noise is high, V_0 has several similar peaks, then the right one might be confused with others; in this case, the estimate of the Z map may be wrong.

For points outside objects (or on object boundaries):
V_0 presents a low peak then the confidence S is low but non-zero, as the beamforming process give rises to artefacts (halos) on object boundaries (due to side lobes [5]). For these pixels the depth estimation is wrong, as the detected peak does not correspond to any object in the direction considered.

From these considerations, one can deduce that a depth value is correctly estimated only when the confidence value is high, but a low confidence value cannot confirm the depth measure, as the Z estimate may be correct or incorrect in dependence of noise effects.

A possible solution is to consider only those Z points whose confidence value are above a certain threshold. In this case side lobes can be removed, but many correct points inside objects are erroneous rejected as their confidence values are below the threshold. Then, a punctual analysis of the Z and S maps is not sufficient, and a coupled restoration of such maps is required.

3. The MRF Approach

Image restoration is considered as the problem to find the optimal configuration of a lattice field Y of random variables. To define an MRF is necessary to specify a probability distribution on a discrete field, where the probability of a random variable y_i depends only on a neighbouring set N_i:

$$p(y_i \,/\, Y) = p(y_i \,/\, y_j) \quad \forall \, y_j \in N_i$$

When an a-priori image model $p(Y)$ is of MRF type, and the observation model satisfies weak properties [6], an estimation of the solution Y is given by maximizing the a-posteriori probability of the field given observations X, which is known to be a Gibbs distribution (Hammersley-Clifford equivalence):

$$p(Y \,/\, X) = \frac{1}{Z_p} \cdot \exp\left[-\frac{E(Y \,/\, X)}{T} \right]$$

where $E(Y/X)$ is the energy function to be minimized, Z_p is the normalizing partition function, and T is the temperature of the field. The energy function $E(Y/X)$ can be written as a sum of local energies $E_p(Y/X) = \sum_c E_c(Y/X)$, where each energy $E_c(Y/X)$ depends only on the neighbouring system. To compute the MAP estimate, we need only to minimize $E(Y/X)$ by means of an optimization algorithm. We use simulated annealing to perform energy minimization, as the energy function may have many local minima. The energy functional is so composed by several terms linked to two types of information: the prior model and the sensor model of the depth and confidence maps Z and S. The global function embedding this information is:

$$E(Z, S \,/\, G_z, G_s) = E_p(Z) \cdot V_m^{(1)}(S) + E_p(S) + E_{obs}(Z \,/\, G_z) \cdot V_m^{(2)}(S) + E_{obs}(S \,/\, G_s)$$

where:

G_s and G_z are the observation fields of S and Z, respectively;

$E_p(Z)$, $E_p(S)$ are the functions related to the a-priori model of the fields Z and S;

$V_m^{(1)}(S)$, $V_m^{(2)}(S)$ are the modulating functions of S that reinforce or weaken $E_p(Z)$ and $E_{obs}(Z/G_z)$ functions;

$E_{obs}(Z/G_z)$, $E_{obs}(Z/G_s)$ derive from the models of the observations.

The components of the energy functional related to the Z field are those used in classical MRF algorithm for restoration and segmentation problems [7]. Assuming that g_{zi} is a generic pixel i of the observations, z_i is a generic pixel i of the depth lattice Z, s_i is a generic pixel i of the confidence lattice S, the energy functions are:

$$E_p(z_i, z_j) = (z_i - z_j)^2 \quad \forall \, z_j \in N_i \quad \text{and} \quad E_{obs}(z_i \,/\, g_{zi}) = \frac{1}{2} \cdot \sigma_z^{-2} \cdot (z_i - g_{zi})^2$$

where z_j is a generic pixel j of the neighboring set N_i.

In the proposed approach these terms are weighted by two functions of S.

The first function, modulating the a-priori model function of Z is:

$$V_m^{(1)}(s_i, s_j) = -K^{-2}(s_i \cdot s_j) + 1 \quad \forall \, s_j \in N_i$$

this function stresses the smoothness constraint if a generic pixel z_j, belonging to N_i, is low confident; in this case the observed image is unreliable and the only useful information is the prior model of the image.

The normalization value K is needed to restrict the function value between {0, 1}. The second function modulating the sensor model is:

$$V_m^{(2)}(s_i) = \sqrt{\frac{s_i}{K}}$$

this function reaches the maximum value, equal to 1, if the confidence of the pixel z_i is 1, stressing the observation constraint. For the function related to the prior model of S field, should be noted that the information on Z lattice is not useful for weighting the prior and sensor knowledge of confidence. In practice, the pixels of S are changed independently from the Z field, while the pixels of Z are varying taking into account the S information. The prior knowledge on S field should force the observed confidence map to become binary: 1 (maximum confidence) where an object is present and 0 otherwise. The function related to the prior model is:

$$E_p(s_i) = (s_i - s^*)^2$$

where the value s^* is calculated using a simple majority rule, with reference to a fixed threshold: s^* is equal to 1 if most of the pixels in the neighborhood of s_i are above this threshold, and is equal to 0 otherwise. A critical point of this algorithm is the choice of the confidence threshold: this value should be chosen carefully so as to preserve S values inside objects and to remove side lobe effects outside objects. For the sensor model of S, a Gaussian model is used, as previously done for the Z lattice.

4. Results and Discussion

The proposed algorithm was tested on simulated acoustic images provided by an active sonar simulator whose output signals had been processed by a beamforming algorithm. The goal of the algorithm was to restore the acoustic images in order to obtain an image similar to the ideal one shown in Fig.1a, starting from the noisy images in Fig.2a and Fig.2b (for depth and confidence, respectively). Fig.1a represents the ideal 2D projection of three objects located under the sea at different distance from the antenna. The three objects were a rectangular plate (20×30 cm) placed 5 meters far from the array and parallel to it; a square plate (20 cm side) placed 3.5 meters far from the array and inclined with respect to it; a spherical cap (30-cm diameter) placed 6.5 meters far from the array. The images were formed by means of 128×128 beams obtained by broad-band focused beamforming in quadrature in the time domain. The depth of each object is easily estimated according to the gray levels of the pixels: darker pixels correspond to shorter distances, and vice versa. Fig.1b shows the ideal confidence levels of the image; the confidence is high inside the objects and is null outside them. Fig.2a and 2b present real images obtained by the beamforming process; they are corrupted by speckle noise and the confidence map shows some "holes" inside the objects and side-lobe effects outside them. Fig.2c is an image obtained by keeping only the pixels of the Z image whose confidence values were above a fixed threshold. The image show many holes inside the objects. This proves that a punctual restoration is not sufficient. The proposed MRF algorithm uses the following parameters for the restoration process: size of input images: 128×128; 256 gray levels; second-order neighboring system;

logarithmic scheduling for temperature; confidence threshold level: 35. Fig.3a and 3b show the reconstructed depth image and the restored confidence image. The result is twofold: first, the algorithm removes holes and side-lobe effects inside and outside objects respectively; secondly, the final confidence map is a binary image that separates confident points from non-confident ones. After reconstructing the image, it was necessary to perform a further step. This processing was simply an AND operation on the restored confidence image in Fig. 3b. As a result, only confident points are present in the 3D image, and information about the object shapes can be simply derived from the restored confidence image. The final output image is shown in Fig.3c: one can notice a notable improvement over the image shown in Fig.2c. This result represents a good approximation to the ideal image shown in Fig.1a.

5. Conclusions

In this paper, we have presented a method that uses confidence information to restore underwater acoustic images. A-priori confidence information allows one to obtain a more accurate estimate of a depth image affected by speckle noise (modelled as AWGN) and to extract information about object shapes from the restored confidence image. To model both images, an MRF approach has been used, for which a single energy functional has been used. By minimizing this functional, the optimal estimate (in the MAP sense) has been found. We have shown that at each iteration, confidence values drive the restoration of noisy pixels of depth image, and, at the same time, confidence pixels, too, are restored; therefore both the 3D and confidence images are improved. Special attention has been given to the choice of the threshold for the confidence level.

References

1. A. Macovsky, "Ultrasonic Imaging using Acoustic Arrays", *Proc. IEEE,* Vol. 67, No.4, April 1979, pp. 484-495.
2. G. L. Sutton, "Underwater Acoustic Imaging", *Proc. IEEE,* Vol. 67, No.4, April 1979, pp. 554-566.
3. V.Murino, A.Trucco, "Acoustic Image Improvement by Confidence Levels", *IEEE Ultrasonics Symposium*, Cannes (F), November 1994, pp. 1367-1370.
4. J. H. Hockland, T. Taxt, "Ultrasound Speckle Reduction using Harmonic Oscillator Models", *IEEE Trans. Ultrasonic Ferroelectrics, and Frequency Control,* Vol. 41 No.2, March 1994, pp 215-224.
5. V. Murino, C.S. Regazzoni, A. Trucco, G. Vernazza, "A Non-Coherent Correlation Technique and Focused Beamforming for Ultrasonic Underwater Imaging: a Comparative Analysis", *IEEE Trans. Ultrasonics, Ferroelectrics, and Frequency Control,* vol. 41, September 1994, pp. 621-630.
6. J.L. Marroquin, *Probabilistic Solution of Inverse Problems.* Ph.D. thesis, Massachusetts Institute of Technology, 1985.
7. S. Geman, D. Geman, "Stochastic relaxation, Gibbs distribution, and Bayesian restoration of images". *IEEE Trans. Pattern Anal. Machine Intell.,* PAMI-6, November 1984, pp 721-741.

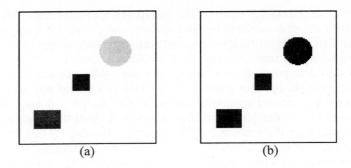

Fig. 1 (a) Ideal depth image; (b) ideal confidence image.

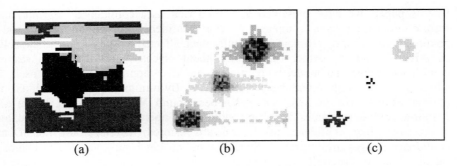

Fig. 2 Images resulting from the beamforming process and peak detection: (a) noisy depth image; (b) noisy confidence image; (c) AND between the depth and confidence images.

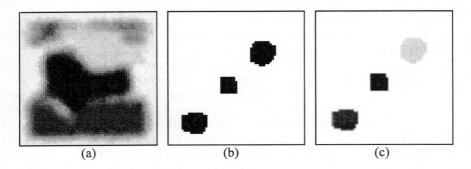

Fig. 3 Restored images: (a) restored depth image; (b) restored confidence image; (c) AND between the restored depth and confidence images.

Plausibilistic Preprocessing
of Sparse Range Images

Björn Krebs, Bernd Korn and Friedrich M. Wahl

Institute for Robotics and Computer Control,
Technical University Braunschweig,
Hamburger Str. 267, D-38114 Braunschweig, F.R.G.

Abstract. Range image interpretation often suffers from contaminating noise and sparseness of the input data. Non-Gaussian errors occur if the physical conditions in the scene violate sensor restrictions. To deal with such drawbacks we present a new approach for range image preprocessing. To provide dense range information initial sparse data is augmented via appropriate interpolation. Furthermore, we propose a measure of plausibility which depends on the density of the initial data to judge the result of the interpolation.

Range image preprocessing – perspective distortions – non-Gaussian errors – sparse range data – plausibility

1 Introduction

In many fields, e g. advanced robot applications, 3d scene interpretation becomes more and more important. Range images are an appropriate interface between sensor systems and scene interpretation. Usually, the first step in scene analysis is to extract 3d primitives representing surface or edge descriptions (e. g. planar patches [4], [5] or simple 3d primitive shapes such as spheres or cylinders [2]). The use of fixed models restricts the scene interpretation to a specific class of detectable objects. To allow more flexible object models some authors propose burdensome direct matching of range data with freeform-surfaces (e. g. [3], [7]). In contrast to these well-known image analysis algorithms, our approach preprocesses range images without any a priori knowledge of the object surfaces. The preprocessing (subdivided into four layers) serves as an interface between the sensor dependent level and the more abstract level of 3d scene interpretation:
The first step is *Range Image Acquisition*; a subsequent *Error Reduction* has to rely on the used scanner (section 2). In the third layer the perspective distorted range information is transfered to functional data by *Orthogonalization* with simultaneous *View Integration* (section 3). To get dense range information an *Augmentation* in layer four interpolates the data (section 4). The interpolation has to obey the following criteria:
1. The original data must not be changed. 2. The geometrical properties of the object surfaces, i. e. smoothness and curvature, must be preserved. 3. The interpolated data need to be judged with respect to its reliability. Hence, the result of the interpolation has to be judged by a measure of *plausibility*.

2 Range Image Acquisition and Error Reduction

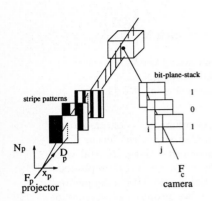

Fig. 1. Time-space encoding of the measurement area by means of n stripe patterns.

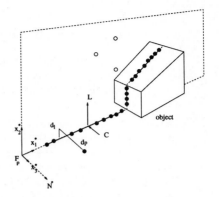

Fig. 2. All points which do not form a cut with an object surface are eliminated.

The used range sensor is based on the coded light approach which is a well-known fast and robust active triangulation technique (e. g. [9]). The system contains a camera, projectors and dedicated image processing hardware. A projector illuminates the scene with a sequence of n stripe patterns, allowing a distinction of 2^n different *illumination wedges* and the extraction of the corresponding *illumination planes* (see figure 1). The 3d coordinates of the observed scene points can be computed by triangulation, i. e. the intersection between the view line and the illumination plane.

In our experimental setup we use a depth-eye-in-hand configuration where the camera is mounted on a robot's hand while the projectors are installed rigidly around the work space of the robot. To supply different views, efficient recalibration and triangulation has been implemented in special image processing hardware[1]([8]). However, raw sensor data is often sparse and contaminated with noise and non-Gaussian errors. The type of errors depends on the physics of the used range sensor. Lens distortions are eliminated in the triangulation step itself [8]. The non-random errors occur very often in exclusive parts of the image[2]. An appropriate approach is to select only samples of data points which fulfill specific geometrical constraints. By sorting all point with respect to their distance to the plane origin a sample of points $\forall P \in S = \{P_0, \ldots, P_n\}$ with n elements in the illumination plane (defined by the normal vector N^* and the plane origin F_p) can be judged if they define a cut with an object surface (see figure 2).

$$n > \nu \qquad (1)$$

[1] DATACUBE MAX-VIDEO 10
[2] especially near jump edges

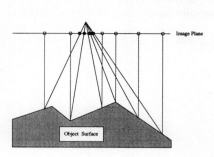 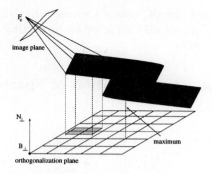

Fig. 3. Perspective and orthographic projections.

$$d_p = |x_3^*| < \theta \tag{2}$$
$$d_l = |((x_1^* l_1 + x_2^* l_2) - (c_1 l_1 + c_2 l_2)| < \zeta \tag{3}$$

$L = (l_1, l_2)$ is the normal vector and $C = (c_1, c_2)$ the centroid of a straight line through the sample (coordinates with respect to the local coordinate frame of the illumination plane). The error reduction is controlled by the thresholds ν, θ and ζ.

3   Perspective Distortions and Multiple View Integration

Usually, range images suffer from perspective distortions. Generally, this is true for any type of range image because the data is generated via a perspective map (see left image of figure 3). Nevertheless, many applications assume long focal distances with almost parallel view-lines and therefore no perspective effects (e. g. [1]). But in real world ranging systems the operational distance, and therefore the focal distance as well, is short. To apply homogeneous filter operations[3] without any further geometric distortions the perspective distorted range data has to be transferred to functional data by *orthogonalization* onto a *orthogonalization plane*. The orthogonalization restricts the observed surfaces to *Monge surfaces* which preserve their geometrical properties under orthogonalization ([6]). If a part of the object surface is non-Monge we select the *maximum distance* from the 3d point to the orthogonalization plane P with the normal vector $N_\perp = (n_1, n_2, n_3)$ and the origin $B_\perp = (b_1, b_2, b_3)$ (see right image of figure 3).

$$\pi_{max}(x_1, x_2, d_{max}) : d_{max} = \max |\sum_{i=0}^{2} x_i n_i - \sum_{i=0}^{2} b_i n_i| \tag{4}$$

The loss of information by orthogonalization of non-Monge surfaces can be handled by using more than one orthographic image: e. g. , orthographic maps in the three Cartesian coordinate system planes will supply all surface information.

---

[3] e. g. to extract gradients or surface normals

But not all parts of a surface are visible. View integration solves this problem and can be easily achieved by projecting all data samples onto one orthogonalization plane.

4 Augmentation of Sparse Data

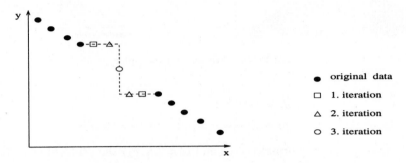

Fig. 4. Artefact produced by iterative sparse data averaging or median filtering.

The orthogonalization described above enables us to process a range image with similar homogeneous operations like greyscale images[4]. So, we can easily adapt well-known techniques from 2d image processing like median or average filtering. Let $I = \{(x, y) \mid 0 \leq x \leq N, 0 \leq y \leq M\}$ be the set of pixels of an orthogonalized range image after error reduction. The range image is a function on an array of pixels $r_0 : [0, \ldots, N] \times [0, \ldots, M] \mapsto \Re \cup \{\bot\}$ of valid range values $r_0(x, y) \in \Re$ or of unknown depth information $r_0(x, y) = \bot$. So, we can divide the range image into two disjoint sets of pixels:

$$S_0 = \{(x, y) \in I \mid r_o(x, y) \neq \bot\} \tag{5}$$

$$\bar{S}_0 = I - S_0 \tag{6}$$

A *sparse data averaging* filter can be realized with an $n \times n$ kernel $k_n(x, y)$ as follows:

$$r_{i+1}(x, y) = \begin{cases} r_i(x, y) & : & (x, y) \in S_i \\ \sum_Q \frac{r_i(\xi, \eta)}{|S_i \cap k_n(x, y)|} & : & (x, y) \notin S_i \end{cases} \tag{7}$$

and $Q = (\xi, \eta) \in (S_i \cap k_n(x, y))$
and $S_{i+1} = \{(x, y) \in I \mid r_{i-1}(x, y) \neq \bot\}$.
The averaging operation tends to generate smooth edges. To preserve steep edges, we propose a *sparse data median* filter:

$$r_{i+1}(x, y) = \begin{cases} r_i(x, y) & : & (x, y) \in S_i \\ m(S_i \cap k_n(x, y)) & : & (x, y) \notin S_i \end{cases} \tag{8}$$

---

[4] Thus, common image processing hardware allows efficient computation

with m(S) to be the median of the Set S of range values.

The update of $r_i(x,y)$ satisfies the first requirement of section 1 and the algorithm terminates already after a few iterations with $S_i = I$, depending on the cardinality of $\bar{S}_0$. However, in general the filter operation doesn't satisfy the second requirement (see figure 4).

Therefore we propose a kind of relaxation algorithm and modify the update in (7) to:

$$r_{i+1}(x,y) = \begin{cases} r_i(x,y) & : \quad (x,y) \in S_0 \\ \sum_{Q_1} \frac{r_i(\xi,\eta)}{n^2} & : \quad (x,y) \notin S_0 \end{cases} \tag{9}$$

with $Q_1 = (\xi,\eta) \in k_n(x,y)$.

The pixel $(x,y) \in \bar{S}_0$ must be initialized with an appropriate value $r(x,y) \in \Re$ and the algorithm stops when:

$$\forall (x,y) \in I \qquad |r_{i+1}(x,y) - r_i(x,y)| \leq \varepsilon \quad , \quad \varepsilon \in \Re^+ \tag{10}$$

is satisfied.

The third criterion we proposed in section 1 requires a measure of plausibility for the interpolated data. The plausibility of interpolated data $r(x,y)$ must increase with the number of measured data within the neighborhood of pixel (x,y). Therefore we choose the initial plausibility

$$pl_0 = \begin{cases} 1 & : \quad (x,y) \in S_0 \\ 0 & : \quad (x,y) \notin S_0 \end{cases} \tag{11}$$

and the update

$$pl_{i+1} = \begin{cases} pl_i & : \quad (x,y) \in S_i \\ \frac{1}{n^2-1} \sum_{Q_1} pl_i(\xi,\eta) & : \quad (x,y) \notin S_i \end{cases} \tag{12}$$

with S_0, S_i, S_{i+1} as in (7)
and Q_1 as in (9).

5 Experimental Results and Conclusions

In our experiments the range images consist of 512×512 of 2byte values. Only data near the borders of the illumination wedges provide proper values. Furthermore, highly erroneous data occur in some scene parts, e. g. due to occlusion. After the error reduction only 20%- 30% of all pixels have depth values assigned. To supply complete surface information, six views of the whole scene are integrated; the orthogonalized image and the result of the preprocessing is shown in figure 5.

In this paper we have proposed a surface and object model independent preprocessing of range images. Different images from several views can be integrated before scene analysis. With the introduced concept of geometrical constraints it is possible to select valid 3d data. A subsequent augmentation provides dense range information judged by a plausibility measure. First solutions and results

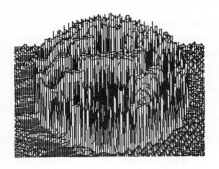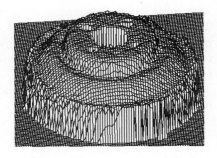

Fig. 5. The left image shows orthogonalized data of six integrated views. The right image shows the resulting 3d information of the preprocessed range data with a plausibility greater than 0.1. For a better visualization $r_0(x, y) = \perp$ was replaced by $r_0(x, y) = -1$.

have been outlined and the proposed algorithm has been validated with real world range images. In future, the preprocessing will be fully integrated in our scene interpretation system to support more complex surface models. Furthermore, the range image acquisition and preprocessing will support sensor guided grasping and assembly currently investigated and developed at our institute.

References

1. P. J. Besl and R. C. Jain. Segmentation through variable-order surface fitting. *IEEE Transactions on Pattern Analysis and Machine Intelligence*, 10(2):167–192, 1988.
2. R. Bolles and M. A. Fischer. A ransac-based approach to model fitting and its application finding cylinders in range data. In *7th Int. Conf. on Art. Int. , Vancouver*, pages 637–643, 1981.
3. K. Higuchi, M. Hebert, and K. Ikeuchi. Bulding 3-d models from unregisterd range images. In *Proc. IEEE Conference of Robotics and Automation, San Diego, CA*, pages 2248–2253, 1994.
4. J. M. Jolion, P. Meer, and S. Bataouche. Robust clustering with applications in computer vision. *Pattern recognition*, 13(8):791–802, 1991.
5. R. Krishnapuram and C. P. Freg. Fitting an unknown number of lines and planes to image data combatible cluster merging. *Pattern Recognition*, 25(4):385–400, 1992.
6. M. A. Penna and S. Chen. *Image Understanding in Unstructured Enviroment*, volume 2 of *Series in Automation*, chapter Sperical Analysis in Computer Vision and Image Understanding, pages 126–177. World Scientific, 1994.
7. D. A. Simon, M. Hebert, and T. Kanade. Real-time 3-d pose estimation using a high-speed range sensor. In *Proc. IEEE Int. Conf. on Robotics and Automation, San Diego, CA*, volume 3, pages 2235–2240, 1994.
8. T. Stahs and F. M. Wahl. Fast and Versatile Range Data Acquisition in a Robot Work Cell. In *Proc. IEEE Int. Conf. on on Intelligent Robots and Systems (IROS) 1992, Raleigh, North Carolina*, 1992.
9. F. M. Wahl. A Coded Light Approach for 3D Vision. In *IBM Research Report*, RZ 1452, 1984.

Segmentation/reconstruction of range images based on piecewise-linear approximation

Enrico Puppo

Istituto per la Matematica Applicata – Consiglio Nazionale delle Ricerche
Via De Marini, 6 (Torre di Francia) - 16149 Genova - ITALY
Email: puppo@ima.ge.cnr.it

Abstract. A technique for segmenting and reconstructing range images of polyhedral objects is presented. A piecewise-linear approximation based on a triangulation of the image is computed first. Optimization techniques are applied next to detect jump and crease edges, to eliminate false boundaries, and to merge nearly coplanar triangles into larger polygonal patches.

1   Introduction

The visible portion of a real world scene framed by a range image is a piecewise-smooth surface defined by the graph of a bivariate function, with discontinuities of various order along characteristic lines. The range image is a noisy, sampled version of the surface. In order to extract useful information, the image plane can be segmented into coherent regions, corresponding to homogeneous smooth surfaces in the scene, and each surface patch can be reconstructed through a suitable function that fits data.

Several techniques for segmenting and reconstructing range images have been developed in the literature (e.g., [1, 3, 4, 6]). Here, we propose a surface-oriented method that combines refinement and aggregation to provide a segmentation of a range image, and its representation through a piecewise-linear surface.

An important characteristic of this method is the use of irregular triangulations to segment the image. Triangulations result more adaptive than recursive decomposition schemes based on regular shapes (e.g., quadtrees), which are traditionally used in split-and-merge methods. Triangulations have been used extensively to reconstruct continuous surfaces, but they have been seldom applied to image processing. A recent triangulation-based reconstruction technique, which allows to deal with surface discontinuities has been proposed in [3]. Such a technique uses traditional contour-oriented segmentation and labeling techniques to detect characteristic lines in the image, which are used next to compute a constrained triangulation: much work is performed at pixel level, while limiting the use of triangulations to a final reconstruction stage. A peculiarity of the method proposed here is that image pixels are considered only during an initial refinement stage to build a fragmentation of the image into triangles. The rest of processing performs aggregation on a subdivision of the image into polygonal regions (initially trianlges), while disregarding the underlying pixel grid.

2 Problem definition

An *orthographic range signal* of a scene is a piecewise-smooth function $f : \mathbb{R}^2 \longrightarrow \mathbb{R}$ defined as follows: for each point X in a *frame* (i.e., a rectangular window on a plane) $D \subset \mathbb{R}^2$, $f(X)$ is the distance between X and the first point of an object in the framed scene that is hit by a ray through X and perpendicular to the frame plane. Assuming that the framed scene is composed of objects bounded by piecewise-smooth surfaces, f is describable by a partition of D into a finite number of regions $\mathcal{R} = \{R_1, \ldots, R_k\}$, such that f is represented over each region R_i by a smooth function ϕ_i belonging to family $\mathcal{F} = \{\phi_1, \ldots, \phi_k\}$. The pair $(\mathcal{R}, \mathcal{F})$ is called a *representation* of f. $\mathcal{R}$ is called a *segmentation* of f if and only if it is composed of a minimum number of maximal regions, otherwise it is called a *fragmentation* of f. Note that f can have discontinuities of various order along the boundaries of regions of $\mathcal{R}$: a discontinuity of the 0th order is called a *jump*, while a discontinuity of the 1st order is called a *crease*.

Given a zero-mean random noise process n, we call $\tilde{f} = f + n$ a *noisy signal* of f, and we consider a range image as a sampled version of $\tilde{f}$. A *regular grid* over a frame D is a subdivision of D into rectangular regions of uniform size, called *pixels*; the points at the corners of pixels are called *vertices*. A *range image* I of $\tilde{f}$ is a restriction of $\tilde{f}$ to the centers of pixels, and it is represented as a two dimensional array of $N \times N$ values, where $I_{ij} = \tilde{f}(x(i), y(j))$, and the vector function $(x(), y())$ gives the centers of all pixels for $i, j \in \{1, \ldots, N\}$.

Let $\hat{f}$ be a real-valued function described on D by $(\hat{\mathcal{R}}, \hat{\mathcal{F}})$. The error in approximating $\tilde{f}$ by $\hat{f}$ on the basis of I is defined as

$$E(\hat{f}) = max_{i,j=1 \ldots N} |\hat{f}(x(i), y(j)) - I_{ij}|.$$

The *approximate segmentation/reconstruction problem* for I, at precision $\varepsilon > 0$ is to find a minimal representation $(\hat{\mathcal{R}}, \hat{\mathcal{F}})$, such that $E(\hat{f}) < \varepsilon$.

Ideally, all discontinuities of f - and only such lines - should appear as boundaries of regions of $\hat{\mathcal{R}}$. In practice, the precision threshold will influence the quality of the result. The value of ε can be set according to the expected variance of the noise affecting I: in general, the smaller its value, the larger the number of regions in $\mathcal{T}$, and the smaller their size.

In the following, polyhedral scenes, i.e., signals that can be represented in terms of polygonal subdivisions and linear functions, will be considered. The method we propose is also useful ato obtain approximate polygonal segmentatations, and piecewise-linear reconstructions of scenes with curved objects.

3 Adaptive image triangulation

The first stage of the method (refinement phase) consists of an algorithm to compute a fragmentation $\mathcal{T}$ of I into triangular regions having their vertices at vertices of the pixel grid. $\mathcal{T}$ is called a *triangulation* of I. In order to be effective, $\mathcal{T}$ should be formed by a restricted number of triangles, whose edges form chains that approximate all edges of the desired segmentation $\mathcal{R}$.

Given $\mathcal{T}$, a piecewise-linear fragmented surface $f_{\mathcal{T}}$ is defined as follows. For each triangle t of $\mathcal{T}$, for each vertex v_i of t, $i = 0, 1, 2$, consider the pixel X_i incident into v_i and whose intersection with t has maximum area. $f_{\mathcal{T}}$ is defined over t as the linear function interpolating range values at the centers of pixels X_i, $i = 0, 1, 2$, restricted to t. Note that different triangles incident into a vertex v can have different range values at v, thus yielding a possible jump at every edge of $\mathcal{T}$ (see Figure 1a).

We are interested in finding $\mathcal{T}$ such that $E(f_{\mathcal{T}}) \leq \varepsilon$ for a given $\varepsilon \geq 0$. $\mathcal{T}$ is built based on a simple heuristic technique that iteratively refines an initial triangulation covering D by inserting one new vertex at a time, until the required precision is reached. This algorithm is reminiscent of a technique proposed first in [5] to reconstruct continuous surfaces.

This refinement technique naturally tends to select points along characteristic lines of the image, hence approximating such lines through edges of the triangulation. Such good behavior, which had already been demonstrated in the literature for continuous surfaces, is confirmed in our experiments for the non-continuous case of range data.

4 Elimination of false boundaries

Although characteristic lines of I are approximated by chains of edges of $\mathcal{T}$, many other edges of $\mathcal{T}$ do not correspond to line features of I, and many false jumps are present in the reconstructed surface. False jumps are eliminated through local analysis of range values of the reconstructed surface at the vertices of $\mathcal{T}$.

Let $t_{i_1}, \ldots, t_{i_h}$ be the set of triangles incident into a vertex v_i of $\mathcal{T}$, in counterclockwise order around v_i, and let $\phi_{i_1}, \ldots, \phi_{i_h}$ be the linear restrictions of $f_{\mathcal{T}}$ to such triangles. The values $P_{i_j} = \phi_{i_j}(v_i)$, $j = 1, \ldots, h$, are in general all different (see Figure 2a): if $P_{i_j} \neq P_{i_{j+1}}$ then $f_{\mathcal{T}}$ is necessarily discontinuous along the edge common to t_{i_j} and $t_{i_{j+1}}$. We wish to maintain such a discontinuity only if it corresponds to a jump. In order to eliminate a discontinuity, we need to make P_{i_j} and $P_{i_{j+1}}$ equal. We consider the sequence $P_{i_0}, P_{i_1}, \ldots, P_{i_h}$, where $P_{i_0} = P_{i_h}$, as a noisy sample of a piecewise-constant univariate signal at points $0, 1, \ldots, h$: each constant segment of the signal spans all P_{i_j}s that should share the same range value. We reconstruct the signal in the interval $[0, h]$ through a function u_i that fits data, and that satisfies the constraint $u_i(0) = u_i(h)$.

The technique used to reconstruct u_i is a modification of the *weak elastic string* described in [2]: u_i is modeled as an elastic string under weak continuity constraints; breaks are placed in the string where the continuity constraint is violated. An energy is associated to the string, which is composed of three terms:

P: the sum of penalties α for each discontinuity $P = \sum_{j=1}^{h} \alpha l_{i_j}$, where $l_{i_j} = 1$ if there is a discontinuity between $P_{i_{j-1}}$ and P_{i_j}, and $l_{i_j} = 0$ otherwise.

D: a weighted measure of faithfulness to data $D = \sum_{j=0}^{h} w_{i_j}(u_i(j) - P_{i_j})^2$, where w_{i_j} is a weight proportional to the area of triangle t_{i_j}.

S: a measure of slant $S = \lambda^2 \sum_{j=1}^{h}(u_i(j) - u_i(j-1))^2(1 - l_{i_j})$.

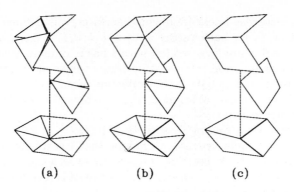

(a)   (b)   (c)

Fig. 1. Patches incident into a vertex (a) before optimization, (b) after jump detection: thick edges are classified as jumps, (c) after crease detection: dummy edges have been eliminated.

Parameter α is a coefficient of strength of the string: the higher ts value, the less discontinuities we get. Weights w_i tends to maintain the result faithful to range values of large triangular patches. Parameter λ is aimed to prevent continuous segments of u_i to be slanted.

The non-convex energy of the string $E = D + S + P$ must be minimized. To this aim, we have used an iterative algorithm that implements the Graduated Non-Convexity technique proposed in [2], whose details are omitted here for brevity.

Jump edges are detected by applying such operation to all vertices of the triangulation (see Figure 2b). Remaining edges of $\mathcal{T}$ must be classified next into crease edges, and *dummy* edges, i.e. edges separating nearly coplanar patches.

Detection of creases is performed through a similar optimization technique, although this time an energy is computed over the whole surface. We consider the piecewise linear function $f_{\mathcal{T}}$ currently defined on the triangulation, and we fit to such function another piecewise-linear function $\hat{f}$ that minimizes an energy depending on its distance from $f_{\mathcal{T}}$, on its "smoothness", and on the amount of creases in it. Function $\hat{f}$ is obtained by modifying the range values assumed by $f_{\mathcal{T}}$ at the vertices of triangles, while maintaining the continuity along non-jump edges. Details are omitted here for brevity.

After identifying creases, patches adjacent along dummy edges are merged. The resulting structure is a planar subdivision of the image frame, made of polygonal regions (see Figure 1c).

Each tiny region, whose area is smaller than a given threshold, is merged next into one of its adjacent region, based on maximum likelihood of surfaces. Finally, contours of regions in the subdivision are simplified through a simple line simplification algorithm.

Fig. 2. The input image image rendered as a grey level map.

5 Experimental results

The algorithm proposed has been tested on several range images. We show results obtained on a popular test image representing a frame of a drilled cube, containing both plane and curved surfaces (see Figure 2). The image is composed of 128×128 integer values in the range $[0, 255]$.

The fragmentation of the image obtained after the triangulation stage with $\varepsilon = 3.84$ (i.e., 1.5% of the input range) has 971 vertices, i.e., nearly 6% of the input grid. Most edges at this stage appear as jumps. The main lineal features of the image are approximated by chains of edges: edges of the cube and borders of the holes are clearly visible. After detection of jumps and creases, and deletion of dummy edges, all important features have been retained, although some of them are approximated by jagged chains; some spurious creases have been detected. Figure 3 shows the different stages of reconstruction.

6 Concluding remarks

The results presented are relative to an initial study, which is limited to piecewise-linear surfaces, and thus it is suitable to handle polyhedral scenes. Approximate reconstructions of scenes containing objects with curved surfaces can be obtained as well. An immediate extension consists in using higher order polynomial patches over triangles. Curved patches permit to handle a broader class of range images containing curved surfaces.

The method proposed is promising, though we believe that its performances can be highly improved, both in the quality of the results, and in terms of running time.

A further possibility is to adapt a hierarchical triangulation methods [7] to non-continuous surfaces, in order to obtain a multi-scale pyramidal model based on layers of triangulations at different levels of resolution.

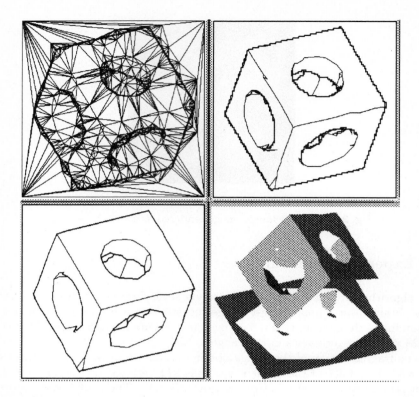

Fig. 3. Different stages of the segmentation: (a) fragmentation after triangulation; (b) segmentation after the detection of jumps and creases (jumps are depicted as bold lines; dummy edges have been removed); (c) final segmentation; (d) a view of the final reconstruction.

References

1. Besl, P.J., Jain, R.C., Segmentation Through Variable-Order Surface Fitting, IEEE-PAMI, **10**, 2 (1988) 167-192.
2. Blake, A., Zisserman, A., Visual Reconstruction, The MIT Press, Cambridge, MA (1987).
3. Chen, X., Schmitt, F., Adaptive range data approximation by constrained surface triangulation, in Modeling in Computer Graphics, B. Falcidieno, T.L. Kunii (Eds.), Spriger-Verlag, Berlin Heidelberg (1993) pp.95-114.
4. Fan, T., Medioni, G., Nevatia, R., Recognizing 3-D Objects Using Surface Descriptions, IEEE-PAMI, **11**, 11 (1989) pp.1140-1157.
5. Fowler, R.J., Little, J.J., Automatic extraction of irregular network digital terrain models, ACM Computer Graphics, **13**, 3 (1979) pp.199-207.
6. Hoffman, R., Jain, A.K., Segmentation and Classification of Range Images, IEEE-PAMI, **9**, 5 (1987) pp.608-620.
7. Scarlatos, L., Pavlidis, T., Hierarchical triangulation using cartographic coherence, CVGIP: Graphical Models and Image Processing, **54**, 2 (1992) pp.147-161.

Low-level Image Processing

In-Place Covariance Operators for Computer Vision

Terry Caelli and Mark Ollila

School of Computing, Curtin University of Technology,
Perth, Western Australia, Box U 1987, Email:tmc@cs.mu.oz.au
Australia

Abstract. Perhaps one of the most common low-level operations in computer vision is feature extraction. Indeed, there is already a large number of specific feature extraction techniques available involving transforms, convolutions, filtering or relaxation-type operators. Albeit, in this paper we explore a different approach to these more classical methods based on non-parametric in-place covariance operators and a geometric model of image data. We explore these operators as they apply to both range and intensity data and show how many of the classical features can be redefined and extracted using this approach and in more robust ways. In particular, we explore how, for range data, surface types, jumps and creases have a natural expression using first and second-order covariance operators and how these measures relate to the well-known Weingarten map. For intensity data, we show how edges, lines corners and textures also can be extracted by observing the eigenstructures of similar first- and second-order covariance operators. Finally, robustness, limitations and the non-parametric nature of this operator are also discussed and example range and intensity image results are shown.

1 Introduction

We have explored the use of covariance operators for the computation of local shape measures relevant to range and intensity (surface) image data. The covariance approach dispenses with surface parameterization and provides invariant descriptions of shape via the eigenvalues and eigenvectors of covariance matrices of different orders. The aim of this paper is to summarize the properties and results of the covariance approach as they can are applied to feature extraction in both types of surface image data.

Following Liang and Todhunter [1] we define the *local* (first–order) surface covariance matrix (C_I) as:

$$C_I = \frac{1}{n} \sum_{i=1}^{n} (\tilde{x}_i - \tilde{x}_m) \cdot (\tilde{x}_i - \tilde{x}_m)^T \, , \tag{1}$$

where $\tilde{x}_i = (x_i, y_i, z_i)$ correspond to the image projection plane (x, y) and (z_i) corresponds to depth or intensity values at position i; $\tilde{x}_m = 1/n \sum_{i=1}^{n} \tilde{x}_i$ to the mean position vector; n to the total number of pixels in the neighborhood of $\tilde{x}_i$ used to compute Eqn.(1).

The eigenvalues and eigenvectors of C_I determine three orthogonal vectors two of which define a plane whose orientation is such that it minimizes, in the least square sense, the (squared distance) orthogonal projections of all the points (n) onto the plane. Liang and Todhunter[1] originally proposed this plane as a reasonable approximation to the surface tangent plane and, so long as the two eigenvectors are chosen to preserve the "sidedness" of the surface points, it can be viewed as an analogue to this. That is, the eigenvectors corresponding to the two largest eigenvalues (λ_1 and λ_2) form this "tangent plane" if "sidedness" is preserved, otherwise the plane must be formed by the eigenvectors corresponding to eigenvalues λ_2 and λ_3. This "sidedness"–preserving or *topology preserving principle*, can be seen as a replacement of surface parameterization. It is important to note that the eigenvalues of C_I are already invariant to rigid motions and they also define the aspect rations of oriented spheres which determine the strengths of directional information in the neighborhood of a pixel. As we will see, from a geometric perspective such eigenvalues and vectors define surface types while, from an intensity image perspective, they can be used to infer edge, corners and other local shape structures in an image. First, the geometric interpretation.

2 Local geometries.

From a Differential Geometry perspective, C_I defines local tangent planes and surface normals (defining the first fundamental form) in the least-squares sense. From this an analogous definition of the second fundamental form follows. We define, about a point $\tilde{x}_0$ on a surface, a two-dimensional covariance matrix in the following manner:

$$C_{II} = \frac{1}{n} \sum_{i=1}^{n} (\tilde{y}_i - \tilde{y}_m) \cdot (\tilde{y}_i - \tilde{y}_m)^T, \qquad (2)$$

with the two dimensional vectors $\mathbf{y}_i$ defined by the projection:

$$\tilde{y}_i = s_i \cdot (\tilde{x}_i - \tilde{x}_0)_{proj}, \qquad (3)$$

where $\mathbf{x}_i$ is in a small neighborhood of $\tilde{x}_0$. We project the difference vector which points from $\tilde{x}_0$ to $\tilde{x}_i$ onto the "tangent plane" as determined by Eqn.(1) and weight the resulting two–dimensional vector by distance s_i which measures the orthogonal distance from the "tangent plane" to point $\tilde{x}_i$. Given both tangent vectors $\tilde{t}_1$ and $\mathbf{t}_2$ and the normal vector $\tilde{n}$ at point $\tilde{x}_0$ derived from Eqn.(2) we may write y_i explicitly as:

$$\tilde{y}_i = \underbrace{\left[(\tilde{x}_i - \tilde{x}_0)^T \cdot \tilde{n} \right]}_{s_i} \cdot \begin{pmatrix} (\tilde{x}_i - \tilde{x}_0)^T \cdot \tilde{t}_1 \\ (\tilde{x}_i - \tilde{x}_0)^T \cdot \tilde{t}_2 \end{pmatrix} \qquad (4)$$

We can then define the quadratic form

$$II_C = \tilde{v}^T \cdot C_{II} \cdot \tilde{v} \tag{5}$$

as a "second fundamental form" based on covariance methods with the defined covariance matrix according to Eqn. (2) and a chosen unit vector v in the tangent plane. Analogous to classical computations of surface geometry we define the **principal directions** as those directions which are given by the eigenvectors of this covariance matrix. The eigenstructures of C_{II} capture how the surface points, in the neighbourhood of a pixel, depart from the estimation of the tangent plane.

Further to this, we have an analogous operator to the Gauss map of classical Differential Geometry[2]. That is, we can determine how the estimated surface normals, in the neighborhood of a pixel, project onto the estimated tangent plane by the two-dimensional covariance matrix:

$$C_P = \frac{1}{n} \sum_{i=1}^{n} (\tilde{v}_i - \tilde{v}_m) \cdot (\tilde{v}_i - \tilde{v}_m)^T, \tag{6}$$

with the two dimensional vectors v_i being defined by the projection:

$$\tilde{v}_i = (\tilde{n}_i)_{proj} = \begin{pmatrix} \tilde{n}_i^T \cdot \tilde{t}_1 \\ \tilde{n}_i^T \cdot \tilde{t}_2 \end{pmatrix}, \tag{7}$$

where $\tilde{t}_1, \tilde{t}_2$ are the estimated tangent vectors obtained from Eqn.(1) and assigned to the point (x, y, z) at the center of the current window and $\tilde{n}_i$ is the normal vector obtained by Eqn.(1) at a position (x_i, y_i, z_i) in the neighborhood of (x, y, z).

In all, then, the operators C_I, C_{II} and C_p determine local geometric characteristics of surfaces without the use of Calculus and with the constraint that such eignestructures correspond to least squares estimates of different order orientation fields. They can be used in identifying different surface types - including discontinuities - in the following ways (using appropriate thresholds, see [2]:

Jump-edge detection: The covariance matrix C_I, according to Eqn.(1), is calculated in a 5 × 5 pixel neighborhood at each pixel of the range image. Pixels with values of the maximal eigenvalue larger than a certain threshold are labeled as *jump*. The eigenvectors calculated in this stage are used for the next step.

Crease-edge detection: The covariance matrix C_p, according to Eqn.(6), is calculated in a 5 × 5 pixel neighborhood at each pixel of the range image. The maximal eigenvalue of this covariance matrix C_p is utilized as a crease-edge detector. Pixels not already labeled as *jump* and with values of the maximal eigenvalue larger than a certain threshold are labeled as *crease*.

Region segmentation: Pixels neither being labeled as *jump* nor *crease* are labeled as *planar, parabolic* (developable) or *curved*. In order to do so the covariance matrix C_I according to Eqn.(1) is calculated in a 7 × 7 pixel neighborhood at each pixel of the range image. The eigenvectors of this covariance matrix are again used as input for the next stage which calculates the covariance matrix C_p

according to Eqn.(3) from the projected normal vectors in a 7×7 pixel neighborhood where only pixels which have not been labeled so far as *jump* or *crease* are taken into account. Two thresholds for each of the two eigenvalues of this covariance matrix are applied. Pixels with smaller values of both eigenvalues than the thresholds are labeled as *planar*. Pixels with a smaller value of the smaller eigenvalue than the threshold but with values of the larger value beyond the corresponding threshold are labeled as *parabolic*. Pixels not meeting above conditions are assigned the label *curved*.

We have obtained experimental results for the proposed segmentation technique using many synthetic range images - one of which is shown in Figure 1a. The range images have neither been filtered nor preprocessed in any way. However, we have computed surface Mean and Gaussian curvatures using the bi-quadratic surface algebraic form proposed in [2]. We have also used the jump–and crease–edge detection technique presented in [3]for comparison. Together these region and boundary detection methods are termed the Smoothed Differential Geometry, or SDG method.

3 Image features

Viewing intensity images as surfaces is not that common but is natural to this covariance approach. Indeed, form this perspective, "edges", "corners" and "linear segments" all can be described by different types of surface types [5] including jumps, creases and non-planar region types. Further, we can even provide first- and second-order descriptions of "textural" features in terms of the associated eigenstructures of the covariance operators (Figure 1c). Figure 1b and c shows examples of such interpretations. Lines were determined by first-order eigenvalue ratios, corners by second-order eigenvalue strengths.

4 Discussion

The aim of this paper was to simple expose and demonstrate the covariance approach to feature extraction. It contrasts with more classical methods - and, in particular, differential and filter-based techniques, in two ways. One, there is no use of parameterization or differentiation directly. Two, there is an implicit optimization criterion: orthogonal least squares approximations of linear structures of different orders. Furthermore, we have been able to define analogous operators to the classical Weingarten map (second fundamental form) and the Gauss map on the basis of covariance matrices. The eigenvalues of the covariance matrices are invariant to rigid motion as well - since the computation operates in the tangent plane only. In addition, we have shown how the covariance method treats discontinuities in a very natural manner.

Finally, since covariance methods do not rely on a consistent local surface parameterization, the spectrum of the covariance matrix (the full set of eigenvalues) provides us with a type of smooth transformation between lines, surfaces and volumes. One non-zero eigenvalue defines merely linear components while a solution of three equal eigenvalues corresponds to data of uniform density in a volume about a point – analogous to fractile-dimensionality.

It should also be noted that covariance methods provide ideal ways of treating signals embedded in additive Gaussian or white noise.In these cases the total covariance matrix decomposes into the sum of the signal and noise covariance matrices. These covariance models also provide linear approximations to the local correlations or dependencies between pixels. Consequently, they are related to techniques for modeling the observed (local) pixel "cliques" or "support" kernels defined in Markov Random Fields and determined using Hidden Markov Models. However, space prohibits more detailed analysis of these connections.

5   References

[1] Liang, P. and Todhunter, J.(1990) Representation and recognition of surface shapes in range images. *Computer Vision, Graphics and Image Processing*, 52, 78-109.

[2] Berkman, J. and Caelli, T.(1994) Computation of Surface Geometry and Segmentation using Covarinace Techniques.*IEEE: Trans. Patt. Anal. and Machine Intell.* (In Press)

[3] Besl, P. and Jain, R.(1986) Invariant surface characteristics for 3D object recognition in range images. *Computer Vision, Graphics and Image Processing*, 33, 33-80.

[4] Yokine, N., and Levine, M.(1989) Range IMage Segmentation Based on Differential Geometry: A Hybrid Approach.*IEEE: Trans. Patt. Anal. and Machine Intell.*, 11, 6, 643-649.

[5] Barth, E., Caelli, T. and Zetzsche. C (1993) Image Encoding, Labeling, and Reconstruction from Differential Geometry. *Computer Vision, Graphics and Image Processing: Graphical Models and Image Processing*, 55, 6, 428-446.

Figure 1 - Following Page

Top: Shows input range image; second row:jumps(white), creases(grey) for SDG(left) and covariance(right) methods; third row: region segments from SDG(left) and covariance(right) methods.

Middle: Input textures(left) and rotation invariant segmentation based on first-order covariance eigenvalues(right) and 2-class K-Means clustering.

Bottom: Input intensity image(left) lines(centre), and corners(right) using first and second-order covariance eigenvalues respectively.

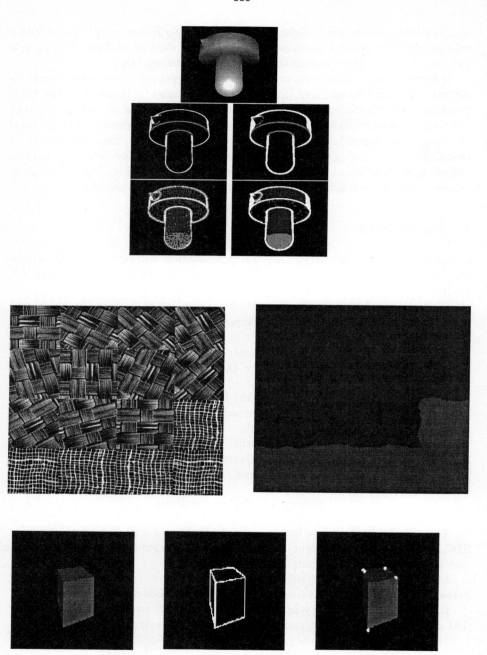

A New Heuristic Search for Boundary Detection

Marc Salotti and Mostafa Hatimi

Groupe VISIA, Centre de Mathématiques et Calculs Scientifiques,
Faculté des Sciences, BP 52, 20250 Corté (FRANCE)
e-mail: salotti@univ-corse.fr, hatimi@univ-corse.fr

Abstract

Using the A* algorithm, we propose a new heuristic search strategy to find object boundaries. We show that a cost function with a Gaussian curvature is more appropriate to develop only the best paths. An application to the detection of cells boundaries is presented.

Keywords

Boundary detection, Edge following, Heuristic search

1 Introduction

Heuristic search strategies have been developed for many applications of artificial intelligence. Martelli applied the A* algorithm to boundary detection in 1972 [6]. Other similar approaches have been proposed, most of them for the detection of boundaries in biomedical applications. Encouraging results have been presented, but some difficulties were not successfully overcome.

In the first part, we propose a short overview of edge following techniques using the A* algorithm or variations of it. Then, we present new ideas to improve the search and we give the implementation details for the detection of an arbitrary number of cells with specific region properties. Some results are presented.

2 Heuristic Search Strategies

2.1 The Basic Algorithm

In 1972, Martelli showed that the problem of boundary detection can be brought back to the problem of finding the minimal cost path in a weighted and directed graph, with positive costs [6]. Although different modelisations can be made, we consider that nodes of the graph correspond to pixels, and arcs exist between two connected pixels P_i, P_j if the direction of vector (P_i, P_j) is the same as the tangent to the boundary given by vector (P_{i-1}, P_i), with a maximal deviation of one pixel. Then, the key of the problem is the choice of the evaluation function used to define the quality of the path. This function is called f. The main stages of the A algorithm are described below [8], [2] :

> 1. Expand the start node (put the successors on a list called OPEN with pointers back to the start node).
> 2. Remove the node P_i of minimum f from OPEN. If P_i is a goal node, then stop. Trace back through pointers to find the optimal path. If OPEN is empty, then fail.

3. Else expand node Pi, putting successors on OPEN, or eventually updating them, with pointers back to Pi. Go to step 2.

The nodes are selected at each step according to an evaluation function f(Pi), which is an estimate of the cost of a minimal cost path from s the starting node to a goal node constrained to go through node Pi. This function f(Pi) can be expressed as :

$$f(Pi) = g(Pi) + h(Pi)$$

g(Pi) is the cost of the path from s to Pi given by summing the arc costs encountered while following the pointers from Pi to s (this path is the lowest cost path from s to Pi found so far). h(Pi) is an estimate of the cost of a minimal cost path from Pi to a goal node. If no information is available on the shape to be found, it is preferable to avoid estimating the cost from the current node to a goal node and therefore to set h(Pi) = 0 for all Pi. In this case, algorithm A* coincides with Dijkstra's algorithm [7], [9].

2.2 Choice of Heuristics

Martelli suggested to use an expression roughly equivalent to the following :

$$cost(Pi,Pi+1) = M - gradient(Pi+1), \text{ where } M = \max_x \{gradient(x)\} \qquad (1)$$

The problem of the heuristic search is that it must keep track of a set of current best paths, and this set may become very large if using heuristics like expression (1): a good but long path may eventually look expensive compared to small undesirable paths. Several ideas have been proposed to solve the problem. The selection of the next node in the OPEN list can be made with a depth-first strategy or using a rating function [3], [10], [11]. Lester and *al* suggest to take the maximum cost arc of the path as the value of g. The advantage is that g does not build up continuously with depth, so that good paths can be followed for a long time [5]. Since the value of g necessary increases with depth if the costs are positive, Ashkar and Modestino proposed a cost function that takes negative values if the edge has a good evaluation [1].

However, if some interesting ideas have been proposed, it seems that it is not possible to keep looking for the path with minimum cost and at the same time to have a depth first strategy, avoiding small undesirable paths.

2.3. Proposition for a Suitable Cost Function

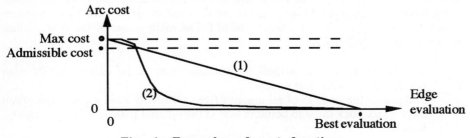

Fig. 1. Examples of cost functions.

The problem is that the sum of arc costs of a good and long path may be equal to the sum of arc costs of a small and undesirable path. In order to overcome this difficulty, it is necessary to assign a very high cost when the evaluation is bad and a very small cost if the evaluation is good. Two cost functions are displayed figure 1. Clearly, a linear expression like $cost(Pi, Pj) = M - gradient(Pi)$ is not appropriate. Function (2) is better, it could be generated by $cost(Pi, Pj) = e^{-gradient^2(Pi)}$. In a practical standpoint, if all gradient values of the best path are significant and little superior to the gradient values of other paths, this best path is developed first. On the contrary, if low gradient values are present, the path is not sure, the cost is high and other paths are explored. Since a Gaussian curvature is suitable to favor a depth first strategy, we propose a generalization, taking simultaneously into account several features:

$$cost(Pi, Pi+1) = \Sigma_k \ \alpha_k \ . e^{-x_k^2} \qquad (3)$$

For k features, x_k is an expression of feature k, and α_k determines the weight of each feature, such that $\Sigma_k \ \alpha_k = 1$.

3 Application

We present an application of heuristic search to determine the boundaries of specific objects. We assume that regions corresponding to theses objets have homogeneous properties. Then we propose to specify the cost function (3) by the following expressions:

• Gradient criterion:
$\quad x_1 = \sigma_1.gradient(Pi+1)$
σ_1 specifies the shape of the function. A gradient map is obtained with Deriche's operator at the beginning of the program [4].

• Region criterion:
The goal is to find a cycle by turning around the object. We assume that the object is on the right side of the contour when expanding the nodes. In order to control the stability of the region properties, we propose to observe the gray level variations between the pixels located close to Pi+1 and the pixels located close to the edge pixels Pi, Pi-1...Pi-n:
Let Pi-1, Pi, Pi+1 be 3 connected pixels of the contour, if θ is the angle determined by:
$\quad \theta = angle(vector \ P_{i-1}.P_{i+1}) - \pi/2$
and Pj is the pixel determined by:
$\quad X_{pj} = X_{pi} + round(2.cos \ (\theta) \ )$
$\quad Y_{pj} = Y_{pi} + round(2.sin \ (\theta) \ )$
then we can reasonably assume that Pj belongs to the region corresponding to the object. Moreover, in order to avoid errors due to noise, a 3x3 median filter is applied to Pj.
Finally, with $\quad Gray1 = Gray\_Level\_Average \ (P_{j-10}, P_{j-9}, ..., P_j)$
$\quad$ and $\quad Gray2 = Gray\_Level \ (P_{j+1})$
we propose to characterize the region criterion with the following expression:

if (| Gray2 - Gray1 | > maxdiff) then x2 = 0
else $x_2 = \sigma_2$.(maxdiff - | Gray2 - Gray1 |)

where maxdiff is a threshold for the maximum difference of the Gray level variations.

Other constraints have been added to the search:

1) if all objects are convex, the convexity of the shape can be controlled by measuring the curvature of the contour;

2) small closed contours with length inferior to a threshold are rejected;

3) since the exploration should stop if the quality of any feature is too weak (for example, the gradient value is 0), we propose avoiding the creation of a new arc if cost(Pi, Pi+1) is superior to a threshold admcost (admcost corresponds to the admissible cost of figure 1);

4) only the 3 first pixels of the search are declared goal nodes.

In order to find all objects, the image is first split into 10x10 square windows and a list of edge seeds is determined by selecting the pixels with maximum gradient value in each window.

4 Results

We propose to evaluate our method with a synthetic image and a real image. The same thresholds have been used in both cases:

$\alpha_1 = \alpha_2 = 0.5$, $\sigma_1 = 0.01$, $\sigma_2 = 0.1$, maxdiff=25, minlength=30, admcost=0.45.

Each time a contour is closed, all pixels are marked and the corresponding region is filled. It takes approximately 2 minutes on a Sparc 2 to process a 256x256 image.

In the first image "test1", the Gray value of the regions corresponding to the objects is 100 and the Gray value of the background is 130. In order to add some difficulties, a small part of each boundary (a 10x10 window) has been blurred. All objects have been correctly detected. The second image is a noisy version of "test1": Gaussian noise with $\sigma = 10$ has been added. All objects have been found, but the boundaries do not always correspond to the exact frontiers.

In the second image "test2", we propose to try finding the contours of black cells uniquely. An edge map obtained with Deriche's detector and hysteresis thresholding (15,8) is presented to illustrate the difficulties: these cells are lowly contrasted and sometimes merged with other dark cells. The convexity criterion has been used to constrain our heuristic search. All cells detected have been correctly located without any error. Few cells too close from the border of the image have not been completely detected (these missing contours should not be considered as errors) and very few are missed, their contour being indistinguishable. Note that a constraint on the Gray level of regions has been used to avoid the search around light Gray cells.

5 Conclusion

Our conclusion is that heuristic search strategies are powerful tools in pattern recognition: few errors are made and good results can be obtained even with complex images. The advantage of the heuristic search approach holds in the possibility of adding many contextual constraints to the detection, thus making the search adaptive to the application. The major problem is to find a suitable expression of the different criteria and our proposition is finally a step toward an acceptable solution.

References

[1] Ashkar,G.P. and Modestino,J.W. "The contour extraction problem with biomedical applications", CGIP, vol. 7, 331-355, 1978.

[2] Ballard, D. H. and Brown, C. "Computer Vision", Prentice Hall Inc., Englewood Cliffs, New Jersey, USA, 1982.

[3] Ballard,D.H. and Slansky,J. "A ladder-structured decision tree for recognizing tumors in chest radiographs", IEEE Trans. Computers, vol. 25, 503-513, 1976.

[4] Deriche, R. "Using Canny's criteria to derive a recursively implemented optimal edge detector", International Journal of Computer Vision, vol. 1, 167-187, 1987.

[5] Lester, J. M., Williams, H.A., Weintraub, B.A., and Brenner J.F., "Two graph searching techniques for boundary finding in white blood cell images", Computers in Biology and Medicine, col. 8, 293-308, 1978.

[6] Martelli, A. "Edge detection using heuristic search methods", CGIP, vol. 1, n°2, 169-182, 1972.

[7] Martelli, A. "An application of heuristic search methods to edge and contour detection", Comm. of the ACM, vol. 19, 73-83, 1976.

[8] Nilsson,N.J. "Principles of artificial intelligence", Palo Alto, CA:Tioga, 1980.

[9] Pearl, J. "Heuristics: intelligent search strategies for computer problem solving", Addisson-Wesley, 1984.

[10] Persoon, E. "A new edge detection algorithm and its applications in picture processing", CGIP, vol. 5, n°4, 425-446, 1976.

[11] Weschler, H. and Sklansky J. "Finding the rib cage in chest radiographs", Pattern Recognition, vol. 9, 21-30, 1977.

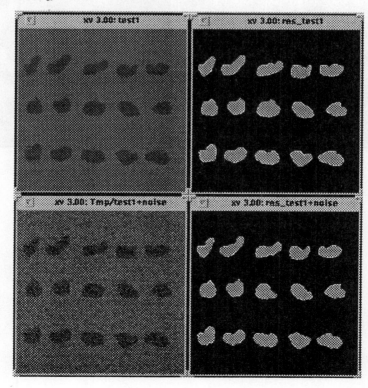

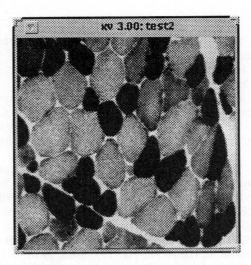

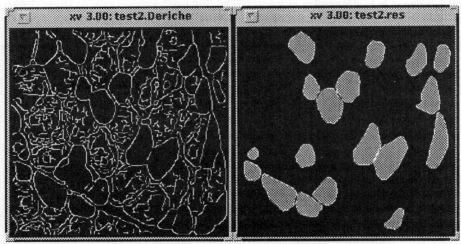

Sub-Pixel Registration for Super High Resolution Image Acquisition Based on Temporal Integration

Yuji Nakazawa, Takashi Komatsu and Takahiro Saito

Department of Electrical Engineering, Kanagawa University
Yokohama, 221, Japan

Abstract. Towards the development of super high resolution image acquisition, we present a temporal integration imaging method. The image processing algorithm for a generic temporal integration imaging method consists of the three stages: segmentation, sub-pixel registration and reconstruction. The segmentation stage and the sub-pixel registration stage are interdependent and extremely difficult to construct completely. Instead, aiming at a particular type of application where a user indicates a region of interest (ROI) on an observed image, we construct a prototypal temporal integration imaging method which does not involve the segmentation stage at all. Moreover, we develop a new quadrilateral-based sub-pixel registration algorithm, the key idea of which is to cover a ROI with deformable quadrilateral patches whose spatial fineness is automatically changed in accord with the curvature of the object's surface and then to describe an image warp between two image frames as deformation of the quadrilateral patches and finally to perform a sub-pixel registration by warping an observed image to a temporally integrated image with the recovered warping function.

1 Introduction

Some research institutes are elaborating plans to develop super high definition (SHD) image media beyond the HDTV. The SHD images should have high spatial and temporal resolutions at least equivalent to those of 35/60 mm films and/or motion pictures. Imaging is a principal problem for handling SHD images. CCD camera technology is considered promising as a high resolution imaging device. Although 1 inch CCD imagers with two million pixels have been developed for HDTV, spatial resolution should be increased further to acquire SHD images. Reducing the pixel size is the most straightforward way to increase spatial resolution, which makes a CCD imager much more sensitive to shot noise. To keep shot noise invisible on a monitor, there needs to be a limitation in the pixel size reduction. Current CCD technology has almost reached this limit. Therefore, a new approach is required to increase spatial resolution further beyond the resolution bounds qualified by shot noise physically.

One promising approach towards improving spatial resolution is to incorporate image processing techniques into the imaging process [1]. Imaging methods based on this approach can be classified into two main categories: a spatial integration imaging method and a temporal integration imaging method. As a spatial integration imaging method, recently we have presented a method for producing an improved-resolution image with sufficiently high SNR by integrating multiple images taken simultaneously with multiple different cameras [1].

This paper presents a temporal integration imaging method. A temporal integration imaging method uses a single camera with insufficiently low spatial resolution, and produces an improved-resolution image by integrating multiple consecutive image frames taken with the camera. The principle of a generic temporal integration imaging method is that we reach an increase in the sampling rate by integrating more samples of the imaged object from an image sequence where the object appears moving.

The image processing algorithm for a generic temporal integration imaging method consists of the three stages: segmentation, sub-pixel registration and reconstruction. In the segmentation stage and the sub-pixel registration stage, while dividing a moving area into multiple small distinct patterns undergoing coherent motion with a certain motion segmentation algorithm, we make precise sub-pixel interframe correspondence between pixels that appear in given two image frames. In the reconstruction stage, an improved-resolution and/or improved-SNR moving image sequence with uniformly-spaced samples is reconstructed from a given moving image sequence by integrating the nonuniformly-accumulated samples composed of samples showing the sub-pixel interframe correspondence.

It seems that there is no unsolved problem about the reconstruction stage, and a number of algorithms are available for it [2, 3]. As described in [3], we can construct an interpolation algorithm that takes several degradation factors into account by extending the image reconstruction algorithm of projections onto convex sets (POCS). Presently, we believe that the POCS-based iterative interpolation method is fairly flexible, suited to the temporal integration imaging method.

On the other hand, the segmentation problem and the sub-pixel registration problem are highly interdependent and extremely difficult to solve completely. It seems that there is little prospect of a general solution of the highly interdependent problem in the near future and hence for the time being we cannot construct a generic temporal integration imaging method completely. To escape the predicament and then to form a prototypal temporal integration imaging method, for the present, we should relinquish a generic approach and confine its applications in a particular type of application such as surveillance, where a user indicates a region of interest (ROI) on an observed low-resolution image before applying the temporal integration technique to an observed image sequence [4].

Aiming at such a limited application, we construct a prototypal specific temporal integration imaging method which does not involve the segmentation stage at all. Moreover, we develop a new quadrilateral-based sub-pixel registration algorithm which describes an image motion within the ROI with sub-pixel accuracy as deformation of quadrilateral patches covering the ROI and then performs a sub-pixel registration by warping an observed image in accord with a temporally integrated image with the warping function recovered from the deformed quadrilateral patch. Before processing an observed image sequence, we should fix the size of quadrilateral patches properly in advance; hence it is certain that the algorithm works well for a plane surface of object, while it does not necessarily work satisfactorily for a curved surface of an object. In order to solve this problem and then to enhance the applicability of the prototypal specific temporal integration imaging method, furthermore we introduce a hierarchical patch splitting algorithm for controlling spatial fineness of quadrilateral patches in accord with the curvature of an object surface corresponding to the ROI.

2   Hierarchical Quadrilateral-Based Sub-pixel Registration

2.1 Computational Framework

Fig. 1 illustrates the computational framework of our proposed hierarchical quadrilateral-based sub-pixel registration algorithm. The key idea of the new hierarchical quadrilateral-based algorithm is to cover a ROI, indicated by a user on an observed low-resolution image frame, with deformable quadrilateral patches whose spatial fineness is automatically changed in accord with the curvature of the object's surface corresponding to the ROI and then to describe an image warp between two image frames within the ROI as deformation of the quadrilateral patches and finally to perform a sub-pixel registration by warping an observed image to a temporally integrated image with the recovered warping function.

2.2 Estimation of Grid Points' Displacement Vectors

We estimate displacement vectors of all the grid points within the ROI, describe an image warp with sub-pixel accuracy inside the ROI as deformation of the quadrilateral patches determined by the grid points shifted by an estimated sub-pixel displacement vector and thus track interframe correspondence of image points inside the ROI with sub-pixel accuracy over consecutive image frames.

The quadrilateral-based sub-pixel registration, of course, depends on accurate estimation of the grid points' displacement vectors. The displacement of each grid point is closely linked with the deformation of the patches. To determine the deformation of each quadrilateral patch, we should compute displacement vectors of all four grid points. Simultaneously, these grid points influence the deformation of four contiguous quadrilateral patches. We iteratively refine the displacement vector estimates of the grid points so that the matching errors for the four contiguous quadrilateral patches decrease after iteration. We refer to this algorithm as the iterative refinement algorithm.

2.3 Image Warping   - Perspective Transformation -

In most applications, the ROI corresponds to a two-dimensional plane surface or a smoothly curved surface of an object, and a smoothly curved surface can be well approximated by a two-dimensional plane locally in the vicinity of a point within the ROI. It was shown that the mapping of an arbitrary two-dimensional plane surface of an object onto the image plane of the arbitrary moved camera is described by two-

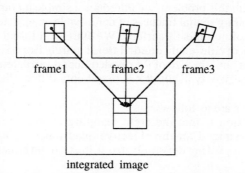

Fig. 1. Computational framework of the hierarchical quadrilateral-based sub-pixel registration.

dimensional perspective transformations [5]. For perspective transformations, the forward mapping functions are

$$x' = \mathbf{T}_1(x, y) = \frac{a_1 \cdot x + a_2 \cdot y + a_3}{a_7 \cdot x + a_8 \cdot y + 1} \qquad (1)$$

$$y' = \mathbf{T}_2(x, y) = \frac{a_4 \cdot x + a_5 \cdot y + a_6}{a_7 \cdot x + a_8 \cdot y + 1} \qquad (2)$$

The eight mapping coefficients can be determined by establishing correspondence between four points in two images. Perspective transformations facilitate planar quadrilateral-to-quadrilateral mappings. Because of this property, we employ quadrilateral patches. When the three-dimensional structure of an object within a quadrilateral patch can be well approximated as a rigid two-dimensional plane, a good approximation of the image warp is provided.

2.4 Global Displacement Estimation

If an image warp inside the ROI is very large, we cannot obtain an accurate image warp estimate only with the iterative refinement algorithm. To cope with the problem, we use a global displacement estimation algorithm for shifting the entire ROI as a pre-process of the iterative refinement algorithm. The global displacement estimation algorithm operates on the entire ROI in the same manner as the iterative refinement algorithm.

2.5 Control of Quadrilateral Patches' Fineness

The size of deformable quadrilateral patches is an important factor. Larger quadrilateral patches are required to sufficiently constrain the solution of the problem of recovering an image warp from an observed image sequence and to provide some insensitivity to noise. The larger the quadrilateral patches are made, the less likely the local plane surface assumption about an object's local surface will be valid within each quadrilateral patch; the local plane surface assumption is that an objects' local surface corresponding to a quadrilateral patch defined in the observed image is well approximated by a two-dimensional plane.

The appropriate size of quadrilateral patches depends on the curvature of an object's surface, and hence it is extremely difficult to determine the proper size of quadrilateral patches. To cope with this problem, we introduce a simple hierarchical patch splitting algorithm for controlling spatial fineness of deformable quadrilateral patches in accord with the curvature of an object's surface. We start with initial coarse quadrilateral patches, and then divide each quadrilateral patch iteratively into four finer quadrilateral patches until the size of the divided finer quadrilateral patches reaches the prescribed minimum size, typically 8×8, if and only if the following two conditions are satisfied.

The two conditions are as follows:
(1) The first condition is that the eight mapping coefficients of the perspective transformation determined for the objective quadrilateral patch with the iterative refinement algorithm differ definitely from those determined for its contiguous

quadrilateral patches sharing a side. We define the distance between a couple of the eight mapping coefficients, on the Lie algebra associated with the Lie group of perspective transformations corresponding to all possible patterns of the eight mapping coefficients. The distance $D(a,b)$ between two patterns of eight mapping coefficients, $a=(a_1,a_2,...,a_7)^t$, $b=(b_1,b_2,...,b_7)^t$, are defined by the equation:

$$D(a,b) = \iint_R \|v(a) - v(b)\|^2 \, dx \cdot dy \qquad (3)$$

where the vector, $v(a)$, corresponds to the optical flow field determined by the perspective transformation with the eight mapping coefficients a.

(2) The second condition is that the edge structure included in the objective quadrilateral patch is complex enough. If the objective quadrilateral patch satisfies the inequality,

$$\frac{1}{S(P)} \cdot \iint_P \left[\frac{\partial^2 I(x,y)}{\partial x^2} + \frac{\partial^2 I(x,y)}{\partial y^2} \right]^2 dx \cdot dy \geq Th \qquad (4)$$

$I(x,y)$: image intensity $\qquad Th$: threshold value
P : quadrilateral patch $\qquad S(P)$: area of P

then we regard the patch as a textured region containing salient local image features. On the other hand, once the objective quadrilateral patch is judged not to be a textured region, we will never divide the patch and thereafter we will exclude the patch from the region where the matching errors are computed in the iterative refinement algorithm.

3   Experimental Simulations

We conduct experimental simulations on a real test image sequence which we take with a normal 8-mm video camera for home use while moving the camera before a NESCAFE bottle. The test image sequence consists of 50 consecutive image frames. Fig. 2(a) shows the first image frame of the test image sequence, where a ROI is superimposed. We assume that the objects of interest are small characters and figures shown on a NESCAFE bottle. Both the vertical size and the horizontal size of the reconstructed improved-resolution image are made five times as large as that of the observed image frame. Fig. 2 contrasts the reconstructed improved-resolution image with the first observed image frame, but in Fig. 2(b) for ease of comparison the first observed image frame is interpolated and magnified by five times in the vertical and horizontal directions with the normal bilinear interpolation technique. As shown in Fig. 2, the hierarchical quadrilateral-based algorithm works well for a smoothly curved surface of an object, and temporal integration of pixels within the tracked ROIs enhances spatial resolution to such a degree that in the reconstructed improved-resolution image we can read small characters and figures which we can scarcely make out in the original observed image frame.

4   Conclusion

We develop a new hierarchical quadrilateral-based sub-pixel registration algorithm which is equipped with the capability of automatically controlling spatial fineness of quadrilateral patches in accord with the curvature of an object surface, thus organizing a prototypal specific temporal integration imaging method oriented to specific applications. The experimental simulations demonstrate that the temporal integration imaging method works well when the local geometrical structure of an object's surface corresponding to a quadrilateral patch can be well approximated as a rigid two-dimensional plane. To extend the applicability of the temporal integration imaging method further, we should develop more flexible algorithms for segmentation and sub-pixel registration.

References

1. T. Komatsu, T. Igarashi, K. Aizawa, T. Saito: Very high resolution imaging scheme with multiple different-aperture cameras, Signal Processing: Image Communication 5, 511-526(1993)
2. R.Y. Tsai, T.S. Huang: Multiframe image restoration and registration, Advances in Computer Vision and Image Processing, JAI Press Inc.(1984)
3. A.J. Patti, M.I. Sezan, A.M. Telekap: High resolution image reconstruction from a low-resolution image sequence in the presence of time-varying motion blur, Proc. IEEE 1994 Int. Conf. Image Process., 343-347(1994)
4. Y. Nakazawa, T. Saito, T. Komatsu, T. Sekimori, K. Aizawa: Two approaches for image-processing based high resolution image acquisition, Proc. IEEE 1994 Int. Conf. Image Process., 1147-1151(1994)
5. G. Wolberg: Digital image warping, Los Alamitos, CA: IEEE Computer Society 1990

(a) ROI in the 1st image frame

(b) 1st observed low-resolution image frame   (c) Reconstructed improved-resolution image
(bilinear interpolation technique)

Fig. 2. The tracked regions of interest (ROIs) and spatial resolution improvements.

Robust features for textures in additive noise

C.Ottonello[1], S.Pagnan[2] and V. Murino[1]

[1]*Dip. di Ingegneria Biofisica ed Elettronica-University of Genoa*
Via all'Opera Pia 11A, 16145 Genova, Italy
[2]*Istituto di Automazione Navale- National Research Council of Italy*
Torre di Francia, Via De Marini 1, 16149 Genova, Italy

Abstract. The paper describes a method for texture classification in noise by using third-order cumulants as discriminating features. The problem is formulated as a test on K hypotheses and solved by a Maximum Likelihood (ML) criterium applied in the third-order cumulant domain. Since in the case of image processing complete third-order cumulant computation is not feasible, we reduced the estimation to a limited number of cumulant slices and lags. This reduction makes the classification algorithm suboptimal. Thus, a criterion for the choice of cumulant samples to be computed is introduced in order to guarantee the selection of those lags which better identify the different textures in the training phase of the classifier.
Experimental tests are carried out to evaluate third-order cumulant performances on noisy textures and the importance of lags selection.

1 Introduction

The aim of texture classification algorithms is to produce a set of measures that make it possible to discriminate between different classes of textures so that each class may be described by parameters that can be used by a segmentation algorithm to partition an image into homogeneous regions. When images are affected by noise, this purpose can be strongly compromised and it is necessary to produce robust measures for each class of texture. In the paper, a Maximum Likelihood (ML) classifier working in the domain of Higher Order Statistics (HOS) [1] is proposed; in particular, third-order cumulants are estimated as texture features, thanks to their insensitivity to symmetric and independent, identically distributed (i.i.d.) noises. In order to reduce computation efforts and to determine an appropriate set of cumulant samples to be included in the feature vector, the number of outliers for each class is evaluated during the training phase. Cumulant samples inconsistent with the expectation produced by each class population are ranked out by assigning them small efficiency weights.

In section 2, the classifier scheme is presented, and, in subsection 2.1, the adopted criterion for lag selection is discussed. Section 3 describes the experiments performed on natural textures corrupted by i.i.d. as well as coloured Gaussian noise; in this section, a comparison with an autocorrelation based classifier is also made. Some conclusions are carried out in section 4.

2 The Classifier

The problem of classifying signals corrupted by noise can be reduced to a test of K hypotheses, and, for additive noise, can be expressed as:

$$H_k: \quad x(i) = n(i) + s_k(i) \qquad i = (i_1, i_2) = 1...R; \; k = 1...K \qquad (1)$$

where n(i) denotes noise samples, $s_k(i)$ is the image template belonging to the class k, x(i) are observation samples, and i=(i_1, i_2) are the pixel indexes.

As alternative to energy classifiers/detectors, schemes based on higher-order statistics have been proposed, which are insensitive to additive Gaussian noise and have the characteristic of preserving phase information (i.e., they exhibit sensitivity to signal shape) [2] [3]. In (1) the observation vector x(i) can be replaced by the Nx1 vector $\hat{f}_N$, which contains all estimated HOS samples and the ML classifier can be implemented, as follows:

$$\ln p(\hat{f}_N / H_k) \underset{<H_l}{\overset{>H_k}{\gtrless}} \ln p(\hat{f}_N / H_l) \quad (2)$$

From [4], it derives that such probabilities are normal.

In the proposed classification scheme, in order to obtain a classifier robust to a large class of noises, we replace $\hat{f}_N$ with N lags of the third-order cumulant, that is defined for zero-mean signals as:

$$\hat{c}_3^x(\tau_1, \tau_2) = E\{x(i)x(i+\tau_1)x(i+\tau_2)\} \quad (3)$$

where we assume $\tau_1 = (\tau_{11}, \tau_{12})$ and $\tau_2 = (\tau_{21}, \tau_{22})$ as the spatial lags in the two image directions, and reduce test (2) to a minimum-distance or minimum-HOS-energy classifier (by setting the covariance matrix $\hat{\Sigma}_{(H_k)} = I$). We choose the

hypothesis H_k $\qquad$ iff $\qquad \left| \hat{c}_3^x - \bar{c}_3^{s_k} \right|^2 < \left| \hat{c}_3^x - \bar{c}_3^{s_l} \right|^2$ $\quad$ (4)

Due to computational efforts, only some cumulant lags can be computed in practice, thus exploiting the HOS potentiality only partially. These drawbacks make HOS performances comparable to those of energy-based classifiers, in some practical cases where few samples are available.

Within the context, a considerable improvement can be obtained by performing the classification process after selecting the most effective lags for discrimination [3].

2.1 Feature Selection

From an initial set of lag samples, estimated on noise-free images, N lags can be selected to obtain an appropriate feature vector by assigning larger penalties to lags with larger numbers of outliers and interclass-overlapping regions, according to the following scheme:

$$W(lag_i) = \sum_{k=1}^{K} ouliers(lag_i, class_k) + \sum_{k=1}^{K} \sum_{j=k+1}^{K} \delta\{C_k^i \cap C_j^i\}$$

(5)

where outlier(lag_i,class$_k$) is a cumulant sample outside the range $\left[m_k^i - 2 \cdot \sigma_k^i, m_k^i + 2 \cdot \sigma_k^i \right]$ for cumulant sample i of class k, C_k^i is the set including all estimated lags i for class k, and $\delta\{C_k^i \cap C_j^i\}=1$ if $C_k^i \cap C_j^i \neq \varnothing$ and it is equal to 0, otherwise.

3 Experimental Tests

Test Case #1: The classification algorithm defined by relation (4) was applied to the first image set (Fig.1). Each 256x256 image was divided into several sub-images. I.i.d. Gaussian and coloured Gaussian noises were superimposed upon each sub-image. Coloured noise was obtained as the output of a linear filter driven by i.i.d. Gaussian noise. The leave-one-out strategy was adopted for the classification process. Each sub-image was presented to the classifier, after training it by using the remaining noise-free images. The classification probability was evaluated by using 50 noise realizations for different SNR values (calculated as $10 \cdot \log 10\left\{ E\left[s^2 \middle/ n^2 \right] \right\}$) for each test. The classification results obtained by the third-order cumulants, according to the minimum-distance criterion, were compared with autocorrelation results. In the absence of noise, 100% classification results were obtained by both classifiers. $\hat{c}_3^x$ yielded better results than $\hat{c}_2^x$ on noisy images, thanks to the insensitivity of the former cumulant to noise with zero skewness (Tables 1 and 2).

In both experiments, a limited number of lags were used. For $\hat{c}_3^x$, the feature vector was the set $\{c_3^x(\tau_{11}, \tau_{12}; \tau_{21}, \tau_{22}): \tau_{11} = 0; \tau_{12} = 0,1; 0 \leq \tau_{21}, \tau_{22} \leq 3\}$.

For $\hat{c}_2^x$, the feature vector included $\{c_2^x(\tau): \tau = 0,...,5\}$.

Then lag selection for $\hat{c}_3^x$ was performed to reduce the number of parameters necessary for an efficient classification. According to (5), the first N lags were extracted from the initial set:

$$\{c_3^x(\tau_{11}, \tau_{12}; \tau_{21}, \tau_{22}): \tau_{11} = 0; \tau_{12} = 0; 0 \leq \tau_{21}, \tau_{22} \leq 20\}$$

estimated on clean images. Improvements obtained in classification results are shown in Table 3.

Test Case #2: A new image set (Fig. 2) was tested to evaluate $\hat{c}_3^x$ insensitivity when i.i.d. is replaced with coloured noise. Unlike the previous test, lag selection was also performed for $\hat{c}_2^x$ samples in the set $\left\{c_2^x(\tau): \tau = 0,\ldots,19\right\}$. Although $\hat{c}_2^x$ performances were strongly improved by eliminating lag outliers, $\hat{c}_2^x$ sensitivity to coloured noise was verified. Figure 3 compares the $\hat{c}_2^x$ and $\hat{c}_3^x$ probability of correct classification in the presence of i.i.d. Gaussian noise and coloured Gaussian noise, for different SNRs and different numbers of features (N). $\hat{c}_2^x$ classifier resultsed completely inneficient in the presence of coloured noise, whereas $\hat{c}_3^x$ performances remained unchanged.

 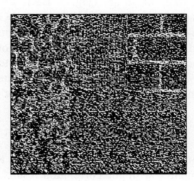

Fig.1: On the right: five original textures of test set #1: coffee, cloth, wall, naphtha, tweed, and synthetic noise. On the left: test set #1 corrupted by coloured Gaussian noise (SNR=-10 dB).

4 Conclusions

An image classification scheme that exploits cumulant features has been proposed. The HOS reduced sensitivity to i.i.d. and coloured Gaussian noise and the capability of characterizing non-Gaussian signals justify the use of HOS in classification problems. The adopted cumulant-based method, even though suboptimal, is computationally efficient and yields good results, especially for very noisy images. A considerable improvement in classification results can be obtained by performing a cumulant sample selection, i.e. by removing from the feature set all samples that may lead to class overlapping.

Results show that the proposed classification features are robust to very critical signal-to-noise ratios (SNRs); in particular, a correct classification is invariant to both i.i.d. and coloured Gaussian noise. Moreover, combination of third and second-order statistics can provide effective parameters for a noise-robust segmentation algorithm.

Table 1: Probabilities of correct classification (32 lags of $\hat{c}_3^x$)

SNR (dB)	Gaussian Noise	Coloured Gaussian Noise
-15	.611	.481
-10	.859	.748
-5	.961	.934
0	.977	.974
5	.992	.986

Table 2: Probabilities of correct classification (36 lags of $\hat{c}_2^x$)

SNR (dB)	Gaussian Noise	Coloured Gaussian Noise
-15	.200	.200
-10	.304	.220
-5	.989	.950
0	1	1
5	1	1

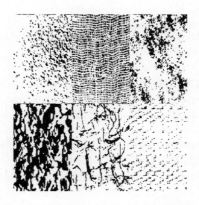

Fig. 2: Test set #2. From left to right From left to right and from top to bottom: cork, gauze, ice, mica, ricepaper and wicker.

Table 3: Probabilities of correct classification (N selected lags of $\hat{c}_3^x$)

SNR	Gaussian Noise			Coloured Gaussian Noise		
dB	N = 10	N = 20	N = 30	N = 10	N = 20	N = 30
-15	.445	.5	.64	.39	.51	.59
-10	.775	.83	.9	.725	.85	.87
-5	.94	.97	.955	.915	.96	.96
0	.995	.985	.975	.995	.985	.975
5	1	.995	.98	1	1	.98

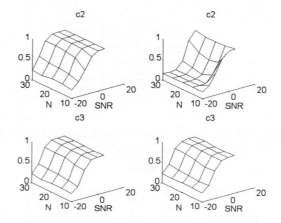

Fig. 3: Classification probabilities in the presence of i.i.d. (left) and coloured (right) Gaussian noises vs. SNR and number of lags, for $\hat{c}_2^x$ and $\hat{c}_3^x$.

References

[1] C.L. Nikias and J.M. Mendel: Signal processing with higher-order spectra. IEEE Signal Processing Magazine, pp. 10-37 (1993).

[2] G.B. Giannakis and M.K. Tsatsanis: A unifying maximum-likelihood view of cumulant and polyspectral measures for non-Gaussian signal classification and estimation. IEEE Trans. on Information Theory, 38, pp. 386-406 (1992).

[3] A. Makovec, and G. Ramponi: Supervised Discrimination of noisy textures using third-order moments. Fourth COST WG.1 and 2nd Workshop on Adaptive Methods and Emerging Techniques for Signal Processing and Communications, Slovenia, April 1994.

[4] D.R. Brillinger and M. Rosenblatt: Asymptotic theory of estimates of kth-order spectra. In: Spectral Analysis of Time Series, B.Harris, ed., Wiley, NY, pp.153-188, (1967).

Decomposing Contours into Curves of Different Families

Vincenzo Caglioti

Artificial Intelligence and Robotics Project, Dipartimento di Elettronica e
Informazione, Politecnico di Milano, P.za Leonardo da Vinci 32, 20133 Milano, Italy

Abstract. Many contour segmentation methods search for a decomposition into curves of a same family (as, e.g., the polygonal approximation presented in [1]). However, the contours of real objects often are constituted by curves of different families. In this paper an entropic criterion is presented for the decomposition of a given contour using curves of different families. The entropic criterion accounts both for the information cost associated to the use of complex curves and the uncertainty cost associated to the larger curve fitting error deriving from the use of simple curves. Experimental results on real images are reported to show the application of the criterion to two different curve families, namely the lienar segments and the circular arcs.

1 Introduction

Many contour segmentation methods decompose a contour into curves of the same family (see, e.g., [2], [4]). Saund [3] presented a scale-invariant criterion given by the ratio between the lenght of the fitted curvilinear arc and the mean distance between the arc and the points assigned to it. However, this criterion can not be applied to the case of overlapping curve families such as, e.g., the family of the line segments and the family of the circular arcs. In fact, no criterion based on the fitting error would prefer a line segment to the best fitting circular arc, since the latter curve is characterized by a smaller fitting error.

In this paper an entropic criterion is proposed for the decomposition of a set of points into curves belonging to different, but possibly overlapping, families. According to this criterion, the larger fitting error resulting from fitting a set of points to a simpler curve is traded for the lower information cost associated to the representation of the simpler curve. The main issue in this criterion concerns how to combine, in a nonparametric way, the information cost involved in representing a curve and the fitting error related to associating a set of points to the curve.

The determination of a globally optimum decomposition, according to the adopted criterion, would involve a combinatorially explosing complexity. Therefore a method is proposed to determine a suboptimal decomposition. The performance of the suboptimal method are demonstrated on contours extracted from real images.

In Sect. 2, the entropic criterion is introduced and the decomposition problem is formulated. In Sect. 3, the suboptimal method is presented. In Sect. 4, the performance of the suboptimal method on contours extracted from real images is illustrated. Section 5 contains some concluding remarks.

2 Entropic criterion for contour decomposition

Let us consider a set $\mathcal{P}$ of N points P_i $(i = 1, \ldots, N)$, which are known to lie within a rectangular planar region of dimensions $R \times R$. A *decomposition* of the point set $\mathcal{P}$ is characterized by a set $\mathcal{C} = \{C_1, \ldots, C_n\}$ of curves, and it specifies an assignment of any point $P_i \in \mathcal{P}$ to a curve $C_j \in \mathcal{C}$.

One way to represent the position of the points $P_i \in \mathcal{P}$ in the planar domain $R \times R$ up to a resolution δr_o is to represent their coordinates (x_i, y_i) directly: this involves an information quantity related to both the domain dimension R and the resolution δr_o.

An alternative way to represent the position of the points is through the set $\mathcal{C}$ of curves: any point $P_i \in \mathcal{P}$ is assigned to one of the curves of $\mathcal{C}$. In this case the involved information includes: *(i)* the information needed to specify the curves C_j with $j = 1, \ldots n$, and, for any point $P_i \in \mathcal{P}$, *(ii)* the information needed to specify that the point P_i is associated to the curve C_j (and not, e.g., to other curves), and *(iii)* the information required to represent the coordinates of the point P_i up to a resolution δr_o, once P_i is known to be assigned to the curve C_j.

The proposed criterion is given by the difference between the information associated to representing the set of points through the assigned curves and the information associated to representing the coordinates of the points directly.

To represent the information quantity, the *relative information content* is used. In particular, suppose the a priori and the a posteriori distribution of a random variable U have the same shape: if the standard deviations of the a priori and a posteriori distributions are, respectively, A and δu_o, then the relative information content associated to U is given by $H(U) = \ln(A/\delta u_o)$.

2.1 Direct representation of contour points

In this representation, the coordinates of the contour points are directly specified up to a resolution δr_o.

We chose to represent a point through its distance from a reference curve. In this way, we represent the "lateral" coordinate of the point, while neglecting to specify its "longitudinal" coordinate (i.e., the its coordinate along a direction which is locally parallel to the reference curve). Thus, the relative information content associated to representing a set of contour points is

$$H_d = \sum_{i=1}^{N} \ln \frac{R}{\delta r_o} = N \ln \frac{R}{\delta r_o}.$$

2.2 Contour points represented through a set of curves

To represent the contour points through a set of curves, first the curves are represented, and then the lateral coordinate of each contour point with respect to its assigned curve is specified up to a resolution δr_o.

Suppose that the decomposition of the contour is through a set of n curves. Each curve is represented by a set of parameters. For instance, straight lines are represented by the polar parameters ρ and θ, which represent the distance of the origin from the line, and the angle formed by the X axis and the normal to the line. A circular arc is represented by its radius and by the cartesian coordinates of the center of the circle (r_j, x_j, y_j).

If both the a priori and the a posteriori distribution of the parameters of the curve C_j are Gaussians, whith covariance matrixes, respectively, Λ_{j_o} and Λ_j, the relative information content of the curve representation is

$$H(C_j) = (1/2)\ln(|\Lambda_{j_o}|/|\Lambda_j|)$$

Here Λ_{j_o} is relative the dimension R of the image screen, while Λ_j is relative to the parameters of the curve C_j fitted to the contour points assigned to it.

Now that the parameters of the curves have been specified, the single contour points have to be represented. For each contour point, first the assigned curve has to be specified, and then the lateral coordinate, relative to the associated curve, must be provided. Suppose that N_j, out of the N contour points, are associated to the curve C_j: then the information quantity needed to specify the curve associated to a generic contour point P_i is

$$H_i = -\sum_{j=1}^{n} \frac{N_j}{N} \ln \frac{N_j}{N}.$$

Once a contour point P_i is known to be associated to a curve C_j, the the lateral dispersion of the point about the curve has to be represented (neglecting the "longitudinal" coordinate of the point) up to a resolution δr_o.

The a priori uncertainty of the position of the point P_i is now related to the variance σ_j^2 of the lateral dispersion about the curve C_j assigned to the point: σ_j^2 is estimated, e.g., by a least squares method. To specify the lateral coordinate of P_i with an accuracy δr_o, the needed information quantity is

$$H_{ij(i)} = \ln(\sigma_{j(i)}/\delta r_o),$$

where $j(i)$ indicates the index of the curve assigned to the point P_i. For a given decomposition of the set of contour points, the relative information content is

$$H_c = \sum_{j=1}^{n} H(C_j) + N \cdot H_i + \sum_{i=1}^{N} H_{ij(i)}.$$

2.3   Global criterion

The global criterion proposed in this paper, is the difference between the relative information content of the curve representation of the contour points and the relative information content of their direct representation Notice that the parameter δr_o vanishes from the global criterion.

$$H = \frac{1}{2}\sum_{j=1}^{n} \ln \frac{|A_{j_o}|}{|A_j|} - \sum_{j=1}^{n} N_j \ln \frac{N_j}{N} + \sum_{i=1}^{N} \ln \frac{\sigma_{ij(i)}}{R} \tag{1}$$

2.4   Expression of the determinants

First let us consider a straight line C_j fitted to a set of points $\{P_1, \ldots, P_{N_j}\}$. Suppose that the least square estimate of the line parameters are (ρ_j, θ_j), which represent, respectively, the distance from the origin to the line and the orientation of its normal. Let $\bar{x}_j = (\sum_i x_i)/N_j$, $\bar{y}_j = (\sum_i y_i)/N_j$, and $L_j^2 = \sum_i((x_i - \bar{x}_j)\sin\theta_j - (y_i - \bar{y}_j)\cos\theta_j)^2/N_j$. The a posteriori covariance matrix referred to an origin placed at the center $(\bar{x}_j, \bar{y}_j)$ of the line is:

$$A_j = \frac{1}{N_j}\begin{bmatrix} \sigma^2 & 0 \\ 0 & \sigma^2/L_j^2 \end{bmatrix}$$

Its determinant is $|A_j| = \sigma^4/(N_j^2 L_j^2)$. It is supposed $|A_{j_o}| = R^2$.

Now let us consider a circular arc C_j fitted to a set of points P_i $(i = 1, \ldots, N_j)$ characterized by their polar coordinates (ρ_i, θ_i). (For the fitting of a circular arc to a set of points see [5].) Suppose that these polar coordinates are referred to an origin which coincides with the estimated center of the cirle. Let σ^2 be the variance of the lateral dispersion of the points about the circular arc (σ^2 can be approximated by the mean squared distances of the points from the best fitting circle). The inverse of the a posteriori covariance matrix of the circle parameters is:

$$A_j^{-1} = \frac{1}{\sigma^2}\begin{bmatrix} N_j & \sum_i \cos\theta_i & \sum_i \sin\theta_i \\ \sum_i \cos\theta_i & \sum_i \cos^2\theta_i & \sum_i \cos\theta_i\sin\theta_i \\ \sum_i \sin\theta_i & \sum_i \cos\theta_i\sin\theta_i & \sum_i \sin^2\theta_i \end{bmatrix}.$$

The form of $|A_j|$ is particularly simple if the reference frame is rotated such that $\sum_i \sin(2\theta_i) = 0$. In this case is

$$|A_j| = \frac{(\sigma^6/N_j)}{(\sum_i \cos^2\theta_i)(\sum_i \sin^2\theta_i) - (\sum_i \cos^2\theta_i)(\sum_i \sin\theta_i)^2 - (\sum_i \sin^2\theta_i)(\sum_i \cos\theta_i)^2}$$

For the a priori covariance matix, it is assumed that $|A_{j_o}| = R^6$.

3   Criterion minimization

Finding the decomposition which minimizes the criterion (1) would require a time exponential in the number N of points. Therefore in this work a suboptimal method is adopted. A set of threshold values are considered. For each threshold value, the following procedure is executed.

The contour is decomposed into curves of the broadest family (in our case, into the circular arcs) by means of an edge-following technique [2]. The considered threshold value is adopted to compare the distance between the currently analyzed point and the currently followed edge. After the decomposition, the entropic criterion is calculated and a series of local refinements are attempted.

A first local refinement, called *merging*, consists in substituting two neighboring curves by a single curve, by fitting one curve to the set of points associated to the two above curves. The curve may belong to any of the considered families. A second local refinement, called *splitting* consists in substituting a single curve with two curves belonging to any of the considered families.

First merging steps are attempted. After each refinement, the new criterion value is calculated and compared with the previous one. If the criterion value decreases, then a further refinement step is attempted on the curve obtained at the previous step. If the criterion value does not decrease, the situation previous to the currently attempted step is recovered, and splitting steps are iterated on the last modified contour segments, until the criterion value does not decrease anymore. Once the process ends, the same process is carried out starting from a different set of neighboring segments.

The above operations are carried out for each of the considered threshold values. Once all the threshold values of the set have been considered, the best decomposition is retained (among those obtained in the previous steps).

4   Experimental results

The suboptimal method described in Sect. 3 has been applied to the contours, extracted from real images, shown in Figs. 1a-2a. In our case the dimension of the image is $R = 256$. The threshold values used in the process range from 1 to 6 pixel. The obtained contour decompositions are shown in Figs. 1b-2b. The obtained contour decomposition shown in Figs. 1b and 2b capture the correct decomposition of the model of the reproduced objects.

5   Conclusions

An information-theoretic criterion has been presented for driving the decomposition of contours into curves of different families. The criterion allows to evaluate both the fitting error, deriving from fitting a curve to a set of points, and the information cost involved in representing the fitted curve. The reported experimental results seem promising, in that they capture the correct contour decomposition of the objects present in the original images. Future work will address the introduction of further curve families, the search for better optimization strategies, and the investigation of the applicability of the criterion to the fitting of 3D data.

References

1. J.G.Dunham, – "Optimum uniform piecewise linear approximation of planar curves" – IEEE Transactions on Pattern Analysis and Machine Intelligence **8** (1986)
2. D.S.Chen, – "A data-driven intermediate level feature extraction algorithm" – IEEE Transactions on Pattern Analysis and Machine Intelligence **11** (1989)
3. E. Saund, – "Identifying salient circular arcs on curves" – GVGIP: Image Understanding **58** (1993)
4. B.K.Ray, K.S.Ray, – "A non-parametric sequential method for polygonal approximation of digital curves" – Pattern Recognition Letters **15** (1994)
5. S.Yi, R.M.Haralick, L.G.Shapiro, – "Error propagation in machine vision" – Machine Vision and Applications **7** (1994)

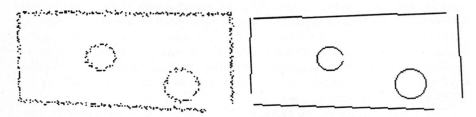

Fig. 1: A contour (a) and its decomposition (b)

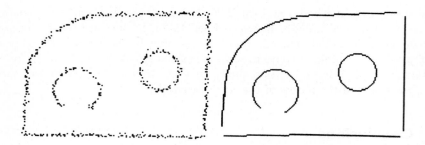

Fig. 2: A contour (a) and its decomposition (b)

A proposal on local and adaptive determination of filter scale for edge detection

Domenico G. Sorrenti

AI and Robotics Lab., Dept. Elettronica e Informazione
Politecnico di Milano, Pzza Leonardo da Vinci 32, Milano, I-20133, Italy
fax +39 2 2399 3411; email: sorrenti@elet.polimi.it

Abstract. In this work a proposal about local and adaptive scale selection in edge detection is presented. Such a proposal follows Canny's optimality criterion, i.e., the smallest scale that provides a minimum value for the signal to noise ratio should be selected for each edge pixel. The proposal exploits a local regularity, i.e. SNR, measure that is based on a simplified version of the Kitchen-Rosenfeld edge quality evaluator. This measure makes possible a local definition of the scale. The work has been carried out with non-directional gaussian smoothing operators. The integration of the results, obtained at different scales on a pixel basis, into a single edge map is also tackled. Experimental work on real images is presented.

1. Introduction

The edge detection problem has been regarded so far as the problem of defining a filter whose convolution with the image is easily searched for the edge points, e.g., the peaks of the convoluted image. The need for multi-scale edge detection algorithms has been pointed out first by Rosenfeld and Thurston [1]. Such a need stems from admitting that images are collections of different edges, each one characterized by its own contrast and noise energy. Therefore the edge detection task turns into treating each single edge at its best, according to some optimality criterion. An important approach in the edge detection field, introducing the scale-space concept, is presented by Witkin in [2]; in this work the behavior of edges in scale-space is exploited in order to define their perceptual saliency. Basing on empirical observations Witkin defines relevant edges, i.e. to be detected, as those edges whose stability across the scales is high. On the edge positioning side of the problem, a coarse-to-fine tracking (not really well defined) try to avoid the typical errors arising in edge localization performed at large scales. The edge detection task has been clearly stated by Canny [3]. He defined criteria for the quantitative evaluation of the performance of a filter, namely *detection* and *localization*. From the criteria Canny determined a class of optimal filters for different types of edges, e.g., step edges. A specific filter in the class is defined by the scale parameter, that is the parameter defining the bandwidth of the filter, and therefore the signal to noise ratio in the convoluted image. The image is modeled as a signal corrupted by gaussian white noise. A small scale results in a nearly unaffected frequency content of the convoluted image, on the other hand a large scale attenuates the higher frequencies. A larger scale provides a larger signal to noise ratio, given the signal amplitude. A major point in Canny's work is that a trade-off between *detection* and *localization* emerged; as the scale parameter increases the *detection* increases, whilst *localization* decreases. An optimal compromise between the two criteria has to be defined; Canny proposes to select the smallest scale, provided that a minimum value for the signal to noise ratio is obtained.

work partially funded by the C. N. R. Targeted Project on Robotics, sub-project URMAD

Being SNR linked to error probabilities (both marking false edges, and missing true ones), Canny's compromise means that we do not like to risk more than a certain amount. On the other hand, the selection of the minimum SNR satisfying the *detection* limit comes from the maximization of *localization* performance. In other words Canny set a priority between the criteria: first *detection*, then *localization*. In this work the take the same optimality definition.

In order to set the appropriate value for the scale parameter it is required to know the noise energy. Unfortunately it is not an easy task to locally measure the noise energy because both noise and signal affect any local measure, i.e., the signal corrupts the noise measure, as pointed out in [3]. Therefore only global measures can affordably be obtained, hence resulting in nearly single scale approaches.

Different approaches do not locally determine an optimal value for the scale parameter, but rather concentrate on the integration of results obtained at different scales. A first proposal is Canny's *feature synthesis* [3]; where, in a fine-to-coarse scale computation, edges not marked at smaller scales are cumulatively added to the edge map. Another relevant contribution is Bergholm's *edge focusing* [4] where, in a coarse-to-fine direction, edges are first detected (at the coarsest scale) and then tracked to the smallest scale. In this work integration means tracking, i.e. coarser scale is exploited to forecast edge location at the smaller scale. The algorithm generates its output at the smallest scale, therefore introducing artifacts like clouds of edge points, around the edges detected at the coarsest scale. This is a consequence of handling edges at the smallest scale of the range where edge focusing takes place, and it is unlikely that the smallest scale coincide with their optimal scale.

Differently from the above mentioned approaches, Jeong and Kim [5] aim at locally selecting the scale, on a single pixel basis. Jeong and Kim define an energy, function of the space-variant scale parameter, to be minimized under some "natural constraints". As an example of constraint they cite limiting abrupt variation of the scale parameter itself, in order to discourage fragmented edges due to noise. In the author opinion such constraint should not be *wired* in the algorithm whose aim is the determination of the scale parameter, but should rather emerge as the algorithm output. Moreover, in correspondence to edges between regions with different noisiness (e.g., due to different machining of the object surface) an abrupt variation of the scale parameter is desired. Jeong and Kim energy function comes from the desired behavior of the scale parameter (it should be small where intensity changes significantly and large in uniform areas), but the scale parameter generated by the algorithm is not the consequence of any optimality criterion. In particular a small error probability, i.e. a sufficiently large scale parameter, sometime is not achieved, as shown by Jeong and Kim themselves in their experimental section. Moreover, it is unclear whether a single threshold can be used to mark the edge pixels, given that the pixels are filtered at different scales.

The work here presented operates in the same mainstream as Jeong and Kim, i.e. to solve the basic conflict between *detection* and *localization* on a pixel basis. The conflict is solved according to Canny's proposal (namely, the smallest scale granting a sufficiently small error probability). Being not possible to directly measure the noise energy, a direct estimation of SNR is attempted. Integration of output at different scales is also performed in a consistent way and results are provided.

In the next section the algorithm and the local SNR estimator is presented. Experimental results and conclusions follows.

2. The proposed algorithm

In order to accomplish the above defined task we cannot follow the proved-to-fail approach of noise estimation; we rather observe that a local estimate of SNR could suffice. Such a local measure can be obtained by adapting the *local edge coherence* introduced by Kitchen and Rosenfeld [6]. It is briefly sketched in the following. Kitchen and Rosenfeld aim was a quantitative evaluation of the performance of different edge detectors. They defined a local measure, differently from previous approaches, and then averaged it on the whole image to give the performance figure. The local measure is built around two components, one measuring *edge thinness*, the second *edge continuity*. The two components are intuitively supposed to capture the most relevant qualities of well detected edges.

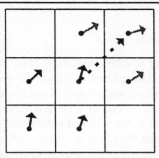

Fig. 1. Edge continuity (Left and Right) according to Kitchen and Rosenfeld; it is a function of i) central pixel edge direction (black), ii) edge direction in the neighboring pixels (right: light gray, left: dark gray), iii) the direction toward the neighbor (dotted, only the one toward the $\pi/4$-neighbor is indicated)

In our work *edge thinness* is meaningless because current edge detectors use to have an explicit thinning phase (NonMaximaSuppression), moreover we aim at an estimate of SNR, that has no impact on edge thinness. On the other hand *edge continuity* reveals to be strictly tied to SNR. We recall that Kitchen and Rosenfeld evaluation takes place after edge detection, when the whole edge map is available. *Edge continuity* rates how well formed is the edge centered on a pixel. The maximum is reached when the two neighboring pixels in the edge direction (8-conn) are edge points and their edge directions are similar to the edge direction in the central pixel (Fig. 1). On the other hand, the minimum is reached when no one of the two neighboring pixels are edge points or their edge directions are dissimilar from the edge direction in the central pixel. Kitchen and Rosenfeld expression, normalized to the [0, 1] interval, is computed for the three neighboring pixels in the edge direction, provided they are edge pixel. The maximum between the three value is taken as the continuity, e.g. the left continuity. Similar operations are performed for the opposite three neighbors. The average between the left and the right continuity is then taken as the local continuity figure.

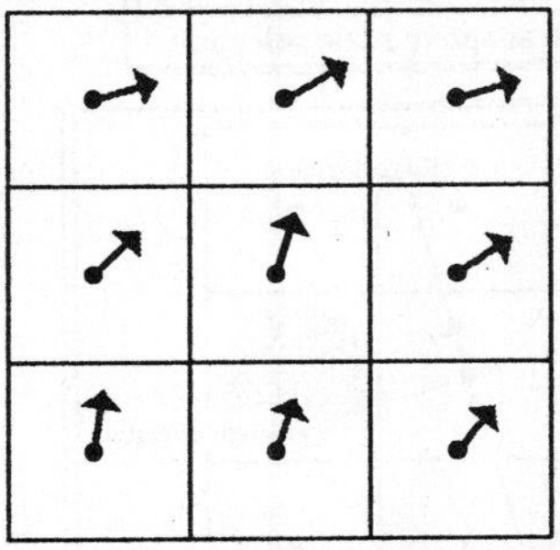

We claim that, in order to give an estimate of SNR, it is enough to rely on just the continuity component of Kitchen and Rosenfeld edge evaluator, provided that the measure is monotonically increasing with SNR. Moreover, the continuity measure proposed by Kitchen and Rosenfeld can be further simplified considering the differences between the direction in the central edge and in the 8-conn neighboring pixels. In our opinion wecan build different estimators, but they can merely modify the convergence speed of the algorithm proposed in the sequel (provided they give an

Fig. 2. SNR estimate, based on the edge direction around each pixel, central pixel direction (black), the $neighbor_i$ (gray); i=0, 1, ..., 7

increasing output with SNR), i.e. they cannot affect its convergence. Therefore we just take, as the SNR estimate we are looking for, a very simple function of the 8 differences, namely the minimum.

$$\text{SNR estimate} = \min_i (\text{CentralPixelDir.} - \text{Dir.Neighbor}_i)$$

The SNR estimator makes possible to set up the following adaptive and multi-scale algorithm. The uniqueness of the optimum value of the scale parameter could be exploited for speeding up the process, but in the actual implementation this is not done. The scale parameter is allowed to have a finite number of values in a predetermined range. At the start the scale parameter is set for each pixel at the smallest extremum in the range; it is then increased as long as its SNR is less than the required value. Different search strategies of the best scale can only affect the speed performance of the algorithm. The algorithm resembles a trivial control system, where the user sets the desired minimum value of SNR, and the system is in charge of minimizing the difference between its desired and actual values by acting on the scale parameter.

At the end of the adaptive smoothing phase we get both the gradient module and phase. The problem now is how to generate the final edge map. Actually this is the point where Jeong and Kim's contribution gets unclear.

Adaptive smoothing smoothes out the effects of the noise, but the module of the gradient image is still a non-thinned image; therefore a thinning phase is required. Such a computation involves comparison between adjacent

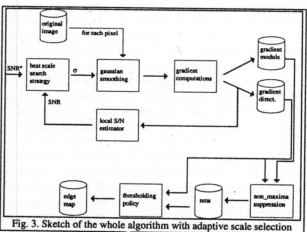

Fig. 3. Sketch of the whole algorithm with adaptive scale selection

pixels (Fig. 4) that could have been filtered at different scales and therefore cannot be straightforwardly compared. Our proposal is to compare pixels at the same scale; where the central pixel, the one whose suppression is under consideration, impose its scale to the others. The central pixel has the right to be considered at its very best, i.e. at its own optimal scale. The module of the gradient of the neighboring pixels at the desired scale could be already available or not, according to the algorithm implementation. In order to avoid

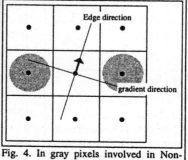

Fig. 4. In gray pixels involved in Non-MaximaSuppression

cumbersome computations and/or storage requirements, we tried to *synthesize* it starting from the behavior (determined off-line) of a normalized ideal step edge in the considered range of scales. We could not succeed in this approach, the attempt resulted in many kind of errors; in conclusion we propose to compute ex-novo the module of the neighboring pixels whenever it is not already available. The whole edge detection algorithm therefore includes the local adaptive smoothing phase followed by a thinning phase similar to the standard NonMaximaSuppression.

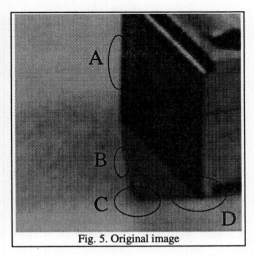

Fig. 5. Original image

3. Experimental results

Relevant areas of the image used for the presentation of the experiments (Fig. 5, the object is a small piece of iron):

area A, well contrasted edge, we expect this edge to be handled at a small scale

areas B and C, not contrasted edges; a straight line in area B, a curve/corner in area C. The intensity is not monothonic like it could look at first glance (see the edges detected at the smaller scales in Fig. 6). We expect these edges to be handled at a quite large scale

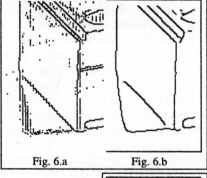

Fig. 6.a   Fig. 6.b

area D, junctions of different edges

In Fig. 6.a: the edge map detected at a small scale with the single scale Canny's simple detector (σ=0.3). 6.b: the same algorithm at a larger scale (σ=2.5). 6.c: the edge map detected by the proposed algorithm (σ is in the range [0.3, 2.5]). Fig. 7. shows the output of Canny's single scale superimposed to the output of the proposed algorithm. Light gray: single scale, dark gray: multi-scale, black: pixel edges detected by both the algorithms. The mislocalization and distortion inherent in the output of the single scale are magnified by using a scale (σ=5) higher than the maximum in the range of the multi-scale (σ=2.5) in order to ease observation of mislo-

Fig. 6.c

calization errors. In area A most edges are handled at $\sigma \cong 0.7$; in areas B and C: most edges are handled at σ=2.5; in area C a few would require to be handled at a scale larger than the largest in the range (they did not reach the SNR threshold); in area D: most edges are handled at relatively small scales (with the exception of those on the shadow edge)

4. Conclusions

A local and adaptive scale selection algorithm for edge detection is presented. It aims at the selection of the best scale for each pixel, where best is defined as in [3] as the smallest scale that provides a minimum value for the signal to noise ratio. The algorithm exploits a local regularity, i.e. SNR, measure that is based on a simplified version of the Kitchen-Rosenfeld edge quality evaluator. This makes possible a definition of the scale on a basis as local as the measure support. The results, obtained at different scales for each pixel, are integrated into a single edge map by means of a Non-MaximaSuppression procedure similar the standard one. Experimental work confirms that the approach provides a good answer to the local and adaptive edge detection problem. The limits of the algorithm, in the author opinion, are a consequence of the fact that adaptive smoothing is performed with a standard non-directional smoothing filter; therefore, where the signal amplitude is small compared to the noise and the edge

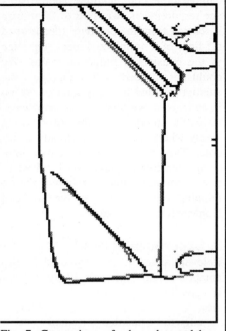

Fig. 7. Comparison of edges detected by a (large) fixed scale and by the proposed multiscale algorithm.

is short interaction with neighboring edges arises. Integration of a discontinuity-preserving approach like [7] is under investigation, as well as recursive implementation of linear filtering [8]. The algorithm is well suited for parallel implementation on loosely-coupled MIMD machines because it involves highly local computations.

References

[1] Rosenfeld A., Thurston M., "Edge and curve detection for visual scene analysis", IEEE Trans. on Computers, May 1971

[2] Witkin A. P., "Scale-space filtering", Proc. Int. Joint Conf. Artificial Intelligence 1983

[3] Canny J., "A computational approach to edge detection", IEEE Trans. on PAMI, November 1986

[4] Bergholm F., "Edge focusing", IEEE Trans. on PAMI November 1987

[5] Jeong H., Kim C. I., "Adaptive determination of filter scale for edge detection", IEEE Trans. on PAMI, May 1992

[6] Kitchen L., Rosenfeld A., "Edge evaluation using local edge coherence", IEEE Trans. on SMC, September 1981

[7] Perona P., Malik J., "Scale-space and edge detection using anisotropic diffusion", IEEE Trans. on PAMI, July 1990

[8] Deriche R., "Fast algorithms for low-level vision", IEEE Trans. on PAMI, January 1990

On the Detection of Step Edges in Algorithms Based on Gradient Vector Analysis

A. Larré, E. Montseny

Computer Engineering Dept. Universitat Rovira i Virgili
Carretera de Salou s/n 43006 Tarragona, Spain
Email: alarre@etse.urv.es

Abstract. The study presented in this paper is an analysis for finding what is the minimum signal-to-noise ratio that must have a contour in order to allow its detection. For finding this minimum, all the study has been based on the analysis of the gradient vector, because this kind of algorithms are those which obtain best results. It can be concluded that: a) The analysis of the gradient vector argument is more robust than the module; b) A contour must have a signal-to-noise ratio greater than 4/5 in order to be detected, if small operators are used to obtain the gradient vector, and so a good contour location be assured.

1. Introduction

Since the beginning of 60's, many contour extractor algorithms have been developed. A contour in an image allows to obtain a very important feature of the objects in the scene: the shape of these objects. This feature is used to get an object description for its later recognition [10].

At this moment, there is not a general purpose edge detector algorithm useful for any kind of images independently of the scene characteristics. There are many factors who affect the contour obtaining process. Now, algorithms based on the 1st derivative are the most important.

In [6], Haralick states that one edge extractor is better than another if reducing the information of the image (that is, adding noise) preserves its behavior. Haralick states that any edge extractor obtains good results in absence of noise. The processing of noise is the main problem in modern edge extraction algorithms. So, Davis in [4] lists the main factors adding noise in the images. These factors are photon noise, defocusing and textural structure of the objects. For this reason the noise immunity of the contour extractor algorithms must be one of its most important features.

At present, there are some algorithms that permit a good contour extraction in noisy images. These algorithms are Canny [3], Deriche [5] and Marr-Hildreth [8], all of them based on the gradient vector and 2nd directional derivative analysis.

This work studies when an edge can be detected in function of its signal-to-noise ratio (SNR) with different algorithms based on the gradient vector analysis. The section 2 of the paper gives the problem of edge detection depending on the noise and the contrast. In section 3 the influence of noise in the gradient vector is studied. Finally, in function of this noise, the minimum contrast to be detected is analyzed.

2. Problem Definition

V.S. Nalwa [9] defines an edge as a local discontinuity in the illumination function defining a scene. A discontinuity of the nth order is defined as a function whose nth

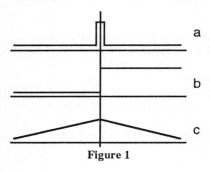

Figure 1

derivative contains a delta function. Davis [4] classified the edges of a scene in three classes according to their profile. These are: lines (discontinuity of order 0), figure 1a; steps (discontinuity of order 1), figure 1b; and roofs (discontinuity of order 2), figure 1c. The discontinuities of greater orders are not relevant. The majority of algorithms attempt to solve the problem of extracting step edges as these are the ones that appear most frequently in any scene.

In real images, there are some factors that produce discontinuities in the illumination function, and they are not real contours. In the image of figure 2-a appear discontinuities associated to the noise, there are many discontinuities but they are very small. In figure 2-b appears discontinuities due to changes in the illumination of the scene. In figure 2-c, shows discontinuities due to real contours.

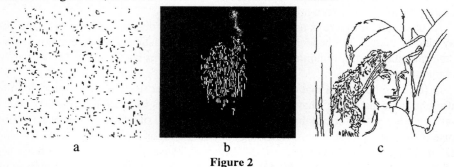

a   b   c

Figure 2

Haralick states that in images with a very low level of noise, all the edges can be detected with any algorithm. In this case contours with a very low contrast, can be detected. An example of this situation appears in figure 3-a. It has a circle at the center of the image with a very low contrast. Figure 3-b shows the contour extracted with Canny algorithm. But when the same image has a higher level of noise, like figure 3-c, contours with low contrast are more difficult to detect. Figure 3-d shows that the circle has not been detected. This example explains that the detection of a contour depends on its SNR.

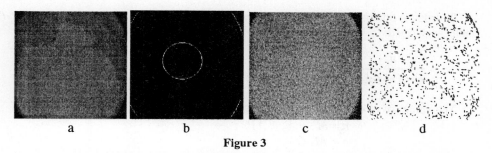

a   b   c   d

Figure 3

This work analyzes problems introduced by the noise, trying to find the minimum SNR that a real contour must have, in order to be detected.

3. Analysis of the contour detection

A contour in an image defines a discontinuity of the illumination function. Those arguments of the gradient vectors near of the contour change continuously along the contour and their modules have higher values than the rest of the image. For this reason, a contour can be detected with: a) the analysis of the gradient vector arguments, b) the analysis of the gradient vector modules, or c) the analysis of both, modules and arguments.

Because of the objective of this work is looking for the minimum SNR, it studies only straight step edges. For curved edges the SNR will be equal or higher.

In order to assure a good location of the contour, the operators for the gradient computation must be small. In the analysis made in this paper, the gradient vectors are obtained with two pairs of operators: the first pair are 3x3 sized, that locates the gradient vectors at the center of each pixel, and the second pair are 4x4 sized, that locates them at the intersection of every 4 pixels. So, a more dense map of gradient vectors is computed.

Elliptical neighborhoods of different sizes have been considered for this analysis. These class of neighborhoods used in the edge detection algorithm in [7] have 7, 13 and 19 gradient vectors inside them that are oriented depending on the direction of the central gradient vector of the straight edges used for this analysis. Figure 4 shows these elliptical neighborhoods for horizontal contours.

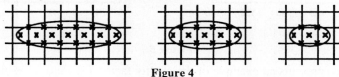

Figure 4

Different values of the SNR has been considered for this study. The standard deviation of the noise is fixed to 20, and the contrasts of the contours analyzed are 1, 2, 4, 8, 12 and 16 gray levels per pixel. So, the SNR are defined as 1/20, 1/10, 1/5, 2/5, 3/5 and 4/5 respectively.

In order to compare the different analysis made in this work, the following quality factor is defined:

$$\eta = \frac{n}{n'+(n''-n)}$$

where n: Number of contour pixels detected.
n':   Number of contour pixels in the image.
n'':   Number of pixels detected as a contour (false detection included).

The values of n, n' and n'' are obtained as follows

$$n = M \cdot p_c(i, N, t)$$
$$n' = M \cdot C$$
$$n'' = M \cdot \left[C \cdot p_c(i, N, t) + (1-C)p_{nc}(i, N, t) \right]$$

where M is the number of pixels in the image, and C is the ratio of contour pixels in the image. The value of $p_c(i, N, t)$ depends on the analysis. In the case of the argument analysis $p_c(i, N, \alpha)$ is the probability that at least i arguments of the N gradients vectors in the ellipse fulfill $\left| \alpha_j - A \right| \le \alpha$ when the central point of the ellipse is a contour pixel. α_j represents the set of the gradient vector arguments inside the ellipse. In the case of the module analysis $p_c(i, N, m)$ is the probability that at least i modules of the N gradients vectors in the ellipse fulfill $m_j \ge m$ when the central point of the ellipse is a contour pixel. m_j represents the set of the gradient vector modules inside the ellipse. The value of $p_{nc}(i, N, \alpha)$ is similar to $p_c(i, N, \alpha)$ but it corresponds to the case when the central point of the ellipse is a non-contour pixel. These values are different for different values of SNR.

An example of these functions are shown in figure 5. Figure 5-a represents the probability function $p_c(13, 13, \alpha)$ for the argument analysis, and figure 5-b a represents the probability function $p_c(13, 13, m)$ for the module analysis. In both cases the SNR is 3/5 and they are obtained in a heuristic analysis. The X axis represents the threshold α and m respectively.

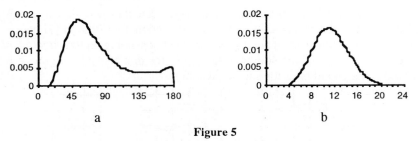

a   b

Figure 5

Figure 6 represents an example of the quality factor. Figure 6-a for the argument condition and figure 6-b for the module condition. It corresponds to the same case as figure 5. The Y axis represents the quality factor and the X axis represents the threshold. A 10% of contour pixels in the image is assumed. So, $C=0.1$.

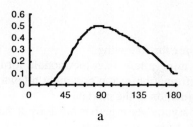 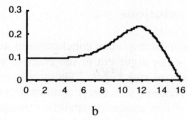

a   b

Figure 6

The most important value of these functions is the maximum and their thresholds. So, in the case of the argument the best threshold is 86° where the quality factor is 0.506, and the best threshold for the module is 12 where its quality factor is 0.232. These are not the best quality factors for the elliptical neighborhood of size 13. For the argument the best quality factor is when at least 12 of the 13 gradient vectors $\left(p_c(12,13,\alpha)\right)$ fulfill the condition. In this case the quality factor is 0.536 with a threshold of 72°. With the module, the best quality factor is when at least 5 of the 13 gradient vectors $\left(p_c(5,13,m)\right)$ fulfill the condition. Here, the quality factor is 0.399 with a threshold of 6.

Tables 1 shows the bests quality factors for different SNR, and different sizes of the elliptical neighborhood for the argument analysis, the module analysis, and the module and argument analysis. In each cell of the table there is the best quality factor that belongs to the best threshold, and in the case of the argument analysis and the module analysis how many gradient vectors must fulfill at least, the condition.

	Ellipse Size	Contrast = 4 SNR = 1/5	Contrast = 8 SNR = 2/5	Contrast = 12 SNR = 3/5	Contrast = 16 SNR = 4/5
Argument Analysis	7	0.161 (6 from 7)	0.281 (6 from 7)	0.421 (7 from 7)	0.561 (7 from 7)
	13	0.179 (9 from 13)	0.343 (11 from 13)	0.536 (12from 13)	0.705 (12 from 13)
	19	0.202 (14 from 19)	0.409 (15 from 19)	0.636 (15 from 19)	0.805 (16 from 19)
Module Analysis	7	0.104 (5 from 7)	0.170 (2 from 7)	0.311 (2 from 7)	0.483 (3 from 7)
	13	0.109 (5 from 13)	0.204 (5 from 13)	0.399 (5 from 13)	0.625 (6 from 13)
	19	0.113 (11 from 19)	0.235 (9 from 19)	0.478 (8 from 19)	0.732 (9 from 19)
Argument and Module Analysis	7	0.162	0.298	0.475	0.652
	13	0.180	0.356	0.580	0.776
	19	0.203	0.427	0.682	0.866

Table 1

The conclusions of the results of this table are discussed in the next section.

4. Conclusions

The most important conclusions of this work are the following:
- The argument of the gradient vector is more robust than the module for the detection of step edges. This means that the information of the argument allows to detect contours with lower contrast than the information of the module.
- The results of analyzing module and arguments together are only a little better, but the algorithm would be more complex than the other cases.
- The detection improves with the number of gradient vectors analyzed. The number of gradient vectors is limited because of the region to be analyzed must be over the contour. In curved lines the region to be analyzed must be curved too, in order to follow the same direction of the contour.
- Larger operators for the gradient vector obtention improve the detection of the contour, but it must be small to assure good location.
- Contours with SNR of 2/5 are very difficult to detect. Contours with a SNR of 4/5 can be detected with good results with the argument or the module information and an elliptical neighborhood of size 13.

References

[1]   Kim L. Boyer, Sudeep Sarkar, "On the Localization Performance Measure and Optimal Edge Detection," IEEE Trans. on PAMI Vol. 16 No. 1 (106-110) January 1994.

[2]   J. Brian Burns, Allen R. Hanson, Edward M. Riseman, "Extracting Straight Lines," Computer and Information Science Dept. Tech. Rep. 84-29, Univ. of Massachusetts, December 1984.

[3]   John F. Canny, "Finding Edges and Lines in Images," MIT., Cambridge, Tech. Rep. 720, June 1983.

[4]   L. S. Davis, "A Survey of Edge Detection Techniques," Comp. Graph. and Image Proc., 4 (248-270) 1975.

[5]   R. Deriche, "Using Canny's Criteria to Derive a Recursive Implemented Optimal Edge Detector," The Int. Jour. of Comp. Vision, 1 (167-187), 1987.

[6]   Robert M. Haralick, "Digital Step Edges from Zero Crossing of Second Directional Derivatives," IEEE Trans. on PAMI Vol. 6 No.1 (58-68) Jan.1984.

[7]   Albert Larré, Eduard Montseny, "A Step Edge Detector Algorithm Based On Symbolic Analysis, " 12th IAPR Inter. Conf. on Pattern Recognition, Vol. 1 (6-10) October 1994.

[8]   D. C. Marr, E. Hildreth, "Theory of Edge Detection," Proc. of the Royal Society of London, Series B, Vol. 207 1980.

[9]   V. S. Nalwa, T. O. Binford, "On Detecting Edges," IEEE Trans. on PAMI Vol. 8 No.6 (699-714) November 1986.

[10] Azriel Rosenfeld, M. Thurston, Y. Lee, "Edge and Curve Detection: Further Experiments," IEEE Trans. on Comp., Vol.C-21 No.7 (677-710) July 1972.

[11] Vincent Torre, Tomaso Poggio "On edge detection," IEEE Trans.PAMI Vol. 8 No.2 (147-163) February.1986.

Directions analysis in grey-level images

Stéphane Bres, Hubert Emptoz

Equipe Reconnaissance de Formes et Vision,
INSA de Lyon, Bat. 403,
69621 Villeurbanne, France.

Abstract. This paper presents a study of anisotropy in digital images, i.e. the detection of main directions and the quantification of their occurrence rate. Human vision is usually very powerful for such a feature-based analysis, because it simultaneously performs a multi-level analysis, from the local inspection of details to the more global analysis of spatial distribution of patterns. We show that our method is able to perform the anisotropy feature analysis using this global approach, unlike classic methods of directions analysis. Moreover, our method directly processes grey level images, uses the inner part of the patterns instead of their contours and is able to inspect all the directions of a picture. These specifications eliminate most of the limitations of usual methods.

1 Introduction

When we observe a picture, or a scene of our every day life, and focus on the different directions this picture contains, we usually see that some of them are highlighted by our visual inspection and some are not, or even completely disappear. Our visual perception system is very powerful for such an inspection. Unfortunately, all we can do is a qualitative inspection. Yet, the quantification of the results of such an inspection can be an interesting analysis tool. It has direct practical applications in different fields, like the study of materials, textures or physical phenomena.

In this paper, we will study what we can call the *Anisotropy* feature of an image. The more this distribution is non-homogeneous, the more the anisotropy level of the picture is high. First of all, we will describe the way we feel the presence of an orientation in a picture. This orientation can have local or global characters. Then, we will present some of the most famous methods of directions detection (mostly based on contour inspection in binary images) and their limitations. Then we will present a new method of anisotropy inspection and quantification, which can be used directly on grey level images, and performs a very complete analysis of the orientations distribution in the picture.

2 The Notion of Anisotropy

2.1 The human feeling of orientation

Our first aim, at the beginning of this study on anisotropy, was to reproduce the conclusions of the human visual impression of orientation, with the possibility to give a quantitative and objective result. The following examples of patterns shows the different kind of interpretations we can give in term of anisotropy, i.e. what characterises the anisotropy for us. Figure 1 gives the examples of isolated patterns : an anisotropic pattern (figure 1.a) and an isotropic pattern (figure 1.b). Anisotropy is directly the consequence of the lengthening of each pattern. A circle is the most possible isotropic pattern. A segment is the most anisotropic pattern and highlights its own direction.

(a) (b)

Fig. 1. Isolated patterns : (a) anisotropic pattern, (b) isotropic pattern.

The following examples present a set of patterns (figure 2). Each of these patterns have their own isotropic or anisotropic properties. But their relative dispositions generate new characteristics of anisotropy, which seem interesting to quantify also.

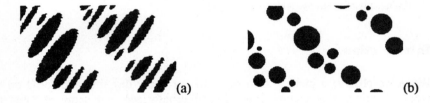

(a) (b)

Fig. 2. Patterns with isotropic or anisotropic properties, which spatial configurations influence the *Anisotropy* feature of the picture.

2.2   Anisotropy : a global feature

These examples show that the anisotropy feature has local and global manifestations. Our visual inspection of orientations presence in images takes both of them into account [1, 2, 5, 6] and both of them are important to quantify. As a matter of fact, the notion of anisotropy is a combination of local and global characters. For a global interpretation of the anisotropy feature, we consider the relative spatial distribution of each pattern. It highlights notions like groups, arrangements, *influence zone* and *interaction distance*. These notions correspond to the maximum distance we accept between two patterns so that we take into account the direction of the line which connect them. The shorter this distance is, the closer we are to a local anisotropy analysis.

3 Some Common Methods of Directions Detection

In the literature, we can find some methods of directions detection, which can lead to an anisotropy analysis. Here are some of the most common.

3.1　Methods

- The intercepts roses and Féret diameters [4] :
The principle is to determine the diametrical variation evolution of the patterns as a function of an angle.

- The directions rose [9] :
This method realises the anisotropy analysis through the analysis of the contours orientations. The final result is given on a polar diagram which represents the level of presence in the image contours of the different directions we can detect.

- The Hough transform :
The Hough transform [8] is able to detect pixels which belong to a given type of curves. These curves, described by their equation, can be lines, circles, ellipses or any curves [10]. The Hough transform allows to perform an Anisotropy analysis throught a lines detection. The directions distribution can be shown on a polar diagram (a kind of directions rose).

3.2.　Limitations of these common methods

- The first limitation of these methods is the obligation of using a binary image as an input. This pre-requested binarisation causes an important loss of information.

- A second limitation comes from the principle of these detections which are based on the hypothesis that contours of patterns contain the anisotropy information. This seems to be wrong, as we can see on the following figure, which presents an anisotropic pattern with an isotropic contour.

Fig. 3. An anisotropic pattern with an isotropic contour [4].

- Another limitation is the number of directions that each method is able to inspect. Even if this number can be parametrable, we never inspect all the possible directions of the image. This limitation has great consequences in cases where this number is too low.

- And finally, all these methods perform a local anisotropy analysis. They never take the different spatial dispositions into account this is very important for real cases analysis.

4. Orientations Analysis on Grey Level Images

Now, we are going to present a new method of orientations analysis on grey level images. This new method [3] was developed as a response to the preceding limitations, and is able to give a result as a rose which is a polar diagram that shows the presence levels of each orientations in the image and that we call orientations rose to differentiate it from the preceding directions rose.

4.1 Anisotropy, translations and autocorrelation

Our method is based on the autocorrelation function C_{xx} which is defined by [7] :

$$C_{xx}(a,b) = \sum_{i=-\infty}^{+\infty} \sum_{j=-\infty}^{+\infty} X(i,j).X(i+a,j+b)$$

for an image X. This function is able to give information on the recovery rates of the image by itself for each translation or the vector (a,b). Figure 4 presents the relations between translations and preferential orientations in an image, and then relations between autocorrelation function and anisotropy. When an isolated pattern presents a lengthening in a given direction, translations in this direction of the pattern in comparison with itself will give important recovery rates, even for long translations.

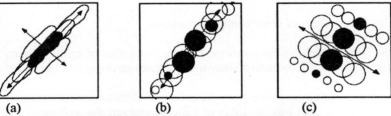

(a)   (b)   (c)

Fig. 4. Detection of patterns lengthening by autocorrelation.
Example of an isolated pattern (a). Example of a set of patterns (b) and (c).

These recovery rates will be much lower for translations in non-privileged directions (figure 4.a). The phenomenon is exactly the same with a set of patterns (figure 4.b and 4.c). This way of orientations detection uses the inner part of the patterns and not their contours.

4.2 Computation of the orientations rose

The orientations rose is a polar diagram which shows for each angle the presence rate of the corresponding orientation. These presence rates can be directly deduced from the recovery rates which are present on the autocorrelation function. For a direction θ_i, the corresponding presence rate $R(\theta_i)$ is computed by the sum of the autocorrelation levels in that direction θ_i :

$$R(\theta_i) = \sum_{D_i} C_{II}(a,b)$$

where D_i is the line of orientation θ_i and containing the origin of the autocorrelation function. The points (a,b) belong to that line D_i. Then, every directions of the input image are inspectable. From these values $R(\theta_i)$, it is possible to display relative variations of the contribution of each direction on new roses. These new roses can be defined, for example, by :

$$r_1(\theta_i) = \frac{R(\theta_i) - R_{min}}{R_{max} - R_{min}} \quad \text{et} \quad r_2(\theta_i) = \frac{R(\theta_i) - \frac{1}{2}R_{min}}{R_{max} - \frac{1}{2}R_{min}}$$

4.3   Examples of orientations roses

The following figure present examples of orientations roses of the given images. They are very close to the visual impression the corresponding images give. Figure 5.d and 5.e present the roses $R(\theta)$ and $r_1(\theta)$ of figure 5.c. The global anisotropy of this image is visible on these roses.

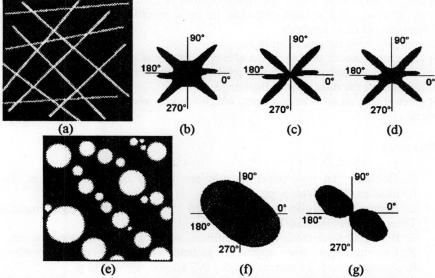

Fig. 5. Orientations rose. Orientations roses of figure (a) : rose $R(\theta)$ (b), rose $r_1(\theta)$ (c) and rose $r_2(\theta)$ (d). Orientations roses of figure (e) : rose $R(\theta)$ (f), rose $r_1(\theta)$ (g).

The part of the line Di on which this sum $R(\theta_i)$ is realised is parametrisable and corresponds to the notions of *influence zone* and *interaction distance*. With this choice, we choose only certain translations among the possible translations, i.e. we can choose the relative spatial positions of the patterns in interaction.

4.4   Computation of a anisotropy rate

From these roses, it is possible to extract a anisotropy rate. It is computed as a function of the differences between the current rose and an isotropic rose, which will give a rose as a circle of radius M. Then, this anisotropy rate A (with 0 < A < 1) will be :

$$A = \sum_{i=0}^{N-1} f\left(\left|R(\theta_i) - M\right|\right)$$

Each function f will give special characteristics of discrimination and dynamic. We can choose, for example, the functions identity, square or square root.

The anisotropy rates computed on rose 5.b and for the different functions we proposed are : 15% with the identity function, 0% with the function square, and 48% with the square root function..

5. Conclusion

This new method of anisotropy quantification is much more powerful than the common methods of directions detection. Its main specifications give a solution to the limitations of other methods :
- it uses grey level images as an input,
- it takes into account the inner part of each pattern and not its contours,
- it computes an analysis at different levels, from the local to the global aspect,
- it inspects all the directions of the image,
- it gives results on orientations roses, which is a very visual way to summarise them.

References

1. C. Bonnet : La perception visuelle des formes et du mouvement. In : Intellectica, 1988, Vol. 5, pp. 57-87.
2. C. Bonnet, R. Ghiglione et J.F. Richard : Traité de psychologie cognitive. Tome 1, Paris : Bordas, 1989. 266 p.
3. S. Bres : Contributions à la quantification des critères de transparence et d'anisotropie par une approche globale. Thèse de doctorat, Equipe Reconnaissance de Formes et Vision, INSA Lyon, 1994. 234 p.
4. M. Coster et J.L. Chermant : Précis d'analyse d'images. Presses du CNRS, 1989. 560 p.
5. M. Denis : Image et cognition. Paris : Presses Universitaires de France, 1989. 290 p.
6. M. Imbert : La vision naturelle. In : Intellectica, 1988, Vol. 1, No. 5, pp. 3-31.
7. M. Kunt : Traitement numérique des signaux. Lausanne : Editions Presses polytechniques romandes, Paris : DUNOD, 1981, 402 p.
8. H. Maitre : Un panorama de la transformation de Hough. In : Traitement du Signal, 1985, Vol. 2, No 4, pp. 305-317.
9. J. Serra : Image analysis and mathematical morphology. London : Academic Press Inc. (London) Ltd. 1982. 610 p.
10. J. Sheng-Ching et T. Wen-Hsiang : Scale- and orientation-invariant generalized Hough transform - A new approach. In : Pattern Recognition, 1991, Vol. 24, No. 11, pp. 1037-1051.

An Unsupervised and Non-parametric Bayesian Image Segmentation

M. Zribi and F. Ghorbel
Groupe Recherche Images et Forme
Institut National des Télécommunications, ENIC - INT
cité scientifique, rue G. Marconi, 59650
Villeneuve d'Ascq FRANCE.

Abstract. We propose here an unsupervised Bayesian image segmentation based on a non-parametric Expectation-Maximisation (EM) algorithm. The non-parametric aspect comes from the use of the orthogonal probability density function (pdf) estimation, which is reduced to the estimation of the first Fourier coefficients (FC's) of the pdf with respect to a given orthogonal basis. So, the mixture identification step based on the maximisation of the likelihood can be realised without hypothesis on the distribution of the conditional pdf. This means that we do not need some assumption for the gray level image pixels distribution. The generalisation to the multivariate case can be obtained by considering the multidimensional orthogonal function basis. In this paper, we intend to give some simulation results for the determination of the smoothing parameter. This algorithm is applied to a contextual image segmentation. Such method conjugated with Bootstrap Sampling allows as the exploration of a large neighbourhood context.

1   Introduction

It is well known that the Bayesian classification rule is the best one in the mean of the a posterior probability of miss-classification. However, the application of this rule in an unsupervised context needs some parametric hypothesis for the conditional probability distributions of each class. Using the orthogonal pdf estimator, we intend to introduce an original non-parametric and unsupervised Bayesian classifier of which we will prove the robustness in the image segmentation. The unsupervised Bayesian classification consists of the two following important steps : The first one is the *mixture identification*. It can be achieved by the Expectation-Maximisation (EM) algorithm which was proposed in 1977 [1] for parametric unsupervised Bayesian classification. It is an iterative procedure which assumes that the mixture pdf of data is a linear combination of a finite number of Gaussian distributions where the coefficients of this combination are the a priori probability of each class. The second step consists of the application of the *Bayesian classification rule*. Using the orthogonal pdf estimator [4], a non-parametric segmentation can be realised. Effectively, this estimator is reduced to the estimation of a finite number of parameters. These features are the FC's of the conditional pdf of each class. During iterations, the number of FC's to estimate changes since it depends on the seize of the sample belonging to each class.

In this article, we propose to apply this algorithm to the segmentation of real images. We use the Bootstrap sampling [2, 3] in order to reduce the dependence between the pixel sample and the computing time without destroying the quality of the segmented image. The reduction of computing time allows us the exploration of the high dimensional contextual segmentation of images.

2   The classical EM algorithm

The classical EM algorithm assumes that the observed image is a realisation of a mixture distributions, so that its pdf can be written as:

$$f(x)=\sum_{j=1}^{K} \pi_j f(x / \theta_j), \text{ with } 0 \leq \pi_j \leq 1 \text{ and } \sum_{j=1}^{K} \pi_j = 1$$

where $f(x / \theta_j)$ is the conditional pdf and π_j is the probability a priori of class j. This algorithm is iterative and has the three following steps. Gaussian hypothesis are generally taken (i.e: $\theta_j = (\mu_j, \sigma_j^2)$ where μ_j represents the mean and σ_j^2 is the variance of the class j).

- *Initialization step*: The number of classes K is assumed to be known. An initial solution of the parameters of the mixture are extracted from the histogram.

- *Expectation step*: This consists on the estimation of the a posterior probability $\hat{\pi}_j^n(x_i)$ for the pixel x_i belonging to the class j at the n-th iteration:

$$\hat{\pi}_j^n(x_i) = \frac{\hat{\pi}_j^n f(x_i / \hat{\theta}_j^n)}{\sum_{l=1}^{K} \hat{\pi}_l^n f(x_i / \hat{\theta}_l^n)}$$

- *Maximization step*: The parameters needed for the next step are constructed, in the following way:

$$\hat{\pi}_j^{n+1} = \frac{1}{N}\sum_{i=1}^{N} \hat{\pi}_j^n(x_i), \quad \hat{\mu}_j^{n+1} = \frac{\sum_{i=1}^{N} x_i \hat{\pi}_j^n(x_i)}{\sum_{i=1}^{N} \hat{\pi}_j^n(x_i)} \quad \text{and} \quad \left[\hat{\sigma}_j^2\right]^{n+1} = \frac{\sum_{i=1}^{N} \hat{\pi}_j^n(x_i)(x_i - \hat{\mu}_j^{n+1})(x_i - \hat{\mu}_j^{n+1})^T}{\sum_{i=1}^{N} \hat{\pi}_j^n(x_i)}$$

3   The proposed non-parametric EM

3.1   Estimation based on orthogonal expansions

The estimation of the pdf based on Fourier analysis methods is suitable for this context. Then an estimator of the pdf based on a independent sample $X_1, \ldots, X_N$ with density f is given by:

$$\hat{f}_{K_N}(x) = \sum_{m=0}^{K_N} \hat{a}_{m,N} e_m(x) \quad \text{where} \quad \hat{a}_{m,N} = \frac{1}{N}\sum_{i=1}^{N} e_m(X_i), \quad \left\{e_m(x)\right\}_{m\in\mathbb{N}} \text{ is an orthonormal}$$

complete basis of the Hilbert space $L_2([a,b])$ and K_N is called truncation point.

3.2 Description of the non-parametric EM Segmentation

The kind of the conditional pdf's of a gray level of image pixels do not need to be known with this approach, since we propose to define θ_j of section 2 as:

$$\theta_j = (a_{0,j},\ldots\ldots,a_{k_{N_j},j}). \text{ So we denote by } \hat{f}_{k_N}(x) = f(x/\hat{\theta}_j)$$

In the following, we intend to describe the proposed algorithm. It consists of three steps:

a. *Initialization step* : Using the histogram, the observed sample are arranged in K groups. So, the parameters of mixture can be initialised as following:

$$\hat{\pi}_j^0 = \frac{\hat{N}_j^0}{N}, \quad K_{\hat{N}_j^0} = \text{int}\left[(\hat{N}_j^0)^{1/s}\right] \quad \text{and} \quad \hat{a}_{m,j}^0 = \frac{1}{\hat{N}_j^0}\sum_{l=1}^{\hat{N}_j^0} e_m(X_l)$$

where int [x] is the biggest integer is inferior to the real number x.

b. *Expectation step* : In this step, we estimate the a posterior probability $\hat{\pi}_j^n(x_i)$ for the pixel x_i belonging to the class j at the n-th iteration :

$$\hat{\pi}_j^n(x_i) = \frac{\hat{\pi}_j^n f(x_i/\hat{\theta}_j^n)}{\sum_{l=1}^{K} \hat{\pi}_l^n f(x_i/\hat{\theta}_j^n)}$$

c. *Maximization step* : The a posterior probability $\hat{\pi}_j^n(x_i)$ of each pixel x_i is computed. So that, at (n+1)-th iteration, we have:

$$\hat{\pi}_j^{n+1} = \frac{1}{N}\sum_{l=1}^{N} \hat{\pi}_j^n(x_l), \quad K_{\hat{N}_j^{n+1}} = \text{int}\left[(\hat{N}_j^{n+1})^{1/s}\right] \quad \text{where} \quad \hat{N}_j^{n+1} = N\hat{\pi}_j^{n+1}$$

$$\hat{a}_{m,j}^{n+1} = \frac{\sum_{l=1}^{N} e_m(x_l)\hat{\pi}_j^n(x_l)}{\sum_{l=1}^{N} \hat{\pi}_j^n(x_l)} \quad \text{for} \quad m = 0,\ldots\ldots,K_{\hat{N}_j^{n+1}}$$

The Bayesian rule: After the mixture identification, the Bayesian rule is applied in order to classify the pixels according their gray level x:

$$j(x) = \text{Arg}\left[\max_{1 \le j \le K} \left\{\pi_j f(x/\theta_j)\right\}\right]$$

where j(x) represents the label of the class of the pixel x.

4   Smoothing parameters estimation

It is well known that the orthogonal estimation method depends on the smoothing parameter which we notice 1/s ($k_N \approx N^{1/s}$). Convergence theorems are well studied for finding the optimal values of 1/s [4, 5, 6]. However, in the mixture identification, such problem appears to be more difficult. In order to give an idea on the values of 1/s, we propose here some simulation studies. We randomly generate a Gaussian mixture distribution. So, the proposed pdf estimation of the mixture can be written as:

$$\hat{f}_{1/s}(x) = \sum_{j=1}^{K} \hat{\pi}_j f(x/\hat{\theta}_j) \text{ with } \hat{\theta}_j = (\hat{a}_{0,j},\ldots\ldots,\hat{a}_{K_{Nj},j}).$$

Where $\hat{\theta}_j = (\hat{a}_{0,j},\ldots\ldots,\hat{a}_{K_{Nj},j})$ are the estimated FC's of the conditional pdfs and K_{Nj} is $\text{int}\left[N_i^{1/s} \right]$ for some values of $s \geq 1$. This asymptotic study is realised in the sense of mean integrated square error (MISE) between the theoretical pdf f and $\hat{f}_{1/s}$:

$$E\left[d_2(f,\hat{f}_{1/s})\right] = g(1/s).$$

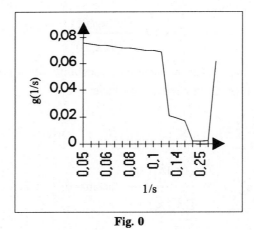

Fig. 0

Fig. 0. represents the MISE according 1/s for a Gaussian mixture. We can observe that the optimal values of 1/s are about 0.25. This result can be shown for kind of mixtures distribution. So, the optimal values of 1/s are similar to those the classical theoretical results for the non mixture case [4, 5, 6].

5 Non-parametric image segmentation

The non-parametric and unsupervised Bayesian classifier is applied to Retina image, and traffic signs image. The original images are represented by 256 gray levels. We consider three classes. Trigonometric basis are used for the non-parametric mixture identification step. However, Gaussian mixture is assumed for the classical EM segmentation procedure. It improves the segmentation result with respect the classical EM procedure for the traffic images (see Fig 2, 4 , 6).

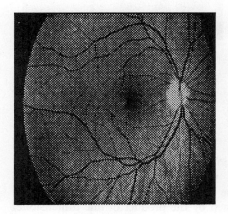

Fig. 1. Original Image

Fig. 2. Original Image

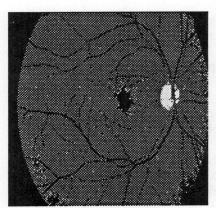

Fig. 3. Segmented Image with
Gaussian mixture

Fig. 4. Segmented Image with
Gaussian mixture

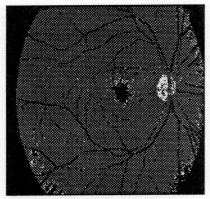

Fig. 5. Segmented Image with non-parametric EM

Fig. 6. Segmented Image with non-parametric EM

Perspectives:

The extension to multivariate case of the classical EM algorithm for non Gaussian multivariate pdf's is not trivial. However, the generalisation of the proposed algorithm to the contextual case can be done easily since the orthogonal function basis in the multivariate case could be obtained by the product of one dimensional orthogonal system functions.

References

1. A. P. Dempester et al: Maximum Likelihood from Incomplete Data via the E.M Algorithm. Journal of the Royal Statistical Society, Series B, 39:1, 38 (1977).
2. C. Banga et al: Optimal Bootstrap Sampling for fast image Segmentation, Application to Retina Image. IEEE-ICASSP 1993, April 27-30 Minneapolis USA.
3. F. Ghorbel et al: Bootstrap sampling approach in pattern recognition, The special session, the Bootstrap and Its Application. ICASSP April 94 South of Australia.
4. R. Kronmal et al: The estimation of probability and cumulatives by Fourier series methods. J.Amer Statist. Ass, Vol-63, pp 925-952, 1968.
5. P. Hall: Measuring the efficiency of trigonometrique series estimates of density. J.Multivarate Anal, in press, 1981 .
6. P. Hall: Comparison of two orthogonal series methods of estimating a density and its derivatives on a interval. Journal of multivariate analysis, 12, 432-449, 1982.

A.I. Based Image Segmentation

S.Vitulano, C.Di Ruberto, M.Nappi

Faculty of Medicine, University of Cagliari, ITALY
E-Mail vitulano@facmed.unica.it, fax (39) 70-663651

Abstract. The paper will show a possible model of the human perceptive process. With the aim to implement the model we have introduced a well know technique of Problem Solving.

The most important roles of our model are played by the Evaluation Function and the control strategy. The Evaluation Function is related to the ratio between the entropy of one region or zone of the picture and the entropy of the entire picture.

The model has been widely experimented on different kinds of pictures: natural scenes, CT Scan, MR and mammographies.

1 Introduction

Segmentation is a significant issue in the field of image processing and image understanding. The segmentation is the process, both human and automatic, that individuates in a pictorial scene zones or regions showing some characteristics with respect to a certain uniformity predicate (UP). In our case, the UP is determined by means of evaluation function (*EF*).

It is possible to choose as *EF* local measures as: the grey tone, the colour, the application of a local operator (gradient, laplacian, average, Sobel, etc,) filters or global measures as Fourier transform, histogram, fractal dimension, entropy and so on. In the computer vision literature, the whole segmentation process is often shown by two opposite but related phases: Merge-and-Split.

The merge phase tries to obtain the maximum possible organisation of the picture, enhancing the transition zones between different regions.

We consider as regions all those zones of the picture whose entropy is negligible and for which it is possible to determine a statistical function or mathematical structure describing them. On the contrary, for those regions that present a high value of entropy it is not possible to determine a statistical function or a mathematical structure but, perhaps, a catastrophe. We say that the zones of the picture where entropy is high correspond to the contour or silhouette of the object.

The purpose of the split phase is to divide each region, individuated by the merge phase, in subregions which can be described by means of different mathematical structures; for these subregions the entropy changes are very small.

The rest of this paper is organised as follows: in section II we introduce a segmentation algorithm based on A.I. technique; experimental results are discussed and shown in section III, while final conclusions are given in section IV.

2 The Proposed Algorithm

The task of this phase is to obtain information about the different regions of the picture: the number of existing regions, their areas and topological positions in the domain of the image, their features and their contours.

For the choice of the characteristics, of each region, and their quantification we have resorted to a typical A.I. strategy, Problem Solving, and we define:

Physical space. The domain of the picture will be the physical space of the problem and will usually be a square matrix of 512 x 512 pixels;

Research tree. A hierarchical structure to represent the state of the problem during its various phases is utilised, for instance the quad-tree.

Evaluation function. The histogram can give some useful information about the picture under investigation, such as entropy, grey level distribution and so on. For example, an image that is totally-homogeneous consisting of a single grey level, has a histogram with only one peak; we can associate with the peak a Gaussian distribution that has zero standard deviation, σ. Therefore, the entropy of the whole picture is zero. In other words, the entire scene is homogeneous and it is composed by only one region characterised by a single grey level, then, we can state that the entropies of both the region and the pictorial scene are zero. On the contrary, if we consider a picture with only two grey levels, white and black for instance, its histogram has two peaks. If we associate a Gaussian function with each one of these peaks, its standard deviation is different from zero. The entropy of the whole picture is different from zero and it is related to $\sigma_1 + \sigma_2$, i.e. the sum of the two standard deviations. Such an image could be a regular structure (chess-board) or a chaotic scene (black and white noise). In this picture two or more homogeneous regions could exist, with entropy equal to zero, but the entropy of the whole picture is different from zero.

Usually, a real picture histogram is very complicated to read, even if it gives us some information about the whole image. This information is not enough to characterise the regions contained inside the image. It is possible to make a remark: the entropy of each region is small since it represents a function of the whole image entropy. For this reason *EF* as entropy measure of each child domain with respect to the entropy of the whole of the domain of the father has been chosen. So the *EF* is:

$$EF_c = \frac{f_c(\sigma_c)}{f_f(\sigma_f)}$$

where: the function $f_c(\sigma_c)$ is a measure of the entropy of the child and where σ_c is the sum of the standard deviations of the peaks in the histogram of the child c

and $f_f(\sigma_f)$ is a measure of the entropy of the father and where σ_f is the sum of the standard deviations of the peaks in the histogram of the father.

The lists. In order to store the quad-tree's nodes visited during the various search phases some lists are utilised; they are:

the F-*Open list* contains the nodes generated along an optimal path but not yet expanded, they are ordered in the list with respect to the *EF* values;

the *F-closed list* contains the nodes expanded along an optimal path;

the *B-closed list* contains the nodes whose associated domain covers a region already individuated.

The rules. The choice of the rules plays a significant role in the field of Problem Solving; in fact, it can influence the number of operations necessary to reach the problem solution.

The control strategy. The purpose we intend to realise in this phase is to highlight, inside the pictorial scene, those zones whose entropy is zero or quasi-zero. Better, the zones individuated as regions are those ones whose *EF* is stable or gradually decreasing. So, we can state:

if Q is the domain of a region then there exists a set H= {H_1, H_2,..., H_n } of subdomains of Q such that $H_n \subseteq H_{n-1} \subseteq ... \subseteq H_1 \subseteq Q$ and whose EF value is less than or equal to the EF value of Q.

Such a set H has a subdomain whose EF value is higher than the EF value of Q as lower bound. We call this subdomain piece of the region.

The Backward strategy. We have said that along an optimal path the *EF* assumes constant or decreasing values. Along such a path, when the dimensions of the subdomain relative to a level of the search tree are less than the dimensions of the domain of the piece, then a sudden variation of the *EF* value occurs. This means that the dimensions of these nodes are smaller than the dimensions of the piece of the region. The backward strategy allows to determine exactly the dimensions of the piece.

The split phase. The aim of the splitting phase is to distinguish the structures associated with each region whatever their statistics be.

The merging phase divides the image into regions whose elements satisfy a predicate of uniformity, but the fact that inside the same region two elements satisfy its uniformity predicate does not imply that they have the same texture or structure. The structure associated with an element is a description of the mutual relation among the pixels contained in the partition element. The basic task of the split phase is to associate a structure with the region under investigation or to split the region in more regions.

3 Experimental Results

Let's consider, as a first example, a *theoric* picture where all the grey tones contained in its histogram are uniformly distributed in the picture domain.

432

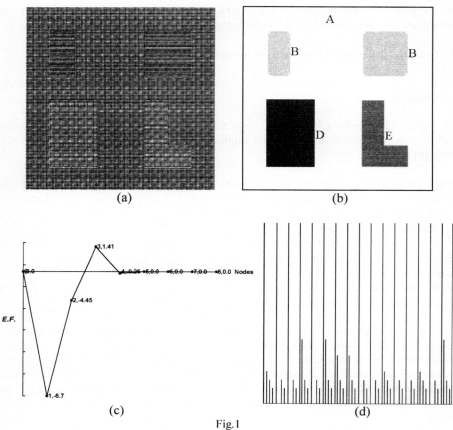

Fig.1

In fig 1a) it is shown the input picture; in (b) it is shown the result obtained applying the segmentation algorithm proposed; in (c) is shown the behaviour of the *E.F.* and in (d) the histogram of the picture in (a).

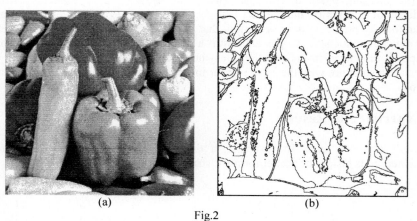

Fig.2

In (a) it's shown a well known picture: peppers and in (b) are shown the contours obtained applying our method.

Many authors propose local operator in order to obtain the contour of the objects contained in a scene. In fig 2 it's shown the contours obtained applying our method to a well known picture.

Let us show the results obtained of our method to a real picture: a CT picture of a human brain.

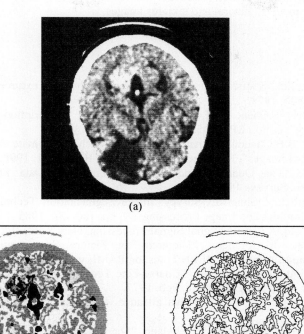

(a)

(b)   (c)

Fig.3

In (a) it is shown a real picture: CT picture of a human brain; in (b) it is shown the segmented picture and where each grey-tone represents a region; in (c) it is shown the contours between the different regions.

4 Conclusion

The wide experimentation of our model on different kinds of image allows us to formulate the following considerations:

the method proposed here gives information not only about the regions contained in the scene (small entropy) but also about the contours of the regions themselves. In fact contours are the transition zones between regions (high entropy). A comparison between the method here proposed and local operators strategy allows us to make the following considerations: it is not necessary to choose the size of local operators, it is not necessary introduce *thresholds* ad hoc, it is not time consuming (our algorithm is $O(n\log n)$) and, finally, are not necessary preliminary iterated smoothings: smoothing procedure is absolutely forbidden for medical pictures. We wish, in the next future, to show the results obtained comparing our method with the ones proposed in (15).

We have widely experimented the method here proposed on different types of pictures, i.e. medical, natural and textures images.

References

1. R.M.Haralick, Statistical and Structural Approaches to Textures, Proc. IEEE 67, pp.786- 804, 1979
2. J.M.Keller, S.Chen, R.M. Crownover, Texture "Description through Fractal Geometry", C.V.G.I.P., v.45, pp.150-166, 1989.
3. N.Sarkar, B.B.Chaudhuri, "An Efficient Approach to Estimate Fractal Dimension of Textural Images", Patt. Recog., v.25, n.9, pp.1035-1041, 1992.
4. H.Samet, "The Quadtree and Related Hierarchical Data Structures", ACM Computers Surveys 1984.
5. R.M.Haralick, L.Shapiro, "Survey: Image Segmentation Techniques", Computer Vision Graphics and Image Processing, 29, pp. 100-132, 1985.
6. A.Rosenfeld, A.C.Kak, Digital Signal Processing, New York, 1982.
7. J.Beaulieu, M.Goldberg, "Hierarchy in Picture Segmentation: a Stepwise Optimization Approach", IEEE Trans. on PAMI, v.11, n.2, 1989.
8. A.Rosenfeld, G.J.Vanderbrug, "Coarse-Fine Template Matching", IEEE Trans. on System, Man and Cybernetic, Feb. 1977.
9. I.Prigogine, I.Stengers, La nuova alleanza. Metamorfosi della Scienza, Einaudi, 1981.
10. C.Di Ruberto, R.Prudente, S.Vitulano, "Segmentation of Biomedical Images: CT and MNR", Digital Signal Processing '91, Ed.: V.Cappellini e A.G. Constantidines, Elsevier Science Publishers, 1991.
11. C.Di Ruberto, M.Nappi, S.Vitulano, "Problem Solving in a Vision Model", Computer Application in Industry, Ed. M.H.Hamza, ACTA Press, .
12. P.Brodatz, Textures: A Photographic Album for Artists and Designers, New York: Dover, 1966.
13. P.J.Burt, T.Hong, A.Rosenfeld, "Segmentation and Estimation of Image Region Properties Through Cooperative Hierarchical Computation", IEEE Trans. on System, Man and Cybernetics, vol.11, no.12, 1981.
14. C.Di Ruberto, N.Di Ruocco, S.Vitulano "A Segmentation Algorithm based on AI Techniques", Pattern Recognition in Practice IV, ED. E.S. Gelsema and L.N. Kanal, Elsevier Science Publishers pp 153- 164, 1994.
15. W.E. Higgins, C. Hsu, "Edge Detection Using Two-Dimensional Local Structure Information", Pattern Recognition, vol. 27, n.2, pp. 277-294, 1994.
16. R.M. Haralick, "Digital step edges from zero crossing of second directional derivatives", IEEE trans. pattern Analysis Mach. Intell., Pami 6 58- 68 (1984)
17. F Tomita and S. Tsuyi, "Extraction of multiple regions by smoothing in selected neighbourhoods", IEEE Trans, Syst. Man. Cybernetics SMC-7,107 109 (1977).

Quality Enhancement in Image Enlargement

Francescomaria Marino, Giuseppe Mastronardi

Dipartimento di Elettrotecnica ed Elettronica, Politecnico di Bari
Via Re David, 200 - 70125 Bari (Italy)

Abstract. Interpolation methods, usually employed in image enlargement, cause a degradation of the image quality in proximity to the edges. In fact, the best interpolation leads to fuzziness effects often due to the image features ignored during the process. In order to have a contained definition loss after the enlargement operation, it is necessary to consider the original image as a subsampled image of the enlarged one. It follows that the image needs to be represented by its low and high frequency components together. A method able to consider and provide the above mentioned image components is the bidimensional Fast Fourier Transform (2D-FFT). In the paper a strategy of partitioning & overlapping is proposed to reduce the computational complexity, conditioned by the number and the accuracy of the operations, when the 2D-FFT is applied to the whole image. So, the image enlargement achieves the goal with low definition loss and without expensivecomputational load. The quality enhancement is evaluated in terms of extracted edges.

1 Introduction

The expansion operators are very often used in the field of image processing, mainly as a preliminary approach to the local spatial processing. Interpolation methods are usually employed for this aim; nevertheless, they cause a degradation of the image quality in proximity to the edges. The interpolation, however applied, leads to fuzziness effects mainly due to the contrast smoothing, since it produces the absent information (for instance: 3 new pixels calculated on 4) starting only from the luminance values of the original image.

In order to avoid the definition loss after the image enlargement, it is necessary to consider the original image as a subsampled image (for instance: 1 pixel on 4) of the enlarged one. It follows that the image needs to be represented by its low and high frequency components together. The low frequency components are obtained by the luminance of the window, selected in the source image. The other components are inserted in order to achieve a better definition by generating an artificial texture.

A method able to provide the above mentioned image components is the bidimensional Fast Fourier Transform (2D-FFT). The enlargement operation by Fourier transform is usually achieved by expanding the image transform with zero values and then appling the inverse transform to the matrix so obtained. During an enlargement x4 of NxN pixel images, this algorithm requires

$$N^2 \lg_2(N) + 4N^2 \lg_2(2N) = N^2 \left(5\lg_2(N) + 4 \right)$$

complex multiplications. Moreover, in the case of large value of N, the angles in the goniometric circle are similar, so an high precision in the representation of the Fourier coefficients is necessary: consequently, the computational complexity increases. From the hardware point of view, a very large size of memory is required, too [1].

2 The Algorithm

To spare the high computational complexity due to 2D-FFT, when applied to the whole image, a first strategy could be adopted partitioning the original image in 2x2 pixel windows and using them as source image within the classic 2D-FFT approach. Nevertheless the set of inverse FFTs generates an enlarged image with an unseemly "mosaic", caused by the boundary effect of the contiguous antitransforms (evaluations about this method and other ones are reported in [2]).

To overcome the above mentioned chequered effect, an overlapping strategy is proposed. This one has been approached in [3] by using the Hadamard transform in the same enlargement operation, with results good about the computational load, but not excellent concerning the image quality.

If we call $P_{r,c}$ the grey level of the pixel belonging to the r th raw ($r=0, ..., M-1$) and to the c th column ($c=0, ..., N-1$) of the source image, our algorithm (o.a.) is composed by the phases sketched and explained in the following:

$$B_{r,c} \equiv \begin{bmatrix} P_{r-1,c-1} & P_{r-1,c-1} \\ P_{r,c-1} & P_{r,c} \end{bmatrix} \equiv \begin{bmatrix} P_1 & P_2 \\ P_4 & P_3 \end{bmatrix} \xrightarrow{2x2\,FFT} T_{r,c} \equiv \begin{bmatrix} T_1 & T_2 \\ T_4 & T_3 \end{bmatrix} \rightarrow$$

$$\xrightarrow{R_i = K_R \cdot T_i \quad S_i = K_S \cdot T_i} \begin{bmatrix} R_1 & S_1 & S_2 & R_2 \\ S_1 & T_1 & T_2 & S_2 \\ S_4 & T_4 & T_3 & S_3 \\ R_4 & S_4 & S_3 & R_3 \end{bmatrix} \xrightarrow{4x4\,IFFT}$$

$$\rightarrow \begin{bmatrix} \tilde{E}_1 & \tilde{E}_2 & - & - \\ \tilde{E}_4 & \tilde{E}_3 & - & - \\ - & - & - & - \\ - & - & - & - \end{bmatrix} \xrightarrow{Norm.} E_{2r,2c} \equiv \begin{bmatrix} E_1 & E_2 \\ E_4 & E_3 \end{bmatrix} \equiv \begin{bmatrix} E_{2r-1,2c-1} & E_{2r-1,2c} \\ E_{2r,2c-1} & E_{2r,2c} \end{bmatrix}$$

1) to consider the 2x2 block $B_{r,c}$ for each $P_{r,c}$ ($r=1, ..., N-1$; $c=1, ..., N-1$);
2) to compute for each block $B_{r,c}$ its Fast Fourier Transform $T_{r,c}$;
3) to extrapolate, for each term T_i, the high frequency components R_i and S_i by the formulas $R_i = K_R \cdot T_i$ and $S_i = K_S \cdot T_i$ using suitable value for K_R and K_S;
4) to compute for each block so obtained the four lowest components of its Inverse Fast Fourier Transform;
5) to normalize these components and adopt them as new pixels for the $E_{2r,2c}$ block;
6) to mosaic these blocks in order to achieve an enlargement x4 of the selected window.

This enlargement x4 is obtained because for each $P_{r,c}$ pixel in the original image, one block $E_{2r,2c}$ (composed by four pixels) is generated. The enlarged image will be carried out by mosaiking the blocks $E_{2r,2c}$.

Moreover, the propagation error is strongly reduced, because the 2D-FFT required by this approach are computed on $(N-1)^2$ windows of 2x2 (direct transform) and 4x4 (inverse transform) pixels by adding a little number of terms: in these cases, the Fourier coefficients require a very short precision.

The number of complex multiplications, required in order to compute $(N-1)^2$ 2x2 (full) and 4x4 (partial) 2D-FFT, is

$$4(N-1)^2 + 8(N-1)^2 = 12(N-1)^2$$

this one is less than that one required for a non partitioned 2D-FFT processing, above all, without high accuracy requiring or large size of memory employing.

At last, the partitioning of the whole image reaches two others fundamental advantages respect to the classic 2D-FFT approach:
a) to implement the algorithm in parallel way;
b) to consider images not needfully having power-of-2 dimensions.

Fig. 1. The portrait of "Lenna" with the selected window to be enlarged.

3 Conclusions and Remarks

We have used as source image a 128x128 pixel window of the famous portrait of "Lenna" shown in Fig. 1. In Fig. 2 some enlargements x16 are shown in order to compare the results of the respective elaborations. They are obtained iterating twice the base process to make more evident the effects of the implemented solutions.

The enlargements in Fig. 2 are respectively obtained: 2a) by bicubic interpolation; 2b) by o.a. with $K_R = K_S = 1$; 2c) by o.a. with $K_R = 0.1$, $K_S = \sqrt{2} K_R$; 2d) by o.a. with $K_R = 0.05$, $K_S = \sqrt{2} K_R$; 2e) by o.a. with $K_R = 0.025$, $K_S = \sqrt{2} K_R$; 2f) by o.a. with $K_R = K_R = 0$. The ratio K_S/K_R is obtained in experimental way.

The quality enhancement is evaluated in terms of extracted edges: they are shown in Fig. 3. The algorithm adopted to extract the edges is a spatially adaptive filter reported in [4].

The bicubic law, used in the interpolation for the enlargement in Fig. 2a, produces a very low definition loss [5]; the image, so generated, appears similar to that one shown in Fig. 2f, that is obtained by o.a. with null values of both parameters K_R and K_S (see Fig. 3a vs. Fig. 3f).

A proper use of the K-parameters generates an artificial texture. This effect, proportional to the K-parameters, produces a better definition of particulars in no strongly contrasted areas.

Taking care of the texture does not become preponderant (as in the Figg. 3b and 3c), significant results can be obtained. In fact, by a parametric use, it is possible to adapt the enlargement quality to the image features in order to thin the fuzziness effect. For the image under consideration, the Figg. 3d and 3e give prominence that values of $K_R=0.025 \div 0.05$ can be suitably used.

4 References

1. A. V. Oppenheim, R.W. Schafer: Digital Signal Processing. (Prentice Hall International Inc., New Jersey, 1975).
2. F. Marino, G. Mastronardi: Techniques of Image Enlargement. Invited paper at the 18th International Conference MIPRO'95, sponsored by IEEE (Croatian Section), Rijeka, Croatia, May 22-25, 1995, MIS 2, pp. 1-13.
3. G. Mastronardi, F. Spilotros: Image Expansion Operators based on the Hadamard Pyramidal Coding. Progress in Image Analysis and Processing III, (Edited by S. Impedovo, World Scientific, 1993), pp. 111-114.
4. G. Mastronardi, S. Sangiovanni: An Adaptive Filter for Image Processing in OCCAM2. Proc. of the ISMM Int. Symp. on Industrial, Vehicular and Space Applications of Microcomputers, New York, 1990, pp. 69-72.
5. M. Sonka, V. Hlavac and R. Boyle: Image Processing, Analysis and Machine Vision. (Chapman & Hall, Cambridge, 1994).

Fig. 2. Enlargements x16 of 128x128 pixels of "Lenna": a) by bicubic interpolation; b) by o.a. with $K_R=K_S=1$; c), d), e) by o.a. respectively with $K_R=0.1$, 0.05, 0.025 and $K_S=\sqrt{2}\,K_R$; f) by o.a. with $K_R=K_S=0$.

Fig. 3. Edge extractions of the enlargements shown in Fig. 2.

Synthesising Objects and Scenes
Using the Reverse Distance Transformation
in 2D and 3D

Ingela Nyström[1] and Gunilla Borgefors[2]

[1] Centre for Image Analysis, Uppsala University,
ingela@cb.uu.se
[2] Centre for Image Analysis, Swedish University of Agricultural Sciences,
gunilla@cb.uu.se

[1,2] Lägerhyddvägen 17
S–752 37 Uppsala
Sweden

Abstract. The reverse distance transformation has proved useful in image synthesis. This paper describes how digital objects are created from a number of seed labels in an image. The shape of the obtained objects depends on the metric used. In 2D the Euclidean and the 3-4 metrics are mentioned, and in 3D the D^6, the D^{26}, and the 3-4-5 metrics are discussed. The proposed method has no need of expensive CAD systems. It is an excellent image synthesising tool when developing image processing algorithms, i. e. shape quantification, visualisation, scene analysis and range imaging, as the obtained objects are well-defined in the image. The method is most advantageous in 3D, as there is an increasing need for volume images, but synthesising objects in 2D can also be useful.

1 Background

The approach described in this paper is an aid in obtaining synthetic volume objects without using expensive and complex CAD systems.

When implementing algorithms for the 3D image processing tool box it is of importance to have a number of mathematically well-defined objects for algorithm evaluation. With a "known" object you know what result to expect and can debug your algorithms accordingly. This also gives the possibility to perform quantitative comparisons between algorithms.

Often one can see generalisations of the world into blocks, spheres, cylinders, cones, etc. Anyone who has been involved in applications will agree with the fact that the world does not consist of any of these geometrical models (especially not cylinders!). It is therefore desirable to use more general objects.

In a biomedical application it can also be the case that a sufficient number of images can not be provided fast enough; some tomography methods are still tedious and expensive.

The objects resulting from our method have smooth surfaces (depending on the metric used), which is consistent with many biomedical applications.

2 Reverse Distance Transformation

This section will give an overview of how to obtain a *reverse* distance transform.

A distance transformation (DT) is an operation that gives every object pixel a label that approximates the distance to the closest background pixel, or, *vice versa*, every background pixel the distance to the closest object pixel. In a *weighted* DT a mask, with weights corresponding to the distance between neighbouring pixels, is used [2]. Each type of neighbour is given a different local distance from the central pixel. When applying a reverse DT the same mask can be used. Similarly to the algorithm for computing the weighted DT, the sequential algorithm to compute the reverse DT requires a forward and a backward scan, during which half of the neighbours of every pixel are taken into account.

It is also possible to make a parallel implementation of the algorithm, similarly to the weighted DT case [3], but we have only implemented the sequential algorithm so far.

The input to the reverse DT is an image with seed labels in different positions. The algorithm performs a propagation with decreasing labels. Every pixel is assigned the maximum label of the current pixel and the neighbouring pixels subtracted by the local distances to the current pixel.

The reverse DT, using different metrics, has been used earlier in the literature when reconstructing the shape of an object from its local maxima (skeleton). Examples are the 3-4 weighted DT [1] and the Euclidean DT [4].

When extending to 3D the principle is the same as in the 2D case, but the algorithm now uses a 3D mask [2]. The weights in the mask are chosen according to the metric used. The D^6 metric is obtained by setting the distance to the 6 face-neighbours to 1 and omitting the other neighbours. The D^{26} metric is obtained by setting the distance to the closest 26 neighbours to 1. The 3-4-5 metric, where the distance to the face-, the edge-, and the point-neighbours is set to 3, 4, and 5, respectively, is a reasonably good integer approximation of the Euclidean metric. Pseudo code for forward pass of the 3-4-5 reverse DT:

```
/*------------------FORWARD PASS--------------------------------*/
LOOPZ(image) {
 LOOPY(image) {
  LOOPX(image) {
   if (x<xLO || x>xHI || y<yLO || y>yHI || z<zLO || z>zHI)
   image = 0; /* image border omitted */
   else
   image = MAX(I(x-1,y-1,z-1)-5, I(x,y-1,z-1)-4, I(x+1,y-1,z-1)-5,
               I(x-1,y,z-1)-4,   I(x,y,z-1)-3,   I(x+1,y,z-1)-4,
               I(x-1,y+1,z-1)-5, I(x,y+1,z-1)-4, I(x+1,y+1,z-1)-5,
               I(x-1,y-1,z)-4,   I(x,y-1,z)-3,   I(x+1,y-1,z)-4,
               I(x-1,y,z)-3,     I)
  }
 }
}
/*------------------END--FORWARD PASS--------------------------*/
```

Extending the Euclidean DT to 3D when running on a sequential computer requires a large amount of calculation and data storage; it is quite possible though, if desired. Even a parallel algorithm uses four sub-iterations in the image with large masks [5]. Also, the reverse Euclidean DT is not simply the reversal of the Euclidean DT. Unexpected difficulties occur, that can be overcome, but requires special care in designing the algorithm already in 2D, see [4]. The 3-4-5 DT should be a fair approximation in many cases.

3   Synthesis Algorithm

The two-step algorithm is very straightforward.

First of all, seed labels are placed in the image, randomly or manually. Placing and labeling the seeds randomly are done according to an appropriate probability distribution. The seed label equals the radius of the corresponding "sphere".

In the second step a reverse DT of some metric is applied. The metric used decides what the associated "sphere" looks like. In 3D, the D^6, the D^{26}, and the 3-4-5 reverse DT generate octahedrons, cubes, and 24-sided polyhedrons, respectively.

It might be tricky to place the seeds manually and really obtaining the planned object, as there are so many degrees of freedom. How close the seed labels are positioned decides how "grown together" the spheres are. An isolated seed label with a low value generates a single "sphere". If the seed labels are placed along a line an elongated object is generated, e. g., a block when using the D^{26}.

If the obtained 3D object has internal cavities, not apparent from simply viewing the object, they may bias quantitative measurements. But as the process of placing the seeds may be done in a manual manner, this situation can be avoided by a careful planning of the labelling and positioning of the seeds. To be certain that there are no internal cavities, a 3D connected component labelling [7] can be applied to the object and the background.

4   Examples

The shown examples are 3D objects synthesised in $128 \times 128 \times 128$ images. For the visualization we are using $ANALYZE^{TM}$, developed at the Mayo Foundation Clinic, USA [6].

Fig. 1 shows a cluster of five partially fused "spheres" with different radii obtained by using the D^6, the D^{26}, and the 3-4-5 reverse DT, respectively. By choosing the seed positions and labels manually a (more or less) planned object is created. In the three cases the same seed labels have been used, except for a rescaling of the labels for the D^6 image.

Another way of creating an object, is to place seeds of a special or a random label at random in the image, and then performing the reverse DT. In Fig. 2 randomly chosen seed labels are placed at random positions and the 3-4-5 reverse

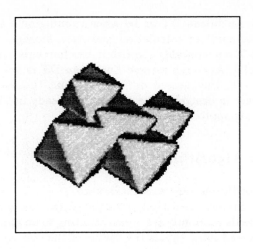

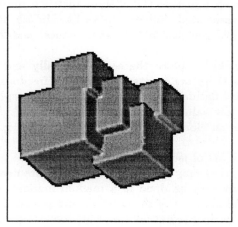

Fig. 1. A cluster of partially fused "spheres". The metrics are D^6 (top), D^{26} (middle), and 3-4-5 (bottom).

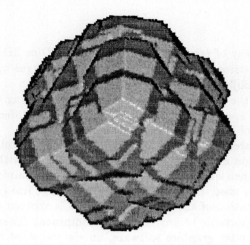

Fig. 2. A random 3D "blob" object.

DT is used. The positions and labels are chosen according to the rectangular probability distribution.

The method can also be used to obtain volume images suitable for scene analysis. A "blocks world" is generated with the D^{26} reverse DT on some seed labels. Volume rendering techniques may then produce 2D projections in desired directions. An example is shown in Fig. 3. Range images may also be obtained from the volume images.

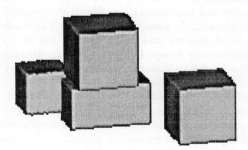

Fig. 3. A blocks world generated with the D^{26} reverse DT on a few seed labels.

5 Conclusion

We have found a way of creating useful objects in 2D or 3D by using a reverse DT in different metrics. After some practice, seed images can be created that generate complex objects and scenes.

At our Centre the method has been useful when testing tools for 3D shape quantification, i. e. approximation of the convex hull and different skeletonisation methods. Various objects have been synthesised.

An additional area where "known" objects may be useful is during development of visualization algorithms. You know what the algorithm should produce and if it does not, then something is wrong. There are certainly more areas where this method can be used, i. e. scene analysis and range imaging.

The implementation of the described method is written in C, so the image synthesis is easily portable to any computer equipment. There is really no need for expensive computer graphics software, unless you want to do some visualisations.

6 Acknowledgments

A special thanks to Dr. Gabriella Sanniti di Baja, who came up with the idea of using "small bunches of grapes" as almost convex test objects, when we were working on a 3D concavity filling algorithm.

References

1. Arcelli, C., Sanniti di Baja, G.: Finding local maxima in a pseudo-Euclidean distance transform. Computer Vision, Graphics, and Image Processing. **48** (1988) 361–367
2. Borgefors, G.: Distance transformations in arbitrary dimensions. Computer Vision, Graphics, and Image Processing. **27** (1984) 321–345
3. Borgefors, G., Hartmann, T., Tanimoto, S. L.: Parallel distance transforms on pyramid machines: theory and implementation. Signal Processing. **21** (1990) 61–86
4. Borgefors, G., Ragnemalm, I., Sanniti di Baja, G.: The Euclidean distance transform: finding the local maxima and reconstructing the shape. Proceedings of 7th Scandinavian Conference on Image Analysis, Aalborg, Denmark. (1991) 974–981
5. Ragnemalm, I.: The Euclidean distance transform in arbitrary dimensions. Pattern Recognition Letters. **14** (1993) 883–888
6. Robb, R. A., Barillot, C.: Interactive display and analysis of 3-D medical images. IEEE Transactions on Medical Imaging **8(3)** (1989) 217–226
7. Thurfjell L., Bengtsson E., Nordin B.: A new three-dimensional connected component labelling and object feature extraction algorithm. Computer Vision, Graphics, and Image Processing **54** (1992) 357–364

Texture Segmentation Using Local Phase Differences in Gabor Filtered Images

Anne M. LANDRAUD and Suk OH YUM

Laboratoire L3I, Faculté des Sciences, Av. Marillac, 17000-La Rochelle, France

Abstract. The present study concerns the phase information in the complex image resulting from the application of frequency-and-orientation-selective Gabor filters. The image phase is made up of a global linear part and a local phase. A strong variation of the phase occurs in the image when a boundary between two different regions is crossed. Two methods using local phase information for segmenting textures have been recently advanced. In the first one, the local phase is obtained by unwrapping the phase of the filtered image and by computing the global linear component. The second method derives the local phase from the filtered image without a phase unwrapping. Both methods are rapid but cannot be implemented with filters other than horizontal and vertical ones. Another method, which uses the phase-gradient information and allows filtering in any direction, achieves a more accurate segmentation. This new method does not require phase unwrapping since the phase gradient is obtained from the convolution of the input image with the filter gradient. We propose a method, both robust and rapid, which is a combination of the first and the third approaches, allowing use of filters with any orientation.

1 The phase components in a Gabor filtered image

The transformation which is supposed to be locally performed by cortical visual cells is the following. The Fourier transform $I_0(u,v)$ of the input image $i_0(x,y)$ is filtered by a function $G(u,v)$ giving the simple product $I(u,v)$. The filter $G(u,v)$, chosen here, is the Fourier transform of a complex Gabor function. The inverse Fourier transform of $I(u,v)$ is the output image $i(x,y)$ obtained by convolving $i_0(x,y)$ and the point spread function $g(x,y)$ of the filter in the spatial domain. The output image is a complex-valued function. This fact allows us to take into account two important parameters: the modulus and the phase. The complex output image can be written:

$$i(x,y) = i_0(x,y) * g(x,y) = Re(x,y) + jIm(x,y), \qquad (1)$$

where $*$ indicates convolution and $j = \sqrt{-1}$, or:

$$i(x,y) = |i_0(x,y)| \exp[\varphi(x,y)]. \qquad (2)$$

where $\phi(x)$ is the phase. The phase of the image representation (1) can be found by:

$$\varphi_{p.v.}(x,y) = \arctan[Im(x,y)/Re(x,y)]. \qquad (3)$$

The phase $\varphi_{p.v.}(x,y)$, obtained by Eq.(3), is always expressed, modulo π, with a value comprised between $-\pi/2$ and $\pi/2$. It is the so-called principal value of the phase. Thus, there are discontinuities in $\pm\pi$. In order to estimate the phase values from $-\pi$ to

$+\pi$, using $\varphi_{p.v.}(x,y)$, we take into account the numerator and denominator signs in formula (4). The four possibilities, corresponding to the four quadrants, are:

$$\arctan(a/b) = \varphi_{p.v.}(x,y) \qquad \text{if } a \geq 0, \ b > 0 \qquad (4\text{ a})$$
$$\arctan(a/b) = \pi - \varphi_{p.v.}(x,y) \qquad \text{if } a \geq 0, \ b < 0 \qquad (4\text{ b})$$
$$\arctan(a/b) = \pi + \varphi_{p.v.}(x,y) \qquad \text{if } a < 0, \ b < 0 \qquad (4\text{ c})$$
$$\arctan(a/b) = 2\pi - \varphi_{p.v.}(x,y) \qquad \text{if } a < 0, \ b > 0. \qquad (4\text{ d})$$

On the other hand, the "true value" of the phase $\varphi(x,y)$, as represented in Eq.(2), is comprised between $-\infty$ and $+\infty$. We have to change it into a continuous function by a procedure which is called the phase unwrapping. A solution of the unwrapping problem is generally to add or to subtract 2π from the part of the function which lies just after a discontinuity:

$$\varphi(x,y) = \varphi_{p.v.}(x,y) + 2\pi n(x,y) \qquad (5)$$

where $n(x,y)$ is given integer values.

This algorithm requires only the knowledge of the principal value of the phase but it does not allow fluctuations greater than π between different samples. Thus, if the phase difference between two samples is less than $-\pi$ (greater than $+\pi$), we just add 2π (subtract 2π) to the phase of the first sample [1]. From the point of view of information volume, the phase unwrapping is strictly equivalent to the determination of the $n(x,y)$ factor which is necessary to get a continuous phase as in Eq.(5).

The position of a point (x,y) can be inferred from the phase of the filtered image at that point. The knowledge of the phase also allows us to estimate the distance which separates a point from another one and the amount of shift between two or several points. It has been shown [1] that the general form of the unwrapped phase is made of a global linear phase component and a local phase component. The former is a periodic one, while the latter corresponds to noise acting on the periodical part. For texture analysis, the local phase component bears an important information about the place where two different regions meet and it plays a very important role for segmenting images of textures [1-3]. The local phase is theoretically computed by subtracting the linear component from the unwrapped phase. However this would assume a solution exists to the "zero-point" problems - discussed below.

2. Other methods of texture-image segmentation using the information contained in the phase

When the phase is computed from Eq.(3), it is assumed that the numerator and denominator values are not simultaneously equal to zero. A point such that $Re(x,y)=Im(x,y)=0$, is called a zero point. It has been shown that images of the real world contain a great number of zero points which confound the application of the phase-unwrapping procedure.

The first stage of an algorithm proposed by Nicoulin [1] consists in locating the zero points in the filtered image by integrating the phase gradient over a closed contour having a minimum length, made of four neighbouring samples which are on the corners of a square: (x,y), $(x+1,y)$, $(x+1,y+1)$ and $(x,y+1)$. The zero points with the corresponding discontinuities $\pm 2\pi$ are then shifted to the edges of the image. The second stage consists in unwrapping the phase to get $\varphi(x,y)$, while the local phase $\Delta\varphi(x,y)$ is obtained during the third stage by:

$$\Delta\varphi(x,y) = \varphi(x,y) - \varphi_{lin}(x,y), \qquad (6)$$

where the linear component $\varphi_{lin}(x,y)$ is mathematically represented by an inclined plane, the orientation of the maximum gradient of $\varphi_{lin}(x,y)$ being the same as the orientation of the filter. This method allows rapid computation of the unwrapped phase. However, the algorithm cannot be applied if the filtered image includes too many zero points. Moreover, the linear component cannot be computed with other filters than horizontal or vertical ones.

Another method, proposed by Du Buf [2], avoids the problem of zero points because it computes the local phase directly from the output image (2) without unwrapping the phase. The local phase can be computed even in the presence of a great number of zero points. This method is also very fast since it requires only one image convolution, but the oriented filters ($0<\theta<\pi/2$, $\pi/2<\theta<\pi$) cannot be used.

Bovik et al. [4] use information concerning both modulus and phase for segmenting textures. They compute a function of the phase gradient directly from the input image previously convolved with the filter gradient. This method avoids the problem of zero points. Another advantage is that the phase gradient can be obtained with oriented filters. But, at least five image convolutions have to be computed for each channel, i.e. for a frequency-orientation pair. Other researchers [5-7] have recently investigated the phase information in computer vision.

3. A new method for segmenting texture images by using the phase derivative with oriented filters

It is well known that the phase value varies abruptly at places where different textures meet. Therefore, if we use an adequate filtering process, we expect a stable phase everywhere except for the boundaries of textures. In principle, the phase gradient is more sensitive to a local variation than the local phase itself and the derivative of the phase will take an absolute value much greater on a boundary than in the other places. A straightforward method would be to estimate the derivative by means of simple differences between the phases of two neighbouring points. But, with that technique, we risk of getting a very bad approximation of the phase derivative for two reasons: (1) we have to know the unwrapped phase as a continuous function - that is not a simple problem in the presence of many zero points - and (2) an awkward noise is associated with that operation. In this section, we propose a more precise method based on some mathematical developments. Let us recall the classical formula:

$$\frac{d}{dx}\left[\text{arctg}(x)\right] = \frac{1}{1+x^2}. \tag{7}$$

We suppose that a derivative of the phase exists everywhere. Under this assumption it can be shown by simple algebra that the phase derivative with respect to x at any point (x, y) is:

$$\varphi_x(x,y) = \frac{\partial}{\partial x}\left[\varphi(x,y)\right] = \frac{\text{Im}\left[i^*(x,y)i_x(x,y)\right]}{|i(x,y)|^2}, \tag{8}$$

where $|i(x,y)|$ and $i^*(x,y)$ are the modulus and the conjugate complex quantity, respectively, of the complex filtered output image $i(x, y)$ and $i_x(x,y)$ is the derivative with respect to x of the complex output image:

$$i_x(x,y) = \frac{\partial}{\partial x}\text{Re}(x,y) + j\frac{\partial}{\partial x}\text{Im}(x,y). \tag{9}$$

The phase derivative with respect to y is computed in a similar manner. Notice that the above formulas, which describe the phase derivatives, are computed along two directions: the horizontal and the vertical ones. For any other direction, corresponding to an angle θ at which the filter is applied, we can readily deduce the derivative:

$$D_\theta[\varphi(x,y)] = \cos\theta \frac{\partial}{\partial x}\varphi(x,y) + \sin\theta \frac{\partial}{\partial y}\varphi(x,y) = \varphi_x(x,y)\cos\theta + \varphi_y(x,y)\sin\theta$$

$$= \frac{1}{|i(x,y)|^2}\left\{ \text{Im}\left[i^*(x,y)i_x(x,y)\right]\cos\theta + \text{Im}\left[i^*(x,y)i_y(x,y)\right]\sin\theta \right\}. \quad (10)$$

The feature $D_\theta[\varphi(x,y)]$ thus provides information for extracting contours of texture images. However as $D_\theta[\varphi(x,y)]$ is arbitrar negative or positive, we have used its absolute value. This measure allows us to clearly distinguish the points where that value is high from the other points where it is practically constant. Before computing the final derived phase, we perform another stage which consists in subtracting the mean of the derived phase. Indeed, Nicoulin [1] has shown that the significant phase information intended to segment an image is contained in the local phase $\varphi_{loc}(x,y)$. This is obtained by subtracting the linear phase $\varphi_{lin}(x,y)$ from $\varphi(x,y)$:

$$\varphi_{loc}(x,y) = \varphi(x,y) - \varphi_{lin}(x,y), \quad (11)$$

where:

$$\varphi_{lin}(x,y) = f_x x + f_y y, \quad (12)$$

f_x and f_y being two constants to be determined.

The derivative of the local phase is then:

$$\frac{\partial}{\partial x}\varphi_{loc}(x,y) = \frac{\partial}{\partial x}\varphi(x,y) - \frac{\partial}{\partial x}\varphi_{lin}(x,y) = \varphi_x(x,y) - f_x. \quad (13)$$

The values of f_x and f_y are chosen so that the greatest oscillations of the local phase be located principally near the contour, in other words so that we obtain high values on the contours and approximately zero values elsewhere. We estimate f_x as the mean of $\varphi_x(x,y)$.

When we apply a multichannel Gabor filtering, we have to integrate the texture characteristics which come from each channel. To do so, we determine the maximum value of the feature-vector components at each pixel, as we list in more details below:

. $D_k(x,y)$ is the value of the derived phase at point (x,y), i.e. the considered feature, which is obtained by applying the kth filter. $k = 1, ..., P$, with $P = M \times N$.

. C_k is the mean of the values $D_k(x,y)$ which were computed in all points of the kth filter output.

. The P filters are applied and the quantity $D_k(x,y)-C_k$ is computed for $k=1,...,P$.

. The characteristic value attached to (x,y) is found as $\max_k[D_k(x,y)-C_k]$.

Then, a classical method for detecting contours is achieved by using an empirically determined threshold. For this purpose we use the following family of separable filters [8, 9]:

$$F_{\rho_{m_i},\theta_{m_j}}(\rho,\theta) = F\rho_{m_i}(\rho)F_{\theta_{m_j}}(\theta), \quad (14)$$

ρ being the radial frequency, θ the orientation, ρ_{m_i} the ith preferential frequency and θ_{m_j} the jth preferential orientation of the filter (i,j), $0 \le i \le M-1$, $0 \le j \le N-1$, M and N being the considered numbers of frequencies and orientations, respectively, and $P = M \times N$. $F_{\rho_{m_i}}(\rho)$ is the expression in polar coordinates of the Fourier transform (where $K_i = (2\pi\sigma_i)^{-1}$):

$$F_{u_{m_i}}(u) = \exp\left[-\frac{(u - u_{m_i})^2}{2K_i^2}\right] \tag{15}$$

of the complex Gabor function:

$$f_{u_{m_i}}(x) = \frac{1}{\sigma_i\sqrt{2\pi}}\exp\left(-\frac{x^2}{2\sigma_i^2}\right)\exp\left(j2\pi u_{m_i}x\right). \tag{16}$$

The second function of Eq.(14) is the gaussian directional filter:

$$F_{\theta_{m_j}}(\theta) = \exp\left[-\left(\theta - \theta_{m_j}\right)^2 / \left(2\sigma_\theta^2\right)\right]. \tag{17}$$

Our algorithm was tested on both natural and artificial textures. Fig.1 shows an example of segmentation in the case of an image made of two parts of the same texture which were lightly shifted. Other examples can be seen in Fig.2, 3 and 4. In some places, especially in Fig.2, the phase value is close to the background and we get broken edges. If we lessen the threshold value, the noise is increased. A solution might be to calculate the phase more accurately while increasing computation time.

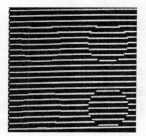 

Fig. 1. An example of texture image and the result of the segmentation process using phase information. The frequency of the filter is 22 u_{m_0} and the orientation: 90°.

Fig. 2. Another example of artificial texture image containing circles with different shiftings and the result. The frequency of the filter is 10 u_{m_0} and the orientation: 90°.

452

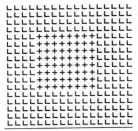
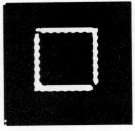

Fig. 3. An example ofsynthetic image made of "L" and "+" and the result of the segmentation process. The frequency of the filter is 24 u_{m_0} and the orientation: 135°.

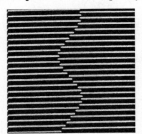
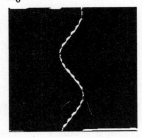

Fig. 4. An artificial texture image showing an illusory sinuous contour and the result of the segmentation process. The frequency of the filter is 21 u_{m_0} and the orientation: 90°.

References

1. A. Nicoulin, "Analyse d'images par spectre local de phase", Presses Polytechniques et Universitaires Romandes (Mars 1990).

2. J.M.H. Du Buf, "Gabor phase in texture discrimination", Signal Processing, Vol. 21 (janvier 1990), pp. 221-240.

3. J.M.H. Du Buf and P. Heitkämper, "Texture features based on Gabor phase", Signal Processing, Vol. 23 (1991), pp. 225-244.

4. A.C. Bovik, M. Clark and W.S. Geisler, "Multichannel texture analysis using localized spatial filters", IEEE Transactions on Pattern Analysis and Machine Intelligence, Vol. 12 (janvier 1990), pp. 53-73.

5. K. Langley, T.J. Atherton, R.G. Wilson and M.H.E. Lacombe, "Vertical and horizontal disparities from phase", Proc. of 1st E.C.C.V., Antibes, New York, Springer-Verlag, 1990, pp. 315-325.

6. K. Langley, D.J. Fleet and T. Atherton, "Multiple motions from instantaneous frequency", Proc. of IEEE CVPR, Champain, Il., 1992, pp. 846-849.

7. D.J. Fleet and A.D. Jepson, "Stability of phase information", I.E.E.E. P.A.M.I., vol. 15, 1993,pp. 1253-1267.

8. A.M. Landraud, "Vision-based model of artificial texture perception", SPI/SPSE Symposium on Electronic Imaging Science & Technology, San Jose, Ca, USA (February 24 - March 1, 1991).

9. S. Oh Yum, A.M. Landraud and G. Stamon, "Une méthode de classification des textures utilisant l'information dérivée de la phase dans un système de filtrage multicanaux", Colloque AGI'94, Poitiers (1994), pp. 349-352.

Disparity Estimation for Stereo Sequences Based on Adaptive Size Hierarchical Block Matching

Marco Accame and Francesco De Natale

University of Genova, Dept. of Biophysical and Electronic Engineering
Via Opera Pia 11a, Genova 16135 Italy

Abstract. The paper focuses on a new method for block based disparity estimation (BBDE) that is specially suited for real time coding of stereo sequences. The estimation is performed by exploiting a preliminary disparity field, obtained from the coarsest level of a multiresolution pyramid, and its successive refinements in the finer ones. A strong correction strategy for wrong estimates has been implemented by using an appropriate propagation of disparity vectors from one level to the next one and an adaptive search range. Finally, a variable-resolution disparity field is achieved stopping the propagation at an intermediate level of resolution, if a vector allows a good reconstruction quality of its block.

1 Introduction

Stereo sequences greatly enhance the effect of presence in visual communication, providing realness by the perception of the third dimension. So their integration in a variety of video applications (such as video-conferencing, video-telephony, digital TV and so on), seems very interesting. A great problem to be solved, is the definition of suitable coding strategies that reduce the large bandwidth required by stereo transmission. Usual video coding techniques, exploiting spatial and temporal redundancy of the scene, may be used to code each sequence independently. However stereovision properties permit considering even cross redundancy (i.e., the similarity between right and left), since two cross-frames result to be locally shifted of an amount called disparity, which is related to the depth of the scene.

This information can be extracted, and then exploited for an efficient crossframe predictive coding (disparity estimation and compensation). In block based approaches (BBDE, BBDC), to find out the disparity vector which will be used for the compensation, a block of pixels from the current right frame is compared with blocks that belong to the related left frame and that fall within a search window along the epipolar line [1, 2, 3]. Usually, a minimum mean square error (MSE) match criterion is used, performed with a full search within the window. In spite of its algorithmic simplicity, the method requires a high computational load that can be hardly managed on line. BBDE presents another major drawback: when the block size is great if compared to objects in the scene, in general it is impossible to achieve a good estimation, for one block may contain several objects at different depth (i.e., different disparity values). This results in false matchings that produce annoying block artifacts and evident reconstruction errors. And by reducing block sizes, the computational load is increased and compression efficiency greatly reduced. In the authors' opinion, these intrinsic limitations of fixed size BBDE can be overcome only by using adaptive criteria able to better exploit local characteristics as in [4]; moreover, they have to be used jointly with a multiresolution representation, which allows speeding up the disparity estimation process (see [5, 6] for applications to motion estimation).

In a multiresolution context, efficient strategies for propagating and updating disparity vectors permit to recovery from wrong estimates, and to increase the estimation quality, without the addition of redundant disparity field information. In the following, a BBDE technique using a multiresolution representation is described; after, a novel and efficient propagation strategy is presented, which is aimed at defining: (i) if a finer estimation of the disparity inside a block has to be performed; (ii) the value of the relative vector from which the update has to start; (iii) the adaptive searching-window size.

Results achieved by applying our algorithm to a standard test stereo sequence (aqua) are reported at the end, together with a comparative example of BBDEs.

2 Multiresolution Block-based Disparity Estimation

A multiresolution BBDE is helpful as, in general, in a scene there are objects covering different scales. Consequently, a matching criterion processing same scale objects together, succeeds in eliminating false matches and achieves a better computational efficiency. There is, indeed, just a small correction of the already processed objects at coarser scales.

For the multiresolution approach, we use a L-levels Gaussian pyramid [7], which is a set including the original image (*(0)-th* level) and its *L-1* approximations at a dyadic sequence of resolutions. At the *(L-1)-th* level (lowest resolution), a fixed size BBDE is carried out using a full search in a reduced window along the epipolar line. Each disparity vector is then mapped, on the *(L-2)-th* level, into four new vectors, each one pointing to the starting position of a more accurate estimation. The new vectors describe the disparity of the four B-sized blocks, originated by the dyadic two-dimensional decimation at this level of the B-sized block of *(L-1)-th* level. A coarse-to-fine propagation and estimation are then iterated until the *(0)-th* level is reached. Due to the corrective nature at finer levels, a much smaller searching window is needed for matching, so allowing for a considerable reduction in computational load even at these levels. If the previous vector is consistent with the exact disparity, the little correction is satisfactory, otherwise the small window cannot recover from false matching. A specific strategy is then needed, both for the choice of initial vectors and for setting adaptively the width of the updating window. Moreover, if a block covers completely an area with constant disparity and the compensation is satisfactory, it is not necessary to generate other blocks from it.

3 Propagation Strategy

The propagation strategy classifies a generic block $b_{B,L-1}(i,j)$ (size B and *(L-1)-th* level) on the basis on the reconstruction error and then:

- decides if the block needs a more accurate compensation (that is, it requires more than one vector in its inside);
- finds out the four initial vectors for the further disparity estimation phase, together with the relative search window.

In the following, a detailed explanation of the modus operandi is given.

3.1 Adaptive Size Block Generation

Each block $b_{B,L-1}(i,j)$ is formed by four sub-blocks of size $B/2$ at the same $(L-1)$-th level: $b_{B/2,L-1}(2i,2j)$, $b_{B/2,L-1}(2i+1,2j)$, $b_{B/2,L-1}(2i,2j+1)$, and $b_{B/2,L-1}(2i+1,2j+1)$. The compensation is evaluated for each one of the sub-blocks; they are compensated with the same disparity vector $d_{B,L-1}(i,j)$ of the block $b_{B,L-1}(i,j)$; if at least one of them has a bad reconstruction, the block $b_{B,L-1}(i,j)$ is split and generates four new son-blocks at the $(L-2)$-th level, where a new estimation is to be carried out for each block starting from an adequate initial vector. That is, by splitting the father-block $b_{B,L-1}(i,j)$, the four son-blocks are: $b_{B,L-2}(2i,2j)$, $b_{B,L-2}(2i+1,j)$, $b_{B,L-2}(2i,2j+1)$, $b_{B,L-2}(2i+1,2j+1)$ with the associated initial vectors $d_{B,L-2,\text{init}}(2i,2j)$, $d_{B,L-2,\text{init}}(2i+1,2j)$, $d_{B,L-2,\text{init}}(2i,2j+1)$, $d_{B,L-2,\text{init}}(2i+1,2j+1)$. Their values are set using the strategy reported in sub-paragraph 3.2.

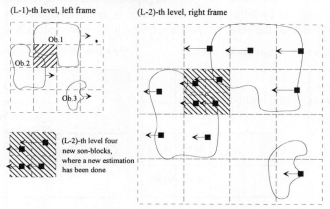

Fig.1. An example of the adaptive size BBDE.

The split and update process is iterated across levels $l\in\{L-1,\cdots,1\}$, until the reconstruction is satisfactory for every sub-block. If that is reached for e.g., the (l)-th level, then it is checked the compensation of the block $b_{2^l B,0}(i,j)$ ((0)-th level, block size $2^l B$), done with the best vector $d_{2^l B,0}(i,j)$ within the search window pointed by $d_{2^l B,0,\text{init}}(i,j)$, which is the vector $d_{B,l}(i,j)$ expanded by a factor 2^l. The level-dependent width of the window is 2^l+1 and gives subpixel accuracy for the (l)-th level (see Fig.2). If the compensation is satisfactory the block $b_{B,l}(i,j)$ at the (l)-th level is not split anymore in the propagation, and just one vector is used for its disparity description. The mechanism permits to create blocks of different dimensions, which adaptively cover the right frame, and the vector concentration is higher just where a denser disparity field is needed. This kind of block generation is equivalent to a quadtree growing process, whose root is the block $b_{B,L-1}(i,j)$. That helps to code very compactly the side information needed to place vectors on the frame. Indeed, it is just necessary to code with one bit each split decision. With this information, it is possible to reconstruct the quadtree and place the blocks on the frame with the relative vectors

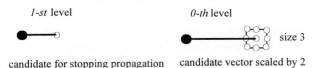

1-st level   *0-th* level

size 3

candidate for stopping propagation   candidate vector scaled by 2

Fig.2. Example of subpixel search window for testing a propagation stop hypothesis.

3.2 Vector Propagation Strategy

The propagation of a vector from the *(l)-th* is different whether the related compensation is satisfactory or not. In the first case the vector $d_{2^l B,0}(i,j)$, associated to the block $b_{2^l B,0}(i,j)$, is taken directly on the *(0)-th* level, as seen before. In the other case, four new vectors are propagated to the *(l-1)-th* level in this way (Fig.4).

- At first each initial vector takes the best scaled vector among the set of its neighbors at the *(l)-th* level (see [8], and [7] for an interesting application).
- Then, during the updating phase, the vector is again compared to its neighbor ones, but those on the same *(l-1)-th* level, which have been already updated.

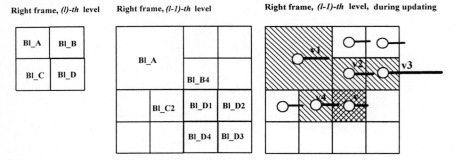

Right frame, *(l)-th* level   Right frame, *(l-1)-th* level   Right frame, *(l-1)-th* level, during updating

Fig.3. At first the initial vector *v* for Bl_D1 in the *(l-1)-th* level is chosen as the one related to the block set *{Bl_A, Bl_B, Bl_C, Bl_D}* giving the best compensation. In the next step, the already refined neighbor vectors *{v1, v2, v3, v4}* and *v* are checked to get the best one.

3.3 Adaptive Size Setting for the Search Window

The initialization of a vector onto the *(l-1)-th* level, needs to be followed by the search for the one which minimize the MSE of the compensation at that level. If the initial MSE is not too large, but above a fixed threshold (LOW_THR, see Fig.4), it is not necessary to set a large window width; on the contrary, a high MSE means that the initial estimate is not satisfactory, and a new search with large scale is needed. In this work, the width of the search window is set as a piecewise linear function of the initial MSE. The width depends even on the resolution level, so a normalization factor of 2^{-l} is applied when the search is done on the *(l)-th* level. Moreover, the window is not square and depends on the slope of the epipolar line, but the shape factor is set fixed across every level.

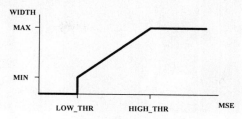

Fig.4. Normalized width function

4 Experimental Results

The results achieved by the proposed adaptive BBDE for stereoscopic video coding are satisfactory. Our approach can be used in a wide variety of application, from very low bitrates (where only the compensation scheme can be used, with a sparse field), to mid-range bitrates (in this case the field needs to be more accurate, and coding of the compensated frame difference (CFD) may be used). In any case, the adaptive BBDE performs better than the full search BBDE, in terms of computational load and better reconstruction quality, with the same number of disparity vectors. A comparison is given in Fig.5 for the stereo sequence *"aqua"* (256×256 pels).

References

1. D. H. Ballard and C. M. Brown, *Computer Vision*, Prentice-Hall, Englewood Cliffs, New Jersey 1982.
2. S. T. Barnard and M. A. Fischler, "Computational stereo", *ACM Computing Surveys*, Vol. 14, No. 4, pp. 553-572, December 1982.
3. F. Chassaing, B. Choquet and D. Pelé, "A stereoscopic television system (3D-TV) and compatible transmission on a MAC channel (3D-MAC)", *Signal Processing: Image Communication*, Vol. 4, No. 1, pp. 33-43, November 1991.
4. W. Li and E. Salari, "Efficient coding method for stereo image pairs", *Visual Communications and Image Processing '93, Proc. SPIE 2094*, pp. 1470-1476.
5. M. Bierling, "Displacement estimation by hierarchical block matching", *Visual Communications and Image Processing '88, Proc. SPIE 1001*, pp. 942-951.
6. F. Dufaux and M. Kunt, "Multigrid block matching motion estimation with an adaptive local mesh refinement", *Visual Communications and Image Processing '92, Proc. SPIE 1818*, pp. 97-109.
7. P.J. Burt and E. H. Adelson, "The Laplacian pyramid as a compact image code," *IEEE Trans. Communications*, Vol. 31, No. 4, pp. 532-540, April 1983.
8. P. Anandan, "A computational framework and an algorithm for the measurement of visual motion", *International Journal of Computer Vision*, Vol. 2, pp. 283-310, 1989.

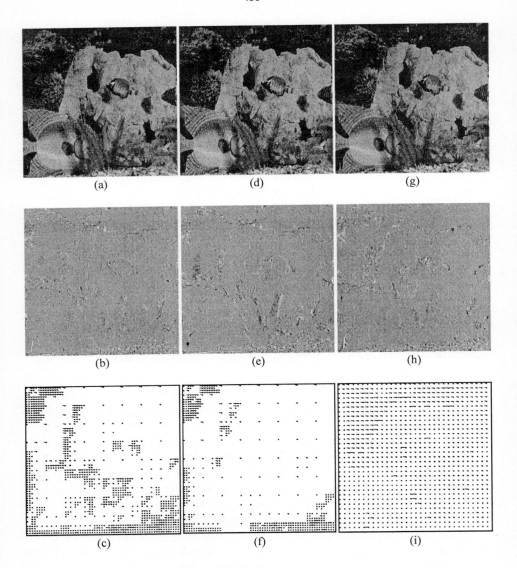

(a) (d) (g)

(b) (e) (h)

(c) (f) (i)

Fig.5. Experimental results:

(a, b, c): right frame reconstruction, CFD and needle diagram with adaptive
BBDE (B=4, MSE=198, 1024 vectors);

(d, e, f): right frame reconstruction, CFD and needle diagram with adaptive
BBDE (B=4, MSE=247, 514 vectors);

(g, h, i): right frame reconstruction, CFD and needle diagram with full-search
fixed size BBDE (B=8, MSE=248, 1024 vectors).

Matching of Stereo Curves – A Closed Form Solution

Yaonan Zhang and Jan J. Gerbrands*

Information Theory Group
Department of Electrical Engineering, Delft University of Technology
P.O. Box 5031, 2600 GA Delft, The Netherlands

Abstract. A method is described in this article to solve the stereo matching problem of general closed planar curves under possible imperfect segmentation and occlusions, provided that camera parameters are known. The method decomposes the parameters related to an object plane, i.e. slant, tilt and scale factor, and uses a histogram technique to estimate these parameters. The parameter estimation is based on the disparity information of the stereo curves. Point correspondence plays an important role in the method. We solve this problem in a dynamic programming style. The final matching is assessed by applying a distance transformation. The method has been applied successfully to several practical examples.

1 Introduction

In this paper, we present a method to solve the stereo matching problem of general closed planar curves, provided that camera parameters are known. The stereo curves are assumed to be the output of some segmentation procedures and represent the boundaries of certain objects. Usually, segmentation procedures are not perfect and occlusions may occur. That means that some portions of the stereo curves may be detected incorrectly or may be missing altogether. The general solution of this problem is not available. There are quite a number of publications dealing with curve-based matching [1–5]. Relevant surveys can also be found in [6, 7]. The existing curve matching methods have the following problems: 1) some methods can only deal with two dimensional rotation, translation, and scaling, but cannot deal with perspective transformation; 2) some methods are sensitive to noise and cannot tackle errors due to bad segmentation and occlusion; 3) some methods have strict constraints with respect to the shape of the curves concerned, etc. In this article, we develop a curve matching method which is capable of dealing with perspective transformation and of handling possible errors due to bad segmentation and occlusion without making assumptions on the shape of the curves. The assumptions we do make in this paper are: 1) the stereo curves are the perspective projections of 3-D planar curves in an object space; 2) the stereo curves are closed; 3) any curve concerned

---

* This research is partially supported by the NWO/SION project "Model Based Image Sequence Analysis"

is a simple curve (if a curve representing a region boundary stems from the output of a region growing algorithm, this requirement is alway guaranteed; if the curve is produced by other kinds of segmentation techniques, some operations may be needed in order to meet this requirement). The method we propose in this paper decomposes parameters related to an object plane, i.e. slant, tilt and scale factor, and uses a histogram technique to estimate these parameters based on disparity information. The point correspondence problem is solved in a dynamic programming style. The final matching is assessed by applying a distance transformation.

2 The Method

The merit of the overall matching between the stereo curves can be judged by the difference between the two normalized curves. By normalization in the stereo case, we mean that one curve is transferred before matching. Suppose we transfer the left curve into the right image and denote the transferred x' coordinate as x^* (note that there is no difference on y coordinates on a correctly matched point pair [2]).

We use the following formula for the geometric transformation:

$$\begin{cases} x^* = x' + S_c(f_c + \tan(\Omega)\cos(\Gamma)x' + \tan(\Omega)\sin(\Gamma)y') \\ \qquad \text{if} \quad \Omega < 45^o \\ x^* = x' + S_c'(\tan(\Omega')f_c - \cos(\Gamma)x' - \sin(\Gamma)y') \\ \qquad \text{if} \quad \Omega' \le 45^o. \end{cases} \tag{1}$$

In this equation, x', y' are x, y coordinates of points along the left curve, f_c is the principal distance of the camera, the slant $\Omega(\Omega' = 90^o - \Omega)$, tilt Γ, and scale factor S_c determine the orientation and distance of object plane corresponding to the stereo curves. The normal vector of the object plane is determined by three points P_1, P_2, P_3 or two vectors $\mathbf{V_1}$ and $\mathbf{V_2}$ on the plane. Let $\mathbf{n} = \{n_x, n_y, n_z\}$ be the normal vector, then Ω and Γ can be calculated by

$$\Omega = \arccos\left(\frac{|n_z|}{\sqrt{n_x^2 + n_y^2 + n_z^2}}\right) \tag{2}$$

$$\Gamma = \arccos\left(\frac{|n_x|}{\sqrt{n_x^2 + n_y^2}}\right). \tag{3}$$

Where

$$n_x = \{(y_1'd_3 - y_3'd_1)(d_2 - d_3)f_c - (y_2'd_3 - y_3'd_2)(d_1 - d_3)f_c\}L/d_1d_2d_3^2$$

---

[2] We assume in this article that the stereo curves are rectified.

$$n_y = \{(x'_2 d_3 - x'_3 d_2)(d_1 - d_3)f_c - \\ -(x'_1 d_3 - x'_3 d_1)(d_2 - d_3)f_c\}L/d_1 d_2 d_3^2$$

$$n_z = \{(x'_1 d_3 - x'_3 d_1)(y'_2 d_3 - y'_3 d_2) - \\ -(x'_2 d_3 - x'_3 d_2)(y'_1 d_3 - y'_3 d_1)\}L/d_1 d_2 d_3^2.$$

In the above equations, L is the length of the baseline; (x'_i, y'_i) and (x''_i, y''_i) are the photo coordinates of corresponding left and right image points on the stereo curves, respectively; and d_i is the disparity value for a pair of points over the stereo curves.

S_c is solved in the near-horizontal case ($\Omega < 45^o$) by

$$S_c = -\frac{d}{f_c + \tan(\Omega)\cos(\Gamma)x' + \tan(\Omega)\sin(\Gamma)y'} \tag{4}$$

In the case of near-vertical ($\Omega \geq 45^o$), the scale factor is defined as S'_c, and is calculated by

$$S'_c = -\frac{d}{\tan(\Omega')f_c - \cos(\Gamma)x' - \sin(\Gamma)y'} \tag{5}$$

3   The Algorithm

The first step in our algorithm is to establish the point correspondence. Based on the disparity information, we can estimate the parameters for curve normalization for each three points along the stereo curves. The most likely values for the parameters are calculated by a histogram technique. After curve normalization, the differences between the two curves are measured using a distance transformation. The details of the algorithm are described below.

3.1   Point correspondences

The aim of the point correspondence is to establish the corresponding relationship between the points on the stereo curves. The algorithm we used is developed in a dynamic-programming style. Each point on the left curve is tested against a number of candidate points on the right curve. The final matches are the unique matches which minimize the sum of the disparity values.

3.2   Weighting the curve direction

The accuracy of the disparity measurement along the curve is quite dependent on the direction of the curve at a specific position. Obviously, the disparities of near-horizontal segments are less reliable than the ones on near-vertical segments. They should be treated differently in the parameter estimation procedure.

3.3 Histogram analysis

Histogram analysis is a useful technique for image processing to measure the relative frequency of occurrence of signal values within a certain range. The technique is used here as a kind of voting mechanism to select the most populous value for the estimated parameter. In our algorithm, a hierarchical histogram is actually adopted, which means that for each parameter, several levels of histograms are used, with the number of bins at each level fixed. The calculated parameter values are first put into corresponding bins at level 1. After all values have been computed, the peak position at that level is detected. At the next level, the same parameter values are used, but the bin width is reduced. After the computations at several levels have been finished, the peak position at the final level is used to calculate the final parameter value. The reason for using this multi-level histogram is to avoid possible multiple peaks on the histogram and to get a more accurate peak position through the refinement of the levels.

3.4 Distance transformation

The final step of the algorithm uses the distance transformation and distance measurement techniques to measure the similarity between two shapes. The distance transformation converts the pixels of a curve into a gray-level image where all pixels have a value corresponding to the minimum distance to the curve.

3.5 Matching measure

In order to get a proper match measure, the searching curve is transferred into a distance image, the left normalized curve is then superimposed on this image. Defining the distance value at point j on the reference curve as D_j, the good match points are detected by thresholding the D_j, i.e.
if

$$D_j < \epsilon_d \tag{6}$$

then point j is a good match point. Here ϵ_d is the distance threshold.

Let N_c be the total number of points on the reference curve, N_c' be the total number of points having good matching decided by Eq. (6), the match measure between the stereo curves are expressed by

$$m_s = \frac{N_c'}{N_c}. \tag{7}$$

Another way of defining matching measure is to consider the average value of D_j. But in case there are some errors caused by bad matches, the average distance value becomes unreliable. The definition by Eq. (7) reflects what fraction of the reference curve has been matched reliably.

4 Experiment

We first show the results carried out on the stereo curves shown in Fig. 1 (the curves are highlighted in the figure). These stereo curves result from region-based segmentation. Given these two curves (the left curve has 149 pixels and the right one has 199), our task is to match them based on the method described in the previous sections. The result for point correspondences shows that the big errors occur along near-horizontal segments. Consequently, the weight function is calculated based on the curve direction. To estimate the parameters slant and tilt, a three-level hierarchical histogram analysis was used. At each level, 8 bins are used, the population of a certain value is put into the corresponding bin. A peak is detected and is refined to the next level where the width of bin is reduced by half. The angle ranges of slant and tilt are both 90° (estimated Γ value from the histogram is in the first quadrant, and the real quadrant Γ belongs to is decided by the signs of n_x, n_y, the finest bin on the last level of the histogram has width of $90/32 = 2.81250^o$. The width of a bin should not be too small otherwise there will be too many local peaks on the histograms. The final position of a peak can be improved by averaging the data positions within the bin of the detected peak. When calculating slant and tilt, a four-point interval is used to sample points on the disparity curve. In order to evaluate the match, the left curve is transferred into the right image by formula (1).

Another example is shown in Fig. 2. In this figure, there are some differences between the stereo curves produced by bad segmentation. From our many experiments, it follows that our method produces acceptable results.

5 Conclusion

In this paper, we described a closed form solution to the general stereo curve matching problem. Experiments have shown that the method is quite promising and is robust against imperfect segmentation and occlusion. The extension of this work is to consider other kinds of curves, for example, the curves on polynomial surfaces.

References

1. B. Bhanu and O. Faugeras, "Shape matching of two-dimensional objects," *IEEE Transactions on Pattern Analysis and Machine Intelligence*, vol. 6, pp. 137–156, 1984.
2. E. Salari and S. Balaji, "Recognition of partially occluded objects using B-spline representation," *Pattern Recognition*, vol. 24, pp. 653–660, 1991.
3. D. Sherman and S. Peleg, "Stereo by incremental matching of contours," *IEEE Transactions on Pattern Analysis and Machine Intelligence*, vol. 12, pp. 1102–1106, Nov. 1990.
4. A. Bruckstein, R. Holt, A. Netraval, and T. Richardson, "Invariant signatures for planar shape recognition under partial occlusion," *CVGIP: Image Understanding*, vol. 58, pp. 49–65, July 1993.

5. Z. Pizlo, "Recognition of planar shapes from perspective images using contour-based invariants," *CVGIP: Image Understanding*, vol. 56, pp. 330–350, 1992.

6. B. Kamgar-Parsi, A. Margalit, and A. Rosenfeld, "Matching general polygonal arcs," *CVGIP: Image Understanding*, vol. 53, pp. 227–234, 1991.

7. E. Arkin, L. Chew, D. Huttenlocher, K. Kedem, and J. Mitchell, "A efficiently computable metric for comparing polygonal shapes," *IEEE Transactions on Pattern Analysis and Machine Intelligence*, vol. 13, pp. 209–216, 1991.

(a)   (b)   (c)

Fig.1. In this example, the left curve (a) has 149 points, and the right curve (b) has 199 points; (c) shows the normalized stereo curve. The final match measure is 0.46.

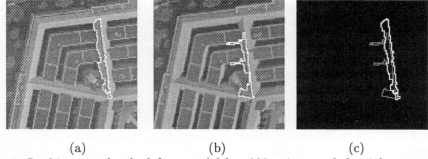

(a)   (b)   (c)

Fig.2. In this example, the left curve (a) has 223 points, and the right curve (b) has 259 points; (c) shows the normalized stereo curves. The final match measure is 0.42.

Multi-Polygonal Object Tracking

Guido Tascini, Paolo Puliti and Primo Zingaretti

Istituto di Informatica, Facoltà di Ingegneria, Università di Ancona, Italy
e-mail: zinga@anvax2.cineca.it

Abstract. In this paper we consider the problem of the recognition and tracking of moving objects from monocular sequences of images. In particular, we describe a new strategy for 2D intelligent temporal tracking of objects. The approach is feature-based and resolves the correspondence between points in two consecutive images with geometric tracking. While in literature often they are used edges or corners here we use an advanced segmented image, that may be represented with a series of polygons, whose correspondence in consecutive images constitutes the basis of the tracking. The architecture of our vision system is structured on two sub-systems: the 'Static Recognizer', which allows to verify the presence of an object in a scene, and the 'Dynamic Recognizer', which allows to track an object after the relative translation and/or rotation.

1 Introduction

Correspondence detection and motion estimation are two main topics in image sequence analysis and tracking.

The motion estimation from a series of images is a well described problem in literature [3, 5, 6]. Normally there are two types of approaches: pixel based and feature based. The *pixel based*, or flow based, methods use local changes in light intensity to compute optical flow at each image point and then derive 3D motion parameters [1]. The *feature based* methods extract the features (corners, point of curvature, lines, etc.), and use them to estimate the motion [7].

The motion correspondence problem concerns the features in two consecutive frames. Given a frame sequence and m points in each frame, the *motion correspondence* maps points between frames such that no two points of a frame map on the same point of the other frame. The problem is combinatorial and may be constrained by some assumptions as the 'proximal uniformity constraint' proposed by Rangarajan and Sah [8].

In this paper we consider the problem of the recognition and tracking of moving objects from monocular sequences of images. In particular, we describe a new strategy for 2D intelligent temporal tracking of objects. This strategy, adopted in the active vision system developed at our Institute, speeds up the whole object recognition process, and represents a fundamental step towards a real time high level symbolic scene interpretation. Tracking intelligence mainly relies on the integration of 2D motion estimation results (dynamic knowledge), updated at every frame, directly into the feature-correspondence detection-process: the use of dynamic knowledge greatly facilitates the recognition by reducing both the search space and the algorithm computational complexity.

Our approach is feature based. The polygonal representation of the regions obtained from the segmentation is used as a basis for the extraction of features, as edges, corners or whole-regions attributes, which can be easily retrieved in the successive frames, forming a correspondence chain. We adopt a *geometric matching* approach which uses parametrized geometric models for the matching of two point patterns from two consecutive images, and may be viewed as a parametrized optimization problem.

The architecture of our active vision system is structured on two sub-systems: the 'Static Recognizer', which allows to verify the presence of an object in a scene by analysing the information contained in a single frame of a sequence, and the 'Dynamic Recognizer', which allows to track an object after the relative translation and/or rotation produced by its movements in the visual field and/or by a camera moving through the environment.

The paper is organized as follows. In the next section we adduce reasons for a multi-polygonal approach to object tracking. Section 3 describes how acquired images are processed and transformed to obtain the input data for the recognizers, which are described in section 4. Results and conclusions are finally reported in section 5.

2 Polygonal Approach

As we will see, our approach to object tracking needs of well segmented images. So fundamental are the choices adopted concerning the environment, the techniques of image pre-processing and the geometric knowledge representation.

Preliminary to object tracking is object extraction. Our approach intends to resolve the problem of object extraction from images by a process of segmentation through multithresholding, and therefore represented from homogeneous regions in intensity. In particular, our vision system adopts a *viewer centred representation* both for the data image and for the models in the knowledge base. Objects are represented by a number of views (aspects) and are partitioned into regions from each of which the visual features are visible.

This choice involves the object extraction process to be characterized by data transformations in order to arrive at the process of matching among homogeneous representations. In the next section we will describe the structure and the characteristics of a vision system that supports this representation, specifying the transformations of low and intermediate level in the processes of vision.

Despite using regions as primitives at a low level knowledge-representation, we adopt a boundary description at an intermediate-level of knowledge-representation for the regions resulting from the segmentation process [4]. But, for their genesis such contours are different from those obtained through edge representations as there are neither problems of closing or thickness (we easily obtain one-pixel width contours without application of thinning algorithms or similar processing). The principal reason of such choice is related to the semantic meaning of the entities resulting from the segmentation: in the case of edge representation the segmentation process halts to a level of organization before the region representation, which, in a sense, furnishes both edges and regions.

In the edge representation the edges obtained from the segmentation process characterize the separation parts among regions with certainty, but they do not allow to characterize regions exactly; this involves a great effort of elaboration in the transformation processes that will conduct to the interpretation of the image.

In the region representation the regions obtained from the segmentation process denote an almost certain clustering of the pixels, even if the certainty decreases toward the boundaries of the regions in the cases of uniform changes of light intensity. The greater computational complexity of this process and the lower reliability of the obtained results are greatly balanced from the simplification that is obtained in the following phases of matching. Obviously the capability of integrating the two representations (data fusion) improves the throughputs of the system simplifying the high level processes.

We perform the segmentation to obtain a region representation, but we also adopt image transformations (image reduction [4]) that conduct to the revision of the primitives at intermediate level (generalized polygonals [9]) in terms of low level primitives (edges). This exploits the analogy with the perceptive abilities of the natural vision systems, which do great use of the edges and of the relations (parallelism, co-linearity, co-ending, etc.) among them.

Again, very important is the influence of the semantic content assigned to an entity in a given representation and, above all, as this semantic value is used in the transformation. For example, in the transformation from a generalized polygonal to a description of the contour through "asperities" [9], assigning the attribute of "gulf" or "peninsula" involves a semantic choice that will be held in consideration in the following processes.

3 Image Acquisition and Pre-processing

The implemented algorithms operate on images acquired from three-dimensional scenes in external environment with natural light. This involves treating noisy input data as reflections and diffusions of the natural light on the object surfaces.

The images, acquired from a camera connected to a frame grabber, have a resolution of 512*512 pixels and 256 grey levels. A real time processor performs pre-processing operations: noise removal, spatial filtering, and extraction of image characteristics as intensity histograms and profiles.

The images are segmented using a multithresholding process that selects grey-level intensity-ranges (thresholds) on the basis of local image characteristics. An analysis of these homogeneous connected components is then performed and the resulting labelled regions are 'polygonalized', that is described by a polygonal representation.

In general the polygonals obtained from the segmentation will have a correspondence n to 1 with (the sub-parts of) an object, and in particular they could be in correspondence one to one with (the sub-parts of) an object. In presence of noise, as in our case, also the correspondence n to 1 could be broken and a polygonal could correspond to more than one (sub-part of an) object.

In these situations it becomes important to characterize robust criteria for a segmentation refinement. For example, by using a geometric processor for assigning a probability of merging / splitting (to obtain the set of the regions to match with the model regions), or by using perceptual criteria for driving the grouping process.

In fig. 1 it is sketched the transformation chain from the grabbed image to the input image furnished to the recognizers.

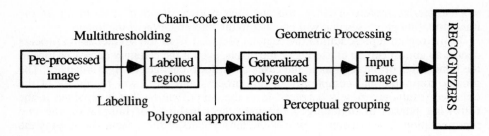

Fig. 1. Transformation chain of the image segmentation process.

4 The Recognizers

The general system architecture contains two sub-systems: the 'Static Recognizer' and the 'Dynamic Recognizer'.

Each input image sequence represents a 2D projection of a 3D scene at a given time. Then the matching among the representations of the models and the input data may happen at 2D level. Even if both the recognizers of the system operate at 2D level they have different tasks.

The 'Static Recognizer' is involved in the recognition of a scene, which will result in the selection of the models corresponding to objects which are present in the scene, and, possibly, if they are static or dynamic objects.

The task of the 'Dynamic Recognizer' is the tracking of a 'known' object in a sequence of frames, and involves both the detection of correspondences between successive frames and the motion estimation of the polygonals (objects). The 'Dynamic Recognizer' use a linear model of motion which will be briefly described.

4.1 The 'Static Recognizer'

The 'Static Recognizer' is activated in the initial frame and in any other frame where a new ('unknown') object appears in the scene or the 'Dynamic Recognizer' has failed in tracking a ('known') object.

At present state of implementation, it is responsibility of the user to furnish the 'Static Recognizer' with the models it should recognize in the actual frame. Then, starting from the input image resulting from the segmentation process, and chosen a distinctive polygon (*guide-polygon*) belonging to the 2D model of the selected object, the 'Static Recognizer' performs a search for similar polygons on the image. The criterion used for establishing the similarity among polygons has been based on the algorithm of Arkin et al. [2]. This search results in a set of capabilities that will be reviewed by considering the knowledge on the complete model. In this phase there are further couplings among polygons (domain dependent grouping), derived both from the model and the input data, which confirm the local matching.

It follows a matching phase which furnishes the scaling, rotation, and translation parameters that allow to superimpose the model to the image data. The aim of this phase is to improve the solution previously found using a simple iterative algorithm.

4.2 The Linear Model of Motion

The motion information is inferred directly from the sequence of images. The object motion is extracted from two successive frames making possible to anticipate the object position in the next frame. The changes of position and of appearance of the regions are due to the combined effect of the relative motion of the object respect to the image plane and of the projection performed by the camera. We use a linear affine model for the motion estimation.

In general we can assume that a region R in the image at time t+1 is the result of an affine transformation $f(\phi, u)$ of the image at time t, where $\phi(t)$ is the rotation matrix and $u(t)$ is the displacement vector. The points of R at time t+1 are located at $(x(t+1), y(t+1))$ and satisfy the relation:

$$\begin{pmatrix} x \\ y \end{pmatrix} (t+1) = \phi(t) \begin{pmatrix} x \\ y \end{pmatrix} (t) + u(t)$$

Let p the position of a point of the object in the frame at time t and p' the correspondent position in the frame at time t+1; we could write: $p' = f(p)$ where f is a vectorial linear function. Then to estimate the polygon modifications it will be used the following linear relation, which is characterized from 6 independent parameters (4 for the S matrix and 2 for the u vector):

$$p' = S\,p + u \qquad \text{where} \quad S = \begin{bmatrix} S_{11} S_{12} \\ S_{21} S_{22} \end{bmatrix} \quad \text{and} \quad u = \begin{bmatrix} u_x \\ u_y \end{bmatrix}.$$

4.3 The 'Dynamic Recognizer'

The dynamic analysis of the frames is directed to moving objects, which are hypothesized as previously detected in the static phase and then must be tracked in the dynamic phase.

The 'Dynamic Recognizer' is constituted by three modules: 'Correspondence Detector', 'Motion Estimator' and 'Tracking Controller'.

'Correspondence Detector'. The 'Correspondence Detector' analyses the polygons in the ROIs of two consecutive frames i and i+1, and formulates a set of hypothezes about plausible correspondences. The correspondence reliability will be evaluated in terms of indexes (*translation index, index of dimension, index of shape*, etc.) combined in the aggregate *index of plausibility*.

'Motion Estimator'. The 'Motion Estimator', using a linear affine model, evaluates the motion parameters by an iterative matching process, giving the values of the 6 parameters that minimize a certain index of disagreement (*mismatch index*) between the corresponding polygons. This index is constituted by the total area of the two polygons that does not belong to the zone of overlap, after their barycentres has been superposed. The calculus of the geometric intersections among polygons is performed by a 'Geometric Processor'.

'Tracking Controller'. The 'Tracking Controller' is the module supervising the complete tracking operation. It uses the results obtained at frame i from the 'Motion

Estimator' for a correct positioning and sizing of the ROI on which the 'Correspondence Detector' will perform the search. Finally it decides for a missing or successful recognition, respectively invoking the 'Static Recognizer' or giving the polygons in the frame i+1 that should be tracked in the i+2 frame.

5 Results and Conclusions

The tracking of multi-polygonal objects was tested on different real frame sequences. A statistical analysis has not been performed yet. Anyway the observations have confirmed as the region representation by generalized polygonals allows an efficient tracking. In particular, our region based tracking allows a very easy matching process, a more accurate estimation of motion parameter (assigned to a dense region) than in optical-flow point based methods, and a good tolerance to noise and partial occlusions.

As the frame acquisition speed, while the 'Static Recognizer' does not require the usual constraints of small motion and smooth intensity profile (it only requires well segmented images), the tracking strategy ('Dynamic Recognizer') requires no too high rate, sufficient to keep limited the changes between two consecutive frames.

Furthermore, this strategy may efficiently be implemented on parallel machines, assigning the tracking of each object part (or multiple target objects) to independent processes and the overall control to another process.

6 References

1. Adiv G., "Determining Three-Dimensional Motion and Structure from Optical Flow Generated by Several Moving Objects", *IEEE Trans on PAMI*, vol. 7, no. 4, pp. 384-401, 1985.
2. Arkin E.M., Chew L.P., Huttenlocher D.P., Kedem K., Mitchell J.S.B., "An Efficiently Computable Metric for Comparing Polygonal Shapes", *IEEE Trans. on PAMI*, vol. 13, no. 3, pp. 209-216, 1991.
3. Broida T.J., Chellappa R, "Estimation of object motion parameters from noisy images", *IEEE Trans. on PAMI*, vol. 8, no. 1, pp. 90-99, 1986.
4. Giretti A., Puliti P., Tascini G., Zingaretti P., "Segmented Image Reduction", Intelligent Perceptual Systems, V.Roberto ed., Lectures Notes in Artificial Intelligence, Springer-Verlag, pp. 181-194, 1993.
5. Jain R., "Segmentation of frame sequences obtained by a moving observer", *IEEE Transaction on PAMI*, vol. 6, pp. 624-629, 1984.
6. Kalivas D.S., Sawchuk A.A., "A Region Matching Motion Estimation Algorithm", *CVGIP*, vol. 54, no. 2, pp. 275-288, 1991.
7. Lee C.H., "Interpreting image curve from multiframes", *Artificial Intelligence*, vol. 53, pp. 145-163, 1988.
8. Rangarajan K., Shah M., "Establishing Motion Correspondence", *CVGIP: Image Understanding*, vol. 54, no. 1, pp. 56-73, 1991.
9. Tascini G., Puliti P., Zingaretti P., "Segmentation Suggested Geometric Scheme", *Proc. SPIE 1832*, pp. 280-286, 1992.

Optimal Parameter Estimation of Ellipses

Yuntao Cui[1], John Weng[1] and Herbert Reynolds[2]

[1] Computer Science Department, Michigan State University,
East Lansing, MI 48823, USA
[2] Ergonomics Research Lab, 742 Merrill Street,
Lansing, MI 48912, USA

Abstract. In this paper, we propose an unbiased minimum variance estimator to estimate the parameters of an ellipse. The objective of the optimization is to compute a minimum variance estimator. The experimental results show the dramatic improvement over existed weighted least sum of squares approach especially when the ellipse is occluded.

1 Introduction

The methods of estimation of the parameters of quadratic curve can be classified into two categories, the least squares curve fitting [1, 5, 7, 8], and the Kalman filtering techniques [3, 6]. The general quadratic curve can be written as follows:

$$Q(X, Y) = aX^2 + bXY + cY^2 + dX + eY + f = 0, \qquad (1)$$

with $b^2 < 4ac$ corresponding to the ellipses. Suppose that points $(x_i, y_i), i = 1, 2, ..., n$ are the detected elliptical points, then the least sum of squares fitting method finds the ellipse parameters (a, b, c, d, e, f) by minimizing following objective function:

$$\sum_{i=1}^{n} \varepsilon_i = \sum_{i=1}^{n} (ax_i^2 + bx_iy_i + cy_i^2 + dx_i + ey_i + f). \qquad (2)$$

The data points have a non-uniform contribution to the above objective function [1]. In order to achieve better performance, the weighted least squares approach has been used [7]. However, the optimal weights would be highly involved, complex and computationally expensive.

Kalman filtering is a sequential technique in the sense that the observation data are sequentially fed into the algorithm, and new estimates are recursively computed from previous estimates and the current new observation. The performance of the sequential techniques is relatively poor for nonlinear problems. Typically, the Kalman filter requires many data points to converge to an acceptable solution.

---

* This work has been supported by Delphi Interior & Lighting System, General Motor Corporation

In this paper, we propose an optimal unbiased minimum variance estimator using the objective function based on the normal distance of a data point to the ellipse to estimate parameters of the ellipse. The error function is non-linear. We use a parameter space decomposition technique to reduce the computation costs.

2 Estimation Criteria

Suppose that an observation vector y is related to a parameter vector m by an equation $y = Am + \delta_y$, where δ_y is a random vector with zero mean, $E\delta_y = 0$, and covariance matrix $\Gamma_y = E\delta_y \delta_y^t$. The unbiased, minimum variance estimator of m (i.e., that minimizes $E\|\hat{m} - m\|$) is also the one that minimizes

$$(y - Am)^t \, \Gamma_y^{-1}(y - Am). \tag{3}$$

(see, e.g.,[4]). The resulting estimator is

$$\hat{m} = (A^t \, \Gamma_y^{-1} A)^{-1} A^t \, \Gamma_y^{-1} y \tag{4}$$

with an error covariance matrix

$$\Gamma_{\hat{m}} = E(\hat{m} - m)(\hat{m} - m)^t = (A^t \Gamma_y^{-1} A)^{-1}. \tag{5}$$

With a nonlinear problem, the observation equation becomes

$$y = f(m) + \delta_y \tag{6}$$

where $f(m)$ is a nonlinear function. As an extension from the linear model, we minimize

$$(y - f(m))^t \, \Gamma_y^{-1}(y - f(m)). \tag{7}$$

In other words, the optimal parameter vector m is the one that minimizes the matrix-weighted discrepancy between the computed observation $f(m)$ and the actual observation y. At the solution that minimize (3), the estimated $\hat{m}$ has a covariance matrix

$$\Gamma_{\hat{m}} = E(\hat{m} - m)(\hat{m} - m)^t$$
$$\simeq \{\frac{\partial f(\hat{m})^t}{\partial m}\Gamma_y^{-1}\frac{\partial f(\hat{m})}{\partial m}\}^{-1}. \tag{8}$$

One of the advantages of this minimum variance criterion is that we do not need to know the exact noise distribution, which is very difficult to obtain in most applications. The above discussion does not require knowledge of more than second-order statistics of the noise distribution, which often, in practice, can be estimated.

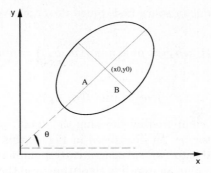

Fig. 1. Illustraion of five ellipse parameters (x_0, y_0, A, B, θ).

3   The Objective Function

The ellipse can be represented by the following equation:

$$\frac{(x\cos\theta + y\sin\theta - x_0\cos\theta - y_0\sin\theta)^2}{A^2} + \frac{(-x\sin\theta + y\cos\theta + x_0\sin\theta - y_0\cos\theta)^2}{B^2} = 1,$$

$$(9)$$

with $m = (A, B, \theta, x_0, y_0)$ as five free parameters. Fig. 1 illustrates the meaning of the parameters. Let $u_{i,j}$ denote the jth component of 2D vectors of the ith point ($j = 1$ for X coordinate and $j = 2$ for Y coordinate). Let the true two-dimensional position of the ith point P_i be defined by parameter α_i, where $P_i(\alpha_i) = (A\cos\alpha_i + x_0, B\sin\alpha_i + y_0)$, and the collection of all such parameters be $\alpha_\bullet$. The set of direct observation pairs consists of all noise corrupted versions $\hat{u}_{i,\bullet} = (x_i, y_i)$ of u_i with i from 1 to n. Suppose that the noises between different observation points are uncorrelated and that the correlation between the errors in different components of the image coordinates is negligible. Without loss of generality, we also assume the same error variance in the different components of image points, which has the same variance σ^2. According to the criteria discussed in the above subsection, the objective function to minimize is

$$f(m, \alpha_\bullet) = \sum_{i=1}^{n} \sum_{j=1}^{2} \sigma^{-2} \|u_{i,j}(m, \alpha_i) - \hat{u}_{i,j}\|^2 \qquad (10)$$

where $u_{i,j}(m, \alpha_\bullet)$ is the noise-free projection computed from m and $\alpha_\bullet$.

4   Minimizing the Objective Function

The objective function in equation (10) is neither linear nor quadratic in m and $\alpha_\bullet$ and an iterative algorithm is required to get a solution: m and $\alpha_\bullet$. Instead of performing a computationally expensive direct optimization, we reduce the dimension of the parameter space first. Since the objective functions are continuous, we have

$$\min_{m,\alpha_\bullet} f(m, \alpha_\bullet) = \min_m \{\min_{\alpha_\bullet} f(m, \alpha_\bullet)\} = \min_m g(m) \tag{11}$$

where

$$g(m) = \min_{\alpha_\bullet} f(m, \alpha_\bullet) \tag{12}$$

is the smallest "cost" computed by choosing the "best" points $\alpha_\bullet$, with a given ellipse parameter vector m. As illustrated in Fig. 2, this means that the space $(m, \alpha_\bullet)$ is decomposed into two subspaces corresponding to m and $\alpha_\bullet$, respectively. In the subspace of m, an iterative algorithm (e.g.,the Levenberg-Marguardt method or conjugate gradient method) is used. The subspace $\alpha_\bullet$ can be further decomposed. Each individual α_i can be computed noniteratively for any given m. According to the decomposition shown in equation (12), the search space in $\min_m g(m)$ is just five-dimensional.

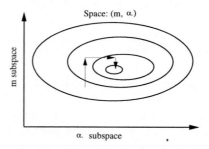

Fig. 2. Decomposition of parameter space. Iterative algorithm is used only in a small subspace corresponding to m. Given each m, the best α_i is computed indiviually. Since the dimension of $\alpha_\bullet$ is very large (n-dimensional with n points), this decomposition significantly reduces computational cost.

Now we consider how to compute the best α_i given m. The problem can be defined as follows: Given a point $P = (x_i, y_i)$ and an ellipse with parameters $m = (A, B, \theta, x_0, y_0)$, find the point P' on the ellipse which minimizes the distance $\|PP'\|$. For the details of computing P', the reader is referred to [2].

5   Simulation Results

The simulation experiments carry out a comparative study of three different estimation approaches, which are the least sum of squares fitting approach, Safaee-Rad etc.'s weighted least sum of squares fitting approach [7] and our unbiased minimum variance estimator. The measure for "Goodness" of fit is defined as "the sum of normal distances of all the data points to the optimal ellipse", which is considered as an objective and independent measurement [7].

A perfect ellipse is defined by given the ellipse parameters, (A, B, θ, x_0, y_0). The two-dimensional edge points were generated randomly for each trial. The noise was simulated as the digitization error. Three kinds of experiments were conducted during the simulation. The first set of experimental data was based on the edge points which are uniformly distributed on the entire curve of the ellipse. In the second and third set of experiments, the given edge points only covered one-half or one-quarter of the ellipse respectively. During each set of experiments, different numbers of data points were used.

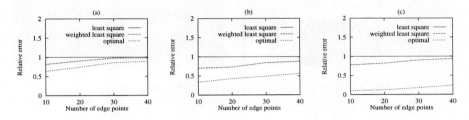

Fig. 3. Relative error of the sum of normal distances of all the data points to the optimal ellipse. (a) Edge points are uniformly distributed on the entire ellipse, (b) Edge points are uniformly distributed on the half of the ellipse, (c) Edge points are uniformly distributed on the quarter of the ellipse.

Fig. 3 shows the average error of the sum of normal distances of all the data points to the optimal ellipse based on 100 trials. The error is plotted as a relative ratio, where the error of the least sum of squares fitting is treated as the base. Fig. 3(c) presents the simulation results when the edge points are coming from one quarter of the ellipse. The improvement of the optimal approach is dramatic. This is because that least squares fitting is statistically biased, the estimation results of these approaches based on the edge points from partial ellipse are far from accurate.

6 Experiments with Real Images

The optimal unbiased minimum variance estimation approach has been applied to a project investigating posture in automobile seats. The body posture of the human driver is estimated by recovering the three dimensional structure of natural body feature points using multiple calibrated cameras. The body feature points are marked with round targets, which appear to be ellipses due to perspective projection. The approach proposed by Wu and Wang [8] was used to detect the boundary points.

After the boundary detection, the optimal estimation approach was used to obtain (x_0, y_0, θ, A, B). Fig. 4 shows the results of two input images. The results are quite good despite the facts: 1) very noisy boundary since the nonplanar feature of the surface of the human body makes the surface of the tag no longer planar; 2) preprocessing errors, such as quantization errors, the edge detection errors and boundary detection errors.

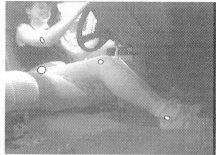

Fig. 4. Detected ellipses are highlighted with dark curves.

7 Conclusions

In this paper, we propose a new unbiased minimum variance estimator for the ellipse parameters (x_0, y_0, θ, A, B). The comparative study shows that this approach can achieve much better performance than least sum of squares fitting and weighted least sum of squares fitting especially when the edge points are coming from the partial boundary of the ellipse and the number of the edge points is relatively small.

References

1. F.L. Bookstein, "Fitting conic sections to scatter data", in *Comput. Graphics Image Processing*, vol. 9, pp. 56-71, 1979.
2. Y. Cui, J. Weng and H. Reynolds, "Optimal parameter estimation of ellipses", Tech. Report, Ergo-95-001, Ergonomics Research Lab, 1995.
3. T. Ellis, A. Abbood and B. Brillault, "Ellipse detection and matching with uncertainty", in *Image Vision Comput.*, vol 10, no. 5, pp.271-276, June, 1992.
4. D.G. Luenberger, *Optimization by Vector Space Methods*, New York: Wiley, 1969.
5. Y. Nakagawa and A. Rosenfeld, "A note on polygonal and elliptical approximation of mechanical parts, in *Pattern Recogn.*", vol. 11, pp.133-142, 1979.
6. J. Porrill, "Fitting ellipses and predicting confidence envelopes using a bias corrected Kalman filter", in *Image Vision Comput.*, vol. 8, no. 1, pp. 37-41, 1990.
7. R. Safaee-Rad, I. Tchoukanov, B. Benhabib, and K. C. Smith, "Accurate parameter estimation of quadratic curves from grey-level images", in *CVGIP: Image Understanding*, vol. 19, no. 8, pp. 532-548, 1992.
8. W. Wu and M. Wang, "Elliptical object detection by using its geometric properties", in *Pattern Recognition*, vol. 26, no. 10, pp. 1499-1509, 1993

A Linear Discriminator of Width

José A.F. Leite* and Edwin R. Hancock

University of York,Computer Science Department
YO1 5DD,York,England

Abstract. Our aim in this paper is to present a methodology for linearly combining multi-channel filter information, using principal components analysis. We take as our application vehicle the detection and classification according to width of line-features. Here we operate on the raw data with oriented multi-scale line detection kernels. The objective in this study is to develop a linear discriminator of line width which can be used to classify the detected features. By performing principal components analysis on the response of the multi-scale filter bank to a set of example line-profiles, we transform the filter-outputs. Under this transformation, the principal component of maximum variance is demonstrated to have linear dependance upon line-width. We demonstrate the utility of our line-classification process under conditions of controlled noise.

1 Introduction

Over the past decade, the rapidly emergent paradigm of multichannel analysis has proved to be an overwhelmingly powerful technique for extracting information concerning the properties of intensity features [1, 5, 6, 7]. Basic to the approach is the idea that intensity structure can be represented by the output of a filter bank that has channels specialized to the selection of different orientation [1], scale [5] or symmetry [6, 7]. Examples include the use of oriented quadrature filter pairs to detect lines or edges [1, 7] and the use of Gaussian wavelets for multiscale edge detection [5].

Whatever the nature of the filter-bank being used to characterise the different types of intensity structure, a framework is required for combining the individual channel responses and making decisions concerning the identity of the underlying image features. In the quadrature filtering approach [6] this is done by adding the squares of the even and odd symmetry filter responses to construct a local energy measure and using their relative phase [1] to separately label edge and line artifacts. In the wavelet approach the multiscale persistence of features can be modelled by the use of Lipschitz exponents [5] or by coupled membranes [4]. Adopting a statistical standpoint, Leite and Hancock [7] have shown how the Mahalanobis distance can be used as a fusional device for integrating together the information represented by a multiscale filter-bank for subsequent processing by

* This work is being sponsored in part by the Brazilian Research Council(CNPq)

non-linear relaxation operations. Taking these ideas one stage further, Hancock [2] has demonstrated that by adopting a phase representation of sine and cosine-phase S-Gabor filter responses, edges and lines can be simultaneously detected and discriminated using relaxation operations.

Central to each of the channel combination models described above is the requirement for highly non-linear processing operations. Although these existing methods are certainly effective, they may be regarded as somewhat premature in their rejection of simpler linear processing operations.

Our aim in this paper is to describe how the channel responses can be combined in a linear way using principal component analysis [8] to discriminate between different classes of intensity features. Although we have selected the task of discriminating line-features of differing width to demonstrate the utility of our ideas, the framework is potentially applicable to a multitude of feature interpretation problems. In essence, principal components analysis allows the multichannel feature-vector to be linearly transformed so that it is alligned with the direction of maximum variance for a set of training examples. Components of insignificant variance may then be discarded. In our line-discrimination example, the training patterns are provided by idealised examples of known width, while the filter bank consists of second-derivative of Gaussian operators of varying scale. Here the principal component of maximum variance is shown to be linearly dependant on line-width. Moreover, the resulting width discrimination process is shown to be robust to severe levels of noise.

The outline of this paper is as follows. In Sect. 2 we describe the ingredients of our multi-scale filtering approach. Section 3 describes the statistical framework and principal components analysis. Experiments are presented in Sect. 4 and conclusions are offered in Sect. 5.

2 Multichannel Filtering

We are interested in identifying a filter basis consisting of oriented even symmetry kernels that can be used to characterise directional line features at a variety of different spatial scales. There are several alternatives available to us including directional second derivatives of the Gaussian and the cosine phase S-Gabor functions [2]. However, since our main concern here is proof of concept, we will not be interested in optimal choice of filter basis and we therefore choose to work with the second-derivative of Gaussian due to its parametric simplicity. The kernel of scale s and orientation θ is as follows

$$L_{\theta,s}(x,y) = \left(1 - \frac{(x\cos\theta + y\sin\theta)^2}{s^2}\right)\exp\left[-\frac{x^2 + y^2}{2s^2}\right] \qquad (1)$$

It is readily verifiable that this filter is of zero d.c. bias, i.e. that the result of the double integration $\int_{-\infty}^{\infty}\int_{-\infty}^{\infty} L_{\theta,s}(x,y)dx.dy$ is equal to zero.

The channel responses are obtained in the usual way, by convolving multiscale filter kernels with the available image data $I(x,y)$

$$R_{\theta,s}(x,y) = \int_{-\infty}^{\infty}\int_{-\infty}^{\infty} L_{\theta,s}(u,v)I(x-u,y-v)du.dv \qquad (2)$$

We are interested in constructing feature-vectors from the output of oriented multiscale filters which form a dyadic basis. This is in line with the physiology of natural vision systems, which are known to employ three different scales each separated by one octave. At each of the orientation states of the filter basis the filter responses of different scale are stacked to form the following feature vector for the lines in the image data

$$\mathbf{F}_\theta(x,y) = \begin{pmatrix} R_{\theta,s}(x,y) \\ R_{\theta,2s}(x,y) \\ R_{\theta,4s}(x,y) \end{pmatrix} \tag{3}$$

In the work reported here we use two orientation states aligned along the vertical and horizontal axes of the pixel lattice. The filter stacks of different orientation can be amalgamated to give the combined multiscale feature-vector $\mathbf{W}(x,y)^T = (\mathbf{F}_0(x,y) \ \mathbf{F}_{\frac{\pi}{2}}(x,y))^T$.

3 Principal Components Analysis

In order to establish a set of training examples for principal components analysis, we construct a series of idealised line-profiles of known width w. For each such profile, we compute the set of responses for the distinct kernels of the filter bank and such responses are stacked to form a normalised pattern-vector $\mathbf{U}_w$.

Our basic aim in performing principal components analysis on the output of the multichannel filter bank, is to identify the linear transformation of the individual filter reponses that has maximum variance with respect to the set of example line-profiles [8]. Underpinning this analysis is the observation that if $F_{\theta,s}(x,y)$ is a random-vector with partially correlated components, then we can perform a linear transformation of the individual instances into a new coordinate space. The axes of this new feature-space are the eigenvectors of the covariance matrix for the example profiles. The associated eigenvalues are the fractions of total variance residing in the corresponding components of the transformed vectors. Since the different components of the transformed feature-vector are naturally graded according to their degree of intrinsic variation, they may be systematically examined to determine their dependance on line-width.

The basic computational ingredient of principal components analysis is the covariance matrix for the components of the feature-vector evaluated over the set of example profiles. If $\mu = E[\mathbf{U}_w]$ is the mean feature vector, then the covariance matrix is equal to $\Sigma = E[(\mathbf{U}_w - \mu)(\mathbf{U}_w - \mu)^T]$.

The principal components transformation is obtained by solving the eigenvalue equation $|\Sigma - \lambda I| = 0$, where I is the 3×3 identity matrix. The eigenvalues λ_1, λ_2 and λ_3 are the variances of the components of the transformed feature-vectors for the example line profiles. Associated with each of these three eigenvalues is an eigenvector $\mathbf{V}_\lambda$ whose components satisfy the system of linear equations given by $\Sigma \mathbf{V}_\lambda = \lambda \mathbf{V}_\lambda$.

These eigenvectors are the axes of a new orthonormal co-ordinate system. Specifically, the three eigenvectors are the columns of the transformation matrix between the original feature-vectors and the principal components representation. If we denote the transformation matrix by $\Theta = (\mathbf{V}_{\lambda_1}, \mathbf{V}_{\lambda_2}, \mathbf{V}_{\lambda_3})$, then the transformed feature-vectors are

$$\begin{pmatrix} C_{\theta,1}(x,y) \\ C_{\theta,2}(x,y) \\ C_{\theta,3}(x,y) \end{pmatrix} = \Theta^T . \mathbf{F}_\theta(x,y) - \boldsymbol{\mu} \tag{4}$$

Figure 1 shows the three principal components extracted from the example line-profiles as a function of line-width. The striking feature of these plots is the near linear dependance of the first principal component upon line-width. Superimposed on the extracted values of the first principal component is the result of a simple linear fit. It is this linear interpolation of the principal component data that we use as our discriminator of line-width. The fraction of variance residing in this leading component is 96%. By contrast, in the second principal component, the associated eigenvalue accounts for less than 4% of the total variance of the example line profiles. The fractional variance in the third component is negligible.

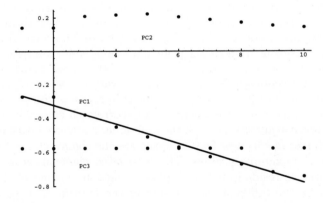

Fig. 1. First,second and third principal components

4 Experiments

In order to demonstrate the effectiveness of our width discrimination technique under conditions of controlled noise, we have generated a series of synthetic images. Each image consists of 7 concentric circles with widths varying from 2 to 8 pixels. The inner circle has a radius of 30 pixels and for the others, this values is increased by 20 pixels plus the width. The grey-level of the circles is 157 while that of the background is 57. To each image we have added Gaussian noise of zero mean and fixed variance. Figure 2 shows the results of applying our technique to noise-free data and in Fig 3 the same results to the same image in which the signal-to-noise ratio (SNR) is 25:10. From top to bottom and from left to right, the subimages show the original picture, the combined line-map and the six width-maps for $w \in \{2, 3, 4, 5, 6, 7\}$ pixels.

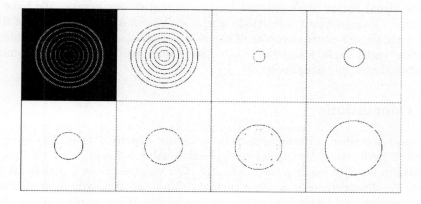

Fig. 2. Width discrimination on a noise-free image

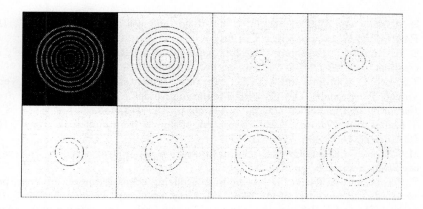

Fig. 3. Width discrimination on an SNR = 25:10 image

These results are very encouraging. Under conditions of zero noise the discrimination is almost perfect. At intermediate noise levels when SNR=25:6, some

75.85% of pixels are correctly classified according to width; the degree of contaminating error comes only from the two neighbouring width-bins. At the highest noise level when SNR=25:10, although the fraction of contamination has increased to 36.78%, the bulk of the classification errors are still only a single width bin away from the ideal result.

We have chosen to use filters of three different scales (i.e. $s = 1$, $s = 2$ and $s = 4$) on the basis of physiological evidence that suggests that the *channels* in the visual cortex are separated by approximately one octave. We have also repeated the experiments described above with both a larger number of filters and with a more dense coverage of the range of scale. Although the results are in some cases slightly superior, their use is not justifiable on the grounds of additional computational cost.

5   Conclusions

We have described a linear approach to texture discrimination. The technique relies on performing a principal components analysis of a series of example line profiles to identify linear combinations of filter response from a multiscale filter bank. We demonstrate that the principal component of maximum variance has an approximately linear dependance of line-width. When applied to the classification of variable-width line structure subjected to controlled levels of noise, the resulting discriminator is demonstrated to operate robustly provided that the SNR does not fall below 25:10.

References

1. E. Adelson and W.T. Freeman, "The design and use of steerable filters", *IEEE PAMI*, **PAMI 13**, pp. 891–906. 1991.
2. E. R. Hancock, "Resolving Edge-line Ambiguities by Relaxation Labelling", *Proceedings of IEEE CVPR Conference*, pp. 300-306, 1993.
3. T.S. Lee, D Mumford and A Yuille, "Texture Segmentation by Minimising Vector-Valued Functionals: The Coupled Membrane Model", *Proceedings of the Second European Conference on Computer Vision*, pp 165–173, 1992.
4. S. Mallat and S Zhong, "Characterisation of signals from multiscale edges", *IEEE PAMI*, **PAMI 14**, pp. 710–732. 1992.
5. M C Morrone and R Owens, "Feature detection using local energy", *Pattern Recognition Letters*, **6**, pp 303–313, 1987.
6. P. Perona and J. Malik, "Detecting and localizing edges composed of steps, peaks and roofs", *Proceedings of the Third Int. Conf. on Comp. Vision*, pp. 52–57, 1990.
7. J.A.F.Leite and E.R.Hancock, "Statistically combining and refining multi-channel information", *Progress in Image Analysis III: Editor S.Impedovo*, pp. 193–200, 1993.
8. P. Devijver and J.Kittler, "Pattern Recognition: A Statistical Approach", *Prentice-Hall*, 1982.

Binarization of Inhomogeneously Illuminated Images

Volodymyr V. Kindratenko, Boris A. Treiger, Piet J.M. Van Espen

Department of Chemistry, University of Antwerp (UIA),
Universiteitsplein 1, B-2610 Antwerpen-Wilrijk, Belgium

Abstract. Most grey level thresholding methods produce good results only when the illumination of the image is homogeneous. An automatic binarization technique suitable for images containing regions of different brightness is presented. It does not use image's grey level histogram as a source of information. Rather, the method is based on the transformation of the raster image in such a way that the transformed image can be easy thresholded. A comparison with histogram based threshold selection technique is given.

1 Introduction

Image analysis often involves the discrimination between "objects" and the background. Grey level thresholding (also known as binarization) technique has been proven a suitable tool for this purpose. Various approaches for the automatic selection of a threshold have been proposed. A survey of them can be found in [1].

In general, thresholding can be defined as mapping the grey scale into the binary set $\{0, 1\}$:

$$S_{xy} = \begin{cases} 0, & if \quad g_{xy} < T_{xy} \\ 1, & if \quad g_{xy} \geq T_{xy} \end{cases} \tag{1}$$

where S_{xy} is the value of the segmented image, g_{xy} is the grey level of the pixel and T_{xy} is the threshold value at the coordinates (x, y). In the simplest cases T_{xy} is coordinate independent and a constant for the whole image. It can be selected, for instance, on the base of the analysis of the grey level histogram (GLH). When the GLH has two pronounced maximum, which reflect grey levels of object(s) and background, it is possible to select a single threshold for the whole image. However, sometimes GLH has only one maximum. This can be caused, e.g., by inhomogeneous illumination of various regions of the image. In such cases it is impossible to select a single thresholding value for the whole image and local binarization technique [2] must be applied. General methods to solve the problem of binarization of inhomogeneously illuminated images, however, are not available.

The proposed technique, which works well in case of inhomogeneously illuminated images (similar to given in Figs 1a and 4a), is based on the application of discrete convolution filtering technique that produces a transformed image, which can be easy thresholded. The threshold value T_{xy} in equation (1) for the transformed image is coordinate independent and equals to 1.

2 Discrete Convolution Filtering Technique

The descrete convolution of image f with mask h produces an image g:

$$g(x, y) = \sum_{i=-v}^{v} \sum_{j=-u}^{u} h(i, j) * f(x+i, \quad y+j)$$

(2)

where h is a $2v+1$ by $2u+1$ matrix. The output $g(x, y)$ at a point (x, y) is given by a weighted sum of input pixels around (x, y) where the weights are given by the $h(i, j)$. Equation (2) can be implemented by a series of shift-multiply-sum operations. The values of h are referred to as the filter kernel [3]. The effect of convolution on the original image depends on the type of filter kernel used. It can have a smoothing effect (e.g., weighted mean filter), an edge enhancing effect (e.g., the Laplacian added back to the original image), etc. [4]. We use the descrete convolution technique to produce a "binarization" effect.

3 Description of Algorithm

The main idea of the proposed technique, as it was mentioned above, is to transform the grey level image in such a way that it will be possible to select a single threshold value T_{xy} for the whole transformed image. The operation of transformation of the grey level image can be represented as the descrete convolution of grey level image with the filter kernel of a special form. This filter kernel is constructed according to the following rule: the elements of the kernel are equal to -1 if they belong to the first or last rows or the first or last columns, -2 if they are not defined yet and belong to the second or last but one rows or the second or last but one columns, and so on. The central element of the kernel equals minus the sum of all elements of the kernel plus an additional parameter p. We used filter kernels of size 5 by 5 and 7 by 7. The filter kernel of the size 3 by 3 can be used, but results are slightly worse. Increasing of the size of the filter kernel does not lead to much better results and only increases computational time. Two examples of such kernels are given in Figs 2a and 5a.

The problem remains, however, how to define the parameter p. The higher it is, the more details will be lost. With small p a lot of details in the image may not be enhanced. With $p = 0$ the proposed filter produces edge enhancement. For decision making about a suitable numerical value of parameter p, the coefficient of correlation between original and convoluted images was used. It is calculated as follows:

$$r = \frac{Cov(X, \ Y)}{\sqrt{Var(X) * Var(Y)}}$$

(3)

where X and Y are the original grey level image and the image convolved using a value p for the parameter, respectively. As shown in Figs 2b and 5b, the correlation coefficient reaches a maximum at a certain p. The parameter p corresponding to the maximal correlation can be used as a suitable one for the current image.

4 Results

Figs 1a and 4a represent two images of the surface of material with a cell structure. Both images have inhomogeneously illuminated regions. The upper left corners of both images are much darker than the centres. The grey level histograms of both images (Figs 1b and 4b, respectively) have no pronounced maxima related to "objects" and/or background. A single threshold value selection for the whole image does not allow to obtain a binary image (see Figs 3a and 6a, respectively). The proposed technique gives binary images shown in Figs 3b and 6b. Dark spots can be observed on the binarized images. They appear as a result of noise and can be removed by noise filtering techniques [4].

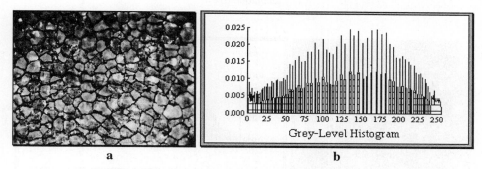

a   b

Fig. 1. Original image (a) and its grey level histogram (b).

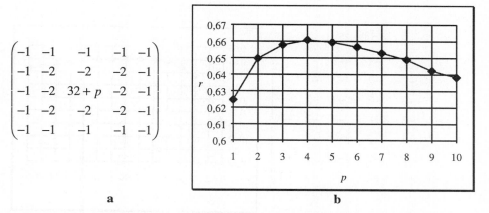

$$\begin{pmatrix} -1 & -1 & -1 & -1 & -1 \\ -1 & -2 & -2 & -2 & -1 \\ -1 & -2 & 32+p & -2 & -1 \\ -1 & -2 & -2 & -2 & -1 \\ -1 & -1 & -1 & -1 & -1 \end{pmatrix}$$

a   b

Fig. 2. Convolution kernel (a) and graph with the coefficient of correlation between original (Fig. 1a) and convoluted (with kernel from Fig. 2a for $p=1,2,...,10$) images vs. parameter p (b).

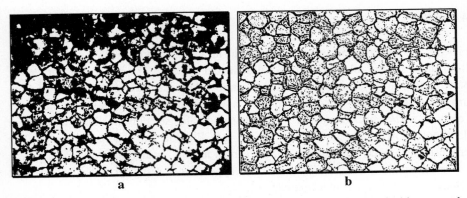

Fig. 3. Image from Fig. 1a binarized based on grey level histogram [5] (a) and with proposed technique (b).

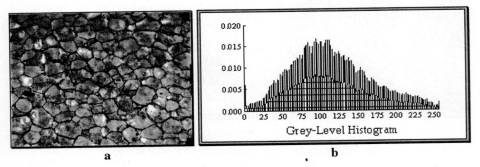

Fig. 4. Original image (a) and its grey level histogram (b).

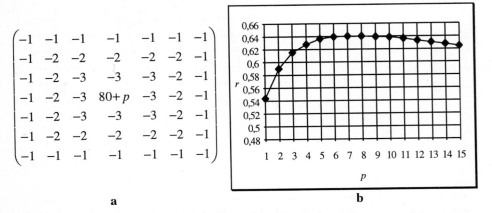

Fig. 5. Convolution kernel (a) and graph with the coefficient of correlation between original (Fig. 4a) and convoluted (with kernel from Fig. 5a for $p=1,2,...,15$) images vs. parameter p (b).

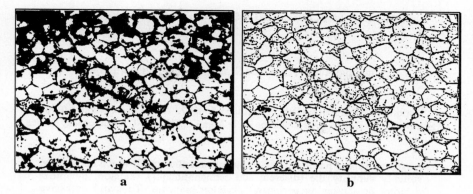

a   b

Fig. 6. Image from Fig. 4a binarized based on grey level histogram [5] (a) and with proposed technique (b).

5 Conclusion

A new binarization method based on discrete convolution filtering is developed. The algorithm is relatively simple to implement with low computational cost.

Examples of its application for binarization of images with inhomogeneous illumination are given. A comparison with histogram based threshold selection method is presented.

References

1. P.K. Sahoo, S. Soltani, A.K.C. Wong: A survey of thresholding techniques. Computer Vision, Graphics, and Image Processing 41, 233-260 (1988)
2. J.R. Parker: Gray level thresholding in badly illuminated images. IEEE Transactions on Pattern Analysis and Machine Intelligence 13, 813-819 (1991)
3. W. Niblack: An Introduction to Digital Image Processing. Englewood Cliffs, N.J.: Prentice-Hall International 1986
4. Anil K. Jain: Fundamentals of Digital Image Processing. London: Prentice-Hall International 1989
5. A.D. Brink: Gray-level thresholding of images using a correlation criterion. Pattern Recognition Letters 9, 335-341 (1989)

Pruning Discrete and Semicontinuous Skeletons

Dominique Attali[+], Gabriella Sanniti di Baja[*], Edouard Thiel[+]

[*]Istituto di Cibernetica, CNR, Via Toiano 6, 80072 Arco Felice, Naples, Italy
[+]Equipe TIMC-IMAG, IAB-Domaine de la Merci, 38706 La Tronche cedex, France

Abstract. In this paper pruning techniques are illustrated, which allow us to suitably simplify the (discrete and semicontinuous) skeleton, by either deleting or shortening peripheral skeleton branches. To avoid excessive shortening, which might reduce the representative power of the skeleton, the relevance of the figure regions mapped in the skeleton branches is used to decide on pruning. Different definitions of relevance are introduced and features allowing the quantitative evaluation of the relevance are suggested.

1 Introduction

The skeleton is a stick-like representation of a figure, which accounts for different figure properties. It is a curvilinear set consisting of branches and, in case of multiply connected figures, of loops. Each skeleton component is placed in the medial regions of a figure subset, and is oriented along the directions of the main symmetry axes of the corresponding figure subset. Each element of the skeleton can be interpreted as the centre of a disc fitting the figure, and is labelled with the corresponding radius; thus, the length of a skeleton branch gives an evaluation of the elongation of the represented figure subset and the labels of the skeleton elements provide a measure of the local thickness of the figure.

The literature includes a relevant number of papers dealing with skeletonization. Most of them refer to the computation of discrete skeletons. More recently, algorithms using the Voronoi graph have become of interest to compute semicontinuous skeletons. In fact the computation cost of the Voronoi graph is no longer prohibitive; moreover vertices approximating the figure contour in the continuous plane are often available, which can be used directly to guide skeletonization without performing any shape digitisation. A problem affecting both discrete and continuous skeletons is the presence of a number of peripheral skeleton branches, originated in correspondence with figure protrusions having no perceptual relevance. This makes the skeleton structure complex and limits the possibility to use the skeleton for shape analysis. This paper provides different pruning criteria for the discrete and the continuous skeleton, which allow us to eliminate unwanted branches without significantly altering the topological and representative power of the skeleton.

2 Discrete Skeleton

On the discrete plane, the identification of the skeletal pixels can be conveniently accomplished on the distance map of the figure (e.g., [1,2]). The distances most commonly used to compute the distance map are the city-block distance, the chessboard distance, the

(3,4)-weighted distance and the (5,7,11)-weighted distance. The latter two distances, introduced in [3], provide a better approximation to the Euclidean distance and allow to originate skeletons almost stable under figure rotation. On the distance map, one can identify a nearly thin set of skeletal pixels (i.e., the centres of the maximal discs, the saddle pixels, and the linking pixels). This set is then reduced to the unit wide skeleton, by employing removal operations, which are topology-and-end-point preserving.

The discrete skeleton cannot be perfectly centred within the figure, wherever the thickness of the figure is given by an even number of pixels. In these regions, the set of the centres of maximal discs is 2-pixel wide. The unit wide skeleton includes almost all the centres of the maximal discs, so that the figure can be nearly completely recovered by applying to its skeleton the reverse distance transformation. Complete recovery is not compatible with skeleton unit thickness. Each skeleton pixel is the centre of a disc fitting the shape. The discs are polygons approximating the Euclidean circle to a different extent, depending on the adopted distance function. Discs obtained by the city-block distance and chessboard distance are 4-side polygons, while those obtained via the (3,4)-weighted distance and the (5,7,11)-distance are 8-side and 16-side polygons, respectively. Maximal discs, associated with skeleton pixels sufficiently close to each other, partially overlap so that the set of the maximal discs does not provide a partition of the figure. The contour of a maximal disc and the contour of the figure share one, two or more connected subsets, each of which may include more than one pixel.

3 Semicontinuous Skeleton

Semicontinuous skeletonization does not require image digitisation. The skeleton is a graph, computed starting from a polygonal approximation of the continuous shape (provided, for instance, by segmentation methods using a deformable curve model). The vertices of the polygonal approximation in the continuous plane sample the boundary of the continuous shape, and are called the sampling points. A measure of the quality of the approximation is given by bounding the greatest distance between two neighbouring sampling points on the boundary. The more numerous those sampling points, the more accurate the approximation.

In recent papers [4-6], the skeleton of continuous shapes is approximated by using the Voronoi graph of the sampling points [7]. A partition of the polygonal shape is computed by using the Delaunay triangulation; then, the approximated skeleton is defined as the dual of this partition. For sufficiently regular shapes, due to the convergence theorem [8], the Voronoi vertices of the sampling points tend to the complete skeleton (i.e., the endoskeleton and the exoskeleton of the shape) when the sampling points tend to the shape boundary. The semicontinuous skeleton consists of the Voronoi vertices associated with the Delaunay triangles contained inside the shape, and of the straight line segments connecting the vertices. Two vertices are connected by a segment if their associated triangles are adjacent. To reconstruct the continuous shape, one can use either the Delaunay triangles associated with the Voronoi vertices, or the Delaunay discs (i.e., discs circumscribed to Delaunay triangles). Triangles provide a partition of the region of the plane enclosed by the initial polygonal approximation. In turn, the discs partially overlap. Their union tends to the continuous shape, when the density of the sampling points sufficiently increases.

4 Pruning

To avoid topology modifications, pruning always concerns with peripheral branches, i.e., branches delimited by an end point. In the discrete skeleton, an end point is a skeleton pixel having only one neighbour in the skeleton. In the semicontinuous case, an end point is a Voronoi vertex having only one neighbouring vertex in the skeleton. The second extreme of a skeleton branch is generally a branch point. In the discrete case, this is a pixel with more than two neighbours in the skeleton; in the continuous case it's a Voronoi vertex having exactly three neighbouring vertices in the skeleton. All the other pixels of a discrete skeleton branch (Voronoi vertices of a semicontinuous skeleton branch) have exactly two neighbours (neighbouring vertices) in the skeleton, and are called normal points. In the following, the term skeleton will be used to equivalently refer to the discrete or the continuous skeleton; the term element will denote a skeletal pixel or a Voronoi vertex.

The discrete skeleton [1] computed by using the (5,7,11)-weighted distance and the semicontinuous skeleton [6] of a test shape are respectively shown in Fig.1a and Fig.1b, as they result before applying any pruning. Pruning may involve either partial shortening or complete deletion of a peripheral branch. The elements of the branch are checked one after the other against a given pruning condition. Pruning is accomplished as far as the pruning condition is satisfied. Generally, the pruning condition should prevent excessive shortening of skeleton branches, as this may result in loss of skeleton representation power. Thus, pruning should be based on a measure of protrusion relevance and the only branches to be pruned are those associated with protrusions regarded as non meaningful according to the relevance measure.

When all the branches sharing a branch point are totally deleted, new peripheral branches are possibly originated in the modified skeleton. These branches can be furthermore subjected to pruning, provided that the protrusion whose relevance is evaluated is the protrusion mapped in the union of the current peripheral skeleton branch with the neighbouring, already pruned, skeleton branches.

4.1 Pruning Criteria

Branch length. The length of a peripheral branch can be computed in terms of the number of elements constituting the branch. A peripheral branch can be entirely removed if its length is below an a priori fixed threshold. This criterion has not general applicability, as the length of a noisy branch depends also on the thickness of the region from which the noisy protrusion sticks out. A length based criterion can be used to remove very short branches, say 1 or 2 elements, or for particular classes of figures (e.g., alphanumerics, where any figure is the superposition of elongated narrow strokes having constant thickness).

In the discrete case, the distribution and number of centres of maximal discs along the branch can provide some more information on the relevance of the represented protrusion. Generally, a small percentage of pixels of a noisy branch are centres of maximal discs. In the continuous skeleton, by using the Euler constant one can prove that the number of skeleton vertices is related to the number of sampling points. If the sampling points are regularly spaced, then, computing the number of vertices is equivalent to computing the length of the boundary of the protrusion associated with the branch. Pruning methods based on the computation of the length of the boundary have been proposed in [5].

Intuitively, a portion of a branch can be safely pruned if a negligible difference exists between the two regions corresponding to the entire skeleton branch and to the pruned skeleton branch, respectively. The difference in elongation or in area between the previous regions can be used to decide on pruning.

Elongation. Let r and R be the radii of the discs associated with the end point p of a peripheral skeleton branch and a more internal element q, along the same skeleton branch. Let d be the distance between the two elements p and q. The quantity $(r-R+d)$ measure the distance between the contour of the two regions, respectively associated with the entire

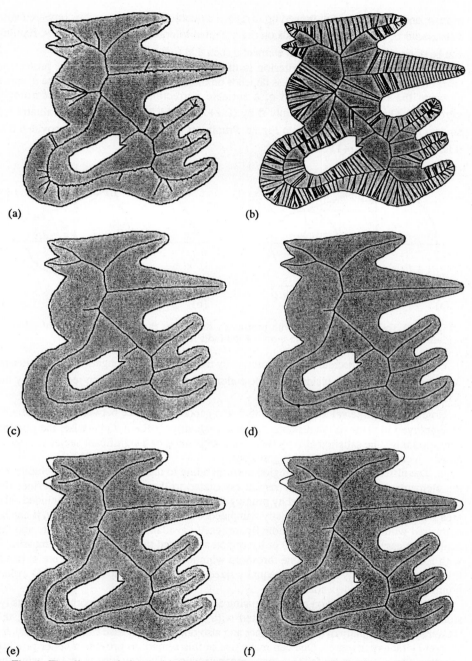

(a)

(b)

(c)

(d)

(e)

(f)

Fig. 1. The discrete skeleton, (a), and the continuous skeleton, (b), before pruning. Effect of elongation-based pruning, (c), and area-based pruning, (e), on the discrete skeleton. Effect of area-based pruning on the continuous skeleton: the protrusion area is compared with the area of a single disc in (d), and with the area of the entire shape in (f).

branch, and with the branch pruned up to q (q excluded). This value can be compared with a threshold ϑ, whose value depends on the accepted tolerance in figure recovery. Pruning can be done up to the most internal element q such that $(r-R+d) \leq \vartheta$.

A disadvantage of the above criterion is that it does not take into account protrusion sharpness. Since protrusion sharpness depends on the difference in radii and on the distance between the two elements p and q, a suitable correction factor can be accordingly introduced. Our choice is to multiply ϑ by $(R-r+1)/d$, which approximately evaluates the tangent of the angle β, as shown in Fig.2a. Pruning is accomplished up to q, provided that it is: $(r-R+d) \leq \vartheta \times (R-r+1)/d$.

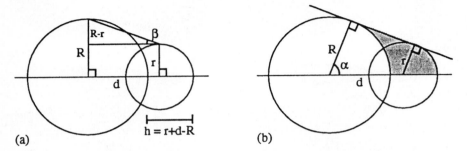

(a)   (b)

Fig. 2. (a) The angle β changes with protrusion sharpness; (b) the area of the protrusion (dark region) can be computed in terms of the radii R and r, and of the angle α.

Fig.1c shows the discrete skeleton, pruned by using the above elongation-based criterion with $\vartheta=2$. White regions inside the shape contour identify the pixels non recovered by the pruned skeleton.

Note that if a large value is assigned to ϑ, the geodesic distance between p and q should be employed in place of the distance d, when evaluating $(r-R+d)$. Otherwise, the pruning condition might be satisfied also by two elements p and q of a significant skeleton branch, along which relevant curvature changes occur.

Area. The difference between the region corresponding to the skeleton branch including all the elements from p to q, and the region associated to the element q alone, defines the protrusion that would be flattened by pruning the skeleton branch up to q, q excluded. The area P of the protrusion can be directly compared with a threshold or, preferably, it can be compared with the area F of the whole figure (easily available in both the discrete and the continuous case) or with the area D of the region associated to the element q. Comparing P with F allows one to use the same threshold whichever is the size of the figure at hand. Pruning will be equally effective on equally sized protrusions. Comparing P with D makes pruning more context dependent.

In the continuous skeleton case, protrusion area evaluation can be accomplished easily, since the skeleton vertices are associated with non overlapping triangles. In turn, the Delaunay discs partially overlap each other and also one Delaunay disc can overlap partially several Delaunay triangles. One can make use of this remark to provide a better pruning criterion as well as a more faithful shape reconstruction. The key idea is that of using the Delaunay triangles while evaluating the contributions provided by the Voronoi vertices that are going to be removed by pruning (i.e., the vertices from p to q, q excluded), and to use the Delaunay discs when evaluating the area of the region associated to q and, in general, for shape recovery. If the current protrusion mapped in the skeleton branch from p to q is significantly overlapped by the Delaunay disc associated to q, then the branch can be safely removed; a rather faithful recovery is still possible, provided that the shape is reconstructed by using the union of the Delaunay discs rather than the union of the triangles. Let P be the area obtained by adding the area of the Delaunay triangles associated with the skeleton vertices from p to q, q excluded. Let D be the area of the Delaunay disc associated to q. The branch is shortened up to q if it results: $P \leq \vartheta \times D$. Alternatively, if F denotes the area of the shape, pruning can be accomplished provided that $P \leq \vartheta \times F$. In both cases, the value of ϑ depends on the tolerance in figure recovery.

Fig.1 d,f shows the continuous skeleton after applying the area-based pruning; shape reconstruction is done by employing the Delaunay discs. Both the polygonal approximation of the initial shape and the reconstructed shape (grey region) are illustrated. In Fig.1d, the area P of the protrusion is compared with D and it is $\vartheta=0,25$; in Fig.1f, P is compared with the area of the entire shape and it is $\vartheta=0,005$.

In the discrete case, since the maximal discs partially overlap, generally the computation of the area of the protrusion is not straightforward. Only for skeletons driven by the city-block and the chessboard distances (i.e., in case of square-shaped discs), convenient algorithms have been introduced to compute the area of the union of the maximal discs [9]. For the general case of skeletons driven by weighted distances, providing more rounded discs, an approximated evaluation of the protrusion area can be computed as $(R^2-r^2) \times (\tan \alpha - \alpha)$, where α is the angle shown in Fig.2b.

As in the continuous case, the area of the protrusion can be compared with the area F of the whole figure, or the area D of the disc associated with the element q. Fig.1e, shows the effect of the area-based criterion on the discrete skeleton, when the area of the protrusion is compared with F and the threshold is $\vartheta=0,03$. As before, both the initial shape and the reconstructed shape (grey region) are illustrated.

Acknowledgements

This work has been partially supported by the "Galileo Program for the cooperation between France and Italy".

Reference

1. G. Sanniti di Baja, E. Thiel: Computing and comparing distance-driven skeletons. In: C. Arcelli et al. (eds.): Aspects of Visual Form Processing. Singapore: World Scientific, 1994, pp. 475-486.
2. E. Thiel: Les distances de chanfrein en analyse d'images: fondements et applications. PhD Thesis. Grenoble 1994. University J. Fourier.
3. G. Borgefors: Distance transformation in digital images. CVGIP, 34, 344-371 (1986).
4. J.W. Brandt, V.R. Algazi: Continuous skeleton computation by Voronoi diagram. CVGIP: Image Understanding. 55, 329-337 (1992).
5. R.L. Ogniewicz: Discrete Voronoi skeletons. Konstanz: Hartung-Gorre Verlag 1993.
6. D. Attali, A. Montanvert: Semicontinuous skeletons of 2D and 3D shapes. In C.Arcelli et al. (eds.): Aspects of Visual Form Processing. Singapore: World Scientific 1994, pp. 32-41.
7. M. Schmitt: Some examples of algorithms analysis in computational geometry by means of mathematical morphology techniques. In J.D. Boissonat et al. (eds.): Geometry and Robotics. Lecture Notes in Computer Science. Berlin: Springer-Verlag 1989, pp. 225-246.
8. F.P. Preparata, M.I. Shamos: Computational Geometry: an Introduction. Text and Monographs in Computer Science. Berlin: Springer-Verlag 1988.
9. L.P. Cordella, G. Sanniti di Baja: Geometric properties of the union of maximal neighborhoods. IEEE Trans. on PAMI, 11-2, pp. 214-217 (1989).

On Commutative Properties of Halftoning and Their Applications

Y. B. Karasik

School of Computer Science, Carleton University,
1125 Colonel By Drive, Ottawa, Canada K1S 5B6
and
Computer Science Department, Tel Aviv University, Israel
e-mail: karasik@turing.scs.carleton.ca or karasik@math.tau.ac.il

Abstract. We investigate relations between halftoning and the binary image algebra which includes set and morphological operations. We show that halftoning has important commutative properties with respect to unions, intersections, Minkowski additions and subtractions. Based on this fact a novel approach to implementation of 3-dimensional optical image processing through manipulating 2-dimensional images is proposed.

1 Introduction

The halftoning technique, which converts a gray scale image into a binary image, is an important tool for image encoding and processing.

First, it is aimed at rendering continuous tone images on devices such as laser printers where only two levels (usually black and white) can be represented at any point. In such devices, a continuous gray scale image, whose gray tone varies across the plane as $gray(x, y)$, is encoded as a discrete binary image consisting of black dots of varying sizes located at the nodes $\{(ih, jh)\}_{i,j=-N}^{N}$ of a lattice having a spacing h, so that the size of a dot at a point (x, y) is proportional to the value $gray(x, y)$. Such a halftone image when viewed at normal reading distance gives the effect of a continuous tone image.

Besides image encoding, halftoning has also turned out to be an important technique for nonlinear image processing. For example, in conjuction with a coherent optical system, it was used for analog–to–digital conversion [L], and for a variety of other nonlinear transformations on images [KG, DS].

However, there exists another powerful tool for image processing, namely: image algebra [HJS] whose connection with halftoning has not yet been properly investigated. The purpose of the present paper is to bridge this gap and to show that halftoning has important commutative properties with respect to basic set and morphological operations such as union ($\bigcup$), intersection ($\bigcap$), dilation ($+$), and erosion ($-$). It is investigated more closely in the section that follows.

2 Commutative properties of halftoning

Let $f_X(x, y)$ be the gray tone distribution of an image X across the plane. Then we assume that

$$f_{A \cup B}(x, y) = \max\{f_A(x, y), f_B(x, y)\},$$

and

$$f_{A \cap B}(x, y) = \min\{f_A(x, y), f_B(x, y)\}$$

respectively.

Based on these assumptions, we can now state the following two lemmas:

Lemma 1. *Let $HT(X)$ be the halftone representation of an image X.*
Then the following relationships hold:

$$HT(A \cup B) = HT(A) \cup HT(B);$$

$$HT(A \cap B) = HT(A) \cap HT(B).$$

Proof. Let $d[x, y, R]$ be a dot of the size R located at the point (x, y) which satisfies the relationship

$$d[x, y, R_1] + d[x, y, R_2] = d[x, y, R_1 + R_2].$$

For example, it can be a circular or rectangular dot.

Obviously,

$$d[x, y, R_1] \cup d[x, y, R_2] = d[x, y, \max\{R_1, R_2\}];$$

$$d[x, y, R_1] \cap d[x, y, R_2] = d[x, y, \min\{R_1, R_2\}].$$

Since,

$$HT(A) = \bigcup_{ij} d[ih, jh, f_A(ih, jh)],$$

and

$$HT(B) = \bigcup_{ij} d[ih, jh, f_B(ih, jh)],$$

then we have:

$$HT(A) \cup HT(B) = \bigcup_{ij} \{d[ih, jh, f_A(ih, jh)] \cup d[ih, jh, f_B(ih, jh)]\} =$$

$$\bigcup_{ij} d[ih, jh, \max\{f_A(ih, jh), f_B(ih, jh)\}].$$

Let $z = f_{A \cup B}(x, y)$ be the equation of the surface bounding $A \cup B$. Obviously,

$$f_{A \cup B}(x, y) = \begin{cases} f_A(x, y) & \text{if } f_A(x, y) \geq f_B(x, y); \\ f_B(x, y) & \text{otherwise.} \end{cases}$$

Hence,

$$f_{A \cup B}(x, y) = \max\{f_A(x, y), f_B(x, y)\}$$

and we can conclude that

$$\bigcup_{ij} d[ih, jh, \max\{f_A(ih, jh), f_B(ih, jh)\}] =$$

$$\bigcup_{ij} d[ih, jh, f_{A \cup B}(x, y)] = HT(A \cup B).$$

Thus, we obtain that

$$HT(A) \cup HT(B) = HT(A \cup B).$$

The proof that

$$HT(A \cap B) = HT(A) \cap HT(B)$$

is analogous.

Lemma 2. *Let $\tilde{\tilde{A}}$ be a step approximation of the image A, i.e. $f_{\underset{A}{\approx}}(x, y) = f_A(ih, jh)$ if $x \in (ih - h/2, ih + h/2]$ and $y \in (jh - h/2, jh + h/2]$. Then the following relationship holds:*

$$HT(\tilde{\tilde{A}} + \tilde{\tilde{B}}) = HT(\tilde{\tilde{A}}) + HT(\tilde{\tilde{B}}) = HT(A) + HT(B).$$

Proof. Due to the definitions introduced in the proof of the previous lemma, we have:

$$HT(\tilde{\tilde{A}} + \tilde{\tilde{B}}) = \bigcup_{mn} d[mh, nh, \max_{\substack{x_1 + x_2 = mh \\ y_1 + y_2 = nh}} (f_{\underset{A}{\approx}}(x_1, y_1) + f_{\underset{B}{\approx}}(x_2, y_2))].$$

Let $x_1 \in (ih - h/2, ih + h/2]$, $y_1 \in (jh - h/2, jh + h/2]$, $x_2 \in (kh - h/2, kh + h/2]$, and $y_2 \in (lh - h/2, lh + h/2]$ for some i, j, k, l. Then

$$f_{\underset{A}{\approx}}(x_1, y_1) + f_{\underset{B}{\approx}}(x_2, y_2) = f_{\underset{A}{\approx}}(ih, jh) + f_{\underset{B}{\approx}}(kh, lh).$$

Hence,

$$\max_{\substack{x_1 + x_2 = mh \\ y_1 + y_2 = nh}} (f_{\underset{A}{\approx}}(x_1, y_1) + f_{\underset{B}{\approx}}(x_2, y_2)) = \max_{\substack{i+k=m \\ j+l=n}} (f_{\underset{A}{\approx}}(ih, jh) + f_{\underset{B}{\approx}}(kh, lh)).$$

On other side,

$$HT(\tilde{\tilde{A}}) + HT(\tilde{\tilde{B}}) = \bigcup_{ij} d[ih, jh, f_{\underset{A}{\approx}}(ih, jh)] + \bigcup_{kl} d[kh, lh, f_{\underset{B}{\approx}}(kh, lh)] =$$

$$\bigcup_{ijkl} (d[[(i+k)h, (j+l)h, f_{\underset{A}{\approx}}(ih, jh) + f_{\underset{B}{\approx}}(kh, lh)]) = \bigcup_{mn} d[mh, nh, \max_{\substack{i+k=m \\ j+l=n}} (f_{\underset{A}{\approx}}(ih, jh) + f_{\underset{B}{\approx}}(kh, lh))].$$

Hence,

$$HT(\tilde{\tilde{A}}) + HT(\tilde{\tilde{B}}) = HT(\tilde{\tilde{A}} + \tilde{\tilde{B}}).$$

It follows from the lemma proved that

$$\lim_{h \to 0} HT(A + B) = \lim_{h \to 0}(HT(A) + HT(B)).$$

It can also be shown that

$$HT(\widetilde{\widetilde{A}} - \widetilde{\widetilde{B}}) = (HT(\widetilde{\widetilde{A}})^c + HT(\widetilde{\widetilde{B}}))^c,$$

provided $f_A(x,y) \geq f_B(x,y)$, where A^c means the complement of the figure A and $X - Y$ means Minkowski difference.

3   On possible applications of the commutative properties of halftoning

As is known, manipulating 3-dimensional objects is an important problem in many practical applications ranging from solid modeling in Computer Aided Design to mobile vehicle motion planning amidst obstacles and simulation of moving through virtual worlds. Such manipulations of 3-D objects are often reduced to performing set operations on them and to computing their Minkowski sums and differences (i.e. dilations and erosions respectively).

However, the latter operations are very time consuming and have, for example, $O(NM)$ time complexity even for 3-D polyhedrons in vector representation, where N and M are numbers of vertices of polyhedrons to be intersected, united, added, or subtracted [P].

That is why during the last decade there was significant interest in implementation of these operations optically because these operations on 2-D figures can be performed in constant time [CB, CAK, LKKE]. However, it was not clear how to implement these operations on 3-D figures optically.

The commutative properties of halftoning, proven above, show a way to optical implementation of set operations and Minkowski additions and subtractions on 3-D objects. Specifically, an algorithm for manipulating 3-D objects optically in constant time looks as follows:

Step 1. Encode 3-D objects A and B to be manipulated as 2-D gray scale images. Specifically, an object $A = \{(x,y,z)|z = f_A(x,y)\}$ is encoded as a planar gray scale image whose gray level at a point (x,y) is equal to $f_A(x,y)$. In what follows we denote such a gray scale representation of objects A and B as $GS(A)$ and $GS(B)$ repectively.

Step 2. Perform halftoning of $GS(A)$ and $GS(B)$ and thereby obtain images $HT(A)$ and $HT(B)$ respectively.

Step 3. Compute either $HT(A) \bigcup HT(B)$, or $HT(A) \bigcap HT(B)$, or $HT(A) \pm \overline{HT(B)}$ depending on the request, and thereby obtain either $HT(A \bigcup B)$, or $HT(A \bigcap B)$, or some approximation of $HT(A \pm B)$ respectively, so that approximation of $HT(A \pm B)$ can be obtained to within any choosen in advance degree of accuracy.

Step 4. Convert $HT(A \bigcup B)$, $HT(A \bigcap B)$, or $HT(A \pm B)$ into $GS(A \bigcup B)$, $\overline{GS(A \bigcap B)}$, or $GS(A \pm B)$ respectively and thereby obtain a planar gray scale representation of the results of the above operations on 3-D objects A and B.

Since, the gray scale representation can be converted at Step 2 into the halftone representation optically in constant time [B], and, since, the operations $\bigcup$, $\bigcap$, $+$, and $-$ on 2-dimensional images at Step 3 can be also performed optically in constant time [CB], we obtain:

Theorem 3. *Union, intersection, dilation (Minkowski sum), and erosion (Minkowski difference) of two 3-dimensional images can be (approximately) computed optically in constant time to within any degree of accuracy chosen in advance.*

The theorem just stated paves the way to implementation of the 3-D binary image algebra optically.

The commutative properties of halftoning with respect to set and morphological operations may also have other applications.

4   Conclusion

We investigated the interaction between two important tools for image processing – halftoning and the binary image algebra. We found that halftoning commutes with such basic operations of the binary image algebra as union, intersection, Minkowski sum and difference. It immediately implied the possibility of manipulating 3-D images optically in constant time.

I belive that practical consequences of this marriage of halftoning and the binary image algebra will not be too long in coming.

References

[B] O. Bryngdahl, Halftone images: spatial resolution and tone reproduction, *Journal of Optical Society of America*, **68**, pp. 416 (1978).

[CAK] A. K. Cherri, A. A. S. Awwal, M. A. Karim, Morphological transformations based on optical symbolic substitution and polarization-encoded optical shadow-casting systems, *Optics Communications*, 82(5,6), 441 – 445 (1991).

[CB] D. Casasent, E. C. Botha, Optical symbolic substitution for morphological transformations, *Applied Optics*, 27(18), 3806 – 3810 (1988).

[DS] S. R. Dashiell, A. A. Sawchuk, Nonlinear optical processing: nonmonotonic halftone cells and phase halftones, *Applied Optics*, 16(7), 1936– 1943 (1977).

[HJS] K. S. Huang, B. K. Jenkins, A. A. Sawchuk, Binary image algebra and optical cellular logic processor design, *Computer Vision, Graphics, and Image Processing*, **45**, 295–345 (1989).

[KG] H. Kato, J. W. Goodman, Nonlinear filtering in coherent optical systems through halftone screen processes, *Applied Optics*, 14(8), 1813– 1824 (1975).

[L] H. K. Liu, Coherent optical analog–to–digital conversion using a single halftone photograph, *Applied Optics*, 17(14), 2181– 2185 (1978).

[LKKE] Y. Li, A. Kostrzewski, D. H. Kim, G. Eichmann, Compact parallel real-time programmable optical morphological image processor, *Optics Letters*, 14(18), 981 − 983 (1989).

[P] F. Preparata, M. Shamos, Computational geometry: An Introduction, *Springer-Verlag*, 1985.

Pattern Recognition and Document Processing

Combining Resolution and Granularity for Pattern Recognition

P. Bottoni, L. Cinque, S. Levialdi, L. Lombardi[+], P. Mussio

Dipartimento di Scienze dell'Informazione, Universita' di Roma,
Via Salaria 113, 00198 Roma, Italy
[+] Dipartimento di Informatica e Sistemistica, Universita' di Pavia,
Via Abbiategrasso 209, 27100 Pavia, Italy

Abstract. This paper combines two approaches for shape recognition to reduce the memory needed to store the salient features of the image and the time to describe images. A first approach uses pyramidal multiresolution to provide successively condensed representations of the information of input images. The second approach describes shapes by typical characteristic patterns at different levels of abstraction granularity, corresponding to different levels of detail at which an image can be studied. Since different patterns of interest are visible in a restricted range of resolution levels, the combination of the two allows the recognition of such patterns to be performed at the minimum sufficient level, thus saving costly description of irrelevant details.

1. Introduction

This paper addresses the problem of defining the efficient strategies for recognizing objects in digital images, based on the structural description of shape.

Shape description can be considered as a first step in pictorial object recognition. It has been shown that humans can often recognize object in line drawings as quickly and accurately as in color photographs [1]. This means that recognition can be essentially achieved from the contour shape and therefore it is often unnecessary to provide the gray-level intensity, color, texture, etc. of the objects in an image. This is advantageous for a computer vision system since processing contour alone reduces complexity and increases efficiency. If recognition is to be based on contours a method is required to represent the contour shape.

A multiresolution approach, to shape detection and coding, has been proposed, based on pyramids [2]. Pyramids provide successively condensed representations of information present in input images [3]. This approach has proved useful in supporting planning strategies for edge and contour detection: a coarse-to-fine search is performed at a coarse scale to quickly locate potentially interesting regions; further search is restricted to the detected regions at a finer resolution to verify the presence of specific features.

A shortcoming of this approach is that the contour code may be influenced by noise and object rotation.

On the other hand, an approach was proposed to describe shapes by detecting typical characteristic patterns at different levels of granularity in abstraction, corresponding to the different levels of detail at which an image can be studied [4]. Descriptions at each level are based on those at the previous level, providing more and more synthetic descriptions of the same image at a given resolution level. A drawback of this approach is that several irrelevant details present in the image have however to be taken into consideration and described.

This paper combines the two techniques to derive descriptions at different levels of granularity for each level of resolution in the pyramid, so reducing the mentioned drawbacks. Since different patterns of interest are visible in a restricted range of resolution levels, the combination of the two techniques allows the search for such patterns to be performed at the minimum sufficient level, thus saving the costly description of irrelevant details. Pyramidal resolution is taken as analogous of optical resolution, granularity levels as analogous of the different levels of detail at which an image can be studied. Their combination allows a reduction of the amount of memory needed to store salient features of the image and a reduction of time in the retrieval of images with particular features. In fact, since different patterns of interest are visible in a certain range of resolution levels, the search for such patterns can be performed at the minimum sufficient level, thus saving the costly description of irrelevant details.

Applications can be envisaged in the field of retrieval of images presenting some patterns from image databases.

2.   Shape characterization through resolution levels

An image may be subdivided into subimages in a number of different ways: in particular, if quadrants are considered and this operation is recursively performed until a single pixel is reached, then a set of hierarchical levels is created each one having a power of 2 number of elements: these levels correspond to versions of the image at different resolution (thus using multiresolution systems) [2]. A pyramidal structure is particularly suited to implement this approach, each plane of the pyramid contains an image at a different resolution: higher at the base and coarser as raising towards the apex.

In order to extract the different images corresponding to the pyramidal levels we apply a technique known as the Gaussian pyramid [5], and to characterize the object shape present in the image at the base of the pyramid we will use a labeling approach.

Our technique is based on the analogy with a heat diffusion process on the object by computing numerical values for every pixel of the object in order to obtain the object contour temperature. Initially, the contour of the object is heated at time t_0 at a given energy value g; at each iteration, new neighbouring pixels will be contaminated provided they lie in the interior of the object.

After a time t_f, the simulated diffusion process is halted, and the temperature on the contour is measured; the temperature value of the contour pixels can be used to characterise a shape-related code [6]. The contour elements that have local maxima correspond to local convexities while those having local minima correspond to local concavities: we may see that the values distributed along the contour depend on the shape local configuration.

As shown in Fig.1, we can obtain a contour shape characterisation, at each level of resolution, based on the results of the diffusion process: these characterizations are given in terms of labels attached to different contour segments.

We define an alphabet of contour labels CL = {w, c, s, x, y} where, the symbols w, c, s, x, y denote, respectively the geometrical primitives very concave, concave, straight, convex and very convex.

Labeling provides the working image for deriving descriptions of different granularity.

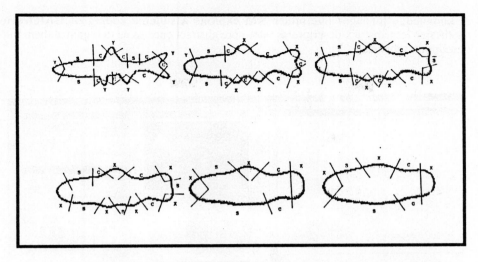

Fig.1 A picture resuming the results of contour shape characterisation on six pyramidal layers

3. Shape characterization through granularity levels

An *image description* is a vector (tuple) of values (the features of the image) assumed by a set of variables - the attributes considered useful to classify the image at hand [7]. A global description is a tuple consisting of a name denoting the image and of the set of features an observer exploits to describe the image as a whole. Some features are names of the image characteristic patterns, i.e. of those shapes that an observer identifies as relevant for the image interpretation. In turn, characteristic patterns are also described by tuples , comprising the name of the characteristic pattern and the names of its sub-patterns - whenever the characteristic pattern can be decomposed by the relation "part-of". This decomposition process is rooted in the descriptions of non decomposable, elementary patterns.

As a result, the image is described at different levels of *abstraction granularity* by a set of tuples named *description scheme* (ds). The most abstract description is the global one, the less abstract (also called the *primal description*) is the set of tuples describing the elementary patterns. Descriptions at two subsequent levels of granularity are related by the part-of relation.

This process is automated exploiting the concept of *attributed symbol* and *Conditional Attributed Rewriting System* (CARW). The finite set of names used in image description is assumed as an alphabet and the attributes/features as attributes/values of the symbols. Descriptions become attributed symbols and description schemes are organized as strings of attributed symbols. CARWs formalize the relation "part-of" by mapping a description d_{g-1} (at level of granularity $g-1$) into d_g at the next level.

A knowledge manager interpreter KM exploits a suitable coding of CARWs to synthesize less abstract descriptions into more abstract ones so as to organize them into description schemes.

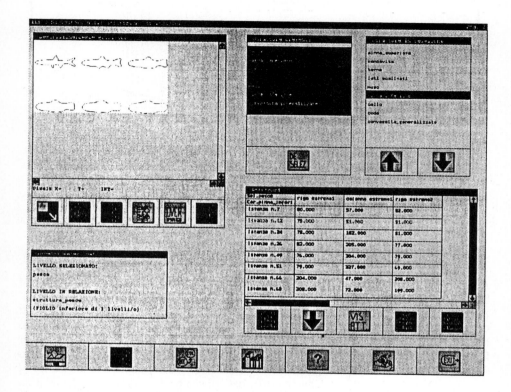

Fig. 2 A screen dump resuming the results of contour description

For instance, binary images can be described by this process assuming as elementary patterns those black pixels in which the contour changes its direction locally. Each such pixel is described by a code resuming the pixel 8-neighborhood information, and by its coordinates. In [8] it is demonstrated that this primal description allows the reconstruction of the original binary image; in the same paper an algorithm is also described which maps an arbitrary binary image into its primal description.

Figure 2 shows the result of applying the interpreter KM to the binary images obtained at the different resolution levels. Five different windows appear in the figure, displaying different views of the obtained description schemes. The graphical window in the upper left corner contains the silhouettes of the recognised features at different levels of resolution. In particular, the nose, upper and lower fins, and tails in the original image and at the first level of resolution appear.

4. Combining Resolution and Granularity

Fig.3 shows how the technique for the characterisation of a shape at different resolution and at different levels of granularity are combined .

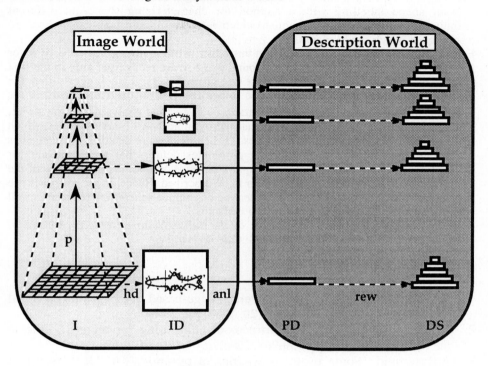

Fig.3 Combining the two techniques

On the left hand side the Image World schematizes the multiresolution approach. Each level of the pyramid contains an image at a different resolution (the base with the original image $i \in I$ and progressively other resolution levels until the apex by the repeated application of the function p), the heat diffusion functions (hd) map every image in I into a contour labeled shape $id \in ID$ as shown inside the squares. The right hand side represents the Description World where, for each resolution a primal description $pd \in PD$ is obtained by applying a function anl to the corresponding id.

The functions p and hd are computed by programs implementing the algorithms mentioned in section 2 and the function rew by the interpreter KM.

The function anl can be computed by extending the pre-existing tools for translating a binary image into a primal description. More precisely, each image id is translated into a binary image, memorizing in a table T the label and coordinates of each contour pixel in id. Then the algorithm to extract a primal description pd' is applied to the binary image. Finally for each element in pd' (i.e. a description of a single pixel belonging to the contour where it changes direction) the coordinates (x,y) are extracted. This couple is used as an index to find in T the label l that was associated with the described pixel in the image id. l is added as a fourth feature to the element at hand. The overall primal description pd is thus obtained.

5. Conclusion

We have presented a method which, by combining a multiresolution approach for contour shape labelling with a method for characterizing shapes at different granularities, allows one to establish which is the minimum convenient level of resolution at which can be classified.

For the particular example considered in this paper, when applying a same CARW for fish description and recognition it was found that fish are recognised only in the first four levels of resolution (the first four shown in Fig.1). This suggests that the detection of fish can be obtained on the image resolved at the fourth level in the pyramid, thus reducing the amount of data to be stored and analysed by a factor k^3, where k is the reduction factor. Note that the pyramid has a blurring effect, which reduces the details in the contour of shapes. This elimination of details is advantageous up to the point in which the blurring eliminates details that are irrelevant for recognition. The size of the primal description depends on both the dimensions of the image and on the roughness of the contour to be described. Hence, in general, even the sizes of the primal description and of the derived description scheme are reduced as well as the complexity of the strategies for their deduction.

This work has been partially supported by the Italian National Reasearch Council, Progetto Finalizzato "Sistemi Informatici e Calcolo Parallelo".

6. References

1. I.Biederman and G. Ju "Surface versus edge-based determinants of visual recognition", *Cognitive Psychol.* 20, 1988, pp. 38-64.
2. V.Cantoni, S.Levialdi, "Multiprocessor computing for images", *Proc. of the IEEE*, vol.76, n.8, pp.959-969, 1988.
3. A.Rosenfeld, "Some useful properties of pyramids", in *Multiresolution Image Processing and Analysis*, A. Rosenfeld (Ed.), Springer-Verlag, 1984, pp. 2-5
4. P. Bottoni, P.Mussio, M. Protti, "Metareasoning in the determination of image interpretation strategies",*PRL* , vol.15, n.2, 1993, pp.177,190.
5. P.J.Burt, "Smart sensing within a pyramid vision machine", *Proc. of the IEEE*, vol.76, n.8, pp.1006-1015, 1988.
6. L.Cinque, L.Lombardi, "Shape description and recognition by a multiresolution approach", to appear in *Image and Vision Computing*
7. R.M.Haralick, L.G.Shapiro, "Glossary of Computer Vision Terms", *Pattern Recognition*, vol.24, n.1, pp.69-93, 1991
8. P. Mussio, D. Merelli, M. Padula."An approach to the Definition, Description and Extraction of structures in Binary Digital Images"*CVGIP*, vol. 31, pp. 19-49, New York, Academic Press, 1985.

Line and Cell Searching in Tables or Forms

E. TUROLLA, Y. BELAÏD, A. BELAÏD

CRIN-CNRS, Bât. Loria, BP 239,
54506 Vandoeuvre-lès-Nancy Cedex, France
Tel: (33) 83.59.20.83 - Fax: (33) 83.41.30.79
E-mail: {turolla, ybelaid, abelaid}@loria.fr

Abstract. Item searching is an important step in form analysis. The goal of this paper is to describe a robust method to locate the items whose boundaries are black lines, and an algorithm to classify those items. Our method is based on detection of lines by Hough transform and on searching of cycles in a graph that represents the cells (boundaries of the items). We will show that our approach is robust, skew independent and can be applied to several kind of lines such as continuous, dashed, doubled, etc. We classify the items in four categories : blank, black, gray and meaningful.

Keywords : Hough Transform, French Tax Forms, Line Searching, Item Extraction, Item Classification

1. Introduction

According to Doermann *et all* ([1]), there are three kinds of form processing systems, depending on the a priori knowledge they use. The most general system treats *unknown forms*, by using general knowledge about the form domain; their performance are limited by their lack of knowledge. In opposite, the systems that deal with *known forms* use a detailed model for each member of a small group of specific forms; they are often efficient in their field and are widely used ([2]); but they can be hardly applied to other kind of forms. In half way between these two classes, the systems which deal with *known classes of forms* have logical information but no information about the exact form layout; they seem to be the good compromise between reliability and flexibility ([3]).

In our case, we treat unknown forms, in such a way that results can be used for systems designed for known forms or known classes of forms. The physical and logical structure of a form is based on the neighboring relationships between its items (alignment, juxtaposition, etc.). We want to prove that is possible to approach the structure of a specific form by using general knowledge such as the existence of boundaries between items, the meaning of alignments, etc. We limited this paper to expose the item extraction, based on the search of the horizontal and vertical segments, and the item classification.

We have applied our system on French tax forms (see Figure 1) and on tables. The items are separated by continuous lines, vertical alignments of brackets or edges of black strips. The tables we have tested, come from scientific publications. Their items are separated by continuous, dashed or double lines.

In the two following sections, we describe our line and cell searching. In the third

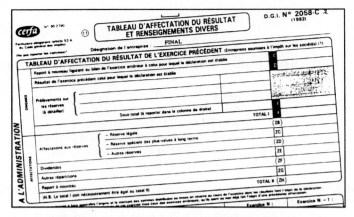

Figure 1. An extract of a French tax form page.

section, we expose our classification algorithm. We discuss the results in the last section.

2. Line searching

Lines are diverse and their quality can be poor (cut by bad scanerization, closed to text, skewed, etc.). Projective methods are unsuited when the lines are slant or too close to the text. Various line following methods have been developed to analyze mechanical drawings ([4]). But they need a lot of thresholds which closely depend on the kind of the document and they have difficulties with crosses. So we have chosen the *Hough transform*. It detects globally all the straight lines that exist in an image. Although this method is time consuming, it is insensitive to noise and slope, is not disturbed by crosses and can detect all kind of lines among text blocks.

In order to obtain a regular distribution probability of the voting points, we have chosen the Hough parameter space (α, ρ) ([5]). α is the slope of the line with respect to the x-axis, and ρ is the distance between the line and the origin of the cartesian space. The Hough space is represented by two dimensions matrix (one for horizontal lines, one for vertical lines).

To avoid finding several straight lines within one thick line and to increase the system speed, only the points that can belong to the edges of a horizontal (resp. vertical) line, vote. The black pixels that vote in the vertical Hough matrix, must have the configuration «001 » (white, white, black) for the left side and "100" (black, white, white) for the right side. The transpositions of those configurations are used to fill the horizontal Hough matrix. These filters are enough to get rid of points inside black strips and keep working on slanted lines

We adapted the algorithm described in [6] to look for the significant clusters of the filled Hough space. After removing the accumulator cells whose value is very low, the system makes a recursive cutting up of each matrix; it stops when the size of the cluster is insufficient or its voting counter becomes smaller than a threshold T. This threshold is calculated to avoid to find the very short line or the alignments of letters in the image; T is the multiplication of the average of the value of the accumulator items, with a given weight (> 1).

In the chosen parameter space, the shape of the area which corresponds to a line, is like a butterfly. So, the clusters which are adjacent, or not too far and in the good direction, are grouped together. Finally, the Hough lines are the barycenters of each remaining group of clusters

Then, the searching of black segments consists in following each Hough line on the image and merging together into a segment, the black consecutive points which are the nearest of the Hough line. The very short segments are suppressed. By this way, the continuous line and the brackets are found. To take into account the dashed lines, the system links the segments that belongs to the same Hough line and are not too far.

3. Cell searching

We use a graph to extract the form items. The nodes correspond to the intersection points between the horizontal and vertical lines of the image. Arcs correspond to segments joining two neighboring nodes. The cells are the minima cycles of this graph. The graph is represented by a two-dimensional table, whose rows correspond to the horizontal Hough lines and columns to the vertical ones. The horizontal (resp. vertical) Hough lines are sorted in an ascending vertical (resp. horizontal) order. Each element (i, j) of the table represents the intersection point between the ith horizontal line and the jth vertical line. A node (i,j) can be linked to his four neighboring nodes $(i-1,j)$, $(i+1,j)$, $(i,j+1)$ and $(i,j-1)$.

The graph analysis operates as follows : after suppressing recursively the nodes that have only one arc, it obtains the set E_{ca} by extracting all the nodes that may be a left up vertex. It calculates $Min_{\angle}(E_{ca})$, which is the minimum element of E_{ca} for the relation $\angle$:

$$(i_1, j_1)\angle(i_2, j_2) \Leftrightarrow (i_1 \leq i_2) \vee ((i_1 = i_2)\wedge(j_1 < j_2))$$

The search starts from the node $N_0 = Min_{\angle}(E_{ca})$, then proceeds to the right. When a node N is reached from the direction d, it proceeds by the direction where an arc exists and whose priority is the lower (see Figure 2).A partial failure is met when the current node is prior to the cycle or above N_0; then the search makes a back tracking. If it backtracks till N_0, then it is a complete failure and it choose a new start node. To avoid finding several times the same cycle, the start node N_0 and all the nodes that are reached from bottom ($\uparrow$) and are left by the right ($\rightarrow$) are suppressed from E_{ca}. All the cycles of the graph are found when E_{ca} is empty. To accelerate the algorithm, the nodes that have lead to a partial failure for the current cycle are stored; if any of these nodes is reached again, then it is a complete failure.

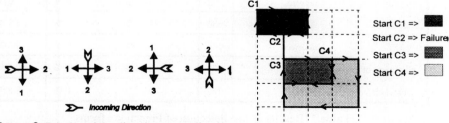

Figure 2. Priority of the direction, according to the incoming direction.

Figure 3. Examples of cycle searching.

4. Item classification

The classification identifies four classes: *blank* (this item contains no information), *black* (its background is black and its foreground is white), *gray* and *« meaningful » item* (it contains information: preprinted or filled by the user).

First, the density D_1 of black pixels is calculated. A high density ($D_1 > T_{1b}$) means a black item, whereas a low density ($D_1 < T_{1e}$) denotes a blank item. An average density implies a gray or a meaningful item. In this case, the cell is divided into little squares of five pixels length and the density D_2 of *« black squares »* (they contain at least one black pixel) is calculated. A very low density ($D_2 < T_{2m}$) denotes a meaningful item, with few text. In order to definitely discriminate gray between meaningful item, the regularity of black squares is studied by calculating the standard-deviation SD of black squares. If SD is low ($SD < T_{3g}$), then the item is classified as gray, else it is classified as meaningful. The best thresholds we have fixed by experiments are: $T_{1b} = 75\%$, $T_{1e} = 1\%$, $T_{2m} = 50\%$ and $T_{3g} = 0.35$.

5. Experiments and discussion

We have applied this item searching method on 41 images of French tax forms and 92 tables, scanned at a resolution of 300 dpi. All these images have been analyzed with the same set of parameters. For skewed documents, it was only necessary to increase the size of the Hough matrices. The treatment of an image of a French tax form takes about 2 minutes on a SUN SPARC station IPX.

The results of line detection are described in Table 1 and Table 2. The segments whose length is smaller than 8 % of the longest line in the image and very thick lines are often ignored. It is due to the way the threshold T is calculated for Hough analysis.

Line type	Quality	Total number	Whole found lines	Partially found lines	Missed lines
Continuous	Unskewed	4553	95.02 %	0.02 %	4.96 %
	Skewed	814	96.2 %	0.12 %	3.68 %
	Merged characters	22	100 %	0 %	0 %
	Total	5389	95.21 %	0.037 %	4.75 %
Boundaries	Unskewed	544	99.3 %	0 %	0.7 %
of	Skewed	92	100 %	0 %	0 %
black strips	Total	636	99.37 %	0 %	0.63 %
Vertical	Unskewed	144	81.7 %	0 %	8.3 %
alignment of	Skewed	38	89.4 %	0 %	10.6 %
brackets	Total	182	91.2 %	0 %	8.8 %

Table 1. Results of line detection of French tax forms.

Line type	Quality	Total number	Whole found lines
	Unskewed	656	96.95%
Continuous	Skewed	143	100%
	Total	799	97.49%
Dashed	Unskewed	575	97.73%
Double	Unskewed	266	100%

Table 2. Results of line detection of tables.

Cells	Total	Well found cell	Under-segmented cells	Over-segmented cells	Forgotten cells
French Tax Forms	7377	88.59%	4.58%	2.72%	4.11%
Tables	3915	92.87%	4.90%	0.18%	2.04%

Table 3. Results of cell extraction.

Item type	Total number	Well classified items	Bad classified items			
			Blank	Black	Gray	Meaningful
Blank	2281	92.9 %	/////////	0 %	0 %	7.1 %
Black	371	99.7 %	0 %	/////////	0 %	0.3 %
Gray	80	41.3 %	13.7 %	3.75 %	/////////	41.2 %
Meaningful	3582	97.5 %	2.48 %	0 %	0 %	/////////
Total	6314	95.2 %	1.58 %	0.475 %	0 %	3.10 %

Table 4. Results of item classification.

The results of cell extraction are described in Table 3. Under-segmentation comes from missed lines. Over-segmentation comes from black items crossed by a Hough line or from characters inside an item which are lined up with a Hough line and very close to two other lines. Forgotten cells are due to rounded vertices which are bad taken into account. All these errors can be corrected by specific treatments.

We have classified the items of the French tax forms. The results of item classification are described in Table 4. The classification of blank or meaningful items can be easily improved by removing the black pixels of their boundaries. But the classification of gray items appears to be a complex problem. Indeed, it is difficult to model the binary image of a gray area; this image is very dependent of the scanner and the quality of the original document. We have realized that the regularity of a gray area is very theoretical !.

6. Conclusion

We have exposed a general method for item extraction by line searching. Based on Hough transform, it can deal with medium quality documents, even skewed, and can extract several kind of lines (continuous, dashed, vertical alignment of brackets, boundaries of blacks strips). The boundaries of items are represented in a graph. The minima cycles of this graph represent the items. We have proved the robustness of our method by applying it on French tax forms and on tables.

We will improve our line following module by dealing with pointed lines and classifying the found lines. We will complete the line searching by a module that searches the white bands.

We have begun to classify the found items in French tax forms; the results are promising but need to be improved for gray items.

The next step consists in building a specific structure of the document, that contains the physical and logical relationship we can extract from the items.

References

[1] D. S. DOERMANN, A. ROSENFELD, *"the Processing of Form Documents"*, ICDAR 1993, pp. 497-501

[2] J. YUAN, L. XU, C.Y. SUEN, *"Form Items Extraction by Model matching"*, ICDAR 1991, pp.210-218.

[3] G. MADERLECHNER, *"'Symbolic subtraction' of fixed formatted graphics and text from filled in forms"*, MVA'90, IAPR Workshop on Machine Vision Applications, Nov 28-30 1990, Tokyo

[4] R. KASTURI, R. RAMAN, C. CHENNUBHOTLA, L. O'GORMAN *"Document image analysis, an overview of techniques for graphics recognition"*, SSPR 1990.

[5] T. RISSE, *"Hough Transform for Line Recognition : Complexity of Evidence Accumulation and Cluster Detection"*, Computer Vision, Graphics, and Image Processing, vol. 46, pp. 327-345, 1989

[6] Y. MULLER-BELAÏD, R. MOHR *"Planes and quadrics detection using Hough transform"*, 7th International Conference on Pattern Recognition; Montreal, Canada, July 30 - August 2, 1984

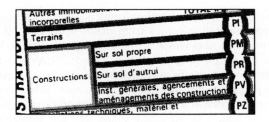

Figure 4. Extract of the found cells in a form.

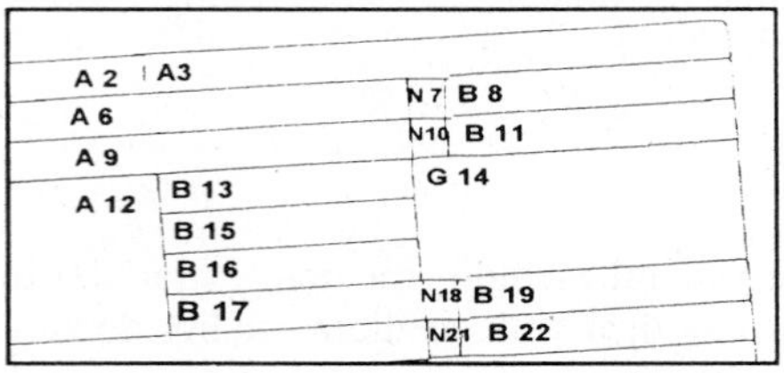

Figure 5. An extract of the found and classified items of Figure 1; .'A' indicates a meaningful item, 'G', a gray item, 'N', a black item and 'B', a blank item.

Describing Words by Graphs

G. Congedo, G. Dimauro, S. Impedovo, G. Pirlo

Dipartimento di Informatica - Università di Bari
Via Amendola 173 - 70126 Bari - Italy

Abstract. In this paper a new technique for handwritten word description is presented. It is based on the development of a description graph which results little sensitive to handwriting variability. The experimental results show the effectiveness of the description graph in retaining the basic information for the recognition of the word image.

1 Introduction

Research on cursive word recognition has several fundamental applications in the development of automatic reading systems for bankchecks, tax forms, insurance forms and so on [1]. Unfortunately, it is commonly believed that cursive word recognition is a very complex task. This is due to the existence of many writing styles and to personal variability which depends on the physical and physiological condition of the writer as well as on the type of the supports used for writing (type of paper, pen or pencil etc.) [2]. .

Up to date, two different kinds of approaches have been considered for cursive word recognition [2,3]: analytical and wholistic. In both cases the recognition process strictly depends on the capability to obtain a suitable description of the word image: a very detailed description of the image can convey useless details which make the recognition process slower and less accurate; a poor description generally lacks of information useful for recognition.

In this paper a simple procedure for the description of handwritten words by graphs is proposed. The graphs, which retain fundamental morphological information of the word image, like oriented segments, concavities/convexities, loops and near-loops, are also used as input for a word recognizer based on an elastic pattern matching procedure. The experimental results demonstrate the effectiveness of the new technique in describing handwritten words for recognition aims.

2 The Word Description Graph

A cursive word is here considered as consisting of regular parts and singularities [2]. Regular parts are the segments of the word while singularities are the discontinuity regions. Three kinds of discontinuity regions are here considered: *cross regions*, *bend regions* and *end regions*. The *cross regions* are the intersection regions of segments; the *bend regions* are the adjacency regions of two segments with different orientation, the *end regions* are the extremity regions of a segment.

ACKNOWLEDGEMENT: This work was supported by MURST 60% - 1994, Italy.

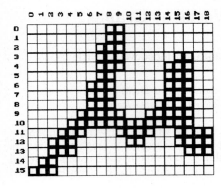

Figure 1

The description of a word can be accomplished by a description graph in which regular parts are represented by arcs while singularities are represented by nodes. The graph is realized by an automatic procedure which performs four different scanning on the word image according to four basic directions: horizontal, vertical, -45° and +45°. The scanning in a specific direction produces a data structure containing the information about groups of adjacent pixels in the perpendicular direction. Specifically, the data structure is a list of sequences of nodes in which each node contains the information of the group of

⇓ ⇒(8,2)
   ⇓⇑

⇓ ⇒(8,2)
   ⇓⇑

⇓ ⇒(7,3)
   ⇓⇑

⇓ ⇒(7,3) ⇒(15,2)
   ⇓⇑   ⇓⇑

⇓ ⇒(7,3) ⇒(14,3)
   ⇓⇑   ⇓⇑

⇓ ⇒(7,2) ⇒(14,3)
   ⇓⇑   ⇓⇑

⇓ ⇒(6,3) ⇒(14,3)
   ⇓⇑   ⇓⇑

⇓ ⇒(6,3) ⇒(14,3)
   ⇓⇑   ⇓⇑

⇓ ⇒(6,3) ⇒(13,4)
   ⇓⇑   ⇓⇑

⇓ ⇒(5,5) ⇒(12,5)

⇓ ⇒(4,13)

⇓ ⇒(3,3) ⇒(9,4) ⇒(15,4)
   ⇓⇑   ⇓⇑   ⇓⇑

⇓ ⇒(2,3) ⇒(10,2) ⇒(10,4)
   ⇓⇑   ⇓⇑

⇓ ⇒(2,3) ⇒(16,3)
   ⇓⇑

⇓ ⇒(1,2)
   ⇓⇑

⇓ ⇒(0,3)

Figure 2

adjacent pixels found during the scanning process and the adjacency between groups on different rows is represented by pointers between the nodes. In each node the following information is stored: the first column of the group, the width of the group, the right pointer to the next group in the same row (if it exists), the upper pointer to the adjacent group in the upper row (if it exists), the lower pointer to the adjacent group in the lower row (if it exists). The upper and lower pointers are assigned according to simple rules: (1) two different nodes in the same row cannot point to the same node in the lower row; (2) the lower pointer of a node of a

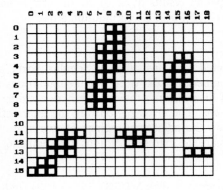

Figure 3

group which is adjacent to two or more groups in the lower row must be null. Figure 1 shows a simple example of a pattern. Figure 2 shows the corresponding data structure obtained by the horizontal scanning of the image. Each couple contains the column and the width of the corresponding group. The average width w of the groups is then computed and each group whose width is greater than w is removed as shown in figure 3. Successively the writing units are identified. A

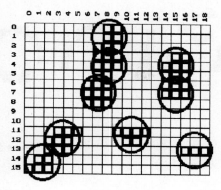

Figure 4

writing unit is here defined as each set of n groups linked together in the vertical direction (by upper and lower pointers) for which $\left| \underline{w} - n \right| \leq 1$, where $\underline{w}$ is the average width of the groups. Figure 4 shows that nine writing units are detected in the pattern in figure 1. A sequence of adjacent writing units constitutes a segment. Blocks of one writing unit are removed and the direction of the remaining segments is computed in order to extract segments with vertical direction. Of course, the same process is repeated with other scanning directions in order to detect other segments. Figure 5A shows an example of handwritten word, figures 5B,C,D, and E show respectively the detected segments with vertical, horizontal, +45° and -45° orientation.

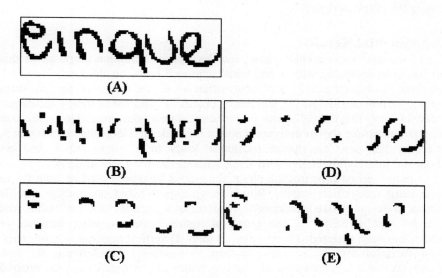

Figure 5

Figure 6

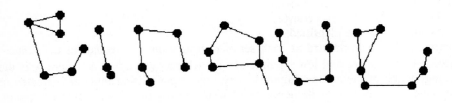

Figure 7

After that all the oriented segments have been detected, they are superimposed as figure 6 shows, and the adjacencies between segments are detected, obtaining the graph in figure 7.

3 Experimental Results

In the first experimental phase, the description technique proposed in this paper has been compared with a traditional approach based on thinning. Figures 8 and 9 show two typical examples of handwritten words, the results of the Safe Point Thinning Technique (SPTA) [4], the results obtain by and the proposed technique. It is easy to verify that the SPTA can yeld to unsatisfactory outcomes. This is the case of the spurious stroke shown in figure 8 and of the undesired loop shown in figure 9. Conversely, the new description technique leads to a graph which conveys fundamental information on the word shape without retaining useless details.

In the second experimental phase, the graphs have been used as input for an off-line word recognition system. Word recognition is based on finding specific structures in the graph like ascenders/descenders, loops , concavities/convexities, and simple oriented segments. In the graph of figure 7 the following sequence of structures has been detected: 1) convexity, 2) loop, 3) vertical segment, 4) concavity, 5) loop, 6) descender, 7) convexity, 8) loop, 9) convexity. Furthermore, for each structure the horizontal position of the barycentre is computed and the ordered sequence of structures is used for recognition. Word matching is carried out by an

519

elastic matching procedure according to a suitable distance measure defined by an a-priori analysis of the confusing structures.

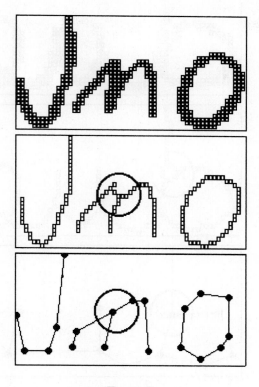

Figure 8

The experimental tests have been carried out in the domain of the 50 basic words of the worded amounts of italian bankchecks [5]. For this purpose, a reference database of 2000 words written by 40 writers and a test database of 2000 words written by further 40 writers have been used. For each test word a ranked list of candidates has been produced by the recognition system. On the entire test database the following results have been obtained: 69% of correct recognition in the first position of the list; 78% of correct recognition in the first two positions of the list; 79% of correct recognition in the first three positions of the list.

4 Conclusion
In this paper a useful technique is proposed for the description of handwritten word by graph. The experimental results show that the new technique is able to retain morphological information on the word shape without retaining usefulness details.

520

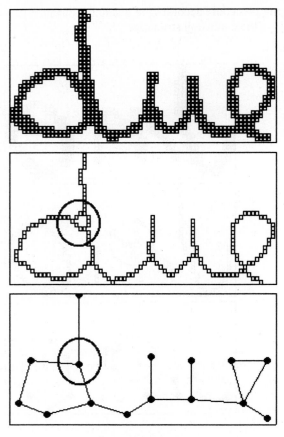

Figure 9

References

1. K. Sakai, H. Asami, Y. Tanabe, "Advanced application systems for handwritten character recognition", *in From Pixels to Features III - Frontiers in Handwriting Recognition*, S. Impedovo, J.C. Simon, (eds.), Elsevier Publ., 1992, pp. 175-186.
2. E. Lecolinet, O. Baret, "Cursive Word Recognition: Methods and Strategies", in *Fundamentals in Handwriting Recognition*, S. Impedovo (ed.), NATO ASI Series, Springer Verlag 1994, pp. 235-263.
3. J.Hull, T.-K. Ho, J. Favata, V. Govindaraju, S. Srihari,"Combination of Segmentation-based and Wholistic Handwritten Word Recognition Algorithms", in *From Pixels to Features III - Frontiers in Handwriting Recognition*, S. Impedovo, J.C. Simon, (eds.), Elsevier Publ., 1992, pp.229-240.
4. N.J. Naccache, R. Shingal, "SPTA:A proposed algorithm for thinning binary Patterns", *IEEE Trans. Syst. Man, Cybern.*, no. 3, 1984.
5. G. Dimauro, M.R. Grattagliano, S. Impedovo, G. Pirlo, "A System for Banckchecks Processing", IEEE Computer Society Press, Vol. 4960-02, ICDAR '93, Tsukuba Science City -Japan, Oct. 1993, pp. 454-459.

Selecting Reference Signatures
for On-Line Signature Verification

G.Congedo, G.Dimauro, A.M. Forte, S.Impedovo, G.Pirlo

Dipartimento di Informatica - Università di Bari
Via Amendola 173, 70123 Bari, Italy

Abstract. Personal stability in signing in one of the key factors for high-accuracy signature verification. In this paper the local stability of a dynamic signature is measured. The acccuracy of an on-line signature verification system is improved by selecting, among a set of specimens available, the subset of reference signatures which is near-optimal for different signature representation domains in terms of impostor-acceptance error rate.

1 Introduction

Signature is the costumary way of identifying an individual in our society and it is rightly considered as one of the best means for personal identification; in fact signature cannot be lost, stolen or forgotten. Because of this, with the diffusion of remote data-banks and widely distributed computer networks there is a growing need of effective systems for automatic signature verification as the renewed interest in this field demonstrates [1,2].

Up to now, a lot of efforts have been carried out by many research groups in order to develop high-accuracy systems for automatic signature verification but it is now generally accepted that the performances of such systems can vary strongly depending on several factors. It is quite obvious that a short, commonly-written signature should be more easy to imitate than a long specimen with many pictorial strokes. Recently an attempt has been proposed to estimate quantitatively and a-priori the difficulty in reproducing the signature using a functional model of a typical forger [3]. Unfortunately, signature more than other kinds of writings is written from habits and therefore is not possible to modify it in order to obtain higher level of accuracy in automatic verification. Conversely, the detection of stable regions in the signatures of a writer allows to tune the system according to personal characteristics and specific application requirements [4].

In this paper, the local stability in on-line signatures is measured. From the analysis of high/low-stability regions in the signatures of a writer, the near-optimal subset of reference signatures in terms of low impostor acceptance error rate is selected for different representation domains.

ACKNOWLEDGEMENT: This work was supported by MURST 40% - Italy.

2 Local Stability in On-Line Signatures.

Local stability in an on-line signature can be computed in automatic way by an iterative procedure based on multiple matchings among the specimens in the set of genuine data [4]. Let be

$$S^g : z^g(i) = (x^g(i), y^g(i)) \qquad g=1,2,...,n; \; i=1,2,...,M^g$$

the set of n on-line genuine signatures of a writer, where the coordinates $x^g(i)$ and $y^g(i)$ describe the position of the pen with respect to the plane of the graphic tablet at time i of the writing process of the g-th signature. A normalization process in the space domain, allows to have for each signature S^g :

$$0 \le x_i^g \le 1 \qquad \text{and} \qquad 0 \le y_i^g \le 1 \qquad \text{for } i=1,2,...,M^g .$$

In order to evaluate the local stability of the r-th specimen in the set ($r \in \{1,2,...,n\}$), a two-steps procedure must be accomplished. The first step deals with the detection of the Direct Matching Points (DMPs) between signature S^r and each other signature S^v in the set ($v=1,2,...,n$, $v \ne r$). A DMP of the signature S^r with respect to signature S^v is a point which has a direct (one-to-one) coupling with a point of signature S^v when the two signatures are matched togheter by an elastic matching procedure. More specifically the point $z^r(p)$ of the r-th signature coupled with the point $z^v(q)$, $q \in \{1,2,..., M^v\}$, is a DMP of the r-th signature with respect to the v-th signature if and only if:

$$\forall \; p = 1,2,...,M^r , \; p \ne p, \; z^r(p) \text{ is not coupled with } z^v(q)$$

and

$$\forall \; q = 1,2,...,M^v , \; q \ne q, \; z^v(q) \text{ is not coupled with } z^r(p).$$

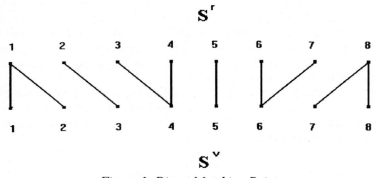

Figure 1: Direct Matching Points

Figure 1 shows some typical couplings between samples of S^r and S^v. In this case the DMPs in the sequence S^r are the points 2 and 5. In the second step the local index of stability of the signature S^r is computed by a simple avaraging procedure. In fact a DMP indicates the existence of a region of the r-th signature which is roughly stable when compared to the corresponding region of the v-th signature. For each point of the r-th signature a score can be introduced according to to its type of coupling with respect to the points of the v-th signture:

$$\text{Score}^v \, (z^r \, (p)) = \begin{cases} 1 & \text{if } z^r(p) \text{ is a DMP} \\ 0 & \text{otherwise} \end{cases}.$$

Therefore, the index of local stability of the point $z^r(p)$ is given by the frequency of direct matchings of the point when signature S^r is matched against the other genuine signatures available:

$$I(z^r \, (p)) = \frac{1}{n-1} \sum_{\substack{v=1 \\ v \neq r}}^{n} \text{Score}^v \, (z^r \, (p)).$$

3 Selection of the Near-Optimal Set of Reference Signatures

It is widely accepted that the performance of a signature verification system depends on the characteristics of the set of reference specimens [1]. In this field, beside intrinsic difficulty in imitating a signature, the instability in the set of reference signatures is one of the most common cause of deterioration in accuracy of the verification [1]. However, in many real applications more genuine signatures are available than those required by the system for reference. Therefore, suitable rules are necessary to select the subset of specimens to be used as reference according to some near-optimality criteria. A powerful tool to select such subset is the stability index described previuosly. From the consideration that the improvement of the impostor-acceptance error rate (type II error rate) is generally considered more important than the reduction of the false-rejection error rate (type I error rate), in the following an application of such tool is presented in order to select the subset of signatures for which the importor-acceptance error rate is as lower as possible.

For this purpose, let be $R^1, R^2, ..., R^n$ the set of genuine signatures available, $z^i = R^i$ for i=1,2,...,n, and $I(z^1 (.)), I(z^2 (.)), ..., I(z^n (.))$ the respective indexes of local stability. The selection of the near-optimal subset of m reference signatures (m<n) is accomplished by selecting specimens which show wide stable regions in corresponding parts of the trace. This criterion descends from the well-known assumption that the imitation of highly stable pattern in a wide region is more difficult than the imitation of minor, variable patterns. Thus, the near-optimal m-tupla of reference signatures is the m-tupla:

$$R^{i_1},..,R^{i_j},..R^{i_m}$$

(where $\forall j=1,2,...,m$, $i\in\{1,2,..,n\}$; and $\forall j,t=1,2,...,m$, $j\neq t \Rightarrow i_j\neq i_t$) for which it follows:

$$\underset{m-tuple}{Max} \quad \underset{i_j}{Max} \quad \sum_p \left[\prod_{\substack{a=i_1,i_2,...,i_m \\ a\neq i_j}} I_a(z^{i_j}(p)) \right]$$

where

$$I_a(z^{i_j}(p)) = \frac{1}{card(K_a^{i_j}(p))} \sum_{q\in K_a^{i_j}(p)} I(z^a(q))$$

and

$$K^{i_j}{}_a(p)= \{q \mid point \; z^a(q) \; is \; coupled \; with \; z^{i_j}(p)\} \; .$$

4 Experimental Results

The stability index proposed in this paper has been used to select the near-optimal subset of reference specimens for the verification system presented in ref. [5].The test has been carried out using a graphic tablet of 500 dpi and 110 Hz, connected to a PS/2 (80486 - 33Mhz). For the experimental test, two databases of on-line signatures have been used for each of the ten signers considered. The first database contains fifty-seven genuine signatures, the second database contains fifty forgeries written by ten forgers in controlled writing sessions. Each forger has had about ten minutes to practice himself with the electronic tablet and five minutes to affix five signatures.

For each signer, seven genuine signatures have been randomly selected and used as the set of available reference signature. Among this set, the near-optimal subset of three signatures has been selected and used as reference for the verification system. In the test sessions, the standard dissimilarity measure has been considered both for the selection of the reference signatures and for the verification process:

$$D(S^r,S^v) = \sum_{k=1}^{K} d(c_k)$$

where the measure $d(c_k)$ is defined for each coupling $c_k=(i_k,j_k)$, $k=1,2,...,K$, provided by the elastic matching procedure. Moreover such measure has been defined with respect to different domains of representation of the signatures:

(1) position

$$d(c_k) = d(z^r(i_k), z^v(j_k)) = \sqrt{(x^r(i_k) - x^v(j_k))^2 + (y^r(i_k) - y^v(j_k))^2}$$

(2) velocity

$$d(c_k) = d(v^r(i_k), v^v(j_k)) = \sqrt{(v_x^r(i_k) - v_x^v(j_k))^2 + (v_y^r(i_k) - v_y^v(j_k))^2}$$

(3) acceleration

$$d(c_k) = d(a^r(i_k), a^v(j_k)) = \sqrt{(a_x^r(i_k) - a_x^v(j_k))^2 + (a_y^r(i_k) - a_y^v(j_k))^2}$$

where for each specimen the velocity and acceleration functions have been derived by the position signals detected directly by the graphic tabled during the apposition process.

When the subset of reference signatures is selected following the criterion presented in this work, the importor acceptance rate of the system is on average three times lower than using other 3-tuple of genuine signatures for reference. Moreover, the possibility to detect for each writer the representation space in which his own signatures are more stable allows to evaluate in a fast way the domain in which the matching procedure is more profitable for each writer. Specifically, for the set of signers under consideration, it has been shown that the more stable domain for signature representation is the velocity domain (5 signers), followed by the acceleration (3 signers) and position domain (2 signers). Figure 2 and 3 shows respectively a genuine signature and its high/low stability regions detected in the velocity domain.

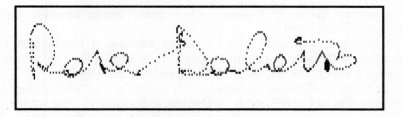

Figure 2. On-line input signature

High-Stability regions

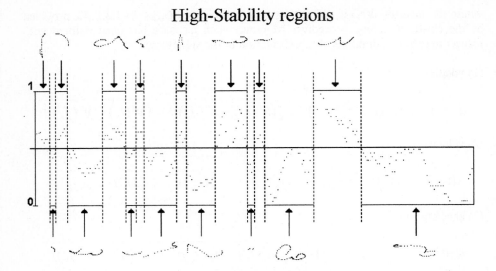

Low-Stability regions

Figure 3. Detection of High/Low stability regions in the velocity domain

5 Conclusion

In this paper the local stability in dynamic signatures is measured. Local stability has been applied to select the near-optimal set of reference signatures for on-line signature verification. The experimental results demonstrate the usefulness of this information for the development of high-accuracy signature verification systems.

References

[1] G. Pirlo, "Algorithm for Signature Verification", in *Fundamentals in Handwriting Recognition*, NATO-ASI Series, Springer-Verlag, pp.139-152, 1994.

[2] *Progress in Automatic Signature Verification*, World Scientific Publ., R. Plamondon (ed.), 1994.

[3] J.-J. Brault, R. Plamondon, "A complexity Measure of Handwritten Curves: Mideling of Dynamic Signature Forgery", *IEEE Transactions on System, Man, and Cybernetics*, Vol. 23, no. 2, March/April 1993, pp. 400-413.

[4] G. Congedo, G. Dimauro, S. Impedovo, G. Pirlo, "A New Methodology for the Measurement of Local Stability in Dynamical Signatures", Proc. of the *Fourth International Workshop on Frontiers in Handwriting Recognition*, Taipei, Taiwan, Dec. 7-9, 1994, pp. 135-144.

[5] G. Dimauro, S. Impedovo, G. Pirlo, " On-line Signature Verification through a Dynamical Segmentation Technique", Proc. of the *Third International Workshop on Frontiers in Handwriting Recognition*,Buffalo-NY, USA, 1993, pp. 262-271.

ApOFIS: an A priori Optical Font Identification System

Abdelwahab Zramdini and Rolf Ingold

Institute of Informatics, University of Fribourg
Chemin du Musée 3, CH-1700 Fribourg (Switzerland)
E-mail: abdelwahab.zramdini@unifr.ch   http://www.-iiuf.unifr.ch/GIRAF/giraf.html

Abstract. The detection of the font style, point size, etc. of a text is an obvious way to improve the capabilities of text recognition algorithms. The *ApOFIS* system has been designed in order to satisfy such a requirement. It adopts an a priori font identification approach where the recognition of a text font is done without considering the characters that appear in the text. In *ApOFIS*, a font is characterized especially by its family, weight, slope and size. Features used in the system represent global aspects of text line images. They have been extracted essentially from projection profiles and from connected components bounding boxes. Statistical tests have revealed that these features follow approximately normal laws so that parameter estimation is used in learning.

A multivariate Bayesian classifier, based on these features, has been designed for font recognition and applied on a base of 240 font models created from a training set of texts written with these fonts. On text lines having the same length as those used for learning, the system allows to discriminate fonts with an average accuracy of 96.5% for top choice and 98.3% within the two top choices.

Keywords: *Document Analysis, a priori Font Recognition, Bayesian Classifier*

1   Introduction

The SSPR'90 working group on Character Recognition stated that [1]:

The detection of the font style, point size, etc. of a text is an obvious way to improve the capabilities of text recognition algorithms. This would allow for hundreds of fonts to be used for training but retain the recognition accuracy and potential speed of a system that uses a small number of fonts. This appears to be a promising but hitherto almost neglected topic.

In this context, the Optical Font Recognition (OFR) become an important topic in optical reading. Furthermore, OFR is gaining some interest during the last few years [1, 2, 3]. Indeed, there are two possible approaches for font recognition:

1. *a priori font recognition* where fonts are identified without considering the characters that appear in the text. Thus, features have to be extracted from global aspects of the text images. These aspects are generally detected by non experts in typography such as the text density, letters size, orientation and spacing, and the presence of serifs. Features may be computed from different text entities such as words or lines.
2. *a posteriori font recognition* taking a substantial benefit from the knowledge of the letter classes. Thus, features can be based on local aspects of individual letters or their particularities such as the shape of serifs (cove, squared, triangular, etc.) and the representation of particular letters like g and g, a and *a*, etc.

In the context of a structured document recognition system developed in our laboratory [4], font identification plays an important role. For instance, an *a priori font recognition approach* has been adopted and evaluated through the *ApOFIS* system.

In our OFR system, a font is defined by five attributes: *family* (Times, Helvetica, etc.)[1], *size* expressed in typographical points, *weight* (regular, bold), *slope* (roman, italic) and *width* (normal, expanded, condensed).

The goal of the present paper is to present the *ApOFIS* system and an evaluation of its performances.

2 ApOFIS Architecture

ApOFIS is based on a Bayesian classification model and uses nine global features extracted from text line images.

2.1 Selected features

Three features are extracted from the vertical projection profile V_p (see Fig. 1):

- *hwp*: height of the whole vertical profile;
- *hup*: height of the upper part of the vertical profile;
- *hcp*: height of the central part of the vertical profile.

These heights simulate the typographical notions of *font-height*, *X-height* and *x-height*. Two other features are extracted from the horizontal projection profile H_p (see Fig. 1):

- *dn*: the density of black pixels in a rectangular box around the considered text line. Indeed, the weight of a font is reflected by the density of black surfaces on the white background. If n is the size of H_p, then this density is defined by

$$dn = \frac{1}{n} \sum_{i=1}^{n} H_p[i];$$

- pd^2: the means of the squared values of the horizontal profile first derivative. One can observe from H_p that roman texts are characterized by a set of upright and tall peaks. For italic texts the peaks are less tall, rounded and boarder. Let n be the size of H_p, this feature is defined by

$$pd^2 = \frac{1}{n-1} \sum_{i=1}^{n-1} (H_p[i+1] - H_p[i])^2.$$

Two other features are extracted from connected components and especially from their bounding boxes (see Fig. 2). Let $d(e_i, e_{i+1})$ be the horizontal distance in pixels between two consecutive connected components bounding boxes and let $w(e_i)$ be the width of the connected component i bounding box. Let us suppose a text line with n connected components, the following features have been defined:

---

[1] Commonly, the family attribute corresponds to the typographical notion of "typeface".

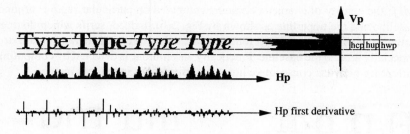

Fig. 1. *Projection profiles and first derivative of the horizontal profile.*

– *ccsp*: the means of connected components spacing within text words, defined by

$$ccsp = \frac{1}{n-1} \sum_{i=1}^{n-1} d(e_i, e_{i+1}).$$

As shown by Fig. 2, characters spacing within the same word changes from one font to another. One notices that this spacing is very important and variable for *fixed pitch* fonts. It is also more important for *sans-serif* fonts than for *seriffed* ones;

– *ccw*: the average width of connected components bounding boxes in the text line, defined by

$$ccw = \frac{1}{n} \sum_{i=1}^{n} w(e_i).$$

This feature is relevant especially for connected components with squared forms corresponding to main characters in the central zone of the vertical profile such as characters *a, c, n* of the word *Spacing*. Indeed, these character widths depend on the font family, weight and size.

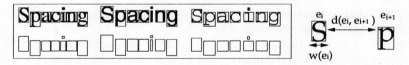

Fig. 2. *Connected components bounding boxes spacing and widths.*

Finally, two other features are extracted from special zones in the text image:

– *hrl*: the average width of the vertical stems within characters which depends on the font family, weight and size (see Fig. 3(a)). The stem widths are estimated by computing the horizontal run-lengths within the text image.
Let hrl_i be the length of a horizontal run i and n be the amount of runs in the text, this feature is defined by

$$hrl = \frac{1}{n} \sum_{i=1}^{n} hrl_i;$$

Since it considers all characters, hrl includes other information than stems such as arcs of characters o and c.

- $pdif$: the density of character contours computed on particular regions around the baseline and the upperline as shown by Fig. 3(b). Indeed, serifs which distinguish between seriffed and sans-serif fonts, are located at the end of character main strokes, especially around the baseline. Practically, the contour is detected by computing the difference between consecutive horizontal pixel lines in the text image.

Fig. 3. *(a) Stems width for various styles*   *(b) pixel lines difference.*

2.2 Classification model

ApOFIS is based on a Bayesian classification model where the font models base used in the classification is created from a training set of text line images. We have shown in [5] that we can assume that features follow normal laws, thus the learning process consists in estimating the multivariate normal density parameters (μ, Σ) where μ expresses the means features vector and Σ the corresponding covariance matrix.

Learning The learning allows to create a font models base from samples of the considered fonts. Learning a font (f_i) consists in:

1. generating an image by scanning a text page printed with f_i;
2. performing some preprocessing on the image in order (1) to detect and correct possible skew and (2) to filter noise, punctuation and diacritical signs which do not carry any pertinent information;
3. segmenting the image into lines;
4. extracting a features vector $x = (x_1, \ldots, x_n)$ from each line (in our case $n = 9$);
5. collecting these vectors to estimate μ and Σ by the maximum likelihood estimator;
6. finally, adding the font model (μ, Σ, f_i) to the base.

The current font models base of *ApOFIS* system contains 240 font models corresponding to 15 typefaces (see Table 1), four sizes (10, 11, 12, 14pt) and four styles (regular, italic, bold, bold-italic) often used in documents. Therefore, 16 fonts have been considered for each typeface. For each font, μ and Σ have been computed on features extracted from 100 English text lines of about 6 cm length. Texts have been arbitrarily extracted from existing documents and produced by a laser printer. Binary images have been produced from these texts by a scanning at 400 dpi.

Classification The goal of the classification process is to assign a font to a text line modeled by a features vector. The classification consists in traversing the font models base and associating to this vector the font model which maximizes a certain function. This function measures the probability that this vector corresponds to the font model[2].

3   Evaluation of the System Performances

3.1   Methodology

For each font, the classification has been done on 100 French text lines with almost the same length as those used for learning. Texts have been produced under the same conditions as during learning: they have been printed with the same printer and scanned using the same scanning parameters (thresholds, resolution, etc.). The classification has been performed by a multivariate Bayesian classifier using the base of 240 font models.

The algorithm is applied on each image (corresponding to a text written with one font) such that for each text line, the classifier returns in a decreasing order the most probable font identifiers. A font is considered to be correct if its family, weight, slope and size are correct. A simple averaging is then performed and a font recognition rate is retained for the whole image. Other average recognition rates are computed for individual font attributes, i.e. the family, weight, slope and size.

3.2   Results

Table 1 shows the average recognition rates for fonts and font attributes. Fonts have been quite well recognized with an average rate of 96.5% for top choice, 98.3% within the two top choices. These rates reach an accuracy of 99% within the four top choices. Weights and slopes have been accurately recognized with rates above 99%, while families and sizes recognition has been less accurate with an average rate of 97%.

The experience has shown the presence of three typeface clusters (*seriffed, sans-serif* and *fixed pitch*), such that family misclassifications mainly occur within the same cluster. In practice, a 99% discrimination rate between *seriffed* and *sans-serif* fonts has been reached, which can be very helpful for OCR systems.

Table 1. *Average recognition rates of fonts and font attributes performed by a Bayesian classifier applied on 240 fonts using a base of 240 font models [(1): top choice, (2): two top choices].*

	Font (2)	Font (1)	Family(1)	Weight(1)	Slope(1)	Size(1)
Avant Garde	0.996	0.990	0.991	0.999	0.998	0.996
Bookman	0.985	0.971	0.973	0.999	0.999	0.989
Computer Modern	0.957	0.913	0.981	0.999	0.999	0.934
Courier	0.970	0.945	0.991	0.999	0.997	0.929
Franklin Gothic	0.997	0.980	0.980	0.999	0.997	0.991
Garamond	0.995	0.992	0.994	0.999	0.999	0.995
Helvetica	0.982	0.967	0.967	0.999	0.999	0.990
Lubalin Graph	0.966	0.960	0.964	0.999	0.995	0.974
Lucida Bright	0.996	0.987	0.991	0.999	0.997	0.994
Lucida Sans	0.996	0.964	0.967	0.999	0.998	0.991
Lucida Sans TypeW.	0.995	0.983	0.985	0.999	0.998	0.993
New Century Schlbk	0.997	0.980	0.984	0.998	0.999	0.991
Palatino	0.980	0.959	0.960	0.997	0.998	0.983
Souvenir	0.982	0.955	0.957	0.996	0.997	0.986
Times	0.959	0.924	0.959	0.998	0.993	0.956
average	0.983	0.965	0.976	0.998	0.997	0.979

The approach seems very promising in the context of real document structures recognition because:

- generally, the fonts used in the document is known a priori, which allows to reduce the size of the font models base;
- typographers use few and very distinct typefaces within the same document.

4 Conclusions

We have shown in this paper the reliability of an *a priori font recognition* approach based on statistical analysis of text line images. Some evaluations have shown the system efficiency when images used in the tests are produced under the same conditions as those used for training. Indeed, the system allows an accurate identification of fonts and their individual attributes. In order to improve the system usability, other quantitative evaluations have to be carried out.

The first one consists in evaluating the system behavior in the context of degraded images. We have shown in [6] that the system is relatively robust against natural degradations generated by scanners and photocopiers but its performances drop dramatically in presence of very degraded images. In order to improve the system tolerance to degradations, an approach allowing an automatic adaptation of the system to modeled degradations, has been adopted and provided some promising results.

The second one aims at the study of the effects of the text length and content on the system performances. Preliminary evaluations of *ApOFIS* on single words have shown that it performs an accurate font style (weight, slope) recognition on words of more than 3 letters. We are currently, analyzing how well font models computed from text lines (set of words) can be adapted to recognize single words.

References

1. T. Bayer, J. Hull, and G. Nagy, 'Character Recognition: SSPR'90 working group report', in *Structured Document Image Analysis*, eds., H.S. Baird, H. Bunke, and K. Yamamoto, 567–567, Springer Verlag, (1992).
2. A. Zramdini and R. Ingold, 'Optical font recognition from projection profiles', in *RIDT'94: Third International Conference on Raster Imaging and Digital Typography*, pp. 249–260, Darmstadt, Germany, (4 1994).
3. R. A. Morris, 'Classification of digital typefaces using spectral signatures', *Pattern Recognition*, 25(8), 869–876, (1988).
4. Tao Hu, *New Methods for Robust and Efficient Recognition of the Logical Structures in Documents*, Ph.D. dissertation, University of Fribourg, 1994.
5. A. Zramdini and R. Ingold, 'A priori font recognition using a Bayesain classifier', Technical report, IIUF, University of Fribourg, Switzerland, (2 1994).
6. A. Zramdini and R. Ingold, 'A Study of Document Image Degradation Effects on Font Recognition', to be published in *ICDAR'95: Third International Conference on Document Analysis and Recognition*, Montreal, Canada, (8 1995).

Extracting Words and Multi-part Symbols in Graphics Rich Documents

Mark Burge* and Gladys Monagan

Institut für Informationssysteme
Swiss Federal Institute of Technology (ETH)
ETH-Zentrum, CH-8092 Zurich, Switzerland

Abstract. We present an algorithm for grouping multipart symbols, dashed lines, and character strings for extraction from line drawings. The image undergoes a lossless raster-to-vector conversion creating as its vector representation an undirected graph, a so-called *run graph*. Next, the image elements of the run graph are extracted and classified probabilistically based upon their geometric features using a decision tree. An *area Voronoi tessellation* of the members of the sets is constructed, from which a *neighborhood graph* is derived, which is guaranteed to be minimal and complete. The graph is then traversed to group the members of the various sets for extraction and input to different recognition modules. No a priori font or other domain specific information is required for the grouping, and no special geometrical relationships among the elements are assumed. Results are presented with example images taken from those used by our Swiss cadastral map understanding system.

1 Introduction

Line drawings contain semantically meaningful objects which are often made up of several visually disjoint parts, or image elements. Analyzing the elements of these objects in isolation may lead to an incorrect interpretation of the object. Such is the case when analyzing a single dash of a dashed line, and the dash in the letter "i", instead of considering the entire dashed line or the complete glyph for the letter "i". We propose an algorithm which uses a neighborhood graph to represent explicitly the spatial relations between all the elements of an image. Using this graph and simple geometric features, we show that it is possible to group image elements into semantically meaningful objects without recourse to application specific knowledge.

The grouping for extraction of symbols, dashed lines, and character strings which appear in a document is a problem encountered in interpreting line images such as maps and engineering drawings [1]. Nakamura et al. [2] give five reasons why character string extraction is difficult in topographic maps: characters often touch background figures, existence of many character like figures, various orientations of strings, intra-character spacing is different from string to string, and character strings are often close together. These problems are present in many types of line drawings, not just topographic maps. The algorithm presented in this paper deals with all but one of the problems listed by

---

* Now with Systemtheorie, Johannes Kepler Universität, A-4040 Linz, Austria

Nakamura et al., that of characters which touch background figures. To resolve this, high level knowledge is required, which our algorithm specifically avoids using.

We present a two phase algorithm, the first phase we call *localization*, and the second phase *grouping*. (Another name for localization is segmentation, namely, the separation of text and graphics.) The algorithm does not require a priori knowledge of the font, point size, or intra-character spacing of the character strings. Descriptions of the symbols used or of the dashed line patterns are also unnecessary. The only information necessary is the resolution at which the image was scanned, and the feature values which can distinguish between the image elements which are to be localized.

2 Localization

Input is a raster image scanned at any resolution, which undergoes lossless raster-to-vector conversion. The result of this vectorization is a graph, the run graph [3], with subgraphs which correspond to the connected components, or image elements. Using the run graph representation as the basis, geometric features are computed which are used by a decision tree to classify probabilistically each element of the image into one of five sets: dots, circles, dashes, symbols, and graphics. The features used are based on a convex hull formed from the nodes of the run graph. Features used are the area of this convex hull, the longest side and the complexity, i.e. the number of nodes in the run graph of the image element.

3 Grouping

Conceptually we wish to group together image elements which when combined form an object which has some semantic meaning. An example would be glyphs of various fonts which when taken together form a character string, or dashes which when grouped together form a dashed line or dashed curve. In fact, if the grouping is not done at this level, serious irreparable errors could occur. Such is the case when multi-part symbols are not grouped. A circle within a circle in a cadastral map signifies a "main point." If the circles are not grouped into one symbol, they will be interpreted independently as "marking rocks" which is one of the interpretation of single circles.

The ambiguity which arises during localization, for example between dashes, ones, and "i"'s is resolved by referring to the neighborhood graph for contextual information. In our implementation, the neighborhood graph is constructed from the area Voronoi tessellation.

The point based Voronoi tessellation is unsuited for line image understanding since the useful primitives (in the context of document understanding) are not individual pixels corresponding to geometric points, but groups of pixels corresponding to segments, arcs, ellipses, or polygonal objects. To produce the neighborhood graph, we have developed an approximated area based Voronoi tessellation which can be calculated efficiently [4].

Given that $A_1, A_2, \ldots A_n$ are image elements and that p and q are particular image elements, then we can define the distance, $d_a(p, A_i)$, from p to A_i as: $d_a(p, A_i) = \min_{q \in A_i} d(p, q)$ where this represents the minimum Euclidean distance from p to any

Fig. 1. Area Voronoi tessellation.

location in A_i. Using this definition [5], one can formulate the area based Voronoi tessellation as $V_a(A_i)$, where this represents the region in which the distance to A_i is less than or equal to the distance to any other area:

$$V_a(A_i) = p|d_a(p, A_i) \leq d_a(p, A_j), j \neq i, j = 1, \ldots, n \ . \tag{1}$$

In the neighborhood graph, each node represents an element of the image, and each edge a connection to its neighboring element, as defined in Equation 1. Since the neighborhood graph is derived from the Voronoi tessellation, the neighborhoods represented by the graph can be shown to be both *minimal* in the number of neighbors and *complete*. Minimal in the sense that only those image elements which are closest are included, and complete in the sense of including all of the closest image elements.

This guarantee of the neighborhood graph being minimal and complete is important so as not to lose information or introduce false information. Other methods like radial search may result in incomplete neighborhoods, and methods like Wahl's [6] run length smearing and Meng's [7] bounding box extents may result in neighborhoods which are neither complete nor minimal. These methods can still group because they rely on additional knowledge (assumptions) about the relative positions of the elements.

The goal of the grouping process is to search the neighborhood graph to generate subgraphs so that the members of the set of dots, set of symbols, and set of dashes can be combined into candidate character strings, candidate multipart symbols, and dashed lines. When grouping, the proximity of the image elements is important but not a sufficient condition to enforce grouping. For example, in a labeled graphic, a given character may be closer to the graphic then it is to the other characters in the label; yet, this character should be grouped with the other characters into a label and not with the graphic. We want to group image elements that are similar to each other; thus we consider as local features for the grouping, both descriptive features of the image elements, as well as proximity.

The search for the grouping is as follows:

1. traverse the neighborhood graph in a breadth first manner
   - calculate the local neighborhood features
2. traverse the graph in a depth first manner
   - join edges into a subgraph, depending on the local neighborhood features
3. traverse the subgraphs in a breadth first search
   - adaptively recalculate the local thresholds to try to extend the subgraph
   - repeat until no more changes occur

The candidate dash lines are then passed on to the vectorization module, whereas the candidate multipart symbols and the candidate strings need to be interpreted by the symbol recognizer and/or word recognizer.

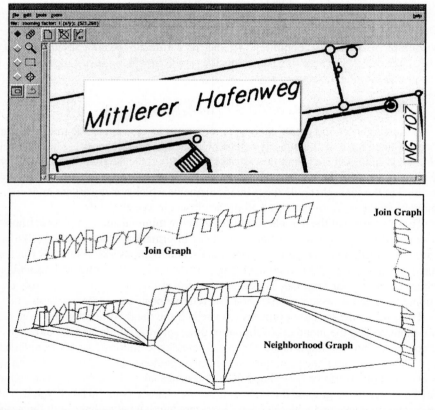

Fig. 2. In the top part of the figure, two strings which were grouped: they are written on bitmaps which are placed on top of the original raster image. Underneath, the two subgraphs resulting from the search graph algorithm. Underneath these, the neighborhood graph from which the subgraphs were derived.

4 Experimental Results

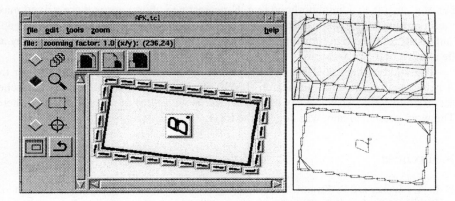

Fig. 3. On the left, a grouped dashed line and a grouped character string. On the top right the corresponding neighborhood graph and on the bottom right its subgraphs.

Figure 2 and Figure 3 show the results of our algorithm as used in our system for understanding Swiss cadastral maps. Table 1 below summarizes performance times for some sample images. The entry *components* is the number of image elements in the image, *image size* is given in pixels, *localization* is the number of seconds for the first phase including: reading the file, raster-to-vector conversion, feature calculation, and CART classification. *Grouping* includes the time for constructing the area Voronoi tessellation and the graph search for grouping. All tests were conducted on a Sparc 10 with 48 MB of real memory using algorithms implemented in C++.

name	components	image size	localization	grouping
Duden [8] document image	155	447 × 271	1.3s	7.3s
CAD image	47	1386 × 1741	3.4s	13.3s
Mittlerer example	153	3556 × 1052	5.7s	30.7s
cadastral map A0	2829	13584 × 19019	86.1s	844.1s

Table 1. Performance of localization and grouping.

5 Conclusions

After an image has been scanned, the image elements have to be localized and grouped so that they can be processed by higher level recognition modules. The main advantages that we see for the presented method of localization is that no a priori font or other domain specific information is required, and when grouping image elements, minimal and complete neighborhoods are examined. To state again, there is no predetermined number of elements in a neighborhood (e.g. K-nearest neighbors) nor is there a fixed maximum or minimum distance between the elements of a group (e.g. inter-character gap), and lastly, no fixed requirements about the geometry of the elements (e.g. that they all lie along a line) are assumed.

The authors thank the financial support of the Swiss Federal Commission for the Advancement of Scientific Research (KWF), project 2540.1, and of the Aargauisches Elektrizitätswerk (AEW). AEW also provided us with most of the test data used here. The work presented here was done at the ETH, Zürich.

References

1. L. A. Fletcher and R. Kasturi, "A Robust Algorithm for Text String Separation from Mixed Text/Graphics Images," *IEEE Transactions of Pattern Analysis and Machine Intelligence*, vol. 10, pp. 910–918, Nov. 1988.
2. A. Nakamura, O. Shiku, M. Anegawa, C. Nakamura, and H. Kuroda, " A Method for Recognizing Character Strings from Maps Using Linguistic Knowledge ," in *Proc. of the Second International Conf. on Document Analysis and Recognition*, (Tsukuba, Japan), pp. 561–564, IEEE Computer Society Press, Oct. 20-22 1993.
3. L. Boatto et al., "An Interpretation System for Land Register Maps," *IEEE Computer*, vol. 25, pp. 25–34, July 1992.
4. M. Burge and G. Monagan, " Using the Voronoi tessellation for grouping words and multi-part symbols in documents ," in *Proc. VISION GEOMETRY IV, SPIE's International Symposium on Optics, Imaging and Instrumentation*, Vol. 2573, (San Diego, California), July 9-14 1995.
5. A. Okabe, B. Boots, and K. Sugihara, "Nearest neighbourhood operations with generalized Voronoi diagrams: a review," *International Journal of Geographical Information Systems*, vol. 8, pp. 43–71, January-February 1994.
6. F. Wahl et al., "Block Segmentation and Text Extraction in Mixed Text/Image Documents," *Computer Vision, Graphics and Image Processing*, vol. 20, pp. 375–390, 1982.
7. L. Meng, "Toward the Automatic Digitization of Map Text," in *Mustererkennung 1991, Proc. 13. DAGM Symposium* (B. Radig, ed.), pp. 361–366, Springer Verlag, Oct. 1991.
8. W. R. der Dudenredaktion: G. Drosdowski, ed., *Der Duden: in 12 Bänden; das Standardwerk zur deutschen Sprache.* Dudenverlag, 1992.

A Semi-Automatic Method
for
Form Layout Description

Silvia Bussi Fulvia Mangili

Elsag Bailey, R&D Department
Via Puccini 2, I-16154 Genova

Abstract. The paper refers on some recent research results in document processing. A semi-automatic processing tool is described, in order to produce the fixed layout description of a preprinted form. A hierarchical collection of geometric primitives is the basis for a robust description of the layout, which is opened to incorporate further extended and richer representations in the future. Such geometric structures can be solid, dashed, dotted and text lines. The description is only geometric and not semantic, that is, for example, a text description contains only information about position and dimension but not the semantic content of the text. This module takes as input an empty deskewed form, in order to compensate for irregular orientations due to the acquisition process. The method is anyway very general and allows also the definition of more complex structures obtained from arbitrary combinations of the basic structures mentioned above. Many of these research results are already partially integrated and in use in a document processing machine.

1 Introduction

Intelligent document processing and recognition is a challenge of modern systems for office and business automation. There has been recently an increasing interest in the scientific literature with different proposed approaches for document description and modelling, as in [2] and [3]. The overall system for document reading and management is a collection of different technological components, from high speed image enhancement and processing, pattern recognition, network communication, data base management and archival, MMI for video coding, high-level context analysis and postprocessing. Typical applications are document processing for banks, insurance companies and various administrations.

In the following we will focus our attention to a particular stage of the document processing chain, i.e. the modelling of the static form layout which has to be properly identified and described in order to provide the necessary information to the following modules in the chain. These modules will be responsible for removal of this structure and to perform the following stages of character and text recognition (both typewritten and handwritten). For this layout description a model based approach is used, with extremely promising results in a variety of form processing applications. The main goal of a form image processing system is to produce a symbolic description of these features, suitable for the following automatic interpretation of the contained information.

In order to simplify the filling of the form, the majority of forms are organized in a complex structure which often represents an obstacle to information retrieval expecially when the applications cannot readily use drop-out colored forms, cannot control the form they process, or have to process faxes and copies of original documents. For this reason, the form layout has to be properly identified, described and dinamically removed from the document itself, to simplify the on-line stages of character and text recognition and to allow, at the same time, the compression of the data to be processed by the optical character recognizer. A system capable to give a form description by means of the basic structures of the form layout is then required. In general, the form description is obtained during an off-line step of the analysis process dealing with empty forms and the time complexity does not represent a serious problem. This form description is then used by the following stage of on-line form removal which has much more severe processing time constraints.

It's necessary to distinguish geometric analysis from semantic analysis. The geometric analysis gives a set of topological information about every object contained in the form. The semantic analysis gives information also about the meaning of the objects. The process of geometric analysis produces a symbolic description of the form which concerns the geometric structure and the classification of the objects contained on it.

The necessity of an automatic description of the form layout raises from some considerations about the most common manual description process. Also with the availability of powerful graphical tools, the manual layout definition is a long and boring process for a user. Moreover the user must know very deeply the on-line process to produce a "good" layout description. In fact, any user's error during the definition of the layout may cause erroneous behaviour of the on-line process. These considerations are all in favour of an automatic description process. The interactivity with the user can be saved, but must be strictly constrained in order to avoid mismatching between off-line and on-line process.

2   Structure of the Description Process

In this section, the main steps of the semi-automatic description process will be considered. Three main steps can be identified in the form layout description process:

1. input interactivity;
2. automatic description;
3. **output interactivity.**

The first step consists in the selection of the form image to be processed. The description process can be activated on the whole form or only on interesting parts of it. The user can also modify some parameters used by the automatic process to produce the description. The second part is the activation of the automatic description process on the previously selected portions of the form with the parameters chosen by the user or with the default parameters. It produces the description of the basic structures contained in the selected areas of the form. The third part of the process has the purpose of results visualization and allows the user the analysis and the correction of them.

2.1 Input Interactivity

When the form image to be processed has been selected, some input interactivity with the user is necessary in order to obtain a description which can be used by the form removal process. In fact, the user may need to treat different areas present on the form in a particular way. In general, independent rectangular regions located in a form image can be classified in two main groups:

1. regions that must be described as rectangles, regardless of which kind of topological entities they contain;
2. regions on which the automatic process must be runned to find the basic structures belonging to them;

The user should, before activation of the completely automatic description part, identify and select such regions. Moreover, once the regions to be described in detail have been chosen, some other decisions may influence the behaviour of the automatic process. The automatic recognition of structures is based on some default parameters that guide the classification of the different structures. If the form image contains regions that present different characteristics from one region to another as character dimension or average line thickness, it is useful to have the possibility to adapt the parameters to the characteristics of the various regions in order to improve the quality of the description. This is done by allowing the automatic process to stop while passing from one region to another and asking the user for parameter settings confirmation. Obviously, the information required must be easily understood by the user and not proper of the computation, and must be translated in a second time in a set of values that will then be used during the description process. For example, the user may be asked if the form is noisy or not, but the user will never be able to set the maximal dimension of a noise component due to the scanning process.

2.2 Automatic Process

The basic structures identified by the automatic process are horizontal and vertical solid, dotted, dashed and text lines. According to the different characterization of the solid lines with respect to the dotted, dashed and text lines, different approaches have been applied for each class. Dotted, dashed and text lines are obtained by grouping collinear adjacent small components of the image that have similar dimensional characteristics such as characters, dots or dashes. This step is performed by computing all the connected components of the form image whose width and height can be compared with a character, a dot or a dash, but that are greater than a threshold established for noise components. These connected components are then classified in dots, dashes and characters on the basis of dimensional thresholds and of the context in which they are placed. These connected components are then grouped according to their collinearity and proximity properties in order to obtain composite structures such as text lines and so on.

All the conncted components computed in this first step of the automatic process are then removed and a quite different approach is adopted for the description of the structures still present on the form image. An auxiliary structure is used to represent endpoints, intersections or curvature change of the lines. The auxiliary structure is a graph whose nodes represent intersections or curvature changes and whose edges are horizontal or vertical line segments [1] [4]. An image representation by mean of a graph appears as a good model for a form layout description because it results in a reduction of the number of entities to treat with respect to the bitmap.and gives more information about the topology of the structures contained in the image. This structure can be used to identify both isolated or intersecting solid lines contained in the empty form. Horizontal and vertical lines are identified by visiting the graph starting from every node and merging all the horizontal and vertical line segments incident in the node.

2.3   Output Interactivity

As no information about the form type is required, a totally automatic form description algorithm may be very difficult to implement. At a certain step, the user knowledge must be employed. For this reason a third interactive step is allowed. In this phase, the user can examine the results obtained in the previous step and evaluate their quality. Moreover, the user must be able to correct some results obtained by the automatic procedure. The three main actions which allow a complete correction of possible errors which may be detected at this step are:

1. complete definition of a non-detected structure;
2. modification of attributes of type and dimension of the detected structures;
3. suppression of a detected structure.

The quality of the obtained results and, as a consequence, the quantity of structures to be added or modified during this step depends on the parameters set at the input interaction step. The algorithm has been tested on about a hundred of different forms, an example of which is showed in Fig. 1, and, for each of them, a quite good description of the basic structures belonging to its layout has been obtained even with the default parameters. In general, only particular

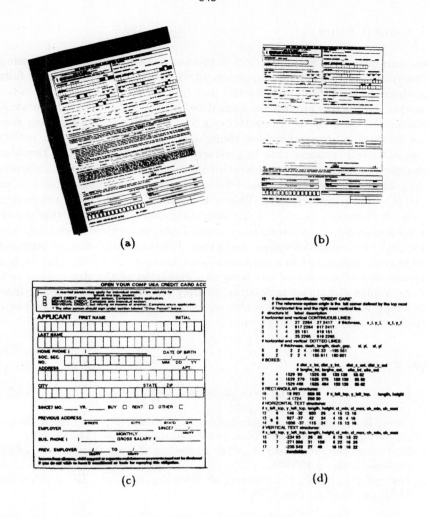

Fig. 1. The main steps of the form layout description tool: image acquisition (a), image deskew (b), field selection (c), output description file (d)

structures like single character strings or very short lines must be manually introduced by the user. Another important feature which makes more realiable a semi-automatic tool description is the possibility of complete or partial rejection of the description obtained. In this case, the repetition of the second part of the computation on some portions of the form with different parameters may be more efficient. Once the geometric process and output interactivity have produced the correct description, all the data necessary to the description of the layout structures are saved in a file that is used by the on-line form removal process.

3 Conclusions

The semi-automatic form description process has been implemented using the C language and a Windows environment. The user interface is used both for the visualization process and for the user interaction. The tool is already integrated in a Windows application named Definition Utility, which is the standard environment used by the Elsag Bailey's SLAMNET product for form reading. This description process has been thought as a first step towards the realization of an automatic process able to produce not only the geometric but also the semantic description of a form or other more generic document. This ambitious final goal is suggested by the powerful algorithm methodologies, too specialized to stop at the present results. This is the reason why the best automatic tool is not an objective, in fact, in the operating context in which this product is placed is convenient, and sometime unavoidable, to preview a level over the one an expert user must determine the system behaviour and tuning. Anyway, it's sure that the possibility of a semi-automatic description of a form layout is a real improvement of the actual situation. An extension of the description to other more complex structures is possible, as they are all obtained starting from the basic type whose description is already realized, as sequences of boxes or comb structures, check boxes and so on.

References

1. Boatto, L. et al., "An Interpretation System for Land Register Maps", *IEEE Computer*, vol.25, pp. 25-34, July 1992.
2. Casey, R., Ferguson, D., Mohiudiuddin, K., Walach, E., "Intelligent Forms Processing System", *Machine Vision and Applications*, 5, pp. 143-155, 1992.
3. Doermann, D.S., Rosenfeld, A., "The Processing of Form Documents", *IEEE Proceedings*, 1993, pp.497-501.
4. Monagan, G., Röösli, M., "Appropriate Base Representation Using a Run Graph", *IEEE*, pp. 623-626, July 1993.

Improving the Use of Contours for Off-Line Cursive Script Segmentation

G. Boccignone, M. De Santo

DIIIE
Università Degli Studi di Salerno
84084 Fisciano (Sa) - Italy
{desanto, boccig}@dia.unisa.it

A. Chianese, A. Picariello

DIS
Università Degli Studi Federico II
Via Claudio, 21 80125 Napoli - Italy
chianese@nadis.dis.unina.it

Abstract. In this paper we show a new method for efficiently extracting information from contour analysis and demonstrate their use to face the Cursive Script Segmentation challenge. Expressly, our method improves the use of contour to detect the so-called baselines and to use them for a "zone" analysis of the text rows. Furthermore, we show how it is possible to use our method to solve critical situations due to slant and character overhanging. A final discussion about experimental results is also reported.

1. Introduction

Cursive Script Recognition (CSR) is a great challenge in the field of Optical Character Recognition. Several methods have been proposed in the literature for *off-line* CSR, but they can be collected into two main classes: analytical and global methods [1]. While global methods attempt to recognise an entire word, the analytical ones attempt to recognize the characters which form a word. In analytical methods, the character *segmentation,* i.e. the detection of characters within a word, is considered particularly critical. Many authors have used contour derived information for segmentation, due to the fact that it is relatively inexpensive to be obtained and moreover it can be extracted during other pre-processing phases, which are considered mandatory ones by a number of authors. Maier [2] introduced a scanning strategy for a classification of neighboring pixels which brings to achieve vertical cuts in the local minima of connecting strokes of the text. Lecolinet and Crettez [1] made a distinction between significant components of the script related to the characters in a word -'*graphemes*'- and unessential components related to the so called '*ligatures*' among characters. They showed that most of the ligatures are the valleys of the upper outline of a given word. Kahan et Al. [3] detected the cutting points of the touching characters analyzing a function based on vertical pixels projection. Other authors [4,5] improved the previous method adding the use of a function H(x) defined as the difference between the top and the bottom profiles of external contour. In spite of their interesting characteristics, last quoted methods have been principally applied for segmenting *printed* characters, where main problems derive from the skew of text lines in the document. On the contrary, cursive script shows very significative and pervasive slant and overhanging problems and, what is worst, text lines that are not linear or horizontal. So, for CSR various techniques have been introduced to recover these problems. Bozinovic and

Srihari [6] used contour to formulate some heuristic considerations (e.g., presence of peaks and valleys, dots over i, holes, ...). Kimura et al. [7] proposed a methodology based on recursive classification/segmentation, using a 'similarity function' based on contours. Finally, there are other approaches [8] where the slant and overhanging problems have been faced using concavity and convessity analysis. As we noted, it is important to remark that all mentioned methods make a considerable amount of work to avoid slant and overhanging effects and, what is worst, this work is frequently lost for successive elaboration. Starting from these considerations, in this paper we introduce a new method which uses contour derived information in detecting cut points through "base-lines" determination which can be usefully utilized in the classification phase (see for ex. [5, 6]). The paper is organized as follows. In section, 2 we propose a preliminary analysis useful for base-lines extraction. In the same section, we furnish a new method for cut-points detection based on the idea that ligatures mostly occur in middle-zone of the text line. Finally, in section 3 we discuss some experimental results and in section 4 some concluding remarks.

2. The Method: Base-lines Detection and Segmentation through Contour Analysis

The first step of our method consists in detecting zones in which is possible to simply and reliably recognize the ligatures among characters. As a starting point, a ligature can be detected considering the histograms of the vertical thickness of the contour, described by the function called $H(x)$ that, as reported in [4], is defined as the distance between the maximum and minimum y-coordinate of the contour of a word, for each x-coordinate. So, the ligatures could be found in the region in which $H(x)$ has valleys surrounded by peaks. However, it is easy to note that $H(x)$ gives wrong information in the case of characters with overhanging boxes. We can avoid these drawbacks considering that ligatures among characters occur in the *middle zone* of the text line and defining a new function based on the histograms calculated in the middle zone itself. In this work we start from the hypothesis that the input text has been previously organized in text lines. A text line, in occidental languages, is a collection of proximate successive words which span from left to right position in the document and may be segmented into three regions or zones: upper, lower and middle zone; in the literature this activity is known as 'base-line determination', and refers to the detection of the main body of a cursive line [6]. The upper zone refers to the ascender components of characters, the lower zone to the descender components, and the middle zone to the main body. Different methods have been used in the literature for base lines individuation, mostly based on histograms analysis [6, 7]: however, these methods are strongly influenced by the presence of writing and digitalization skew of the text line and by the quoted fact that they are not strictly horizontal or linear. We avoid the problems generated by histograms, using contour features that will be reused in successive phases. For this aim, it is useful to introduce some preliminary definitions. Let us consider a discrete plane $\{x_n, y_n\}$ and a closed region whose border is detected by Γ, described by the function $\Gamma: f(x_n, y_n) = 0$ where $(x_n, y_n) \in I \times I$, being I the integer set. Let $R_x(x^*_n)$ and $R_y(y^*_n)$ be subranges of $\mathbf{I}$ defined as

$$R_x(x^*_n) = [x^*_n - \Delta x_n, x^*_n + \Delta x_n] \qquad R_y(y^*_n) = [y^*_n - \Delta y_n, y^*_n + \Delta y_n]$$

We define *extremes* of Γ those points (x^*_n, y^*_n) for which one of the following condition occur:

(1)	$f(x^*_n, y^*n) > f(x^*n+k, y^*n)$	(local maximum)
(2)	$f(x^*n, y^*n) < f(x^*n+k, y^*n)$	(local minimum)
(3)	$f(x^*n, y^*n) > f(x^*n, y^*n+r)$	(local left extreme)
(4)	$f(x^*n, y^*n) < f(x^*n, y^*n+r)$	(local right extreme)

for each $k \hat{I} Rx(x^*n)$ and for each $r \hat{I} Ry(y^*n)$

Using geometrical considerations, we also define ascender contour a set of points of G (i.e. and arc) which starts from a local minimum and finishes to the first successive local maximum, and descender contour an arc which starts from a local maximum and finishes to the first successive local minimum, upper contour the arc which starts from left extreme minimum and finishes to the first successive right extreme, and lower contour an arc which starts from right extreme and finishes to the first successive left extreme (figure 1).

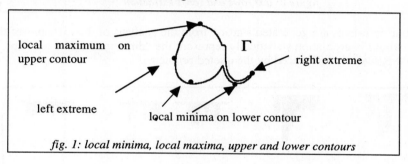

fig. 1: local minima, local maxima, upper and lower contours

Given a text line, let us define Σ as the set of local maxima detected on the upper contours of the line and σ as the set of local minima of lower contours. By means of a least square interpolation of Σ and σ, we may usefully approximate the upper-baseline and the lower base-line of our text. Note that our algorithm dinamically chooses the Δx_n and Δy_n ranges of the previous formulas taking care of avoiding the inclusion in Σ and σ of points which are near to "pen-up" and "pen-down" areas. In fact, these points do not represent significative information for base-lines determination but, on the contrary, may bring to distortions. Fig. 2 shows the steps for base-lines detection.

While it is clear that the accuracy of our approximation depends on the number of points considered, i.e. it is better if the text-line is long, we have found that experimental results show how, even if considering short lines (e.g., a single word), our method does not reduce the number of right cut points and does not introduce wrong cut points.

Furthermore, it is important to note that the use of this method to detect the middle zone avoid the effects of the skew and minimize those due to the unperfect linearity of text.

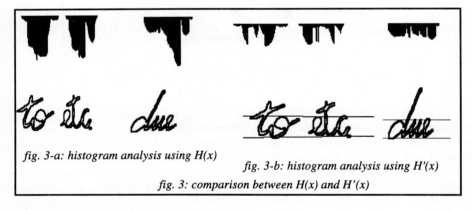

2.a : the input bitmap

2.b: upper contour

2.c: lower contour

2.d: local minima and local maxima

2.e: base-lines detection

figure 2: text rows and zones extraction

In fact, our base-lines are generated starting from the analysis of the *extremes* of the contour, whose interpolation statistically represent the "direction" of the text-line in a way that is scarcely influenced by the quoted problems.

fig. 3-a: histogram analysis using H(x)

fig. 3-b: histogram analysis using H'(x)

fig. 3: comparison between H(x) and H'(x)

Now, using the base-lines, we can usefully introduce a new function for cut points detection. This function, that we call H'(x), is defined as the distance along y'-axis between the upper and the lower contours in the middle zone for each x'-coordinate. The new y'axis is obtained considering the new coordinate system which assumes the lower base line as its x'-axis. In figure 3 we show the application of H(x) and H'(x) functions to some word samples. It is easy to see that H(x) function fails in presence of horizontal strokes which span over successive characters in the word. In the same figure we illustrate an equivalent critical situation caused by 'slant'. An excessive slant, in fact, causes the overhanging among boundary boxes of 'd' and 'u'. Using the function H'(x), we can avoid this kind of problems.

3. Experimental results and discussion

We tested our Cursive Script Segmentation method on a database of some hundreds of handwritten mail addresses collected by students of our University (see fig. 4 for an example). The aim of our experiments was an analysis of safe cut points detected by our method by means of a comparison between the results obtained using H and H' functions and the average of points detected by human observers [9]. In particular we want to find: a) the number of undetected cut points; b) the number of intercharacter cut points detected within a single letter (typical of "m", "n", "u", etc.); c) the number of superfluous cut points detected within the same ligature. In tab. 1 we show the results of the described experimental work.

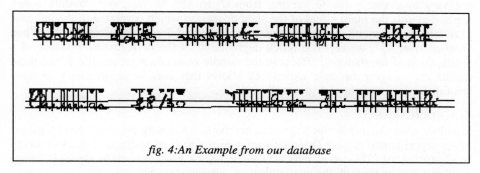

fig. 4:An Example from our database

	undetected cut points (%)	inter-character cut points (% on Total)	superfluous cut points - (% on Total)
H(x)	13	28	12
H'(x)	5	31	14

tab. 1: experimental results

The use of H' reduces the number of undetected cut points from 13% to 5%. This first positive result confirms the hypothesis that a large number of ligatures is located in the middle zone of the text line. Let us now consider the results relative to intercharacter cut points. Both H and H' functions exhibit an high number of wrong cut points due to the fact that the human observer uses semantical information to individuate true ligatures. Our approach slightly amplifies the number of inter character cut points because "m", "n", "u" characters are mainly located in the middle zone. For the same reasons, similar results can be observed for superflous cut points. However, it is important to remark that intercharacter cut points and superflous cut points can be recovered in an easier way than undetected cut points in successive steps of the recognition process. In a successive phase of our experiments, we start to quantitatively test the effectiveness of our method with respect to its sensitivity to the slant. To this aim, about one hundred of mail addresses samples have been transformed by means of a bit map rotation (see fig. 5), using graphical facilities.

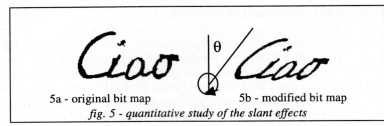

5a - original bit map   5b - modified bit map

fig. 5 - quantitative study of the slant effects

Let us call Θ the rotation angle. We found that the performance of the method does not vary appreciably for Θ varying from 0° to 45°. In fact, our "middle zone" method avoids the main effect of the slant distorsion which is related to the increase of overhanging among characters due to ascender and descender components. When Θ spans over 45°, we report a quick degradation of the performance because of a propagation of overhanging effects to the middle zone components. We found these results encouraging because experience shows that cursive script rarely exhibits slant distortions with angles greater than 45°.

4. Conclusions

The discussed Cursive Script Segmentation method has been implemented by means of an experimental Object Oriented Environment System realized at the University of Salerno [10]. This environment allows a C++ programmer to develop image processing tasks through the availability of suitable class libraries. We are currently integrating the method with some heuristic knowledge to be applied in the character detection and we are starting a massive testing of the method itself.

References

1.   E.Lecolinet and J.Crettez, "A Grapheme-based Segmentation Technique for Cursive Script Recognition", *ICDAR 1991*, Saint-Malo, 1991.

2.   M. Maier, "Separating Characters in Scripted Documents", *ICPR 1986*, Paris 1986

3.   S. Kahan, T. Pavlidis, H. S. Baird "On the recognition of printed characters of any font and size", IEEE Trans. Pattern Anal. and Machine Int., PAMI-9, 2, 1987.

4.   F. Arcelli, A. Chianese, M. De Santo, A. Picariello, "Approaching Char. Segmentation Using Cont./Skel. Analysis",. *SCIA 1993, Tromso, 1993*.

5.   S. Liand, M. Shridhar, M, Ahmadi, "Segmentation of touching char. in printed document recognition", Pattern Recognition, 27, 6, 1994

6.   R.M. Bozinovic, S.N. Srihari, "Off-line Cursive Script Word Recognition", IEEE trans. on PAMI, PAMI-11, 1989.

7.   Kimura, Shridhar, Chen "Improvements of a Lex. Dir. Algorithm for the Recogn. of Uncostrained Handw. Words", *ICDAR 1993*, Tokyo, 1993.

8.   B. Plessis, et. al., "A Multiclassifier Combination Strategy for the Recognition of Handwritten Cursive Words",. *ICDAR, Tokyo* 1993.

9.   A. Fisher, R. Bolles, "Perceptual organization and curve partitioning ", IEEE Trans. Pattern Anal. and Machine Int., PAMI-8, 1, 1986.

10.   G. Boccignone, A. Chianese, M. De Santo, A. Picariello, "Building an O.O Env. for Image Analysis and Processing", *ICDAR 1993*, Tokyo, 1993.

A Method for Determining Address Format in the Automated Sorting of Japanese Mail

T. Tsuchiya, N. Nakajima and T. Kamimura

Information Technology Research Laboratories, NEC Corporation
4-1-1 Miyazaki, Miyamae-ku, Kawasaki, Kanagawa 216 JAPAN

Abstract. This paper presents a new method for determining the address format used on mail that has been hand-written in Japanese, a task made particularly difficult by the variety of formats possible in Japanese. We classify possible formats into six types, and identify the distinguishing features of each. In the proposed method, features characterizing any of the six types are identified for a given address, and from this a list of format-candidates is generated. Character lines are then determined for any format candidate, and one candidate is subsequently selected on the basis of the statistical likelihood of any address being written in that particular way, given the location and size of the character lines. We also demonstrate the effectiveness of the new method experimentally.

1 Introduction

A lot of research has been done related to ways of automating mail sorting work at post offices to handle greater volumes of mail smoothly and to improve the postal service. In particular, there is a strong need to automize delivery sorting, i.e. sorting letters and postcards to delivery zones which are assigned to individual carriers. Many ideas for delivery sorting have been proposed[1][2].

Delivery sorting includes four processes, Address Block Finding (ABF), character segmentation, character recognition, and address recognition. Because ABF is the first step in the delivery sorting process, and because it is not easy to recover from ABF errors, it should be as stable and accurate as possible.

ABF detects macroscopic address information, i.e. the address format and the location of the address lines. Here, we define "address format" as the direction of address lines (vertical or horizontal) and the direction of the written characters. The location of an address line may be expressed in the form of the coordinates of a rectangle circumscribing that address line. All of this macroscopic address information is necessary for delivery sorting of Japanese mail because without it individual characters cannot be segmented out and recognized.

It is difficult to determine the address format being used, because Japanese mail has many format variations. One of the most common methods for determining format does so on the basis of the location and size of detected address lines. However, some classes of address formats are hard to determine by this method.

We propose an address format determination method that uses features that characterize Japanese address formats. First, we define specialized terms, then point out the difficulties involved in format determination, and then descirbe our method. Finally, we present details of experiments with mail images which were carried out to demonstrate the effectiveness of our method.

2 Japanese Address Formats and The Difficulty in Determining Them

2.1 Japanese Address Formats

In an address recognition machine, letters and postcards are conveyed by a belt and their images are obtained when they run past scanner line sensors. The origin and the coordinates for a mail image are defined according to the running direction, as shown in Fig. 1. The figure also shows the top, bottom, forward and back positions relative to the mail image.

Addresses written in Japanese have many more variations than ones written in English. (For example, English is almost never written vertically.)

Japanese address formats are broadly classified in terms of two items, the postcard or envelope format (portrait or landscape) and the address line direction (vertical or horizontal) (See Fig. 2).

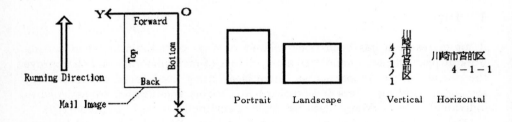

Fig. 1. Origin and Coordinates   Fig. 2. Classification of Japanese mail

We thus defined six formats as shown in Fig. 3. The "vertical plus landscape" combination is not considered, because this combination is almost never used in actual mail. The term "normal" means that the characters are written toward the bottom or back edge and the term "reverse" means that they are written toward the top or forward edge.

2.2 The Difficulty in Determining Address Formats

In one conventional method, the address format is determined based on the location and size of the detected address lines. Specifically, it assumes that there are several formats as format candidates, and detects address lines for each format candidate and compares the likelihood scores of the detected lines. The score is calculated based on rank tables where approximate probability densities for the location and size of the address lines are listed. The tables are determined

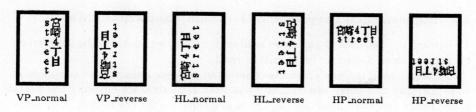

Fig. 3. Mail Format Definition

in advance based on mail image databases for each format[3]. The method is effective when the locations and sizes of address lines are for the most part fixed in each format.

However, some classes of address formats are hard to determine with this method. Figure 4 shows two examples of format determination errors that occur with this method.

Figure 4(a) shows VP_normal format mail and Fig. 4(b) shows HL_normal format mail. In Fig. 4(a), the sender's name was detected wrongly as an address line when the format candidate was HL_normal format. In a case like this, it is difficult to distinguish VP_normal format from HL_normal format with this method because both the location and size of the detected address line are almost identical.

This indicates that there are some types of mail for which it is difficult to determine address formats by using only the location and size of address lines.

In order to develop a better format determination process, it is necessary to use other information that is present in mail images.

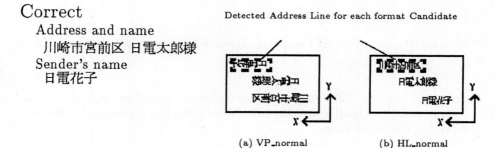

Fig. 4. Example of Difficult-to-Determine Format Using Only Detected Address Lines

2.3 Supplemental Properties of Japanese Handwritten Mail

To extract the features of Japanese address formats, we consider the supplemental properties of Japanese handwritten mail.

Japanese mail include elements other than address lines. There is almost always a stamp and there are frequently printed red boxes for entering the postal code. There are sometimes line segments as guidelines for the addrss. And sometimes there are printed character lines, for example the corporation name and address, near the edge. These may be printed within a box frame. Detecting special elements such as these is one effective way to determine the address format.

3 Format Determination Process Using Mail Features

3.1 Outline of Format Determination Process

The format determination is composed of two processes, Format Candidates Selection (FCS) and Final Determination.

FCS detects specific features on a mail image and generates format candidates. If FCS generates several format candidates, Final Determination selects the final format from the candidates by means of a conventional method using the location and size of the address lines.

To solve the problem described in Section 2, we introduce FCS into the format determination. FCS helps to make format determination more accurate because it makes it possible to determine address formats that cannot easily be determined by the conventional method. How this is done is explained in the next section.

3.2 Format Candidates Selection

FCS is mainly composed of two parts, Feature Detection and Selection Using Rules.

Feature Detection This consists of several feature detectors, such as stamps, red boxes, mail item length, direction of address lines, line segments, box frames, and printed character lines.

Selection Using Rules This generates format candidates using detected features. It has three rule-based processes as follows.

A. Conversion

This process converts detected features into attribute values. Examples of attributes are "Existence of stamps", "Existence of red boxes", and "Existence of pattern on forward-back position".

For example, the attribute "Existence of pattern on forward-back position" has five values: "box frame on forward position", "box frame on back position", "printed character line on forward position", "printed character line on back position" or "not clear".

B. Matching

This process selects address formats that match each attribute value. For example, the value "printed character line on back position" can match the VP_normal and HP_normal formats in Japanese mail.

All the formats that match all the attribute values become format candidates as the result of the matching process.

C. Adjustment

This process adjusts conflicts in matching rules. Conflicts occur if no format candidates are left after applying the matching rules.

4 Experiment

We have developed a simulator of a handwritten address recognition system on EWS4800 (NEC workstation with MIPS risc chips). The proposed method was applied to 455 mail images by the simulator. We compared these simulator results with the correct results, which were prepared manually in advance. The format determination results are shown in Table 1. The total rate of correct format determination was 93.4%, confirming that the proposed method is effective.

Table 1. Results of Format Determination

True		Results						Total	Correct rate
		VP_n	VP_r	HL_n	HL_r	HP_n	HP_r		
	VP_n	274		6	1	4	·	285	96.1%
	VP_r							0	
	HL_n	7	1	49	3			60	81.7%
	HL_r	2			7	1		10	70.0%
	HP_n	5				95		100	95.0%
	HP_r							0	
								455	93.4%

Table 2 shows the rate of correct format determination that was calculated from the number of format candidates. The number is 1 or 2 and the two cases are mutually irreconcilable. Format determination is regarded as correct when the macroscopic address information, which is the ABF output, includes the correct address format.

The rate of correct determination was 95.8% when the number of format candidates was 1 and 91.4% when the number of format candidates was 2. This compares with 87.9% and 87.1% respectively for the conventional method.

Thus the proposed method yields better results than the conventional method. Its performance is significantly better when the number of format candidates is 1, in which case the format is determined only by FCS. Hence, it is clear that FCS is effective.

Table 2. Rate of Correct Format Determination by Number of Format Candidates

Number of Format Candidates	Number of Mail [percent]	Rate of Correct Format Determination by Proposed Method	Rate of Correct Format Determination by Conventional Method
1	306 [67.3%]	95.8%	87.9%
2	139 [30.5%]	91.4%	87.1%

5 Conclusion

Our address format determination method for mail sorting is more effective than the conventional method. It introduces Format Candidates Selection (FCS), which detects features that characterize each format, and generates format candidates using these features, before address lines are detected. This makes it possible to determine address formats that are hard to determine using only results of address line detection.

When the number of format candidates is 1, the rate of correct format determination is especially improved, proving that FCS is effective.

We would like to improve the ABF process by using the results of other processes such as character recognition and address recognition.

References

1. A. Gardin (CGA-HBS): A Real Time Address Block Location System for Handwritten Letter and Flat Mail. USPS Advanced Technology Conference 4 1992
2. Sargur N. Srihari, Ching-Huei Wang, Paul W. Palumbo and Jonathan J. Hull (State University of New York): Recognizing Address Blocks on Mail Pieces: Specialized Tools and Problem-Solving Architecture. AI Magazine, Winter (1987) 25–40
3. T. Ishikawa, Y. Nishijima, Y. Tsuji, I. Kaneko and T. Bashomatsu: Address Block Location and Format Identification of Japanese Address Reading Letter Sorting Machine. NEC RESEARCH & DEVELOPMENT Vol.33, No.2, April (1992) 217–225,
4. T. Bashomatsu: Address Block Location and Format Identification. USPS Postal Service Advanced Conference 5 (1992) 1295–1303

Structural Features by MCR Expression for Printed Arabic Character Recognition

A.B.C. ZIDOURI, S. CHINVEERAPHAN, and M. SATO

Precision and Intelligence Laboratory, Tokyo Institute of Technology
4259 Nagatsuta-cho, Midori-ku, Yokohama, JAPAN 226

Abstract. This paper discribes how stroke features in document images are extracted and used for the recognition of printed Arabic characters. It is of importance to provide a good base representation that facilitate analysis and processing of document images. The strokes are extracted by a method called Minimum Covering Runs (MCR)[1]. This method of representing binary images by a minimum number of horizontal and vertical runs is used as a preprocessing step. The strokes are labeled and ordered, a feature space for the 100 shapes of the 28 Arabic characters is build. The system is under developement but the recognition rate obtained at this stage, 95.5% is encouraging.

1 Introduction

Text is a major part of information in document images, and automatically recognising the characters is not an easy problem. This is true especially for languages like Arabic which is written cursively (connected) even when printed. It is of importance to provide a good base representation that facilitate the processing and analysis of such huge amounts of information. Recognising well formed and neat characters in many languages have been solved for, and research is being focussed on more challenging problems of poor quality print omnifont or unconstraint handwriting. However, cursive writing like Arabic, where the segmentation problem of text into distinguishable characters is the source of many errors of several OCR, is still not yet fully explored. Structural features of basic patterns in document images such as characters or tables are mainly horizontal and vertical stroke components.

Our approach to Arabic cursive writing, using stroke components, is a novel one. It considers Arabic like any other stroke-like languages. This has been made possible by the structural features that MCR expression offers to represent text and tabular components in binary document images[2].

Several researchers have attempted to solve the problem of cursive writing with success for On-line recognition, however Off-line cursive writing, where the order of the strokes made by the writer is lost, has not been satisfactorily solved[3]. A great amount of further effort is required in this domain.

Our method is similar to that of El-Dabi *et al.*,[4] in the sense it deals with the problem of separating the characters after they are recognised. This is because the segmentation process is not an aim by itself. However they proposed a system

Name	ALIF	BAA	TAA	THAA	JEEM	HAA	KHAA
Form_I	ا	ب	ت	ث	ج	ح	خ
Form_B		ﺑ	ﺗ	ﺛ	ﺟ	ﺣ	ﺧ
Form_M		ﺒ	ﺘ	ﺜ	ﺠ	ﺤ	ﺨ
Form_E	ا	ﺐ	ﺖ	ﺚ	ﺞ	ﺢ	ﺦ
Name	DAAL	DHAL	RAA	ZAY	SAAD	DHAD	TTAA
Form_I	د	ذ	ر	ز	ص	ض	ط
Form_B					ﺻ	ﺿ	ﻃ
Form_M					ﺼ	ﻀ	ﻄ
Form_E	ﺪ	ﺬ	ﺮ	ﺰ	ﺺ	ﺾ	ﻂ
Name	ZHAA	KAAF	LAAM	MEEM	NOON	AYN	GHAYN
Form_I	ظ	ك	ل	م	ن	ع	غ
Form_B	ﻇ	ﻛ	ﻟ	ﻣ	ﻧ	ﻋ	ﻏ
Form_M	ﻈ	ﻜ	ﻠ	ﻤ	ﻨ	ﻌ	ﻐ
Form_E	ﻆ	ﻚ	ﻞ	ﻢ	ﻦ	ﻊ	ﻎ
Name	FAA	QAAF	SEEN	SHEEN	HAH	WAW	YAA
Form_I	ف	ق	س	ش	ه	و	ي
Form_B	ﻓ	ﻗ	ﺳ	ﺷ	ﻫ		ﻳ
Form_M	ﻔ	ﻘ	ﺴ	ﺸ	ﻬ		ﻴ
Form_E	ﻒ	ﻖ	ﺲ	ﺶ	ﻪ	ﻮ	ﻲ

Fig. 1. The 100 shapes of the 28 Arabic character set in their different forms: Beginning, Middle, End or Isolated Form.

for typed Arabic text, which involves a statistical approach using accumulative invariant moments as identifier. Invariant moments are very sensitive to slight changes in a character shape. They reported a 94% recognition rate and a speed of 10.6 characters/minute. In our case the characters are segmented once their composing parts are successfully recognised. Although our system is also font dependant, nevertheless we achieved a better recognition rate, at a speed of recognition faster of at least one order of magnitude. Our test set is composed of 10 documents of 500 characters each.

2 Review of MCR Expression

The MCR expression method was developed to express binary document images by a minimum number of both types horizontal and vertical runs. It represents binary images with no redundancy and without any loss of information. Some of the horizontal and vertical runs called covering runs are suitably selected to represent the image with a minimum number of runs. In binary images, no runs from the same direction cross each other and every black pixel can be considered as a crossing point of one horizontal run and one vertical run. Using this fact, it has been shown that horizontal and vertical runs of binary image can be thought of as *partite* sets of a *bipartite* graph. From this correspondence between the

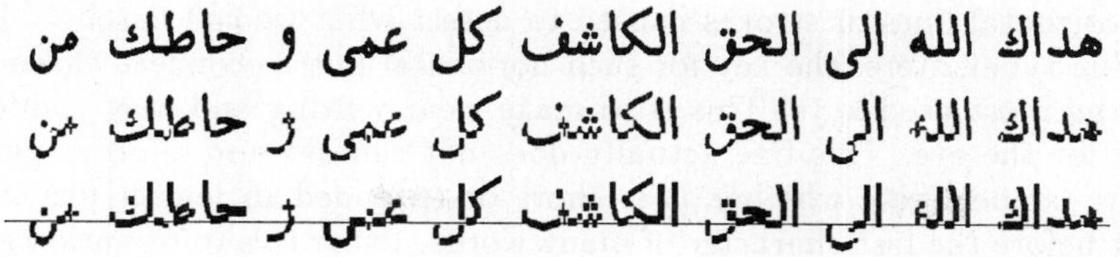

هداك الله الى الحق الكاشف كل عمى و حاطك من

هداك اللء الى الحق الكاشف كل عمى ز حاطك من

هداك اللء الى الحق الكاشف كل عنى ز حاطك من

Fig. 2. A part of Text with detected baseline, its non-overlapping strokes extracted by modified MCR expression, and approximate strokes for visualisation.

binary image and the bipartite graph, where runs correspond to partite sets and edges of the graph correspond to pixels in the image, finding the MCR expression amounts to constructing a minimum covering in the corresponding bipartite graph. This in turn is the same as finding the maximum matching, which has been solved for by graph theoritical algorithms. The modified version of MCR that we are using, which extracts strokes more accurately, is based on 2 essential procedures, a local stroke analysis used to find elongated segments and stroke like patterns in a binary image and a maximum matching constructing algorithm in the corresponding graph to find the remaining parts. The technique can partition characters and lines in images into horizontal and vertical segments. The use of horizontal and vertical terms in our case is very flexible as all the binary patterns are classified either as horizontal or vertical strokes according to a stroke decision criteria. Horizontal strokes are represented by sets of adjacent horizontal covering runs while vertical strokes are represented by vertical covering runs.

Then a description of characters in forms of strokes (parts) connections is constructed. An example is shown in fig.2 of a part of an Arabic text with detected baseline, the non-overlapping parts extracted by modified MCR, and approximate strokes for visualisation.

3   Characteristics of Arabic Writing

The characteristics of the Arabic language do not allow direct implementation of many algorithms used for other languages. Segmentation of words into characters to be recognised is still the most difficult and source of errors of OCR systems that are developed for cursive writing.

Arabic is written cursively whether printed or handwritten, from right to left, with an alphabet set of 28 characters, which take up to four shapes each, depending on their position. Fig.2 shows the set of the Arabic characters in their different forms within a word: beginning form, middle form, end form or isolated form. Seven characters do not allow for connection from the left side and have only two shapes each.

One of the main characteristics of Arabic writing is that it is written following a baseline. This is where almost all characters are connected to each other for most of the fonts and writing styles. This fact of character connection by some short horizontal lines or strokes is used to detect what we call baseline. In fact, on Arabic typewriters, the key for such horizontal line to connect characters is one of the most used keys. This is to make clear writing and easy comfortable reading for the eye. This line actually does not contain any information apart from the connectivity whether it is short or extended in length like in some titles or before the last character of many words. Using this information and the fact that most Arabic characters themselves have their horizontal part written on this baseline, we detect the line containing the largest number of horizontal strokes and label it as baseline fig.3.

4 Features Extraction and Character Reference Building

Modified MCR gives a good structural information to provide the dynamic information from a static image, needed for character recognition. After the MCR expression is obtained for a binary image the non overlapping parts of a pattern or a character are labelled and ordered in a top down, left to right priority, to follow the Arabic way of writing to allow for future eventual connection to a speech synthesiser machine after the recognition is accomplished. The writing line or baseline is detected and a feature vector is then associated with each character, it takes into account the following parameters:

1. Number of parts constituting the character.
2. Size of the parts (length).
3. Position with respect to the baseline.
4. Sequence or order information.
5. Type of strokes (horizontal or vertical).
6. Direction or angle information.
7. Width of parts.

It must be noted that not all these parameters are needed at the same time to identify a specific character. These are used only in some ambiguous cases of multi-response. Fig.4 shows an Arabic word and the different parts parameters.

The baseline is used as a reference to divide the text line into four horizontal zones where the zone zero is that which contain the baseline. The three others are a lower zone below, and a middle and upper zones above the baseline. Two levels have been chosen above the baseline for the reason that the main information of the Arabic text is contained in the upper part of the main body of each character[5]. Also most stress marks like dots are above the main body of the character (four times as many as those which have dots below).

5 Segmentation And Recognition

In the framework of structural approaches to OCR most methods are based on representation, feature extraction, description and classification. For cursive writing it is necessary to overcome the complicated problem of letter separation.

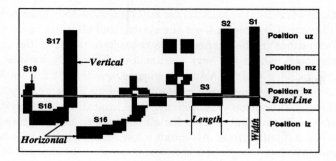

Fig. 3. An Arabic word and its decomposition into parts showing the features used

In our novel approach with MCR representation we are able to segment words or subwords into characters at the same time the recognition is achieved. The document image is first scanned at 216 by 216 resolution and input to the system in the CCITT format size. Then the MCR expression of image is found and the features extracted. The recognition is performed by matching to the reference prototypes build for this purpose. A reference model is built for the 100 shapes of the Arabic characters in their different positions, in the beginning of, within, or at the end of a word or subword, or when it is in its isolated form. Some additional prototypes were necessary for the special ligatures and special characters or stress marks like LAAM-ALIF (ﻻ), TAA-MARBOOTA (ة) or HAMZA (ء). The prototypes are made flexible to account for variations in the character shape due to noise from printing processes or quantisation noise of the scanning device. These reference characters were build from observation of the patterns obtained by MCR expression for Arabic printed text documents.

The feature vector built from the character pseudo-segmentation into parts provide the dynamic information from the static image, needed for characters recognition. The recognition is performed from left to right to be in a natural way to allow for an eventual link to a speech synthesiser machine. The recognition rate for the model document is more than 99%. The recognition drops however to just above 95% for the test documents at this early stage, but most of the errors are due to characters being rejected not misrecognised. These being mainly connected characters, as most isolated characters are recognised successfuly. We use a threshold as a unit length that is the size of a dot. From experimental results we found this threshold to be $th = 1/6$ the length of character ALIF. We call this a *baselength*. This is used for zoning of lines of text into 4 horizontal zones, as mentioned previously, and for the parameter size of the strokes. The text is segmented into words or subwords and this in turn separated into characters. Arabic characters take a special shape at the end of a word or a subword. This is used to segment text into words or subwords. Unfortunately there is no way to tell a word from a subword which could be composed of as few as one character unless high level recognition is sought as the use of a dictionary or a spell checker.

The segmentation of words into characters is automatically done after a character is recognised. There is still room to train the system and allow for the prototypes to cover more variations in the shapes of the characters. Most of the errors are now due to characters which are very similar and differ only by one dot like FAA and QAAF or TAA and THAA, and when character MEEM or AYN are within a word or subword. These are represented in some cases by only 2 strokes which makes the discrimination very difficult. The limitation of the system now is that it assumes the same font is used for all the text to be recognised and to be the same as the modelling font. Although a low level recognition scheme, we achieved a reasonably fast recognition speed of more than 10 characters/sec, for a test set of more than 5000 characters. There will be always a trade off between accuracy and effectiveness of a system and its cost and complexity. Up to now we have trained our system only to one type of font. We believe that there is plenty of room for improvement and work to be done.

6 Conclusion

In this paper we have shown that MCR provides a good representation of stroke components in text. As an application we used it to printed Arabic cursive writing to overcome the difficult problem of segmentation inherent to all cursive writings. We labelled the extracted strokes and rendered them meaningful as structural features for recognition and segmentation. The system is in progress and at this stage we achieved a minimum recognition rate of 95.5% for a test set of 10 documents of about 500 character each of the same font as the training set.

References

1. Chinveeraphan, S., Douniwa, K., and Sato, M., *"Minimum Covering Run Expression of Document Images Based on Matching of Bipartite Graph."* IEICE *Trans. Inf. & Syst.*, vol.E76-D, no.4, pp.462–469, Apr. 1993.
2. Chinveeraphan, S., Zidouri, A., and Sato, M., *"Stroke Representation by Modified MCR Expression as a Structural Feature for Recognition,"* IWFHR-IV, (Taipei), pp. 11–19, Dec. 7-9, 1994.
3. Senior A. W., and Fallside F., *"Using Constrained Snakes for Feature Spotting in Off-line Cursive Script,"* Proceedings of the Second Int. Conf. on Document Analysis and Recognition (Tsukuba Japan), pp.305–310, Oct. 20-22, 1993.
4. El-Dabi, S. S., Ramsis, R., and Kamel, A., *"Arabic Character Recognition System: A Statistical Approach for Recognising Cursive Typewritten Text,"* Pattern Recognition, vol. 23, no.5, pp. 485–495, 1990.
5. Margner, V., *"SARAT - A System for the Recognition of Arabic Printed Text,"* 11th IAPR, vol. 2, (The Hague), pp. 561–564, Aug. 30-Sep. 3 1992.

Genetic Algorithm for Thinning Gray-Scale Images of Characters

Tadayoshi Shioyama, Akira Okumura and Yoshikazu Aoki

Department of Mechanical and System Engineering
Kyoto Institute of Technology, Sakyo − ku, Kyoto 606

Abstract. In this paper, we propose a new method for thinning digital gray-scale pictures by fitting piece-wise linear line to skelton. The method is not based on the point-wise gray-value information as Salari and Siy method, but based on an objective function robust for noise contaminated pictures. To optimize the function, we search the solution of the optimization problem with genetic algorithm. The procedure of the method is as follows. We allocate small regions so that they cover the neighborhood of skelton of object. In each small region, we extract a linear line which approximates the skelton in the small region, and represent the line by a couple of label numbers of points with nonzero gray values through which the line passes. A string is constructed by collecting all the label pairs. The fitness of a string is given by an objective function which is constructed so as to evaluate the proximity to skelton and to be robust for noise. After a genetic algorithm generates successive populations of strings, a skelton is given by the string with the greatest fitness.

1   Introduction

The transformation of images of linelike objects such as printed or handprinted characters and chromosomes to their skeltons is a problem of great importance in pattern recognition. In pattern recognition, the thickness of the lines which constitute the objects, is not effective, but the skeltons are used to extract topological features of the objects.

In methods for thinning objects in black and white (binary) pictures, the skelton is obtained from the boundary of the binary picture. However, the boundary is not well defined for many applications such as handwritten characters and chromosomes. Salari and Siy [1] have proposed a method for thinning digital gray-scale pictures. In their method, the skelton is obtained by seeking the ridge (high gray value) areas to position the skelton. In their thinning process, the boundary point of object area which is a set of connected elements with nonzero gray values, is removed using the gray value information of its eight neighbors if the conditions for retaining structure information and topological properties are satisfied. Hence their method which is based on the point-wise gray value information, is considered to have a drawback that it is not robust for noise contaminated images of objects.

In the present paper, we propose a new method for thinning digital gray-scale pictures by fitting piece-wise linear line to the ridge i.e. skelton. The method is not based on the point-wise gray value information but based on the noise resistive objective function. To optimize the function, we search the solution of the optimization problem by genetic algorithm [2] which surpasses traditional search procedures in the quest for robustness due to parallel flavor. The procedure of the present method is as follows. We allocate small regions so that they cover the neighborhood of skelton of object. In each small region, we extract a linear line which approximates the skelton in the small region, and represent the line by a couple of label numbers of points with nonzero gray values determining the line. A string is constructed by collecting all the label pairs. The fitness of a string is given by an objective function which is constructed so as to evaluate the proximity to skelton and to be robust for noise. After a genetic algorithm composed of three operators: reproduction, crossover and mutation, generates successive populations of strings, a skelton is obtained from the string with the greatest fitness. To evaluate the present method, the resulting skeltons are shown for noise contaminated original pictures.

2 The Piece-Wise Linear Line Fitting Method

2.1 Elementary Objective Function

Let $f(x,y)$ be the gray value at a point with image coordinate (x,y). Denote by m_{pq} the moment of order $p+q$ of a gray value function $f(x,y)$ around a point (x_0, y_0): $m_{pq} = \int (x - x_0)^p (y - y_0)^q f(x,y) dx dy$, $p + q \geq 0$, by $\tilde{m}_{pq}$ the normalized moment: $\tilde{m}_{pq} = m_{pq}/m_{00}$. Define an elementary objective function ϕ as $\phi = \sqrt{\tilde{m}_{10}^2 + \tilde{m}_{01}^2}$. The function ϕ takes a small value at the ridge (high gray value) areas. Then, the function ϕ can be used to evaluate the proximity of skelton. The function ϕ is considered to be robust for noise contaminated images because it is based on moments.

In practical computation, the function ϕ is obtained by computing the moments in the 5 × 5 mask with the center at the considered point (x_0, y_0). Figure 1 shows the perspective of the value of ϕ for character "A", where for convenience, ϕ is set as zero at (x_0, y_0) such that $f(x_0, y_0)=0$.

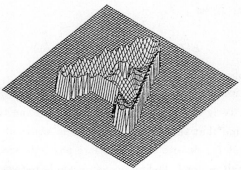

Fig.1 The perspective of ϕ for character "A".

The function ϕ is used for two purposes. First, it is used to determine whether the considered point (x_0, y_0) is in the neighborhood of skelton or not. That is, the point (x_0, y_0) is decided to be in the neighborhood of skelton if the function ϕ is less than a threshold θ. The threshold θ is determined by P-tile method in which the area of histogram of values of ϕ less than θ is t-percent of the area of total histogram of values of ϕ. In the sequel, t is set to be 50 percent. Then, θ is determined to be 1.06 from values of ϕ for 26 alphabetical capital letters.

Second, the function ϕ is used to construct an evaluation(objective function) of string as described in the later section.

2.2 Allocating Small Regions

Small regions of size $w \times w$ are allocated by raster scanning without overlapping each other so that each small region contains more points than a threshold, which are in the neighborhood of skelton.

In each small region, a linear line is extracted by choosing two points with nozero gray values in the small region such that the extracted line passing through the two points is the best approximation to skelton in the small region.

2.3 Thinning with GA

2.3.1 String

Let N denote the number of allocated small regions, and k_i, i=1,...,N, the number of points with nonzero gray values in the i-th small region, where each point is labelled by a number in a set $\{1,2,...,k_i\}$. Denote by L_i^1 and L_i^2, labels of two points which specify a linear line in the i-th small region. A string is given by collecting the label couple (L_i^1, L_i^2), i=1,...,N, i.e. $(L_1^1 L_1^2 L_2^1 L_2^2 ... L_i^1 L_i^2 ... L_N^1 L_N^2)$.

2.3.2 Evaluation of String

Let a linear line P_i, which is extracted in the i-th small region, be composed of a set of n_i points as follows: $P_i = \{p_i^1, p_i^2, ..., p_i^{n_i}\}$, $i = 1, ..., N$, where the points p_i^1 and $p_i^{n_i}$ are on the boundary of i-th small region. Define the objective function

$E(P_i)$ of a linear line P_i as

$$E(P_i) \equiv \frac{1}{n_i} \sum_{j=1}^{n_i} \phi(p_i^j). \qquad (1)$$

The objective function E_ϕ of a string is defined as $E_\phi \equiv \frac{1}{N} \sum_{i=1}^{N} E(P_i)$. To evaluate the connectivity of the thinning results, we define the objective function E_c as follows. Assume that the j-th small region is adjacent to the considered i-th small region. Let P_{ij} denote the linear line passing through two centers of the i-th and j-th small regions. Define θ_{ij} as an angle between P_i and P_j, and ρ_{ij} as an angle between P_i and P_{ij}. Define x_{ij} as

$$x_{ij} = 0 \quad for \quad 0 \leq min(\rho_{ij}, \pi - \rho_{ij}) < \pi/4,$$
$$x_{ij} = 1 \quad for \quad \pi/4 \leq min(\rho_{ij}, \pi - \rho_{ij}) \leq \pi/2.$$

Then, we define the evaluation V_{ij} for connectivity between the i-th and j-th small regions as $V_{ij} \equiv x_{ij}(1 + cos\theta_{ij}) + (1 - x_{ij})(1 - cos\theta_{ij})$. We define the objective function E_c for connectivity of string as $E_c \equiv \frac{1}{N} \sum_{i=1}^{N} \sum_{j=1}^{m_i} V_{ij}$, where m_i denotes the number of small regions adjacent to the i-th small region. Hence, the objective function E of a string is defined as $E \equiv E_\phi + \alpha E_c$, where a constant α is a weighting coefficient. The small E implies the greater fitness of string.

2.3.3 Reproduction

Strings with lower fitness are selected and cannot have ability to contribute offspring in the next generation. The ratio of the number of such inferior strings to the population size is called generation gap. The generation gap of strings are supplemented by copying strings with higher fitness to retain the population size. In reproduction process, strings are copied according to their objective function values.

2.3.4 Crossover and Mutation

Uniform crossover proceeds on mated string couple among a population. That is, exchange is performed in uniform probability on a label-by-label basis. The subsequent operator, mutation, is performed on a label-by-label basis on the created two new strings.

After a genetic algorithm composed of three operators: reproduction, crossover and mutation, generates successive populations of strings, a thinning result is obtained from the string with the greatest fitness.

2.4 Postprocessing

Postprocessing is performed in the following steps to guarantee a connectivity of thinning result.

(step 1) Find an end point P in the thinning result. Set a 2w×2w mask with center at the end point.

(step 2)In the mask, find end points other than P which satisfy the following two conditions. If exist, choose an end point Q among them such that the linear line PQ has the least objective function value defined as (1). Then, connect the end points Q and P by a linear line.

Condition 1. The point is not connected with P in the 2w×2w mask.

Condition 2. The linear line determined by the point and P, has an objective function value given by (1) less than a threshold.

If there is not such end point other than P, go to step 3.

(step 3) In the image of thinning result, choose the black point Q such that it satisfies the above two conditions and moreover the linear line PQ has the least objective function value given by (1). Connect P and Q by a linear line.

These steps are performed for every end point P.

3   Results

The present thinning method is evaluated using 64×64 original image with 16 gray levels with additive gaussian noise of mean 0 and variance 1. In the experiment, the size of small region is set as 6×6, and the following parameters in GA are used: *The population size* = 100, *The number of generations* = 500, *The probability of crossover* = 0.8, *The probability of mutation* = 0.003, and *The generation gap* = 0.4. Figure 2 shows original images. Figure 3 shows the results by the present thinning method. For comparison, the results by the Salari and Siy method are shown in Fig.4. Here, the procedure where noisy isolated points are removed, is performed only in Fig.4. In the results by Salari and Siy method, there are improper protrusions. From these results, it is found that the present thinning method is robust even for noise contaminated images.

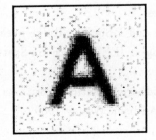

Fig.2 Original images.

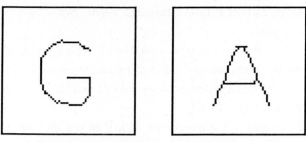

Fig.3 The thinning results by the present method.

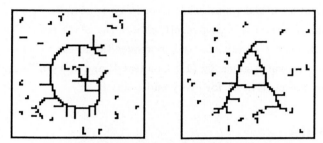

Fig.4 The thinning results by Salari and Siy method.

4 Conclusion

We have proposed a new method with GA for thinning digital gray-scale pictures by fitting piece-wise linear line to skelton. The method is not based on point-wise gray value information but based on a noise resistive objective function. To optimize the function, we have searched the solution with GA. In comparison with the Salari and Siy method which is a typical conventional method, the present method is found to be robust even for noise contaminated pictures.

The computational burden required by the present algorithm is heavy. It is the problem to overcome the drawback.

References

[1] E.Salari and P.Siy:The Ridge-Seeking Method for Obtaining the Skelton of Digital Images. IEEE Trans. on Systems, Man, and Cybernetics, SMC-14, 524-528 (1984)

[2] D.E.Goldberg:Genetic Algorithms in Search, Optimization, and Machine Learning. Addison-Wesley 1989

Blue-Print Document Analysis for Color Classification

Gabriel Marcu   and   Satoshi Abe*

Array Corporation, 3-32-11 Yoyogi, Shibuya-ku, Tokyo-151, Japan
* Department of Information Science, Fac. of Science, Univ. of Tokyo, Japan

Abstract. The blue print copies result in common designing process by copying the half-transparent white paper hand drawn in black ink. Over the blue print papers, very often, the designer marks in different colors some area or symbols of interest. This paper presents the particularities of analysis of such kind of documents in order to extract the colored lines as well as the original blue lines resulted during the copying process. The procedure is based on processing for clusterization of the colormap of the original image, and on post processing of the resulted run-length files for each classified color. It runs for any size of the scanned image and is independent on the contents of the image. For A1 size document, the color classification and bynary coding takes 20 min for 400MB image data.

1. Introduction

The blue print copies result in common designing process by copying the half-transparent white paper hand drawn in black ink. Over the blue print papers, very often, the designer marks in different colors some area or symbols of interest. The drawn colors on a very bluish and not uniform background are different from the drawn colors on a white background. A small sample of a blue print image is presented in figure 1. The images to be processed for color classification have the following particularities.

First, the blue-print scanned images are very large, usually A0 and A1, and are scanned with 16 dots/mm and 8 bits/color components, resulting in a large files. An A1 image stored in RGB raw format is 9000 x 12000 pixels and it conducts to about 400 MBytes image data, if each R,G,B channel is coded with 8 bits/component. A high speed processing algorithm is required in order to implement a practically usable procedure. Second, the background of the blue-print copy is bluish and locally not uniform. The bluish background color varies in a wide range from sample to sample, from very light to dark blue. Additionally, in time, the paper color changes from bluish to grey-yellowish, dependent on the exposure of paper to the sun light. Third, the color range of the document is narrow, enabling quantization of the image in hundred of colors.

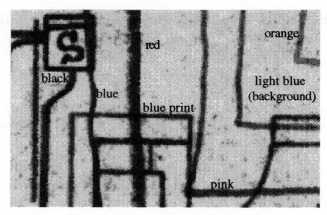

Fig.1. Sample of the original image

Over the blue print papers, the designer marks in different colors some area or symbols of interest. Pens or pencils are used for manual drawing. The colors non-uniformities and color deviations must be compensated in the color classification algorithm.

This paper presents the particularities of analysis of such kind of documents in order to extract the colored lines as well as the original blue lines resulted during the copying process. The paper proposes a fast classification method for color images, that can be effectively used for particular blue-print images.

For a high speed color classification, processing the image using spatial information of pixels [1,2] results in unacceptable time performance. As an example, the segmentation method for color images proposed by Yan and Hedley[1], can process a 336x336 pixels image in 6 seconds, but for 9000x12000 image size, the algorithm takes about 2 hours. In case of classification method proposed by Tominaga[2], the processing time increases more due to the color transformations required to perform an iterative process of detection of significant peaks and valleys of the histogram of main color components computed in Lab color space. Other methods based on spatial pixel information processing are not applicable due to limitation of processing time.

The procedure proposed in this paper is based on processing for clusterization of the colormap of the original image, and on post processing of the resulted run-length files for each classified color. The color classification method based on colormap clusterization has proved to be a very fast procedure, limited only by the access time to image data, and with acceptable results. The clusterization procedure runs optimally in Lab color space in order to compute the color distances between clusters, based on perceptually color distances between colors. The procedure runs for any size of the scanned image and is independent on the contents of the image. Results and diagrams illustrating the procedure are included.

2. Global Processing Procedure

The procedure proposed in the paper is based on few fast steps, as it is illustrated in figure 2.

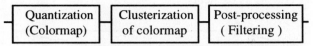

Fig.2. Diagram of fast color classification procedure

First step consists of extraction of the colormap of original image by a color quantization procedure. We used a modified version of Wu algorithm [3], due to time performance, but in case when more accurate results are required, the algorithms oriented on color properties can be selected [4,5]. The original image is transformed from true color to pseudo color (pixelmap image).

The second step consists of clusterization of the colormap. The clusterization process enables to derive a classification color table for each color class. The advantage of the an individual classification color table for each class is that in a single class can be included more that a single cluster. This can be usefully when in a class are included also the clusters resulted by overlapping the class color with other color classes.

The third step consists of derivation of binary color classes files, by passing the pixelmap image through the LUTs defined by each classification table.

The last step consists of post-processing the classes files for elimination of noise and transition colors.

3. Colormap Clusterization Procedure

The clusterization procedure uses an agglomerative algorithm based on nearest neighbor principle[6,7]. It starts assigning the colormap colors in different clusters. The clusterization is performed iteratively. For each clusterization step, the algorithm finds the closest colors, or clusters, to be merged, in order to reduce the clusters number with 1. The algorithm stops at a threshold number of clusters. At one step, the algorithm selects a pair of two elements of the current classified colormap that have to be merged. The pair of elements have the minimum distance over the current classified colormap. The elements can be two colormap colors, a colormap color and a cluster or two clusters. The distance, D_{clust}, between two elements, referred as 1 and 2 (and named as colormap color or cluster), is: $D_{clust} = D - R_1 - R_2$, where D is the euclidean distance between the elements center (color components or cluster mean value components) and R_1, R_2 have the significance dependent on the elements involved in distance computation. For a color, R_i is:

$R = A.\log((a.x+b)/x_{max})+B$, if $R_{min} < R < R_{max}$,

$R = R_{max}$, if $R > R_{max}$, and $R = R_{min}$, if $R < R_{min}$,

where A, B, a, b, R_{min}, R_{min} are constants, and x_{max} is the maximum value of the histogram. For a cluster, R_i represents the radius of cluster, that is maximum Mahalanobis distance[8] of the cluster elements to its center, such that the cluster volume includes 80% of its elements.

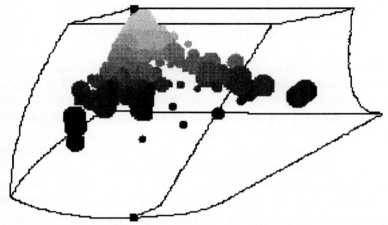

(a) 3-D visualization of color map including information about histogram values

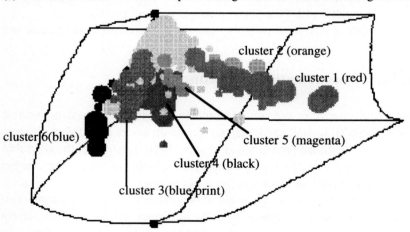

(b) result of clusterization procedure for 6 clusters

Fig.3. The clusterization procedure takes advantage of the 3-D visualization parameters for the definition of the cluster distance criteria

We select for clusterization the Lab color space, determined by the color transformation described bellow. The RGB color components of each colormap, based on scannr calibration, are transformed to XYZ device independent coordinates. The XYZ device independent components are then transformed in L*a*b* components, corresponding to a more uniform color space. Using other color transformations, also other color spaces can be investigated. The L*a*b* space was selected in order to take advantage of the color difference formula that models better than other spaces the particularities of the human observer. Figure 3 illustrates the colormap clusterization procedure in Lab color space.

The detection of the clusters can be investigated over specific region of the color space, in order to increase the accuracy of classification algorithm, and in the same time to detect overlapped colors.

The clusterization of the colormap enables to build the color look-up table that is used to associate to each pixel value a color class. The image file is passed through the look-up table and the files resulted for each color are saved in run-length format. Effective clusterization of colormap it takes seconds but processing of the large image file is determined by the access time to image data on storage media. Figure 3 illustrates the clusterization procedure. Figure 3a represents the 3-D colormap and histogram visualization. Figure 3b depicts the 6 clusters identified by the algorithm.

4. Post-Processing the Classes Files

The colormap classification procedure may conduct to transition colors, represented as thin lines describing the edge of some hand drawn lines, dependent on the color and their relative position in the color space. For these cases, a post-processing procedure is required. The post-processing procedure takes advantage of the binary format of the files coding each color class. The post-processing procedure filters all the segments having the width smaller than a threshold value. The procedure is very fast since it uses only boolean operations on a binary image, avoiding the manipulation of 3 components pixels that is more time consuming procedure. The size of run length files depends on the contents of the image, but it was observed that can vary from 500KB to 5-6 MB for very complex diagrams and it takes less than 1 minute for A1 document size. For improved efficiency, the postprocessing procedure is applied simultaneously with run-length coding procedure.

5. Results and Conclusions

The procedure was implemented and tested on IBM PC with Pentium processor, at 90 Mhz. The procedure requires about 20 minutes for extraction of 7 colors from an A1 size document. Figure 4 depicts three samples of classified colors from image in figure 1.

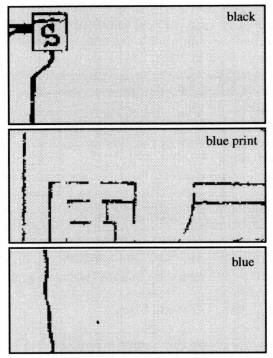

Fig.4. Samples of 3 classified colors

References

[1].M.Hedley, H.Yan, "Segmentation of Color Images using Spatial and Color Space Information", *Journal of Electronic Imaging*, Vol.1, N.4, October 1992.

[2].S.Tominaga, Color Classification of Natural Images, *Color Research and Applications*, V17, N4, 1992.

[3]. X.Wu, Efficient Statistical Computations for Optimal Color Quantization, *Graphics Gems III*, edited by D.Kirk, Academic Press, 1992.

[4].T.Orchard, A.Bouman, Color Quantization of Images, *IEEE Trans. on Signal Processing*, V39, N12,1991.

[5].R.Balasubramanian, C.Bouman, J.Allebach, Sequential Scalar Quantization of Color Images, *Journal of Electronic Imaging*, V3., N1., January 1994.

[6]. R.O.Duda, P.E.Hart, *Pattern Classification and Scene Analysis*, John Wiley & Sons, 1973.

[7].B.V.Dasarathy, Nearest Neighbor Pattern Classification Techniques, IEEE Comp. Society Press, 1990.

[8].J.M.Jolion, P.Meer, S.Batauche, Robust Clustering with Applications in Computer Vision, *IEEE Trans. on PAMI*, V13, N8, 1991.

Optical Character Recognition

Document Image Analysis: What Is Missing?

George Nagy

ECSE, RPI, Troy, NY, USA 12180-3590, nagy@ecse.rpi.edu

Abstract

The conversion of documents into electronic form has proved more difficult than anticipated. Document image analysis still accounts for only a small fraction of the rapidly-expanding document imaging market. Nevertheless, the optimism manifested over the last thirty years has not dissipated. Driven partly by document distribution on CD-ROM and via the World Wide Web, there is more interest in the preservation of layout and format attributes to increase legibility (sometimes called "page reconstruction") rather than just text/non-text separation. The realization that accurate document image analysis requires fairly specific pre-stored information has resulted in the investigation of new data structures for knowledge bases and for the representation of the results of partial analysis. At the same time, the requirements of downstream software, such as word processing, information retrieval and computer-aided design applications, favor turning the results of the analysis and recognition into some standard format like SGML or DXF. There is increased emphasis on large-scale, automated comparative evaluation, using laboriously compiled test databases. The cost of generating these databases has stimulated new research on synthetic noise models. According to recent publications, the accurate conversion of business letters, technical reports, large typeset repositories like patents, postal addresses, specialized line drawings, and office forms containing a mix of handprinted, handwritten and printed material, is finally on the verge of success.

1 Introduction

As a thirty-five year veteran, I feel entitled to air some insights or prejudices nurtured in the course of years of patient labor in the vineyards of Optical Character Recognition and Digital Image Analysis. I make unsupported statements and unfounded claims without apology. I focus heedlessly on picayune issues like representative samples at the expense of important contributions like new features. I vent my indignation over fuzzy propaganda and wax irascible about persistent failures. To avoid misattribution of inflammatory ideas to innocent parties, I dispense altogether with citations and references, even to my own rambling publications.

I propose to explore continuity and change in research paradigms and real-world applications in OCR and DIA. To this end, I try to differentiate innovative solutions that make me exclaim "I wish I had thought of that!" from the application of new technology to old problems. Of course, in many instances, someone did think of *that* long ago, but lacked the means to test the idea, or the time to write it up. As in other fields, many good ideas are forgotten and reinvented again and again.

Commercial OCR and DIA (the two have always been inseparable) have just celebrated their 40th birthday and may therefore be rightfully suspected of nearing a mid-life crisis. Complaints abound. Researchers in other disciplines, even those in other areas of pattern recognition, consider OCR a long-solved problem. Many regard document image analysis as *infra dig* compared with medical, astronomical, and satellite image analysis and computer vision. Scholars, bureaucrats, lawyers, engineers or geographers seldom make use of OCR and DIA. While document imaging already represents a huge market, only a tiny fraction of the current applications requires techniques beyond the "trivial" processes of acquiring, indexing, compressing, storing, transmitting, retrieving, and displaying thousands of full-page images. Potentially, these files are all grist to our mill.

I first review the earliest applications of pattern recognition to printed matter, and trace the gradual shift to more and more demanding tasks. Then I list some old ideas that have stood the test of time before mentioning techniques that have been adopted or considered seriously only recently. Because the evaluation of OCR and DIA systems has recently become such an active area, I dedicate a short section to this topic. I conclude with the mandatory expressions of faith, optimism, and belief in progress.

2 Old Applications

The easy problems were solved in the sixties and early seventies, but the economics and methods of implementation have changed. Systems that used to require expensive electromechanical or CRT scanners and hardwired special purpose processors now run on PCs using scanners and paper transports that are first cousins to inexpensive fax machines. Nevertheless, the most successful automated data-entry applications still involve specially designed fonts and rigidly formatted documents, where the document preparation is controlled by the organization that must eventually read it (*turn-around forms*). OCR-A and OCR-B fonts, which have both undergone several rounds of revision and expansion by standards committees, are still alive. The phrase *document reader,* which used to refer to reading forms, now means a machine that reads entire pages of text.

The multi-million dollar IBM 1975 that was installed at the Social Security Administration headquarters in Baltimore in 1965 read full-page, "omni-font" reports of social security numbers, names, and earnings. It chugged along for more than twenty years, reading millions of typed and line-printer generated pages. The IBM 1975 had thresholds for rejecting both individual lines and whole pages, and it rejected plenty. The resulting low error rate was improved further downstream by automated cross-checking against the SSA's master-file of one million names.

From the point of view of pattern recognition, the essential aspect of form reading is that most fields contain at most a dozen characters of interest (e.g. name, address, account number, amount). Thus the system can run at a relatively high character reject rate and still recognize all the characters on most fields. At a high reject rate, expensive substitution errors are very rare. Any field on which even a single character is rejected is submitted to manual handling. In the last decade, the availability of sufficient digital storage to retain and display the entire field has increased the efficiency of operator intervention enormously. The form that contains the error no longer needs to be physically accessed: scanning, recognition, and post-processing can be separated in time and space. Current form readers can process printed, imprinted (stamped), and handprinted characters on the same form, as is typical on credit card slips.

Another important application introduced in the sixties was postal address reading. The first machines read only *outgoing mail* (city, country and postal code). In Europe and the United States, most of the addresses were typed or printed, but in Japan the postal codes were handprinted in preprinted boxes. Printed and handwritten mail were separated before the machines took a crack at it. Some machines were later modified to read *incoming mail* (street addresses). The current machines find and read the entire address block and spray a bar-code on the envelope (including, in the US, two extra digits, for a total of eleven, for delivery sequencing). Accuracy is enhanced by reference to a postal address database. The US database takes about 300 megabytes and is updated weekly. As in the case of form readers, postal readers pay their way even if they reject 30% of the mail. Postal research, spearheaded in the US by CEDAR, has also produced many innovative byproducts for other applications. However, reading 95% of the first-class mail is still five years away, just as it was in the sixties.

A bane of early postal readers, blotchy Addressograph plates,* has virtually disappeared. Another application that was displaced in the seventies (by key-to-disk) was retyping documents, on a typewriter with an OCR font, as a substitute for keypunching. More importantly, the replacement of office typewriters and chain printers by 300 dpi laser printers, and of thermal copiers by intelligent reprographics, did wonders for OCR error rates on office documents. (But typewriters and dot-matrix printers are still important in some form-processing applications.) We must note also the remarkable difference between the image quality of CCD scanners and that of the flying-spot scanners of yore.

3 New Applications

Among the new DIA applications, the most compelling is the World Wide Web. As anyone who has surfed the Internet can testify, it is an ocean of meta-information and pointers to documents still on paper. Some current authors do post their latest drafts, converted with minimal human intervention from TeX to HTML (the Web *lingua franca*), but the bulk of the technical material is reference information: bibliographies, directories, and catalogs (some already obsolete). Housekeeping is rudimentary. The cost of keying-in archival technical information, such as the contents of the last ten years' accumulation of the 2000-3000 largest-circulation periodicals and conference proceedings, is staggering.

Nevertheless, the WWW is likely to host the first large-scale digital libraries. A recent call for consortium proposals by the National Science Foundation, the Advanced Research Projects Agency, and the National Aeronautical and Space Administration drew so many applications that NSF was hard-pressed to find knowledgeable referees who were not affiliated with any applicant. Five projects, led by major research universities, were funded at about five million dollars each. However, only a small part of this funding addresses conversion from paper to electronic form.

Other important DIA targets are the repositories of the world's bureaucracies. Government agencies yearn to convert their bulging files to computer readable form, ready to be searched, indexed, annotated, and shuffled from civil servant to civil servant. We cite three such applications in the United States that have triggered significant DIA activity. The first is the Department of Energy's Licensing Support System, which contains all the documents relevant to the disposition of spent nuclear fuel. Some of these papers may be subpoenaed in lawsuits in the year 3000! The legal requirements to retain access to the information - recently estimated at forty million pages - spawned the Information Science Research Institute at the University of Nevada.

Increasing demands on the Bureau of Census for timely demographic information (primarily for marketing purposes) triggered the first and second competitions on HCR (handprinted character recognition). The tests were conducted by the National Institute of Standards and Technology whose competence in OCR can be traced at least to Jacob Rabinow's 1952 "Rapid Selector" for upper-case typewritten characters. The training material for the first contest consisted of forms written by Census employees. The contestants were tested on similar forms filled out by students. The second test was based on three fields related to employment from 1980 census forms (paper and microfilm). Many of the entries were from Europe, but the volume of data and the rigid reporting requirements precluded "amateur" participation. The analysis of the results of these massive tests brought to light many issues related to sampling, error metrics, recognition confidence markers, reject/error trade-offs, and the use of subject-specific context in the form of lexicons.

The third example is the current Internal Revenue Service's initiative to automate processing of individual income tax forms. In a pilot project, many forms are already scanned and stored in compressed bitmap form, which reduces to a few weeks the delay in making them available to local IRS bureaus for detailed audits. The OCR effort is concurrent with the continued encouragement of electronic filing, so fewer forms will be processed than are now manually entered. The system development, which is scheduled to last about five years, is to be shared by large military and aero-space contractors. The need for accurate form classification, form parsing, printed and handprinted character recognition, internal and external contextual constraints, and tight quality control constitute a challenge that eclipses in scale the mammoth SSA system of the sixties.

Most of the large legal reference libraries have already been key-entered for on-line access (*Lexis, WestLaw*), but systems including OCR components are marketed for trial-document management. Besides large, concentrated projects, there are a number of successful mom-and-pop operations. Examples include systems to scan and file visiting cards, read classified advertisements to obtain accurate price ranges for used cars, and convert music notation for publishing new arrangements.

Inexpensive desk-top scanners and OCR programs are reviewed regularly and enthusiastically in personal computing magazines. These systems interface directly with the most popular word processors, so errors can be readily corrected. On well-spaced, high-contrast natural-language text in a normal point size, the error rate is no higher than that of an average typist. It increases significantly on low-contrast, crowded copy, on tables and equations, and non-text alphanumeric strings (because most systems make use of letter n-gram tables or lexicons). Recently several OCR vendors have added fax-reading capability: as expected, the error rate at the normal fax resolution is much higher than at the usual 300 dpi. Many OCR systems offer versions for several languages. According to the current wisdom, individual OCR users don't want to train their machines for new typefaces, but provisions must be included for incorporating special symbols. The number of OCR systems sold annually is still insignificant compared to the number of PCs, and in the United States most of the leading OCR vendors are small, specialized companies.

Among the few examples of line-drawing systems in actual use, a shining example is the cadastral system developed by the late IBM Scientific Center in Rome. In other applications, tracing drawings and maps on a digitizing tablet is gradually being replaced by tracing the lines, using a mouse or "electronic ink," on a raster-digitized version of the drawing. Some systems offer a vectorizing capability: any errors can be easily corrected by the operator using computer-aided drafting commands. To the best of our knowledge, none of these systems have a useful OCR component, and labels, captions and dimensions must be keyed in. However, the most time-consuming aspect of label entry, tying the label to an appropriate graphic entity or coordinate position, has been greatly accelerated by point-and-click operation or automatic high-lighting of consecutive fields.

4 Old Ideas

In this section I list some of my pet aversions as a referee. Many of the techniques mentioned have a lively and productive past and perhaps an honorable and secure future, but they tend to make me nod off within a few paragraphs. Because most are quite central to DIA and OCR, they are revisited again and again. The graduate students who implement them feel compelled to write up each new wrinkle, and thesis advisors (like me) complacently affix their names. The ultimate blame lies with promotion and tenure committees who would rather count than read.

Among preprocessing methods, I single out global thresholding. Almost any binarization method works for high-contrast documents, and no *global* thresholding can be effective on faded or colored copy. A good scanner is worth three times its weight in preprocessing algorithms.

Text-nontext filters based on the spatial spectrum and related orthogonal expansions, morphological operations, and fractal dimensions are solutions in search of a problem. Commercial zoning algorithms do a surprisingly good job of locating text. Methods that do not take into account typesetting rules and layout conventions gerrymander each page into a patchwork guaranteed to bedevil any downstream program.

I am tired of ad hoc format analysis with built-in rules for narrow, specific types of documents. After the first few demonstrations that one can write a program to separate the titles of articles from photo captions, or subheadings from citations, the novelty fades. Writing the rules in a new "high-level" syntax instead of Lisp or Prolog or C does not warrant sacrificing yet another graduate student.

Preprocessing followed by character segmentation followed by feature extraction followed by classification followed by context correction makes for a nice flow chart, but there must be more interesting ways of arranging these blocks. Isn't it time to introduce a few feedback loops?

While speaking of features, please don't send me any more papers about hand-crafted "topological" features or "new" orthogonal expansions, no matter how many hundreds of characters you offer that demonstrate their palpable superiority. For a specific set of shapes, any hacker can think of dozens of workable, easily programmed features. They won't be any better than the hundreds that have been tried (although they may not be much worse either). If they don't work as well as expected, just add a few more features to take care of the errors, or find more suitable data. On the other hand, orthogonal expansions - Zernike moments, Fourier, Haar, Schwartzenegger - generate coefficients that are sensitive to any difference between shapes, whether essential or incidental. If we used them to recognize people, they would make a mistake every time they changed their hat. We need automatically generated features that are based on the observable difference between classes yet resist common, predictable sources of character distortion.

Enjoy your insight that optimal classifiers based on analytically tractable parametric distributions don't work very well except on artificially generated data, and that in hyperspace the boundaries between classes of live data aren't planes, paraboloids, or much of anything else that can be described by a neat equation. The realization that non-parametric classifiers with lots of adjustable parameters work better will surely follow. Then you can opt either for nearest neighbors and Parzen-windows based directly on the training samples, or for neural networks that internalize the sample vectors in a mere few thousand passes through the data. Whichever you pick, be assured that your initial algorithms for training *and* classification can be (and most likely already have been) improved upon. But please don't write about it.

Skeletons have many interesting quirks. Unmasking each quirk and then getting rid of it is an entire industry, but not the *OCR* industry. Although there is little evidence that *thinning* is a natural human activity or that alphabetic characters grow from the spine out, feature extraction and evaluation methods based on these ideas are popular. The hand-written U-V discrimination, first thoroughly studied twenty years ago, will keep the skeleton crews busy. Interestingly enough, in the sixties printed characters were sometimes *fattened*. Times change.

My list has many more items, but I don't want to sound grumpy. Let me mention instead some topics that turn me on.

5 New Ideas

Combining automated zoning, general-purpose layout analysis, font recognition and character classification leads to *page reconstruction*. Page reconstruction requires extracting sufficient information from a printed page image to generate the source code for generating a computer-editable facsimile version of the page. The target can be either a word-processor representation or a platform-independent page-layout language. The result can be searched, windowed, scaled, and manipulated like any computer-generated document. Although surprisingly good results have already been demonstrated, plenty of challenging problems remain. Among them are the identification of the best matches among the fonts available for reconstruction; imperceptible format perturbations to compensate for minor changes in font geometry; half-tone to dither conversion; table representation; and the generation of sensible code for formulas and equations.

One step beyond page reconstruction is *functional description* of arbitrary documents in a format such as SGML or ODA. While page reconstruction requires the identification of only physical attributes such as blank lines, indentations and page breaks, logical markup requires the recognition of functional components like subtitles, footnotes, paragraphs, and emphasis. Long term goals include partial automation of link-insertion. Most of the functional document image analysis reported to date stops short of converting the results to a widely usable format. This is not a trivial step. As already mentioned, the major impetus for functional description is the rapid growth of the World Wide Web, but conventional text retrieval will also benefit.

Complete systems of the type just mentioned cannot be developed by isolated researchers. Therefore the question of *intermediate representations* and data structures takes on additional importance. We hope that some of the current endeavors will lead to flexible, widely accepted representations for zoning, layout, typeface, segmentation, classification, context, and functional component identification for both test documents and reference data. Such representations must, of course, be compatible with current document interchange standards.

There is innovative research on *segmentation-free classification*, especially for hand-printed and cursive writing. But I would like to see an operational definition of the difference between segmentation-free classification and iterative or recursive segmentation-and-recognition. Much of this work is modeled on recent work on speech recognition. Linguistic constraints are often invoked, but I believe that there is hope for recognizing touching or overlapping characters even without them. I welcome the gradual erosion of the distinction between word and character recognition. These two processes surely work best in synergy.

Classifier combination remains an exciting topic. Voting methods have long been used for isolated character recognition, but robust string comparison algorithms have been only recently applied to align the output of "black box" OCR devices with the reference text for accurate counting of errors. It is possible that new OCR devices that provide word or character coordinate output will reduce the need for string matching, but this requires precise geometric registration of the page. For systems which also produce confidence measures, there are new methods based on rank-ordering. But will classifier combination survive if someone discovers how to combine disparate features into a single classifier?

The development of completely automated conversion systems for engineering drawings and maps is, in my view, a very long-term proposition. We should start with an interactive system such as AUTOCAD or CAD-OVERLAY that allows complete conversion of *any* drawing, instead of aiming at complete automation and relegating error correction to an afterthought. The human functions should gradually be taken over by the computer, starting with the easiest ones. The operator should use only AutoCad-type commands to enter parts of the drawing. Both the original bitmap and already converted portions can be shown as screen overlays. The parameters necessary for automating the process must be extracted from this type of user interaction only. For example, after the operator labels a resistor or two on the first drawing of a family, almost every resistor of the same size and shape should be automatically recognized. The converted documents should, of course, be in some standard format such as DXF.

More generally, it is time to concentrate on *systems that improve with use.* Now that even commercial OCR systems run entirely in software, nothing prevents continued fine-tuning of system parameters. We need more attempts to introduce feedback from down-stream programs (information retrieval, accounting, or schematic analysis) to alter layout analysis or classification parameters rather than simply correct individual errors. Current systems don't even exploit human post-editing in a manner analogous to adding "custom" words to ordinary spell-checkers. While these methods require only supervised adaptation, we should not rule out unsupervised learning. The basic premise is that the mass of data processed by any DIA or OCR systems in daily operation is more representative of future data than anything available at the factory.

6 Automated Evaluation

Automated benchmarking is not all that new. In the fifties and sixties, robot typewriters and printers pounded out hundreds of thousands of documents for testing MICR and OCR font readers. The IBM 1975 was tested on 1.3 million *pages.* Academic researchers reported results mainly on isolated characters. Although many different public test sets were produced and distributed through the Computer Society Repository, none attained the popularity of the hand-digitized data released by Highleyman in 1963. Competitive large-scale tests were conducted for address reading and, more recently, for census forms. Several large Japanese character sets were compiled. As the size of the test data sets grew, they migrated to CD-ROM. In the last few years, there has been renewed emphasis, and significant progress, on testing OCR and zoning accuracy on complete pages.

As has been painfully demonstrated over and over, large data sets don't guarantee accurate prediction of performance in the field. The data must also be *representative.* Sampling schemes reported to date have been based more on expediency than on sound statistics: we need to develop stratified and sequential sampling designs for large document populations. I am not aware of any organization or agency that has reported a well-planned *document census* centered on OCR and DIA variables.

In addition to the nature of the sample, we must consider the *sample size.* Even if the errors are statistically independent, it takes a sample size 1000 times greater to achieve a given level of significance at an error rate of 0.1% than at 1.0%. (Most researchers don't, however, report *any* confidence intervals). Underlying every statistical test there is some independence assumption that must be carefully examined.

Regardless of the specific performance aspects (*"metrics"*) being evaluated, I find it convenient to divide methods of automated evaluation into five major paradigms. The first is based on *manually labeled samples*. The accuracy of the labels can be ensured by comparing the results of data entry by two independent operators in the spirit of key-punch verification. Isolated characters are identified by a serial or accession number, and errors can be readily counted and classified. For text data, sophisticated string-matching algorithms are used to align the OCR output with the reference data. Character errors, word errors, error-reject trade-offs, and zoning accuracy have all been reported on large, manually-labeled document data bases.

The second paradigm is based on the *cost of post-processing*. In some applications, including layout analysis and line-drawing conversion, there exists no truly satisfactory classification of errors. In these cases, the cost (or time) of correcting the errors manually may be used as a performance measure. Even for text, character errors do not reflect post-editing costs accurately. However, the components of the edit-distance that are a by-product of string matching may yield a more satisfactory approximation.

The third paradigm measures performance on *pseudo-random defect models* that generate bitmaps directly. Elaborate defect generators have been constructed for both isolated characters and printed pages. Such models are useful only to the extent that they mirror the distortions found in some real population of documents. The validation of defect models is the objective of vigorous research.

The fourth paradigm, based on *synthetic documents*, must be distinguished from the third. It too is based on bitmaps generated by a word-processor or layout language that mimic real documents. These bitmaps are, however, "ideal" bitmaps that are not perturbed by artificial noise. The reference data required to evaluate performance is the source code of the synthetic document, and inherently contains every typographic or layout detail that might be of interest. The noise component is modeled by printing, copying, and scanning the synthetic documents. Regardless of the faithfulness of the format and typeface reproduction, such synthetic documents are useful only to the extent that the printers, copiers and scanners used to produced them are representative of some actual application. This method seems restricted to printed documents and computer-generated maps and line-drawings. (However, hand-print classification research could take advantage of writing test samples with both real and "electronic" ink, and retaining the temporal trace of the stylus for comparison with off-line classification.)

The last paradigm, *goal-directed evaluation*, may eventually supersede all others. Except for digital library applications, OCR and DIA are not, after all, intended to produce output for human consumption. Converted documents will be processed by some downstream program for information retrieval, accounting, market analysis, or circuit simulation. Most errors can be detected, at least in principle, by these programs. Goal directed evaluation requires, however, the availability of down-stream programs and large databases beyond the reach of most researchers.

Because there has been only very limited success in analytically modeling DIA and OCR systems, sound, objective and reproducible experimental evaluation is of the utmost importance. Therefore the current interest in large-scale automated evaluation of a range of performance measures is most welcome. In the United States, extensive DIA and OCR evaluation activities have been recently reported by CEDAR, ISRI, NIST, and the University of Washington.

7 Conclusion

Did I say that DIA and OCR may be approaching a mid-life crisis? Perhaps it is only puberty. They are surely on the threshold of success and prosperity. The best pickings still await us in this fertile patch at the intersection of pattern recognition, image analysis, and artificial intelligence.

Acknowledgment

The author gratefully acknowledges the influence of knowledgeable colleagues and of excellent students, past and present. Some of them commented kindly (and some less kindly) on drafts of this paper. He is also appreciative of the sustained support of the Central Research Laboratory, Hitachi, Ltd. and of the Northern-Telecom/BNR Education and Research Networking Program. This work has been conducted in the New York State Center for Advanced Technology (CAT) in Manufacturing, Automation and Robotics, which is partially funded by a block grant from the New York, State Science and Technology Foundation.

Invariant Features for HMM Based On-Line Handwriting Recognition

Jianying Hu, Michael K. Brown and William Turin

AT&T Bell Laboratories, Murray Hill, NJ 07974, USA

Abstract. In this paper we address the problem variability in handwriting due to geometric distortion of letters and words by rotation, scale and translation. In general, translation has not been a problem because it is easy to choose features that are invariant with respect to translation. It is more difficult to find features that are invariant with respect to all three types of geometric distortion. We introduce two new features for HMM based handwriting recognition that are invariant with respect to translation, rotation and scale changes. These are termed *ratio of tangents* and *normalized curvature*. Writer-independent recognition error in our system is reduced by a factor of over 50% by employing these features.

1 Introduction

The principal difficulty in the recognition of patterns by computer is dealing with the variability of measurements, or features, extracted from the patterns. There are several sources of variability, depending on the type of pattern data being processed. In on-line handwriting recognition these sources include input device noise, temporal and spatial quantization error, and variability in the rendering of input by the writer.

In this paper we address the spatial component of the last source of variability, that is, the geometric distortion of letters and words by rotation, scale and translation. We introduce two new features for handwriting recognition that are invariant with respect to all three factors of geometric distortion. These are termed *ratio of tangents* and *normalized curvature*. Similar features have appeared previously in the literature for planar shape recognition under partial occlusion [1].

2 Invariant Features

The concept of invariant features arises frequently in various machine vision tasks. Depending on the specific task, the geometric transformation ranges from simple rigid plane motion to general affine transformation, to perspective mapping, etc. [1]. In the case of handwriting recognition, the transformation of interest is *similitude* transformation, which is a combination of translation, rotation and scaling [1]. In our basic HMM based handwriting recognition system [2], we used the tangent slope feature which is invariant under translation and scaling, but not rotation. In this section we consider features that are invariant under arbitrary similitude transformation.

A similitude transformation of the Euclidean plane $\mathbb{R}^2 \to \mathbb{R}^2$ is defined by $\mathbf{w} = c\mathbf{U}\mathbf{r} + \mathbf{v}$, where c is a positive scalar, $\mathbf{U} = \begin{bmatrix} \cos\omega & -\sin\omega \\ \sin\omega & \cos\omega \end{bmatrix}$, $\mathbf{v} = [v_x \; v_y]^T$, representing a transformation that includes scaling by c, rotation by angle ω and translation by $\mathbf{v}$. We regard two curves as equivalent if they can be obtained from each other through a similitude transformation. Invariant features are features that have the same value at corresponding points on different equivalent curves.

Suppose that a smooth planar curve $\mathbf{P}(t) = (x(t), y(t))$ is mapped into $\tilde{\mathbf{P}}(\tilde{t}) = (\tilde{x}(\tilde{t}), \tilde{y}(\tilde{t}))$ by a reparametrization $t(\tilde{t})$ and a similitude transformation, i.e.

$$\tilde{\mathbf{P}}(\tilde{t}) = c\,\mathbf{U}\,\mathbf{P}(t(\tilde{t})) + \mathbf{v} \; . \tag{1}$$

Without loss of generality, assume that both curves are parametrized by arc length (natural parameter), i.e. $t = s$ and $\tilde{t} = \tilde{s}$. Obviously, $d\tilde{s} = c\,ds$. It can be shown [1] that curvature (the reciprocal of radius) at the corresponding points of the two curves is scaled by $1/c$, i.e. $\tilde{\kappa}(\tilde{s}) = \kappa\left((\tilde{s} - \tilde{s})/c\right)/c$. It follows that:

$$\frac{\tilde{\kappa}'(\tilde{s})}{(\tilde{\kappa}(\tilde{s}))^2} = \frac{\kappa'\left((\tilde{s} - \tilde{s}_0)/c\right)}{(\kappa((\tilde{s} - \tilde{s}_0)/c))^2} \; , \tag{2}$$

where $\tilde{\kappa}' = d\tilde{\kappa}/d\tilde{s}$ and $\kappa' = d\kappa/ds$, thus eliminating the scale factor from the value of the ratio. Equation (2) defines an invariant feature which we shall refer to as *normalized curvature*.

The computation of the normalized curvature defined above involves derivative estimation of up to the third order. Another set of invariants that require lower orders of derivatives can be obtained by using the invariance of distance ratios between corresponding points. Consider again the two equivalent curves $\mathbf{P}(t)$ and $\tilde{\mathbf{P}}(\tilde{t})$ defined above. Suppose P_1 and P_2 are two points on $\mathbf{P}(t)$ whose tangent slope angles differ by θ; $\tilde{P}_1$ and $\tilde{P}_2$ are two points on $\tilde{\mathbf{P}}(\tilde{t})$ with the same tangent slope angle difference. P and $\tilde{P}$ are the intersections of the two tangents on $\mathbf{P}(t)$ and $\tilde{\mathbf{P}}(\tilde{t})$, respectively (Fig. 1). Since angles and hence turns of the curve are invariant under the similitude transformation, it can be shown that if point $\tilde{P}_1$ corresponds to point P_1, then points $\tilde{P}_2$ and $\tilde{P}$ correspond to points P_2 and P respectively [1]. It follows from (1) that:

$$\frac{\left|\tilde{P}\tilde{P}_2\right|}{\left|\tilde{P}_1\tilde{P}\right|} = \frac{|PP_2|}{|P_1P|} \; . \tag{3}$$

Equation (3) defines another invariant feature which we shall refer to as *ratio of tangents*.

---

[1] The same transformation was referred to as *similarity* transformation by Bruckstein *et. al.* [1]. We have chosen another term to avoid confusion with the well known similarity transformation of linear algebra.

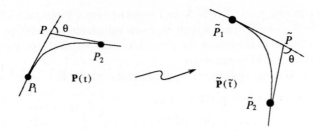

Fig. 1. Ratio of tangents

3 Implementation Issues

3.1 Derivative Estimation

To evaluate accurately the invariant features described above, high quality derivative estimates up to the third order have to be obtained from the sample points. In the following we describe how to use spline smoothing operators for derivative estimation.

Let $\mathbf{y} = (y_1, y_2, ..., y_n)^T$ be the noisy input vector obtained by sampling a "smooth" function $g(t)$ at $\mathbf{t} = (t_1, t_2, \ldots, t_n)$. Given an integer $m \geq 1$, the *natural polynomial smoothing spline approximation of order* $2m$ to $g(t)$ is the unique solution to the problem of finding $f(t)$ with $m - 1$ absolutely continuous derivatives, and square integrable mth derivative, which minimizes [3]: $\frac{1}{n} \sum_{i=1}^{n} (f(t_i) - y_i)^2 + \lambda \int_{t_1}^{t_n} (f^{(m)}(u))^2 du$. Let $\mathbf{g}$ be the output vector obtained as a result of evaluating the smoothing spline approximation at values of t given by $\mathbf{t}$. Since the smoothing spline approximation is linear we have $\mathbf{g} = \mathbf{A}(\lambda)\mathbf{y}$, where $\mathbf{A}(\lambda)$ is independent of the input vector. In a typical application to data lengths greater than n, the row $k = (n+1)/2$ of $\mathbf{A}(\lambda)$ is convolved with the data to produce the set of filtered data. Operators for estimating the νth derivative of $g(t)$ at the sample points can be constructed by evaluating the derivative of the spline approximation at the sample points. Since this is a linear functional, the result can also be expressed by a matrix-vector multiplication, i.e. $\mathbf{g}^{\nu} = \mathbf{A}^{\nu}(\lambda)\mathbf{y}$. Efficient algorithms for the construction of the $\mathbf{A}^{\nu}(\lambda)$'s exist [3].

To obtain estimates of third order derivatives, m has to be at least 3, yielding a spline of degree 5. The length of the operator is constrained by: $n \geq 4m - 1$. Given the degree of the spline, a wider operator provides better support for the spline estimation but also involves more computation. We chose to use a spline filter of degree 5 ($m = 3$) and length $n = 15$. The smoothness parameter λ controls the cut-off frequency f_c if the spline smoothing operator is viewed as a low pass filter. Since the handwriting signal (with the exception of cusps) consists predominantly of low frequencies and the predominant noise sources (mostly quantization error and jitter) are of high frequency content, it is easy to choose λ so that the spline filter cuts off most of the noise without causing significant distortion of the signal. In our system the handwritten scripts are parameterized in terms of arc length by resampling at 0.2 mm intervals before feature extraction and $\lambda = 20$ ($f_c \simeq 0.425\text{mm}^{-1}$) is used for all our experiments. Cusps tend to be smoothed out when such derivative operators are applied. However this does not pose a severe problem as long as the resulted ratio of tangents and normalized curvature features are used along with the tangent slope feature, since information related to cusps can be captured by the last feature [2].

3.2   Calculation of Ratio of Tangents

First we describe how to compute the ratio of tangents with an arbitrary tangent slope angle difference. Suppose P_1 and P_2 are two points along a script whose tangent slope angles differ by θ, as shown in Fig. (2). P is the intersection of the two tangents. The ratio of tangents at P_1 is defined as: $Rt_\theta(P_1) = |PP_2| / |P_1P|$. Suppose $\mathbf{u}_1$ and $\mathbf{u}_2$ are unit normal vectors at P_1 and P_2 respectively, using the law of sines we get: $Rt_\theta(P_1) = \sin\alpha / \sin\beta = |P_1P_2 \cdot \mathbf{u}_1| / |P_1P_2 \cdot \mathbf{u}_2|$. For convenience, we shall call P_2 the θ *boundary* of P_1.

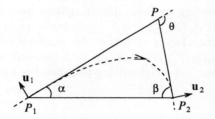

Fig. 2. Calculation of ratio of tangents

In order to use ratio of tangents as an invariant feature in handwriting recognition, a fixed angle difference $\theta = \theta_0$ has to be used for all sample points in all scripts. In real applications we normally have only scattered sample points instead of the continuous script and, in general, we can not find two sample points whose slope angle difference is equal to θ_0. Suppose sample point P_i's θ_0 boundary is between points P_j and P_{j+1}, i.e. P_j is P_i's θ_1 boundary and P_{j+1} is its θ_2 boundary where $\theta_1 < \theta_0 < \theta_2$, $Rt_{\theta_0}(P_i)$ is estimated from $Rt_{\theta_1}(P_i)$ and $Rt_{\theta_2}(P_i)$ using linear interpolation.

Obviously the choice of θ_0 greatly affects the tangent ratio values. If θ_0 is too small, the feature tends to be too sensitive to noise. On the other hand, if θ_0 is too large, the feature becomes too global, missing important local shape characteristics. Currently this value is chosen heuristically and $\theta_0 = 10°$ is used in all our experiments.

In order to enhance the distinctive power of the feature, we augment the ratio of tangents at each point by the sign of the curvature at that point. The resulted feature is referred to as *signed ratio of tangents*, and is used instead of ratio of tangents in the experiments described later.

3.3 Combined Scores

In our discrete HMM system, each feature is quantized into a fixed number of bins. To simplify our models, we chose to treat the features as being independent from each other. When the three features are used together, the joint probability of observing symbol vector $S_{k_1 k_2 k_3} = [k_1, k_2, k_3]$ in state j is: $b_j(S_{k_1 k_2 k_3}) = \prod_{i=1}^{3} b_{ji}(k_i)$, where $b_{ji}(k_i)$ is the probability of observing symbol k_i in state j according to the probability distribution of the ith feature. It follows that the corresponding log-likelihood at state j is: $L_j(S_{k_1 k_2 k_3}) = \sum_{i=1}^{3} \log[b_{ji}(k_i)]$.

In a conventional HMM implementation with Viterbi scoring, the likelihood defined above is used directly in training and recognition. In this case, each of the three features contributes equally to the combined log-likelihood and therefore has equal influence over the accumulated scores and the optimal path. However, our experiments with the three features show that when used alone, the tangent slope feature gives far better recognition performance than each of the two invariant features. This suggests that the tangent slope is a more discriminative feature and therefore should have more influence on decision making than the other two features. In order to adjust the influence of different features according to their discriminative power, we use instead the *weighted log-likelihood*: $\hat{L}_j(S_{k_1 k_2 k_3}) = \sum_{i=1}^{3} w_i \log(b_{ji}(k_i)) - \log(N_j)$, where N_j is the state normalization factor such that the *weighted probabilities*, defined by $\hat{b}_j(S_{k_1 k_2 k_3}) = \prod_{i=1}^{3} [b_{ji}(k_i)]^{w_i} / N_j$, sum up to 1 for each state. The normalization factor ensures that the weighted log-likelihood is not biased towards any particular state.

4 Experimental Results

The detailed description of our recognition system can be found in a previous paper [2]. A brief review is provided here.

The handwriting data was collected using a newly developed graphics input tablet [4] at 200 samples per second. Writers were asked to write on a lined sheet of paper, without any constraints on speed or style. The preprocessing steps include cusp detection, smoothing, deskewing and finally resampling at 0.2 mm intervals. Features are then extracted at each resampled point.

Sub-character models called *nebulous stroke models* are used as the basic model units. Currently each stroke is modeled by a single HMM state. A letter model is a left to right HMM with no state skipping, constructed at run time by concatenating the corresponding stroke models. These HMM's are embedded in a stochastic language model which describes the vocabulary. The current vocabulary contains 32 words, targeting an underlying application of a pen driven graphics editor.

8595 samples have been collected from 18 writers, with each word in the vocabulary written 15 times by each writer. 10 writers were chosen (after data collection) to be the "training writers". The training set is composed of about 10 samples of each word from each training writer, a total of 3180 samples. The test set is composed of all samples not used for training, divided into two groups. Group A contains 1592 samples from the 10 training writers, group B contains 3823 samples from the 8 other writers not used for training.

Table 1 compares the error rates of the system when a single feature – tangent slope was used and when the two invariant features – signed ratio of tangents and normalized curvature were added. As shown in the table, by adding the two invariant features we have achieved an error rate reduction of 46% for training writers, and 54% for non-training writers.

Table 1. Comparison of error rates

	A (Training Writers)	B (Non-training Writers)
single feature	10.3%	23.4%
invariant features added	5.6%	10.7%

5 Concluding Remarks

We have introduced two new features for handwriting recognition that are invariant under translation, rotation and scaling. The use of these features in an HMM based handwriting recognition system has been demonstrated and significant improvement in recognition performance has been obtained. Invariant features have been discussed extensively in computer vision literature. However, they have been rarely used in real applications due to the difficulty involved in estimating high order derivatives. We have demonstrated that high order invariant features can indeed be made useful with careful filtering in derivative estimation.

References

1. Bruckstein, A. M., Holt, R. J., Netravali, A. N., Richardson, T. J.: Invariant signatures for planar shape recognition under partial occlusion. CVGIP: Image Understanding **58** (1993) 49–65
2. Hu, J., Brown, M. K., Turin, W.: Handwriting recognition with Hidden Markov Models and grammatical constraints. Fourth Int. Workshop on Frontiers in Handwriting Recognition, Taipei, Taiwan (1994) 195–205
3. Lyche, T., Schumaker, L. L., Computation of smoothing and interpolating natural splines via local bases. SIAM J. Numer. Anal. **10(6)** (1973) 1027–1038
4. Boie, R., Ruedisueli, L., Wagner E.: Capacitive position sensor. U.S. patent filed (1993)

A Robust Analytical Approach for Handwritten Word Recognition

G. Congedo, G. Dimauro, S. Impedovo, G. Pirlo, D. Sfregola

Dipartimento di Informatica - Università di Bari
Via Amendola 173 - 70126 Bari - Italy

Abstract. This paper presents a system for handwritten word recognition based on an analytical approach. Through a fast segmentation procedure, basic strokes are detected and used to recognize the word by a bottom-up approach. The system has been tested on the worded amounts on bankchecks. The experimental results show the effectiveness of the approach.

1 Introduction

Handwritten documents are generally massively used to transfer information in administrative and technical offices. In fact, from tax offices to banks, from insurance companies to health offices, and everywhere handwriting is used, handwriting recognition is required. In these environments a wide number of activities could receive a significant support by systems able to handle handwritten data forms [1,2].

Unfortutely automatic document analysis and processing involves a wide range of tasks, ranging from intelligent layout analysis to typewritten and handwritten character and word recognition, and for many of them the scientific community in still far to obtain conclusive solutions. Among the others, handwritten word recognition involves the integration of several different aspects in the field of pattern recognition as well as in the field of the exploitation of knowledge at several levels [3].

In this paper a new system for handwritten word recognition is presented. The system uses an analytical approach based on the consideration that each character consists of a sequence of basic strokes: humps, cusps, oriented lines, circles, semicircles, etc.; therefore each word can rightly be considered as a sequence of basic strokes [4]. Therefore the process of word recognition here proposed follows four main phases: preprocessing, segmentation of the word into basic strokes, feature extraction and stroke classification, and recognition of the word. The experimental results have been carried out testing the system on the basic words extracted from italian bankchecks. The results point out the robustness of the approach and demonstrate the effectiveness of the design strategies used for this system. These strategies mainly consist in performing recognition at the last classification stages in a closed-loop procedure and using the complete a-priori lexical knowledge available in the specific domain of the application.

ACKNOWLEDGEMENT: This work was supported by MURST 60% - 1994, Italy.

2 Preprocessing

The preprocessing phase involves three successive steps: smoothing, word zones detection and noise removal. Smoothing is carried out by the traditional

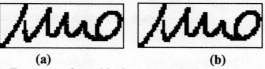

(a) **(b)**

Fig. 1. Exampes of word before (a) and after (b) smoothing.

algorithms which work on the contour of the image [5]. Fig. 1 shows an example of handwritten word before and after smoothing. Word zones identification is accomplished by detecting the Upper Basic Line (UBL) and the Lower Basic Line (LBL) by the analysis of the horizontal histogram [6]. This approach allows to detect three basic zones of the word: the "middle zone", the "upper zone" and the "lower zone". The "middle zone" is

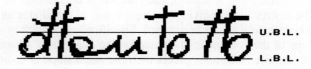

Fig. 2. Word zone detection.

delimited by the UBL and the LBL and is always present. The "upper zone" is the part of the word above the UBL and indicates the existence of ascending pulls (like those of characters "l", "d", "t") or dots of the letters "i". The "lower zone" is the part of the word below the LBL and indicates the existence of descending pulls (like those of character "q"). Figure 2 shows the "middle zone", the "upper zone", and the "lower zone" detected by this approach on an input word. Noise removal is carried out by detecting noisy patterns like spurious points in the image. At this purpose, a connected component detector is used and isolated components of few pixels are removed in an automatic way [5].

3 Segmentation

The objective of the segmentation phase is to obtain a representation of the word as a sequence of basic strokes. A stroke is here defined as a pattern between two consecutive local minima in the upper contour of the word. Therefore, stroke detection starts from the detection of the local maxima and minima in the vertical direction of the word. Successively the cutting line r is defined according to the direction of the bisecting line of s and t as shown in figure 3, where: M is the local minimum candidate for splitting, P is the

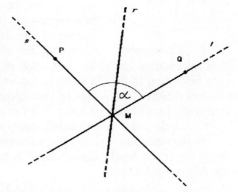

Fig. 3. Computation of the cutting angle.

local maximum preceding M in the upper contour, Q is the local maximum following M in the upper contour, s is the straight line connecting P and M, t the

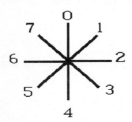

Fig. 4. A segmented word.

straight line connecting Q and M. Fig. 4 shows a typical result of the segmentation procedure. For each stroke of the word some features like position, dimension (height and width), and zone of the stroke. The positions are: *lower, middle- lower, middle, middle-upper, upper.*

4 Feature extraction and stroke classification

After segmentation, each stroke is normalized in a matrix of 45x45 pixels. The matrix is then divided in nine zones of equal size as fig. 5 shows and for each zone a feature vector is detected. Specifically, for each pixel of the contour, the local slope of the contour is computed by the analysis of adjacent contour pixels. Finally, the number of contour pixels with the same slope is counted. Eight basic slopes are considered as fig. 6 shows.

(a)   (b)   (c)

*Fig. 5. (a) A stroke, (b) The stroke after normalization,
(c) The nine zones of the normalization matrix.*

Consequently, for each zone a feature vector of 8 elements is extracted and the feature vector v of 72-elements for the entire stroke is obtained by connecting the feature vectors of each zone.

In the learning phase, the stroke feature vectors are used, for the selection of the classes of basic strokes by a supervised classification process. For each class, the vectors of the averages and standard deviations of the feature of the elements in the class are computed. A

Fig. 6. Basic slopes.

further information stored for each class in the feature vector is the position of the strokes belonging to the class inside the word. Successively, a suitable grouping of the classes detected into groups of basic shapes (humps, cusps, circles,etc.) is carried out manually by a human operator. The use of the groups allows to simplify the word description code and to improve the recognition process.

In the recognition phase, the test word is splitted into strokes and each stroke is firstly classified. Stroke classification is based on the similarity measure provided by the following distance:

$$d(v, \mu^j, \sigma^j) = \frac{1}{72} \sum_{i=1}^{72} \frac{|v_i - \mu_i^j|}{\sigma_i^j},$$

where: $v = (v_1, v_2, ..., v_{72})$ is the feature vector of the stroke that must be classified, $\mu^j = (\mu_1^j, \mu_2^j, ..., \mu_{72}^j)$ is the mean feature vector and $\sigma^j = (\sigma_1^j, \sigma_2^j, ..., \sigma_{72}^j)$ is the vector of the standard deviations of the strokes of the j-th class.

According to the similarity measure and from the analysis of the position of the stroke, the classification process returns a ranked list of classes to which the stroke could belong. Successively, each class is replaced with the corresponding group and successive occurences of the same group is the ranked list are removed.

5 Recognition

The recognizer works according to a simple iterative strategy: at each step of the recognition phase, the system checks if the sequence of groups under examination is a lawful character. At this purpose, a list of sequences of groups of strokes is associated to each character and for each different sequence of groups the zones of affiliation of the strokes (lower, middle-lower, middle, middle-upper, upper) are also stored.

When a character is identified, an entry in the trie structure of the data dictionary is gained and the recognition procedure continues for another lawful character until a complete word is identified. The data dictionary contains the description of each basic word. It is organized as a trie structure in which each node is a character and each sequence of characters obtained beginning from the root of

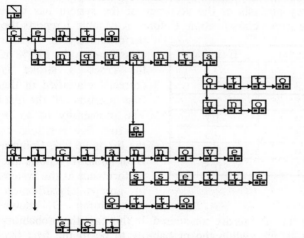

the trie structure to one of its terminal nodes represents a basic word. Figure 7 shows the structure of a small part of the data dictionary used in the system.

In some cases the code contains strokes useless to the aim of the recognition or it lacks of entire characters due to very poor writing quality. The recognizer recovers these conditions by cashing from the code a boundless number of

Fig. 7. The trie structure representing the data dictionary.

groups retained bindings or adding a finite number of character supposed to be lacking. Of course, this recovery procedure stresses the normal operation of the system and therefore this condition adds a penalty to the similarity index used for the final classification. Specifically, the recognition index used to select the candidate words is computed in the following way:

$$Ind = \frac{1}{n}\left(\sum_{j=1}^{n} pos_j + b\_penality \cdot \sum_{i=1}^{nb} i + l\_penality \cdot \sum_{k=1}^{nl} k\right)$$

where: n is the number of code groups used to compose the word, pos_j is the position of the selected group in the group list, nb is the number of bindings, $b\_penality$ is the penality when a binding is found, nl is the number of characters not found in the code, $l\_penality$ is the penality when a character is not found in the code.

6 Experimental Results

The system presented in this paper has been realized in TURBO C++ 3.00 for DOS and it actually runs on a PC - 80486DX/33Mhz. Data acquisition has been accomplished by a HP scanjet IIc with a resolution of 150 dpi.

It has been tested on the set of fifty basic words extracted from the worded amounts of italian bankchecks [7]: *uno, due, tre, quattro, cinque, sei, sette, otto, nove, dieci, undici, dodici, tredici, quattordici, quindici, sedici, diciassette, diciotto, diciannove, venti, ventuno, ventotto, trenta, trentuno, trentotto, quaranta, quarantuno, quarantotto, cinquanta, cinquantuno, cinquantotto, sessanta, sessantuno, sessantotto, settanta, settantuno, settantotto, ottanta, ottantuno, ottantotto, novanta, novantuno, novantotto, cento, mille, mila, unmilione, milioni, unmiliardo, miliardi.* The experimental results have been obtained using two suitably defined databases. The *reference database* has been used by a human operator for the definition of the basic strokes and of the sequences of strokes composing each character. It consists of 200 reference words written by four different writers. The test *database consists* of the 4050 words, written by 81 different writers.

A preliminary analysis of the accuracy of the system has been carried out with respect to the stroke classification. Column A of table 1 reports the performance in stroke classification. In 79.1% of the cases a stroke is correctly classified in the first position of the list. The probability to be in the first five positions of the list is 96.9%. Column B of table 1 reports the performance of the system in groups classification. As column B shows,

Position	A (in %)	B (in %)
1	79.1	87.5
2	9.3	7.7
3	4.8	2.1
4	2.3	1.2
5	1.4	0.3
more than 5	2.6	0.7
Error	0.5	0.5

Table 1.

misclassification goes down when groups are considered. In this case the probability to be in the first position is 87.5%, while the probability to be in the first five positions of the list is about 98.8%.

The complete test about the performance of the system in recognizing basic words is reported in column A of table 2. The recognition in the first position has been achieved in 70.7% of the cases, in the first ten positions in 93.9% of the cases. The recognition error, due to the absence of the correct word in the list of hypotheses of the recognizer, is 5.4%. A manual segmentation procedure has been carried out on the word images in order to demonstrate how uncorrect segmentation can lead to misclassification.

Position	A (in %)	B (in %)
1	70.7	82.6
2	12.8	8.5
3	4.0	2.6
4	2.2	1.2
5	1.5	0.6
6	0.9	0.2
7	0.6	0.2
8	0.4	0.2
9	0.5	0.1
10	0.3	0.1
more than 10	0.7	0.2
Error	5.4	3.5

Table 2.

Column B of Table 2 reports the performance of the system when manual segmentation is performed.

Currently, the system performs word recognition with a speed of about 5 words/second when complete ranked lists are considered for stroke classification. Faster output can be enforced if reduced ranked lists are considered.

The system has been also tested on complete worded amounts written by different writers also by using the syntactic rules of the italian worded amounts [7]. On a test database of 750 amounts, the system obtains a correct recognition the 60.8% of the cases when automatic segmentation is accomplished. When manual segmentation is carried out, the correct recognition reaches the 72.2% of the cases.

7 Conclusion

In this paper an analytical approach for handwritten word recognition is presented. The approach uses a simple scheme to identify shapes in the word image and a recursive procedure to verify each hypothesis generated at the lower level from the analysis of the basic shapes. The recognition results points out the effectiveness of the approach in domains with a reduced dictionary.

References

1. K.Sakai, H.Asami and Y.Tanabe, "Advanced application systems for handwritten character recognition", in *From Pixels to Features III - Frontiers in Handwriting Recognition*, S. Impedovo and J.C.Simon eds., Elsevier Publ.,pp.371-384, 1992.
2. R. Casey, D. Ferguson, K. Mohiuddin, E. Walach, "Intelligent Forms Processing System", *Machine Vision and Applications*, Vol. 5, N. 3, pp. 143-155, 1992.
3. L.J. Evett, C.J. Wells, F.G. Keenan, T. Rose, R.J. Whitrow, "Using linguistic information to aid handwriting recognition" in *From Pixels to Features III- Frontiers in Handwriting Recognition*, S. Impedovo, J.C. Simon, (eds.), Elsevier Publ., pp.339-348, 1992.
4. C.A. Higgins, D.M. Ford, "A new segmentation method for cursive script recognition", in *From Pixels to Features III- Frontiers in Handwriting Recognition*, S. Impedovo, J.C. Simon, (eds.), Elsevier Publ., pp. 75-86, 1992.
5. T.Pavlidis, *Algorithms for Graphics and Image Processing*,Berlin-Heidelberg,Springer-Verlag, 1982.
6. J. Hull, Tin Kam Ho, J. Favata, V. Govindaraju, S. Srihari, "Combination of Segmentation-based and Wholistic Handwritten Word Recognition Algorithms", in *From Pixels to Features III- Frontiers in Handwriting Recognition*, S. Impedovo, J.C. Simon, (eds.), Elsevier Publ., pp. 261-272, 1992.
7. G. Dimauro, M.R. Grattagliano, S. Impedovo, G. Pirlo, "A System for Banckchecks Processing", IEEE Computer Society Press, Vol. 4960-02, ICDAR '93, Tsukuba Science City -Japan, Oct. 1993, pp. 454-459.

Recognition of Rotated Characters
by Inexact Matching

L. Cinque, S. Di Zenzo, S. Levialdi

Dipartimento di Scienze dell'Informazione, Universita' "La Sapienza" di Roma
Via Salaria 113, 00198 Roma, Italy

Abstract. In this note we address the recognition of rotated hand-printed characters. We define the sides and the lids of a plane figure and outline an inexact matching process using such features based on the edit distance between circular words of different lengths.

1   Introduction

Numerous object recognition tasks can be seen as a case of inexact matching [1]. In order to solve this problems we must find a correspondence between two relational structures. Given a candidate object to be recognized and a model object called the prototype, we seek a match between such structures which is not necessary a perfect one, but only good enough. Often these structures can be represented by graphs whose nodes stand for subparts and whose arcs represent relationships between the nodes they connect [2]. When matching graphs representing two different objects, an optimal permutation of a matrix that minimizes a similarity distance measure between the graphs must be found.

In order to improve the efficiency of the matching process we must consider: (1) the choice a suitable distance between the structural descriptions to be matched, (2) the reduction of the number of permutations of the candidate structure needed to compute that distance, (3) matching the candidate structure with a subset of the stored prototypes. Regarding point (3), we note that, if a distance measure is available, the search for the best D_k matching D is actually a search for a nearest neighbor (NN). Hence we know that, with suitable preprocessing in storing the $D_1, \ldots, D_N$, it is possible, in principle, to achieve a more efficient search in the recognition process [3].

In this note we focus on points (1) and (2) with reference to a specific problem, namely, the recognition of rotated characters. In this particular OCR problem, current efforts are far below the performances of the standard OCR commercial systems. There are expectations that inexact matching can have a key role to improve the recognition rate. In section 2 we give an overview of the recognition of rotated characters, and in section 3 we define the sides and the lids of a plane figure and outline an inexact matching process using such features.

2   Recognizing rotated characters

Our research is motivated by a computer application that has received much attention over the past few years, namely, the automatic acquisition of line drawings (engineering drawings, land-use and land-register maps, utility maps).

The raster-to-vector conversion of line drawings is a non trivial computer task. Effective techniques for the automatic input of drawings into a database are difficult to implement, and many efforts in this direction over the past twenty years have found limited success. Solutions that meet the requirements of the users to a reasonable extent have been developed only recently [4]. The automatic computer acquisition of line drawings is performed through the basic steps of vectorization of the line-structure and recognition of the text data interspersed in the document. The latter processing step is a particularly difficult OCR problem: indeed, the characters and symbols in a drawing are usually handprinted, and can be however rotated and rescaled. Besides, characters and symbols are often isolated, offering no contextual information; symbols and lines may overlap, and the separation of symbols arranged into strings may be imperfect.

A portion of a digitized drawing is shown in Figure 1. It can be seen that the separation of text data from the rest of the image is usually a critical computing step. In most raster-to-vector conversion systems text is separated from graphics by extracting and classifying the connected components in the image. This procedure fails, however, when symbols touch or overlap lines, something that occurs often when the originals are blurred or degraded.

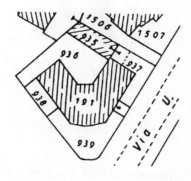

Fig. 1. Portion of a land-register map

We use SEADT[4] a feature-based system, where all the features are invariant under 'shifts, rotations, and scaling. These invariants are rather obvious requirements, of course: the true criteria in the formation of a feature set has been robustness to distortion and noise. Various reliable features are derived from *lids* and *sides*, as defined in the following. The patterns formed by lids and sides convey much information about the shape of a figure. Figures 2 show some examples of possible lid/side configurations [5].

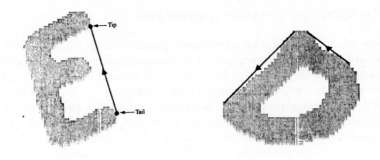

Fig. 2. Example of a lid and sides

SEADT is a rule-based recognition system. That means that each class of symbols has a set of requirements that a shape should satisfy to be considered as a candidate for that class. To have an example, consider symbol T in Figure 4: a very simplified set of requirements that a symbol should satisfy to be considered as a candidate T is the following: (1) there are two lids, (2) neither lids is twice as long as the other, (3) the angle between AB and CD is greater than 110 degrees.

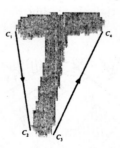

Fig. 3. Shape of T

Each requirement is so formulated that it satisfied with probability 1 by every shape actually belonging to the class. This strategy allows adding as many requirements as needed for actual discrimination without performance degradation. Our concern in this paper is actually that of implementing techniques of inexact and approximate matching within the framework of the SEADT system. While in this paper we just discuss ideas and proposals, we do have an actual system which seems very suited to the addition of the algorithms discussed below.

3 Inexact matching based on sides and lids

In this section we present and discuss a inexact matching procedure for the OCR of rotated characters. The computational cost is considerably lower. We shall use the notions of side and lid introduced in the cited paper. This section is subdivided in two subsections: in the first we define sides and lids, while in the second we specify a method of inexact matching relying upon those features.

3.1 The lids and sides of a planar figure

Since we are concerned with the recognition of characters of any size, position, and orientation, we need features that do not change value under the similarity transformations of the plane, (motions of the plane that are combinations of shifts, rotations, and scalings). Another requirement for a feature to be of use in this context is the following: it should be as insensitive as possible to certain transformations of the plane such as stretching along one direction and rubber-sheet distortions. These more general motions of the plane are usually present in hand-written characters. Of course, to be of actual use in the classification of shape, a feature should also exhibit certain basic properties: it should have discriminating capability, be rather insensitive to the sampling rate, be computationally cheap, have small variance on each of the subpopulations it helps to discriminate from one another, have high overall variance.

In this section we concentrate on the two features lids and sides. They seem to meet the above wishlist for features to a reasonable extent. However, the use done of these features in that recognizer is somewhat straightforward. It is well known that the number of concavities (or bays) is somewhat unreliable as a feature; indeed, to use it reliably, we should need a threshold to separate noise concavities from true concavities, which obviously does not exist. However, bays can be extremely useful in the recognition of rotated characters: the key idea here is not to use the bays themselves, but their lids.

By a lid of a plane figure we mean a maximal portion of the perimeter of the convex hull of the figure not belonging to the figure itself. The arrow in figure 5 represents a lid. Lids are vectors, each equipped with a tail, a tip, and an orientation. We stipulate that orientation is chosen such that each lid leaves the figure on the left (anticlockwise orientation). To define the sides of a figure, let us first note that the boundary of the convex hull of a figure always consists of a sequence of segments. We normalize the lengths of these segments by dividing them by their sum. Segments whose normalized length is greater than a fixed threshold, typically 0.15, are called sides. A side is a vector, hence it has a tail, a tip, and an orientation. For orientation we stipulate the same convention as for lids.

3.2 Matching strings of lids and sides

Let F be any plane figure. In general, if we traverse the boundary of the convex hull of F, we will encounter certain lids and sides. We may associate to F a string over a two-letter alphabet $\{s, l\}$ to represent the sequence of sides and lids encountered in the traversal. For example, a string such as *sls* will denote a sequence composed by a side, followed by a lid, followed by a side. Obviously, the strings *sls*, *ssl*, and *lss* will represent the same sequence of sides and lids.

Pursuing this approach a bit further, we may replace the strings over $\{s, l\}$ with strings of complex numbers in exponential form $re^{i\alpha}$. We stipulate that r represents the length of the vector (side or lid), while the amplitude α represents the angle formed with the subsequent vector ($\alpha = 0$ if there is no subsequent vector).

To distinguish between sides and lids we store $-re^{i\alpha}$ when the vector is a side. This amounts to taking a negative imaginary part for sides (note that the angle formed by a vector (side or lid) with the subsequent vector does not exceed 180 degrees. Thus, we represent the salient features along the boundary of the convex hull by a sequence of complex numbers. As with strings over $\{s, l\}$, two strings that can be obtained from one another by a circular permutation represent the same sequence of features.

Let us now examine how matching can be implemented. In the training phase we will store a string S_k in correspondence of each prototype figure. S_k illustrates the sequence of lids and sides along the border of the convex hull of the k-th figure used to train the recognizer. A similar string S is associated with the query figure F. To reduce to a search for NN (or kNN) we need a notion of distance between strings of complex numbers that possibly can be of different length.

We propose a generalization of the so called *edit distance* between strings over an alphabet. Before we get into the details, however, we find it convenient to note that the number of permutations of the query string S needed to find the best match is not greater than to the length of S. Indeed, we only have to permute S circularly.

The edit distance is connected with the string-to-string correction problem. This problem is to determine the distance between two strings as measured by the minimum cost sequence of edit operations needed to change one string into the other. The edit operations allow changing one symbol in a string, deleting one symbol from a string, inserting a symbol into a string. The cost of each of these edit operations is set equal to a costant C (usually, $C = 1$). There are various algorithms that compute the edit distance between two strings in time proportional to the product of the lengths of the strings [6]. Fortunately enough, a generalization of the edit distance to circular strings has already been studied, and we also have an algorithm that computes the distance between circular strings in O($nm \log m$) time [7].

To generalize the edit distance to strings of numbers we simply replace the constant C with the absolute value of the number that is inserted or deleted, or the absolute value of the difference of the two numbers, when a number is replaced with another. It is straightforward to show that the quantity so computed satisfies the axioms of a distance measure. The overall complexity of this matching is thus O($n^2 \log n$), where n is the average length of our strings.

References

1. L. G. Shapiro and R. M. Haralick, "Structural descriptions and inexact matching", *IEEE Trans. on Pattern. Analysis. Machine. Intell.*, **5**, pp. 504-519, (1981).
2. A. Sanfelieu and K. S. Fu, "A distance measure between attributed relational graphs for pattern recognition", *IEEE Trans. on System, Man and Cybernetics*, **13**, pp. 353-362, (1983).
3. F. P. Preparata and M. I. Shamos, *Computational Geometry − An Introduction*, Springer-Verlag, 1985.
4. L. Boatto, V. Consorti, M. Del Buono, S. Di Zenzo, V. Eramo, A. Esposito, F. Malcarne, M. Meucci, A. Morelli, M. Mosciatti, S. Scarci, M. Tucci, "An Interpretation System for Land Register Maps," *IEEE Computer.* **25**, 25–33 (1992).
5. S. Di Zenzo, M. Del Buono, M. Meucci, A. Spirito, "On the recognition of hand-printed characters of any size, position, and orientation," *IBM J. Res. Develop.* **36**, 487–501 (1992).
6. R. Wagner and M. Fisher, "The string-to-string correction problem", *Journal of the ACM.*, **1**, pp. 168-173, (1974).
7. M. Maes, "On a cyclic string-to-string correction problem", *Infornation Processing Letters*, **35**, pp. 73-78, (1990).

Biomedical Applications II

Bayesian Image Reconstruction
Using a High-Order Interacting MRF Model*

Michael Chan[1], Gabor T. Herman[2] and Emanuel Levitan[3]

[1] Dept. of Computer and Information Science, University of Pennsylvania, Philadelphia, USA.
[2] Dept. of Radiology, University of Pennsylvania, Philadelphia, PA 19104, USA.
[3] Dept. of Physiology and Biophysics, Technion, Haifa 31096, Israel.

Abstract. We propose a Markov Random Field image model formulated using both low-order and high-order interactions of the pixels for piecewise smooth image reconstruction. We discuss and compare the applications of two Bayes estimators, namely the Maximum a Posteriori (MAP) estimator and the Minimum Mean Squares Error (MMSE) estimator, to Positron Emission Tomography (PET) imaging. The usefulness of the high-order interactions is demonstrated especially in the reconstruction of noisy image regions with small supports. The MAP and MMSE estimates are obtained in an efficient way by making use of the noise properties of the images obtained using the Filtered Backprojection (FBP) algorithm. We illustrate the efficacy of our overall approach by simulation experiments.

1 Introduction

The use of *Markov Random Field* (MRF) or *Gibbs* image priors in the Bayesian framework of image analysis has attracted much attention in the past in many different areas since the appearance of [2, 7]. Many of the previously proposed Gibbs priors typically include only nearest neighbor correlational information. The use of higher order interactions has not been very much investigated. Most pairwise interacting models have mainly a smoothing effect on the recovered images. (Sometimes the smoothing is mitigated for sharp intensity changes, as in [8].) It has been observed that randomly sampled images from the distributions specified by such priors have typically a uniform appearance [2, 9] and that the models are often inadequate in a statistical sense [10]. This issue has been recently investigated in [10], where a statistical model with high-order interactions for a particular class of piecewise constant images was proposed. We generalize this type of model in the current paper to model piecewise smooth images, in which we use 3×3 cliques to model the borders between regions. Borders in the images have been modeled by a dual MRF composed of unobservable line elements [7]. The derivation of an equivalent energy function formulated using only the pixel values was found possible in some cases [5], but not when there are interactions between line elements [1, 5].

Using the proposed prior, we discuss and compare the applications of two Bayes estimators, namely the *Maximum a Posteriori* (MAP) estimator and the *Minimum Mean Squares Error* (MMSE) estimator, in *Positron Emission Tomography* (PET). A method to remedy the computational difficulties encountered by previously proposed algorithms for such problems will also be discussed. We illustrate the usefulness of the high-order interactions especially in the reconstruction of noisy image regions with small supports.

* This work was supported by the National Institutes of Health under the grant HL28438.

2 A High-Order Markov Random Field Image Model

A MRF image model is provided by the Gibbs distribution of the general form:

$$\Pi(X) = \frac{1}{Z}e^{-\beta H(X)}. \tag{1}$$

Here $X = (x_1, \cdots, x_N)$ is an image consisting of N pixels, with the gray level of the nth pixel given by x_n. $H(X)$ is an energy function to be expressed as a weighted sum of *clique* potentials. β is a parameter which controls the "peakiness" of the distribution and can be interpreted as the inverse of the *temperature* in statistical mechanics.

Following the development of an image model of a class of piecewise constant images which uses a MRF with interactions not constrained to those between pairs of pixels [4, 10], we present a generalization of the model with a similar neighborhood structure for piecewise smooth images with 256 gray levels.

First, we write the energy function as $H(X) = H_1(X) + H_2(X)$. The first term, which models smoothness, is

$$H_1(X) = -(\sum_{C \in \mathcal{C}_+} \lambda_1 V_C(X) + \sum_{C \in \mathcal{C}_\times} \lambda_2 V_C(X)), \tag{2}$$

with

$$V_{\{j_1, j_2\}}(X) = \exp(-\frac{(x_{j_1} - x_{j_2})^2}{2\delta^2}), \tag{3}$$

where $\mathcal{C}_+$ denotes the set of all horizontal and vertical pair cliques, whereas $\mathcal{C}_\times$ denotes that of the diagonal ones. The coefficients λ_1 and λ_2 control the strengths of each type of interaction and δ controls the degree of smoothing. The second term, which models the borders between regions, is

$$H_2(X) = -\sum_{C \in \mathcal{C}_{3\times3}} (\kappa_1 I_C^1(X) + \kappa_2 I_C^2(X) + \kappa_3 I_C^3(X)), \tag{4}$$

with

$$I_C^1(X) = \begin{cases} 1, & \text{if } |x_k - x_c| \le \delta \ \forall k \in \{w, nw, n, ne, e\} \\ & \text{and } |x_k - x_s| \le \delta \ \forall k \in \{sw, se\} \text{ and } |x_n - x_s| > \Delta, \\ \vdots & \vdots \\ 0, & \text{otherwise.} \end{cases} \tag{5}$$

(I_C^2 and I_C^3 are similarly defined as depicted in Fig. 1.) The subscripts c, n, e, s, w (denoting center, north, east, south and west, respectively) are used to index the pixels in a 3×3 clique, whereas $\mathcal{C}_{3\times3}$ denotes the set of all cliques with 3×3 block of pixels. The case in (5) (three others omitted), for example, is depicted by the left clique labeled "horizontal" in which the pixels labeled by 'x' have gray levels "similar" to those labeled by 'a'. Each term models borders in a particular set of orientations and the coefficients κ_1, κ_2 and κ_3 control the strength of each type of interaction. Δ is a threshold for an edge and δ is defined as before. Note that by using 3×3 cliques, borders at intermediate orientations can also be modeled.

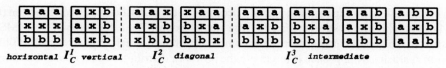

horizontal I_C^1 vertical I_C^2 diagonal I_C^3 intermediate

Fig. 1. Pictorial clique potential definition for borders in different orientations. Pixels labeled by 'a' and 'b' have "dissimilar" gray levels. Pixels labeled by 'x' in a clique all have gray levels "similar" to those labeled by either 'a' or 'b'.

To encourage spatially isotropic neighbor interactions we set $\lambda_2 = \frac{1}{\sqrt{2}}\lambda_1$. For the "border elements", we set $\kappa_1 = \kappa_2 = \frac{1}{2}\kappa_3$ since the vertical, horizontal and diagonal border elements always appear in pairs at the borders between two regions. With properly determined parameters, the model not only encourages the formation of smooth regions, but also sharp transitions at locations which make up the borders between regions.

3 Bayesian Image Reconstruction

3.1 The Posterior Distribution

We discuss the application of the proposed prior in PET image reconstruction. The measurement model in PET imaging is given by the likelihood function of the form [12]:

$$L(Y|X) = \prod_{i=1}^{I} \frac{l_i^{y_i} e^{-l_i}}{y_i!}, \tag{6}$$

where $l_i = \sum_j^J l_{ij} x_j$, J is the total number of pixels and I the number of measurements. Under the Bayesian framework, the posterior distribution of the images is given by

$$P(X|Y) \propto \Pi(X)L(Y|X), \tag{7}$$

where $\Pi(X)$ is the prior distribution given by (1).

3.2 Bayes Estimators

We apply two Bayes estimators to the image reconstruction problem. The MAP estimate is defined by the mode of the posterior distribution, which can be obtained by maximizing (7), or equivalently, by performing the following minimization:

$$\min_X \left[\sum_{i=1}^{I} (l_i - y_i \ln(l_i)) + \beta H(X) \right]. \tag{8}$$

The MMSE estimate is defined by the posterior mean or the expected value (image) over the posterior distribution

$$E[X|Y] = \sum P(X|Y) \cdot X. \tag{9}$$

The MAP estimate is difficult to obtain because of the nonconvexity of $H(X)$. One approach is to use *simulated annealing* (SA) with the Gibbs Sampler [7]. However, this is computationally expensive especially because (7) contains the likelihood function (6) whose neighborhood structure is of a large spatial extent. The calculation of the MMSE estimate is also difficult, because of the complexity of the normalizing factor of the Gibbs prior. A stochastic approach is to sample the posterior distribution using the Gibbs Sampler. The iterates $X^{(k)}$ form a Markov chain with the equilibrium distribution (7). The ergodicity of the chain guarantees that an ergodic average will converge to the MMSE estimate [7], the value of which in practice can be estimated by calculating the average of N samples $\{X^{(k)} : k = K + r \times t$ for $r = 1, \cdots, N\}$ for some large enough values of t and K (the number of *thermalization* steps). Again this sampling process is expensive because of the involvement of the likelihood function (6).

3.3 PET Reconstruction via Restoration

In view of the above computational demands associated with the measurement model (6), instead of using the original PET measurements and this measurement model in the Bayesian estimations, we propose to use the image produced from the PET measurements by the standard *Filtered Backprojection* (FBP) algorithm [11] as input to be restored according to an appropriate model of the combined measuring-and-reconstruction process. We performed a simulation study of the properties (including both blurring and noise) of the images produced by FBP. We model the noise using the Gaussian model:

$$L'(Y|X) = \prod_{j=1}^{J} \exp\left(-\frac{(y_j - x_j - \mu(x_j, \overline{x_{\eta_j}}))^2}{2\sigma^2(x_j)} \right), \tag{10}$$

where x is the original value of a pixel, $\overline{x_\eta}$ the average value of its eight nearest neighbors, y is the value assigned by FBP to the pixel, σ is the standard deviation of the noise, and the bias (which captures the blurring) is empirically estimated using a third order polynomial $\mu(x, \overline{x_\eta}) = \sum_{i=1}^{3} a_i x^i + \sum_{i=1}^{3} b_i (\overline{x_\eta} - x)^i$. The coefficients a_i's and b_i's, and hence σ, can be determined from training data based on non-linear least-squares techniques. The computational demand associated with (7) in sampling using the Gibbs Sampler is much reduced because of the simpler (5×5) neighborhood structure of $L'(Y|X)$. Although this model is only an approximate one [13], we found it very effective in our experiments.

4 Experiments

We simulated the measurement process of PET imaging using a mathematical phantom under the same conditions as those described in [3] (involving 3,000,000 coincidence counts, 300×151 projections with independent Poisson noise). The phantom consists of an elliptical background (gray level equals to 100) with elliptical and rectangular blobby features (the gray level of each equals to either 195 or 200) in it. Fig. 2a displays the digitized phantom on the 95×95 pixel grid on which the reconstructions were performed.

We first apply the FBP algorithm to the PET measurements. We then calculate a Bayes estimate based on (7) with $L(Y|X)$ replaced by the $L'(Y|X)$ of (10). The values of $(a_1, a_2, a_3, b_1, b_2, b_3)$ are estimated from a training data set, which is produced by generating 30 sets of random projection data from the phantom and performing FBP on

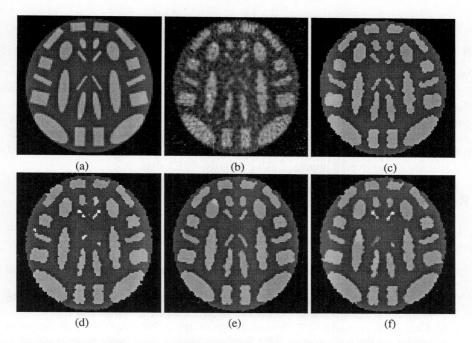

(a)   (b)   (c)

(d)   (e)   (f)

Fig. 2. (a) Original phantom image (b) FBP reconstruction (c) MAP estimate with the proposed prior (d) MAP estimate with the pairwise interacting prior (e) MMSE estimate with the proposed prior. (f) MMSE estimate with the pairwise interacting prior.

each set. The coefficients obtained by a least squares fit were (0.002, -2.36e-05, -6.31e-08, 0.698, 3.39e-04, 2.02e-05).

To obtain the MAP estimate, we employed an approximate version of the Gibbs Sampler algorithm (referred to as the 5-union algorithm in [6]) in the simulated annealing, with an annealing schedule of the form $T_k = 3.5/\log(1+k)$ (k is the number of sweeps), for 500 sweeps. For the MMSE estimate, 250 thermalization steps were used and images subsequently generated (with $T_k = 1.0$) every 25th iterations (for a total of 10 samples) were averaged. The parameters δ and $\triangle$ were set to 3.0 and 75.0, respectively. (These gave consistently good results.) Furthermore, we set $\lambda_1 = 1.0$, $\kappa_1 = 0.94$ and $\beta = 1.56$ in the prior. (These choices produce piecewise smooth image realizations similar to those discussed in [10].)

Fig. 2c shows the final MAP estimate obtained using our prior, whereas Fig. 2d is obtained by replacing the prior with a pairwise interacting prior ($\kappa_i = 0.0$, $\lambda_1 = 1.0$ and $\beta = 1.60$, which produced the best result in terms of the mean squares error). The mean squares errors are 240.1 and 301.0 respectively. It can also be observed that the proposed model did a better job especially at the smaller regions where the spatial supports are small. Fig. 2e and Fig. 2f show the corresponding MMSE estimates and similar observations regarding their relative performances can be made; with the mean squares errors being 186.3 and 204.1 respectively. We also observe in each case that the MMSE estimate is better than the corresponding MAP estimate. It appears that the MMSE estimator performs better at the borders between adjacent regions.

5 Conclusions

We have proposed using a MRF prior with low-order and high-order interactions in Bayesian image reconstruction problems. The desirability of having the higher order terms in the model for modeling the borders has been demonstrated. We have discussed and compared the applications of two Bayes estimators for the PET reconstruction problem. A method which is computationally less expensive than employing stochastic methods directly was proposed for finding both the MAP and MMSE estimates. We have illustrated the efficacy of our overall approach using simulated PET data.

References

1. L. Bedini, I. Gerace, and A. Tonazzini. A deterministic algorithm for reconstructing with interacting discontinuities. *CVGIP: Graphical Models and Image Processing*, 56(2):109–123, 1994.
2. J. Besag. On the statistical analysis of dirty pictures (with discussion). *Journal of the Royal Statistical Society Series B*, 48(3):259–302, 1986.
3. J. Browne and G. T. Herman. Software for evaluating image reconstruction algorithms. In *1994 IEEE Nuclear Science Symposium and Medical Imaging Conference*, to appear.
4. M. Chan, E. Levitan, and G. T. Herman. Image-modeling Gibbs distributions for Bayesian restoration. In *Proceedings of the IEEE Southwest Symposium on Image Analysis and Interpretation*, pages 7–12. IEEE Computer Society Press, 1994.
5. D. Geiger and F. Girosi. Parallel and deterministic algorithms from MRF's: Surface reconstruction. *IEEE Transactions On Pattern Analysis and Machine Intelligence*, 13(5):401–412, 1991.
6. D. Geman, G. Reynolds, and C. Yang. Stochastic algorithms for restricted image spaces and experiments in deblurring. In R. Chellappa and A. Jain, editors, *Markov Random Fields: Theory and Application*, pages 39–68. Academic Press, Inc., San Diego, 1993.
7. S. Geman and D. Geman. Stochastic relaxation, Gibbs distributions, and the Bayesian restoration of images. *IEEE Transactions on Pattern Analysis and Machine Intelligence*, 6(6):721–741, 1984.
8. S. Geman and D. McClure. Statistical methods for tomographic image reconstruction. In *Proceedings of the 46th Session of the ISI, Bulletin of the ISI*, volume 52, pages 5–21, 1987.
9. A. J. Gray, J. W. Kay, and D. M. Titterington. An empirical study of the simulation of various models used for images. *IEEE Transactions on Pattern Analysis and Machine Intelligence*, 16(6):507–512, 1994.
10. E. Levitan, M. Chan, and G. T. Herman. Image-modeling Gibbs priors. *Graphical Models and Image Processing*, 57(2):117–130, 1995.
11. S. W. Rowland. Computer implementation of image reconstruction formulas. In G. T. Herman, editor, *Image Reconstruction from Projections: Implementation and Applications*, pages 9–79. Springer-Verlag, Berlin, 1979.
12. L. Shepp and Y. Vardi. Maximum likelihood reconstruction in emission tomography. *IEEE Transactions on Medical Imaging*, MI-1(2):113–121, 1982.
13. D. W. Wilson and B. M. W. Tsui. Noise properties of filtered-backprojection and ML-EM reconstructed emission tomographic images. *IEEE Transactions on Nuclear Science*, 40(4):1198–1203, 1993.

Minimum Spanning Trees (MST) as a Tool for Describing Tissue Architecture when Grading Bladder Carcinoma

Heung-Kook Choi[1], Ewert Bengtsson[1], Torsten Jarkrans[1], Janos Vasko[2],
Kenneth Wester[2], Per-Uno Malmström[3] and Christer Busch[2]

[1] Centre for Image Analysis, Uppsala University, Läggerhyddvägen 17, S-75237
Department of [2] Pathology and [3] Urology, University Hospital
Uppsala, Sweden

Abstract. In this pilot study we have investigated the possible use of minimum spanning trees, MST, as a way of quantitatively describing the tissue architecture when developing a computer program for malignancy grading of transitional cell bladder carcinoma. The MST was created by connecting the centre points of the nuclei in the tissue section image. These nuclei were found by thresholding the image at an automatically determined threshold followed by a connected component labeling and a watershed algorithm for separation of overlapping nuclei. Clusters were defined in the MST by thresholding the edge lengths. For these clusters geometric and densitometric features were measures. These features were compared by multivariate statistical methods to the subjective grading by the pathologists and the resulting correspondence was 85% on a material of 40 samples.

1 Introduction

Quantitative, computer assisted, analysis of tumor tissues started in the late sixties. Using computerized image analysis, many of the features that are qualitatively used by pathologists can be quantitatively described. Additional features are not readily apparent to the human eye such as high order texture descriptors. Since much of the information used to grade the tissue comes from the relations between the cells, i.e. an assessment of the order vs. disorder of the tissue organization, it is interesting to try to describe these relations. Prewitt proposed to use graph analysis as a mathematical tools for this [11]. Since then several other authors have followed up on the idea and developed graph analysis algorithms for tissue characterization [6, 8, 12]. In particular the MST has proven useful.
A few years ago we started a project the aim of which was to develop an easy-to-use, reproducible, objective, computerized grading system for bladder tumors. The reproducibility of subjective grading seems to indicate considerable prognostic power [3]. In this project we have previously studied parameters obtained after segmentation of the cell nuclei [7], as well as texture parameters [4]. Other groups have presented similar results [10]. In the work reported here we extend this work by investigating the use of the MST to describe the tissue architecture.

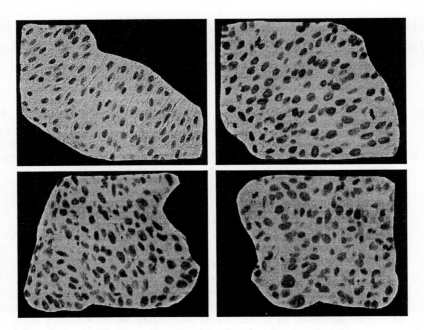

Fig. 1. The digital microscopic images (400X) of the bladder carcinoma cells stained by Feulgen. In each grade the region of interest was chosen by pathologists: (a)grade 1(upper left), (b)grade 2A(upper right), (c)grade 2B(lower left) and (d)grade 3(lower right).

2   Material and Methods

2.1   Material and image acquisition

All patients with newly detected transitional cell carcinoma of the urinary bladder seen at Uppsala University Hospital during the period 1975-1978, were studied retrospectively with an observation time ranging from 5 to 9 years. Our subjective grading divided the material into four groups: grade 1, 2A, 2B and 3. Grade 1 has a slight papillary tendency (Fig. 1a). In grade 2A the cellular pattern shows more variation in nuclear size, shape, internuclear distance and chromatine features (Fig. 1b). In grade 2B the general impression of disorder dominates (Fig. 1c). Grade 3 exhibits an extreme variations in each of the features described (Fig. 1d).

A standard monochrome video CCD camera, giving a data matrix of 512X410 pixels was used on a microscope with an interference filter centered at 550 nm and a 40X lens, making the size of the pixels about 0.5 μm. Digitized images were obtained from 5 μm thick Feulgen stained paraffin wax sections. Pathologists chose a region of interest from each section which best represented the subjective grade. This image acquisition and processing was carried out on an EPSILON workstation using the EGO software system [1].

2.2 Histogram based thresholding

The gray scale image was automatically segmented into a binary image with objects (nuclei) and background. The method, global thresholding, was based on the fact that cell nuclei appear as dark objects on a lighter background. A histogram of a typical image showed a dominant distribution of light pixels, the background, and a smaller distribution of darker pixels, the objects. The two distributions were separated by a weak valley or an infection point. We found the threshold by searching for minima in the first derivative of the histogram.

2.3 Watershed segmentation

The resulting binary image showed the nuclei but also many other spurious objects, holes, as well as touching and overlapping nuclei. The object masks were first cleaned by filling the holes of the objects. Small objects, defined as having an area less than 70 pixels or a maximum width less than 9 pixels, were deleted. The width was obtained by a chamfer 3-4 distance transformation [2]. The distance transform image was also used as a starting point for separating touching objects. The watershed segmentation algorithm [13] can separate irregular *blobs* into more regular parts. We empirically determined that a *waistline* between blobs with a width of less than 90% of the diameter of the smallest side blob should result in a cut. For this cut the watershed line was replaced with the shortest straight line.

2.4 Minimum spanning tree

A minimum spanning tree (MST) connecting the centroids of all nuclei in a tissue section can be used as a basis for extracting features describing the tissue architecture. We first created a linear graph connecting the centre points of adjacent objects. The edges were assigned the *Euclidean* distance between the corresponding vertices as weights. This graph was then reduced to a MST using *Kruskal's* algorithm [9]. In Kruskal's algorithm, edges are first sorted in ascending order of weight and placed in a list structure. This algorithm was implemented here using a *heapsorted* priority queue. As the list is traversed, if an edge is found connecting a vertex that has not been visited before, the edge is included in the tree. When an edge connects two vertices belonging to different clusters, the two clusters are merged, with the new cluster being absorbed into the older clusters. The tree will complete when only one cluster remains i.e. when all the vertices have been visited. For N vertices, the complete tree will have $(N - 1)$ edges. The obtained tree will be the MST i.e. the graph whose weightsum is minimum among all spanning tree graphs. A practical example is shown in Fig. 2a.

2.5 Clustering

In the MST the vertices (nuclei) were clustered based on cutting links longer than a threshold value. Reasonable values for this threshold were obtained by studying

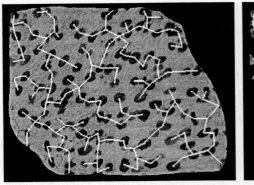 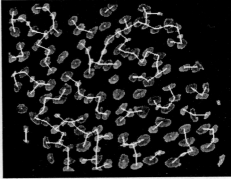

Fig. 2. (a) To the left, the minimum spanning tree of a tissue section with 153 vertices shown superimposed on the grayscale image. (b) To the right, The Euclidean distance between vertices has been thresholded at 16 μm to create the 31 clusters.

the image geometry. Two slightly overlapping nuclei would have an edge distance of about 6 μm and a distance of more than 19 μm is rather unusual. Several distances in this range were used as possible thresholds for creating clusters. The threshold which yielded features with the best discriminating power was finally used. Fig. 2b. depicts the clusters of segmented nuclei.

2.6 Feature extraction

We defined 24 different features. The best of these as determined by their discriminatory power were: The degree of nuclear overlap defined as the number of edges shorter than 6 μm divided by the total number of nuclei (F1), the variation of local orientation of the graph defined as the standard deviation of the directional indices measured in a 2x2 (F2) or 3x3 (F3) neighborhood, the nuclear area to graph size ratio defined as the sum of all nuclear areas divided by the total graph size (F4), the normalized number of clusters defined as the number of clusters divided by the total graph size (F5), the variation in average nuclear size defined as the standard deviation of the average nuclear areas in each cluster, taken over all clusters (F6) and similarly the standard deviation for the average gray values of the nuclei (F7), the variation in inter cluster distance defined as the standard deviation of the distances between the clusters (F8) and finally the cluster area variation defined as the standard deviation of total nuclear area per cluster over all clusters (F9). These features achieved discriminating powers for the different classification phases as shown in Table 1.

2.7 Hierarchical classification

A two stage binary hierarchical classifier using multivariate linear discriminant analysis at each stage was developed. For each stage we applied a *step−wise* linear discriminant analysis, using the BMDP [5] program package. The optimized

Table 1. The following features were selected for each of the three classification stages and training phase using 20 randomly selected images. The corresponding means, standard deviations, variation ranges and the F-values for each of the groups are given.

Features	Mean	Std	Min	Max	Mean	Std	Min	Max	F-dist
Phase I	Grade 1,2A				Grade 2B,3				
F3	14.54	2.31	10.33	18.54	10.95	4.51	2.25	17.10	7.77
F4	11.43	1.94	8.66	14.52	14.93	2.15	12.64	19.17	12.81
F6	61.59	38.23	0.00	105.76	240.95	128.49	83.99	459.75	5.76
Phase II:1	Grade 1				Grade 2A				
F1	0.012	0.01	0.00	0.025	0.007	0.008	0.00	0.02	34.11
F3	15.07	2.50	12.10	18.54	14.01	2.26	10.33	16.49	8.35
F5	0.042	0.004	0.034	0.045	0.042	0.004	0.037	0.047	15.76
F7	14636	2357	11409	16675	16983	4572	10596	22742	44.09
F9	97.77	9.38	86.25	107.91	126.98	26.98	83.52	153.75	92.69
Phase II:2	Grade 2B				Grade 3				
F2	13.63	3.74	8.12	17.66	10.48	4.51	3.92	16.37	4.43
F4	15.98	2.23	14.07	19.17	13.87	1.63	12.64	16.24	9.41
F6	180.2	39.28	150.7	246.5	224.3	70.44	165.3	325.8	7.64
F8	0.445	0.093	0.354	0.596	0.513	0.186	0.298	0.782	6.25

criteria are the ratio between the within group and between group distances using the *Mahalanobis* distance. For each phase in the hierarchical classification this program found the optimal combination of features and the corresponding weights in the discriminant function. The discrimination between grades (1 + 2A) and grades (2B + 3) used clusters with less than 16 μm distance (Phase I) while the threshold 7 μm and 4.5 μm were used for·grade 1 and 2A (Phase II:1) and grade 2B and 3 (Phase II:2) respectively.

3   Results and Discussion

The classifier was trained on a randomly selected subset of 20 out of the 40 images and then tested on all 40 images, 10 images of each grade. The classifier was able to correctly classify 34/40 images (85 % agreement) see Table 2. The method thus showed good correlation with subjective visual grading. This at least indicates that there is some useful information available in these features. These preliminary results need to be verified on a larger test set before we finally can know if these methods can be used as part of such a system. Computing the MST is a computationally demanding task. We used Kruskal's algorithm with a time complexity of $O(eloge)$, where e is the number of edges in the graph.

4   Acknowledgements

Some of the algorithms in this work were implemented in the IMP image processing software system developed by Dr. Bo Nordin. We are grateful for his

Table 2. Confusion matrix for the minimum spanning tree based hierarchical classifier tested on 40 images. The *kappa* value was 0.80 and the *t*-value was 8.83. Visual grade horizontally, computer grade vertically.

grade	grade 1	grade 2A	grade 2B	grade 3	%
1(10)	8	2	0	0	80.0
2A(10)	0	9	1	0	90.0
2B(10)	0	1	9	0	90.0
3(10)	0	1	1	8	80.0
Total(40)	8	13	11	8	85.0

patient support in this work. This research was supported by NUTEK under grant number 90-01178P and by grant 2323-B87-01x from the Swedish Cancer Society and Lion's Association, Uppsala.

References

1. E. Bengtsson, O. Eriksson: The design of an image analysis system. 5th Scan Conf Image Analysis, 217-224 (1987)
2. G. Borgefors: Distance Transforms in Digital Images. Comput Vision Image Process 34, 344-371 (1986)
3. C. Busch, A. Engberg, B.J. Norlen, B. Stenkvist: Malignancy grading of epithelial bladder tumours. Scand J Urol Nephrol 11, 143-148 (1977)
4. H.K. Choi, J. Vasko, E. Bengtsson, T. Jarkrans, P-U. Malmström, K. Wester, C. Busch: Grading of transitional cell bladder carcinoma by texture analysis of histological sections. Anal Cell Pathol 6, 327-343 (1994)
5. W.J. Dixon: BMDP Statistic software manual, 7M. Depart. of biomathematics, University of California Press. Berkeley, Los Angeles, Oxford, 339-358 (1990)
6. C. Dussert, M. Rasigni, J Palmari, G. Rasigni, A. Llebaria, A. Marty: Minimum spanning tree analysis of biological structures. J Theor Biol 125, 317-323 (1987)
7. T. Jarkrans, J. Vasko, E. Bengtsson, H.K. Choi, P-U. Malmström, K. Wester, C. Busch: Grading of transitional cell bladder carcinoma by image analysis of histological sections. Anal Cell Pathol 8, 135-158 (1995).
8. K. Kayser, H. Stute, M. Tacke: Minimum spanning tree, integrated optical density and lymph node metastasis in bronchial carcinoma. Analyt Cell Pathol 5, 225-234 (1993)
9. J.B. Kruskal: On the Shortest Spanning Subtree of a Graph and the Traveling Salesman Problem. Proc. Amer Math Soc 7, 48-50 (1956)
10. U. de Meester, I.T. Young, J. Lindeman, H.C. van der Linden: Towards a quantitative grading of bladder tumors. Cytometry 12, 602-613 (1991)
11. J.M.S. Prewitt: Canonical representations for tissues and textures. IEEE Computer Society:COMPSAC, 470-476, (1979).
12. K. Rodenacker, M. Aubele, B. Jutting, P. Gais, G. Burger, W. Gossner, M. Oberholzer: Image cytometry in histological sections. Acta Stereol 11, 249-254 (1992)
13. L. Vincent, P. Soille: Watersheds in Digital Space: An Efficient Algorithm Based on Immersion Simulations. IEEE Transactions on PAMI 13, 583-598 (1991)

Symbolic Indexing of Cardiological Sequences through Dynamic Curve Representations

M.Baroni[1], G.Congiu[2], A.Del Bimbo[2,3], A.Evangelisti[2], and E.Vicario[2]

[1] Dip. Ingegneria Elettronica, Università di Firenze, Italy
[2] Dip. Sistemi e Informatica, Università di Firenze, Italy
[3] Dip. Elettronica per l'Automazione, Università di Brescia, Italy

Abstract. Digital image analysis supports diagnostic activities by highlighting geometric and temporal features of physiological phenomena that are not perceivable to the human observation. These features can be exploited to build up symbolic representations of visual data in medical reports and to index them within large databases. The comparison of such representations against descriptive queries capturing the properties of significant physiological phenomena supports new diagnostic approaches through the systematic analysis of database reports.

A prototype system is presented which supports the construction of symbolic representations and their comparison against descriptive queries capturing geometric and temporal properties of time-varying 2D shapes deriving from dynamic cardiac analyses. The system is embedded within a visual shell allowing physicians to compose content-oriented queries through iconic interaction.

1 Introduction

Imaging processing and image databases are assuming a growing relevance in medical systems supporting physicians in diagnostic activities through an objective and reproducible exploitation of their experience. Digital image processing and analysis permit the observation of geometric and temporal features of physiological phenomena that are not evident, or even not perceivable, for human observation. These features can be exploited to build up symbolic representations of visual data in medical reports and to index them within large databases. The comparison of symbolic representations against descriptive queries capturing the properties of significant phenomena supports content-oriented analysis of the reports stored in a database and opens the way to systematic diagnostic approaches. On the one hand, in *prospective* diagnosis, a number of standard queries are checked against the indexes of a single report to accomplish a screening of the possible pathologies of a specific patient. On the other hand, in *retrospective* diagnosis, a single query is checked against a large number of reports to identify pathological classes and to investigate them against historical data.

A number of indexing techniques have been proposed which can be exploited to support the symbolic representation of the contents of single static images [7] [4]. But, only a few description techniques have been recently proposed, which can be exploited in the representation of multiple sequenced images [10] [6]. None of these techniques addresses the case of non-rigid imaged shapes, as encountered in most medical applications.

In this paper, a prototype system is presented which supports the construction of symbolic representations and their comparison against descriptive queries for time varying 2D shapes deriving from dynamic cardiac analysis [4]. Qualitative static and dynamic shape descriptors are introduced to provide a symbolic representation of shapes, and a descriptive language is presented which supports the expression of queries on their contents. System interaction is accomplished through a visual language allowing physicians to compose iconic queries on geometric and temporal properties of reports to be retrieved.

2 Visual Data Acquisition and Processing

Cineangiographic analysis produces image sequences presenting the evolution over a complete beating period of the contour of the left ventricle (LV) in the Right Anterior Oblique projection of the heart. Image acquisition is performed on 35 mm films at a rate of 50 frames/second during cardiac catheterization, in mild inspiration. LV contours are traced and digitized frame-to-frame, from the beginning to the end of one cardiac cycle (according to ECG signal), by using a moviola and a graphic tablet (MY-PAD-A3 digitizer, model K-510mk2, 0.1 mm resolution).

At storage time, each cineangiographic report is given as a sequence of sample curves represented in a chain code numerical format. This representation is used to extract the bending function of the individual samples [2], which are then used to build the symbolic representation of the contents of the report according to the description format expounded in Sect.3. The numerical chain code representation of the report is stored in the system along with this symbolic index.

The resulting contour of each frame is processed to derive the normalized curvature function. This comprises an invariant description with respect to rigid transformations and provides local information independent of geometrical assumptions. Curvature values are also correlated with wall stress and evidence local shape characteristics, such as concavity, protrusions and flat segments. To reduce quantization and statistical errors, the curvature function is computed through an adaptive Fourier approximation. The contour is first approximated by a windowed 2D Fourier series which is then differentiated to obtain curvature values in close form. Approximation errors and curvature smoothness are optimized by changing the harmonic number and filter-window along the contour. The benefits of this approach are discussed in [2] with comparison to other commonly used methods. The interobserver and intraobserver repeatability of the entire procedure, from the contour digitalization to the curvature computation, accomplishes a mean deviation lower than 10% and a linear correlation coefficient higher than 90%.

---

[4] The system was developed in cooperation with the Institute of Cardiology of the University of Florence, Italy, and it presently runs on an IBM Risc R6000.

3 Visual Data Representation

Following the results of studies on human perception, the representation of the shape of a curve can be reduced to the representation of a set of sub-curves, each characterized by homogeneous features [11]. Different partitioning criteria have been proposed in the literature to identify subcurves [8] [9]. In particular, in the *codon* approach [12], a curve is divided in subcurves, the *codons*, which are delimited by two subsequent minima of the bending function. This segmentation approach appears suited to cineangiographic reports as it reflects a natural partitionment of the heart body into physiological regions characterized by different values of internal tension forces. This segmentation also prescinds from both the orientation and the zooming factor employed during the imaging process.

Following this approach, each frame of the ventricular contour is represented through a (variable) number of codons, each following one out of 5 possible patterns corresponding to the different combinations of the signs of the bending function in the extrema and by the possible presence of intermediate flexes (see Fig.1). Codons are also associated with three symbolic attributes of *symmetry*, *bending*, and *extension*, accounting for the position of the bending maximum with respect to the codon center, the maximum value of the bending function along the codon, and the linear distance between the extrema of the codon. All the attributes are expressed in *qualitative* terms.

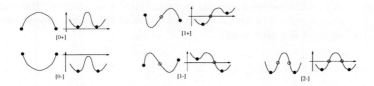

Fig. 1. Codon types corresponding to the five possible combinations of the sign of bending minima and of the number of flex points between the extrema.

The evolution over time of the curve is captured by associating codons belonging to subsequent frames through a *tracking relation*. In the common case, a single codon of sample n is biunivocally associated with a single codon of sample $n+1$, however, due to the appearance or disappearance of bending minima, it may also be the case that a codon of sample n is mapped onto multiple codons of sample $n + 1$ or that multiple codons of sample n are mapped onto a single codon of sample $n + 1$. These merging and splitting phenomena, which are quite frequent in cineangiographic contours, are resolved through an heuristic decision which associates codons in subsequent samples on a minimum displacement basis [5]. An example of tracking is reported in Fig.2.

In general, the tracking relation identifies three disjoint sets of codons, which are never intertwined by mutual splitting or merging. This is a general property of cineangiographic reports, and reflects the fact that the heart is naturally divided into three different anatomical regions, that are usually referred to as *anterior*, *apical* and *inferior*. This result permits the automatic attribution of each part of each sample of the curve to one of these three anatomical regions.

The comparison of subtended areas of codons that are associated to each other by the tracking permits a symbolic characterization of the expansion/contraction trend, which is mainly relevant to the purposes of the diagnostic activity. To capture this trend, the individual codons of each cineangiographic sample are associated with a further attribute of *straining* which is interpreted as *expansive*, *stable* or *contractive*. The straining attribute provides a local representation of the evolution of the curve in that it refers to the individual codons. The simultaneous presence of strongly differentiated straining trends is a condition of main relevance as it is indicative of important pathologies.

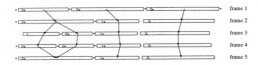

Fig. 2. Codon tracking for six samples of a cineangiographic report. For each sample, the curve is represented in curvilinear abscissa and it is segmented into codons that are tracked along subsequent samples. The first codon of the first sample is tracked onto the first codon of the second sample. In turn, this is split into the first and second codon of the third sample. The first and second codon of the fourth sample are merged into the first codon of the fifth sample.

4   Sequence Querying by Contents

In the retrieval stage, the user expresses content-oriented queries through a visual iconic interface by defining distinctive geometric conditions which characterize sequences to be retrieved and temporal ordering relationships among the lasting intervals of these conditions. Visual queries are automatically interpreted by the system and checked against the descriptions of the sequences in the database.

The basic graphic objects employed in the visual expression of queries are icons and lines. Icons represent significant conditions, such as an expansive straining or the presence of a sharp bending in some of the codons of the cineangiographic curve. Multiple icons can be grouped visually to express Boolean composition of multiple events. Lines denote the lasting time intervals in which conditions and condition compositions occur, and they are annotated with the iconic representation of the condition composition. Each line is also annotated with a special icon denoting the anatomical region in which the event composition occurs and the reference phases of systolic contraction and diastolic expansion.

The mutual positioninig of event lines on the screen defines the temporal ordering among the lasting intervals of distinctive events. To permit the expression of *don't care* conditions in the interpretation of this mutual positioning, each line is associated with one or more colors, and the mutual position between intervals which do not share any common color is ignored in the interpretation of the visual arrangement. The translation of the visual specification occurs through two distinct phases. First, the mutual positioning of event intervals are analyzed and classified into one of the 13 possible mutual ordering relations [1]. Afterwards, icon compositions associated with each interval are parsed and translated into interval formulae. These formulae are checked against the descriptions of the sequences in the database, and matching reports are displayed on the screen.

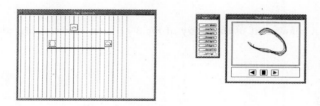

Fig. 3. Fig a: a visual query expressing a contractive strain in the apical region during the systolic phase. Fig b: the list of retrieved reports(left hand side menu) and display (in the right hand side window) frame by frame of one of them.

Fig.3a shows a visual query seeking for reports characterized by an inward motion in the apical region during the diastolic phase. The upper line is annotated with the systolic icon thus denoting the time interval in which the heart should expand. The lower line is annotated with two icons representing a contractive strain (in the middle of the line) and the apical region (in the right hand side of the line) so as to represent a time interval in which part of the apical region exhibit a contractive motion. The two lines are associated with a common color so as to make significant the mutual positioning of the two intervals. The system interpretation of the overall arrangement produces a query statement which causes the retrieval of the reports characterized by a time interval which is contained in the systolic phase and which exhibits a contractive motion in the apical region. When retrieval is completed, reference names of retrieved reports are listed on a menu which permits selection and display on the screen of reports of specific interest for the user (Fig.3ab).

Fig.4a shows a refinement of the query of Fig.3, which adds the request for an outward motion in the apical region during the systolic phase. Two further lines are added (in the bottom part of the screen), one standing for the systolic phase and the other for the lasting interval of the expansive motion in the apical region. The two lines are associated with a common color different from that associating the previous two lines. As a result, the system takes care of the ordering between the two new lines and between the two old lines, but not of the the ordering between new and old lines. Again, after the interpretation of the query, retrieval is carried out and matching reports are listed and displayed on the screen (see Fig.4b).

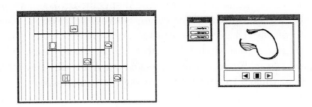

Fig. 4. A refinement of the query of Fig.3 and the corresponing result.

References

1. J.F. Allen, "Maintaining Knowledge about Temporal Intervals," in *Comunications of the ACM*, Vol.26, n.11, Nov. 1983.
2. M.Baroni,G.Barletta, "Digital Curvature Estimation for Left Ventricular Shape Analysis," *Image and Vision Computing*, 1992.
3. E.Binaghi, I.Gagliardi, R.Schettini, "Indexing and Fuzzy Logic-Based Retrieval of Color Images," in *IFIP Trans. Visual Database Systems II*, Knuth, Wegner (Eds.), Elsevier Pub. 1992.
4. S.K.Chang, Q.Y.Shi, C.W.Yan, "Iconic Indexing by 2-D Strings", *IEEE Transactions on Pattern Analysis and Machine Intelligence*, Vol.9, No.3, July 1987.
5. G.Congiu, "Rappresentazione e Ricerca Visuale di Sequenze di Immagini Cardiologiche (In Italian)," *Doctoral Thesis*, Univ. Florence, Italy, July, 1994.
6. A.Del Bimbo, E.Vicario, D.Zingoni, "Symbolic Description and Visual querying of Image Sequences Using Spatio-Temporal Logic," accepted for publication on *IEEE Transactions on Knowledge and Data Engineering*.
7. K.Hirata, T.Kato, "Query by Visual Example: Content-Based Image Retrieval," In *Advances in Database Technology - EDBT'92*, A.Pirotte, C.Delobel, G.Gottlob (Eds.), Lecture Notes on Computer Science, Vol.580, Springer Verlag, Berlin, 1992.
8. M.Leyton, "Shape and Casual History," in *Visual Form*, Plenum Press, 1982.
9. S.Marshall, "Review of Shape Coding Techniques," in *Image and Vision Computing*, Nov.1989.
10. A.Nagasaka, Y.Tanaka, "Automatic Video Indexing and Full Video Search for Object Appearances," in *IFIP Trans. Visual Database Sys. II*, Knuth, Wegner (Eds.), Elsevier Pub. 1992.
11. S.E.Palmer, "The Psychology of Perceptual Organization,: a Transformational Approach," in *Human and Machine Vision*, Academic Press, New York, 1983.
12. W.Richards, D.D.Hoffman, "Codon Constraints on Closed 2D Shapes," in *Computer Vision II*, Natural Computation Group, M.I.T., Cambridge, 1984.

Image Coding II

Visual Communications

Arun N. Netravali

AT&T Bell Laboratories
Murray Hill, NJ 07974

Abstract. Commercial analog television was standardized over fifty years ago, mainly for entertainment, sports, and news, using over the air broadcast. It is only recently that technology of compression, integrated circuits, fiber optic transmission, switching, and storage has advanced to make digital video economically feasible. This is forcing the convergence of diverse industries such as communications, computing, entertainment, and information services. Digitizing video in a cost effective manner and integrating it with computers, telecommunication networks, and consumer products will produce a vast array of new products and services in the future. These will have a long lasting impact on entertainment, education, medicine, and will improve our productivity in daily life.

1 Introduction

Commercial analog television was invented and standardized over fifty years ago mainly for over the air broadcast of entertainment, news, and sports. While some upward compatible changes have been made, such as color, multichannel sound, close captioning, and ghost cancellation, the underlying analog system has survived several generations of technological evolution that has pervaded all other media. After 50 years of refinement, we are poised for a major disruption -- *digital video*. Digitizing video in a cost effective manner and integrating it with computers, telecommunication networks, and consumer products, promises to produce a large array of new products and services. Visual communications is poised to become inexpensive, ubiquitous, flexible, and a necessity for improving productivity in our daily life.

A number of technological advances over the past few decades will enable the growth of visual communications. They are:

1. High quality, standard compression algorithms to reduce the bit rate of the video signal.

2. Inexpensive integrated circuits for processing and storage of video signals.

3. High speed communication networks using fiber optic transmission and wide band switches.

4. High density and fast access storage media (magnetic, optical, or semiconductor).

5. Computer architectures that can handle video-rate pixels with ease.

6. Algorithms and implementations that can rapidly create photo-realistic images by computers.

This technological capability, along with a proven value of visual communications for group discussions, education, entertainment, and medical applications will make visual communications ubiquitous. Consumers will be able to access video programs from multimedia databases populated by a variety of content providers. Video conferencing and video telephones will improve our productivity in business transactions. Prepackaged interactive programs delivered on CD-roms will help medical and educational applications. Most importantly, computers will become friendly by the use of visual interfaces. This, together with falling prices, will lead to consumerization of personal computers resulting in an even richer set of services.

The current state of video processing can be best described as "pixel based." The common paradigm is to scan a video signal, convert it into an array of pixels and process pixels without regard to the semantic meaning of any groups of pixels (spatially and/or temporally). While some progress will continue to be made in pixel based processing, the big frontier to be conquered is: content-based processing. In this, the video signal is described as a set of objects in three dimensions undergoing transformations (rigid or non-rigid) over time. The goal will be to recognize object boundaries and process, store, and transmit objects rather than pixel at a time.

The two current dominant paradigms are:
a. point-to-point video (i.e., video telephone)
b. broadcast video (i.e., terrestrial, satellite, or cable TV.) As visual communications becomes flexible and pervasive, there will be many more. Multipoint video teleconference, as well as interactive television, where a consumer is interacting with the content rather than simply changing channels or connections, will become a common place. Computing, storage, communication networks that enable interactive, distributed and collaborative processing of video signals will give rise to exciting new services.

2 Enabling Technologies

In this section we present an overview of some of the technologies that make digital video economically viable.

1. *Compression* - Analog television signal when digitized produces far too many bits to economically process, store, transmit, or switch. For example, the CCIR-601 television standard results in 216 Mb/sec bitstream for a 6-MHz or 8-MHz analog signal. Digital compression reduces this data rate to manageable sizes consistent with the applications. Standard compression algorithms (JPEG for intra-frame compression; MPEG for inter-frame compression) have been devised. CCIR-601 standard provides "studio quality" at 216 Mb/sec; JPEG is

roughly equivalent at 35-45 Mb/sec; MPEG is equivalent at slightly less than 10 Mb/sec. Similarly, the "VHS quality" requires approximately 1.5 Mb/sec using MPEG and slightly over 4 Mb/sec with JPEG. For video telephone applications, where the motion in scenes is limited (and acceptable quality is lower than VHS), these standards have achieved impressive results with compression ratios of about a thousand. (See Figure 1.)

In addition to producing the highest quality pictures at the lowest bit rate, compression algorithms must satisfy many other requirements. The compressed bits must be robust in the presence of transmission impairments. Abrupt changes, such as scene changes in the video material and channel changes introduced by the viewer, should not create transients that affect picture quality for a long time. If the compressed bit stream is stored in a digital VCR, many of the VCR functions that we are accustomed to TV. (e.g., fast forward - backward searches) should be easy to do. Editing of video and insertion of video within the compressed bitstream should be easy as well. Thus, as the use of the compressed video proliferates, many other requirements become important. While the cost of electronics continues to drop, compression algorithms must be manageably complex in order for them to get wide use. In applications such as digital broadcasting, digital cable, satellite, or interactive television, the cost of decompression done at the consumer end is the primary factor.

Figure 2 shows the approximate relationship between compression efficiency and the complexity of decompression. The compression efficiency is in terms of compressed bits per Nyquist sample, and therefore, pictures with different resolution and bandwidth can be scaled simply by proper multiplication to get the relevant bit rates. MPEG, which uses motion compensated transform coding, is about a thousand times more complex compared to PCM television. The relationship of cost to complexity depends upon the evolving technology. Very soon fast microprocessors will be able to decompress video signals entirely using only software.

Figure 3 shows the computational requirements for encoding and decoding of video at different resolutions. While encoding is still far too complex for even the most powerful processor of today, we are close to decoding done entirely in software.

As mentioned earlier, advances in computer vision and graphics are beginning to allow us to progress from pixel oriented coding to coding of objects in the scene. Techniques of computer vision can be used to identify a variety of objects from a complex scene. Each of these objects may contain hundreds of pixels and may be described by a modeling process available from computer graphics. In order to achieve a high degree of compression, proper models must be used. Since models that can handle arbitrary scenes are not well developed, a highly constrained application domain is usually considered. For example, in video phone applications, most scenes contain a head and shoulders view of a person in front of a static background. Head and shoulders views of people have been modeled as wire frames with some success. This type of work is still in its infancy and faces formidable challenges in segmentation of the scene into objects, building models for each segment, finding specific application domains, and dealing with the complexity of the encoder/decoder. A simpler approach is to segment an image into regions using techniques from computer vision and encode each region separately with coding strategy that is well matched to that region. Regions may not necessarily have semantic meaning and may not be rectangular blocks of pels as in block-transform coding. The principal difficulty of such an approach is the ability of the algorithm to segment images into areas similar statistics and perceptual characteristics. Extensive research is being carried out in model and segment based coding, but the promising results are yet to come.

2. *Multimedia Databases* - Once each of the mono media components of a multimedia signal (e.g., video, audio, data) are digitized, they need to be stored in a database server for a variety of services. Traditional databases store mostly alphanumeric characters as relations or tables. While much research and productization has been done to improve relational databases, they are not suited for storing multimedia signals. Most multimedia applications involving digital video require either simultaneous outputs to multiple subscribers or multiple outputs for a single subscriber. In either case very high bandwidth I/O is required. In addition, audio and video bits can be thought of as bitstreams, whereas computer data is usually bursty. Seamless integration of stream and bursty data presents a new challenge for multimedia databases. If digital video and its corresponding audio samples are stored separately within a database, then precise time synchronization of the video and audio outputs is required in order to create an effective presentation to the viewer. Additional functionality such as pause, restart, rewind, fast-forward, and fast-backward makes synchronization of audio and video even more difficult a challenge. The size of the video objects is usually several magnitudes higher than alphanumeric objects stored in the database. It appears that object oriented database technology in which objects of different types (e.g., video clips, audio clips, data,...) are stored directly is the potential solution. This is being developed in several laboratories and universities.

3. *Bridging* - In distributed and collaborative applications of multimedia services, it is often required to bridge multimedia signals. While bridging an analog audio signal is straight forward (simply add all the component audio signals) bridging video is not simple. In addition, bridging compressed bit streams without having to decompress, bridge, and then compress the bridged signal usually results in loss of quality and is expensive. In the case of video, a variety of options exist. Bridging multiple component video bit streams may be simply done by choosing one of the component video signals depending upon

voice activity. Another possibility is to display some customer-selected component video signals in different parts of the screen simultaneously. A more flexible way is to have different size windows displaying each of the component videos. In the case of data, bridging amounts to either multicasting to a select group of users or broadcasting to a large class of users. The trend in bridging is to allow users to control how each of the monomedia are bridged to create the most effective presentation to the viewer.

3 Applications

A large number of applications of visual communications technology already exist. For example, the concept of video phones has been around for several decades. However, the newer implementations of video phones are quite different from the earlier ones. Person to person video telephony can now be implemented by incorporating a special card in the back plane of a personal computer which may be attached either to a local area network or to a telephone line which can handle different bit rates. On the entertainment front, digital video is making hundreds of channels available on the current cable television systems as well as in the newer generation of high power satellites.. In this section we will describe two emerging applications enabled by digital video.

Interactive Television enables the consumer to exert both coarse and very fine grain control over the contents of the programming being viewed. By contrast, conventional television only allows a viewer to select among a number of programs being broadcast to a large audience. For the most part, each member of the audience for each of these standard programs sees identical contents. With an interactive television program, each viewer can directly control what he is watching. Interactive TV programs can be thought of as a collection of media elements together with software that controls the flow of the program in response to consumer inputs and directs how the media elements are used to create the aural or visual presentation. Media elements, such as audio and video clips, still images, graphics, text, etc., are the primitives of the presentation and are created as part of the process of producing the ITV application. The process and technology for creating ITV applications borrows heavily from computer graphics. Each simultaneous viewer of an interactive TV program can be thought of as totally independent. This requires a separate instance of an application executed by the ITV system for each viewer (although the media elements are shared for economy of storage). In addition, communication bandwidth must be allocated for each viewer to connect computing and storage facilities in the network to customer premises equipment. Overall, delivery system resources must be allocated for each active viewer -- a situation that more closely resembles the telephone network than a conventional broadcast or cable TV system. The development and evolution of interactive television on a large scale presents numerous technical challenges in different areas. However, most of the technical issues are sufficiently developed to allow trials to ascertain consumer interest in this new form of communication.

Distributed, Collaborative Virtual Space - The virtual workspace can be created by connecting several users with a flexible multimedia collaboration between them. Figure 4 shows a configuration for virtual meeting rooms. Virtual Meeting Room 1 involves user A and B as well as shared programs and data from server 1, whereas Virtual Meeting Room 2 involves a multimedia connection between user A and C. Control of these rooms allows users to participate from their own offices. Users may be people or programs. In addition, any of the users may share data that is generated by other users or that can be obtained from a database server. One user may participate in multiple rooms; each room conducting a conversation with a specific context. Interaction with the screen by one user may be seen by designated others. Thus, the attempt is to create a common visual space resembling communication that takes place when all the participants are in the same room.

Summary

In this short paper, we described some key video technology necessary for further evolution of visual communications. Falling prices of hardware and bandwidth, coupled with computer and telecommunication infrastructure will result in faster growth of visual communication than ever before.

References

1. A. N. Netravali, B. G. Haskell, *Digital Pictures: Representation and Compression, and Standards.* Plenum (Second Edition), 1994

2. M. Kunt (Editor), *Digital Television.* Special Issue Proceedings of IEEE, July, 1995

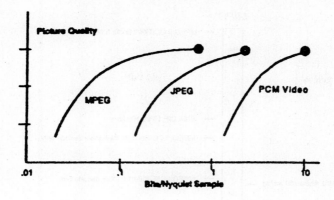

Fig. 1. A rough measure of the rate/quality of picture coders

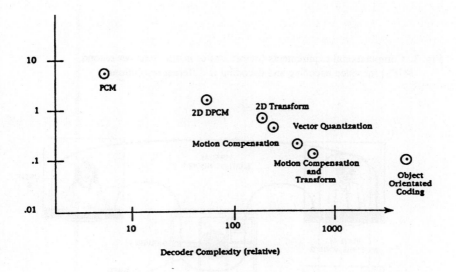

Fig. 2. Compression versus complexity of video decoding

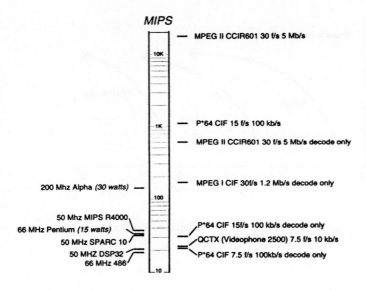

MIPS

— MPEG II CCIR601 30 f/s 5 Mb/s

10K

— P*64 CIF 15 f/s 100 kb/s

1K

— MPEG II CCIR601 30 f/s 5 Mb/s decode only

200 Mhz Alpha *(30 watts)* —   — MPEG I CIF 30f/s 1.2 Mb/s decode only

100

50 Mhz MIPS R4000
66 MHz Pentium *(15 watts)*   P*64 CIF 15f/s 100 kb/s decode only
50 MHz SPARC 10   QCTX (Videophone 2500) 7.5 f/s 10 kb/s
50 MHZ DSP32
66 MHz 486   P*64 CIF 7.5 f/s 100kb/s decode only

10

Fig. 3. Computational requirements (in millions of instructions per second, MIPS) for video encoding and decoding at different resolutions

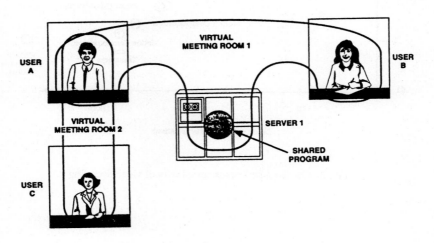

VIRTUAL
MEETING ROOM 1

USER
A

USER
B

VIRTUAL
MEETING ROOM 2

SERVER 1

SHARED
PROGRAM

USER
C

Fig. 4. Virtual Meeting Rooms

Generalized square isometries - an improvement for fractal image coding

Dan C. Popescu[1], Alex Dimca[2] and Hong Yan[1]

[1] Department of Electrical Engineering, Sydney University, Sydney 2006, Australia
[2] Department of Mathematics, Sydney University, Sydney 2006, Australia

Abstract. Most recent advances in fractal image coding have been concentrating on better adaptive coding algorithms, on extending the variety of the blocks and on search strategies to reduce the encoding time. Very little has been done to challenge the linear model of the fractal transformations used so far in practical applications. In this paper we explain why effective non-linear transformations are not easy to find and propose a model based on conformal mappings in the geometric domain that are a natural extension of the affine model. Our compression results show improvements over the linear model and support the hope that a deeper understanding of the notion of self-similarity would further advance fractal image coding.

1 Introduction

Barnsley's iterated function systems (IFS) and the collage theorem have provided the motivation for a class of fractal image compression techniques [1]. However, the collage theorem gives no hint of how to find an IFS that would approximate a given image and in general, a coverage of an image in terms of contracted copies of itself under reasonably simple transforms seems unlikely to exist.

The first practical approach to fractal image compression was proposed by Jacquin [2]. With this method, the image is split in small square blocks that tile the image, called range blocks. A pool of larger size square blocks, called domain blocks, that may be located anywhere in the image is also created a priori. The IFS consists of a set of transformations - one per range block. Every such transformation tries to match the range block with a domain block under a mapping that is a composition of a scaling from domain size to range size, a square isometry and a linear transformation in the gray level. This is a main departure from the spirit of the collage theorem: the image is covered with contracted copies of larger parts of the image, rather than the image itself. In practical terms, this search for a distributed self-similarity at microscopic level increases the chances for finding a good fractal code. It also has the negative effect of leading to long encoding times, to a point where, with any attempt to extend the model, a mechanism for efficient block matching becomes imperative.

Focusing our attention on one domain block - range block transformation, we notice that it is the composition of two linear transformations, one in the gray level domain and the other in the geometric domain. It is normal to ask whether the use of 3 bits only for the geometric transformation (8 isometries of the square) and some 12 to 16 bits for the gray level one represents an optimal code allocation.

In this paper we investigate the virtually unexplored field of non-linear transformations and address the question of whether increasing the number of geometrical transforms would increase the probability of a good match for larger blocks to a point where it would compensate for the extra code and result in higher overall compression ratios.

2 Generalized square isometries

Several attempts with continuous and one-to-one mappings which were not smooth have failed to improve the compression ratio, those transformations being "naturally rejected". We turned our attention to conformal mappings, which produce no angular distorsion.

If the range and domain blocks were disks, not squares, the geometric transformations could be extended by allowing rotations of arbitrary angles, rather than multiples of 90^0. Nevertheless, a square can be mapped into a disk with minimal distortion and preserving angles. This would further allow the construction of transformations that rotate squares into squares with arbitrary angles. It is easier to understand the continuous model of this transformation acting in the complex plane $\mathbb{C}$ as the functions involved have a simpler form as functions of a complex variable. We denote by $\mathbb{Z}$ the set of all integers, by i the imaginary unit ($i^2 = -1$) and also make the following notations for some subsets of $\mathbb{C}$:

$\mathbb{C}_- = \{z | z \in \mathbb{C}, Im z \leq 0\}$, the lower semiplane;
$\mathcal{D} = \{z | z \in \mathbb{C}, |z| \leq 1\}$, the unit disk and
$\mathcal{SQ} = \{z | z \in \mathbb{C}, 0 \leq Re z \leq 1, 0 \leq Im z \leq 1\}$, the unit square.

The *Weierstrass $\wp$ function*:

$$\wp : \mathcal{SQ} \mapsto \mathbb{C}_-$$
$$\wp(z) = \sum_{k,l \in \mathbb{Z}} \frac{1}{(z - 2k - 2l \cdot i)^2} \tag{1}$$

is a conformal one-to-one mapping of the unit square into the lower semiplane [5]. Denote by $z_0 = \wp(0.5 + 0.5i) \approx -1.718796i$ and by $\overline{z_0}$ the complex conjugate of z_0. Then:

$$\omega : \mathbb{C}_- \mapsto \mathcal{D}$$
$$\omega(z) = e^{\frac{3\pi}{4}i} \cdot \frac{z - z_0}{z - \overline{z_0}} \tag{2}$$

is a one to one conformal mapping of $\mathbb{C}_-$ into the unit disk. It follows that

$$\varphi : \mathcal{SQ} \mapsto \mathcal{D}$$
$$\varphi = \omega \circ \wp \tag{3}$$

is a one-to-one conformal mapping of the unit square into the unit disk. The effects of φ are illustrated in Figs. 1 and 2, where the square of Fig. 1 is mapped into the disk of Fig. 2 by φ.

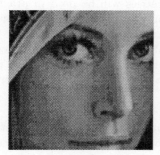

Fig. 1: Initial square   Fig. 2: Mapping of Fig. 1 by φ

For any angle $\theta, \theta \in [0, 360^0)$ we denote by r_θ the usual clockwise rotation of angle θ :

$$r_\theta \ : \ \mathcal{D} \mapsto \mathcal{D}$$
$$r_\theta(z) = e^{-\frac{\pi\theta}{180}i} \cdot z \tag{4}$$

A square rotation of angle θ is defined as the transformation

$$sqr_\theta \ : \ \mathcal{SQ} \mapsto \mathcal{SQ}$$
$$sqr_\theta = \varphi^{-1} \circ r_\theta \circ \varphi \tag{5}$$

Figs. 5-7 show the results of mapping the square of fig. 1 by square rotations of 30^0, 55^0 and 70^0.

Square antirotations are defined similarly by compositions with a flip over the real axis. Finally, we define the set $\mathcal{I}$ of generalized square isometries as the set of all square rotations and antirotations.

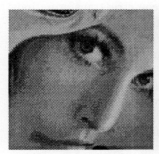
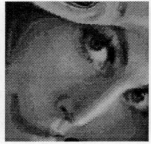
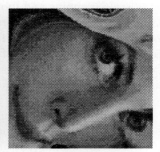

Fig. 3: Mapping of Fig. 1 by rsq_{30^0} Fig. 4: Mapping of Fig. 1 by rsq_{55^0} Fig. 5: Mapping of Fig. 1 by rsq_{70^0}

3 The discrete model

3.1 Discrete generalized square isometries

Only a finite number of generalized square isometries will be used in practice. They are calculated only once according to the previous formulas and the hard-coded patterns of their discrete equivalents are known both at the encoder and the decoder. To understand how these discrete hardcoded patterns are generated without going into unimportant details we shall consider an analogy with classical isometries. A 90^0 rotation of a discrete block of $n \times n$ pixels is given by the mapping

$$(x, y) \mapsto (n - 1 - y, x) \qquad (6)$$

It says that the new pixel at (0,0) is computed from the old pixel at (0,n-1) ,etc. If the mapping is followed by a spatial contraction of (say) factor 2, from a $2n \times 2n$ domain block to a $n \times n$ range block we would have that the gray level of the pixel at (0,0) is obtained by averaging the pixels at (0,2n-1),(0,2n-2),(1,2n-1),(1,2n-2), each with the weight 0.25. It is clear that the transformation can be described either by giving the set of "weighted lists" or by applying formula (6) and both approaches are computationally fast. The "weighted list" approach works with the formulas of chapter 2 as well even if the "real-time formula" approach similar to (6) would be impractical in terms of computational complexity.

3.2 A filtering mechanism

The critical part of this fractal image compression technique is the encoding process. For each range block, it requires a search through the pool of domain blocks for the best match under all available transformations. If we denote by N_r, N_d, N_t the number of range blocks, the number of domain blocks and the number of transformations, then an exhaustive search has a computational complexity $O(N_r \cdot N_d \cdot N_t)$.

However, with the method we propose, moment features can be defined for each block. They are computationally unexpensive and they have to be calculated only once for each block - which means a complexity $O(N_r + N_d)$. The first order moments define the barycenter orientation which is invariant to a linear gray scale mapping and moves consistently with the geometric transformations. Only the best geometric transformations are attempted, typically 2 per range block/domain block pair. If the domain block and the geometric transformation are fixed, the parameters of the linear transformation in the gray level can easily be found by an autoregression formula [3]. Because of this tight control, *the increased number of transformations does not result in a significantly increased computational complexity.* The search complexity is reduced to $O(N_r \cdot N_d)$.

We use a generalization of an idea presented in [5] to further reduce the search for a matching domain block. We define two moment features, $\mathcal{M}_1$ and $\mathcal{M}_2$, associated with (the disk equivalent $\mathcal{D}$ of) each range and domain block as:

$$\mathcal{M}_1(\overline{b}) = \frac{\int \int_{\mathcal{D}} \frac{g(x,y)-m(\overline{b})}{dev(\overline{b})} \cdot (x^2 + y^2) dx dy}{\int \int_{\mathcal{D}} dx dy} \qquad (7)$$

$$\mathcal{M}_2(\overline{b}) = \frac{\int\int_{\mathcal{D}} \frac{g(x,y)-m(\overline{b})}{dev(\overline{b})} \cdot (x^2+y^2)^{-\frac{1}{4}} dxdy}{\int\int_{\mathcal{D}} dxdy} \qquad (8)$$

where $m(\overline{b})$, $dev(\overline{b})$ are the mean and standard deviation of block $\overline{b}$ and $g(x,y)$ is the gray scale level at position (x,y). The absolute values of $\mathcal{M}_1$ and $\mathcal{M}_2$ are invariant under any linear transformation in the gray level and any geometric mapping of $\mathcal{I}$. When looking for a good match for a given range block, only domain blocks with similar features would be investigated. Discrete equivalents of $\mathcal{M}_1$ and $\mathcal{M}_2$ that are computationally unexpensive can be defined.

4 Experimental results

We use an adaptive coding scheme that splits the image into non-overlapping square range blocks, according to a tree structure. This tree structure is more flexible than a quadtree, with smoother transitions in block sizes [5]. It uses 4×4, 6×6, 8×8 and 12×12 size range blocks.

Fig. 6: Decoded image using
8 classical isometries

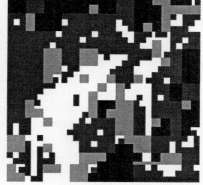

Fig. 7: Range block distribution
for Fig. 6

Fig. 8: Decoded image using
gen. square isometries

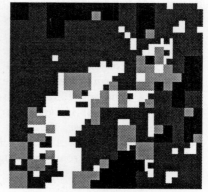

Fig. 9: Range block distribution
for Fig. 8

Figs. 6 and 8 show two compression results of the 256×256 Lena image using this adaptive partitioning scheme. For the reconstruction of Fig. 6 only the 8 classical affine transforms have been used for all range blocks. The reconstruction of Fig. 8 was obtained using 8,16,32 and 32 generalized square isometries for 4×4, 6×6, 8×8 and 12×12 size range blocks, respectively. The PSNR for both reconstructions is 32.8 dB . Since the quality of the reconstruction is pretty high, the PSNR is a good measure of the visual degradation, so we can consider the two reconstructions of equivalent quality. Figs. 7 and 9 display the range block allocation for the reconstructions of Figs. 6 and 8, with 4×4 blocks colored in white, 6×6 blocks in light gray, 8×8 blocks in dark gray and 12×12 blocks in black. Intuitively, the darker the area, the higher is the (local) compression factor. It is noticeable that in Fig. 9 the darker areas "take over" the corresponding lighter ones in Fig. 7. Even if the extra bits necessary to encode more transformation reduce this gain, the compression factor for the reconstruction in Fig. 8 is 9.77 versus 8.97 for the reconstruction in Fig. 6, an 8.9% improvement of the compression factor.

5 Conclusions

In this paper we have developed a non-linear model for fractal image compression based on a set of conformal mappings of the square into itself. A discrete model of these transformations and a filtering mechanism to reduce the encoding time, both of which are easy to implement, have been proposed. Our compression results with an adaptive scheme show improvements over the linear model. In a broader sense, the whole experiment gives ground to hope that a better understanding of the self similarity governing natural patterns could lead to further advances in fractal image compression.

6 References

1. Barnsley M.F.: *Fractals Everywhere*, Academic Press,Inc., 1988
2. Jacquin A.E.: *Image Coding Based on a Fractal Theory of Iterated Contractive Image Transformations*, IEEE Trans. on Image Processing,Vol. 1,No. 1, January 1992, pp. 18-30
3. Fisher Y., Jacobs E.W. and Boss R.D.: *Fractal Image Compression Using Iterated Transforms*, Image and Text Compression, J.A. Storer(ed.), Kluwer Academic Publishers, 1992, pp. 35-61
4. Frances K.: *Complex Algebraic Curves*, London Mathematical Society, Cambridge University Press, 1992
5. Popescu D.C., Yan H.: *A Fractal Based Method for Color Image Compression*, Journal of Electronic Imaging, vol. 4, nr. 1, January 1995, pp. 23-30

A Fast Algorithm for High Quality Vector Quantization Codebook Design

Carlo Braccini, Fabio Cocurullo and Fabio Lavagetto

University of Genova, Via Opera Pia 13, I-16145 Italy
Tel. 39-10-3532983, Fax 39-10-3532948, e-mail Fabio.Cocurullo@dist.unige.it

Abstract. In this paper we present a new theoretical approach to the problem of optimal Vector Quantization. We base, in fact, our method on the a priori explicit analysis of the effects on the MSE distortion introduced by an arbitrary exchange of training vectors among clusters. Even when the theoretical results corresponding to the simplest possible case are used, the proposed algorithm outperforms the GLA method to an impressive extent both in speed and in performance. Experiments on different images from the USC database have proved that the proposed algorithm is 5 to 10 times faster than the GLA method increasing the convergence Peak Signal to Noise Ratio (PSNR) of up to 1.12 dB.

1 Introduction

The term Vector Quantization (VQ) [1] generally indicates a class of widely used techniques for signal compression and coding within applications aiming at storage saving or transmission rate reduction. Since Lloyd and Max [2] optimal scalar quantizer, quite a lot of new proposals have been made, providing innovative algorithms for the design of vector quantizers. Their application initially, limited only to scalar signals and typically to speech, has been progressively extended to multidimensional sources like images and image sequences.

The multidimensional extension of Lloyd's approach, generally known as Lloyd-Generalized-Algorithm (GLA), iteratively modifies an initial quantizer on the basis of a predefined vector pdf or with reference to a vector training sequence: the reconstruction levels migrate to stable configurations which locally minimize a distortion functional usually based on the Mean-Square-Error (MSE). The GLA approach represents a key reference point for any new proposal concerning the design of a vector quantizer and an useful yardstick to compare its performances. The quantizer average distortion, which is locally minimized in correspondence to the GLA convergence configuration, strongly depends on the initial choice of the reconstrution levels and on the reliability of the estimated pdf [3]. Other methods have been recently proposed, including the Pairwise Nearest Neighbor (PNN) [5] and Maximum Descent (MD) algorithm [6]; the mathematical formalism of both of them can be viewed as a particular case of the more general theory presented in this paper.

In this paper we approach the problem of designing the codebook from a different point of view (substantially improving what proposed in [7] and[8]):

after having evaluated a priori the effects produced by any possible single vector redistribution among clusters, we choose and apply the least distortion one. This way of looking at the problem with "new eyes", though quite immediate and simple in its formulation, provides powerful tools for devising a variety of new algorithms and procedures. The proposed algorithm, despite its simplicity and uniformity (there is no speed-up, no combination of different techniques), goes one step forward to solve the classical problems of convergence speed and codebook optimality.

2   Mathematical Formalism

The problem can be formalized as follows: given a *training set TS* of n k-dimensional input vectors

$$TS = \{\boldsymbol{x}_1, \boldsymbol{x}_2 \ldots, \boldsymbol{x}_n\}$$

considered representative of the signal statistics, the task of designing the optimal Quantizer Q^* consists of finding a set $P = \{p1, p2, \ldots, ph\}$ of clusters each containing $n_{p1}, n_{p2}, \ldots, n_{ph}$ vectors:

$$pi = \{\boldsymbol{x}_{pi_1}, \boldsymbol{x}_{pi_2} \ldots, \boldsymbol{x}_{pi_{n_{pi}}}\} \quad i = 1, 2, \ldots, h$$

in a way to minimize a distortion functional. In the following, the Square Error with respect to the training set TS will be considered:

$$D_{TS,P} = \sum_{i=1}^{n} \|\boldsymbol{x}_i - Q(\boldsymbol{x}_i)\|^2$$

where $Q(\boldsymbol{x}_i)$ represents the codeword associated to the cluster which includes $\boldsymbol{x}_i$. These codewords are (according to [9]) the corresponding centroids $\boldsymbol{m}_{p1}, \boldsymbol{m}_{p2}, \ldots, \boldsymbol{m}_{ph}$:

$$\boldsymbol{m}_{pi} = \frac{1}{n_{pi}} \boldsymbol{s}_{pi} \quad \text{where} \quad \boldsymbol{s}_{pi} = \sum_{j=1}^{n_{pi}} \boldsymbol{x}_{pi_j} \quad i = 1, 2, \ldots, h$$

the global quantization distortion D_{TS} can be obtained as the sum of the partial distortions introduced by each cluster:

$$D_{TS,P} = d_{p1} + d_{p2} + \ldots d_{ph} \quad \text{where} \quad d_{pi} = \sum_{j=1}^{n_{pi}} \|\boldsymbol{x}_{pi_j} - \boldsymbol{m}_{pi}\|^2$$

The generic partial distortion d_{pi}, being $\boldsymbol{m}_{pi}$ the centroid of the corresponding cluster pi can be rewritten as:

$$d_{pi} = \sum_{j=1}^{n_{pi}} \|\boldsymbol{x}_{pi_j}\|^2 - \frac{\|\boldsymbol{s}_{pi}\|^2}{n_{pi}} \quad i = 1, 2, \ldots, h$$

The global quantization distortion D_{TS} can be expressed as:

$$D_{TS,P} = \sum_{i=1}^{n} \|x_i\|^2 - \sum_{j=1}^{h} \frac{\|s_{pj}\|^2}{n_{pj}}$$

Being $\sum_{i=1}^{n} \|x_i\|^2$ independent from the specific clusters configuration, the problem of finding the optimal partition P is that of maximizing

$$SQ_{TS,P} = \sum_{j=1}^{h} \frac{\|s_{pj}\|^2}{n_{pj}}$$

Any different configuration of the quantizer can be reached by means of an arbitrary exchange of vectors among the clusters that can be formalized as follows: the generic cluster pi will loose a subset ti^- of vectors and will receive a subset ti^+ of vectors. The new clusters $\hat{p}1, \hat{p}2, \ldots, \hat{p}h$ obtained after the vector redistribution will be $\hat{p}i = (pi \setminus ti^-) \cup ti^+$ Defining the sums and the corresponding centroids of the vectors within each subset $t1^-, t2^-, \ldots, th^-$ and $t1^+, t2^+, \ldots, th^+$ as $s_{t1}^-, s_{t2}^-, \ldots, s_{th}^-$, $s_{t1}^+, s_{t2}^+, \ldots, s_{th}^+$, $m_{t1}^-, m_{t2}^-, \ldots, m_{th}^-$ and $m_{t1}^+, m_{t2}^+, \ldots, m_{th}^+$, the new optimal output point for each cluster $\hat{p}1, \hat{p}2, \ldots, \hat{p}h$ will be:

$$\hat{m}_{pi} = \frac{\hat{s}_{pi}}{\hat{n}_{pi}} \quad i = 1, 2, \ldots, h$$

with

$$\hat{n}_{pi} = n_{pi} - n_{Ti}^- + n_{Ti}^+ \ \hat{s}_{pi} = s_{pi} - s_{Ti}^- + s_{Ti}^+ \ i = 1, 2, \ldots, h$$

The global distortion $\hat{D}_{TS,\hat{P}}$ after the cluster rearrangement will be:

$$\hat{D}_{TS,\hat{P}} = \sum_{j=1}^{n} \|x_j\|^2 - \sum_{j=1}^{h} \frac{\|\hat{s}_{pj}\|^2}{\hat{n}_{pj}}$$

In this way the distortion variation will be:

$$\hat{D}_{TS,\hat{P}} - D_{TS,P} = \sum_{j=1}^{n} \|x_j\|^2 - \sum_{j=1}^{h} \frac{\|\hat{s}_{pj}\|^2}{\hat{n}_{pj}} - \sum_{j=1}^{n} \|x_j\|^2 + \sum_{j=1}^{h} \frac{\|s_{pj}\|^2}{n_{pj}}$$

The **expression of the distortion variation resulting from the cluster reorganization is found to be:**

$$\hat{D}_{TS,\hat{P}} - D_{TS,P} = \sum_{j=1}^{h} \left(\frac{\|s_{pj}\|^2}{n_{pj}} - \frac{\|\hat{s}_{pj}\|^2}{\hat{n}_{pj}} \right) \tag{1}$$

that is to say:

$$\hat{D}_{TS,\hat{P}} - D_{TS,P} = \sum_{j=1}^{h} \left(\frac{\|s_{pj}\|^2}{n_{pj}} - \frac{\|s_{pj} - s_{tj}^- + s_{tj}^+\|^2}{(n_{pj} - n_{tj}^- + n_{tj}^+)} \right) \tag{2}$$

3 A Simple Algorithmic Implementation

Finding the optimal vector redistribution among the clusters, i.e. the subsets $t1^-, t2^-, \ldots, th^-$ and $t1^+, t2^+, \ldots, th^+$ yielding the best value of $\hat{D}_{TS,\hat{P}} - D_{TS,P}$, is not a trivial task. Because of this complexity, the experiments reported in the following have been obtained applying (2) in the simplest case for evaluating the distortion variation obtainable when a single vector is transferred at each time. Given a vector $\boldsymbol{x}$ belonging to cluster pi, reclassifying it into any other cluster pj provides, according to (2), the following distortion variation:

$$\Delta_{i,j} = \frac{\|\boldsymbol{s}_{pi}\|^2}{n_{pi}} - \frac{\|\boldsymbol{s}_{pi} - \boldsymbol{x}\|^2}{(n_{pi} - 1)} + \frac{\|\boldsymbol{s}_{pj}\|^2}{n_{pj}} - \frac{\|\boldsymbol{s}_{pj} + \boldsymbol{x}\|^2}{(n_{pj} + 1)} \tag{3}$$

The GLA algorithm, at each iteration, reassigns each vector $\boldsymbol{x}$ of the training set TS to the cluster P^* whose centroid m^* has minimum distance from $\boldsymbol{x}$. On the contrary, in the proposed approach (which will be referenced in the following as $K - CL$) $\Delta_{i,j}$ (the distortion variation obtainable by transferring each vector $\boldsymbol{x}$ from its originary cluster p_i to any other cluster pj) is evaluated; if none of the distortion variations $\Delta_{i,j}$, $j = 1, 2, \ldots, h$, $j \neq i$ is negative, $\boldsymbol{x}$ is not moved; otherwise it is transferred to $pj*$ corresponding to the maximum (negative) distortion variation $\Delta_{i,j*}$. In this latter case the partitions pi and pj^* are immediately updated:

$$\hat{s}_{pi} = s_{pi} - \boldsymbol{x} \, , \, \hat{n}_{pi} = (n_{pi} - 1) \, , \, \hat{s}_{pj*} = s_{pj*} + \boldsymbol{x} \, , \, \hat{n}_{pj*} = (n_{pj*} + 1)$$

Since any iterative algorithm needs some form of initialization, different approaches have been tried reporting some preliminary results [8] with specific reference to "random guess" technique. Better quality and repeatable codebooks may be obtained using the well-known splitting technique [1] allowing larger codebooks to be produced from smaller ones:

- the optimal partition of the Training Set TS with size $2^{k=0} = 1$ is obviously the single cluster TS;
- given a codebook of size 2^k, an initial codebook of size 2^{k+1} may be obtained by "splitting" each cluster;
- the so obtained initial configuration is improved using an iterative algorithm.

Using the GLA algorithm, each cluster is splitted considering, together with the original centroid, $\boldsymbol{m}$, another codeword $\boldsymbol{m} + \boldsymbol{\epsilon}$ where $\boldsymbol{\epsilon}$ is a vector with small euclidean norm. In the proposed $K - CL$ method, each cluster is splitted by comparing a random selected component of each vector $\boldsymbol{x}$ to the corresponding component of the centroid $\boldsymbol{m}$. A well conceived implementation of the described algorithm provides almost the same complexity of the GLA algorithm for each iteration, since the term $-\frac{\|\boldsymbol{s}_{pi} - \boldsymbol{x}\|^2}{(n_{pi} - 1)}$ from (3) can be computed only once for each vector $\boldsymbol{x}$ and the terms $\frac{\|\boldsymbol{s}_{pi}\|^2}{n_{pi}}$ and $\frac{\|\boldsymbol{s}_{pj}\|^2}{n_{pj}}$ can be memorized for each cluster. In this way only a slight increment in storage requirements is needed.

4   Experimental Results

The proposed method has been tested in the context of image coding where it has shown significant performance improvements combined with faster convergence in comparison to the standard GLA algorithm. The presented results have been obtained on three 512x512 pixels monochromatic (8 bit/pels) images from the USC test set: "lena", "girl" and "f16". The images have been subdivided in 4x4 blocks to form a training set of 16384 vectors clustered around 1024 codewords. A performance comparison between the GLA method and the K-CL algorithm is presented, in terms of MSE, In Table 1. These results have been obtained employing the splitting technique for both methods and keeping fixed the same number of iterations after each splitting operation. Far more impressive than the

MSE	lena: It. # each split				f16: It. # each split				girl: It. # each split			
method	1	4	10	40	1	4	10	40	1	4	10	40
GLA	39.0	24.0	22.6	22.3	48.7	29.1	27.1	25.8	32.7	21.0	19.8	19.3
K-CL	22.3	19.8	19.5	19.4	25.9	22.5	22.3	22.2	19.5	17.6	17.3	17.2

Table 1. Comparison between the GLA and the K-CL performances in terms of MSE on "lena", "f16" and "girl" image (512x512 pixel, 8 bpp, 4x4 blocks, 1024 codevectors)

performance gain in terms of MSE is the convergence speed as it results from Fig. 1, that displays the MSE curves obtained by applying the two methods until similar quality is reached. The quality of the K-CL coded image is noticeable

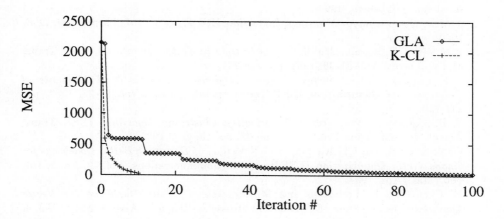

Fig. 1. Comparison of the MSE curves, versus the number of iterations, between GLA method (10 iterations after each splitting - convergence MSE=27.1) and the K-CL (1 iteration afer each splitting - convergence MSE=25.9). Both curves refer to the image "f16" which has been used as training set (512x512 pixel, 8 bpp, 4x4 blocks, 1024 codevectors).

higher in comparison with the GLA coded image expecially in correspondence of the high detail area. In Fig. 2 details of both coded f16 images are presented.

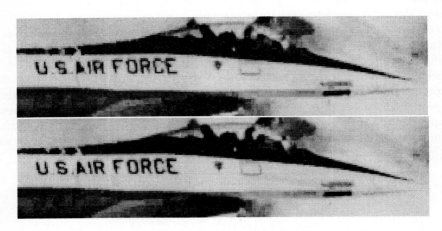

Fig. 2. Details extracted from "f16" coded images (512x512 pixel, 8 bpp, 4x4 blocks, 1024 codevectors 4 iteration after each splitting for both methods) via GLA (top) and K-CL (bottom) techniques

References

1. A. Gersho, R.M. Gray, *Vector Quantization and Signal Compression*, Kluwer Academic Publishers, 1992.
2. J. Max, *Quantizing for Minimum Distortion*, IRE Trans. on Inf. Theory, Vol. IT-6, 1960, pp. 7-12.
3. R.M. Gray, E. Karnin, *Multiple Local Optima in Vector Quantizers*, IEEE Trans. on Inf. Theory, Vol. IT-28, 1982, pp. 256-261.
4. J.H. Conway, N.J.A. Sloane, *Voronoi Regions of Lattices 2-nd Moments of Polytopes and Quantization*, IEEE Trans. on Inf. Theory, Vol. IT-28, 1982, pp. 211-226.
5. W.H. Equitz, *A New Vector Quantization Clustering Algorithm*, IEEE Trans. Acoust. Speech Signal Processing, vol. 37, no. 10, Oct. 1989.
6. Chok-Ki Chan and Chi-Kit Ma, *A Fast Method of Designing Better Codebooks for Image Vector Quantization*, IEEE Trans. on Communications, Vol. 42, No. 2/3/4, Feb./Mar./Apr. 1994.
7. F. Cocurullo, F. Lavagetto and M. Moresco, *Optimal Clustering for Vector Quantizer Design*, Proc. EUSIPCO-92 Brusseles, Begium, August 24-27, Vol. I, pp. 563-566.
8. F. Cocurullo and F. Lavagetto, *A New Algorithm for Vector Quantization*, to appear in Proc. Data Compression Conference - DCC '95 Snowbird, Utah, March 28 - 30, 1995.
9. A. Gersho, *On the Structure of Vector Quantizers*, IEEE Trans. on Inf. Theory, Vol. IT-28, 1982, pp. 157,166.

Fast Fractal Image Coding Using Pyramids

H. Lin and A. N. Venetsanopoulos

Department of Electrical and Computer Engineering,
University of Toronto, Toronto, Ontario, Canada, M5S 1A4

Abstract: In this paper, we present a fast fractal image encoding algorithm which is based on a refinement of the fractal code from an initial coarse level of the pyramid. Assuming that the distribution of the matching error is described by an independent, identically distributed(*i.i.d.*) Laplacian random process, we derive the threshold sequence for the objective function in each pyramidal level. Computational efficiency depends on the depth of the pyramid and the search step size and could be improved up to two orders of magnitude compared with the full search of the original image.

1. Introduction

Fractal image compression is based on the observation that all real-world images are rich in affine redundancy. That is, under suitable affine transformations, large blocks of the image look like smaller ones of the same image. For a given image block, the compression algorithm needs to search through the whole image to find the best matching domain block under an affine transform. This search process is very computationally intensive as compared to the JPEG algorithm. Jacquin[1] used a classification scheme, which restricts the domain block search to the same class as the range block. As the number of the classes is only 3, the computational savings are relatively small. We therefore propose a fast encoding scheme based on pyramidal image representation. The search is first carried out on an initial coarse level of the pyramid. This initial search increases encoding speed significantly, because not only the number of the domain blocks to be searched is reduced, but also the data within each domain block are only $1/4^m$ of those in the finest level, where m is the pyramidal level. Then, only a few numbers of the fractal codes from the promising domain blocks in the coarse level are refined through the pyramid to the finest level with little effect.

2. Fast Pyramidal Domain Block Search Algorithm

Pyramidal image models employ several copies of the same image at different resolutions. Let $f(x,y)$ be the original image of size $2^M \times 2^M$. An image pyramid is a set of image arrays $f_k(x, y)$, $k=0, 1, ..., M$, each having size $2^k \times 2^k$. The pyramid is formed by low pass filtering and resolution subsampling of the original image. The pixel $f_k(x, y)$ at level k is obtained from the average of its four neighbours $f_{k+1}(x', y')$ at level $(k+1)$:

$$f_k(i, j) = \frac{1}{4} \sum_{r=0}^{1} \sum_{s=0}^{1} f_{k+1}(2x+r, 2y+s) \tag{1}$$

The coarsest level (k=0) image has size 1 and represents the average grey level of the original image. The finest level image f_M is the original image of size $2^M \times 2^M$. As the number of the levels decreases, the image details are gradually suppressed and spurious low spatial frequency components are introduced due to the effect of aliasing. Because the pyramidal structures offer an abstraction from image details, they have been proven to be very efficient in certain image analysis and image compression applications[2].

Our encoding process starts with partitioning $f(x,y)$ into a set of nonoverlapping range blocks of size $2^m \times 2^m$. Similarly, the same image is partitioned into a set of overlapping domain blocks that are larger in size than range blocks to meet the contractivity condition. The previous study[3] showed that the general optimization objective function for the best matched domain block search can be written as:

$$E = \frac{1}{4^m} \sum_{x=0}^{2^m-1} \sum_{y=0}^{2^m-1} \left[D(x, y, s, t) - R(x, y) \right]^2 \tag{2}$$

where $D(x, y, s, t) = s f_{M-1}(x, y) + t$ is an affine of the scaled domain block and $R(x, y) = f(x, y)$ is the range block to be encoded. The fractal code is fully specified by parameters: (1)θ_i: the index of rotations/reflections; (2)D_{xi}, D_{yi}: the position of domain block D_i; (3)s_i: contractive factor; (4)t_i: grey level shift.

Instead of a constant contractive factor s, a nonlinear contractive function $s(x, y, a, b, c)$ can be used for fast decoding [3]:

$$s(x, y, a, b, c) = \pm \frac{1}{1 + e^{ax+by+c}} \tag{3}$$

where (a, b, c) are parameters of the contractive function. From the original image a pyramid is created, the depth of which is determined by the range block size. Because the range block is defined in the image, the range block pyramid will be contained in the image pyramid with the k-th level of the range block pyramid corresponding to the (M-m+k)-th level of the image pyramid. Instead of a direct search of the minimum of the objective function at the finest level m, we propose a fast algorithm by introducing a smaller, approximate version of the problem at a coarser level k of the range block pyramid:

$$E^k = \frac{1}{4^k} \sum_{x=0}^{2^k-1} \sum_{y=0}^{2^k-1} \left[D^k(x, y, s^k, t^k) - R^k(x, y) \right]^2 \tag{4}$$

for $k_0 \leq k \leq m$. Therefore, at range block pyramid level k, the encoding amounts to finding the best matching domain block of size $2^k \times 2^k$ in the image of the size $2^{M-m+k} \times 2^{M-m+k}$. For example, for an original image of size 512×512 (M=9) and range block size 64×64 (m=6), the search complexity at k_0=2 is that of the image size 64×64 and the range block of size 4×4. The $k=k_0$ level of the range block pyramid is said to be initial and every location of the image from the $(M-m+k_0)$-th level of the image pyramid needs a test. Now, generate a $2^{k+1}\times 2^{k+1}$ promising location matrix G^{k+1}:

$$(G^{k+1})_{2u,2v} = \begin{cases} 1, & if\ E^{\ k}(u,v) < T^{\ k} \\ 0, & otherwise \end{cases} \tag{5}$$

where (u, v) is the upper left corner coordinates of the domain block and T^k is the threshold at level k. Matrix G^{k+1} is used as a guide in the search of the domain location at the next level $k+1$. Tests are to be performed only at the locations (i, j) for $(G^{k+1})_{i,j} = 1$ and its neighbour locations. Other parameters P^k of the promising locations are also propagated to P^{k+1} for further refining at level $k+1$. For the of affine mapping, we have $\theta^{k+1}=\theta^k$, $D_x^{k+1}=2D_x^k$, $D_y^{k+1}=2D_y^k$, s^{k+1} and t^{k+1} need to reevaluate. In the case of nonlinear contractive functions, the initial parameters at level $(k+1)$ are: $a^{k+1}=\frac{1}{2}a^k$, $b^{k+1}=\frac{1}{2}b^k$, $c^{k+1}=c^k$ and $t^{k+1}=t^k$. The ½ gain before the a^k and b^k is due to the resolution increase in the x and y directions. The algorithm provides a gradual refinement of the fractal code. The process is repeated recursively until the finest level m is reached as shown in Fig. 1. The iterations are over the promising locations. At the finest level, if there exist more than one locations (u, v) such that $(G^M)_{u,v}=1$, select the parameters with the smallest match error as the fractal code. An important feature of the algorithm is estimating of the threshold T^k. The next section shows how to estimate these thresholds under certain assumptions.

3. Determination of Thresholds

Let x_i denote the grey level difference of a pixel between an affine transformed domain block D and a range block R at the finest level m, i.e., $x_i = D_i - R_i$, for $i=0,1,...,(2^m\times 2^m-1)$. At the match location (u^*, v^*), x_i is significantly less correlated. Thus, we may consider x_i as independent, identically distributed($i.i.d.$) random variables with an approximately Laplacian density function:

$$f(x)=\frac{\alpha}{2}e^{-\alpha|x|} \tag{6}$$

$f(x)$ has mean $\mu_0=0$ and variance $\sigma_0^2=2/\alpha^2$. Our experimental data showed a reasonable approximation to the density function. Then, it can be shown that the function of the random variable $y_i=x_i^2$ is exponentially distributed and has mean $\mu_y=\sigma_0^2$ and variance $\sigma_y^2=5\sigma_0^4$. The next step towards the goal is to find the distribution of the mismatch measure as in (2) which can be rewritten as:

$$E=\frac{1}{n}\sum_{i=0}^{n-1} x_i^2=\frac{1}{n}\sum_{i=0}^{n-1} y_i \tag{7}$$

where $n=2^m\times 2^m$. By the central limit theorem, which says that the density of the sum of n independent random variables tends to a normal density as n increases, regardless of the shapes of the densities of the given random variables[4], E is approximately normal with:

$$\mu_E = \sigma_0^2, \quad \sigma_E^2 = \frac{5}{4^m} \sigma_0^4 \tag{8}$$

Let P_α be the probability of finding the best match (u^*, v^*), i.e. $P(E < T^m) = P_\alpha$, then the threshold will be:

$$T^m = \mu_E + x_\alpha \sigma_E = \sigma_0^2 (1 + \frac{\sqrt{5}}{2^m} x_\alpha) \tag{9}$$

where x_α is the P_α point of standard normal distribution. For example, when $P_\alpha = 0.9$, $x_\alpha = 1.28$.

It can be shown[5] that the thresholds at a coarse level k are:

$$T^k = \frac{\sigma_0^2}{4^{m-k}} (1 + \frac{x_\alpha}{2^{(k+\frac{1}{2})}})^2 \tag{10}$$

for $k = k_0, \ldots, m-1$.

4. Computational Efficiency

The computational efficiency of the pyramid algorithm can be evaluated based on the following theoretical considerations. For a given range block, assume each domain block needs the same number of operations to determine the parameters. Then the computational cost is proportional to the product of the number of domain blocks searched and the number of pixel in each block. For an original image of size $2^M \times 2^M$ and range block of size $2^m \times 2^m$, when D_i is chosen in each dimension twice the size of the R_i, the search domain image is $2^{M-1} \times 2^{M-1}$ with the contracted domain block of size $2^m \times 2^m$. This number becomes:

$$C_1 = 8 \left(\frac{2^{M-1} - 2^m}{h} + 1 \right)^2 2^{2m} \tag{11}$$

where h is the step size of the domain block search. When a pyramidal search is applied, the computational resources for the algorithm are determined by the average number of the promising locations n_p on every pyramid level and the number of the shifts n_s around each promising location:

$$C_2 = 8 \left(\frac{2^{(M-m+k_0-1)} - 2^{k_0}}{h(k_0)} + 1 \right)^2 2^{2k_0} + n_p n_s \sum_{i=k_0+1}^{m} 2^{2i} \tag{12}$$

where the first term corresponds to the initial step of the algorithm, testing every domain block on the initial range pyramid level k_0. The search step size $h(k_0)$ is related to th finest level step size h as follows:

$$h(k_0) = \max(1, \frac{h}{2^{(m-k_0)}}) \tag{13}$$

where we assume only the integer search step is used in level k_0, although, in general, the search with sub-pixel accuracy is possible.

The number of operations required to create an image pyramid will be proportional to the number of pixels:

$$C_3 = K_1 \sum_{i=M-m+k_0}^{M-1} 2^{2(M-i)} \tag{14}$$

Compared with the optimization operation during the domain block search, this part can be neglected.

The benefit in computational saving using pyramids relative to the full search of the original image is estimated as:

$$Q = \frac{C_1}{C_2 + C_3} \approx \frac{C_1}{C_2} \tag{15}$$

For a given image and range block size, the value of Q depends on the depth of the pyramid and the search step size. For example, for the image of size 512×512, range block 32×32, when $h=2$, $n_p=20$, $n_s=16$ and $k_0=2$, the computational saving factor will be 194. The actual Q value is expected to be smaller than the theorical one. For example, encoding a 32×32 range block with affine contractive mapping needs 95.89 *CPU* seconds by full search and 0.79 seconds by pyramidal search (Serial implementation on KSR1 parallel computer without optimization of codes), which gives $Q=121$.

5. Experimental Results

Fig. 2 is 512×512×8 bits original Lenna image. Quadtree partition is used for range blocks. The initial range block size is 64×64. The mean square error was determined for each range block. Blocks which had an error exceeding 81 (corresponding rms value 9.0) and were larger than 8×8 in size were split. Fig.3 shows our reconstructed image using nonlinear contractive function by full search algorithm at bit rate 0.2 bpp (compression ratio 40:1) and PSNR=30.2 dB. Fig. 4 is the result of this paper at the same bit rate and PSNR=29.9 dB. Thus, the pyramid search algorithm is quasi-optimal in terms of minimizing the mean square error. The main advantage of the pyramid algorithm is the greatly decreased computational complexity, when compared to full search.

References

1. A. E. Jacquin: Image coding based on a fractal theory of iterated contractive image

transformation. IEEE Trans. image Process. **1** (1992) 18-30

2. P. J. Burt, and E. H. Adelson: The Laplacian pyramid as a compact image code. IEEE Trans. Comm. **3** (1983) 532-540
3. H. Lin and A. N. Venetsanpoulos: Incorporating nonlinear contractive functions into the fractal coding. Proceedings of the International Workshop on Intelligent Signal processing and Communication Systems, Seoul, Korea. (1994) 169-172
4. A. Papoulis: Probability, Random Variables, and Stochastic Processes, McGrawHill, Inc. (1991)
5. H. Lin and A.N. Venetsanpoulos: Fast Fractal Image Compression Using Pyramidal Search. in: Scientific Information Guild (Ed.): Circuits & Systems. India: Research Signpost (to appear)

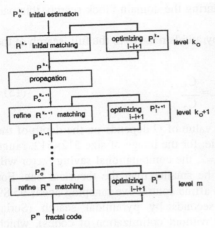

Fig. 1. Refining the fractal code
from coarse to fine level

Fig. 2. Original image

Fig. 3. Full search, 0.2 bpp, 30.2 dB

Fig. 4. Pyramid search, 0.2 bpp, 29.9 dB

Motion

Hidden MRF Detection of Motion of Objects with Uniform Brightness

Adam Kuriański[1] and Mariusz Nieniewski[1,2]

[1] Institute of Fundamental Technological Research, PAS, Warsaw,
[2] Dept. of Fundamental Research in Electrical Engineering, PAS, Warsaw.
e-mail: akurian@ippt.gov.pl   mnieniew@ippt.gov.pl

Abstract. The hidden Markov Random Field (MRF) model for motion detection in image sequences is described. A typical MRF model uses two observations: the difference in brightness between two consecutive images, and the value obtained from the mask of temporal changes. The performance of this model can be improved by including the third observation: the brightness at a given pixel. The paper gives the necessary equations and presents an example of motion detection of the object with uniform brightness.

Notation

$f_k(i,j)$ – brightness value at pixel (i,j) of the k-th image,

$o_k(i,j)$ – brightness difference $o_k(i,j) = f_{k+1}(i,j) - f_k(i,j)$,

$\bar{o}_k(i,j)$ – value at pixel (i,j) of the mask of temporal changes between k-th and $(k+1)$-th image,

$C = \{1,0\}$ – set of values which can be assigned to a particular pixel (i,j) by the mask of temporal changes,

$e_k(i,j)$ – label assigned to pixel (i,j) of the k-th image by the mask of moving objects,

$L = \{a,b\}$ – set of labels which can be assigned to a particular pixel by the mask of the moving object,

Superscript T denotes the transposition of a vector.

1   Introduction

The field of labels assigned to the pixels specifies the mask of moving objects and represents a particular realization of a hidden MRF. The energy of this field includes the energy of the hidden layer itself as well as the energy representing the influence of the observation on the hidden layer. In the MRF model considered, there are three sources of observation: brightness difference o_k at a given pixel (i,j), the value obtained from the mask of temporal changes $\bar{o}_k$ for the same pixel, and brightness f_k at pixel (i,j) in the k-th image. The MRF model with two observation sources is described in [2] and [4], and the authors confirm that their model is not appropriate for motion detection of objects with uniform brightness.

The new model is especially advantageous for such objects. The labels assigned to individual pixels in the mask of moving objects are found by minimization of the total energy W_k of the field of labels of a pair of images, that is of the k-th and $(k + 1)$-th image. This energy can be expressed as

$$W_k = W_{s_1} + W_{s_2} + W_t + W_c, \tag{1}$$

where:

W_{s_1} – spatial energy of the field of labels for k-th image,

W_{s_2} – spatial energy of the field of labels for $(k + 1)$-th image,

W_t – temporal energy for the field of labels of k-th and $(k + 1)$-th image conditioned on the observation $\bar{o}_k$,

W_c – consistency energy of the field of labels of k-th and $(k + 1)$-th image with observation $[o_k, f_k]^T$.

The field of labels assigned to the k-th image is considered final, whereas the field of labels assigned to the $(k + 1)$-th image will be changed subsequently when the next pair of images $(k + 1)$-th and $(k + 2)$-th is analyzed.

Finding the minimum of the energy W_k is described in [1] and [3]. In the current paper the problem of building the appropriate model is considered.

2   A hidden MRF with Three Observation Sources

Instead of dealing directly with total energy it is much easier to consider the local energy of the field ([1] and [3]). Without getting into details, it can be said that by minimization of the individual local energies one can presumably achieve the minimum of the total energy. The local energy at a pixel (i, j) is

$$U_{ij} = U_{ijs_1} + U_{ijs_2} + U_{ijt} + U_{ijc}, \tag{2}$$

where:

U_{ijs_1} – local spatial energy of pixel (i, j) of k-th image,

U_{ijs_2} – local spatial energy of pixel (i, j) of $(k + 1)$-th image,

U_{ijt} – local temporal energy of the realization $[e_k, e_{k+1}]^T$, at pixel (i, j) of k-th and $(k + 1)$-th image,

U_{ijc} – consistency energy of the realization $[e_k, e_{k+1}]^T$ with observation $[o_k, f_k]^T$ at a pixel (i, j).

When calculating the energy according to Eq. (2) one has to know the vector Θ of the parameters of the model. This vector can be written as

$$\Theta^T = [\Theta_s^T, \Theta_t^T, \Theta_c^T], \tag{3}$$

where Θ_s – part of the parameter vector responsible for the spatial energy, Θ_t – part responsible for temporal energy, and Θ_c – part responsible for the consistency energy.

For purposes of calculating the spatial energy, a simple model of the field is assumed, in which only 2-pixel cliques are included. These cliques can have horizontal, vertical, diagonal to the left, or diagonal to the right position. The potential V_{cs} of any 2-pixel clique is assumed to have the form

$$V_{cs} = \begin{cases} \beta_s & \text{if the labels of both pixels are different,} \\ -\beta_s & \text{if the labels of both pixels are the same,} \end{cases} \tag{4}$$

where cs denotes a spatial 2-pixel clique, and β_s is a positive constant.

Each spatial energy U_{ijs_1}, U_{ijs_2} is then equal to the sum of potentials of eight cliques, each of which includes the pixel (i, j), that is

$$U_{ijs} = \sum_{cs \in C'} V_{cs}, \tag{5}$$

where cs denotes a spatial 2-pixel clique, and the summation is over the set C' of all cliques including the pixel (i, j). It is possible to assign a different weight to each of the clique orientations. In the simplest case all the weights are the same, and the vector $\boldsymbol{\Theta}_s$ is

$$\boldsymbol{\Theta}_s = [\beta_s, \beta_s, \beta_s, \beta_s]^T, \tag{6}$$

where β_s is a positive constant.

It is convenient to define the consistency energy U_{ijc} in terms of the auxiliary vector function $\boldsymbol{\Psi}$

$$\boldsymbol{\Psi}[e_k(i,j), e_{k+1}(i,j)] = \begin{cases} [m_{01}, m_{02}, \sigma_{01}^2, \sigma_{02}^2, \rho_0]^T & \text{if } [e_k(i,j), e_{k+1}(i,j)] = (b,b) \\ [m_{11}, m_{12}, \sigma_{11}^2, \sigma_{12}^2, \rho_1]^T & \text{if } [e_k(i,j), e_{k+1}(i,j)] = (a,b) \\ [m_{21}, m_{22}, \sigma_{21}^2, \sigma_{22}^2, \rho_2]^T & \text{if } [e_k(i,j), e_{k+1}(i,j)] = (b,a) \\ [m_{31}, m_{32}, \sigma_{31}^2, \sigma_{32}^2, \rho_3]^T & \text{if } [e_k(i,j), e_{k+1}(i,j)] = (a,a) \end{cases} \tag{7}$$

where:

a – label indicating that the pixel belongs to the moving object,

b – label indicating that the pixel belongs to the background,

$(a,a), \ldots, (b,b)$ – pair of labels assigned to a pixel in two consecutive images,

m_{01} – mean value of the background-to-background brightness difference,

m_{02} – mean value of the brightness of the background,

σ_{01}^2 – variance of the background-to-background brightness difference,

σ_{02}^2 – variance of the brightness of the background,

ρ_0 – correlation coefficient between the brightness of the background-to-background difference and the brightness of the object.

The remaining parameters $m_{11}, \ldots, \rho_3$ in Eq. (7) denote, respectively, the mean values, the variances, and the correlation coefficients for three other situations: object-to-background, background-to-object, and object-to-object, which can occur for two consecutive images.

It is assumed that the consistency of the observation $[o_k, f_k]^T$ at a pixel (i, j) with the realization of the field of labels $[e_k, e_{k+1}]^T$ at the same pixel is characterized by the 2-D Gaussian distribution with parameters $m_{l1}, m_{l2}, \sigma_{l1}^2, \sigma_{l2}^2, \rho_l$, where $l = 0, \ldots, 3$, depending on which realization $(a, a), \ldots, (b, b)$ is considered. The consistency energy at a pixel (i, j) is

$$U_{ijc} = \frac{1}{2(1 - \rho_l^2)} \left[\left(\frac{o_k - m_{l1}}{\sigma_{l1}} \right)^2 - \rho_l \left(\frac{o_k - m_{l1}}{\sigma_{l1}} \right) \left(\frac{f_k - m_{l2}}{\sigma_{l2}} \right) + \left(\frac{f_k - m_{l2}}{\sigma_{l2}} \right)^2 \right] \tag{8}$$

and the vector Θ_c is

$$\Theta_c = [m_{01}, m_{02}, \sigma_{01}^2, \sigma_{02}^2, \rho_0, \ldots, m_{31}, m_{32}, \sigma_{31}^2, \sigma_{32}^2, \rho_3]^T \tag{9}$$

The temporal energy $U_{ijt}(e_k, e_{k+1}, \overline{o}_k)$ at a pixel (i, j) depends on the value at the same pixel of the mask of temporal changes occurring between k-th and $(k + 1)$-th image. In the case of motion of objects with constant brightness there is a high probability of the situation when the value of the mask of temporal changes indicates *no changes* but the pixel should belong to the moving object. The brightness observation f_k allows one to properly describe such a situation. The energy of a temporal clique is found according to the following table

	$\overline{o}_k = 0$	$\overline{o}_k = 1$
(b, b)	$-\beta_t$	β_t
(a, b)	β_t	$-\beta_t$
(b, a)	β_t	$-\beta_t$
(a, a)	$-\beta_t$	$-\beta_t$

where $\overline{o}_k = 1$ indicates that changes at a pixel (i, j) between k-th and $(k + 1)$-th images have been detected, $\overline{o}_k = 0$ – changes have not been detected, and β_t is a positive constant. In accordance with the above table the vector Θ_t is

$$\Theta_t = [-\beta_t, \beta_t, \beta_t, -\beta_t, \beta_t, -\beta_t, -\beta_t, -\beta_t]^T. \tag{10}$$

The assumed MRF model has 32 parameters, some of which are equal. Typically one assumes $\beta_s = 10$ and $\beta_t = 100$. However, there are 20 unknown parameters for consistency energy. The authors estimated these parameters via the "teaching sequence." A sample of a sequence was taken, and one obtained the mask of the moving object by manually indicating pixels of the mask. Then, having both the original sequence and the masks of moving objects one estimated the mean values, variances, and correlation coefficients for o_k and f_k by means of ML estimation.

3   Example of the Use of the MRF with Three Observation Sources

A sequence of images is shown in Fig. 1. The masks of temporal changes for this sequence were obtained by approximating the brightness in the 3×3 window by a linear function and then carrying out the ML test for determining whether the remaining variability of the brightness indicated a change. The mask of temporal changes typically consists of a number of disconnected areas. The hair was not included in this mask except for parts of the contour of the head. In Figs. 2 and 3 contours of the masks of moving objects are shown.

4 Conclusion

The results obtained confirm that the MRF model using three observations offers a sensitive method of motion detection. Experiments showed that the model with two observations gives very low quality masks for sequences such as in Fig. 1. The model with three observations is particularly recommended for moving objects with uniform brightness.

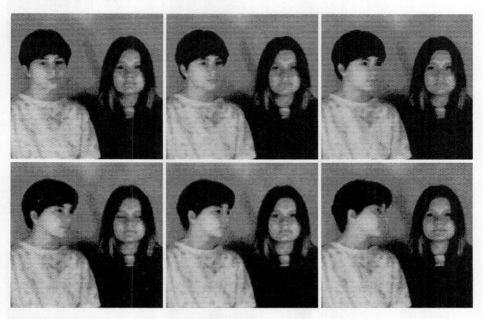

Fig. 1. A sequence showing the rotational motion of the girl on the left around her vertical axis.

References

1. Besag, J.: On the statistical analysis of dirty pictures. Journ. Royal Statist. Soc. **48** series B (1986) 259–302
2. Bouthemy, P., Lalande, P.: Detection and tracking of moving objects based on a statistical regularization method in space and time. Proc. European Conf. Comp. Vision, Antibes (1990) 307–311
3. Geman, S., Geman, D.: Stochastic relaxation, Gibbs distributions and the Bayesian restoration of images. Pattern Anal. Mach. Int. **6** (1984) 721–741
4. Lalande, P., Bouthemy, P.: A statistical approach to the detection and tracking of moving objects in an image sequence. Proc. 5th ESPC EUSIPCO 90, Barcelona (1990)

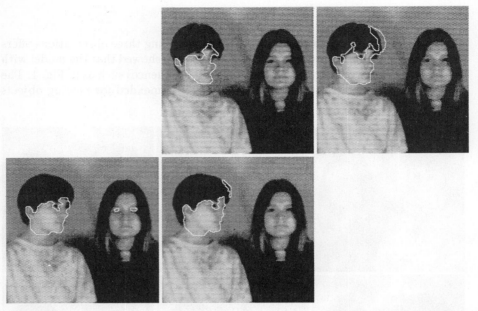

Fig. 2. Contours of the masks of moving objects superimposed on the images of Fig. 1. The model parameters were estimated with the hair included in the background.

Fig. 3. Contours of the masks of moving objects superimposed on the images of Fig. 1. The model parameters were estimated with the hair included in the moving object.

Motion Analysis and Segmentation for Object-Oriented Mid-Level Image Representation

Takahiro Saito and Takashi Komatsu

Department of Electrical Engineering, Kanagawa University
Yokohama, 221, Japan

Abstract. In the object-oriented mid-level image representation, moving images are decomposed into multiple global image regions undergoing different coherent motion, and then each global coherent motion region is described with its texture and region map along with its motion model. With regard to the global motion segmentation, this paper employs the strategy of a gradual migration from a simple local optical flow representation of motion to a compound global segmentation representation of motion, and presents a new matching-based robust local motion estimation algorithm with which we design to cope with the generalized aperture problem. Furthermore, this describes an iterative global motion segmentation algorithm for decomposing moving images into multiple global coherent motion regions, each of which conforms to a different global motion model.

1 Introduction

Image coding and computer vision are closely related domains, and we can apply a hierarchical framework of computer vision, ranging from low-level image representation to high-level image representation, to image coding. Low-level image coding have been well studied so far, but it is said that low-level image coding is now approaching the performance limit. On the other hand, the high-level concept is now considered to be beyond reach except in restricted domains [1].

Object-oriented mid-level image representation is considered to be a promising candidate for the image representation with which a fertile domain of new-generation generic image coding will be opened up. Recently, many research groups are studying mid-level image coding [2, 3]. Mid-level image representations involve such concepts and attributes as surfaces, segmentation, global motion, texture, transparency, and so on. Wang and Adelson have recently advocated a layered image representation [3]. Their layered representation is to describe moving images with sets of overlapping layers, each of which contains three different maps: the intensity map, the alpha map, and the velocity map. The layers are ordered in depth and occlude each other in accord with the rule of composition. They have presented some methods for decomposing image sequences into layers using a global motion segmentation algorithm, and they have employed the standard strategy of a gradual migration from a local optical flow representation of motion to a global segmentation representation of motion. Local motion estimation and motion segmentation are decisively important for generating stable accurate mid-level image representation. An inaccurate local optical flow representation at the first stage of the global motion segmentation will greatly damage quality of the resultant global motion segmentation.

Our work keeps within the framework of mid-level image representation and employs the standard strategy of the gradual migration. In the first stage of the gradual migration strategy, local optical flow estimation is carried out over a local neighborhood, and hence the generalized aperture problem, which means the dilemma surrounding the proper size of an analysis image block, appears to be the most serious problem with local motion estimation. To solve the problem, this paper extends the basic concept of the existing block-matching algorithm, thus forming a matching-based robust local motion estimation algorithm based on the assumption that there may be multiple distinct image patterns undergoing different coherent motion within a given analysis image block. Furthermore, this paper presents an iterative global motion segmentation algorithm which identifies multiple global coherent motion regions by initially fitting global perspective motion models to local optical flow estimates recovered by the extended matching-based robust local motion estimation algorithm and then refining segmentation of multiple global coherent motion regions and their corresponding global perspective motion models alternately iteratively.

2 Extended Matching-Based Algorithm for Estimating Multiple Local Image Motions

We employ the standard strategy of the gradual migration from a simple local optical flow representation of motion to a compound global segmentation representation of motion. In the local optical flow representation of motion, we assume that each local image pattern within an analysis image block undergoes a simple rigid translation, while in the global segmentation representation of motion we employ a perspective transformation as a motion model. The gradual migration strategy circumvents the egg-and-chicken dilemma surrounding interdependence between motion analysis and segmentation, and alleviates the difficulty of recovering compound motion model parameters directly from a given input image sequence. This chapter concentrates on a new-extended matching-based algorithm for recovering the local optical flow representation of motion from a given input image sequence.

2.1 Generalized Aperture Problem in Local Motion Analysis

In local motion estimation, the size of the analysis image block is a critical factor. A large analysis image block is needed to sufficiently constrain the solution and to provide some insensitivity to noise. The larger the analysis image block is made, the less likely the assumption about the motion will be valid over the entire analysis image block. For example, the constant velocity assumption will be violated by an affine and/ or perspective optical flow, motion boundaries, transparency, and so on. The dilemma surrounding the proper size of an analysis image block is referred to as the generalized aperture problem.

The appropriate size of an analysis image block depends on such factors as the size and the velocity of objects in the scene, and hence it is extremely difficult to determine the optimal size of an analysis image block. Therefore, there may be a number of local motion configurations occurring within an analysis image block when its size is arbitrary determined [4, 5, 6].

The existing local motion analysis methods can deal with only the simplest local motion configuration where the block contains a single image pattern undergoing coherent motion. This paper presents a new robust matching-based local motion estimation algorithm which can deal with the more complex local motion configuration where the block consists of multiple distinct image patterns undergoing coherent motion.

Recently, the study group of Bergen and others, and the study group of Shizawa and others, have independently presented the extended gradient-based algorithms which can deal with the other complex local motion configuration of multiple transparent image patterns [4, 5]. Their algorithms are constructed with the second derivatives of image intensity, and hence unfortunately seems to be sensitive to noise. Our algorithm is robust even under real noisy conditions like the existing block matching algorithm, but it requires large computational efforts.

2.2 Prototypal Two-Frame Algorithm [6]

For the analysis of multiple motion vectors, firstly the extended algorithm segments a given analysis image block into multiple distinct image regions undergoing coherent motion, simultaneously applies its matching mechanism to each image region separately, and then provides its multiple matching results as multiple motion vector estimates; the extended algorithm finally provides a single motion estimate for each image region. A segmentation process is usually based on image intensity, but the segmentation process introduced here is based on the assumption that each image region undergoes coherent motion.

We form the extended two-frame algorithm as the following discrete optimization problem, where p motion vectors $(m_1, n_1),......,(m_p, n_p)$ are determined so that the cost function is minimized.

$$\underset{\substack{\{(m_1, n_1),..., (m_p, n_p)\} \\ (m_k, n_k) \in SE}}{Min} \left\{ \underset{i,j \in BL}{\Sigma} \underset{k=1,2,...,p}{Min} (X_{i, j} - Y_{i+mk, j+nk})^2 \right\} \qquad (1)$$

$X_{i, j}$: image intensity of the pixel (i, j) in the present image frame X
$Y_{i, j}$: image intensity of the pixel (i, j) in the next image frame Y
SE : search area,   BL : analysis image block

In addition, we have experimentally derived a good way of determining the proper value of the number of image regions p. The way is as follows: we start with the initial value of 1 for p, and then we increase p by one if and only if the increase of p by one decreases the minimum cost value to less than 20 % of the preceding minimum cost value before the increase of p.

2.3 Improved Algorithms

Three-Frame Algorithm. Generally speaking, the two frame algorithm has a major drawback. The two-frame algorithm cannot handle the image areas covered by other moving image patterns, and it cannot provide correct motion vector estimates in the covered areas, because the covered area does not appear in the next image frame. To solve this problem, we should introduce the concept of a three-frame algorithm which estimates multiple motion vectors by applying the extended segmental matching process to consecutive three image frames. We form the three-frame algorithm as the following discrete optimization problem, where p motion vector estimates $(m_1, n_1),......,(m_p, n_p)$ are determined so that the cost function is minimized.

$$\text{M i n} \left\{ \sum \text{M i n} \left\{ (X_{i,j} - Y_{i-mk,j-nk})^2 , (X_{i,j} - Z_{i+mk,j+nk})^2 \right\} \right\}$$
$$\{(m_1,n_1),...,(m_p,n_p)\} \quad i,j \in BL \quad k=1,2,...,p$$
$$(m_k, n_k) \in SE \tag{2}$$

$X_{i,j}$: image intensity of the pixel (i, j) in the previous image frame X
$Y_{i,j}$: image intensity of the pixel (i, j) in the present image frame Y
$Z_{i,j}$: image intensity of the pixel (i, j) in the next image frame Z
SE : search area,   BL : analysis image block

Introduction of Spatial Coherence Constraint. The prototypal extended algorithm has another defect that it does not take the spatial coherence property of a motion field into consideration. The spatial coherence property means the assumption that surfaces have spatial extent and hence neighboring pixels with similar pixel intensity value in an image are likely to belong to the same surface and likely to have similar motion vectors.

By imposing the spatial coherence constraint to a motion vector estimate assigned to each pixel within an analysis image block, the extended algorithm will probably get to provide correct motion vector estimates even for a pixel in the vicinity of which there are few salient features. The spatial coherence constraint is introduced into the extended three-frame algorithm as follows; firstly we segment an analysis image block into small cells by grouping neighboring pixels having similar pixel intensity into a small cell, and then in the local motion estimation procedure expressed by equation (2) we select an optimal motion vector estimate for each segmented small cell, instead of for each pixel, from among p motion vector candidates, on the assumption that pixels which belong to the same segmented small cell have the same motion vector. The algorithm is organized as follows:
[Spatial Coherence Constraint Algorithm]
(1) If the difference between the maximum pixel intensity value and the minimum pixel intensity value within an analysis image block, typically with 16×16 pixels, exceeds a threshold Th, then we divide an analysis image block into four small square cells equally.
(2) If the condition applied in the step (1) is satisfied in a segmented small cell, then we divide the segmented small cell further into four smaller cells equally. This procedure is repeated until the size of the segmented small cell reaches the prescribed smallest cell size, typically of 2×2 pixels, or further segmentation is not performed.
(3) We solve the optimization problem of equation (2) under the constraint that pixels which belong to the same segmented small cell should be allotted the same motion vector estimate.

2.4 Evaluation

As an instance, this paper presents only the results of the experiment using the natural image sequence where the doll is swinging like a pendulum, but in this image sequence the doll looks as if it were undergoing pure horizontal translation, and the vertical component of the optical flow field is nearly zero for every pixel in the image. In the experimental simulations, the size of an analysis image block and that of a search area are fixed at 16×16 and 13×13 respectively.

Fig. 1 shows the results of the extended matching-based three-frame algorithm with the spatial coherence constraint. The left column shows the horizontal components of the motion estimates and the right column shows the vertical components of the motion estimates. In Fig. 1, pixel intensity is allocated to each pixel according to its estimated motion component. The existing block-matching algorithm incurs estimation errors along the contour of the moving doll. The prototypal extended two-frame algorithm decreases estimation errors along the contour of the moving doll to some extent, but the estimated flow field is rather rugged along the contour. On the other hand, the extended three-frame algorithm with the spatial coherence constraint decreases estimation errors along the moving doll considerably, and recovers the flow field as accurately as we expect.

3   Iterative Global Motion Segmentation

The iterative global motion segmentation algorithm identifies multiple global image regions, each of which undergoes a different coherent motion, by fitting perspective motion models to local optical flow estimates recovered with the local motion estimation algorithm described in the previous chapter.

The iterative global motion segmentation algorithm is organized as follows:

(1) Initial Perspective Motion Model: Beginning with the local optical flow representation of motion, we determine a initial set of global coherent motion regions that have spatially continuous optical fields, by grouping neighboring pixels having spatially continuous optical flows into a global coherent motion region, and then we determine an initial set of perspective motion models each of which well approximates the optical flow field within each global coherent motion region with the standard weighted least square regression algorithm. The perspective motion model is defined by the equations

$$v_x(x,y) = a + bx + cy + dx^2 + exy \qquad (3)$$
$$v_y(x,y) = f + gx + hy + ey^2 + dxy \qquad (4)$$

where v_x and v_y are the x and y components of velocity, (x,y) are the spatial coordinates, and (a,b,c,d,e,f,g,h) are the parameters of the model.

(2) Iterative Refinement of Motion Segmentation: At each iteration step, multiple

horizontal component u   vertical component v

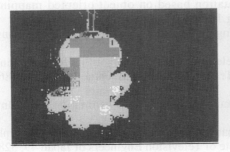
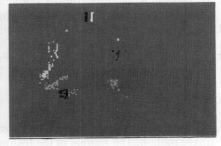

Fig. 1. Motion vector estimates provided by the extended three-frame algorithm with the spatial coherence constraint.

perspective motion models and multiple global motion regions are alternately refined until convergence. At each iteration step, segmentation of multiple global coherent motion regions is updated not only by comparing the local optical flow estimates with the multiple perspective motion models, but also by estimating intensity differences between the observed image frame and its corresponding image frame warped with each of the multiple perspective motion models, and then a proper perspective motion model is newly calculated within each of the refined global coherent motion regions by the standard gradient-based weighted least square motion estimation algorithm based on the perspective motion model.

(3) Interframe Projection of Motion Segmentation (Temporal Integration): The perspective motion models and the segmentation of successive image frames will be similar, because a rigid object shape and motion change slowly from frame to frame. To preserve temporal coherency and continuity, we use the current global motion segmentation results as initial segmentation for the next image frame. Once the global motion segmentation is completed on the entire image sequence, each global coherent motion region will be identified and tracked in the entire image sequence with its corresponding perspective motion model and its spatial location and extent.

We have conducted experiments using the identical natural image sequence to evaluate the performance of the entire process of the iterative global motion segmentation algorithm. The iterative global motion segmentation algorithm works very well for the natural test image sequence.

4 Conclusion

We present a new matching-based robust local motion estimation algorithm and an iterative global motion segmentation algorithm, which are characterized as key technologies for recovering a mid-level image representation from a given input image sequence. We have conducted experiments using natural image sequences to evaluate the performance of the algorithms. The results of the experiments demonstrate that for natural image sequences the algorithms work as well as we expect.

References

1. K. Aizawa, H. Harashima, T. Saito: Model-based analysis synthesis image coding system for person's face, Signal Processing: Image Communication 1, 139-152(1989)
2. M. Hotter, R. Thoma: Image segmentation based on object oriented mapping parameter estimation, Signal Processing 15, 315-334(1988)
3. J.W.A Wang, E.H. Adelson: Representing moving images with layers, IEEE Trans. Image Processing 3, 625-638(1994)
4. B. Bergen, P.J. Burt, R. Hingonari, S. Peleg: A three-frame algorithm for estimating two-component image motion, IEEE Trans. Patt. Anal. Machine Intell. 14, 886-896(1992)
5. M. Shizawa, K. Mase: Simultaneous multiple optical flow estimation, Proc. Int. Conf. Patt. Recog., 274-278(1990)
6. T. Komatsu, T. Saito: Extended block-matching algorithm for estimation of multiple image motion, The Journal of the Institute of Television Engineers of Japan 49, 328-335(1995)

Miscellaneous Applications

Speckle Noise Filtering in SAR Images Using Wavelets

L.Böröczky[1], R.Fioravanti[1], S.Fioravanti[2] and D.D.Giusto[3]

[1]MSZKI-KFKI, Hungarian Academy of Sciences, Budapest, Hungary
boroczky@ecse.rpi.edu
[2]Saclant Undersea Research Center, La Spezia, Italy
steve@saclantc.nato.int
[3]Dept. of Electrical and Electronic Engineering, University of Cagliari, Italy
myrddin@elettro1.unica.it

Abstract. This paper presents a novel multiresolution wavelet-based algorithm for filtering SAR images in order to remove speckle noise. The basic idea is to apply to the wavelet coefficients a size-decreasing half-interpolated median filter. The size of the median filter is adapted to the noise energy reduction between the image pyramid levels and different filter shapes are used in each wavelet subbands according to the dominant frequencies. Experimental results showed, that the proposed algorithm results in significant noise removal while the edges are preserved in the images.

1 Introduction

SAR technology has made possible to obtain images at fairly good resolutions when observing a ground scene from aircrafts or satellites, and it can be used to estimate also features like the dampness of the soil, the thickness of a forest, or the roughness of the sea. However, SAR images are contaminated by speckle noise, a chaotic phenomenon that results from coherent energy imaging, and that obscures the scene content and strongly reduces the possibility to observe objects.

Typical noise-smoothing methods are not well suited to preserving edge structures in speckled images; accordingly, many adaptive filters have been developed. Classical operators are based on the local variance statistics [6,4,1]. Also complex processing has been proposed; as the speckle has a different power spectrum than the signal, they can be separated by an operator in the frequency domain [3]. Some of these operator perform better in smoothing speckle in homogeneous areas, others give superior results at the vicinity of edges, but the research in this area is still open [5,8,10].

Due to the inability of the Fourier theory to represent nonstationary signals, wavelet theory was developed and applied intensively also in image processing [7]. For speckle reduction of SAR images, in [9] a wavelet-based approach has been proposed, where the filtered image is obtained by a weighted linear combination of the inverse transforms and the unfiltered image.

The multiresolution method presented in this paper is also based on the wavelet transform, but a modified median filtering, namely a Size-Decreasing Half-Interpolated Median (SDHIM) is carried out for the wavelet coefficients. The reconstruction procedure using these filtered subbands results in smoothed SAR images with sharp edges.

In Section 2, the principles of image decomposition using wavelets are outlined. Derivation of noise energy reduction between levels of wavelet pyramid, and the SDHIM filtering of wavelet coefficients itself is described in Section 3. Experimental results obtained by the proposed algorithm using different SAR images are given in Section 4.

2 The Wavelet Transform

The Fourier transform, the most useful technique for frequency analysis of signals, has undesirable effects if one deals with signals which are localized in time and/or space, due to the fact that sinusoids have an infinite support.

Wavelet transform decomposes an image into basis functions, which are dilations and translations of a single prototype wavelet function:

$$f(x) = \sum_{m=0}^{\infty} \sum_{n=0}^{\infty} c_n^m \Psi_{m,n}(x) \qquad \text{where} \quad \Psi_{m,n}(x) = 2^{-m/2} \Psi\left(2^{-m}x - n\right)$$

This transform can be implemented using band-splitting Quadrature Mirror Filters (QMF) derived from orthonormal wavelet basis.

Image decomposition by a 2D wavelet transform can be done by splitting an image into its low frequency part and the difference signal, which describes the difference between the lowpass image and the actual one. Due to the correlation which exists in the original image, the difference signal has a histogram that is peaked around the zero. The low frequency subband (SLL) image still contains spatial correlation. Therefore, this decomposition can be applied recursively to the lowpass band, in order to obtain a multiresolution image representation. It results in a decomposition of the signal into subbands of equal bandwidth on a logarithmic scale.

This multiresolution wavelet transform results in a compact non-redundant image representation in contrast to the traditional methods, such as lowpass filtering and Laplacian pyramid transform. The traditional pyramid-type wavelet transform recursively decomposes subsignals in the low frequency channels. However the most significant information regarding textures often appears in the middle frequency channels, at certain levels of the decomposition process [2].

At each decomposition level there are three different signals: SHL, SLH and SHH. The SHL shows scale variations in the x-direction and its high values indicate the presence of vertical edges. Large values of SLH and SHH indicate the presence of horizontal edges and corner points, respectively. The noise is reduced and the edges are kept if the noise-cleaning process takes into consideration both the lowpass noise reduction of SLL and the information about the edges given by SLH, SHL and SHH.

3 SDHIM filtering

The SLL image of the wavelet decomposition is a lowpass image of the original one; consequently the energy of noise contained in SLL is less than the one at the previous decomposition level. The relation between the wavelet coefficients and the noise energy reduction is derived for white Gaussian noise in the following.

Let consider, by hypothesis, $x_i^0 = s_i^0 + n_i^0$, where n^0 is an indipendent stocastic process with $E\{n_i^0 n_j^0\} = \sigma^2 \delta_{i-j}$, δ_i being the Kronecker function.

Let x_i^l being the wavelet transform coefficient obtained through the orthogonal filters c_i such that

$$\sum_{k=-\frac{N}{2}}^{\frac{N}{2}} c_k c_{k+2m} = \delta_{2m} \xi$$

$$x_i^l = \sum_{h=-\frac{N}{2}}^{\frac{N}{2}} c_h x_{i+h}^{l-1} = \sum_{h=-\frac{N}{2}}^{\frac{N}{2}} c_h s_{i+h}^{l-1} + \sum_{h=-\frac{N}{2}}^{\frac{N}{2}} c_h n_{i+h}^{l-1} = s_i^l + n_i^l$$

Thesis:

$$E\{n_i^l n_j^l\} = \sigma^2 \delta_{i-j} \left(\sum_h c_h^2\right)^l = \sigma^2 \delta_{i-j} \xi^l \qquad j = i + 2m$$

Proof (by induction):

$E\{n_i^0 n_j^0\} = \sigma^2 \delta_{i-j}$ by hypothesis.

If $E\{n_i^{l-1} n_j^{l-1}\} = \sigma^2 \xi^{l-1} \delta_{i-j}$ then $E\{n_i^l n_j^l\} = \sigma^2 \xi^l \delta_{i-j}$.

$$E\{n_i^l n_j^l\} = E\left\{ \sum_{h=-\frac{N}{2}}^{\frac{N}{2}} c_h \cdot n_{i+h}^{l-1} \cdot \sum_{k=-\frac{N}{2}}^{\frac{N}{2}} c_k \cdot n_{i+k}^{l-1} \right\} = \sum_{h,k=-\frac{N}{2}}^{\frac{N}{2}} c_h c_k \cdot E\{n_{i+h}^{l-1} \cdot n_{i+k}^{l-1}\} =$$

due to the hypothesis

$$= \sum_{h,k=-\frac{N}{2}}^{\frac{N}{2}} c_h c_k \cdot \sigma^2 \xi^{l-1} \delta_{i+k-(j+h)} = \sum_{h,k=-\frac{N}{2}}^{\frac{N}{2}} c_h c_k \cdot \sigma^2 \xi^{l-1} \delta_{(k-h)-2m} =$$

due to the wavelet orthogonality hypothesis

$$= \sigma^2 \xi^{l-1} \sum_{h=-\frac{N}{2}}^{\frac{N}{2}} c_h c_{h+2m} = \sigma^2 \xi^{l-1} \delta_{i-j} \qquad \text{(QED)}.$$

4 The size-decreasing half-interpolated median (SDHIM)

The above presented derivation shows, that for logarithmic version of the images, the multiplicative noise can be considered as additive one. Furthermore, considering some assumptions, in wavelet subbands the white Gaussian nature of the noise is remained. Taking into account these facts, the SDHIM filtering is applied to the wavelet subbands that are obtained by the wavelet transform of the logarithmic image. The speckle noise energy of a certain pyramid level is decreased relative to the previous decomposition level, if the input image is the logarithmic version of the original one. Therefore, the size of median filter applied at each wavelet decomposition level has to be adapted to this relation.

The initial size of the median is decreased by a factor $\alpha = 2\lambda_{nr}$, where λ_{nr} is the noise reduction factor depending on the wavelet coefficients. The factor 2 is due to the subsampling performed at each decomposition level. In order to preserve edges in the original image, different shapes of median filter (horizontal, vertical, diagonal) are used in each wavelet decomposed image, according to the dominant frequency in that particular subband.

To allow for the use of median filters with size inferior to 1×3 or 3×1 pixels, a linear interpolation of adjacent pixels has been performed. In the case of the interpolated median, the output of the filter is the median between the central pixel and two linearly interpolated pixels computed by the relation:

$$i_p = c_p + \left(n_p - c_p\right)s_m$$

where i_p is the interpolated pixel, c_p the central pixel, n_p the nearest real pixel in the chosen direction, and s_m is the is the half size of the median ($s_m \leq 1$).

The final logarithmic image is obtained as a minimum between the original logarithmic image and the reconstructed one, in order to prevent additional noises originated by the wavelet-based filtering. The exponential function of this reconstructed image gives the final filtered image.

5 Experimental results

Experiments were carried out for the proposed speckle removal algorithm using different SAR images. At first, the natural logarithm was taken for each pixel value of the original image. Then, this "logarithmic" image was decomposed by a 2-level wavelet transform using filter coefficients presented in [7]. The SDHIM filtering was applied only to the highpass subbands (SLH, SHL, SHH) with the initial filter size of 1×5. For the lower resolution subband images the filter size was decreased by $\alpha=5/3$, and an interpolation was carried out by $s_m=0.5$. The exponential function was applied to the reconstructed logarithmic images in order to get back the conventional pixel values of the filtered image.

Two original SAR images are shown in Figure 1 and Figure 2. The filtered images resulted by the proposed speckle reduction process are shown in Figure 3 and Figure 4, respectively. In both filtered images an obvious reduction in the speckle can be seen in homogeneous regions with good edge preservations. Furthermore, the histogram of the filtered image in Figure 4 becomes bimodal in contrast to the unimodal one of the unfiltered image in Figure 2 (Figures 5 and 6).

6 Conclusions

A multiresolution filtering algorithm, based on wavelets and consists of a special median filtering of wavelet coefficients, was proposed for speckle removal of SAR images. This novel approach resulted in a significant smoothing in homogeneous areas while edges are preserved in the images. Theoretical investigations of speckle noise distribution in wavelet subbands and its application to the filtering method are topics of our future research.

Acknowledgments

Lilla Böröczky and Roberto Fioravanti acknowledge the European Community Phare-Accord Mobility Grant H9112-0267. Lilla Böröczky acknowledges also the support by the Hungarian National Science Research Fund F014077.

References

1. S.M.Ali, R.E.Burge: New automatic techniques for smoothing and segmenting SAR images. Signal Processing 14 (4), 333-344 (1988)
2. T.Chang, C.-C.J.Kuo: Texture analysis and classification and tree-structured wavelet transform. IEEE Trans. Image Processing 2 (4), 429-441 (1993)
3. R.A.Cordey, J.T.Macklin: Complex SAR imagery and speckle filtering for wave imaging. IEEE Trans. Geoscience and Remote Sensing 27 (6), 666-673 (1989)
4. J.M.Durand, B.J.Gimonet, J.Perbos: SAR data filtering for classification. IEEE Trans. Geoscience and Remote Sensing 25 (5), 629-637 (1987)
5. T.Hosomura, C.W.Jayasekera: Speckle filtering and texture analysis in SAR images. Proc. IEEE-IGARSS'93, 1423-1425 (1993)
6. J.S.Lee: Speckle suppression and analysis for synthetic aperture radar images. Optical Engineering 25 (5), 636-645 (1986)
7. S.G.Mallat: A theory for multiresolution signal decomposition: The wavelet representation. IEEE Trans. Pattern Analysis and Machine Intelligence 11 (7), 674-693, 1989
8. F.J.Martin, R.W.Turner: SAR speckle reduction by weighted filtering. Int. Journal of Remote Sensing 14 (9), 1759-1774 (1993)
9. M.C.Proença, J.-P.Rudant, G.Flouzat: Using wavelets to get SAR images *free* of speckle. Proc. IEEE-IGARSS'92, 887-889 (1992)
10. M.R.Zaman, C.R.Moloney: A comparison of adaptive filters for edge-preserving smoothing of speckle noise. Proc. IEEE-ICASSP'93, V/77-80 (1993)

Fig. 1. Original SAR image.

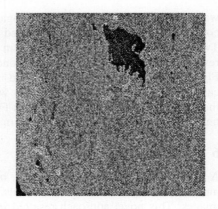

Fig. 2. Original SAR image.

Fig. 3. Filtered SAR image.

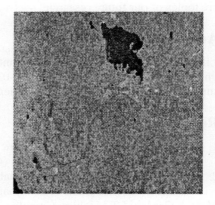

Fig. 4. Filtered SAR image.

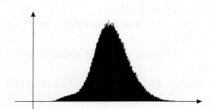

Fig. 5. Histogram of image in Fig. 2.

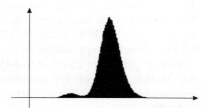

Fig. 6. Histogram of image in Fig. 4.

Prediction of Short-term Evolution of Cloud Formations Based on Meteosat Image Sequences

Raffaele Bolla, Mario Marchese, Carlo Nobile, Sandro Zappatore

Department of Communications, Computer and Systems Science (DIST)
University of Genoa
Via Opera Pia, 13 - 16145 Genova, Italy

Abstract. In this paper an algorithm to predict the short-term evolution of cloud formations with high rainfall probability by using image sequences coming from meteorological satellites (Meteosat images) is described. The proposed algorithm consists of four steps: the first step performs image processing procedures (thresholding and relaxation, edge following, frequency filtering) adapted to a specific environmental application; the second step is dedicated to solve the correspondence problem in different images while the third one deals with the problem of modelling the parameters defining the time evolution of a cloud formation and investigates how these parameters can be estimated. The last step concerns the motion prediction of processed clouds. The main goals of this work are to test the effectiveness of the proposed procedure with different values of the filtering parameters, and to verify the importance and the influence of the mentioned parameters in the prediction mechanism. Some experimental results obtained by using real image sequences coming from Meteosat satellites are shown.

1 Introduction

The framework of this work is the processing of sequences of Meteosat images in order to extract information useful in a decision support system. The goal is to describe the evolution of cloud formations to foresee the rainfall probability in limited areas [1]. Specifically, an algorithm is presented to estimate the evolution parameters of cloud formations characterized by high rainfall probability in a Meteosat sequence. The input to the procedure is represented by the regions of interest, extracted in two or more consecutive images by means of processing steps applied to the infrared data. The proposed technique consists of four steps, as already described in [2]. In the first one specific image processing procedures are applied to the original images in order to extract the regions of interest in each image. Three different types of procedures are used: a first, thresholding and relaxation procedure and an edge follower are applied, then the obtained edges are filtered to detect their relevant features, thus improving the robustness of the following steps. The second step deals with the correspondence problem between regions of interest in different images. The correspondences are therefore determined by minimizing a cost function taking into account the distance among possible regions, as well as their difference in area. In the third step, the parameters defining the time evolution of a cloud formation are estimated. Such parameters are chosen to be those of a linear model, namely a translation vector, a rotation matrix and a deformation matrix. They are estimated by means of an iterative procedure that minimizes the not overlapping surfaces of two corresponding cloud formations. The last step consists of using data obtained from a sequence of images to predict the evolution parameters, i.e. the centre of mass coordinates and the deformation matrix, which are used to compute the rainfall regions of the next two images in the sequence. This task is performed by employing interpolating functions, written as a linear combination of orthogonal functions. The next Section describes the thresholding and relaxation procedures, and the clustering and filtering operations on the available image. In Section 3 the strategy to obtain the corresponding clouds in two consecutive images is described, while the approach to estimate the motion parameters is shown in Section 4. Section 5 presents some experimental results obtained from Meteosat imagery.

2 Meteosat Image Processing

The first step of the image processing concerns thresholding and relaxation. The goal of this step is to extract the regions characterized by high rainfall probability from a sequence of Meteosat images in the infrared band. According with Griffith and Woodley heuristics, the regions whose temperature is below a certain threshold are called clouds. In this way, the so called clouds are associated with a certain rain intensity [3,4]. In order to improve the robustness of the thresholding procedure, a technique, based on a relaxation method, has been applied to obtain a binarized image from the original grey scale IR image. The relaxation is, basically, a stochastic process [5,6] that allows to classify image pixels into two distinct classes (cloud points and background) by taking into account not only the pixel value but also the characteristics of the neighbouring pixels. The aim of the procedure is to correct or to reduce the inherent errors involved in classifying a point according only to its properties. Thus the final binarized image results from removing the isolated cloud points surrounded by background points and from filling the small background areas inside a large set of cloud points. Cloud points identified by the relaxation procedure need to be clustered into some connected regions (called clouds in the following). This operation is performed by using a simple scheme based on the Sobel edge follower algorithm [7]. A blob is classified as a cloud if 90% of inner points is evaluated as belonging to the cloud point set. The last operation of the image processing step is the filtering, aimed at smoothing the contours (by means of low-pass operator in the frequency domain) obtained during the clustering; in fact, a too detailed information about the contours is not necessary for our goal and it could introduce a high computational burden. The filtering employs the Discrete Fourier Transform (DFT) whose mathematical details are described in the following: let us suppose to have N samples of a discrete signal $x(n)$ with $n=0...N-1$; the k-th value $X(k)$ of the DFT of the signal $x(n)$ is

$$X(k) = \sum_{n=0}^{N-1} x(n) e^{-j2\pi kn/N} \qquad k=0...N-1 \qquad (1)$$

In this context the signal to be filtered is the cloud boundary; if the number of values is not a power of 2 (as it has to be to implement a frequency filter), the boundary is oversampled thus obtaining a correct number of samples (let N be this number). A generalized raised cosine function, $F(k)$ in (2), is used to cut-off all the frequencies higher than a fixed threshold f_c.

$$F(k) = \begin{cases} 1 & 0 \le k \le N(1-\alpha)/2\beta \\ \dfrac{1}{2}\left[1 - \sin\left[\dfrac{\pi\beta}{N\alpha}\left(k - \dfrac{N}{2\beta}\right)\right]\right] & \dfrac{N}{2\beta}(1-\alpha) < k < \dfrac{N}{2\beta}(1+\alpha) \\ 0 & k > N(1+\alpha)/2\beta \end{cases} \qquad (2)$$

Then by back-transforming the filtered values of the boundary in the space domain are obtained. The filter allows to choose precisely the cut-off frequency and the smoothing level of the filter. The effect of the frequency filter and his impact on the prediction is extensively described in Section 5.

3 Cloud Identification

After the filtering, a crucial step is the identification of corresponding clouds in two consecutive images. A cloud in two consecutive frames can change position, orientation, scale; moreover, it can split, or merge other clouds generating different cloud formations. More specifically, because of winds and atmospheric currents a cloud can split into two or more parts, thus creating other clouds not existing before; on the other hand, two or more clouds can merge giving birth to one larger cloud. It can be noted that an automatic procedure detecting splitted and merged clouds is very difficult to derive. In our approach the identification is performed in three steps: case of no splitting and merging, case of splitting, case of merging. In the first case the operation is performed by setting a variable side rectangle around the center of mass of a cloud in the first image, then by searching if some centers of mass in the second image fall inside the given rectangle; if more than one center of mass satisfies this condition a cost function is evaluated to choose the 'best' corresponding cloud. The cost function mentioned above is based on: i) the distance between centers of mass, ii) the difference of areas. That is:

$$cost = B * \sqrt{(x_{b,j} - x_{b,i})^2 + (y_{b,j} - y_{b,i})^2} + A * \left| Area_j - Area_i \right| \qquad (3)$$

Being

$x_{b,i}$, $x_{b,j}$,   the x-values of the considered centers of mass in the first and second image, respectively;

$y_{b,i}$, $y_{b,j}$ the y-values of the considered centers of mass in the first and second image, respectively;

$Area_i$, $Area_j$ the area of the considered clouds in the first and second image, respectively

The mechanism is repeated for every cloud inside the first image; if the first step does not identify all the clouds, the case of possible splitting is investigated for not identified clouds. The operation is performed by setting a rectangle around each cloud in the first image; it is important to note that the mentioned rectangle is not set around the center of mass but it contains every point of the cloud. The second operation step is performed by examining if some centers of mass in the second image are inside the rectangle, then every possible combination for all found centers of mass is evaluated and the corresponding cloud is chosen by computing the mentioned cost function for every permutation. After finding the splitted clouds, the last operation is the evaluation of merged blobs. This step is performed by the same mechanism as in the previous case; the difference is that the rectangle is set on the second image and not on the first one.

4 Motion Modelling and Prediction

Before predicting the future position of a cloud it is necessary to model the cloud motion itself. For the sake of simplicity the motion model employed in our approach is linear, meaning that the movement of each point $\underline{x}_i = (x_i, y_i)$ of the cloud can be described [8, 9] as:

$$\underline{x}_i = \underline{d} + F\underline{x}_{i-1} \qquad (4)$$

Being: $\underline{x}_{i-1}$ the position of the point at instant i-1, $\underline{x}_i$ the position of the point at instant i, $\underline{d}$ the translation vector, F the shape matrix. By using the Polar Decomposition Cauchy theorem the shape matrix F can be written as the product of two matrices R and U:

$$F = RU \tag{5}$$

with $R \in O^+$ (positive definite) and $U \in Sym$ (Symmetric matrixes). From a physical point of view, the matrix R can be considered the rotation matrix describing the cloud rotation of an angle θ around an axis through the centre of mass and orthogonal to the plane of the cloud pattern:

$$R = \begin{bmatrix} \cos\theta & \sin\theta \\ -\sin\theta & \cos\theta \end{bmatrix} \tag{6}$$

The matrix U can be interpreted as the matrix containing the information about the deformation of the cloud:

$$U = \begin{bmatrix} u_{11} & u_{12} \\ u_{21} & u_{22} \end{bmatrix} \text{ with } u_{12} = u_{21} \tag{7}$$

The problem is then reduced to the evaluation of the translation vector $\underline{d}$ and the matrix $F=RU$. The vector $\underline{d}$ is computed as the difference of the cloud centers of mass in two consecutive frames, the matrix F is evaluated by a minimization algorithm based on not overlapped surfaces. The estimated parameters associated with the motion and the deformation of a cloud during the transitions between successive frames can be exploited for an efficient prediction of the position and the shape of the cloud in the next unknown frame. The estimation of the center of mass parameters (x_b, y_b), the rotation angle θ and the deformation matrix $U=\{u_{11}, u_{12}, u_{21}=u_{12}, u_{22}\}$ are obtained by using an interpolating function and having as available data: 1) M previous images in successive instants $(t_0, ... t_{M-1})$, 2) M values of (x_b, y_b) for each blob, 3) M-1 values of θ for each blob, 4) M-1 values of U for each blob.

The succession given by the component of the cloud center of mass is interpolated by a suitable function. Once we have the analytical expression of this function we are able to determine the next coordinates of the center of mass by simple substitution of the desired instant. The used strategy involves the use of *orthogonal functions*; for a detailed description of the interpolation method see [10].

5   Experimental Results

In this Section some experimental results are shown, that confirm the effectiveness of the presented prediction strategy. The prediction mechanism has been applied to two sequences of three consecutive images obtained from the same original sequence by using two different cut-off frequencies in the filtering procedure. The sequences contain just a cloud because the aim is to test the prediction mechanism and the presence of other clouds would have made the analysis of the results more difficult. Fig.1 shows the sequence obtained by using the value $\beta=4$ in the filter. Fig.2 and Fig.3 show, respectively, the next predicted image (one step forward prediction), and the image predicted after the next (two steps forward prediction) along with the cloud shape really observed from the METEOSAT images. Results shown in Fig.2 and Fig.3 have been obtained by using the image sequence in Fig.1 as an input. It can be seen that the predicted and the observed shapes are not quite different; little differences can be noted but they derive from the linear model used to describe the motion and deformation of a cloud. It is important to note that the similarity of the shapes is just a qualitative measure of the quality; taking into account the used motion model, for a quantitative evaluation, the analysis of the position of the center of mass and of the rotation angle (reported in the following) are more meaningful. Fig.4 is the image sequence obtained by using the value $\beta=128$ in the filter, that implies a lower cut-off frequency; this sequence is used as an input to get Fig.5 and Fig.6 which have the same meaning as Fig. .2 and Fig. 3. It can be seen that these last images contain a less detailed information then the previous; however ignored details could not be predicted (as can be seen in Fig. 2 and 3) by using the proposed modelling scheme.

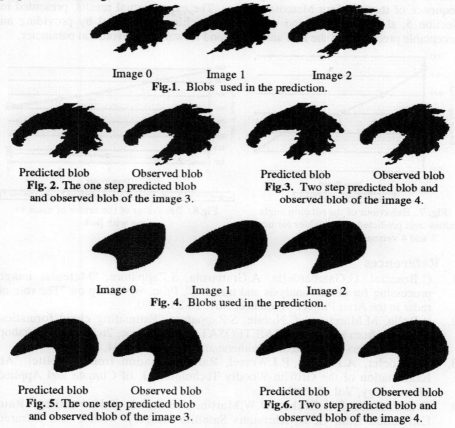

Image 0   Image 1   Image 2
Fig.1. Blobs used in the prediction.

Predicted blob   Observed blob   Predicted blob   Observed blob
Fig. 2. The one step predicted blob   **Fig.3.** Two step predicted blob and
and observed blob of the image 3.   observed blob of the image 4.

Image 0   Image 1   Image 2
Fig. 4. Blobs used in the prediction.

Predicted blob   Observed blob   Predicted blob   Observed blob
Fig. 5. The one step predicted blob   **Fig.6.** Two step predicted blob and
and observed blob of the image 3.   observed blob of the image 4.

In Fig. 7 the behaviour of the observed rotation angle and of the predicted one for images 3 and 4 is depicted versus time; it can be noted that the predicted value is close the observed one, especially in the one step prediction. The same consideration can be made for the behaviour of the center of mass coordinates depicted in Fig. 8. The results shown concern only one situation but they are representative of the average behaviour of the described prediction mechanism.

6   Conclusions

In this paper an algorithm to estimate the evolution parameters of cloud formations characterised by high rainfall probability in a Meteosat sequence has been presented. This algorithm consists of four steps: a first step, described in Section 2, of specific image processing procedures, namely a thresholding, a relaxation, an edge following and a filtering procedure, used to detect the clouds and their relevant features; a second step, described in Section 3, dealing with the correspondence problem; a third step in which the parameters defining the time evolution of a cloud formation are estimated by using a linear model and, finally, the last step in which the prediction of the evolution parameters, i.e. the centre of mass coordinates and the deformation matrix, is computed, as described in Section 4. The aim of this technique was to predict the one or two step forward evolution of the clouds by using a sequence of three or four Meteosat images. The experimental results, presented in Section 5, show that this kind of algorithm achieves this goal by providing an acceptable prediction of the first and the second forward step evolution parameter.

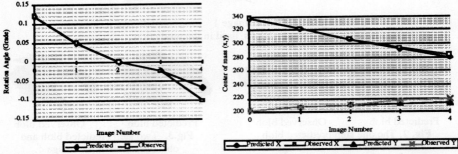

Fig. 7. Behaviour of the rotation angle values and predicted angle values for image 3 and 4 versus time with β=4.

Fig.8. Behaviour of the center of mass vs time with β=4.

References

1. C.Braccini, G.Gambardella, A.Grattarola, S.Zappatore, "Meteosat image processing for clouds analysis and tracking", Proc. Workshop on "The role of radar in the Arno Project", Florence, Nov. 20-23, 1990.
2. R.Bolla, M.Marchese, C.Nobile, S.Zappatore "Estimating cloud formation evolution from sequences of METEOSAT images", Proc. 2nd ACM Workshop on Adv. in Geo. Inf. Systems, Gaithersburg, MD, Dec. 1994.
3. R.F.Adeler, A.J.Negri, P.J.Wetzel, Rain Estimation from Sattelite: "An Examination of the Griffith-Woodly Technique", J. of Climate and Applied Meteorology, Vol.23,1983
4. G.C.Griffith, P.G.Grube, D.W.Martin, J.E.Stoud, D.N.Woodley, Rain Estimation from Geosynchronous Satellite Imagery- visible and Infrared Studies, Mon. Weather Rev. 106,1978
5. A.Touzani, J.G.Postaire "Mode detection by Relaxation" IEEE trans. on Pattern Anal. and Machine Intell. Vol. P.A.M.I. 10, N° 6, 970-978, November 1988.
6. A.Rosenfeld and Russel C.Smith "Thresholding using Relaxation" IEEE trans. on Pattern Anal. and Machine Intell. Vol. P.A.M.I. 3, N° 5, 588-606, Sept. 1981.
7. I.Sobel, "Neighborhood coding of binary images for fast contour following and general binary array processing", Comp. Graph. Im. Proc., vol.8, 1978, pp.127-135.
8. S.Chadhuri, S.Chatterjee, "Motion analysis of a homogeneously deformable object using subset correspondences", Pattern Rec., vol.24, No. 8,1991,pp.739-745.
9. D.Skea at al.,"A control point matching algorithm", Pattern Recognition, vol.26, No. 2, 1993, pp. 269-276.
10. H.F.Harmuth,"Transmission of information by orthogonal functions", Springer-Verlag, 1969, New York, USA.

Prototyping of Interactive Satellite Image Analysis Tools Using a Real-Time Data-Flow Computer

Stéphane Praud[1], Pierre Germain[1] and Justin Plantier[2]

[1] Laboratoire Système de Perception, DGA/ETCA,
16 bis, Av. Prieur de la Côte d'Or, 94114 ARCUEIL Cedex, France
[2] Dpt Sciences Cognitives et Ergonomie, IMASSA-CERMA,
BP 73, 91223 BRETIGNY/ORGE, France

Abstract. Tools for computer-aided satellite image analysis require interactivity, i.e. the capability to modify some parameters and see instantaneously the result of the processing, for efficient work. Due to the amount of data to process that interactivity can only be achieved by parallel architectures. In this paper, we show how a data-flow computer developed in our laboratory can be used to prototype tools for satellite image analysis. Thanks to its high computational power it was possible to implement two complex algorithms without the loss of real-time interactivity : visualization by means of an anamorphosis and image contrast enhancement. Both tools allow to pan across the image and provide smooth and interactive parameters adjustement. The visualization tool enables the image analyst to preserve global information when zooming on a region. Based on human vision characteristics, the contrast enhancement tool lets the image analyst interactively adjust frequency band gains to optimize target perception.

1 Introduction

Computer-aided analysis of satellite image requires interactive tools allowing human analysts to work efficiently. Because of the huge amount of data embedded in satellite images (typical SPOT image size is 6000×6000 pixels), commercially available low-cost workstations perform poorly when faced with low-level processing of such images (visualization, image enhancement, ...). Nevertheless, the fast increase in workstation computational power should enable them to support attractive satellite image analysis tools in the near future.

In order to investigate which algorithms will be relevant for computer-aided satellite image analysis on these future workstations, we propose to implement various tools on a high performance data-flow computer dedicated to real-time image processing and built in our laboratory, the DFFC (see section 2). This massively parallel computer provides real-time interactive tools, because it is able to take into account parameters modification at 25 Hz.

We focused on two satellite image analysis tools: *visualization* by means of anamorphosis and *local contrast enhancement*. The first tool allows to visualize both local and global information in the image by applying an interactively controled anamorphosis operation. Local constrast enhancement is based on human vision and lets the image analyst interactively adjust frequency band gains to optimize target perception.

2 The Data-Flow Functional Computer

The Data-Flow Functional Computer (DFFC) [5] [4] has been developed at ETCA for on the fly processing of digital video streams. The core of the DFFC is a $8 \times 8 \times 16$ mesh-connected network of low-level custom data-flow processors (6×20 Mbytes/s links, 50 millions operations per second). A fine grain task parallelism is exploited by decomposing an application into a a graph of elementary operators that is mapped on the processors network. Additionally 12 "T800" transputers are used for high-level image processing.

A 16×8 data-flow processors plane of the DFFC provides many input/output links. Each processor of this plane provides a 20 Mbytes/s port that is statically configured as an input or an output. These I/O ports are connected to video boards for cameras and monitors, RAM boards, and to the host workstation (via a VME interface). This latter (SPARC 2) is used to configure the DFFC and to enter parameters (threshold, position, images, ...) by means of a graphic interface during the computation.

A programming language called the Data-Flow Functional Programming (DFFP) language, derived from John Backus' FP [1] has been defined as a high level user interface and a complete software environment [6] has been developed. It includes: a DFFP language compiler which translates a DFFP source file into a Data-Flow Graph (DFG) of operators, a mapper that places and routes a DFG onto the physical network and a graphic editor for manual modification or optimization of the physical DFG. A library of operators and macro-functions has been developed for both types of processors. Users may describe their algorithms in a text form or in a graphic form. They may place and route their data-flow graphs themselves using a graphic editor or let the system do it.

The configuration used for implementing image satellite tools on the DFFC comprises a RAM board for storing images to process (image size is currently 1024×1024 pixels) and a video board for displaying processed images on a monitor (image size is currently 512×512 pixels). In this configuration, whatever algorithm is implemented, the DFFC is able to take parameters modification into account and to refresh the screen at 25 Hz.

3 Satellite Image Analysis Tools

Some basic interactive functions are useful and necessary for computer-aided satellite image analysis : pan, image contrast enhancement, zoom, and switching between the initial and processed image.

The first function enables to detect the different interesting sites in the image. The second and third function lead to fit the image to the image analyst's visual characteristics and help him to detect and to identify the sites. The fourth function is useful to compare the two images (initial and processed) and allows the image analyst to avoid analysis errors due to image processing artifacts.

In this paper, two satellite image analysis tools are presented which combine the functions cited before : *visualization* by means of anamorphosis and *local contrast enhancement*. Each tool features a pan facility, i.e. the user can select the region of the image to process at any time.

Thanks to the computational power of the DFFC an image analyst working with these tools does not have to wait when he adjusts some parameters (the visualization screen is refreshed at 25 Hz). The parameters are entered via a graphic interface running on the DFFC host workstation that was designed taking into account ergonomics.

3.1. Visualization by Means of Anamorphosis

Satellite images are so large (typically SPOT : 6000 × 6000 pixels) that current screens cannot display them at high resolution. Most often, visualization tools display only a part of the image depending on the zoom factor and the location selected by the user (pan and zoom functionalities). This approach is not completely adequate as it is known that image analysts need both global and local information. In fact they would like to zoom on a region but continue to see its neighborhood on a large scale. For example, an image analyst examining precisely a crossroad, is also interested in seeing a global view of the network to which it belongs (see figure 1).

We propose to apply an anamorphosis operation to obtain such a morphing effect. It can be seen as the projection of a plane onto a curved surface and results in a continuous decrease of the zoom factor when moving away from the screen center.

The following parameters are available to adjust the anamorphosis effect: the location of the visualization window into the image, the zoom factor (from 1 to 5), and the zoomed region size (from a 20 pixels diameter centered disk to the whole screen).

Algorithm Principle. The location of an output image pixel in the input image is obtained with the following mapping function :
$$map : (x, y) \rightarrow zoom(\sqrt{(x^2 + y^2)}) * (x, y) + (\textit{offsx, offsy})$$
$zoom(d)$ is a continuous function returning a value between 0 and 1, that depends on the zoom parameters entered by the user. This function is designed to obtain replication of pixels by a decreasing factor when moving away from the center of the vizualisation screen [2]. The position of the visualization window in the input image is defined by (*offsx, offsy*).

Finally the pixels, whose size was increased by replication, are smoothed by applying a convolution with a window of the same size . The process of replicating and averaging is similar to a bilinear interpolation.

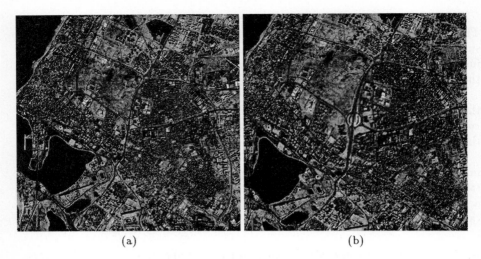

(a)   (b)

Fig. 1. (a): Original image   (b): Anamorphosis with zoom factor 2

Implementation. The implementation uses about 25 % of the DFFC's processors. The algorithm lends itself well to data-flow computation, but requires to compute 16 bits operations (8 bits aren't enough to code the coordinates of pixels) . In fact, its implementation would have been much simpler, if processors were able to compute 16 bits operations.

The host station provides a graphic interface for entering the visualization parameters, and computes the content of two 256 entries look-up tables ($zoom()$, $smooth()$), which are sent to the DFFC. 400 millions of integer operations per second are executed in this application, which prevents from implementing it on a workstation.

3.2. Local Contrast Enhancement

The image contrast enhancement uses a definition of local contrast in complex images. It is based on a spatial and non-linear signal analysis. The original principle has been developped at the IMASSA-CERMA (a french military institute for spatial health) from studies of the human vision [3]. This definition arises from the following visual system properties : logarithmic sensitivity to the luminance and spatial frequency sensitivity.

Algorithm Principle. The wavelet transform provides a multiscale analysis that splits the signal into spatial frequency bands while maintaining the information of localization attached to each component. The local contrast achieves a logarithmic form of this image analysis. As a result of local contrast analysis, such a quantification, modeling visual analysis, leads to determine whether information can be visually perceived, otherwise the information can be fitted to operator's visual characteristics as well as to the task he is involved in.

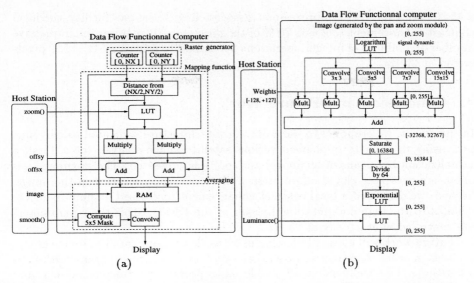

Fig. 2. (a): Anamorphosis Algorithm   (b): Local Contrast Enhancement Algorithm

For the interactive satellite image enhancement, the user can improve selectively the contrast in function of the target size (roads, missile battery, transmission building, ...). For accurate perception of details, it is necessary to amplify specifically high spatial frequencies.

Local contrast enhancement is achieved by applying the following operations to the image. After a logarithmic transformation, a wavelet decomposition is made to obtain images associated with a set of frequency bands. Then a linear combination of the frequency band images, whose coefficients are interactively adjusted by the user, is computed. Finally an exponential transformation is applied, followed by a global luminance and contrast transformation.

Implementation. In our implementation, wavelet decomposition is computed by convolutions rather than by pyramidal wavelet transform. This latter requires less regular data handling (sampling, interpolation) and would be difficult to implement on the DFFC. The wavelets base used are "DOGs" (Differences of Gaussians) functions with 4 different sizes. Therefore it was possible, in order to simplify implementation (only positive weights), to convolve with 4 gaussian kernels of size $3 \times 3, 5 \times 5, 7 \times 7$ and 15×15 and to subtract resulting low-pass filtered images at the linear combination step.

The image analyst can adjust the following parameters to control the local contrast enhancement : the gains applied to 4 spatial frequency bands, and the global luminance and contrast. This kind of algorithm particularly benefits from the implementation on the DFFC, as interactive processing makes easier and faster the adjustment of the set of parameters.

The convolutions represent a major part of the computations and result in a 45 % use of the DFFC's processors. Associated with the pan facility, the local contrast enhancement tool uses 70 % of the processors. It requires the impressive number of 3 billions of integer operations per second to refresh a 512×512 pixels screen at 25 Hz.

4   Conclusion and Future Work

In this paper, we proposed to use a data-flow architecture dedicated to real-time processing in order to implement various satellite image analysis tools. First, a visualization by means of anamorphosis tool was presented. The anamorphosis operation preserves global information while zooming on a region in the image. Second, we described a local contrast enhancement tool based on human vision. Thanks to the high computational power of the DFFC, algorithms parameters are controled interactively at 25 Hz.

Future work will consist in letting users work with our satellite image analysis tools in order to extract some statistics on how they adjust parameters of the algorithms. This feedback could enable us to propose parameters presets, and further to automatically adapt parameters depending on statistical characteristics of the image and on the target of the analysis. We will study the relevance of the different parameters in order to simplify and adapt the algorithms.

Acknowledgements

We are grateful to Véronique Serfaty, Christophe Coutelle and Georges Quénot for their careful reading and suggestions.

References

1. Backus, J.: Can Programming be Liberated from Von Neumann Style ? A Functional Style and its Algebra of Programs. Communications of the ACM **21** (1978)
2. Praud, S., Germain, P., Coutelle, C., Serfaty, V.: Etude et Implantation de Fonctionnalités de P.I.A.O sur le Calculateur Fonctionnel. Etablissement Technique Central de l'Armement (1994), CREA/SP report, n° 94 R 128
3. Plantier, J., Menu, J.P.: Analyse des contrastes locaux de luminance dans les images complexes. Département des Sciences Cognitives et Ergonomie (1993), CERMA n° 93-12
4. Quénot, G., Coutelle, C., Sérot, J., Zavidovique, B.: Implementing image processing applications on a real-time architecture. Computer Architecture for Machine Perception (1993) New Orleans, Lousiana, USA
5. Quénot, G., Coutelle, C., Sérot, J., Zavidovique, B.: A wavefront array processor for on the fly processing of digital video streams. International Conference on Application-Specific Array Processors (1993) Venice, Italy
6. Sérot, J., Quénot, G.M., Zavidovique, B.: A functional data-flow architecture dedicated to real-time image processing. IFIP WGIQ.3 (1993) Orlando, USA

Simulation of Aircraft Images for Airport Surface Traffic Control

Fausto MARTI, Maurizio NALDI

Centro Vito Volterra
TorVergata University of Rome

Abstract. An Air Surface Traffic Control (ASTC) system needs an advanced and integrated surveillance function based on radar sensors for non-cooperative targets. A new generation of Surface Movement Radars is currently being developed that requires the use of tools like radar-image simulators, able to test the performances of such radars. A high resolution radar-image simulator that meets the ASTC requirements is described in this paper.

1. The environment and the application.

Greater safety and efficiency of air transport require an enhanced Air Traffic Management based on a co-ordination among air route traffic control, approach control and surface movement control. In particular, the subsystems, the tools and the operational procedures used for Air Surface Traffic Control (ASTC) have to be integrated, and a better coordination with Air Traffic Control and Management has to be implemented.

ASTC requires an efficient surveillance function to ensure greater safety. There is a growing need for information about precise position and identity of a target, as well as for other target information, which must be reliable and obtained at high rate.

Surveillance is an important function that provides a detailed picture of all the ground traffic (aircraft and vehicles) on the surface area of an airport. The environment of ASTC [2] is divided into a *Movement area* and an *Apron area*. The former includes *approaches*, *runways*, *taxiways* and all *intersections* between taxiways and runways. The latter is a defined area (on a land aerodrome) used to accommodate aircraft to load or unload passengers, mail and for refuelling, parking and maintenance purposes.

The airport environment is characterized by the presence of many tens of aircraft (with dimensions ranging from 10 to 70 m), hundreds of various types of vehicles, (from small cars to large trucks), and numerous obstacles. The number of simultaneously moving targets to be tracked is of the order of five hundreds or more in large airports. Some targets can be cooperative, i.e. carrying equipment for detection and location. Therefore, in the future, the surveillance for ASTC will integrate the use of different types of sensors with data fusion.

A Surface Movement Radar (SMR) is of paramount importance, as it is capable of detecting and locating non-cooperative targets such as obstacles, intruders, etc. The main reason why surveillance is so important for safety is the need for detection of runway incursions and detection of not allowed and dangerous ground movements, such as wrong directions on runways and taxiways. An advanced ASTC system covers a whole aerodrome surface, and provides information about the detection, tracking, classification and orientation of aircraft and vehicles. These functions are used for guidance and control as well as for the detection and management of dangerous movements.

The main characteristics and functions of an SMR to be considered in image simulation can be summarized as follows:

— high tracking precision and class labelling (i.e. aircraft classification depending on dimensions);

— image processing to extract information such as target centroid, target dimensions and target orientation, necessary to implement the tracking algorithms for a large target, like an aircraft, and to improve the performance of the safety logic, which provides alarms or warnings when conflicts are identified on the basis of safety rules.

The image-processing techniques (Moment of Inertia, Hough Transform, Cross-Correlation) that can be applied to digitized radar images of aircraft taxiing on an airport surface are described in [3].

Tests of tracking algorithms and of image-processing techniques require appropriate images resolution and grey-dynamics range, in agreement with the resolution (higher than 3m x 6 m) and the dynamics range (about 80 dB before logarithmic conversion) of the SMR used. In conclusion, a radar-image simulator should provide images of aircraft of an airport surface that are in agreement with the electro-magnetic characteristics of the targets, with the characteristics of next-generation sensors and with the target-sensor positions. The main characteristics of the sensor considered (a millimeter wave SMR) are described in [1] and in [4].

2. Description of the Simulator

This paper describes a simulator of aircraft images for Airport Surface Traffic Control. The simulator aims to test and evaluate:

- radar image processing algorithms;
- plot-extraction algorithms;
- track-while-scan algorithms for extended targets;
- sensor-data fusion.

The motivation for the development of the simulator is the lack of equivalent tools. In fact, some aircraft image simulators exist (e.g., the Radar Imagery Simulator, RIG), by the Technology Service Corporation), but such tools exhibit the following limitations:

a) the electromagnetic model of backscattering is valid up to 16 GHz only, not for the millimeter-wave range;

b) such tools are not easy to integrate with other software for the purpose of research and study.

The software of the proposed SMR implements a simple aircraft electro-magnetic model suitable for the millimeter band (95 GHz). The goal is to provide images of an aircraft like the new images obtained by a new-generation SMR image. The radar simulation is suitable for a monostatic radar, but simple modifications can allow the simulation of a bistatic radar.

It includes computation and graphics routines in C language and utilizes the X Window ToolKit. A simple block diagram is shown in Fig. 1

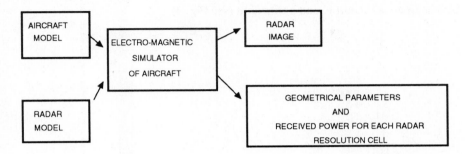

Figure 1 Block Diagram of the simulator.

2.1 Radar Imaging for ASTC

Radar imaging differs from optical (IR, visible, etc.) imaging in two main aspects:

a) Radiation coherence that causes statistical fluctuations (speckle) in time and from pixel to pixel; averaging of independent samples (in terms of time, frequency or polarization) is needed when speckle reduction is required.

b) A polar reference system (range and azimuth coordinates) that makes cross-range resolution vary linearly with the range; in our application, the cross-range resolution is between 0.5 m (at 150 m from the radar) and 5.2 m (at 1500 m from the radar), and the range resolution is 3 m.

2.2 The Radar Model

The features of the radar sensor [1,2] are used to develop the radar model: frequency, antenna rotation speed (r.p.m.), range resolution, azimuth resolution, pulse repetition frequency (PRF), and position of the radar. They are stored in a first data- file at the input to the simulator. Other parameters are derived during simulation.

The transmission and reception of each pulse (or, to reduce the computation load, of 1 out of N pulses) are simulated. The model for simulating the radar while the antenna is rotating is that of a linear system whose impulse response is a sequence of pulses received from a point-like scatterer. Therefore, for each range interval, the imaging process, while the antenna is scanning, corresponds to a convolution of the target reflectivity function over the radar impulse response.

2.3 The Aircraft Model

The second file at the input describes an aircraft (an air carrier or a general aircraft like the Boeing, Airbus, MD, DC and Cesna airplanes). It provides information about its real shapes and dimensions (from AutoCAD files). In order to build up a reasonably simple electromagnetic model of the aircraft, the aircraft surfaces are described by spherical, conical, cylindrical, or flat elements. The aircraft parts that may result in double or triple reflections (e.g., the connections between the wings and the fuselage, the tail wings / rudder, the engine intake inlets) are described by dihedra or trihedra. Therefore, the file is a 3D representation of the aircraft obtained by combining the following geometrical primitives: a cylinder, a truncated cone, a parallelepiped, a dihedron, and a trihedron (see Fig. 2).

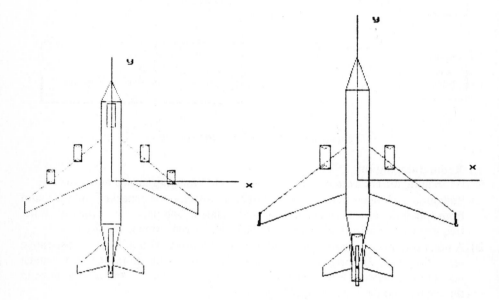

Figure 2 - Section representations of the B747 and MD11 Aircraft

Each geometrical 3D component is described by the coordinates of this component with respect to the reference system of the aircraft. The aircraft (target) is a large and complex object, and it is important that its representation should be easy to the user. The computer code of the simulator automatically exploits the representation of the target and the radar characteristics to identify which components are visible to the radar (at any given aspect angle), by using ray-tracing techniques. In order to describe the interaction between the radiation and the aircraft surface, the latter is modelled as a set of contiguous flat elements (facets) whose dimensions are a trade-off between the need for minimizing the differences with respect to the model surface (i.e., the cylinder, the truncated cone, etc.) and the need for using facet dimensions that are much larger than the wavelength (equal to about 3.3 mm in our application) in order to apply Optical Physics. Typical facet dimensions are about 0.3m, or larger. This information, combined with the aspect angle and with optics-based scattering models, provides the Radar Cross Section, i.e. the backscattering capability (see below) of each scattering center and the power in each radar resolution cell.

For such a target, the main problems lie in representing the model components and in establishing which ones are visible.

The steps to define and build up the aircraft model are the following:

1. analysis of the geometrical structure and of its dimensions;
2. splitting of the aircraft into elementary geometrical 3D components, such as a cylinder, a flat plate, a truncated cone, a parallelepiped, a dihedron, and a trihedron;
3. representation of each component element, with the related scattering law and the related position;
4. definitions of easy data structure files in order not to cause errors on the complex model.

2.4 Output format

The image provided by the simulator measures 64x64 pixels for an area of 128 m x 128 m (i.e., larger than an aircraft). The resolution is 2x2 m /pixel. The output is consistent with the characteristics of the SMR graphic display. In particular, the SMR uses a dedicated window of the work-space to display a surface area of the airport with the aircraft. The output file includes data on the radar position and on the signal power returned from each resolution cell.

3. RCS calculation methods

The Radar Cross Section of a target is a measure of the strength of the scattered field for a given illumination. It can be defined as [5]:

$$RCS = \sigma(\theta, \varphi) = 4\pi \lim_{R \to \infty} R^2 \frac{\left| E^S \right|^2}{\left| E^I \right|^2} \tag{1}$$

where: θ, φ are the azimuth and elevation angles;
            R is the distance between the radar and the target;
            E^S is the scattered electric field;
            E^I is the incident electric field.

A technique that can be applied to a large target is to split it into a collection of smaller scatterers (flat plates). The returns from these can be coherently combined so as to give the total RCS, i.e., the echo of each scattering centre is a complex number that represents both the magnitude and the phase of the return,

$$V_n(\theta, \varphi) = A_n(\theta, \varphi) \, e^{i\psi_n(\theta, \varphi)} \tag{2}$$

with

$$\psi_n(\theta, \varphi) = \frac{4\pi}{\lambda} R_n(\theta, \varphi) \tag{3}$$

The total RCS of each resolution cell is:

$$RCS_{TOT} = \left| \sum_i v_i(\theta, \phi) \right|^2 \tag{4}$$

The model based on physical optics theory is one of the simplest and most robust models for high-frequency scattering. Results on the RCSs of flat and curved surfaces, like cylinders, are available. As an example, the resulting RCS of a flat plate is:

$$RCS_{flat} = \frac{4\pi}{\lambda^2} \left(L_a L_b \cos(\theta) \cos(\phi) \right)^2 \left(\frac{\sin(kL_a \sin\theta)}{kL_a \sin\theta} \right)^2 \left(\frac{\sin(kL_b \sin\phi)}{kL_b \sin\phi} \right)^2 \tag{5}$$

where:

k:	wave number $= 2\pi/\lambda$;
θ:	incidence angle on the horizontal plane;
ϕ:	incidence angle on the vertical plane;
L_a:	dimensions of the flat plate on the horizontal plane;
L_b:	dimension of the flat plate on the vertical plane;
λ:	wave length.

4. Functional structure of the simulator of aircraft images.

Figure 3 describes the fundamental simulation functions. The block diagram (bottom) describes them in more detail.

a) The first step is to define the aircraft with its shape and dimensions, its 3D elementary geometrical components, the type of scattering law for each elementary (a few wavelengths in size) subscatterer (flat plate) in the reference system of the aircraft. This kind of representation is independent of the position of the aircraft with respect to the radar.

b) The second step includes an analysis of the elementary flat plates that are visible to the radar, as a function of the positions of the target and of the radar, and then the calculation of the complex echo $V_n(\theta,\phi)$.

c) Finally, for each position of the radar, the total RCS, (Eq. (4)), and the power of the echo signal for each range and azimuth resolution cell, are calculated. This is accomplished with respect to the reference system of the radar.

The echo power is obtained by:

(I) computation of the RCS of each elementary scatterer inside a resolution cell (Eq. (5) for flat scatterers);

(II) computation of the amplitude of each elementary scattered field as the square root of the RCS, weighted by the 1-way antenna power gain, and computation of its phase by Eq. (3);

(III) sum of all the above computations for each resolution cell to get the total echo power for the assumed geometry.

To simulate the antenna revolution, steps (II) and (III) correspond to a convolution that is made over the reference system of the radar.

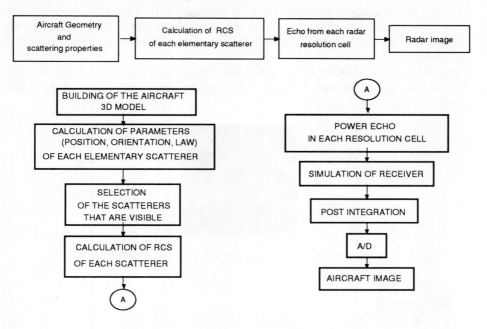

Figure 3 - Fundamental simulation functions.

4.1. Results and computational complexity

For a large aircraft, like the B747 (70 m long and 65 m wide), the number of scattering centres (i.e., the number of facets that form the elementary geometrical 3D components of the aircraft model) is about 70000. The fuselage, the prow, the engines and the tails of the aircraft are divided into flat plates of typical dimensions of 0.3 m.

The 3D model of this aircraft is recorded into a file whose size is about 3.5 Mbytes, as, for each scatterer, it is necessary to record its position with respect to the reference system of the aircraft, the type of scatterer (i.e., the type of scattering law) and the matrix coefficients to pass from the reference system of the aircraft to a local coordinate frame of a facet (local reference system). For this local reference system, the condition of visibility of this scatterer center with respect to the radar sensor is evaluated; this is done by evaluating the possible presence of an intersection between a generic 3D element of the model and the ray that goes from the radar to the scatterer center.

The software includes graphics and independent computation functions in C language and utilizes the Xwindows ToolKit to provide and to manage a Graphical User Interface (GUI): windows suitable for providing the output image, the output data and the description of the input data. The size of the executable software is about 150 Kbytes and the CPU time to obtain an image is less than one second (SUN SPARK CLASSIC 10). Finally, the number of multiplications is about 500 and the number of calls to the C functions (like sin and cos functions) is about 100 for each elementary scattering center. 90 % of the total computation load is due to the resolution of the masking problem, i.e., the calculation of the scattering centers that are visible.

Figure 4 shows an image of the aircraft A310-200 as a simulation output. In this first version, the algorithm used to resolve the masking problem is simplified, so the number of scattering centers is reduced.

Figure 4 - Aircraft image provided by the simulation

5. References

1   G. Galati, M. Ferri, and F.Marti, Advanced Radar Techniques for the Air Transport System: the Surface Movement Miniradar Concept, IEEE National Telesystems Conference, S.Diego, May 1994.

2   G. Galati, M. Ferri, and F.Marti, Distributed advanced surveillance for SMGCS, ECAC-APATSI-EC Workshop on SMGCS, Frankfurt, April 1994.

3   G. L. Foresti, M. Frassinetti, G. Galati, F. Marti, P.F. Pellegrini, C. Regazzoni, Image Processing Applications to Airport Surface Movements Radar Surveillance and Tracking, IECON' 94, Bologna, September, 1994.

4   G. Galati, M. Ferri, P. Mariano, and F. Marti, Advanced Integrated Architecture for Ground Movements Surveillance, IEEE International Radar Conference, Alexandria, Virginia, May 1995.

5   M.I. Skolnik, Introduction To Radar System, McGraw-Hill International Ed., 1993.

Cooperation of Knowledge Based Systems for Galaxy Classification

Jean-Christophe Ossola and Monique Thonnat
INRIA, Sophia Antipolis BP 93, France

Abstract. Astronomical observing systems provide astronomers with a great amount of high quality data. Thus, there is a need for automating data processing. In this paper, we are interested in the galaxy classification problem which is important for universe understanding. The classification consists of two main steps: parameter extraction for object description and data interpretation for object recognition. We present a cooperative architecture based on two knowledge-based systems for automatic classification. The first system is specialized in computer vision and allows parameter extraction. The second is dedicated to galaxy classification. This paper shows our methods and the main components of both knowledge-based system mentioned above.

1 Introduction

This paper presents a system for automatic classification of galaxies based on their images. Most of the current work in data interpretation by computer vision concerns the recognition of man-made objects [4] [5], and takes into account the three-dimensional aspect of objects [6]. In contrast, galaxies are natural complex objects which have no defined geometric models.

Recognizing complex natural objects, requires the cooperation of computer vision techniques, for image description, and data interpretation techniques for object recognition. Artificial Intelligence, with knowledge-based systems (KBS) or neural network [8] may be useful to achieve this task. Neural networks do not reflect knowledge of experts contrary to KBS. Furthermore, KBS can evolve, are flexible and allow knowledge exploitation.

Automation of object classification requires a great amount of knowledge in both computer vision and data interpretation. As this cannot be handled by a single expert, we use a distributed approach to separate different sources of knowledge. Our distributed approach is based on two knowledge-based systems. A schematic view of the global architecture of the system is shown in figure 1. The first stage computes, from the image, numerical parameters describing the galaxy morphology. This phase is performed by PROGAL [2], a knowledge-based system specialized in image processing . The second phase classes the galaxy described by the numerical parameters computed in the first phase, using a taxonomy closed to de Vaucouleurs classification [3]. This phase provides as output the morphological type plus a detailed symbolic description of the galaxy. It is performed by SYGAL [9], a knowledge-based system dedicated to galaxy classification. In the next sections, we will describe PROGAL and SYGAL. Section

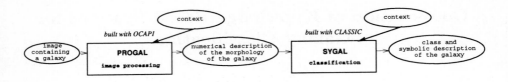

Fig. 1. Synopsis of the global processing of images containing a galaxy

4 gives an overview of the communication protocol and section 5 shows some results. Finally, we conclude with remarks on future works.

2 PROGAL

PROGAL is a KBS built with OCAPI [1], a KBS shell dedicated to program supervision. This means that, given a goal to achieve and a library of programs, pertinent programs are chosen, ordered and their execution is controlled. The PROGAL knowledge base contains several types of useful knowledge described below. A *program description* is associated to each image processing program in the library. It is a description of the program function, input data, output data, tunable input parameters and calling syntax. An important kind of knowledge is also predefined *typical decompositions* of image processing goals into several sequential or parallel sub steps. Rules are explicit descriptions of conditions under which certain decisions are to be taken. Four different types of rules are used. *Choice rules* explicit the way the selection is operated among several programs or sequences of treatment to reach a goal. *Initialization rules* explicit the method used to compute the value of the input parameters of a selected program. *Evaluation rules* explicit how the output data of the programs are assessed. *Adjustment rules* explicit how to modify the value of parameters after a negative evaluation of the results of a program.

The global processing, named *Morphological-description* is described in a typical decomposition into five sub-steps:
(1) Creation and initialization of the file containing the numerical parameters,
(2) Extraction in the image of the object of interest, (3) Computing of global parameters describing the galaxy, (4) Building of 5 iso-intensity contours corresponding to different regions of the galaxy, and (5) For each of these contours, computation of numerical parameters describing them.

Among these steps, two are relatively complex and are decomposed in substeps. *Extraction of the galaxy* is performed as follows: (1) localization of the center of the galaxy, (2) effective isolation of the galaxy and (3) noise removal. *Building of iso-intensity contours* is decomposed into: (1) Creation of thresholded images and (2) Extraction of the contours which are significant.

The knowledge base of PROGAL contains 34 program descriptions, 25 typical decompositions, 12 choice rules, 6 initialization rules, 5 evaluation rules, and 4 adjustment rules.

3 SYGAL

We used CLASSIC [10], a KBS shell, to develop the SYGAL knowledge based system in order to perform classification in the same way experts do, using their usual terminology. CLASSIC facilitates additions of new classes as well as it enables the completion of galaxy descriptions. Models of known object classes are described by *prototypes* which are structured objects organized in a tree reflecting the specialization relations between the different object classes. Prototypes are described in terms of descriptors with symbolic or numerical values. *Inference rules*, organized in rule bases attached to prototypes enable data abstraction. Their role is to allow the passage from numerical input parameters to symbolic values assigned to prototype descriptors. A *fuzzy comparison mechanism* allows to handle noisy, incomplete or imprecise data. To classify an object means to find the closest corresponding prototype or prototypes and to give a correspondence measure.

The inference engine works as follows: At each node of the prototype tree, rules attached to this prototype are activated, the object to be classified is matched against the current prototype. If a prototype is validated, the system selects its sons, if not, the search stops. The SYGAL knowledge base contains 117 rules and 41 prototypes. Examples of a prototype and two rules are shown in the table below:

Prototype SA3		
profile: [0, 10000]	ellipticity: [-.9, 1]	orientation: [-90, 90]
linear-error: [0, 100000]	area: [0, 900000]	shape: spiral
isophotes: smooth, normal distorted	centering: good, average mediocre	resolution: good
bar: absent	arms: incipient	
c1: contour	c2: contour	c3: contour
c4: contour	c5: contour	
RULE 95	**RULE 59**	
Class: contour	Class: galaxy	
if eccentricity < 0.65	if resolution $=$ good	and linear-error > 0.02
then shape $=$ average	then bulge $=$ visible	

4 The communication protocol

In order to make PROGAL and SYGAL cooperate, we use a communication language called CHOOE [7]. It allows to create communication networks between such systems. Communication is made by message sending in different modes (synchronous, asynchronous and deferred). We mostly use the deferred mode: a system waiting for an answer to a message it has sent is still able to answer another message. There is no conflict to manage because each system is highly specialized for a task (image processing or classification) and manages its own control.

An example of message exchanged between PROGAL and SYGAL is shown below:

Name	Classification-request-1	
Communication	mode: deferred	service: object-classification
Content	input-data: NGC7531t.results	
Context	noise: absent	stars: absent
	image-type: density	galaxy-location: unknown

5 Automatic classification

This section presents some results obtained by the global system on several
galaxies, some of which are shown in figure 2.

NGC 7531   NGC 4406   NGC 6936

Fig. 2. Some Galaxies

The external contextual information provided to PROGAL for these images
are different. For example, the first image, galaxy location is unknown, noise
is absent, stars are absent. For the second image, the galaxy is centered in the
image, noise is present, and stars are present. The third galaxy is centered, noise
is present, and stars are present. The classification of galaxies is performed as
follows. Given an image of a galaxy, an external initial request is made on the
image processing goal *morphological-description*. This request triggers the five
previously mentioned sub-steps. (1) The extraction of the object of interest uses
a *choice rule* to choose between two possible decompositions. If we have no a
priori information on the position of the galaxy in the image, we look for the
most important object present in the image. To check that we have actually
detected a galaxy and not a star, an *evaluation rule* compares the size of the
most important object with the size of the maximal star size in the observed
region. If the detection is not correct an *adjustment rule* decreases the threshold
values. (2) Then, we compute five global parameters describing the galaxy: Two
photometric parameters computed using the projected profile along the main
axis (see figure 3) and three geometrical parameters.

(4) The five contours are built (see figure 3). (5) For each contour, six numerical parameters are computed.

Finally, PROGAL provides a set of numerical parameters to SYGAL which
begins the classification process. SYGAL first checks that the input parameters
of each contour are compatible with the description of the prototype *contour*.
Rules attached to the contour prototype are activated and infer new values. Then

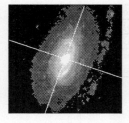 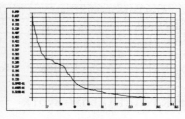 

Extracted galaxy+axes   Projected profile (Main axis)   superimposed contours

Fig. 3. Photometric analysis (NGC7531)

for each of the five contours, the system validates one of the following prototypes: *spiral, average, elliptical.* When the five contours are classified, the galaxy prototype is validated. Once again, the system scans the rule base attached to the current prototype. Rules are activated, infer new values which allow the system to validate one or several prototypes among *spiral, intermediate* and *early.* This process continues until no more rule can be activated or a leaf of the prototype tree is reached.

We have tested this system on two sets of images containing respectively 11 and 8 images. For the first, images set were taken with a low resolution telescope (Schmitt), whereas they were taken with a high resolution one (MacDonald telescope) for the second set. Although the results are quite good, problems remain with estimation of arm size and bar detection. The knowledge base does not contain information about irregular galaxies and about ring pattern structures. Results are shown in the tables below.

Name	Real class	Global result	Best result	Comment
\multicolumn{5}{c|}{Schmitt telescope images}				
NGC 4421	SBS0	lb+ Sb1c	Sb1c	Almost correct
MGC 4540	SA6 SB6	SB3 SB5	SB5	Almost correct
NGC 4595	SA3 SB3	SA3 SB3	SA3 SB3	Correct
NGC 4474	L	la+ lb+ Sa1c Sb1c	la+ lb+	Correct
NGC 4569	SA3 SB3	SB5 SB7	SB5	To improve
NGC 4571	SA7	SA5 SA7	SA5	Correct
NGC 4639	SA4 SB4	SA5 SB5	SA5 SB5	Correct
NGC 4459	la	la-	la-	Correct
NGC 4477	lb	la- lb- la+ lb+	la+ lb+	Correct
NGC 4473	E5	E5 la+ lb+	E5	Correct
NGC 4406	E3	la+	la+	To improve
\multicolumn{5}{c|}{MacDonald telescope images}				
NGC 1433	SB5	SB3 SA3 SB5 SA5	SB3 SB5	Correct
NGC 6300	SB5	SB5	same	Correct
NGC 7531	SA5	SA5	same	Correct
NGC 6946	SA7 SB7	SA7 SB7	same	Correct
NGC 7702	Sa1a	intermediate-b	same	To improve

Name	Real class	Global result	Best result	Comment
MacDonald telescope images				
NGC 3021	SA5	SA5	same	Correct
NGC 4027	SB8	SB5	same	To improve
NGC 4303	SA5 SB5	SA	same	To improve

6 Concluding remarks and future work

We have presented our work on automatic galaxy classification by a distributed system. This system provides the necessary flexibility to process different kinds of images. For example, in [9], all images were taken with a low resolution telescope, and the image shown in section 5 has been taken with a high resolution one. The image processing required in the two cases is different, and our system is able to achieve it in both cases with respect to a certain context. In the future, the system needs to be extended to increase its capability of classification. Owing to our communication mechanism, sending specific requests on image processing programs is possible. For example, there is a crucial need for arm measurement and bar detection algorithms. We are currently working on it. A problem remains for translating quantitative descriptions issued from computer vision algorithms to qualitative description expressed in the data interpretation system. The use of a learning mechanism may provide help in this matter.

References

1. V. Clément and M. Thonnat. Integration of image processing procedures, OCAPI: a knowledge-based approach. *Computer Vision Graphics and Image Processing : Image understanding*, 57(2), March 1993.
2. V. Clément and M. Thonnat. Pilotage de procédures de traitement d'images pour la description morphologique de galaxies. *TS*, 9(5):389–401, January 1993.
3. G. de Vaucouleurs. Classification and morphology of external galaxies. *Handbuch der Physik*, 53:275, March 1957.
4. W. Eric and L. Grimson. *Object Recognition by Computer : The role of Geometric Constraints*. The MIT Press, 1990.
5. G. J. Ettinger. Large hierarchical object recognition using libraries of parameterized model sub-parts. In *Computer Vision and Pattern Recognition*, pages 32–41, 1988.
6. K. Ikeuchi and T. Kanade. Automatic generation of object recognition programs. *IEEE*, 76(8):1016–1035, August 1988.
7. F. Lebastard. CHOOE : un gestionnaire d'environnement distribué. Technical Report 93-22, CERMICS, December 1993.
8. M.C. Storrie-Lombardi, O. Lahav, L. Sodre, and L.J. Storrie-Lombardi. Morphological classification of galaxies by artificial neural networks. *Mon. Not. R. Astronomical Society*, 259:8p–12p, 1992.
9. M. Thonnat. *The world of galaxies*, chapter Toward an automatic classification of galaxies, pages 53–74. Springer-Verlag, 1989.
10. M. Thonnat and A. Bijaoui. *Knowledge-Based Systems in Astronomy*, chapter Knowledge-Based Classification of galaxies, pages 121–159. Springer-Verlag, 1989.

Fuzzy Segmentation and Structural Knowledge for Satellite Image Analysis

Wendling L., Zehana M., Desachy J.

IRIT -UPS, 118 route de Narbonne
33062 Toulouse

Abstract. We present a segmentation method using fuzzy sets theory applied to remote sensing image interpretation. We have developed a fuzzy segmentation system in order to take into account complex spatial knowledge involving topologic attributes and also relative position of searched areas in membership degrees images. A membership degrees image represents the membership degrees of each pixel to a given class and is supposed obtained by a previous classification (involving simple contextual knowledge). To improve this previous classification, we introduce structural rules which allow us to manage with region characteristics. These structural characteristics are obtained by using a fuzzy segmentation technique.

1 Introduction

We present a knowledge-based approach for satellite image interpretation taking into account structural knowledge about searched classes In a first step, we use a locally existing knowledge-based system [1] which gives us membership degrees to each searched class for each pixel. Non-structural expert knowledge describing the favourable context for each class in terms of out-image data (elevation, roads, rivers, types of soils, ...) is at that step already taken into account. This type of knowledge is based on pixel information independently of the neighbourhood.

So we have n images (if we look for n classes) representing the membership function to each class i: $\{\mu_i(x,y)\}$. We have now to introduce expert structural knowledge and out-image data to update the μ_i functions in order to give the final classification. As we have to manage with regions and relations involving different regions or objects and as we have only membership functions μ_i, we produce a fuzzy segmentation of each image $\{\mu_i(x,y)\}$. We have now to produce n sets of fuzzy regions (with their fuzzy geometric attributes) and to compute fuzzy relations involving regions of different classes or objects defined on out-image data.

For each image $\{\mu_i(x,y)\}$ the fuzzy regions are defined by their level-cuts. Note that if a level-cut gives more than one connected crisp region, we define new fuzzy regions. Geometric attributes are computed for fuzzy regions (surface, perimeter, degree of compactness, shape ...). We define relations (inclusion degree, distance, ...) between fuzzy regions of different classes or out-image items (roads, rivers, soils, ...). So our fuzzy segmentation splits the n images $\{\mu_i(x,y)\}$ into hierarchical structures of fuzzy regions with their geometric attributes.

The structural knowledge concerning the searched classes is now introduced in order to update the n membership functions μ_i and finally an improved classification is obtained.

2 Fuzzy Segmentation

2.1 Convex Combination of Sets

Using fuzzy sets theory [7] we can define a fuzzy set by a random sets representation (*also called convex combination of sets*) [2]. This combination is composed of n included crisp sets A_i ($A_1 \supset A_2 \supset A_3 \supset ...$) with $m(A_i)$ corresponding positive weights [5]:

$$\sum_{i=1}^{n} m(A_i) = 1 \qquad (1)$$

Assume n values $\alpha_i \in [0,1]$ and $\alpha_1 < \alpha_2 < ... < \alpha_n$. We can compute $A = \{CUT_{\alpha_1}, CUT_{\alpha_2}, \cdots, CUT_{\alpha_n}\}$ with $CUT_{\alpha_i} = A_i$ is the crisp set obtained with the level-cut α_i. A level-cut is defined as the set of pixels which $\mu(x,y)$ is greater than α_i.

The weight assigned to the set A_i is computed as follows:

$$m(A_i) = \alpha_i - \alpha_{i-1} \text{ and } \alpha_0 = 0 \qquad (2)$$

The membership function of a fuzzy set is obtained from its "convex combination of sets" representation. This property allows us to manage fuzzy regions only by using their level-cuts *(which are crisp regions)*.

2.2 Application on a Membership Degrees Image

Fuzzy Sets Supports. We consider the Euclidean plane of the image as the fuzzy sets referential. Fuzzy sets supports are computed by using a threshold α_1, obtained from the membership degrees image histogram, in order to process only the significant pixels *(fig. 1&2):*

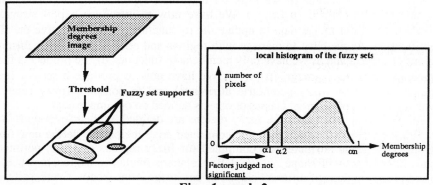

Fig. 1. and 2.

Fuzzy Region. An ordinary region is a set of connected pixels *(with a non-zero membership degrees value)* obtained from a level-cut on the A_1. fuzzy region support. All the crisp sets *(included in the support)* with corresponding weights define a fuzzy region *(fig. 3)*

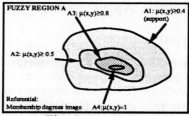

Fig. 3. Fuzzy Region

The level-cuts values can be computed by a constant step or by a thinness threshold as follows *(defining the thinness of each level-cut)*:

$$\alpha_{i+1} = \underset{\alpha > \alpha_i \text{ and } \alpha \leq 1}{Min} \left(\int_{\alpha_i}^{\alpha} H(x)dx \geq Thinness \right) \quad (3)$$

With : α_i: level-cut i; α_{i+1}: next level-cut; H(): local histogram function.

So an image is described as a set of K fuzzy regions and a fuzzy region is defined as a concentric set of crisp regions.

Fuzzy Characteristics. A crisp region A_i is a set of connected pixels. Then we can compute topologic attributes for this region such as: Perimeter, Compactness degree, Moments of order one and two, Surface, ..:

The value of one particular topologic characteristic for each crisp region is computed separately by using a measurable function F. The final measurable function $\tilde{F}$ of the fuzzy region, characterizing the topologic attribute, is obtained by the following formula [3]:

$$\tilde{F}(A) = \sum_{i=1}^{n} m(A_i) F(A_i) \quad (4)$$

We obtain a general aspect of the fuzzy set. Therefore, the image is described as a hierarchical structure of fuzzy and ordinary regions with their geometric attributes.

Fuzzy Relations Between Fuzzy Regions. Moreover we can also determinate fuzzy relations between fuzzy regions such as: distance, at which degree two regions are oriented in the same direction, at which degree one region is included in another one, ...

The realisation degree of a relation between two crisp regions is computed by using a measurable function F. We can also define a measurable function $\tilde{F}$ for the relation between two fuzzy regions [3] A and B:

_ A which is composed by the crisp regions set $\{A_1, A_2, \cdots, A_n\}$

_ B which is composed by the crisp regions set $\{B_1, B_2, \cdots, B_p\}$

- and a function $F(A_i, B_j)$ which links the crisp regions A_i and B_j.

The relation between the two fuzzy regions A and B is computed as follows [3]:

$$\tilde{F}(A,B) = \sum_{i=1}^{n} \sum_{j=1}^{p} m(A_i) m(B_j) F(A_i, B_j) \quad (5)$$

3 Structural Knowledge Application

3.1 Local Rules (Pixels)

Local rules characterize the expert's local knowledge describing the favourable context for each class in terms of out-image data (elevation, roads, rivers, types of soils,...). This type of knowledge [1] is based on pixel information independently of the neighbourhood. For example: "class X: frequently elevation >1000m and on north slope". So we obtain n images (if we look for n classes) representing the membership function to each class i: $\{\mu_i(x,y)\}$ (the membership degrees images).

Let's suppose A is a fuzzy region extracted in membership degrees image for one particular class, we define: $\Pi(A)=\{\alpha_1,\alpha_2, \cdots,\alpha_n\}$ (with $\alpha_0 < \alpha_1 < \cdots < \alpha_n$ and $\alpha_0 = 0$). A level-cut α_i defines a crisp region and the associated weight is computed (section 2.1) to obtain a convex combination of sets.

3.2 Structural Rules

These rules introduce expert structural knowledge and out-image data to update the μ_i functions in order to give the final classification.

For example: "the class X appears *principally* as <u>elongated</u> shapes <u>along</u> <u>big</u> rivers".
We have in this rule three structural information:
    - <u>elongated</u> (characteristic of a region) can be computed from the fuzzy compactness topologic information,
    - <u>along</u> (relation between two regions) can be computed from the distance variations between regions' skeletons (or supports),
    - <u>big</u> (characteristic of a region) can be computed from the fuzzy surface topologic information.
Principally can be considered as the frequency degree for the class to be in this context.

As we have to manage with regions and relations involving different regions or objects and as we have only membership functions $\mu_i(x,y)$, we produce a fuzzy segmentation of each membership degrees image $\{\mu_i(x,y)\}$.

In fact, in our application, after applying the fuzzy segmentation method, we determine topologic characteristics (such as compactness, perimeter,...) and relations (such as distance, adjacency degree...) for two fuzzy regions. Then we introduce the structural information given by the rule concerning the class i in order to update the membership degrees image $\{\mu_i(x,y)\}$. In fact each $\{\mu_i(x,y)\}$ will be modified relatively to the structural knowledge concerning class i.

3.3 Application of the Rules

Assume P is a property (associated to a production rule concerning class C) and $f_p(\ )$ the real function which associates at each crisp region its realization degree.

$\tilde{f}_p(A)$ will be the realization degree for the fuzzy region A.
For example:
    Rule: If <Class C> Then elongated shape
    (Property P: "elongated region").

For all crisp region A_i, $f_p(A_i)$ is the elongated degree of A_i. This degree, which must be a value in the $]0,1]$ set, may be computed from the following ratio: perimeter / surface.

Let's suppose A is a fuzzy region (included in a membership degrees image and characterizing a class C) with $A=\{A_1,A_2,\cdots,A_n\}$ the level-cuts set on A. So, we define the weight function $m_p(\ )$ as follows:

$$\forall\ A_i\subseteq A$$
$$m_p(A_i) = f_p(A_i) \qquad (6)$$

In our method, all the new weights are normalized ($[0,1]$ set). So the combination $m(A_i)$ by $m_p(A_i)$:

- increases the A_i regions' weights which best verify the property P,
- reduces the A_i regions' weights which worst verify the property P.

We modify the weight attached to each crisp region:

$$\forall\ A_i\subseteq A$$

$$m'(A_i) = \frac{m(A_i)\cdot m_p(A_i)}{\sum_{A_j\in A} m(A_j)\cdot m_p(A_j)} = \frac{m(A_i)\cdot f_p(A_i)}{\sum_{A_j\in A} m(A_j)\cdot f_p(A_j)} = \frac{m(A_i)\cdot f_p(A_i)}{\tilde{f}_p(A)} \qquad (7)$$

So, the weight assigned to a crisp region A_i which verify property P will be increased.

In the formula (7), the expert knowledge is considered as sure. But, generally, we have both an expert rule and a frequency degree β for the rule ($-1\leq\beta\leq1$; from -1 for "never" to 1 for "always"). Then we define a new weight function to manage with this frequency degree β:

$$\forall\ A_i\subseteq A$$
$$m_p(A_i) = f_p(A_i)^\beta \qquad (8)$$

We obtain a new fuzzy region:

$$\forall\ A_i\subseteq A$$

$$m'(A_i) = \frac{m(A_i)\cdot m_p(A_i)}{\sum_{A_j\in A} m(A_j)\cdot m_p(A_j)} = \frac{m(A_i)\cdot f_p(A_i)^\beta}{\sum_{A_j\in A} m(A_j)\cdot f_p(A_j)^\beta} \qquad (9)$$

At this step, we use a MYCIN-based combination function [6] F_{mycin} (see annex) to obtain the final fuzzy region:

$$\forall A_i\subseteq A,\ m''(A_i)=\frac{1+F_{mycin}\left(2m'(A_i)-1,\beta\cdot\tilde{f}_p(A)\right)}{2} \qquad (10)$$

The final weights $m''(A_i)$ associated to all crisp regions A_i are defined with the MYCIN-based combination [6] and from the two following values:

- $m'(A_i)$: weight for region $A_i\in[0,1]$

- $\beta\cdot\tilde{f}_p(A)$:   - β frequency degree of the rule.

    - $\tilde{f}_p(A)$ realisation degree of property P for the fuzzy region A.

4 An example of Fuzzy segmentation and structural knowledge application

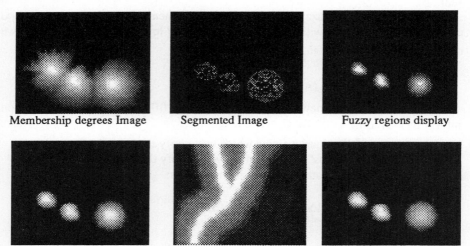

Membership degrees Image   Segmented Image   Fuzzy regions display

If <Class C> Then compacted   Image: "Distance to roads"   New membership Image
shape - A -   If <Class > then roads
(+ property "compacted region").   (combined with - A -)

5 Conclusion

We have proposed a new method to introduce structural knowledge for image interpretation based on fuzzy logic and fuzzy segmentation. Our method splits the membership degrees images in a set of atomic fuzzy regions. In a second step, we use structural rules to update the membership degrees images and then to give final classification. We have applied this method on geocoded images and achieved interesting results. The application to satellite image classification is going on.

References

1. J. Desachy : "A knowledge-based system for satellite image interpretation.", 11th International Conf. on Pattern Recognition. The Hague, The Netherlands, 1992.
2. D. Dubois, M.C. Jaulent : "A general approach to parameter evaluation in fuzzy digital pictures" Pattern Recognition Letters 4, 1987 pp 251-261, 1987.
3. D. Dubois, H. Prade : "Possibility theory, an approach to the computerized processing of uncertainty" Plenum Press - New -York 1988.
4. A. Rosenfeld : "The fuzzy geometry of image subsets" Pattern Recognition Letters 2, pp 311-317, 1984.
5. G. Shafer: "A mathematical Theory of evidence", Princetown Univ Press 1976.
6. E.H. Shortliffe, B. Buchanan : "A model of inexact reasoning in medicine". Mathematical biosciences 23 pp 351-379, 1975.
7. L. A. Zadeh, "Fuzzy sets", *Readings in Fuzzy Sets for Intelligent Systems*, edited by D. Dubois, H. Prade & R.R. Yager, Morgan Kaufmann Publishers, San Mateo, California, pp 27-64, 1993.

Detection of Lofar Lines

J-C Di Martino and Salvatore Tabbone

Crin/Cnrs & Inria-Lorraine, BP 239, 54506 Vandoeuvre-les-Nancy, France
e-mail dimartin@loria.fr, tabbone@loria.fr

Abstract. The problem of extracting spectral lines from sonar images is treated using a complete scheme that combines an edge detector with a line tracking process. That is, the edge detection process allows filtering the sonar image and finding a region (zero-crossings of the second derivative of the sonar image) that includes a spectral line. The founded region is then used to initialize and to limit the extent of the line tracking process. Comparative performance results confirm the efficiency and the reliability of the proposed approach.

1 Introduction

The fundamental objective of a passive sonar system is to detect the presence of target-like signals in underwater acoustic fields. Signals emitted from the acoustic source are recorded by an array of hydrophones, beamformed and spectrum analysed. Narrow-band components of the signals' result provide an image of frequency power versus time, more commonly referred to as a "lofargram". Lofar is an acronym for LOw Frequency Analysis and Recording. Its main characteristics are:

- Signal to noise ratio is low due to the discreteness of the sources.
- A constant frequency tonal produces a darkening over time periods, which appears as a vertical line on the lofargram display.
- Multi-lines can be displayed on the lofargram depending on the spectrum band-width.
- Slanted lines or slightly curved lines can be shown on the lofargram. They are due to the Doppler effect.

By characterizing the spectral lines on a lofargram, one can determine the acoustic source of the sound.

Several methods have been proposed in the literature to extract lines from images. Our review is not intended to be exhaustive, but some of the related works represent important advances in this area, and we have experimentally tested some of them.

Abel *et al.* [1] perform a detection step using a statistical likelihood test. Then mathematical morphology operators are used to extract regions which encompass lines. This method, well tested on sonar images, requires however a priori information regarding the noise environment and the number and type of sinusoids present.

Other contributions in the field of curved line extraction are proposed by introducing perceptual considerations. All these methods rely on a two-step process [12, 10]. Globally, these methods give good results for constant curvature curves but they are not well adapted to fluctuating curves (curvature sign often changes) such as spectral lines.

Neural network approaches are also an important field of methods that have been proposed to deal with the line detection problem. However, these methods [5, 6], based on supervised learning neural networks, need a learning set that reduces their utility in real cases.

Line detectors, based on a differentiation operation [14] or on a local energy computation [4], can also be provided to extract spectral lines. Since the images are very noisy, these approaches require a smoothing stage. High smoothing removes noise and fine structures too. These approaches are not well adapted to lofar images because we need to restore very fine structures.

The work described in this paper is toward automatically processing lofargrams in order to suppress the noise and to detect the spectral lines. This process is very important because the quality of further automatic processing such as classification or tracing acoustic sources relies on the quality of the extracted spectral lines. Thus, we propose to combine an edge detector with a line tracing process. The edge detection process is used to initialize the tracing process.

2 Our approach

The different steps of our approach are described in figure 1. Spectral lines are extracted by the detection process in order to find surrounding regions. Then, a tracing step is executed in each region to find the lines accurately. We describe these three stages more precisely in the following sections.

Fig. 1. Approach scheme.

2.1 Detection process

The detection process consists of detecting and bridging spectral line contours in order to extract regions. The line detection, further described, is achieved in a two-step process: smoothing and differentiation.

Line detection

Smoothing. Noise in lofargram images is essentially characterized by high frequency components which can be reduced by a smoothing step. The Gaussian ensures an optimal compromise between the detection and localization errors:
$g_\sigma(x,y) = \frac{1}{2\pi\sigma^2} e^{-\frac{1}{2}\frac{x^2+y^2}{\sigma^2}}$ where σ is a standard deviation.

Edge detection. The spectral lines in a lofargram can be modelled by peak profiles. To perform the detection, we run a detector based on steerable filters designed in quadrature pairs as described by Freeman and Adelson [4]. The method consists of calculating the response of an arbitrary linear filter by a linear combination of known filter responses, which will in general be significantly faster than computing the convolution of a rotated filter with the image (see [4] for more details).

The detection first computes in each image point a local dominant direction θ_d (the angle of maximum response of $E(\theta)$). This is performed by defining a local energy function $E(\theta) = \left[G^\theta\right]^2 + \left[H^\theta\right]^2$.

Once this is done, we can easily compute the second derivative of the image in order to enhance the lines and also to locate regions encompassing these lines. This is simply achieved by taking the appropriate combinations of the G basis filter outputs, and adaptively steering them along the local direction of dominant orientation. After finding, in the enhanced image, the 2D local maxima in the direction perpendicular to θ_d, the false contours are suppressed using an hysteresis threshold [2]. No additional filtering is required for this step and the whole process (*i.e.*, finding the local dominant direction and the derivation of the image) involves only one pass of the image.

Gap bridging Because of its local nature, the detection process is not able to provide continuity and coherence of image curves. In our case, the major drawbacks of this process result in:

- *gaps*, where weak but significant information is missed due to thresholding errors and intensity discontinuity stemming from various and complex physical phenomena ;
- *deviations in position and uncertainties of orientation* where local estimates of these quantities are adversely affected by noise.

Our aim is to restore continuous curves as long as possible. A linking process is first achieved on the thresholded image in order to obtain elementary curves, called tokens. Each token, defined by a linked edge points lying on the same curve, is characterized by a local dominant orientation at each end of the tokens and its main direction. The grouping step relies on two fundamentals concepts:

1) A strategy of association based on an iterative techniques of grouping by successive refinement [7, 13]. The process is iterated on a set of tokens and ends when no association is achieved at one stage.

2) A set of criteria that ensures a good robustness of the grouping process: proximity, curvature consistency and overlapping (see [13] for more details).

Discussion We have shown a global scheme to efficiently detect spectral lines embedded in a very noisy background. Nevertheless, even if they are well detected, fine structures are blurred by the smoothing step and so are not well localised. We propose to tackle this problem by first locating regions that encompass the contours and then using a tracing algorithm to accurately find the lines in each region.

2.2 Region locating

This process aims to provide the tracing process with a limited search space where we have a high chance of detecting lines. Such an approach enables us to avoid initialization and solution tree overlapping problems inherent in multistage decision processes. The main idea is to determine for each contour the boundaries of the smallest region it encompasses [13]. We first compute the zero-crossings of the second derivative of the image on each side of the contours (zero-crossings are detected in four direction scans). This information cannot be directly used as boundaries of regions because of effect of noise that can produce wrong zero crossings. Our solution consists of estimating each region width by calculating the mean distance from the curve it contains and its zero crossings. A mathematical morphology operator is thus ran to dilate each curve until we obtain the desired width. This method does not insure that a region contains only one curve. It is always possible that some closed spectral lines will be merged by the smoothing step into one contour. In our experiments, we have choosen the smoothing parameter σ equal to 3 in order to have good noise reduction and to be able to distinguish two curved lines 5 pixels from each other.

2.3 Tracing guided by contours

A multistage decision process [8, 3], based on the optimisation of an objective function Φ, is used to extract from each region R the spectral line going through it. The problem consists of finding an optimal path from a set of graph nodes defined by the pixels belonging to a region. This operation is achieved on the original image. The fundamental property underlying this method is that any optimal path between two nodes of a graph has optimal subpaths for any node lying on it. Thus the optimal path between two nodes P_A, P_B can be split into two optimal subpaths $P_A P_i$ and $P_i P_B$ for any P_i lying on the optimal path $P_A P_B$. The function Φ is chosen so that the values for a path increase as the path's total amplitude increases and as its global curvature decreases:

$$\Phi(\mathcal{C}) = \sum_{i=1}^{N} a(P_i) - \alpha \sum_{i=2}^{N-1} |s(P_{i-1}, P_i) - s(P_i, P_{i+1})|$$

where $a(P_i)$ is the amplitude of P_i, $s(P_i, P_j)$ is the slope of the segment $[P_i, P_j]$. The gain coefficient α allows more or less local distortions to be obtained. In our experiments, its value was fixed to 3 in order to allow tracing very fluctuating lines.

3   Experimental results

Our method has been tested on a set of real lofargrams with different signal to noise ratios. The experimental results show that this method is well suited to detect spectral lines accurately, even with a low signal to noise ratio. Figure 2 shows the main steps of our method on a noisy image containing five spectral lines.

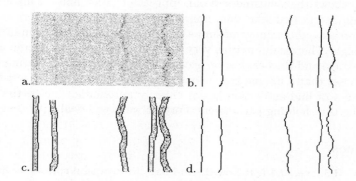

Fig. 2. Main steps of our line extraction method.

a) Real lofargram. b) Result of line detection. c) Regions encompassing the detected lines. d) Extracted spectral lines.

We have compared the location performance of the detection process (edge detector) and the detection combined with the tracing process (tracing guided by contours). Line location accuracy is assessed by the figure of merit proposed by Pratt [11] : $F = \frac{1}{max(I_I, I_A)} \sum_{i=1}^{I_A} \frac{1}{1+\alpha d^2(i)}$ where I_I is the ideal number of edge points, I_A is the actual number of edge points, $d(i)$ is the shortest distance of the ith actual edge point to an ideal edge point, and α is a positive constant. F is maximum (that is, $F = 1$) when $I_I = I_A$ and $d(i) = 0$ for all i. Table 1 shows results obtained from a set of images containing spectral lines with increasing spatial frequency (increasing fluctuation of the curves). We can see that, globally, our approach performs well, especially when the spatial frequency increases.

Spatial frequency	0.10	0.20	0.30	0.40	0.50	0.60
Detection process alone	0.90	0.90	0.89	0.85	0.82	0.81
Tracing process guided by contours	0.94	0.95	0.95	0.94	0.92	0.88

Table 1. Line-location figure of merit as a function of spatial frequency.

4 Conclusion and perspectives

We have proposed a complete scheme of processes to perform the spectral line extraction from lofargram images (sonar images). Our method is built on the cooperation of two techniques which are for our purposes complementary: the detection of contours and the tracing process. Comparative results are presented that show the performance of the tracing guided by contours versus the contour detection alone, especially when fine structures are to be detected (high spatial frequency). A set of experiments performed on images with different signal to noise ratios shows the robustness of our approach against noise. This method is currently applied to real lofargram images in a real-time situation.

At the present, the amount of smoothing, determined experimentally, is chosen to be high. The results is that very near spectral lines can merge after the smoothing stage and thus cannot be accurately localized by our tracing process, which supposes that a unique line belongs to each region. In other respects, our global schema of line extraction is not enterily automatic, due to the manual choice of the thresholding parameters. Further work will deal with these issues.

References

1. J.S. Abel, H.J. Lee, and A.P. Lowell. An Image Processing Approach to Frequency Tracking. In *Proceedings of IEEE ICASSP*, volume 2, San Francisco USA, 1992.
2. J. Canny. A Computational Approach to Edge Detection. In *IEEE Transactions on PAMI*, 8(6):679–698, 1986.
3. J. C. di Martino, J. P. Haton, and A. Laporte. Lofargram line tracking by multi-stage decision process. In *Proceedings of IEEE ICASSP*, Mineapolis USA, 1993.
4. W.T. Freeman and E.H. Adelson. The design and use of steerable filters. In *IEEE Transactions on PAMI*, 13(9), September 1991.
5. A. Khotanzad, J. H. Lu, and M. D. Srinath. Target Detection Using a Neural Network Based Passive Sonar System. In *Proceedings of IJCNN*, 1:335–340, 1989.
6. N. Leeming. The use of Artificial Neural Nets to Detect Lines in Lofargrams. In *Proccedings of UDT*, 1:48–53, 1993.
7. R. Mohan and R. Nevatia. Segmentation and Description Based on Perceptual Organisation. In *Proccedings of IEEE CVPR*, San Diego USA, June 1989.
8. U. Montanari. On the Optimal Detection of Curves in Noisy Pictures. In *Communications of the ACM*, 14:335–345, 1971.
9. R. Nevatia and K.R. Babu. Linear Feature Extraction. In *CVGIP*, 13:257–269, 1980.
10. P. Parent and S.W. Zucker. Trace Inference, Curvature Consistency and Curve Detection. In *IEEE Transactions on PAMI*, 11(8):823–839, 1989.
11. W.K. Pratt. *Digital Image Processing*. Wiley Interscience, second edition, 1991.
12. A. Sha'ashua and S. Ullman. Structural Saliency: The Detection of Globally Salient Structures Using a Locally Connected Network. In *Proceedings ICCV, Tampa, Florida USA*, pages 321–327, 1988.
13. S. Tabbone. Cooperation between edges and junctions for edge grouping. In the *First IEEE ICIP*, Austin, November 1994.
14. D. Ziou. Line Detection Using an Optimal IIR Filter. In *PR*, 24(6):465–478, 1991.

A Fast Object Orientation Estimation and Recognition Technique for Underwater Acoustic Imaging

Alessandra TESEI, Andrea TRUCCO, and Daniele ZAMBONINI

Department of Biophysical and Electronic Engineering (DIBE) - University of Genoa
Via All'Opera Pia 11/A, I - 16145 Genova (Italy)

Abstract. This paper describes a technique for automatic fast orientation estimation and recognition of man-made objects present in underwater acoustic images. A method able (a) to estimate the orientation of the object principal axis, and (b) to recognize the object itself by using information about boundary segments and their angular relations, is presented. This method is based on a simple voting approach applied directly to the edge discontinuities of an image. Often, in underwater acoustic images some undesired effects are present, such as halos around objects, speckle that creates discontinuities, and distortion. Voting approach is very robust with respect to these effects, so allowing good results also when images have a very poor quality. A sequence of real acoustic images is shown for testing the proposed technique validity.

1 Introduction

In modern underwater missions, the use of automatic or autonomous devices is in constant increase, in order to avoid that human operators are directly involved. The analysis of the acoustic images, acquired at the mission site, is a very heavy task for human eyes both for the long duration of the mission and for the poor quality of such images. So, algorithms able to recognize a wanted object on the basis of *a priori* known model and to estimate its orientation are very useful.

We address the problem of fast recognition and orientation estimation of man-made objects present in 2D images particularly acquired by acoustic camera for a wide number of different tasks. One can mention inspection, survey, manipulation and positioning in offshore structures or mineral mining sites, along pipelines or communications cables, relocation of lost objects, identification and removal of toxic wastes, mine counter measures, and so on. For man-made objects we consider objects having a regular and well-defined structure. The term "regular" means the possibility to describe the object by means of geometric surfaces bounded by straight lines.

In literature, some recognizer were suggested for other kind of acoustic images (i.e. side-scan images) [1] but their complexity is high and real-time functioning is very difficult. Other authors [2,3,4] proposed real-time operations on underwater acoustic images, but they limited their goal to object detection.

We present a method able to estimate the orientation of an object by searching for the direction of its principal axis, and to recognize the object itself by using information about boundary segments and their angular relations. This method is based on a simple voting approach [5] applied directly to the edge discontinuities of an image. Two different levels of increasing complexity are used to detect important image features (e.g., straight lines) with different accuracy. The first level is very easy and allows to estimate the object orientation quickly. The second level is computationally more complex as allows to obtain both more accurate orientation estimation and fast object recognition [6]. The proposed method is able to discriminate between different sets of man-made objects, thus making the recognition process invariant to translation and scale, and the orientation estimation process also invariant to rotation.

2 The Voting Approach

In general, voting methods are very robust to noise and are not affected by shape partial occlusions [7]. In underwater images some undesired effects are often present, such as halos around objects (due to poor camera resolution), speckle creating discontinuities in the objects (due to coherent interference among waves back-scattered from the scene), and distortion (due, above all, to inhomogeneities of the medium). The proposed approach is very robust with respect to these effects.

The voting method maps each feature space point $e \in E$ into a set of parameter space points $\pi \in \Pi$ by means of a generating equation $G(e,\pi)=0$, where E and Π represent the feature and the parameter spaces, respectively [8]. Let us consider the edge image, $E(x)=E(x,y)$ as a 2D feature space of x and y coordinates (i.e., $e=x=(x,y)$), obtained by applying a Canny filter [9] to the original image $I(x)=I(x,y)$. An image reference system (X,Y) with origin O in the bottom left corner of the image is chosen. Each edge point x in this space votes for all 2D parameter space points which satisfy the voting equation G. The voting algorithm consists of two processing levels: (a) object orientation estimation, and (b) object recognition.

2.1 Object orientation estimation

The voting process is performed directly from the edge-image space E into a 1D parameter space $\Pi = \{\alpha\}$, composed by all possible angle values α associated with object axes. As shown in Figure 1, α is the angle between the normal to the object axis and the X-axis of the image reference system.

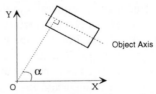

Figure 1. 2D reference system for object orientation estimation.

Let Θ the upper limit of the parameter space (e.g., $\Theta=180°$) and $\Delta\theta$ the quantization step, thus the parameter space dimensions are given by $\Theta \cdot \Delta\theta$. In order to obtain an orientation accuracy equal to 1° it is necessary to choose $\Delta\theta=1$. The relation G is 1D and considers rectilinearity and gradient similarity. Two edge points, x_1 and x_2, belonging the same straight line and with similar gradient orientation $\gamma(x)$ ($|\gamma(x_1) - \gamma(x_2)| \le \gamma_{th}$) vote for the same orientation value α [10]:

$$G(\mathbf{x}, \alpha) = \begin{cases} 1 & \forall\, \alpha \in \Delta\gamma(\mathbf{x}) \\ 0 & otherwise \end{cases}$$

(1)

where $\Delta\gamma(x) = [\gamma(x) - \gamma_{th}, \gamma(x) + \gamma_{th}]$, $\gamma(x)$ is the gradient orientation at the point x and γ_{th} is a local threshold. For each edge pixel x in the image domain, which satisfies the relation $G(x, \alpha)$, an accumulator array $C(\alpha)$ is incremented. At the end of the voting phase, an histogram reports, for each α value, the number of votes of the correspondent cell $C(\alpha)$. The orientations α^*, common to several edge points into the image domain (e.g., object borders), generate multiple peaks in the parameter space. Each peak will be as tall as great the edge point number will be.

A more robust object orientation procedure can be obtained by considering the angles between the object axes. A set of angle relations corresponding to two or more model object axes can be easily computed by:

$$|\alpha_i - \alpha_j| \cong \delta_{ij} \quad \forall\, i, j \in [1, J]$$

(2)

where α_i represents the orientation of the i-th model object axis and J is the number of model object axes. A set of mobile windows $W = \{W_j,\ j=1,..,J\}$ (one for each axis of the object model) of inter-distance $\delta_{ik} = \delta_{ik} = \|\omega_i - \omega_k\|$, where ω_i is the medium point of W_i, and size δ_{th}, are shifted on the histogram to detect candidate peaks, whose position satisfies the angle relation among the 2D object model axes. Candidate peaks related to the same object are grouped into different sets $P_k = \{\alpha_{i,k}:$ i=1,..,J\}, k=1,..,K. In this way, the probability of detecting false maxima because of both false peaks and noise data is reduced. The object orientation will be computed as higher maximum $\alpha_{i,k}^*$ belonging to the group P^*_k whose candidate maxima received the highest number of votes. P^*_k is:

$$P_k^* = \left\{ P_k : \sum_{i=1}^{J} C(\alpha_{i,k}) > \sum_{j=1}^{J} C(\alpha_{j,h}) \quad k,h = 1,..,K\ k \neq h \right\}$$

(3a)

Then, the global maximum α^* (i.e., object orientation) is given by:

$$\alpha^* = \{\alpha_i : C(\alpha_i) > C(\alpha_j)\ \alpha_i \in P^*_k, \alpha_j \in P^*_k, i,j = 1,..,J, i \neq j\}$$

(3b)

2.2 Object recognition

The object recognition process is based on the shape information contained in the object boundaries. Our approach is restricted to regular (man-made) objects, whose boundaries are straight lines. To this end, a complex process able first to extract from the edge image a set $S = \{s_m : m=1,..M\}$ of straight lines and to match them with object model segments contained in a data-base is applied. In order to reduce the process time-consume, a preliminary phase able to focus attention on the most probable object boundaries is performed: a 1D voting step like that used in section

2.1 is applied. A label l(α) is used to mark the edge points corresponding to the histogram peaks. The angles between the boundary segments of the 2D model object are used for selecting the histogram peaks α_p^*, p=1,..P, containing votes coming from segments. P is the number of boundary segments. Then, a 2D voting step, restricted to the edge points characterized by a label l(α_p^*), is performed.

Let (X', Y') a new reference system with origin O' in the center of the image. Each straight line can be parametrically described by polar coordinates (ρ,θ) [8] where ρ is the normal distance of the line from the origin O' and θ is the angle between the line and the x axis. The mapping process is performed from the edge image space E into a 2D a coupled space $\Lambda = \{[\pi, \mathbf{p}_j(\pi)]: \pi \in \Pi, j \in [1..J(\pi)]\}$, where $\pi=(\rho,\theta)$ is a parameter-space point representing a straight line s(π) on the image plane, and $\mathbf{p}_j(\pi)=[x_j^{min}(\pi),x_j^{max}(\pi)]$ is a pair of edge points representing the endpoint coordinates of a straight segment of s(π). J(π) is the number of straight segments belonging to s(π). The 2D parameter space is quantized into cells of size $\Delta\rho$ and $\Delta\theta$ chosen as explained in [11] for reducing the spreading vote effects. This is accomplished by defining a 1D reference system S(π)=S(ρ,θ) for each straight line s(π), whose origin O_S is placed at the foot point of the normal.

The voting method proposed is an improvement over classical voting representation [7], as it allows edge points to be associated with different, adaptively chosen straight segments $s_j(\pi)$. Each collinear segment $s_j(\pi)$ is defined as a set of edge points $\mathbf{x}$ on S(π) satisfying the generating equation and the two constrained endpoints. Each edge point x=(x,y) votes for an image line according with a specific relation which favour collinear segments $s_j(\pi)$ having similar orientation. For each pixel in the image domain, satisfying such relation, the corresponding cell C(ρ,θ) in the 2D parameter space is incremented and the segment endpoints are updated.

At the end of the voting step, the parameter space Π is examined to select higher peaks (ρ^*,θ^*) and recover from these peaks the segments $s_j(\pi^*)$ representing the object boundaries on the image. A set of segments S=$\{s_m : m=1,..M\}$, extracted from the input image is created. Given a set G of model segment groups $GS_q=\{m_{r,q} : r=1,..,R, q=1,..,Q\}$ representing the 2D-projection of the searched object model from different viewpoints, the recognition step must verify whether a group of image segments can match with that set. The matching method proposed follows some basic ideas shown in [6], and uses geometric constraints to reduce the search.

The result of the matching process is object recognition. Recognition can be performed also in the case of incomplete data, due to occlusion or high noise level, if the pair of data-model segments is greater than a fixed threshold. Recognition results allow to give a more accurate orientation estimation. On the basis of the model, the orientation of some segments is used to determine the principal axis direction.

3   Results

Results aim to estimate the validity of the method by testing it on image sequences acquired by an acoustic camera. About orientation estimation, a mean error ε_m is computed as absolute difference between the actual and the estimated object orientation, normalized to the number of images M contained in the sequence:

$$\varepsilon_m = \frac{1}{M} \sum_{i=1}^{M} \left| \alpha_i - \alpha_i^a \right|$$

(4)

where α_i is the estimated and α_i^a the actual object orientation at the i-th image.

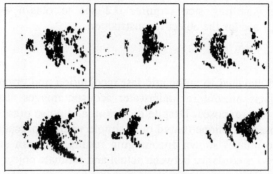

Figure 2. Sequence of actual images acquired by an acoustic camera.

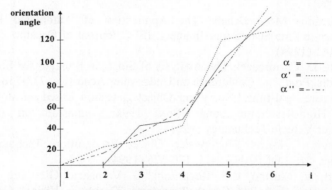

Figure 3. Behaviour of the real and estimated object orientation for the sequence in Fig. 2.

An acoustic camera is able to form both a 3D map of the scene and some interesting 2D projections visualized as images. We show a sequence of orthoscopic images extracted from 3D maps and analogous to optical images. The object that should be recognized is a cylinder floating in the water and having principal axis parallel to the image plane, so, its orthoscopic projection is a rectangle. Figure 2 shows the sequence: the actual orientation of the cylinder is, respectively, $\alpha = 0°$, $\alpha = 0°$, $\alpha = 43°$, $\alpha = 58°$, $\alpha = 110°$, and $\alpha = 145°$.

The cylinder had a diameter equal to 60 mm, a length equal to 150 mm and it was placed about 1 m far from the camera. The carrier frequency used for the scene insonification was equal to 500 kHz and the high distortion of the images is due to low resolution achieved by the acoustic camera. Nevertheless, the algorithm was able to estimate the orientation with a satisfactory mean error value: $\varepsilon_m = 4.1°$ for the coarse voting function described in sub-section 2.1 and $\varepsilon_m = 2.3°$ for the accurate voting function described in sub-section 2.2. The orientation estimation α'' is obtained after object recognition which has been performed successfully on the images in Fig. 2 by matching data with 9 different object models. Figure 3 shows the behaviour of the actual (angle α) and estimate (angle α' for coarse method, angle α'' for accurate method) object orientation. Computation times of about 0.2 sec and 0.8 sec (on a SUN Sparc 10) are required by the 1D and 2D voting functions, respectively. The matching step requires 0.2 sec to obtain a complete object recognition and 0.3 sec for the object reconstruction.

4 Conclusions

A voting approach devoted to automatic fast recognition and orientation estimation of man-made objects present in underwater acoustic images acquired mainly by acoustic cameras was proposed. Despite distortion and noise present in the acoustic images, the proposed technique (based on a simple voting approach applied directly to the edge of the image) shown encouraging results. A mean error measure was used to test the good accordance between actual and estimate object orientation.

References

1. S.G. Johnson, M.A. Deaett: The Application of Automated Recognition Techniques to Side-Scan Sonar Imagery. IEEE Journal of Oceanic Engineering, 19, 138-141 (1994)
2. L. Linnett, D. Carmichael: The Analysis of Sidescan Simages for Seabed Types and Objects. Proc. 2nd Conference on Underwater Acoustics, 733-738 (1994)
3. L. Henriksen: Real-time Underwater Object Detection Based on an Electrically scanned High-Resolution Sonar. Proc. 1994 Symposium on Autonomous Underwater Vehicle Technology (1994)
4. M. Ashraf, J. Lucas: Underwater Object Recognition Techniques Using Ultrasonics. Proc. OCEANS 94, I, 170-175 (1994)
5. G. Foresti, V. Murino, C.S. Regazzoni, G. Vernazza: Grouping of Straight Segment by the Labelled Hough Transform. Computer Vision, Graphics and Image Processing: Image Understanding 58, 22-42 (1994)
6. W.E.L. Grimson: Object Recognition by Computer, The Role of Geometric Constraints, The MIT Press (1990)
7. H.Illingworth and J.Kittler: A survey on the Hough transform. Computer Vision, Graphics and Image Processing, 44, 87-116 (1988)
8. R.O. Duda, P.E.Hart: Use of the HT to detect lines and curves in pictures. Comm ACM 15,1, 11-15 (1972)
9. J.F.Canny: Finding edges and lines in images, Artificial Intelligence Laboratory. Technical Report TR-720, MIT (1983)
10. C.M Brown: Inherent bias and noise in the Hough transform. IEEE Trans. on PAMI 5, 493-505
11. T.M.Van Veen and F.C.A.Groen: Discretization errors in the Hough transform. Pattern Recognition, 14, 137-145 (1981)

Applying Contextual Constraints to Extract Symbolic Representation for Image Understanding

N. Semmar

Institut National des Sciences et Techniques Nucléaires
CE Saclay, 91191 Gif sur Yvette Cedex, France

Abstract. In this paper, we describe a parallel distributed processing model that learns to interpret the content of images using the surface features of their objects. These features are the relative positions of the basic objects in the image and their associated distinguishing attributes. The interpretation starts from the basic objects obtained in the morphological analysis phase and then applies a body of rules for the recursive recognition of the more complex objects contained in the image. The neural model assigns thematic roles to the image objects, disambiguates ambiguous objects, instantiate vague objects, and elaborates implied roles. The image are pre-segmented into a sequence of complex objects. Each object is processed in turn to update an evolving representation of the event described by the image. The model uses the information derived from each object to revise its ongoing interpretation of the image and to anticipate additional objects. The network learns to perform this interpretation through practise on processing example image/event pairs. The learning procedure allows the model to take a statistical approach to determine the concepts corresponding the missing units. The result of the interpretation process is the conceptual representation of the image in terms of the contained objects with their recognition degree.

1 Introduction

Images are very rich in semantics, and are subject to different interpretations according to the human perception or the application domain [1,2]. It is difficult to determine and represent the mutual relationships among the objects contained in an image because they form structures which vary greatly from image to image [3,4].

The automatic interpretation of image content is necessary to support the access to a database of images [5]. The goal of our research has been to develop a model that can learn to convert a simple image into a conceptual representation of the event that the image describes. Specifically, we have been concerned with the later stages of this process: the conversion of a sequence of image objects into a representation of the event. The event constitutes the symbolic representation of the image.

Section 2 presents the description of the image interpretation system. The accomplishments of the model are reported in Section 3. In section 4, we describe the limits of the model and further research efforts.

2 Description of the image interpretation system

Extraction of concepts from an image means the conversion of the content of a simple image into a conceptual representation of the event that the image describes [6,7]. A number of problems make this process difficult. First, the objects of an image may be ambiguous. In the image containing a *man* with *glasses*, the object "man" is ambiguous, it could either refer to a *teacher, bus driver* or a *doctor*. How are the appropriate objects selected so that a single, coherent interpretation of the image is produced? A second problem is the complexity of assigning the correct thematic roles to the objects referred to in an image. In the image containing the objects: "man", "dog" and "leash", either the *dog* follows the *man* or the *man* follows the *dog*, then, if we do not know the spatial relationships between the objects *man* and *dog* we can not determine who is the agent and who is the patient. A third problem for processing content of images is that an image may leave some thematic constituents implicit which are nevertheless present in the event. For example in the image containing a *teacher*, a *schoolgirl*, and a *blackboard*, the *teacher* writes on the *blackboard* with *chalk*. Psychological evidence indicates that missing objects, when strongly related to the action, are inferred and added to the description of the event [8,9].

Our model of the interpretation process centres on viewing the process as a form of constraint satisfaction [10,11,12]. The surface features of an image, its particular objects and their importance provide a rich set of constraints on the image interpretation. Each feature constrains the interpretation in a number of respects. Conjunctions of features, such as objects importance order, provide additional constraints. Together, the constraints lead to a coherent interpretation of the image.

2.1 Architecture of the neural network

The image is presented to the model as a temporal sequence of objects. The information each of these image objects yields is immediately used as evidence to update the model's internal representation of the entire event [13,14]. From this representation, the model can produce, as output, a representation of the event. This event representation consists of a set of pairs. Each pair consists of a thematic role and the concept which fills that role. Together, the pairs describe the event.

2.1.1 Producing the internal representation

To process the objects of an image, we have adapted a neural network architecture with two hidden layers that uses the output of previous processing as input for the next iteration [15,16] (Fig. 1). Each object is processed in turn to update the internal representation. An object, before being processed, it is first represented as a pattern of activation over the current object units. Activation from these units projects to the first hidden layer and combines with the activation from the internal representation created as the result of processing the previous object. The actual implementation of this arrangement is to copy the activation from the internal representation to the previous internal representation units, and allow activation to feed forward from there. Activation in the hidden layer, then, creates a new pattern of activation over the internal representation units. The internal representation, therefore, is not a superimposition of each object. Each new pattern in the image internal representation is computed through two layers of weights and represents the model's new best guess interpretation of the content of the image.

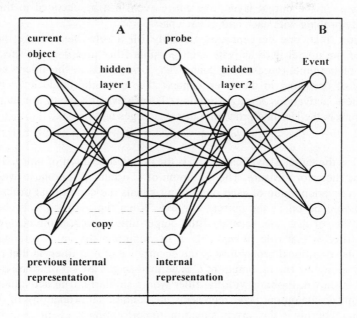

Fig. 1. *The architecture of the neural network. The part A processes the objects into the internal representation, and the part B processes the internal representation into the output representation.*

2.1.2 Producing the output representation

Our model makes the image internal representation a trainable hidden unit layer, which allows the network to create the primitives it needs to represent the content of the image efficiently. We used units that represent the conjunction of semantic features of the object with the semantic features of a concept. To encode an image. the patterns of activity produced for each object/concept conjunction were activated in a single pool of unit that contains every possible conjunction. Instead of having to represent every possible conjunction, only those conjunctions that are useful will be learned and added to the representation. Further, these primitives do not have to be conjunctions between the object and a concept. A hidden layer could learn to represent conjunctions between the concepts themselves or other combinations of information if they were useful for solving its task. Since a layer of hidden units cannot be trained by explicitly specifying its activation values, we decode the internal representation into an output layer. We use back propagation algorithm [17,18] to train the hidden layer. The output layer represents the event as a set of thematic role and filler pairs. For example, the event described by an image of a *man*, an *apple* and a *knife* would be represented as the set {agent/*man*, action/*peel*, patient/*apple*, instrument/*knife*}. The output layer can represent one role/filler pair at a time. To decode a particular role/filler pair, the internal representation is probed with half of the pair, either the role or the filler. Activation from the probe and the internal representation combine in the second hidden layer which in turn activates the entire role/filler pair in the output layer. The entire event can be decoded in this way by successively probing with each half of each pair.

When more than one concept can plausibly fill a role, we assume that correct response of the model is to activate each possible filler to a degree. The degree of activation of the units representing each filler corresponds to the filler's conditional probability of occurring in the given context. The network should learn weights to produce these activations through training. To achieve this goal, we employed an error measure in the learning procedure that converges on this goal:

$$C = -\sum_j \left[T_j \log_2(A_j) + (1 - T_j) \log_2 (1 - A_j) \right]$$

Where T_j is the target activation and A_j is the output activation of unit j. The goal is to minimise the error measure. The minimum of C occurs at a point in weight space where the activation value of each output unit equals the conditional probability that the unit should be on in the current context. When the network is probed with a particular filler of that role, several of the output units represent the occurrence of a particular filler of that role. When C is at its minimum the unit's activation values represent the conditional probability of the occurrence of that filler, in that role, given the current situation. The minimum of C is defined across the set of training examples the model is shown. Probing with the filler works similarly. The activation value of each role unit in the output layer represents the conditional probability of the probed filler playing that role in the current situation. In performing gradient descent in C, the network is searching for weights that allow it to match activations to these conditional probabilities.

2.1.3 Training

Training consists of trials in which the network is presented with an image and event it describes. The learner processes the image and compares the conceptual representation its interpretation mechanism produces to the conceptual representation it obtained from experiencing the event. Discrepancies are used as feedback for the interpretation mechanism. These image/event pairs were generated on-line for each training trial.

While images often include ambiguous objects, the events are always specific and complete: each event consists of a specific action and each thematic role related to this action is filled by some specific concept. Accordingly, each event occurs in particular location, and actions requiring an instrument always have a specific instrument.

During training, the objects of the image are presented to the model sequentially, one at a time. After the model has processed an object, the model is probed with each half of each role/filler pair for the entire event. The error produced on each probe is collected and propagated through the network. The weight changes from each image trial are added together and used to update the weights after every several trials.

2.2 Functionalities and performance of the model

First, we will assess the model's ability to comprehend images generally. Then we will examine the model's ability to fulfil our specific processing goals, and we will examine the development of the model's performance across training trials. Finally, we will discuss the model's ability to generalise.

2.2.1 Overall performance

Correct processing was defined as activating the correct units more strongly than the incorrect units. After 10 000 random image trials, the model began correctly processing the image in the corpus. A set of 100 test image/event pairs were generated randomly from the corpus. These image/event pairs were generated in the same way the training images were generated except that they were generated without regard to their frequency during training, so seldom practised pairs were as likely to appear in the test set as frequently practised pairs. Of these pairs, 45 were set aside for separate analysis because they were ambiguous: at least two different interpretations could be derived from each (e.g. A *man* with *glasses*). Of the remaining image event pairs, every image contained at least one vague or ambiguous object, yet each had only one interpretation. These unambiguous image/event pairs were tested by first allowing the model to process all of the objects of the image. Then the model was probed with each half of each object that was mentioned in the image. The output produced in response to each probe was compared to the target output. For these unambiguous images, the cross-entropy, summed over constituents, averaged 3.9 per image. Another measure of performance is the number of times an output unit that should be on is less active than an output unit should be off. The idea behind this measure is that as long as the correct unit within any set, such as people or gender is the most active, it can win a competition with the other units in that set. Checking that all of the correct units are more active than any of the incorrect units is a quick, and conservative, way of calculating this measure. An incorrect unit was more active in 14 out of the 2000 possible cases, or on 0.7% of the opportunities. The 14 errors were distributed over 8 of the 55 images. In 5 of the 8 images, the error involved the incorrect instantiation of the specific concept, or a feature of that concept, referred to by a vague object. Two other errors involved the incorrect activation of the concept representing a non vague object. In each case, the incorrect concept was similar to the correct concept. Therefore, errors were not random; they involved the misactivation of a similar concept or the misactivation of a feature of a similar concept. The errors in the remaining image involved the incorrect assignment of thematic roles in a image.

Additional practice, of course, improved the model's performance. Improvement is slow, however, because the images processed incorrectly are relatively rare. After a total of 20 000 trials, the number of images having a cross-entropy higher than 15 dropped from 3 to 1. The number of errors dropped from 14 to 11.

2.2.2 Performance on specific tasks

Our specific interest was to develop a processor that could correctly perform several important image interpretation tasks. Three specific images were drawn from the corpus to test each processing task. The categories and one example image for each are presented in Table 1. The parentheses denote the implicit, to be inferred, role.

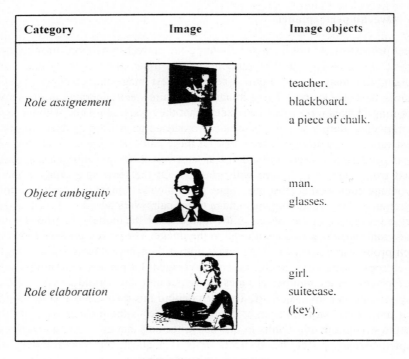

Category	Image	Image objects
Role assignement		teacher, blackboard, a piece of chalk.
Object ambiguity		man, glasses.
Role elaboration		girl, suitecase, (key).

Table 1. Task categories

The first category involves role assignment. In the example containing a *teacher*, a *blackboard* and a *piece of chalk*, of the concepts referred to in the image, only the *teacher* can play the role of an agent of teaching. The network can therefore use that semantic information to assign *teacher* to the agent role. After each image was processed, the internal representation of the image was probed with the half of each role/filler pair. The network then had to complete the pair by filling in the correct thematic role. For each pair, in each image, the unit representing the correct role was the most active. Of course the semantic knowledge necessary to perform this task is never provided in the input or programmed into the network. Instead, it must be developed internally in the internal representation of the image as the network learns to process images.

Surface information does not have to be used in these cases; the semantic constraints suffice. In fact, if the surface location of the objects is removed from the input, the roles are still assigned correctly. Further, if the objects are presented in different orders, the activation values in the output are affected only slightly.

The remaining two categories involve the use of context to help specify the concepts referred to in an image. Images in the object ambiguity category contain one or more ambiguous objects. After processing an image, the network was probed with the role half of each role/filler pair. The output patterns for the fillers were then examined. For all pairs in each test image, the correct filler was the most active.

Desambiguation requires the competition and co-operation of constraints of both the object and its context. While the object itself cues two different interpretations, the context fits only one. In the image containing the objects "man" and "glasses", the object "man" is ambiguous, it could either refer to a *teacher*, *bus driver* or a *doctor*. The context cues both *teacher* and *doctor* because the model has seen images involving both people with *glasses*. All the constraints supporting *doctor* combine, and together they win the competition for the interpretation of the image.

Depending upon the image, however, the context may only partially constrain the interpretation. Such is the case in the image containing a *teacher* and a *girl*. The object "girl" could refer to any of the people found in the corpus. Since, in the network's experience, in a classroom, we can find schoolboys and schoolgirls, the context constrains the interpretation of *girl* to be the *schoolgirl*. Consequently, the model can activate the female and girl features of the patient while leaving the unit representing the *schoolgirl* only partially active. The features adult and child are also partially and equally active because the *teacher* is an adult while the *schoolgirl* is a child. While *schoolgirl* is slightly more active in this example, neither is activated above 0.5. In general, the model is capable of inferring as much information as the evidence permits: the more evidence, the more specific the inference.

Object disambiguation can be seen as one type of this general inference process. The only difference is that for ambiguous objects, both the general concept and the specific features differ between the alternatives, while for vague objects, the general concept is the same and only some of the specific features differ.

Finally, images in the role elaboration category test the model's ability to infer thematic roles not mentioned in the input image. For example, in the image containing the objects: "girl" and "suitcase", no instrument is mentioned, yet a *key* can be inferred. For each test image, after the image was processed, the network was probed with the role half of the to-be-inferred role/filler pair. The correct filler was the most active in each case. For role elaboration, the context alone provides the constraints for making the inference. Extra roles that are very likely will be inferred strongly. When the roles are less likely, or could be filled by more than one concept, they are only weakly inferred.

As it stands, there is nothing to keep the network from generalising to infer extra roles for every image, even for events in which these roles make no sense. For instance, in the image containing a *teacher* and a *blackboard* (Fig. 2.), a *stick* should be inferred, yet the network infers a *piece of chalk* because of its association with the *teacher*. It appears that since the *teacher* uses a *piece of chalk* in many events about teaching, the network generalises to infer *the piece of chalk* as an instrument for this action. However, in events further removed from teaching, instruments are not inferred. For example, in the image containing a *teacher* and a *rose*, no instrument is activated. It appears, then, that generalisation of roles is affected by the degree of similarity between events. When events are distinct, roles do not generalise, and the model has no reason to activate any particular filler for a role.

Fig. 2. An image with the objects: teacher, blackboard and stick.

3 Accomplishments of the model

The extraction of concepts from the image objects is based on conceptual (surface and semantic) constraints. The surface constraints are more difficult for the model to master then the semantic constraints even though we have provided explicit cues to the surface features, in the form of the surface location of the object, in the input. The model masters these constraints as they are exemplified in the set of training images.

The model infers unspecified objects roughly to the extent that they can be reliably predicted from the context. Here, we see very clearly that objects of an event description can be cued without being specifically designed by any object of the image. These inferences are graded to reflect the degree to which they are appropriate given the set of clues provided. The drawing of these inferences is also completely intrinsic to the basic interpretation process: no special separate inference processes must be spawned to make inferences, they simply occur implicitly as the objects of the image processed.

The model demonstrates the capacity to update its representation as each new object is encountered. As each object is encountered, the interpretation of all aspects of the event description is subject to change. If we revert to thinking in terms of meanings of concepts describing particular objects, both prior and subsequent context can influence the interpretation of each object. The ability to exploit subsequent context is an intrinsic part of the process of interpreting each new object. There is no backtracking, the representation of the image is simply updated to reflect the constraints imposed by each object as it is encountered.

While avoiding backtracking, the model also avoids the computational explosion of computing each possible interpretation of an image as it encounters ambiguous objects and thematic role assignments. When an ambiguous object is processed, it initially activates semantic units representing both concepts. The resulting pattern of activation is a combination of the semantic features of both concepts, so the bindings among the features of a concept are lost in the activation pattern. These bindings, however are preserved in the weights, and the model will settle into one interpretation of the object or the other.

Finally the model is able to generalise the processing knowledge it has learned and apply it to new images. To perform this generalisation task, the model learns several types of information. First, it learns the concept referred to by each object in an image. Second, the model learns the order of the objects, which object comes before the main object and which after. This order is important to assigning the agent and patient. Third, the model learns the relevance of relationships among each object and the main object of the image. Fourth, the model learns to integrate the objects order and the information obtained from relationships to correctly assign the thematic roles.

4 Conclusion

The approach described in this article has been experimented on simple images. It will be important to analyse the limitation of the model on the complexity of the images, and the events that they describe [19]. Indeed, it is necessary to characterise the roles and fillers of images with respect to their super ordinate objects. The images are limited in complexity because of the limitations of the event representation. Only one filler can be assigned to a role in a particular image. Also, all the roles are assumed to belong to the image as a whole. Therefore, no embedded actions attached to single objects are possible. Similarly in complex events, there may be more than one actor, each performing an action in a different sub-event of the overall event or action. Representing these structures requires head/role/filler triples instead of simple role/filler pairs.

References

1. S. Zeki, S. Shipp: The functional Logic of Cortical Connections. In nature vol. 335, n°6188. pp.311-317 (1988).
2. P.A. Carpenter. M.A. Just: Reading comprehension as the eyes see it. In M.A. Just and P.A. Carpenter. eds., Cognitive Processes in Comprehension. Erlbaum, Hillsdale, NJ (1977).
3. C. Faloutsos, T. Sellis, N. Roussopoulos: Analysis of object oriented spatial access methods. In Proc. ACM SIGMOD International Conference, pp. 426-439 (1987).
4. F. Rabitti, P. Savino: Automatic indexation to support content-based retrieval: In Information Processing & Management Vol. 28, No. 5, pp. 547-567 (1992).
5. N. Semmar, C. Fluhr: A neural model for image content analysis. In Proceedings of The Third Turkish Symposium on ARTIFICIAL INTELLIGENCE & NEURAL NETWORKS. Ankara METU (1994).
6. N. Semmar, C. Fluhr: Tool kits design for the internationalisation of multimedia courseware. In Tech. Report DELTA programme D2013 project MALIBU, Deliverable 12 WorkPackage 24(3) (1993).
7. N. Semmar, C.Fluhr: Utilisation d'une représentation hybride des bases de données documentaires pour la recherche d'informations. In Actes des journées d'études sur les Systèmes d'Information Elaborée, Ile Rousse Corse (1993).

8. V. Gudivada, V. Raghavan: A system for retrieving images by content. In Intelligent Multimedia Information Retrieval Systems and Management. Rockefeller University, New York (1994).

9. T. -Y. Hou: A content-based indexing technique using relative geometry features. In Storage and Retrieval for Image and Video Databases. SPIE. Vol. 1662 (1992).

10. J. Hertz, A. Krogh, R.G. Palmer: Introduction to the Theory of Neural Computation. Addison-Wesley (1986).

11. G.E. Hinton: Connectionnist Learning Procedures. In Artificial Intelligence vol. 40, n°1-3. pp.185-234 (1988).

12. G.E. Hinton: Learning distributed representations of concepts. In Proceedings of Eighth Annual Conference of the Cognitive Science Society, Amherst. MA (1986).

13. J.L. McClelland, D.E. Rumelhart: An interactive activation model of context effects in letter perception: Part 1. An account of basic findings. In Psychol. Rev.88, pp. 375-407 (1981).

14. M.L. Jordan: Attractor dynamics and parallelism in a connectionist sequential machine. In Proceedings Eighth Annual Conference of the Cognitive Science Society, Amherst, MA (1986).

15. D.E. Rumelhart, J.L. McClelland: Explorations in Parallel Distributed Processing, A Handbook of Models, Programs, and Exercises. MIT Press (1988).

16. D.L. Waltz, J.B. Pollack: Parallel Interpretation of Natural Language. In Proceedings of The International Conference of Fifth Generation Computer Systems, ICOT (1984).

17. D. E. Rumelhart, G. E. Hinton, R. J. Williams: Learning Representations by Back-Propagating Errors. In Nature vol. 323, n°6188, pp.533-536 (1986).

18. D. E. Rumelhart, G.E. Hinton, R. J. Williams: Learning Internal Representations by Error Propagation. In Parallel Distributed Processing-Explorations in the Microstructure of Cognition. MIT Press (1986).

19. C.F. Herot: Spatial management of data. In ACM Trans. Database Syst. 5, pp. 493-513 (1980).

Application of the Hough Transform for Weld Inspection Underwater

Alistair R. Greig

Department of Mechanical Engineering, University College London,
University of London, Torrington Place, LONDON, WC1E 7JE, UK.

Abstract. The inspection of the nodal welds of off-shore steel structures is a necessary but hazardous task at present performed at regular intervals by divers. Automation of this task is an on-going research effort. One requirement for automation is the accurate location of the weld roots and it is this which is considered here. A brief description of the system is provided as well as a summary of the problems of imaging underwater. This is followed by a description of how the relevant segments of the degraded laser stripe can be identified using the Hough transform. The Rainflow method (originally proposed by Tatsuo Endo for analysis of stress histories) is then used to rapidly identify the desired features on the parameter surface and hence locate the weld roots.

1 Introduction

The welds joining the main structural members of steel jacket, off shore platforms can suffer from fatigue cracks due to wave loading on the platform. Inspection is currently done by divers but there is a requirement to automate this [1]. To successfully examine the large, multi-pass welds found at the junction of the tubular structural members it is necessary to locate the weld roots, (the edges of the weld where cracks are most likely to occur). One proposed method for achieving this is by using a manipulator deployed machine vision system.

2 Imaging System

Automatic welding machines which track the weld run are now available from various manufacturers. These use a variety of methods for tracking the weld run, but all rely either on feed back from the weld arc, or the strong visual cues available from the prepared edges of the metal. Neither of these information sources are available when attempting to track an existing weld. The weld and surrounding metal are bland and devoid of useful features, especially if any corrosion has taken place. Marine growth may also be present when inspecting underwater, though most of this will have been removed by grit blasting.

Since no features are readily available a structured light source is introduced. A laser line is projected onto the weld and viewed at an angle by a CCD camera. This produces a profile of the weld and its position can easily be identified by triangulation. In air the laser line projected on to the metal surface is sharp, continuous, clear and easily detected by simple thresholding, see Fig. 1. The weld roots are easily identifiable as the ends of the upper and lower line segments.

The imaging system is mounted on the end of a manipulator which is used to move it along the weld. An initial approximation of the weld path is constructed by assuming that the structural members are cylinders and that the weld occurs at their intersection. As the imaging system profiles the weld and identifies the roots, the results are feed back to the manipulator path planning routine to improve its performance. Unlike land based robotic applications, speed of operation is not a priority for underwater operation.

3 Imaging Underwater

An underwater image will be degraded by the total attenuation due to the transmission of light through the sea water, and also in this case by the debris after cleaning. There are a number of mechanisms which contribute to the degradation of an image captured underwater, fuller descriptions are available elsewhere [2, 3, 4], but in summary;
• Molecular scattering. Pure water will scatter light in all directions, though the distribution is not uniform, the ions in sea water increase scattering (typically 30%).
• Particle scattering. Particles are small (1 - 100μm diameter) pieces of organic or inorganic material held in suspension in the water, these will also scatter light but scattering is predominantly forward. Particle concentration and size distribution depends on geographical location, weather conditions and depth.
• Absorption. Light is absorbed by the water itself and also the organic matter dissolved in it. This is very wave length dependant and minimum absorption occurs in a narrow band (460 - 560nm) depending upon the concentration of organic matter.
    By reducing the operating range to less than 1m and using a green laser the above effects can be reduced. Even so the laser line will not be as sharply defined as in air. Also rigs produce their own micro-marine environments and act like artificial reefs, consequently the biological and suspended matter will be much higher than might be expected than for that particular sea area. Before inspecting the weld area it is necessary to grit blast the soft organic and hard calcareous growth off from the structure, this produces an additional problem of large (diameter >100μm), inorganic particulate matter floating in the viewing volume. These particles are too large to cause significant scattering but depending upon their orientation and position they can either obscure part of the image or create bright spots by reflecting the laser.
    The combination of the above effects produces an image that is blurred and the occlusions will cause the stripe to be discontinuous. After thresholding the line will be surrounded by a number of bright spots due to reflections from the debris. Once the laser stripe is identified within the image it is relatively straight forward to track as it will not vary significantly form one frame to the next as the manipulator's motions are slow. There are two exceptions to this. Large portions of the stripe will occasionally be obscured when a large piece of debris floats (swims) through the viewing volume. As the manipulator moves around the weld it will occasionally have to reconfigure itself to avoid a singularity transit or a joint limitation. This will necessitate moving the laser stripe away from the weld. A reduced version of the Hough transform is used to rapidly identify the line segments for the initial stripe detection or when ever its location is lost.

4 Application of Hough Transform

The Hough transform is a method for identifying sets of co-linear points on an image. Description of the method is available in many standard image processing texts [5, 6]. It is recognised that the method has a number of limitations;
• It is slow to implement
• To achieve good resolution a large matrix is required to represent parameter space.
• The extraction of features from the parameter space can be very time consuming.
    For this application *a priori* knowledge is used to speed implementation of the transform. Location of the local maxima in the parameter plane is simplified by reduction of the problem from three to two dimensions and by subsequent use of the rainflow method to extract features.
    Hough's original implementation of his transform [7] used the gradient/intercept parametrisation of a straight line ($y = mx + c$), subsequently Duda [8] suggested the normal parametrisation ($x\cos\theta + y\sin\theta = r$). The latter form is now generally used for Hough transforms. The normal form has the advantage that it provides a bounded

parameter space, but the disadvantage that it is slower to plot, straight lines are far quicker to compute than sinusoids. By controlling the path of the manipulator around the weld and the pose of the camera relative to the laser it was possible to restrict the orientation of the laser stripe. The gradients of all its significant line segments were within a limited, finite range, $-1.28 \le m \le 1.28$. Constraining the gradient values removes the limitation on using the gradient/intercept parametrisation as a closed solution is now possible. This permits faster generation of the transform.

The Hough transform is generated on the image plane of the image processing board rather than using a large (0.74Mb) matrix within the host computer, each pixel acts as a collection bin. $f(x,y) = h(c,m)$. The value (gray scale) of each pixel is set to zero and the ink colour to one. The parameter plane is then generated by drawing lines on the plane with the logical operator set to AND. The limiting factor on speed is the time taken to interrogate the original image in the frame store, not the time taken to construct the transform. This approach has the advantage that a very large number of bins (512x720) can be used at virtually no extra cost in computing effort.

Two points are sought on the parameter plane, corresponding to the upper and lower line segments. Both points will be local maxima and will have different values of m as the line segments will always have different gradients. As the system is operating in an unstructured environment atypical conditions must be considered. It is possible for the local maximum associated with the weld line segment to be of similar magnitude to that of one of the other line segments, therefore it is also necessary to identify the third most significant maximum on the parameter plane. This will be a rare occurrence as the weld line segment is usually shorter and less regular than the other two line segments. Even when this condition does occur, in the majority of cases the ambiguity can be resolved by considering the indicated positions of the lines on the image plane. If the result is still ambiguous two steps of remedial action are taken before halting the programme. On the assumption that the error was caused by matter within the viewing volume a fresh image is captured and the procedure repeated. If this fails the manipulator repositions the laser stripe in what is hoped to be a better position over the weld. The manipulator is moved so as to reduce the length of the one easily identifiable line segment with the intention to increase the length of the other.

The relative strength of the weld line segment can be reduced by viewing weld from further away, this increases the length of the upper and lower line segments. Increasing the viewing range has three disadvantages; loss of resolution, increased attenuation and scattering, increased line curvature. The upper and lower line segments are not straight lines but arcs of an ellipse, the aspect ratio of the ellipse is such that over short arc lengths the curvature is nominal. The manipulator is programmed to orientate the laser to minimize this effect. The Hough transform provides a best line fit to the arc.

The geometry of the cylinders and the manipulator's path ensure the upper and lower line segments always have different gradients. Hence the local maxima search on the parameter plane can be reduced from considering all the 368,640 cells of $h(c,m)$ to just the 512 elements of $h(c_{Max})$. For each gradient value m the maximum value of c is found. The resulting function is not smooth due to noise on the original image.

The problem is to extract the position of the three distinct maxima. It is not sufficient to find the three highest peaks of the function $h(c_{Max})$ as these are usually clustered around a narrow band and relate to a single line on the image plane. Consider Fig. 2; A, B and C are the peaks of interest, but 1, 2 and 3 are the three largest maxima. A filtering method is required that identifies large amplitude cycles and does not corrupt the original data pertaining to them as is the case with smoothing functions. This was achieved using the rainflow method. This method has the advantage over many filtering and thresholding techniques in that it does not require the values of any coefficients or limit values to be preset. The correct and reliable evaluation of the coefficients and limit values for filters is often a limitation on their effectiveness, especially when operating in environments with poorly controlled conditions.

The rainflow method was developed for the study of stress histories of randomly loaded structures [9]. The method successively filters out small stress reversals to leave only those of large amplitude. It is these that cause the majority of damage to structures and are so important for fracture mechanics and fatigue analysis. The rainflow method preserves amplitude information and operates independently of frequency.

The rainflow method is best explained by means of a physical analogy. The variant presented here is sometimes referred to as the reservoir method. It has the advantage that it identifies the largest rather than the smallest cycles first. Figure 3a shows the simplified function $y = f(x)$. The two local maxima with the greatest magnitude are A and B while the peaks of the two main cycles are at A and C. The function is considered as a continuous loop and rearranged about $x[min]$, Fig. 3b, so as to include any maxima at the limits of x, for example D. Consider a reservoir of buoyant gas trapped beneath the function, find the absolute maximum and let the gas escape at that point, the result is shown in Fig. 3c. The next most significant cycle can be identified by considering the depth of gas trapped beneath it, as can be seen from Fig. 3d. The process is repeated as required, note that B corresponds to only the fourth most significant cycle.

The method is rapid as it requires little computation beyond simple hill climbing and sort routines. It becomes progressively faster as the most significant cycles are progressively identified and removed. The example here is for a two-dimensional system, the principle can be extended to three-dimensions for consideration of the full parameter plane. Work has been done on multi-axial loading of structures [10] which is relevant, the main problem is analysis of the values at the edges of the parameter surface.

5 Results

Tests were conducted on a 1/4 scale tubular Y-joint, the camera-laser system being deployed from a PUMA 560 manipulator. The tests were conducted in air as neither an underwater camera nor an underwater manipulator were available. Access to underwater test facilities is a serious limitation to research in off-shore engineering. To simulate underwater conditions the line optics were defocused and noise was deliberately superimposed onto the image. The image is randomly seeded with two populations of blobs. One population, which covers 5% of the total image is bright (gray scale 255) while the other which covers 10% of the image is dark (gray scale 0).

The system was able to track all parts of the weld around the Y-joint. The example shown in Fig. 4 is a worst case situation taken at the toe of the weld, at this position the angle subtended between the upper and lower line segments is a minimum. The weld is approximately 30mm wide at this point. Note the added complications of reflections from the metal surface (top centre), and the ghost stripe to the right of the main stripe due to internal reflections in the optics.

Figure 5 shows detail of the weld area with the results of the inverse transform and root location superimposed. In this particular case the near vertical, dark grey, line which corresponds to the upper line segment was the third most significant on the parameter surface. The most significant line being the light grey line which includes large proportions of the weld segment and the upper line segment. Despite this the upper weld root (Root 1) has been correctly located as has the lower weld root. The broken, light gray line was the second most significant feature on the parameter plane and corresponds to the lower line segment. Note that the lower root location is not assumed to be at the intersection of the lines, though this is often the case.

6 Conclusions

A system has been constructed and tested that can detect and track the roots of a weld. It is able to operate in an optically harsh environment with serious image disturbances. Suitable design of the image system and deployment strategy permit simplification (and hence acceleration) of image processing. A simplified version of the Hough transform is successfully implemented, using the rainflow (or reservoir) method to assist in feature extraction from the parameter surface. At present the accuracy of the system is compromised by the accuracy of the deploying manipulator.

7 References

1. Bennett, A. and Chadwick, F.J. (1988). Crack detection on subsea structures - a cost effective approach. *IEE Economics of NDE symposium*, London 1988.
2. Duntley, S.Q. (1963). Light in the sea. *J. Opt. Soc. Am.* Vol. 53, pp. 214 - 233.
3. Morel, A. (1974). Optical properties of pure water and pure sea water. In: *Optical aspects of Oceanography* (Jerlov, N. and Nielsen, E.S. eds.), pp. 1 - 24. London: Academic Press. ISBN 0-12-384950-0.
4. Jerlov, N.G. (1976) *Marine Optics*. Elsevier Oceanography Series, 14. Amsterdam: Elsevier Scientific Publishing Company. ISBN 0-444-41490-8.
5. Gonzalez, R.C. and Wintz, P. (1987). *Digital Image Processing* 2nd Ed. pp 130 - 136. Reading MA: Addison-Wesley, ISBN 0-201-11026-1.
6. Leavers, V.F. (1992). *Shape detection in Computer Vision Using the Hough Transform*, London: Springer-Verlag ISBN: 3-540-19723-0.
7. Hough, P.V.C. (1962). Methods and means for recognizing complex patterns. *U.S. Patent 3,069,654.*
8. Duda, R.O. and Hart, P.E. (1972). Use of Hough transforms to detect lines and curves in pictures. *Comm. Association for Computing Machinery 15*, No. 1, pp. 11 - 15.
9. Nelson, D.V. (1991). Rainflow and ordered overall range cycle counting, in *The rainflow method in fatigue, the Tatsuo Endo memorial volume*, Ed. Murakami, Y. pp. 41 - 50. Oxford: Butterworth Heinemann, ISBN 0 7506 0504 9.
10. Beste, A., Dreβler, K., Kötzle, H., Krüger, W. Maier, B. and Petersen, J. (1991). Multiaxial rainflow, a consequent continuation of professor Tatsuo Endo's work, in *The rainflow method in fatigue, the Tatsuo Endo memorial volume*, Ed. Murakami, Y. pp. 31 - 40. Oxford: Butterworth Heinemann, ISBN 0 7506 0504 9.

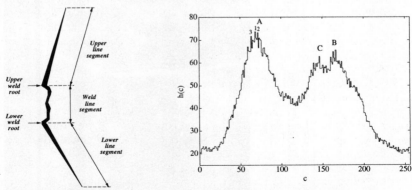

Fig. 1. Typical weld profile.   **Fig. 2.** Reduced parameter data set.

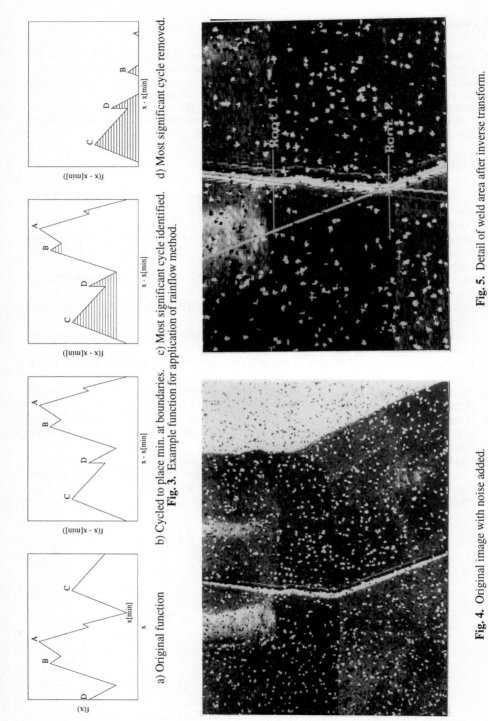

a) Original function

b) Cycled to place min. at boundaries.

c) Most significant cycle identified.

d) Most significant cycle removed.

Fig. 3. Example function for application of rainflow method.

Fig. 4. Original image with noise added.

Fig. 5. Detail of weld area after inverse transform.

Development of an automated bond verification system for advanced electronic packages

Michel Darboux, Jean-Marc Dinten

LETI (CEA - Technologies Avancées) - DSYS/SCSI
CENG, 17 rue des Martyrs
F38054 Grenoble Cedex 9 - FRANCE

Abstract. As the performance of an electronic product improves, the surface density of the board population increases drastically. As a consequence, the technology is moving from conventional components to surface mounted components. The solder joints are more and more difficult to inspect, they are no longer reachable by the standard inspection techniques (visible light or laser). However, X-rays still allow to access information concerning these joints. In this paper, we propose a diagnosis method for characterising solder joints of surface mounted components. This method is based on a comparison between three radiographic images acquired from a mounted component, the corresponding unsoldered component used as a model, and the unpopulated board. Firstly, the radiographies are preprocessed in order to get quantitative information. Then, an angle and X,Y position alignment is performed before the subtraction of the unpopulated board and the model component images from the mounted component image. This leads to the solder joint map, from which quantitative parameters are then extracted. After a description of the image processing algorithms, we present experimental results obtained on a typical component. These results show clearly the possibility to access important parameters characterising the solder joint.

1 Introduction

Compared to visible light or laser, X-rays provide additional information. Radiographies show the internal structures while visible or laser images describe the surface. This property is essential for printed circuit board inspection. X-rays allows the access to information unreachable by other inspection techniques : internal defects, solder hidden by a component (as for BGA or J-lead components for example).

For a complex object as an electronic board, the radiography shows several 3D structures that are superimposed; therefore, direct interpretation is not obvious. Standard ways to deal with this superimposition problem are tomography and laminography. Tomography leads to a 3-D reconstruction of the lead and the solder, but requires a large number of radiographies and powerful computers. This is not possible for "on-line" industrial inspection. Laminography allows to

---

* This work has been achieved within the BRITE-EURAM project "VERBONDS".

reconstruct rapidly a "blurred" image of one slice of the object. It is well suited for detecting defects like a lack or an excess of solder, but it is not adapted for a quantitative characterisation of the solder joint. At LETI, an approach taking into account this superimposition has been developed [1]. This method allows the separation of the contribution of each side of a double sided mounted board, by combining three radiographies taken from different points of view.

Once the separation of the data from each side of the board is achieved, the respective contributions of the solder joint, the lead and the traces, must be identified. To do this, a diagnosis method for solder joint analysis is proposed. It is based on image comparison using a model of the object (the unsoldered electronic component). This approach enables to focus on defect detection and quantitative parameters extraction.

2   General presentation

The proposed X-ray inspection method of printed circuit boards solder joints is based on image comparison. The main steps of this method are presented in figure 1. We use three radiographies : one field from the inspected mounted board, the corresponding field from the unpopulated board, the radiography of the model component. The first step is a preprocessing of each radiography in order to obtain quantitative information. The second step is a spatial alignment (in angle and X,Y position) in order to be able to subtract the unpopulated board and the model component images from the mounted board image. The combination of these three radiographies leads to the solder joint map from which quantitative parameters are then extracted.

For on-line inspection applications, the model components radiographies could be taken from a database previously generated.

3   Preprocessing of the radiographic images

The initial image from the acquisition system shows the remaining flux of X-ray photons after crossing the object to be inspected (see section 5). The initial image pixels have integer values (encoded between 0 and 4095). The purpose of this preprocessing is to transform the pixel values of this initial image into thickness information. To correct the bias introduced by the acquisition system, we use a reference image obtained with a uniform object and a black image obtained when the X-ray tube is off. In the resulting image, the pixel real value is the product of the thickness by the corresponding linear attenuation coefficient, for the different layers composing the printed circuit board.

4   The spatial alignment

4.1   X,Y position alignment

A binary correlation technique is used. The alignment is done for the mounted board and the model component images and then for the mounted board and the unpopulated board ones. It consists first of a gradient computation on the two images to be aligned. The grey levels gradients are then binarized, using a threshold calculated as a percentage of a maximum gradient value. This value is determined after removing the possible false maximum due to noisy information. The correlation function is then computed between the two binarized images. The coordinates of the maximum of the function give the X and Y misalignment.

4.2   Angle alignment

To make a good comparison, it is necessary to have the same orientation for the mounted component, the model component, and the unpopulated board. This is obtained by correcting the orientation of each image. The reference orientation is a vertical line. The correction angle to apply is shown on the right. The angle alignment algorithm is now detailed. Firstly, the gradient of the image is extracted. These grey level gradient values are then projected along the columns, and the standard deviation of the projected signal is measured. The image is rotated for different angles, the misalignment angle is obtained for a rotation that gives the maximum standard deviation. The example given in figure 2 shows how the standard deviation of the projected signal decreases when the image is not correctly aligned along the vertical reference.

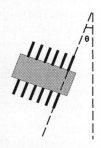

correction angle

5   Experimental conditions

The system used for the radiographic acquisitions is now described (see the figure below). The X-ray source is a continuous emitting microfocus tube with a minimum focus spot size of 10 microns, and a power range from 30 to 160 KV. The inspected board is placed between the microfocus tube and an X-ray image intensifier. The visible light image at the output of the intensifier is then acquired with a CCD camera (512x512 pixels). The resolution depends on the position of the sample between the tube and the intensifier. Concerning the objects to be inspected, an industrial printed circuit board with various SMT (Surface Mount Technology) components is used. The results obtained on a typical J-lead component are presented hereafter.

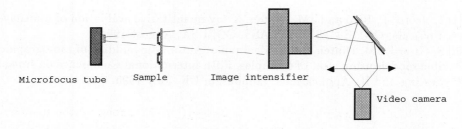

6   Experimental results

A schematic representation of a J-lead component and solder joint is given on the right. Figure 3 shows the images obtained with the microfocus acquisition system for a J-lead component (the pixel size is 17.4 microns for the X axis and 12.3 microns for the Y axis), and the cross sections of one of the solder joints (for these graphical representations, the size of the grid is 100 microns by 100 microns). We can see on these graphs the different parameters characterising the solder joint. The legend for these parameters is, θ : wetting angle, L : joint length, W : joint width, H : heel fillet height. We can note for this example, that this system allows to obtain reliable parameters. These parameters fit well with the real physical shape of the J-lead joint obtained with the scanning electron microscope.

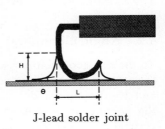

J-lead solder joint

7   Conclusion

The experimental results presented in this paper show clearly the possibility to access important parameters characterising solder joints. Many of these parameters are unreachable by visible or laser inspection techniques.

These parameters are well correlated with the real physical shape of the joints obtained with the scanning electron microscope.

All the algorithms presented have been developed under an interactive image processing environment. At present time their use is highly supervised. The next step would be to automate the processing : the user would just have to define a region of interest on the mounted board and the system would then automatically execute the combination and provide a list of relevant parameters.

The model component may present some differences with the mounted component (tolerances of fabrication and acquisition). LETI is working on the development of methods integrating these kinds of tolerances in the analysis [2].

8   References

[1] C. Icord, P. Rizo and JM. Dinten. X-ray quantitative evaluation of multilayered objects, QNDE94, SNOWMASS, CO, USA, August 1994.

[2] S. Girard, JM. Dinten and B. Chalmond. Automatic building of radiographic flexible models using a set of examples, Fifth International Conference on Image Processing and Its Applications, Edinburgh, UK, July 1995.

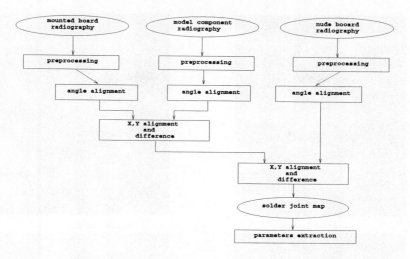

Fig. 1. the main steps of the inspection method

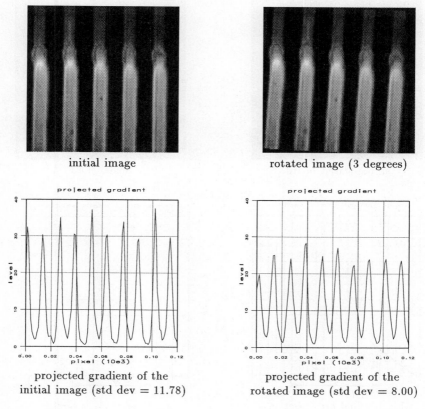

initial image   rotated image (3 degrees)

projected gradient of the
initial image (std dev = 11.78)

projected gradient of the
rotated image (std dev = 8.00)

Fig. 2. angle alignment principle

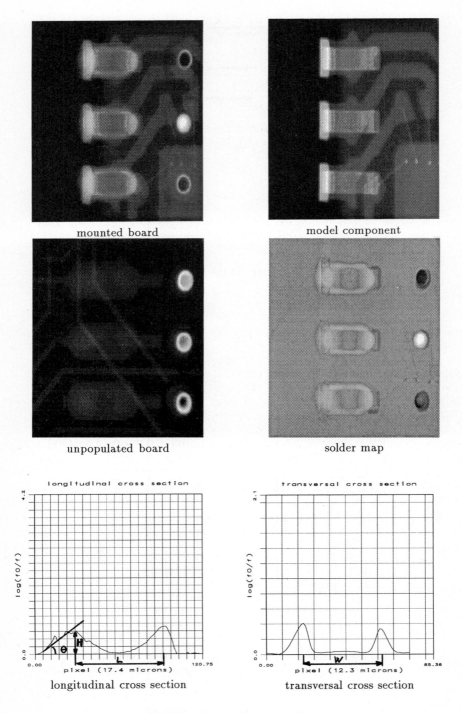

Fig. 3. J-lead component subimages and solder joint cross sections

Crack Detection by a Measure of Texture Anisotropy

L. Bruzzone, F. Roli, and S.B. Serpico

Dept. of Biophysical and Electronic Engineering - University of Genoa
Via all' Opera Pia, 11A, 16145, Genova, ITALY

Abstract. In this paper, the problem of automatic visual inspection of textured surfaces is addressed. In particular, a technique for crack detection on both regularly and randomly textured images is presented. The technique is based on a new measure of texture anisotropy that allows an easy discrimination between defect pixels and defect-free ones. This technique was used to detect cracks on granite slabs. The reported results confirm its effectiveness.

1 Introduction

In manufacturing industries, the inspection of raw materials and end products is a very important task to attain the reliability of products and the customer's confidence. In this context, automatic visual inspection techniques can play a key role.

In the literature, many techniques for detecting defects on regularly textured surfaces, such as textiles, have been proposed [1]. In particular, the detection of large defects has been extensively investigated. By contrast, the problem of detecting very small defects (e.g., cracks or scratches) on randomly textured surfaces has not received much attention [2, 3].

In this paper, we address this problem and describe a new technique developed for the detection of cracks on textured images (both regular and random textures). The technique is aimed at crack detection, as it takes into account explicitly the characteristics of this kind of defect. Cracks usually occupy a very small percentage of a textured surface (1% or less) and exhibit dominant orientations. Therefore, they do not strongly affect "global" texture characteristics, but destroy texture mainly along their dominant orientations. In addition, an inspected texture is not affected along other orientations. Consequently, in order to detect cracks, we propose to analyze the behaviour of a texture along different orientations. Anomalous changes in the inspected texture in a particular direction should give a precise cue for crack detection. Such changes can be detected by measuring a new type of texture anisotropy that we called "conditional texture anisotropy".

2 Conditional Texture Anisotropy

Let us characterize the texture of an image along m different orientations (e.g., $0°$, $45°$, $90°$, and $135°$). Moreover, let us assume to have a set of n textural features for each orientation. Then, the texture of each image pixel can be characterized as follows:

$$X = (\underline{x}_1, \ldots\ldots\ldots, \underline{x}_m) \tag{1}$$

where each $\underline{x}_i$ is defined as:

$$\underline{x}_i = (x_{i1}, \ldots\ldots\ldots, x_{in}) \tag{2}$$

To analyze the directional behaviour of the inspected texture, we use the concept of local texture anisotropy [4]. Anomalous changes in the texture along a particular orientation result in a determinate degree of local anisotropy. However, the classic definition of local anisotropy is not well suited to crack detection. In fact, the texture of a given kind of material can exhibit an intrinsic degree of texture anisotropy that is not related to the presence of a crack. To take into account the anisotropy of the inspected texture, we define a measure of local texture anisotropy "conditioned" by the inspected texture. Let us consider a set $C = \{C_1, \ldots\ldots, C_N\}$ of "N" kinds of texture to be inspected (i.e., a set of "N" classes of texture). Let $p(X/X \in C_i)$ be the conditional density functions of the texture class C_i. As the texture is characterized along different orientations, we can define the conditional density of the texture class "C_i" along the "j" orientation (i.e., the function $p_j(\underline{x}_j/\underline{x}_j \in C_i)$). Then, we define a measure of conditional texture anisotropy (CTA) as follows:

$$CTA(X) = \frac{\underset{j}{Max}\left\{p_j\left(\underline{x}_j/\underline{x}_j \in C_i\right)\right\} - \underset{j}{Min}\left\{p_j\left(\underline{x}_j/\underline{x}_j \in C_i\right)\right\}}{\underset{j}{Max}\left\{p_j\left(\underline{x}_j/\underline{x}_j \in C_i\right)\right\}} \tag{3}$$

The values assumed by $CTA(X)$ are affected by the presence of a crack, as a crack along the dominant "j" orientation strongly changes the function $p_j(x_j/x_j \in C_i)$. In particular, $p_j(x_j/x_j \in C_i)$ decreases, as the crack makes the texture in the "j" direction very different from the texture of the inspected class. The maximum value of $CTA(X)$ (i.e., $CTA(X)=1$) is associated with a pattern containing a crack that causes $p_j(x_j/x_j \in C_i)=0$. On the other hand, when cracks are not present, $CTA(X)$ decreases and its minimum value (i.e., $CTA(X)=0$) is reached when $p_j(x_j/x_j \in C_i)$ is equal in all directions. Therefore, the properties of the CTA measure can be summarized as follows:
a) $CTA(X)$ takes on values in the range [0,1];
b) $CTA(X)=0$ for an "ideal" texture without defects;
c) $CTA(X)=1$ for maximum anisotropy.
To compute $CTA(X)$, we must estimate the conditional probability $p_j(x_j/x_j \in C_i)$ for each "j" orientation by using a training image without defects that belongs to the texture class "C_i" to be inspected. To compute $p_j(x_j/x_j \in C_i)$, we can use both parametric and non-parametric approaches. If an appropriate "parametric form" (e.g., a gaussian form) of $p_j(x_j/x_j \in C_i)$ can be assumed, then parametric methods can be used [5]. Otherwise, non-parametric methods, such as the Parzen windows, have to be adopted [5].

3 Experimental Results

The effectiveness of the *CTA* measure was tested on images of granite slabs affected by cracks. An example of such images is given in Fig. 1. As textural features, we used the ones based on the Fourier Power Spectrum computed in four different spatial-frequency bands. Each feature was computed along four angular orientations ($0°$, $45°$, $90°$, and $135°$). We used a 16×16-pixel window to compute the resulting 16 textural features.

Crack detection was performed in two steps:

a) System training on a defect-free image.

b) Classification of the test image to detect pixels belonging to cracks.

By the first step, we computed $p_j(x_j/x_j \in C_i)$ for each "j" direction by using a defect-free training image of the inspected texture. In particular, the computations were performed by assuming a gaussian form for the distributions. Then, we computed the Conditional Texture Anisotropy *CTA(X)* for each pixel X of the training image, and we derived the *CTA* statistical properties. For the sake of simplicity, we assumed a gaussian form of the *CTA* distribution. Finally, the classification process was carried out on the test image to detect "defective pixels". The Mahalanobis distance $d_m(X)$ between the *CTA(X)* of the pixel X and the ideal *CTA* distribution was computed. Defective pixels were detected by thresholding this distance. We fixed two thresholds, th_1 and th_2 (with $th_2 < th_1$), on the basis of the *CTA* distribution. Each pixel was classified as a "defect pixel" if $d_m(X) > th_1$, or as a "defect-free pixel" if $d_m(X) < th_2$. The pixels with $th_2 \leq d_m(X) \leq th_1$, which were uncertain pixels, were analyzed at the end of the thresholding process. In particular, they were labelled as "defect pixels" if they were connected with other pixels already classified as defective ones; otherwise, they were classified as "defect-free pixels". The result of the crack-detection process is shown in Fig. 2. As can be noticed, the crack was well detected. In order to better assess the effectiveness of our approach, we carried out the interesting experiment of analyzing the behaviour of the *CTA* measure in the presence of a crack. Figure 3 shows the profile of the *CTA* measure as a function of the row number along a fixed column in the test image shown in Fig. 1. The presence of a crack can be deduced from the very steep peak in the graph profile.

4 Conclusions

In this paper, we have presented a new technique aimed at detecting cracks in both regular and random textures. The technique is based on a new measure of texture anisotropy. In our experiments, the effectiveness of the proposed measure in detecting cracks on granite slabs was assessed.

References

1. L.H. Siew, R.M. Hodgson, and E.J. Wood: Texture Measures for Carpet Wear Assessment. IEEE Trans. on Pattern Analysis and Machine Intelligence, Vol.10, No. 1 (1988)

2. G.S. Desoli, S. Fioravanti, R. Fioravanti, and D. Corso: A System for Automated Visual Inspection of Ceramic Tiles. In: Proc. of the International Conference on Industrial Electronics, Control, and Instrumentation. Hawaii, USA, November 15-19, 1993, Vol. 3, pp. 1871-1876
3. K.Y. Song, M. Petrou, and J. Kittler: Texture crack detection. Machine Vision and Applications, Vol. 8, pp 63-76 (1995)
4. D. Chetverikov: GLDH Based Analysis of Texture Anisotropy and Symmetry: an Experimental Study. In: Proc. of Joint Workshop on Mapping Signal and Image Processing Algorithms onto Parallel Processor and Efficient Texture Analysis: Advanced Methods, Application. Hungarian Academy of Sciences, Central Research Institute for Physics, Budapest, October, 1994, pp. 43-56
5. K. Fukunaga: Introduction to Statistical Pattern Recognition. New York: Academic Press Inc. 1990

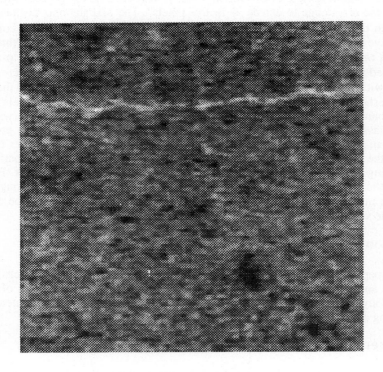

Fig. 1. An inspected granite slab affected by a crack.

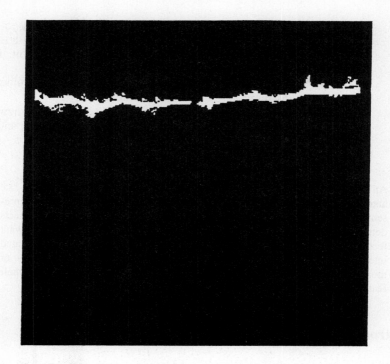

Fig. 2. Result of the crack-detection process.

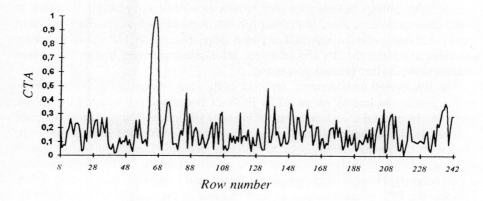

Fig. 3. Values of the CTA measure as function of the row

number along a fixed column in the image shown in Fig. 1.

A Vision System for Automatic Inspection of Meat Quality

M.Barni, V.Cappellini, A.Mecocci

Dipartimento di Ingegneria Elettronica, Università di Firenze
via S.Marta n.3, 50139 Firenze, Italy

An Intelligent Perception System (IPS) is presented which makes full-rate automated inspection of chicken meat feasible. It analyzes RGB images representing the chickens after they have been washed and plucked and detects defects such as burns, hematomas and blisters, together with other relevant features. First the chicken is extracted from the background and it is segmented into its anatomic parts. Then, defective areas are identified by means of morphological reconstruction. Finally, defects are classified by comparing their features against the defect description contained in a reference database.

1 Introduction

Nowadays, quality control of alimentary products is mainly performed by means of manual inspection; however, in many cases inspector's capabilities can not meet the high-speed production rates achieved by means of the modern manufacturing facilities. The introduction of *machine vision* tools can allow a full-rate automated inspection of alimentary products. In this paper an *Intelligent Perception System* (IPS) devoted to *on line* quality control of chicken meat before packing is presented[1]. The IPS analyzes images representing chickens after they have been washed and plucked. Its aim is to detect chicken defects, along with other relevant features such as the chicken size or some parameters related to the chicken global shape. Hematoma, burns and blisters are examples of defects the IPS should detect. Information collected by the IPS is used to address the further product processing.

The IPS overall architecture is reported in figure 1. Images are gathered directly on the production line by means of an RGB CCD camera, providing 760x562 red, green and blue images, with eight bits of intensity per image. The camera is equipped with an electronic shutter, so that images are *frozen* and blurring effects due to the motion of the conveyor belt (about 70 cm/s) are avoided. The Image Vision System is devoted to image analysis; it extracts the chicken from the background, segments it into subparts, i.e. neck, legs, wings and breast, and recognizes any existing defect. Besides, the IVS measures some general features of the sample product under in-

---

[1] The system has been developed in connection with ALINSPEC: a project partially supported by EEC under contract number BRE2-CT92-0132.

spection (e.g. the global color of the chicken skin and the chicken size). The information extracted by the IVS is fed to the Expert System (ES) whose aim is to decide the actions to be performed on the chicken. According to the defects revealed by the IVS, the ES can decide to reject the item, or to use it for a suitable application.

The IVS and the ES are interfaced to the user via the Man Machine Interface (MMI). The ES controls the Manipulation Unit, which is responsible for physically performing the actions the ES decides to undertake.

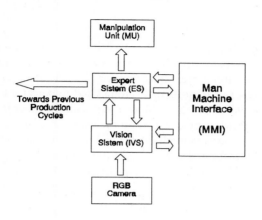

Fig. 1. Overall IPS Architecture.

2 IVS Architecture

The main tasks of the IVS are background removal, chicken segmentation and defect detection. Among them, defect detection is the most complex one, since it is affected by the great variety of defects to be revealed and by the lack of a suitable definition of them. Other requests which increase the IVS complexity are the real-time-functionality as well as the algorithm robustness. In addition to pure defect detection, the IVS also provides information for the global characterization of the hicken. These information (e.g.: the mean skin colour) are used by the Expert System to perform a full quality monitoring of the production.

The IVS architecture is shown in figure 2. The first action to be performed is background removal. To make such a task easier, chickens are framed against a blue rear panel which is clearly distinguishable from the chicken body. Then the chicken is segmented into its anatomic parts. On the segmented image a global analysis is performed, whose task consists of extracting some features characterizing the chicken in its entirety. These features are used by subsequent modules to adapt their behaviour to the sample currently being analyzed. The global analysis gives useful results also for the tracking of the production quality. In many cases, in fact,

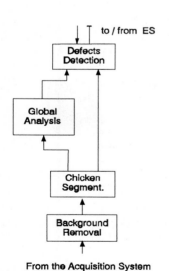

Fig. 2. Image Vision System outline.

global features can be regarded themselves as defects, for example, if a chicken is too little it can not be used for certain applications. The detection of defects constitutes the core of the IVS. Its goal is to identify possibly defective areas and to associate them to one of the defects contained in a given reference list.

3 Background Removal

In order to distinguish the chicken body from the background, a coordinate transformation is first performed to pass from the RGB space to a perceptual representation of colours [1,2]. In fact, it has been demonstrated [2,3] that most of the information needed to extract the chicken from the background and to detect a large class of defects is carried out by the colour hue. More specifically, a coordinate system has been adopted which is obtained by considering the same intensity of the HIS system and by using a cartesian representation for the H and S coordinates, which in the HIS space are expressed in polar form. The coordinate transformation from the RGB to the new system, is expressed by the following equations

$$I = \frac{r+g+b}{3} \; ; \quad C_1 = \frac{b}{r+g+b} \; ; \quad C_2 = \frac{1}{2}\frac{2r+b}{r+g+b}$$

By considering only the two chrominance features (C_1, C_2) and by plotting the scattergram of the chicken images, two well separated clusters appear: a yellowish cluster corresponding to the chicken body and a bluish cluster relative to the background. The area between the clusters is filled by pixels lying on the contour of

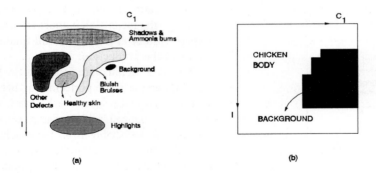

Fig. 3. a) Rough position of the various classes of pixels in the C_1-I space; b) mask used to separate the background from the chicken body.

the chicken. These pixels have intermediate colours between that of the background and those of the chicken skin. It is worthwhile noting that pixels belonging to bluish bruises and shadowy areas can belong to the sparse area between the clusters or even to the blue cluster, so they cannot be distinguished from the background on the basis of chrominance information only. Upon inspection of the scattergrams it also appears that the C_1 feature is sufficient to discriminate between the two clusters. To account for bruise and shadowy pixels, intensity information must be taken into account. In particular, the following points must be considered: pixels relative to healthy skin always belong to the yellowish cluster; pixels relative to bluish bruises belong to the sparse area between the clusters or to the blue cluster, they can be distinguished from the background since they are darker; with regard to pixels belonging to other defects, three cases are possible: they belong to the yellow cluster, they are darker than the background, they are reddish (very small C_1 values); shadowy pixels are darker than the background; pixels belonging to highlights are brighter than the background. The position of the various classes of pixels on a bidimensional feature space, is summarized in figure 3a. To segment the images into chicken and background, the mask reported in figure 3b has been built which splits the features space into two parts: pixels whose features lie inside the white area are classified as chicken pixels, while the others are labeled as background pixels. Background removal has been tested on more than 250 images, and in all the cases it succeeded in distinguishing the chicken from the background.

4 Chicken Segmentation and Global Analysis

Chicken segmentation is achieved by means of mathematical morphology [4]. In particular, an opening is performed with the structuring element chosen in such a way that wings, legs and neck are discarded and only the breast is left. In order to cut legs more sharply, a properly shaped kernel has been used instead of a classical circle. Besides, the kernel size is computed on the basis of the chicken area, thus making the system adaptive with respect to the chicken size.

With regard to global analysis the main task to be performed is the chicken classification according to skin colour. Two different kinds of chicken have been considered: white and yellow chickens. The importance of determining the colour of the chicken skin stems from the fact that skin colour affects the detection of colour-related defects, e.g. hematoma, bruises, livid areas, ammonia burns and so on. Along with white and yellow chickens a third class has been introduced to take into account abnormal skin colour. Defective chickens such as incompletely bled chickens belong to this class. Several approaches have been investigated to achieve a good classification. The best results have been obtained by means of a classical Bayesian classifier with an error probability of 5.3% which, compared to the error probability obtained by means of a manual classification (3.6%), witnesses the effectiveness of the adopted approach.

5 Detection of Defects

So far, the attention has been focused on the detection of colour-related defects, i.e, bruises, ammonia burns, blisters, livid areas and incomplete bleeding. In figure 4 the main steps involved in the defect detection process are summarized. First two sets of points are built: the former comprises pixels whose probability of belonging to defective areas is very high, whereas pixels belonging to the latter set, labeled as *uncertain*, are likely to be defective only to a lesser extent. Classification of defective and uncertain pixels is achieved by carrying out a statistical analysis on the chromatic features of healthy skin. By considering only pixels representing healthy skin, I, C_1 and C_2 can be assumed to be distributed according to a multivariate normal distribution whose parameters have been determined statistically. Then, for each pixel in the images, the Mahalanobis distance from the healthy skin cluster is computed. According to such a distance, pixels are classified as defective, uncertain or healthy.

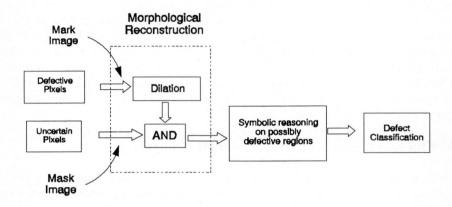

Fig. 4. Processing chain for the detection of defects.

At this point morphological image reconstruction [5] is applied to build possibly defective regions. Let I and J be two binary images, and suppose that $J \subseteq I$, i.e. $J(p) = 1 \Rightarrow I(p) = 1 \ \forall p$. J is called the *marker* image, whereas I is referred to as the *mask* image. The morphological reconstruction $R_J(I)$ of mask I from marker J is defined as the union of the connected components of I which contain at least one pixel of J

$$R_J(I) = \bigcup_{J \cap I_k \neq \varnothing} I_k$$

where $I_1, I_2, ..., I_n$ are the connected components of I. In our case, the marker image contains only defective pixels, while the mask image is composed by defective and uncertain pixels. The next task to be performed is symbolic reasoning on the regions extracted so far. Symbolic reasoning is necessary to minimize the number of false alarms, since morphological reconstruction fails to extract only really defective areas. Thus, regions on chicken borders are discarded as well as regions containing too many highlight pixels or very small regions. Finally, a Bayesian classifier associates each region to a particular defect.

6 Experimental Results and Future Works

Experimental results show the effectiveness of the proposed approach. The IVS has been tested on 150 images; for each of them the background has been removed and the chicken subparts successfully extracted. Classification results are summarized in Table I. For sake of simplicity the table considers only two classes: healthy and defective chickens. In the table the confusion matrix is reported together with the average error probability and the two probabilities of either missing a defect or rising a false alarm. The values along the diagonal represent the percentage of correctly classified images for each class; values along a given row indicate how misclassified chickens are distributed among the classes. The average error probability is 10.68%, with a false alarm rate of 10%. The probability of missing a defect, i.e. to classify as healthy a defective chicken, is 11.7%. Future work includes the analysis of other classes of defects such as shape and texture-related defects. Besides, we will try to increase algorithm robustness: different approaches will be investigated, among them fuzzy techniques are likely to provide good results.

TABLE I
CONFUSION MATRIX

	Mapped Class		
True Class	Healthy	Defective	Number of test images
Healthy	90%	10%	90
Defective	11.7%	88.3%	60

$P_e = 10.68\%$
P(false alarm) = 10%
P(missing a defect) = 11.7%

7 References

1. D.F. Rogers, Procedural Elements for Computer Graphics, McGraw-Hill, 1985.
2. A.Angotti, A.Barducci, M.Barni and A.Mecocci: "Analysis of Signals Retrievable from Standard RGB Tristimulus for Proper Meat Characterization and Classification"; Internal Report No. 940102, Dipartimento di Ingegneria Elettronica, University of Firenze, Jan. 1994.
3. A.Barducci, M.Barni, V.Cappellini, S.Livi and A.Mecocci, ALINSPEC Project: an Intelligent Vision System for Automatic Inspection of Alimentary Products, in "Intelligent Systems, E.A.Yfantis (ed), Kluwer, 1995.
4. P.Maragos and R.W.Schafer, "Morphological systems for multidimensional signal processing", Proceedings of the IEEE, pp. 690-710, Apr. 1990.
5. L.Vincent, Morphological Grayscale Reconstruction in Image Analysis: Applications and Efficient Algorithms, IEEE Trans. Image Processing, Vol.2, No.2, April 1993.

Author Index

Springer-Verlag
and the Environment

We at Springer-Verlag firmly believe that an international science publisher has a special obligation to the environment, and our corporate policies consistently reflect this conviction.

We also expect our business partners – paper mills, printers, packaging manufacturers, etc. – to commit themselves to using environmentally friendly materials and production processes.

The paper in this book is made from low- or no-chlorine pulp and is acid free, in conformance with international standards for paper permanency.

Lecture Notes in Computer Science

For information about Vols. 1–903

please contact your bookseller or Springer-Verlag

Vol. 938: K.P. Birman, F. Mattern, A. Schiper (Eds.), Theory and Practice in Distributed Systems. Proceedings,1994. X, 263 pages. 1995.

Vol. 939: P. Wolper (Ed.), Computer Aided Verification. Proceedings, 1995. X, 451 pages. 1995.

Vol. 940: C. Goble, J. Keane (Eds.), Advances in Databases. Proceedings, 1995. X, 277 pages. 1995.

Vol. 941: M. Cadoli, Tractable Reasoning in Artificial Intelligence. XVII, 247 pages. 1995. (Subseries LNAI).

Vol. 942: G. Böckle, Exploitation of Fine-Grain Parallelism. IX, 188 pages. 1995.

Vol. 943: W. Klas, M. Schrefl, Metaclasses and Their Application. IX, 201 pages. 1995.

Vol. 944: Z. Fülöp, F. Gécseg (Eds.), Automata, Languages and Programming. Proceedings, 1995. XIII, 686 pages. 1995.

Vol. 945: B. Bouchon-Meunier, R.R. Yager, L.A. Zadeh (Eds.), Advances in Intelligent Computing - IPMU '94. Proceedings, 1994. XII, 628 pages.1995.

Vol. 946: C. Froidevaux, J. Kohlas (Eds.), Symbolic and Quantitative Approaches to Reasoning and Uncertainty. Proceedings, 1995. X, 420 pages. 1995. (Subseries LNAI).

Vol. 947: B. Möller (Ed.), Mathematics of Program Construction. Proceedings, 1995. VIII, 472 pages. 1995.

Vol. 948: G. Cohen, M. Giusti, T. Mora (Eds.), Applied Algebra, Algebraic Algorithms and Error-Correcting Codes. Proceedings, 1995. XI, 485 pages. 1995.

Vol. 949: D.G. Feitelson, L. Rudolph (Eds.), Job Scheduling Strategies for Parallel Processing. Proceedings, 1995. VIII, 361 pages. 1995.

Vol. 950: A. De Santis (Ed.), Advances in Cryptology - EUROCRYPT '94. Proceedings, 1994. XIII, 473 pages. 1995.

Vol. 951: M.J. Egenhofer, J.R. Herring (Eds.), Advances in Spatial Databases. Proceedings, 1995. XI, 405 pages. 1995.

Vol. 952: W. Olthoff (Ed.), ECOOP '95 - Object-Oriented Programming. Proceedings, 1995. XI, 471 pages. 1995.

Vol. 953: D. Pitt, D.E. Rydeheard, P. Johnstone (Eds.), Category Theory and Computer Science. Proceedings, 1995. VII, 252 pages. 1995.

Vol. 954: G. Ellis, R. Levinson, W. Rich. J.F. Sowa (Eds.), Conceptual Structures: Applications, Implementation and Theory. Proceedings, 1995. IX, 353 pages. 1995. (Subseries LNAI).

VOL. 955: S.G. Akl, F. Dehne, J.-R. Sack, N. Santoro (Eds.), Algorithms and Data Structures. Proceedings, 1995. IX, 519 pages. 1995.

Vol. 956: X. Yao (Ed.), Progress in Evolutionary Computation. Proceedings, 1993, 1994. VIII, 314 pages. 1995. (Subseries LNAI).

Vol. 957: C. Castelfranchi, J.-P. Müller (Eds.), From Reaction to Cognition. Proceedings, 1993. VI, 252 pages. 1995. (Subseries LNAI).

Vol. 958: J. Calmet, J.A. Campbell (Eds.), Integrating Symbolic Mathematical Computation and Artificial Intelligence. Proceedings, 1994. X, 275 pages. 1995.

Vol. 959: D.-Z. Du, M. Li (Eds.), Computing and Combinatorics. Proceedings, 1995. XIII, 654 pages. 1995.

Vol. 960: D. Leivant (Ed.), Logic and Computational Complexity. Proceedings, 1994. VIII, 514 pages. 1995.

Vol. 961: K.P. Jantke, S. Lange (Eds.), Algorithmic Learning for Knowledge-Based Systems. X, 511 pages. 1995. (Subseries LNAI).

Vol. 962: I. Lee, S.A. Smolka (Eds.), CONCUR '95: Concurrency Theory. Proceedings, 1995. X, 547 pages. 1995.

Vol. 963: D. Coppersmith (Ed.), Advances in Cryptology -CRYPTO '95. Proceedings, 1995. XII, 467 pages. 1995.

Vol. 964: V. Malyshkin (Ed.), Parallel Computing Technologies. Proceedings, 1995. XII, 497 pages. 1995.

Vol. 965: H. Reichel (Ed.), Fundamentals of Computation Theory. Proceedings, 1995. IX, 433 pages. 1995.

Vol. 966: S. Haridi, K. Ali, P. Magnusson (Eds.), EURO-PAR '95 Parallel Processing. Proceedings, 1995. XV, 734 pages. 1995.

Vol. 967: J.P. Bowen, M.G. Hinchey (Eds.), ZUM '95: The Z Formal Specification Notation. Proceedings, 1995. XI, 571 pages. 1995.

Vol. 969: J. Wiedermann, P. Hájek (Eds.), Mathematical Foundations of Computer Science 1995. Proceedings, 1995. XIII, 588 pages. 1995.

Vol. 970: V. Hlaváč, R. Šára (Eds.), Computer Analysis of Images and Patterns. Proceedings, 1995. XVIII, 960 pages. 1995.

Vol. 971: E.T. Schubert, P.J. Windley, J. Alves-Foss (Eds.), Higher Order Logic Theorem Proving and Its Applications. Proceedings, 1995. VIII, 400 pages. 1995.

Vol. 972: J.-M. Hélary, M. Raynal (Eds.), Distributed Algorithms. Proceedings, 1995. XI, 333 pages. 1995.

Vol. 973: H.H. Adelsberger, J. Lažanský, V. Mařík (Eds.), Information Management in Computer Integrated Manufacturing. IX, 665 pages. 1995.

Vol. 974: C. Braccini, L. DeFloriani, G. Vernazza (Eds.), Image Analysis and Processing. Proceedings, 1995. XIX, 757 pages. 1995.

Vol. 975: W. Moore, W. Luk (Eds.), Field-Programming Logic and Applications. Proceedings, 1995. XI, 448 pages. 1995.

Vol. 976: U. Montanari, F. Rossi (Eds.), Principles and Practice of Constraint Programming - CP '95. Proceedings, 1995. XIII, 651 pages. 1995.

Vol. 977: H. Beilner, F. Bause (Eds.), Quantitative Evaluation of Computing and Communication Systems. Proceedings, 1995. X, 415 pages. 1995.

Vol. 978: N. Revell, A M. Tjoa (Eds.), Database and Expert Systems Applications. Proceedings, 1995. XV, 654 pages. 1995.

Vol. 979: P. Spirakis (Ed.), Algorithms – ESA '95. Proceedings, 1995. XII, 598 pages. 1995.

Vol. 980: A. Ferreira, J. Rolim (Eds.), Parallel Algorithms for Irregularly Structured Problems. Proceedings, 1995. IX, 409 pages. 1995.